Advances in Space Exploration

COSPAR Symposium Series

VOLUME 3

X-RAY ASTRONOMY

Advances in Space Exploration
COSPAR Symposium Series

Volume 1	K. A. VAN DER HUCHT and G. S. VAIANA (Editors): New Instrumentation for Space Astronomy
Volume 2	E. A. GODBY and J. OTTERMAN (Editors): The Contribution of Space Observations to Global Food Information Systems
Volume 3	W. A. BAITY and L. E. PETERSON (Editors): X-Ray Astronomy
Volume 4	H.-J. BOLLE (Editor): Remote Sounding of the Atmosphere from Space
Volume 5	W. RIEDLER (Editor): Scientific Ballooning

Forthcoming:
Remote Sensing and Mineral Exploration
Low Latitude Aeronomical Processes
Space and Development
Gamma-Rays

X-RAY ASTRONOMY

Proceedings of a Symposium of the Twenty-first Plenary Meeting of COSPAR
Innsbruck, Austria
29 May to 10 June 1978
including a Catalogue and Bibliography of Galactic X-Ray Sources

Edited by
W. A. BAITY and L. E. PETERSON
University of California at San Diego, La Jolla, California, USA

Published for
THE COMMITTEE ON SPACE RESEARCH
by
PERGAMON PRESS

OXFORD · NEW YORK · TORONTO · SYDNEY · PARIS · FRANKFURT

U.K.	**Pergamon Press Ltd., Headington Hill Hall, Oxford OX3 0BW, England**
U.S.A.	**Pergamon Press Inc., Maxwell House, Fairview Park, Elmsford, New York 10523, U.S.A.**
CANADA	Pergamon of Canada, Suite 104, 150 Consumers Road, Willowdale, Ontario M2J 1P9, Canada
AUSTRALIA	Pergamon Press (Aust.) Pty. Ltd., P.O. Box 544, Potts Point, N.S.W. 2011, Australia
FRANCE	Pergamon Press SARL, 24 rue des Ecoles, 75240 Paris, Cedex 05, France
FEDERAL REPUBLIC OF GERMANY	Pergamon Press GmbH, 6242 Kronberg-Taunus, Pferdstrasse 1, Federal Republic of Germany

First edition 1979

British Library Cataloguing in Publication Data

Symposium on X-Ray Astronomy, *Innsbruck, 1978*
X-ray astronomy. — (Advances in space exploration; vol. 3).
1. X-ray astronomy — Congresses
I. Title II. Baity, W A III. Peterson, L E
IV. COSPAR. Plenary Meeting, 21st, Innsbruck, 1978
V. Series
522'.6 QB472.A1 78-41186

ISBN 0-08-023418-6

In order to make this volume available as economically and as rapidly as possible the authors' typescripts have been reproduced in their original forms. This method unfortunately has its typographical limitations but it is hoped that they in no way distract the reader.

Printed in Great Britain by A. Wheaton & Co., Ltd., Exeter

Symposium on X-Ray Astronomy

Organized by

THE COMMITTEE ON SPACE RESEARCH - COSPAR

and

THE AUSTRIAN ACADEMY OF SCIENCES

through

THE AUSTRIAN SOLAR AND SPACE AGENCY - ASSA

Sponsored by

THE COMMITTEE ON SPACE RESEARCH - COSPAR

and

THE INTERNATIONAL ASTRONOMICAL UNION - IAU

CONTENTS

Low Luminosity Sources and Transients

Binary X-Ray Emitting Systems - Observations

Observations of Transient X- and γ-Ray Cosmic Sources

Large-Scale Distribution of Cosmic X-Ray Emission

PREFACE

The COSPAR/IAU Symposium on "X-Ray Astronomy" held in Innsbruck May 31-June 3, 1978 occurred at a propitious moment in the development of the subject. In addition to synthesis of older data from spacecraft such as Uhuru, Ariel V, SAS-3, and OSO-8, many new results were available from the 1st High Energy Astronomical Observatory (HEAO-1), launched but nine months earlier. Furthermore, theoretical developments in the field, particularly in the area of accretion phenomena around compact objects in binary systems, were at an exciting and fluid state, and many detailed comparisons of models and observations were possible.

One of the outstanding results has been the discovery of many low-luminosity (10^{33} - 10^{35} ergs/sec) galactic sources, and their identification with known optical objects. Also, Nova Ophiuchus, which flared early in the HEAO-1 lifetime was discovered to be an X-ray emitter. From SAS-3, Ariel V and HEAO-1, the positions of perhaps a hundred sources are now known to precisions within ~20 seconds of arc, about two orders of magnitude better in area than previous work. The reader is referred to the Appendix,* which contains a catalog of compact galactic and Magellanic Cloud sources with precise positions. Although not all known sources are identified, of the 87 galactic X-ray error boxes, 57 have possible optical/IR candidates, while 30 are still unidentified. Many of the low luminosity galactic X-ray sources are now identified with binary systems containing a white dwarf such as the RS CVn variables. Even the nearest star, Alpha Centauri, 4.3 light years away, has been detected in the light of X-rays, presumably from a solar-like corona.

Extragalactic research has also taken a large step forward with OSO-8, Ariel V, SAS-3, and now HEAO-1. There are now 41 extragalactic sources associated with clusters of galaxies, whose emission is presumably due to a hot (~10^8 K) intergalactic gas. Fe line emission at 6.7 keV has been detected from many of these clusters. Active galaxies, emitting in the range 10^{42} - 10^{45} ergs/sec are also identified as sources. These include some 31 Seyfert galaxies now known as X-ray emitters, and about 30 others, including radio galaxies (M87 and Cen A), BL Lacertae objects, QSO's, and emission-line galaxies. Models, based mostly on non-thermal phenomena, have been developed for these compact extragalactic sources.

In addition to the X-Ray Symposium, this volume contains the papers presented at the open meeting of COSPAR Working Group 3, Panel 3A, on Astrophysical Observations from Space, held on June 5, 1978. This session was devoted primarily to hard X-ray and γ-ray observations. One of the most unexpected results reported here was obtained from a balloon observation of Her X-1 and confirmed by HEAO-1. A spectral

* See page 22

feature near 58 keV, pulsed at the 1.24 second rotation period of the neutron star, may be due to gravitationally, accreted electrons radiating in the ~10^{12} Gauss field near the neutron star's surface. This discovery has resulted in a wealth of theoretical development, and many planned observations to determine the nature of this cyclotron feature and its characteristics through the phases of the various temporal variations of Her X-1.

Other results covered X-ray transients, the ever-mysterious gamma-ray bursts, and a session on hard X-ray and low energy gamma-ray studies, particularly on the diffuse background and on counterparts for the COS B sources.

The next International X-Ray Symposium will undoubtably report many results from HEAO-B, which will contain the first large area focussing telescope for X-rays in the 0.1 - 3.0 keV range. New γ-ray results are to be expected from HEAO-C by 1980, and eventually from NASA's Gamma-Ray Observatory, now being planned.

Many people contributed to the organization and success of the symposium. L.E. Peterson, Chairman of the Symposium, was ably assisted by an organizing committee consisting of G. Clark, H. Friedman, S. Hayakawa, K. Pounds, I. S. Shklovskii and J. Trümper. The Working Group 3 session was organized by S. Hayakawa, Chairman of Panel 3A. Much gratitude is due to Z. Svestka, Chairman of COSPAR W.G. 3, who shepherded the symposium and panel meetings, and who ably assisted in the program definition in Paris in the months proceeding the COSPAR XX Plenary. The hard work of the COSPAR secretariat, under Z. Niemirowicz, Executive Secretary, and D. Kastel was essential. The authors have all put considerable effort into condensing their papers and producing the camera-ready copy. We especially thank H. Bradt and his co-workers at MIT for the large task of assembling the catalog of X-ray source positions, which should be of lasting value.

Finally the editors much appreciate the assistance of the COSPAR editorial office, headed by M. Mutafchieva, and the patience of the special publications editor of Pergamon Press, J. Denby. The large amount of organization of correspondence and typing at UCSD was ably handled by our secretaries C. Barry and C. Matson.

William A. Baity
Laurence E. Peterson

La Jolla, California, October, 1978

Source Positions, Identifications and Properties

POSITIONS AND IDENTIFICATIONS OF GALACTIC X-RAY SOURCES

H. V. Bradt, R. E. Doxsey and J. G. Jernigan

Department of Physics and Center for Space Research, Massachusetts Institute of Technology, Cambridge, Massachusetts, USA

ABSTRACT

The use of satellite borne modulation collimators during the past few years has led to a dramatic increase in the numbers of identified stellar, >2 keV, X-ray sources associated with the Galaxy. There are now about 65 optical identifications of such sources. These include 40 low galactic latitude ($|b| < 10^o$) stellar objects, 7 globular clusters, up to 7 Magellanic Cloud identifications, and 11 other high latitude stellar objects. These statistics do not include supernova remnants and sources which have been detected only at $\sim 1/4$ keV . Another ~ 30 galactic X-ray sources with positions accurate to $< 60''$ remain unidentified and all but a few of these lie in the galactic bulge region. For comparison there are ~ 130 Uhuru (4U) X-ray sources at $|b| < 10^o$.

We present in the Appendix a Catalog of Compact Galactic X-ray Sources with Precise Positions (CGS Catalog) together with a large bibliography. The object of this catalog is to present in one place the available X-ray, optical, and radio data for the 96 compact galactic sources which are identified or which have precise ($< 60''$) X-ray positions. This catalog contains the most precise of the available optical, radio, or X-ray positions, the maximum and minimum published intensities, references to finding charts, and a comprehensive (as of August 1978) referenced list of X-ray, optical, and radio characteristics.

This paper is organized as follows:

INTRODUCTION

In the past few years, the number of the identifications of galactic X-ray sources with optical, radio, and infrared objects has increased dramatically. The "identification" of a given X-ray source is often a gradual and subjective process. An optical counterpart may be first suggested on the basis of rather tenuous evidence. It may then become generally accepted as more precise X-ray positions become available and as the peculiar nature of the optical object becomes more evident. In some cases, identical periodicities or simultaneous flaring at X-ray and optical wavelengths establish the identification with certainty. Caution is required here, though, because many questionable periodicities have been reported in the literature.

By 1974, the determination of X-ray positions to precisions of 1'-2' by the Uhuru and Copernicus satellites, together with other work with rockets and OSO-7, had led to firm identifications of ~ 10 compact galactic stellar sources (e.g. Sco X-1, Cen X-3, Cyg X-1, Her X-1, SMC X-1) and the suggestion of a number of other candidates. The recent availability of many precise (down to 10") positions obtained with the satellite-borne modulation collimators on Ariel-5, SAS-3 and HEAO-1 (A3 experiment) has accelerated the identification process: new candidates often can be accepted immediately and many previously suggested candidates have been established with high confidence. There are now ~ 65 well-established optical or IR counterparts. X-ray astronomy is now entering an era where detailed modeling of a wide variety of galactic astrophysical systems can be constrained by extensive data at non X-ray wavelengths.

We define a "precise" position as one with an equivalent error radius of $r < 60''$, i.e. with an error region area $A < \pi (60'')^2 = 3.1$ sq. arc min. The numbers of such positions derived from the several satellites are given in Table 1 . The SAS-3 survey of the galactic plane yielded good positions for essentially all Uhuru sources down to an intensity of about 10 Uhuru counts (= 11 μJy for a Crab-like spectrum averaged over 2-11 keV; 1.0 μJy = 0.24×10^{-11} erg s^{-1} cm^{-2} keV^{-1}). In selected regions where longer exposures were obtained, the sensitivity reaches to 1 μ Jy. In contrast, the HEAO A3 experiment will survey the entire sky to a sensitivity of 1.0 μ Jy before the end of the mission.

The role of the several satellites in arriving at identifications is illustrated in Fig. 1. The indicated stars are without doubt the correct identifications. The GX9+9 candidate (Fig. 1a), was proposed [185][1] on the basis of the Copernicus [867] and Uhuru [276] error regions. Only later did the Ariel [871], SAS-3 [211] and HEAO-A3 [322] positions and the detection of λ 4686, 4640-50 [120] lend substantial confidence to this candidate. The candidate [590] for 1735-444 (Fig. 1b) was proposed on the basis of the SAS-3 position because of its pronounced UV excess. Subsequent spectroscopic data [593] show it to be almost identical to the Sco X-1 star. This established without doubt the identification. The B1Vne star proposed [757, 431, 540] for 1145-619[2] (Fig. 1c) on the basis of the Uhuru position [276] gained credence through the preliminary (1S) [78] and final (2S) [205] SAS-3 positions. The faint highly reddened and radio emitting counterpart of Cir X-1 (Fig. 1d), the highly variable black hole candidate, was found [133,853] on the basis of Uhuru [432] and Ariel-5 [870] positions. Correlated X-ray, radio, and infrared variability [853,285] established the identification, and the subsequent SAS-3 and HEAO-A3 positions [77,322] remove any possible residual doubts. For 1538-522 (Fig. 1e), the SAS-3 [6] and HEAO-A3 [742] positions together eliminated all but one (star 12) of ~ 10 suggested candidates [168] in the large region defined by the Ariel-5 [872] and Uhuru [276] positions. Independent optical work [640,820,175] on the two candidates in the SAS-3 circle firmly established the identification. Finally, the weak (~ 5 μ Jy) Orion nebula source, 0532-054 [273] has been shown to be coincident with the ~ 1° nebula by Uhuru, Ariel, and ANS investigators [250,855,162,199]. About 30% of this flux has now been shown with SAS-3 to emanate from the immediate vicinity of the Trapezium [81] (Fig. 1f).

Perusal of a large number of diagrams such as those in Fig. 1 indicates that the Ariel-5 modulation collimator and the Uhuru 4U results apparently include a modest underestimate of the systematic errors. The SAS-3 and HEAO-A3 regions appear to be consistently in agreement within less than about 10". In an analysis of six sources [322], the rms discrepancy between the two sets of positions is 5" . Thus, the intersection of HEAO-1 and SAS-3 error regions should include the true position with a confidence that may exceed the quoted 90%.

We present in the Appendix a Catalog of Compact Galactic X-ray Sources with Precise Positions (CGS). This catalog (see Table 2) includes (1) all unidentified, hard (> 2 keV), compact sources

---

[1] Reference numbers allude to the alphabetical bibliography which follows the source catalog in the Appendix to this paper.

[2] In this text and in tables 3-12 we drop the satellite and catalog prefixes for the source names. We do include them in order of precedence, insofar as possible, in the CGS catalog (Appendix).

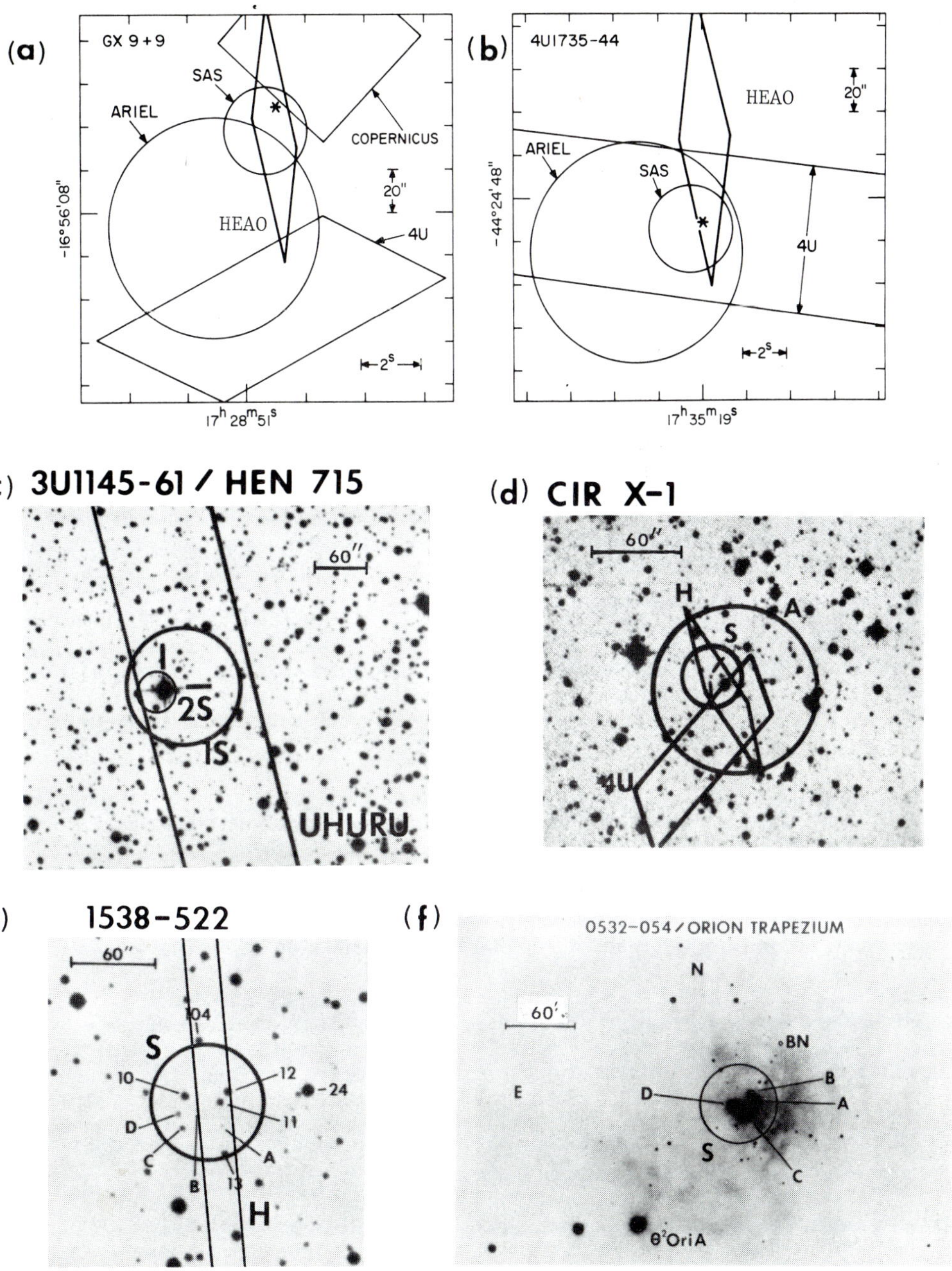

Fig. 1 X-ray error regions for six identified sources from the Uhuru (U), Copernicus, Ariel-5 (A), SAS-3 (S), and HEAO-1 (H, A3 experiment) satellites. Taken or adapted from: (a,b) Gursky et al. [322]; (c) Bradt et al. [78] and Dower et al. [205]; (d) Whelan et al. [853]; (e) Schwartz et al. [742]; and (f) Bradt and Kelley [81]. See text for other references to the x-ray positions and optical counterparts.

with $r_e < 60''$ and $|b| < 10^{\circ}$ and (2) all identified compact sources associated with the Galaxy, independent of the galactic latitude or the precision of the X-ray position that led to the identification. Thus, we include sources in the Magellanic clouds, globular clusters, and two predominantly soft sources with >2 keV emission (SS Cygni and AM Her), but exclude supernova remnants. Two exceptions to the $r_e < 60''$ criterion are the transients 0449-55 and 1118-615. Both are intrinsically interesting and both have quite precise positions.

The celestial position quoted in the CGS catalog is the best available X-ray, optical, or radio position. For an unidentified source, where both SAS-3 and HEAO-A3 positions are published, we give a weighted average position for the intersection of the two error regions. For the identified sources, the optical position to a precision of (1"-3") from the literature or from our own astrometry is given. We also attempt to summarize the presently available X-ray, optical, and radio literature for each of these sources.

There are 96 sources in the CGS catalog of which 65 have good optical, IR, or radio candidates (cf. Table 2). All but a very few (2 or 3?) must be the correct identifications. The vast majority exhibit indisputable spectral (e.g. high excitation emission) or temporal evidence (e.g. simultaneous X-ray/optical novae or identical binary periods). One might be inclined to question a faint UV-excess candidate or an early-type star with no other supporting evidence. However, we estimate that the likelihood of accidentally finding such objects in 30"-60" error regions is extremely small. With very few, if any, exceptions, subsequent work on such candidates has supported the X-ray/optical association. For example, in the summer of 1977, the candidates for 1735-444 [590], 1636-536 [590] and Ser X-1 [183] were known simply as UV excess stars. Now these are quite sound identifications. The spectral evidence [593] mentioned above has firmly established the 1735-444 counterpart, and the recent detection [111] of λ4686 emission in the 1636-536 candidate now confirms this candidate. The Ser X-1 candidate [183], has survived a substantial (factor of 15) reduction in area of the X-ray error region, from that of the early SAS-3 result [215] which led to the proposal of the star to the present joint SAS-3 [210] and HEAO-A3 [336] position. The candidate has now been resolved photographically into two stars, one component of which is quite likely the X-ray source (UV excess, weak λ4686) [792].

The galactic longitude distribution of the CGS sources which lie within 10° of the galactic equator (Fig. 2) shows that the unidentified sources lie almost exclusively in the galactic bulge within 40° of the galactic center. (The higher latitude sources in the CGS catalog are, by definition, identified; see Table 2.) An equivalent statement is that essentially all sources outside the bulge region with a precise (<60") position are now identified, and that new positions with this precision do, in fact, lead to early identifications. The unidentified bulge sources are probably heavily obscured; radio and infrared searches should play an important role in locating counterparts.

The 96 tabulated sources are segregated into 11 overlapping classes in Tables 3-13 [3]. Relevant parameters of the X-ray sources and brief characterizations of the optical counterparts are also given. References to the quoted parameters may be found in the CGS catalog. In the balance of this paper we will comment on the status of identifications within each of the classes and will present X-ray data or results for several specific sources to illustrate recent progress. These classes cut across the entire range of galactic X-ray astronomy much of which will be discussed elsewhere in these proceedings. The optical counterparts of these galactic sources will be discussed in the following paper by Margon [557].

Information acquired since the Symposium, through the September 1978 HEAD/AAS Meeting is also included in Tables 3-13, in the CGS Catalog, and in the following discussion. The authors regret any omissions or misrepresentations which may occur in this work.

---

[3] Tables 3-13 are at the end of the text.

TABLE 1. Precise Galactic Positions: Area $<\pi(60'')^2$.

Instrument	Celestial Region	Number
Rockets	Selected	Few
Uhuru	All Sky	17*
Copernicus	Selected	7^+
Ariel-5 RMC	Selected	23#
SAS-3 RMC	Galactic plane and selected	64** to $\geq 20''$
HEAO-A3 MC	All Sky (1st scan 70% complete - June 1, 1978)	$\sim 50^{++}$ to $\gtrsim 10''$

* Ref. (250)

\+ Refs. (578,686,867)

\# Refs. (116,228,617,680,870,871)

**Refs. (6,7,77,78,81,143,157,205,207,210,211,225,410,412,415, 517,521,700,736,737)

++Refs. (212,213,231,232,295,300,301,322,335,336,422,423,425, 426,742,743,793)

TABLE 2. Catalog of Compact Galactic Sources (CGS) with Precise Positions - Summary.

Criteria for source inclusion

- Identified Sources
  - Exhibits >2 keV flux
  - Associated with the Galaxy
    - Includes: Globular Clusters
      - Magellanic Cloud sources
      - High galactic latitude stellar systems
  - Appears point-like -- excludes supernova remnants
- Unidentified sources
  - Exhibits >2 keV flux
  - $|b^{II}| < 10°$, (exceptions: H0449-55, LMC X-2, H0544-665 (LMC?))
  - Area of x-ray positional error region: $A < \pi\ (60'')^2$
    - (exceptions: transients H0449-55, A1118-61)

Contents: 96 Sources	$\|b^{II}\|<10$	$\|b^{II}\|>10$	Total
Opt/IR Counterparts	45	20	65
Radio only (1742-289)	1	-	1
Unidentified	27	3	30

X-RAY DOPPLER OR ECLIPSING BINARIES (TABLE 3)

The 14 sources in this group are probably all identified correctly. Eleven exhibit well established binary behavior at X-ray wavelenths and 10 of these exhibit the x-ray period at optical or IR wavelengths (Symbol # in Table). These include 6 pulsing Doppler binaries, 2 eclipsing, non-pulsing binaries, one white dwarf system with > 2 keV emisson, and also the Cyg X-1 and Cyg X-3 systems. The other three sources, GX301-2, the recently reported OAO1653-407, and Cyg X-2 exhibit X-ray variability which could well be indicative of binary motion: (1) GX 301-2 may have a 40.8 d X-ray period [580], (2) 1653 -407 exhibited two transitions which were consistent with the binary period of the B01 star, V861 Sco which was in the 9 deg^2 (FWHM) Copernicus field of view [661] and (3) a possible 11.2 d period [380] for Cyg X-2 has been confirmed with 4 years of All-Sky-Monitor data [387]. The ASM group also report a 43.0 d (precessional?) period for Cen X-3 [387]. Reviews of binary X-ray sources by Bahcall [23,25] and by Rappaport and Joss [696,698] are available.

Three recent identifications are 0115+634 [157,422,416], LMC X-4, and 1538-522. The determination [518,862] that LMC X-4 exhibits the same 1.4 d binary period as its proposed optical counterpart [128] establishes its identification. The source 1538-522 was recently found by the MSSL [195] and GSFC [36] groups to be an eclipsing X-ray pulsator. Its X-ray [197] and optical [175] mass functions have been determined. The X-ray mass functions for both 0115+634 [702] (See Fig. 3) and 1538-522 indicate massive optical primaries which independently lend confidence to the B-emission line candidates. In fact, most of the stars in this group are mid to heavy mass objects. The exception are Cyg-X-2, AM Her, and (?) Cyg X-3.

The precise knowledge of the binary orbits and spin periods has led to a deeper understanding of the accretion process through stellar winds, critical potential lobe overflow, X-ray excitation of winds, etc. The relation between the X-ray luminosities and the spin-up rates of the X-ray pulsars indicate that the compact objects are neutron stars and not white dwarfs [697,577, 230]. The orbital parameters yield masses for the neutron stars which are consistent with present theoretical models (1.4-1.8 M)[437,698] .

OTHER X-RAY PULSATORS (TABLE 4)

There are 7 or 8 other X-ray pulsators (excluding the Crab pulsar) which as yet have not exhibited optical or X-ray variability attributable to binary motion. A recent 20" SAS-3 position [517] for 0535+262 confirms the often discussed [762] BOVe star as the candidate. The pulsating transient 1118-615, with r=2'-6' has had two candidates proposed for it, one a Be star [127] and the other a Mira variable [237]. No other X-ray source is known to be of the latter type [see 192, 592] so the latter is probably not the X-ray source. The unidentified source 1239-599 was in the broad field of view of the Ariel-5 RMC detectors when a 191 s periodic signal was detected [388]; the source and the pulsar may not be associated.

It is very likely that X Per, a rapidly rotating, possibly main sequence star, is the X-ray source 0352+309. The nearby (22") fainter companion [603] is now known to be a normal MIII star [606,648] . X Per and γ Cas, establish a class [78] of main sequence rotating Be stars as X-ray emitters, as proposed in [540]. In these stars matter is ejected in the zero gravity equatorial regions. It is possible that some of this matter reaches a companion compact object to produce the relatively low luminosity X-ray fluxes (2-5 x 10^{33} erg s^{-1}), though evidence for such companions in close orbits is lacking [167, cf. 395]. The B1Vne counterpart of 1145-619 may also belong to this class [78,540] ; although its X-ray luminosity can reach as high as 6 x 10^{36} erg s^{-1} [413,80] .

Yet another type of optical counterpart appears in Table 4: a faint star with a UV excess [590] for 1627-673. This is probably a low luminosity star because of its relatively high galactic latitude (-13°). The pulsing does not exhibit a detectable cyclic Doppler shift which indicates a very wide or a very close binary system [695] or a low-mass companion [440]. A model for the latter type of system (Fig. 4) with a very low mass primary (M star) could explain the lack of X-ray eclipses in many of the bulge sources [440,441] .

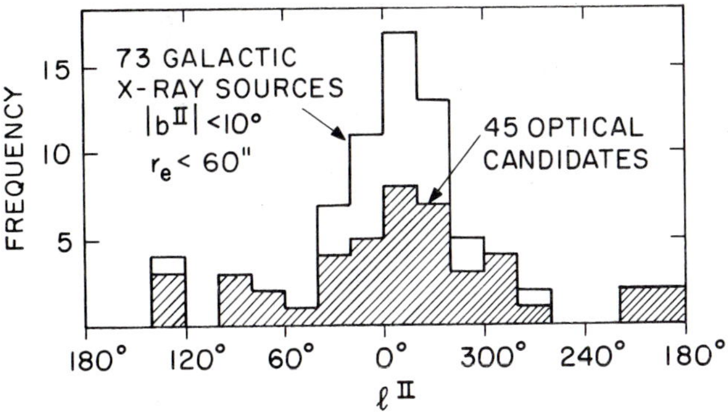

Fig. 2 Galactic longitude distribution of galactic plane sources with precise positions, $r_e < 60''$, i.e. the x-ray error region has area $A < \pi(60'')^2$ or the source has a good optical candidate. The latter are indicated with hatched area.

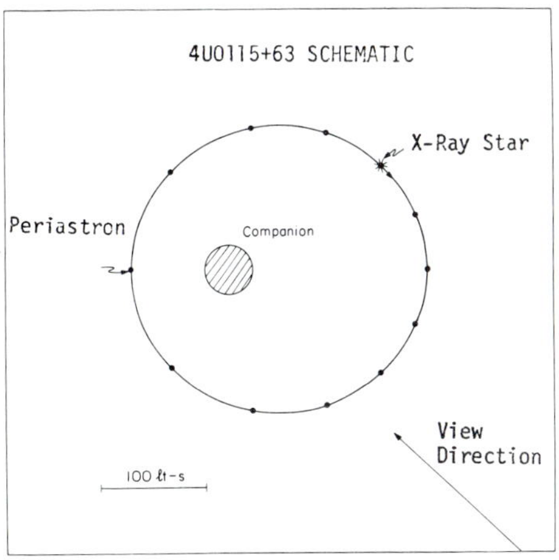

Fig. 3 Schematic of 4U0115+63 orbit; from Rappaport et al. [702].

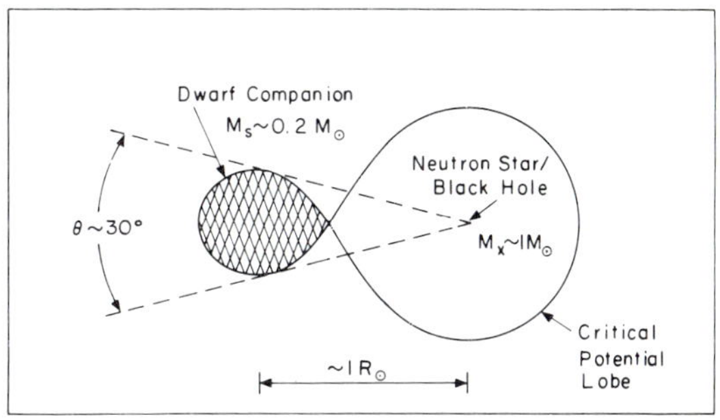

Fig. 4 Compact binary model for galactic bulge sources; from Joss and Rappaport[441].

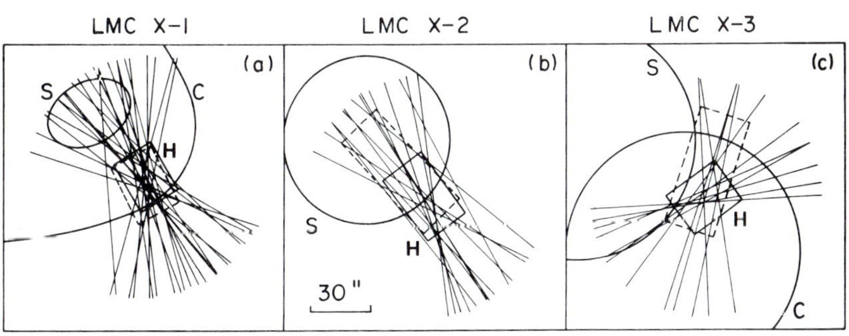

Fig. 5 X-ray lines of position and error regions (H) for LMC X-1, 2,3 from HEAO-A3 [423]. Dashed lines are the previously reported HEAO positions [425]. Copernicus (C) and SAS-3 (S) positions are also shown (see text for references); from Johnston et al. [423].

MAGELLANIC CLOUDS (TABLE 5)

Three SMC sources and up to 6 LMC sources are listed in the CGS catalog. They are included because some and probably all of them are associated with stellar systems. LMC X-5 [562] and the reported LMC X-6 [298] have not yet been detected with the high resolution experiments. Two of the sources (SMC X-1 and LMC X-4) have optical counterparts [525,128] which exhibit the X-ray periodicities (cf. Table 3).

The recent discovery of the transient sources, SMC X-3 and SMC X-2 with the SAS-3 modulation collimators [143] led to immediate suggestions of an OB star candidate for each, the former with emission lines [730,531] and the latter a B1e star [728,612].

Very precise positions for LMC X-1, 2, 3 have been obtained recently from HEAO-A3 [425,423]. Numerous independent lines of position (Fig. 5), each with 10" half-width (90% confidence), were obtained over a period of months as HEAO-1 tracked the sun (the LMC lies at the ecliptic pole). The position for LMC X-1 is in disagreement with that obtained from SAS-3 data [225]. The SAS-3 error regions were defined with a different algorithm than that used for the other SAS-3 galactic positions and the systematic errors may have been underestimated. The HEAO-A3 results also fail to show LMC X-1 as a 1' extended source as inferred [225] from the SAS-3 data. The HEAO-A3 30" FWHM collimator with separated transmission bands [322] is intrinsically more sensitive to such structure than is the 140" FWHM SAS-3 collimator. The present HEAO-A3 positions for LMC X-1 and LMC X-3 again focus attention upon two of the several previously suggested candidates, the B5I star, R148 [431,686] and a BIII-IV star [840], both of which are quite probably the counterparts, although further confirmation is desirable.

Recent additions to Table 5 are (1) the Ariel recurrent transient 0535-668=A0538-66 for which there is a HEAO-A3 position and a B2I candidate [426] and (2) a faint, new, unidentified HEAO-A3 source, 0544 -665, in the direction of the LMC [423].

The LMC/SMC sources as a class have relatively higher luminosities than galactic sources. This has been attributed to the known underabundance of heavy elements in the clouds and the associated higher limiting accretion rates due to X-ray heating of, and scattering in, the infalling material [143]. Optical identifications and the study of the specific optical counterparts could shed light upon this issue.

TRANSIENTS OR X-RAY NOVAE (TABLE 6)

All compact galactic X-ray sources could well be called transients since they are all, to our best knowledge variable. We list in Table 6 only those which are commonly called "transients" in the literature and, of course, are identified or have r_e <60". Most exhibit a temporary increase of intensity by a factor of 10 or more. Other sources, e.g. Cir X-1 and Her X-1, qualify and could well be added. See also a review by Cominsky [158].

The yield of optical candidates for this group is substantially lower (14 out of 23) than that for the pulsators. This class includes examples of the soft-spectrum non-pulsing X-ray novae (e.g. Nova Mon '75 [222] and Nova Oph '77 [300]), and the hard-spectrum pulsing transients, 0535+26 [718,75], 0115+634 [157,386], and 1118-615 [227]. The class includes events with time constants from ~1 month, e.g. 0620-003 [583,456], to events with 2 to 10 min. time constants, i.e. 1715-321[566,375,412], probably more properly called flares. Also there are recurrences with intervals from a year of so, 0535+262 [710,123,145], to 58 years 0620-003 [221]. The so-called high latitude transients, 0042+327 [713,700] and 0449-55 [303] are also present.

The optical counterparts exhibit the same range of types associated with all galactic sources: very low mass and high mass stars. The low mass stars brighten optically (reprocessed X-rays) when the X-ray source enters its bright transient state. There are five examples of these coincident X-ray/optical novae known to us (symbol #, Table 6).

New insight has been obtained into the nature of the hard transients with the demonstration, for the first time, that one of these objects, 0115+634, is in a binary system [702]. The orbit

(Fig. 3) is the largest (24 d) of the six known X-ray Doppler systems. The wide elliptical orbit may be related to the transient behavior because of the relative difficulty of transferring mass over such large distances [702].

Transient behavior must involve episodic mass transfer due to a variable stellar wind (high mass stars), critical lobe overflow (low mass stars), or possibly the slow build up of material in an accretion disk, which occasionally undergoes an instability (see discussion in Griffiths et al., this volume [303]). Nuclear burning is clearly excluded as the source of the X-ray outbursts in some cases because of the absence of flux due to the accretion of the unburned matter between outbursts, e.g. 0620-003 [583]. The study of the quiescent low mass stars, e.g. 0620-003 [636] and Aql X-1 [791,555] is beginning to shed light upon the anatomy of these systems (see Margon, these proceedings [557])

PERSISTENT BRIGHT BULGE SOURCES (TABLE 7)

Perusal of the Uhuru 4U catalog clearly establishes the existence of a set of bright persistent sources within about 35° of the galactic center [e.g. 271,593]. In Table 7, we list 15 such sources which were selected with the following criteria: $I > 100$ Uhuru counts (110 μJy, 2-11 keV), $|b| < 10^{\circ}$, $|l| < 40^{\circ}$, and a variability factor of < 3. All of the 4U sources which meet these criteria are in the CGS catalog. The maximum X-ray luminosity (2-11 keV) of these sources ranges from 6×10^{37} to 4×10^{38} erg s^{-1} at 10 kpc.

Only about one-half of these sources have optical counterparts. A probable Mira variable [592] star with strong IR emission [283] has been noted [881] in the GX349+2 error region. It is not considered a likely candidate [592]. The optical candidate [785] for GX17+2 has the appearance of a normal G or K star [185,551] and is a variable radio emitter [360,367]. Its location within the 20" X-ray error region leaves little doubt about the identification. Five of these bulge sources are burst sources. The galactic longitude distribution of burst sources [507] resembles that of these bright bulge sources, but not that of all low latitude steady sources [507]. A physical association between the bulge sources and globular cluster sources has also been suggested [105] since, in both cases, the X-ray sources are found in dense regions of late-type stars and because their spectra are relatively soft [436] (See refs. in [601]).

The problem of the nature of many of these bright bulge sources is still with us (See discussions in e.g. [593,347,441,601]). They do not exhibit eclipses, periodicities, or fast timing variability other than bursts. Although six stellar-appearing objects have been identified, there is as yet no evidence for binary motion in any of these objects. A close binary system with a very low mass star (e.g. 1627-673, Fig. 4) would explain the absence of eclipses in about 80% of the cases [441]. Another closely related model would restrict X-ray emission to the pole of the orbit [601].

GLOBULAR CLUSTERS (TABLE 8)

Seven X-ray sources have been shown to be associated with globular clusters. One of these, the rapid burster [497], does not exhibit steady X-ray emission. However, positions for the bursts themselves from SAS-3, Ariel-5, ANS, and HEAO-1 (Fig. 6) indicate that the X-ray source lies in the core of the globular cluster Liller 1 [528,475]. This cluster was discovered and proposed as the source on the basis of the initial SAS-3 (4' radius) [497] and Ariel-5[8] positions and the ANS [341] line of position. Five of the six sources which exhibit steady emission have been shown, with SAS-3 positions, to lie within ~ 20" of the cluster centers [415]. The remaining source has been convincingly identified with NGC 6440 [561,252] on the basis of OSO-7 and Uhuru positions. Its transient behavior has precluded more precise measurements.

It is not well understood why such a large proportion, or indeed any, of the ~ 100 known globular clusters with their late-type stellar population are X-ray emitters [137,462]. This has led to several suggestions for the formation of binaries by capture [137,236] in the central dense regions of the cluster. Also, massive black holes accreting stellar debris at the cluster centers have been proposed [22,753]. The recent SAS-3 results indicate that the masses of the X-ray emitting systems are greater than about twice the average stellar mass in the cluster [415].

BURST SOURCES (TABLE 9)

The precise (~30") positions and optical identifications of these objects have come about because relatively precise (5-10') positions have been determined for the bursts themselves, primarily with SAS-3 [503]. These made possible reliable identifications of the burst sources with known, e.g. 4U, steady sources. See reviews by Lewin [503,505,506 and 512 (this volume)]. In turn, modulation collimator experiments have been able to pinpoint the steady sources to high precision. As mentioned above, rapid burster positions (Fig. 6) were obtained from the bursts themselves. Six of the sources on the list (marked "?") have been mentioned in the literature as possible burst-steady source associations and precise positions do exist for the steady components. However, the burst positions are not yet sufficiently precise to insure that these associations are correct [cf 512]. The list in Table 9 constitutes only about 1/3 of the ~30 known burst sources [507].

The optical counterparts of the burst sources include possibly five globular clusters. The six stellar candidates include 5 faint UV excess stars, four of which exhibit high excitation emission lines. The identification of both globular cluster and stellar counterparts, together with the continuing study and modeling of the detailed properties of the bursts themselves is now giving us some insight into their nature [512].

THE GALACTIC CENTER REGION (TABLE 10, FIG. 7)

The 2° region in the vicinity of the galactic center contains an assortment of highly variable X-ray sources (Fig. 7): at least three burst sources [495], three transient sources (GX359+2=4U1735-28 [466,250]; GCX-3=GX+0.2-0.2=A1743-288 [177,680,228]; A1742-289 [226,90,871, 680]), and one persistent but variable source A1742-294=2S1742-294=GCX-1 [10,412,177,680, 228]. Three of these sources have positions precise to better than 60" (Table 10).

The existence of these seven sources is well established; the three listed in Table 10 have been reported by more than one group and their positions have been measured with instruments of high angular resolution (2.3' to 15' FWHM, see footnotes to Table 10). The others (GX359+2 and the bursters) were well above the Uhuru and SAS-3 thresholds respectively. This source complex was first reported as an extended or multiple source, "GCX", region of size ~ 2° (Fig. 7) by Uhuru investigators [466]. At present there is no evidence for diffuse emission. Each of the three sources in Table 10 could be associated with a burster, but the density of sources and the size of the burster error regions make this uncertain [512]. One of the sources, A1742-289, has been identified with a coincident radio-flare source which lies 53" ± 3" SW of the nominal galactic center in Sgr A West [187]. (Note the possible detection of the galactic center in the near infrared, these proceedings [316]). All of the detected sources including the burst sources exhibit hydrogen column densities of ~ 10^{23} cm^{-2} [e.g. 495,680,436].

Some six other X-ray sources have been reported to exist in this region [177,680,228]. In our view, they must be considered with caution because they are near detection threshold or may be confused with other sources. None have yet been confirmed. Also we note the report of 3 small bursts in 6 minutes from the galactic center [680]. If confirmed, this is of great importance. Studies with greater sensitivity and angular resolution, and with sustained or repeated observations, are required to further clarify the highly variable X-ray structure of this region.

OTHER IDENTIFIED SOURCES (TABLE 11)

Table 11 contains 18 objects of which 12 (marked *) are recent identifications. Several which are quite diverse and interesting, viz.: γ Cas [78], a rapidly rotating Be star mentioned above; Algol [736], a flaring radio source, which may yield X-rays through inverse Compton scattering of the radio emitting electrons from the optical photons [3], or from bremsstrahlung of the K star stellar wind accreting onto the B8V star [328]; the Orion source which may arise from shocks created by the stellar winds from the Trapezium stars [676,161,81]; Cir X-1 [542], a black hole candidate; Sco X-1 type systems (see Margon, these proceedings [557]; the dwarf novae SS Cyg [691,342,715,232] and EX Hydrae [846,335,165] which exhibit hard (>2 keV) X-ray emission, as well as soft X-ray emission.

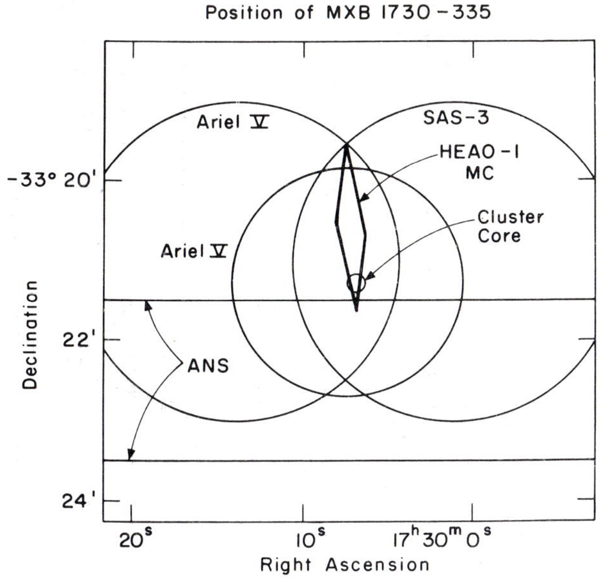

Fig. 6 X-ray error regions for the Rapid Burster [497] from the SAS-3 [411], ANS [341], Ariel-5 [8, 871], and HEAO-A3 [212] satellite experiments. The globular cluster core identified with the burster [528, 475] is also shown; from Doxsey *et al.* [212].

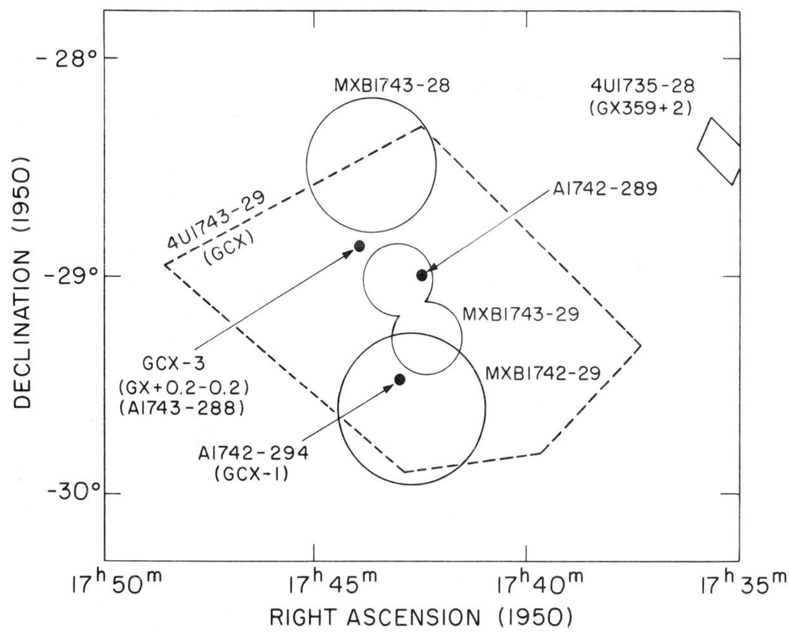

Fig. 7 X-ray sources in the galactic center region. Only well established sources are indicated; see text and Table 10 for discussion and references. The three filled circles indicate positions known to < 60" precision. The northern two of these and 4U1735-28 are transient sources. The open circles represent burst source error regions [495]. The approximate extent of this region, as first measured with Uhuru [466] , is indicated by the dashed line (GCX). The "galactic center" (compact radio source in Sgr A West) lies 53"± 3" NE of the radio flare source identified with A1742-289.

Cir X-1 has been identified with a faint, highly reddened star (R = 16, B = 22) with broad Hα emission and radio emission in a tour de force by a U.K. collaboration [853]. The radio flux flares [853] within a few days after the periodic 16.6 d X-ray transition [446] to lower flux values. The infrared flux also shows systematic variability during the 16.6 d cycle, on one occasion rising 1.5 mag to K=7.3 in 1 hour, 0.5 d after the X-ray transition [285]. Several SAS-3 observations of the source have been carried out around the time of the transitions [203]. One of these transitions was particularly dramatic and complex (Fig. 8). During the 9 hours before the transition, the source was remarkably steady, a state not previously observed. It then exhibited spectrally independent dips while brightening for ~ 3 hours in the lower energy channels. The abrupt transition to the low state then occurred in about 60 s and was immediately followed by dramatic 3 s to 20 s flares, all without substantial spectral variability. A significant flux remained after the transition. The temporal and spectral appearance of the dips, the transition, and the flares suggest that a single process is responsible, e.g. the exhaustion of matter in the accretion disk, albeit with clumpiness, though this may not easily explain the 16.6 d period, which is clearly not a simple eclipse. Any model must also explain the dramatic increase in soft flux just prior to the transition.

OTHER OPTICALLY UNIDENTIFIED SOURCES (TABLE 12)

Table 12 contains those sources in the CGS catalog which do not appear on our previous lists (Tables 3-10). At present they are not known to have any particularly distinguishing X-ray features, or serious optical candidates. One, 1553-542 [6], lies in the 1° radius error region of the COS-B gamma ray source CG327-0 [346].

RADIO EMISSION (TABLE 13):

Radio emission is an important feature of a number of these sources, though it is hardly typical of galactic sources. Table 13 lists sources which have been suggested or demonstrated to have radio emission. Nine of these are certain associations and range from intense dramatic flares (Algol [278,361] and Cyg X-3 [293]) to transient phenomena coincident in time with X-ray novae (0620-003 [638,188,53,532] and 1742-289 [187]), and the post transition radio flares of Cir X-1 [853].

SUMMARY

The several categories of X-ray emission presented here demonstrate the wide variety of X-ray behavior of galactic X-ray sources. The number of optical counterparts (up to 65 out of the 96 sources in the CGS catalog) is quite impressive. These figures do not include the identified supernova remnants or the low luminosity stellar soft X-ray sources discussed by Garmire in these proceedings [263]. Each of the categories of X-ray sources is well represented with optical and in some cases IR and radio counterparts.

The stellar optical counterparts include not only the previously known massive supergiants (e.g. 0900-403 and GX301-2) with stellar winds and the mid mass stars (e.g. Her X-1) which lose mass through Roche lobe overflow (see [23]). They also include fast rotating main sequence stars,"Be stars" [e.g. 78,540, cf 159] (e.g. γ Cas and X Per) which may be transferring the mass ejected in the zero gravity equatorial regions to a compact object. The source 1145-619 with its main sequence classification may also be such a system [540,78]. Also, within the past year, faint low mass stars have been identified with X-ray sources (e.g. 0620-003 [60] and 1735-444 [590, 593]). These, as pointed out by a number of authors (e.g., Margon these proceedings [557]), may simply be distant Sco X-1 type systems.

Essentially all of the listed X-ray sources can be explained as accretion onto a compact object in a binary system. Possible exceptions are Algol [736,328], the Orion Trapezium [81,161], and, of course, the Crab pulsar. However, the natures of the globular clusters, the bulge sources and also the X-ray pulsars without demonstrated Doppler or eclipse behavior are far from settled.

The source GX301-2 with its (infrequent) chaotic and burst-like behavior [465], not unlike that of Cyg X-1 and Cir X-1, is important to our understanding of the significance of rapid

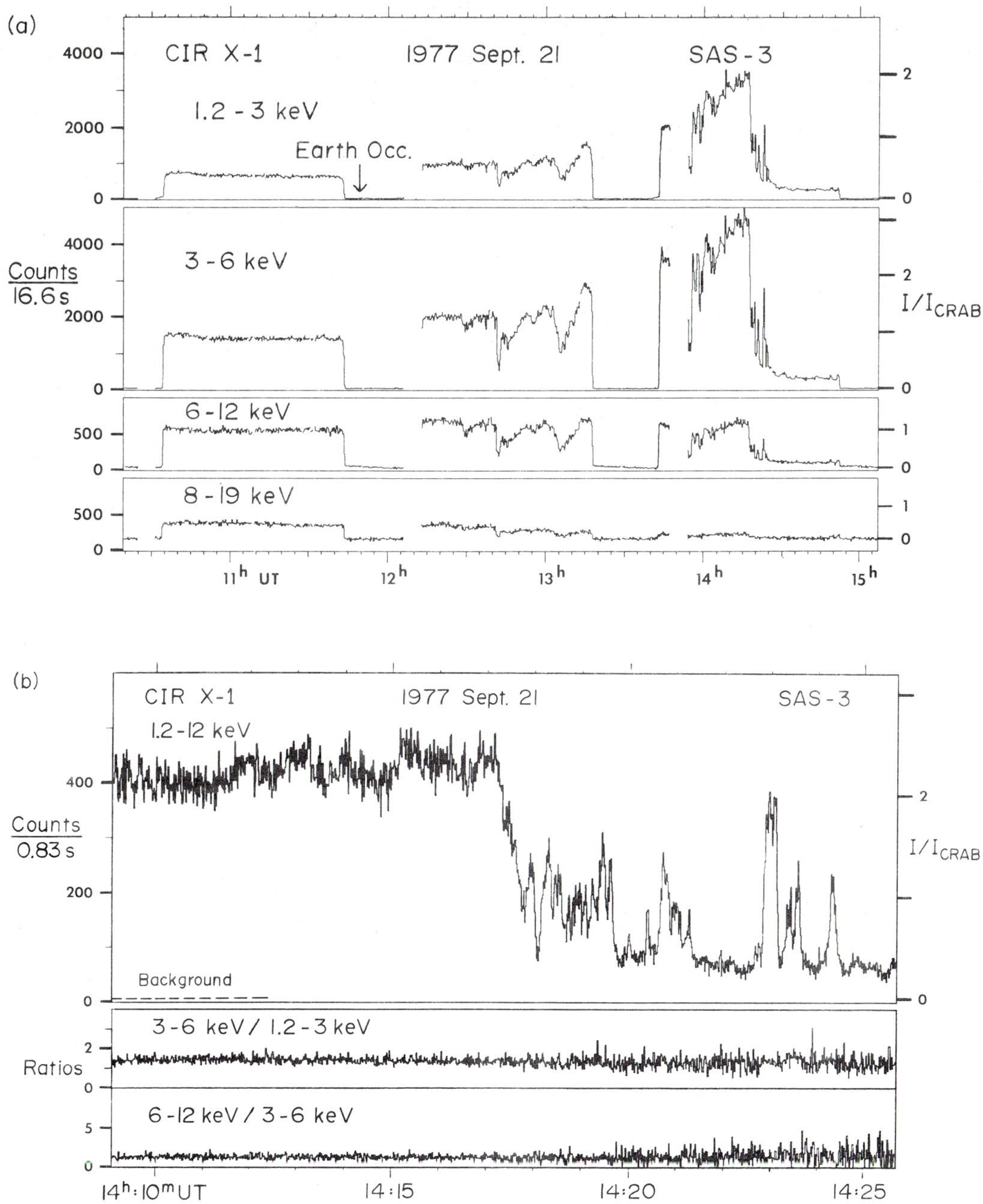

Fig. 8 X-ray transitions of Cir X-1 on September 21, 1977. (a) Three orbits with (1) very steady emission, (2) increasing activity, and finally (3) the transition with pronounced flaring and a substantial residual flux. (b) The transition with 0.83s resolution and intensity ratios which show no pronounced spectral changes during the large intensity variations; from Dower et al. [203].

variability since, unlike these two sources, it is a periodic pulsator, and hence most likely a magnetic neutron star. Another example of such a highly variable source [722] is GX339-4 (1659-487). We also know that Cir X-1 exhibits a wide variety of behavior which is reminiscent at different times of the relatively steady bulge sources, the rapidly variable Cyg X-1, the rapid enhancements of burst sources, and the longer term variability of the transient X-ray sources. The fact that these types of behavior all take place in a single X-ray system should help us to understand the processes taking place in related, but less schizophrenic, sources.

The authors are grateful for helpful conversations with and support from their colleagues associated with the SAS-3 and HEAO-1 modulation collimator experiments, namely K. Apparao, J. Delvaille, R. Dower, A. Epstein, G. Fabbiano, R. Griffiths, H. Gursky, M. Johnston, R. Kelley, J. Kulik, A. Ramsey, H. Schnopper, D. Schwartz, J. Schwarz, G. Spada, and F. Walter. The SAS-3 modulation collimator analysis system was developed, in part, by J. Buff, T. Matilsky, F. Primini, J. Richardson, and J. Sachs. The authors thank C. Canizares, G. Clark, P. Joss, W. Lewin, W. Mayer, J. McClintock, S. Rappaport, and B. Rossi for important contributions to this work. They are also grateful to J.T. Armstrong, M. Blazquez, P. Byrne, K. Jaehnig, A. Johnson, S. Joyce, R. Kelley, C. Reid, T. Sienko, and D. Woodruff for assistance in the preparation of this document. They particularly thank Cliff Reid, who carried the major share of the responsibility for the preparation from its inception to its completion. This work was supported in part by NASA contracts NAS5-11450 and NAS8-27972.

TABLE 3[+]. 11 (14?) X-ray Doppler or Eclipsing Binaries*, 13-14 identified.

Source	Name	X-ray	Optical/Radio
0115-737&	SMC X-1	3.9d, 0.7s, ecl.	B0I*#
0115+634&	transient	24.3d, 3.6s, e=0.34, no ecl.	B-em
0532-664	LMC X-4	1.4d, ecl.	O8III-V#
0900-403&	Vela X-1	9.0d, 283s, ecl.	B0.5Ib#
1119-603&	Cen X-3	2.1d, 4.8s, 43.0d?, ecl.	O6.5II-III#
1223-624?	GX301-2	40.8d?, 699s	B1.5Ia
1538-522&		3.7d, 529s, ecl.	B0I#
1653-407?@	V861 Sco	7.8d (ecl.?)	B0Iae(#?)
1656+354&	Her X-1	35d, 1.7d, 1.24s, ecl.	A-F#
1700-377		3.4d, ecl.	O6.5f, λ4686#
1814+498	AM Her	3.1h, partial ecl.	Mag. White Dwarf#
1956+350	Cyg X-1	5.6d, partial ecl.	O9.7Iab#/Radio
2030+407	Cyg X-3	4.8h, partial ecl., 16.8d?	IR star#/Radio
2142+380?	Cyg X-2	11.2d (partial ecl.?)	V~15.5, uv, var.

& X-ray pulses clearly exhibit Doppler shifts due to binary motion.
* No other such systems known, i.e. all meet CGS catalog criteria (Table 2).
? Binary nature of x-ray variability requires confirmation.
Optical exhibits X-ray periodicity.
@ X-ray error region of 1653-407 is ~20 deg.^{2}
+ FOR TABLES 3-13:
REFS. ARE IN CGS CATALOG (APPENDIX).
SPECTRAL DESIGNATIONS REFER TO EMISSION FEATURES.
UV ≡ UV EXCESS.

TABLE 4. 8 (9?) Other X-ray Pulsators; 7-8 identified.

Source	Name	Period	Optical/Radio
0352+309	X Per	835s, 22h?, 581d?	Be, O9.5(III-V)e
0531+219	Crab Pulsar	0.033s	Pulsar/Radio+Opt.
0535+262	Transient	104s	B0Ve
1118-615*	Transient	405s	Be star*
1145-619		297s+292s(?)	B1Vne
1239-599?		191s	
1258-613	GX304-1	272s	B6-9pe
1627-673		7.7s	uv, V~18, 7.7s#
1728-247	GX1+4	116s	M6III+Hot comp., IR

* The transient A1118-61 (405 s) has a 2'-6' error radius.
? Pulsator (191s period) may not be associated with 1239-599.
Coincident x-ray and optical pulsing and flares.

TABLE 5. 8 (9?) Magellanic Cloud Sources*; Up to 7 identified.

Source	Name	X-ray	Optical/Radio
0050-727	SMC X-3	Transient	OB-em
0052-739	SMC X-2	Transient	Ble
0115-737	SMC X-1	Binary Pulsar	B0I#
0521-720	LMC X-2		
0532-664	LMC X-4	Eclipsing Binary	O8III-V#
0535-668+	A0538-66	Recurrent (P=16.7d?) transient	B2Iab
0538-641	LMC X-3		BIII-IV, var.
0540-697	LMC X-1		B5I@
0544-665	LMC?		

* 1 or 2 others known.
@ Other early type stars in error region.
# Optical exhibits x-ray periodicity.
+ Position derived under assumption that 4 outbursts were from same source.

TABLE 6. 23 Transients*; Up to 14 identified.

Source	Name	X-ray	Optical/Radio
0042+327	b^{II}=-30		Weak λ4686
0050-727	SMC X-3		OB-em
0052-739	SMC X-2		Ble
0115+634		3.6s pulsar	B, Hα
0449-55	b^{II}=-40		(M dwarfs noted)
0535-668	LMC	16.7d(?)	B2Iab@
0535+262		104s pulsar	B0Ve
0620-003	Nova Mon '75	58 yr recurrence	Nova#, B~12, K5-7V/Radio flare
1118-615		405s	Be star@
1524-617	TrA X-1		Nova#, B~17
1543-475			
1608-522		Burster(?)	Nova#, I~18
1630-472		Recurs 600-700d	
1658-298		Burster	V=18.3, uv, λ4686,4640-50
1705-250	Nova Oph '77		Nova#, B=16.5, λ4686
1715-321		Fast 2-10m flares	
1742-289	Near gal. cen.	Burster?	Coinc. Radio flare
1743-322			
1743-288	GX+0.2-0.2	Burster?	
1744-361			
1745-203	NGC 6440		Glob. Cluster
1803-245			
1908+005	Aql X-1		Nova# to B=16.5, K0

* Includes most sources commonly called transients; others could easily qualify, e.g. Her X-1.
# Brightened coincident with x-ray increase.
@ See notes for these sources in Tables 4 and 5.

TABLE 11. 18 Other Identified Sources.

Source	Name	Optical/Radio
0053+604	γ Cas	Rotating Be
*0114+650		B0.5 IIIe, V=11, Broad Hα
0304+407	Algol	B8V+K+Am
*0532-054	Orion	Orion Trapezium
0614+091		B=18.8, λ4640-50
*0921-630		B~17, λ4686
*1052+607		Early KIV, Hα em.
*1249-289	EX Hydrae	Dwarf nova
*1254-690		blue, λ4640-50
1516-569	Cir X-1 (16.6d)	Hα, R~16, B~22, Radio, IR
*1543-624		uv, B~20
1617-155	Sco X-1	Variable, 0.78d, Radio
*1659-487	GX339-4 (fast var.)	V=16.6; λ4640-50,4686, (var.?)
*1755-338		uv, B~19, featureless spectrum
*1822-371		B~16, λ4686,4640-50
1957+115		V=18.7; Like Sco X-1
*2129+470		B~17, λ4640-50,4686, P=5.2h
*2140+443	SS Cyg	U Gem, White Dwarf

* Recent Identifications.

TABLE 12. 8 Other Optically Unidentified.

Source	Comment
0142+614	(Possible λ4650)
0918-549	
1417-624	
1553-542	In 1° radius CG327-0 γ-ray error circle
1556-605	
1624-490	Radio Candidate(?)
1822-000	No POSS stars in error region
1845-024	Slightly peculiar IR colors

TABLE 13. 9-15 Radio emitters.

Source	Name	Radio (1-10 GHz fluxes)
0304+407	Algol	Flares to ~1000 mJy
0531+219	Crab Pulsar	Pulsar
0532-054?	Orion	Orion nebula complex
0540-697?	LMC X-1	MC77 HII region
0620-003	Nova Mon '75	Coinc. flare to 300 mJy
1516-569	Cir X-1	Flares up to 4 Jy after 16.6d x-ray trans.
1617-155	Sco X-1	Flares to 260 mJy
1624-490?		"Weak"
1642-455?	GX340+0	Extended (10')
1742-289	Transient	Coinc. flare to 480 mJy
1744-265?	GX3+1	"Weak"
1758-250?	GX5-1	"Weak"
1813-140	GX17+2	Variable 0.8-3.4 mJy
1956+350	Cyg X-1	Variable to 23 mJy
2030+407	Cyg X-3	Flares to 21,000 mJy

? Radio may not be associated with x-ray source.

APPENDIX

A CATALOG OF COMPACT GALACTIC X-RAY SOURCES WITH PRECISE POSITIONS

COMMENTS

General. The catalog contains 96 sources. The criteria for inclusion (precise (<60") position, > 2keV flux, galactic or LMC/SMC membership) are given in Table 2. The information is largely a compendium of published material, but does include some previously unpublished information, indicated by ref. 80 (this work), e.g. positions of optical candiates accurate to ~ 3" and calculations of ratios of x-ray to optical luminosities.

Column 1: Source

Name: Seven-digit name is based on best known celestial position (Columns 3,4).

X-ray names and prefixes (catalogs): An attempt is made to order these in historical precedence. Names currently in use and the latest catalog prefix from a given group are given. Refs. for common prefixes are: 4U [250], 2A [162], 1M [567], MXB [512], and 2S [6,77,78, 205,207, 210, 211,410,411,412,415,517,700,736]. The latter are collected in a document distributed at the Austin AAS Meeting by H. Bradt. References are given for other names and for other 2S positions. We call attention to the catalog of spectral parameters for 43 Uhuru sources by Jones [436].

Common optical names: Listed here for convenience.

Column 2: Position Type and Reference. Type (e.g. optical) and reference for the position quoted in Columns (3-6). Stellar positions presented for the first time in this work [80] are usually derived from finding charts and astrometry published by the HEAO-A3 and SAS-3 groups (refs. under "F.C." in Column 9). Most of the x-ray positions (unidentified sources) are from SAS-3; the HEAO-3 positions for many of these unidentified sources are not yet published, although their present availability (e.g. for optical observers) is indicated in Column 8 as ref. 336 (i.e. Pos [336]). We also list here the first authors of the references cited for each source.

Columns 3,4: Celestial Position (epoch 1950). Positions of x-ray, optical, or radio object in conventional units and in decimal degrees.

Column 5: LII, BII. Galactic coordinates to nearest 1/10 degree.

Column 6: Error radius. Radius of position error circle (90% confidence) or equivalent radius, r_e, based on the area of the error region, $A = \pi r_e^2$.

Column 7: Flux density. Maximum and minimum flux densities quoted in the literature. Flux densities from the 4U [250], 2A [162] and 1M [567] catalogs are not individually referenced. For the 1M(OSO-7) catalog, we use the argon counter (3-10 keV) fluxes as presented in both the table and in the plots of intensity vs. time. We quote the flux density, averaged over 2-11 keV, that is, the integrated 2-11 keV flux given in the literature divided by the 9 keV band width. We make use of the following relations:

$$1.0\ \mu\text{Jy} = 0.242 \times 10^{-11}\ \text{erg cm}^{-2}\text{s}^{-1}\text{keV}^{-1}$$
$$= 1.51 \times 10^{-3}\ \text{keV cm}^{-2}\text{s}^{-1}\text{keV}^{-1}$$

For an incident Crablike spectrum:

S_{crab}	= 1060	μJy	(at 5.2 keV, or av. over 2-11 keV)
1.0 Uhuru ct s^{-1}	= 1.1	μJy	(" ")
1.0 OSO-7 Argon ct s^{-1}	= 17.6	μJy	(" ")
1.0 Ariel-5 SSI ct s^{-1}	= 2.5	μJy	(" ")

Column 8: X-ray characteristics.

Literature: References are intended to be comprehensive but not exhaustive, though most recent literature (through Aug. 1978) and some talks at the Sept. HEAD/AAS Meeting are included. The later works generally include references to relevant earlier work.

Question mark: The '?' is used liberally to denote the need (in our opinon) for confirmation. The marked items range from those which are quite probably correct to those which quite probably are not.

α: Spectral index for power law energy spectrum: $S \propto E^{-\alpha}$.

L(x): The maximum reported 2-11 keV x-ray luminosity for the distance quoted (usually in Column 9), unless otherwise noted. We use the FORTRAN notation ($3x10^4 \equiv 3E4$) to avoid exponent/reference confusion. Average x-ray flux densities are again quoted (integrated flux ÷ bandwidth).

1.0 μJy → 2.2×10^{-11} erg $cm^{-2}s^{-1}$

1.0 μJy → 2.6×10^{35} erg s^{-1} at 10 kpc (2-11 kev).

L(x)/L(o): The ratio of x-ray to optical luminosities from other literature (references cited) or from this work [80]. For the latter, we use the bolometric optical flux, calculated from V, A(v), and the B.C. [417]. If the spectral type (and hence B.C.) is not known, we calculate the 3000-7000 A flux corrected for interstellar extinction (from V, B-V, U-B , A(v)). The stellar type and photometric quantities are given in Column 9. If any of the latter quantities, e.g. B-V, A(v) are not known to us (i.e. missing in Column 9) they are set to zero. The comment "UV excess" is interpreted as U-B=-1.0. The maximum x-ray and optical fluxes (Columns 7,9) are systematically used unless otherwise indicated.

Pos.: References to published x-ray positions with error regions of area, $A < \pi(60'')^2$. In a few cases, so indicated, the only available position has a somewhat greater area.

Column 9: Optical Characteristics.

General: See first two items under Column 8, above.

Identification references: An attempt is made to describe the identification process with a series of references to the first entry (star name), e.g. (1) the x-ray position or pulse period used to locate the optical counterpart, (2) the first suggestion(s) of the correct counterpart, and (3) significant optical and x-ray confirmations.

"F.C.": References to finding charts are given for each source.

Spectral lines: Lines specified are emission lines.

Column 10: Radio Characteristics. See first two items under Column 8, above.

References. The reference list for the text and CGS catalog follows the catalog. It is arranged in alphabetical and numerical order. Numbers ending with "4" or "9" are deliberately omitted to faciliate late insertions. The topic of each reference is given in the left column.

Index of Source Names. A compact list of source designations and names (x-ray and optical) is provided to facilitate access to the Catalog, and follows the references.

SOURCE	*POSITION TYPE,REF.	RA (1950)	DEC (1950)	LII BII	EQUIVALENT ERROR RADIUS (",90%)	FLUX DENSITY (2-11 keV) MAX (μJy) MIN (μJy)
0042+327	opt[80]	00 42 06.7	32 44 05	121.3	3	55[713]
4U,2A,2S		10.5279	32.7347	-29.8		<0.5
	80 This work 121 Charles 330 Harris 563 Markert 618 Murray 700 Rappaport 713 Ricketts 845 Watson					
0050-727	opt[80]	00 50 19.5	-72 42 24	302.9	3	5[143]
SMC X-3[143]		12.5813	-72.7067	-44.7		≲1[146]
2S[143]	1 Allen 2 Allen 80 This work 143 Clark 146 Clark 173 Crampton 531 Lindsay 730 Sanduleak					
0052-739	opt[80]	00 52 53.0	-73 57 18	302.6	3	7[143]
SMC X-2[143]		13.2208	-73.9550	-43.4		≲1[146]
2S[143]		(Southern Component of Close Pair[612])				
	1 Allen 2 Allen 80 This work 143 Clark 146 Clark 173 Crampton 612 Morton 728 Sanduleak 818 van Paradijs					
0053+604	opt[732]	00 53 40.3	60 26 47	123.6	1	11[572]
2S,4U		13.4179	60.4464	-2.1		5
γ Cas	78 Bradt 80 This work 167 Cowley 205 Dower 392 Hutchings 407 Jernigan 568 Marlborough 572 Mason 603 Moffat 682 Purton 732 SAO Star Cat.					
0114+650	opt[80]	01 14 41.8	65 01 32	125.7	3	4
2S		18.6742	65.0256	2.6		
	80 This work 205 Dower 327 Hardorp 548 Margon					
0115-737	opt[143]	01 15 45.6	-73 42 22	300.4	3	57
SMC X-1[670,488]		18.9400	-73.7061	-43.6		2
4U,2A,1M Sk 160	15 Avni 25 Bahcall 80 This work 143 Clark 162 2A Cat. 196 Davison 250 4U Cat. 343 Henry 397 Hutchings 488 Leong 525 Liller 536 Lucke 655 Petro 670 Price 677 Primini 678 Primini 725 Sanduleak 740 Schreier 802 Tuohy 813 van Genderen 817 van Paradijs 848 Webster 878 Yentis					

*We also list here the first authors of the references cited for each source.

X-RAY	OPTICAL	RADIO	
Mid-latitude (bII=-30°) transient Feb.'77[713,700]; Variable 1971-73[563,618]; P=11.6d?, flat spectrum[845]; L(x)/L(o)=1.8E3 [80]; Pos.[700]	Star 3[700,121]; F.C.[121]; Weak, variable λ4686, V∼19.2, B-V=0.6, A(v)∼0.7[121]	≲20 mJy (610 MHz)[330]	0042+327
Transient in SMC, L(x)=7E37 erg/s[143]; L(x)/L(o)=0.5 [80]; Pos.[143]	Star 4[143,730,2]; F.C. (left hand chart)[143]; OB Star[730]; Emission lines[531]; V∼15, B-V=-0.3, U-B=-1.0 [2]; (Star 198 in[531]); Very weak λ4686, Be type spectrum, O9 III-V[173]; D∼63 kpc [1]		0050-727 SMC X-3
Transient in SMC, L(x)=1.0E38 erg/s[143]; L(x)/L(o)=3, for V=16, type B1 [80]; Pos.[143]	Star 5[143,728,2,612]; F.C. (right hand chart)[143]; OB Star[728]; Highly ultraviolet, V=14.8, B-V=-0.06, U-B=-1.1, variable?, absorption lines[818]; Very weak λ4686, Be type spectrum[173]; Doublet (2") star: southern fainter component is B1e with Hα, Hβ, and is more likely the counterpart, northern component possibly normal O7[612]; D∼63 kpc [1]		0052-739 SMC X-2
L(x)=3E33 erg/s [80]; Variable(?) [78,572]; L(x)/L(o)=6E-6 [80]; Pos [78,205]	γ Cas[407,78,572] =SAO011482; B0.5 (II-V)e, V=1.6-3.0, A(v)=0.16, M(v)∼-4.5, D∼300 pc[603,167]; Rapidly rotating, 0.7d(?)[392]; Coronal emission?[568]; Secondary undetected, <6Mo[167]	<10 mJy (2695, 8085 MHz)[682]	0053+604 γ Cas
L(x)/L(o)=1.5E-4 [80]; Pos[205]	Star 1[205,548] (=LS I +65° 010[327]); F.C.[205]; B0.5IIIe, V=11.0, B-V=1.2, U-B=0.1, A(v)=5.0, broad Hα emission[548]		0114+650
Eclipsing pulsating Doppler binary, Highly variable[250]; Extended lows[162]; P(orb)=3.89d[740,802]; Eclipse 0.6d[740]; P(spin)=0.71s[536,677,678]; F(x,M)=10.8Mo[677,678]; a(x)sin(i)=53.5±0.5 lt-s[678]; $\dot{P}$/P∼-6E-4/y[343,196,878,678]; L(x)=6E38 erg/s[143]; L(x)/L(o)=1.2 [80]; Pos[143]	Sk 160[488,740,848,525]; F.C.[725]; B0I, V=13.3, B-V=-0.14, U-B=-0.98, A(v)=0.1[848]; Ellipsoidal variations[525,655]; Member of SMC, D∼65 kpc[848]; M(opt)=16.2±1.4, M(x)=1.0Mo[397]; 5 color photometry[813,817]; Other refs. in refs. (15,397, 25)		0115-737 SMC X-1 Sk160

SOURCE	POSITION TYPE, REF.	RA (1950)	DEC (1950)	LII BII	EQUIVALENT ERROR RADIUS (",90%)	FLUX DENSITY (2-11 keV) MAX (μJy) MIN (μJy)
0115+634 4U, 2S[157], 1M	opt[422] 80 This work 157 Cominsky 251 Forman 386 Holt 416 Johns 422 Johnston 701 Rappaport 702 Rappaport 851 Wheaton	01 15 13.8 18.8075	63 28 38 63.4772	125.9 1.0	2	250[422] <2
0142+614 4U,2S,1M	x-ray[205] 205 Dower 550 Margon	01 42 55.1 25.7296	61 29 59 61.4997	129.4 -0.4	35	7 4
0304+407 2S Algol β Per	radio[147] 1 Allen 80 This work 147 Clark 257 Frieboes-Conde 278 Gibson 328 Harnden 351 Hill 361 Hjellming 362 Hjellming 736 Schnopper 796 Tomkin 827 Wade	03 04 54.34 46.22642	40 45 52.2 40.76450	149.0 -14.9	0.1	9
0352+309 4U,2A,2S X Per	opt[732] 32 Barbour 40 Becker 52 Bernacca 78 Bradt 80 This work 83 Braes 93 Brucato 100 Campisi 108 Canizares 246 Ferrari-Toniolo 273 2U Cat. 292 Gottlieb 332 Hawkins 336 HEAO A-3 gp. 393 Hutchings 395 Hutchings 545 Margon 547 Margon 573 Mason 603 Moffat 606 Moffett 607 Mook 616 Murdin 620 Mushotzky 648 Penston 732 SAO Star Cat. 808 van den Bergh 857 White 860 White	03 52 15.1 58.0629	30 54 01 30.9003	163.1 -17.1	1	37 11
0449-55 H[303]	x-ray[303] 162 2A Cat. 303 Griffiths 461 Kaluzienski	Multiple Positions		263 -39		140[461,303] ≲5[162]
0512-401 1M,4U,2A, MXB,2S NGC 1851	opt[415] 136 Clark 142 Clark 253 Forman 310 Grindlay 336 HEAO A-3 gp. 415 Jernigan 420 Johnson 652 Peterson 683 Pye 822 Vidal	05 12 27.7 78.1154 (Center of Globular Cluster)	-40 06 03 -40.1008	244.5 -35.0	5	20 1.2

X-RAY	OPTICAL	RADIO	
Transient pulsing Doppler binary; Flareups Jan.'71[251]; Jan.'78[157,386]; P=3.61s[157]; P(orb)=24.3d, ε=0.34, no eclipse, F(x,M)=5.0Mo, a(x)sin(i)=140 lt-s[702]; α=-0.4, L(x)=3E37 erg/s at 5kpc, Pulse shape like Cen X-3[422]; 20 keV cyclotron(?) absorption feature[851]; L(x)/L(o)=2 [80]; Pos.[157,422]	Star 1[157,422,416,701]; F.C.[157,422]; Hα emission, V=15.6, B-V=1.4, U-B=0.3, B type, A(v)$\gtrsim$5 mag.[416]; M(v)$\geq$-4 at D$\leq$7kpc[422]		0115+634
Pos.[205]	F.C.[205]; Suggestion of λ4650 emission, star 3, V$\sim$20[550]		0142+614
L(x)=2E31 erg/s[736]; ¼ keV emission[328]; L(x)/L(o)=5E-5 [80]; Pos.[736]	Algol=β Per[736]; B8V/$\sim$K0IV/A-met[257]; 2.9d, 1.9y[351]; D=32 pc, V=2.2, B-V=-0.1, M(v)=-0.3 [1]; Roche lobe of K star filled.[796]	Variable source[827,362,361]; Flares to 1.0 Jy at 8 GHz[278]; Thermal and non-thermal events[361]; Size$\sim$0.05 A.U.[147]	0304+407 Algol
Variable pulsator; P=835s[857]; 22h(?)[857], not detected[40]; 581d(?)[860]; 7-110 keV flux[620]; L(x)=1.2E34 erg/s, L(x)/L(o)=1E-4 [80]; Pos. [78,336]	X Per[273, 83,808, 93,332, 78]; F.C. [78]; Rapidly rotating $\sim$400 km/s [93]; O9.5(III-V)e, A(v)=1.8, M(v)$\sim$-3.6[603]; V=6.0-6.7[545,607]; D$\sim$350 pc [93]; P(spectros.)=581d(?)[393,395], not in B light[292]; 10-30s dips in U band, 0.1-0.3 mag.[52,100], and at λ4686(?)[108]; V=12, normal MIII companion at 22" [603,606,648]; Period searches[246,292,547]; Flare[616]; Interstellar medium in line of sight[573]; V band polarization variable [32]	<5 mJy at 1.4 GHz [83]	0352+309 X Per
Fast ($\sim$200 min) high galactic latitude transient, Feb.'78 to $\sim$140 μJy[303,461]; Multiple Pos.[303]	F.C.[303]; M dwarfs noted[303]		0449-55
Highly variable; L(x)=5E36[136]; Burster MXB0512-40[253,142]; Pos.[415,336]	NGC 1851 globular cluster[136,683] core[415]; F.C.[415]; UV-bright star(?)[822]; Hα from core[310]; Cluster parameters, r(core)=8", D=9.5 kpc[652]	Radio survey, no source in x-ray error region[420]	0512-401 NGC 1851 MXB

SOURCE	POSITION TYPE,REF.	RA (1950)	DEC (1950)	LII BII	EQUIVALENT ERROR RADIUS (",90%)	FLUX DENSITY (2-11 keV) MAX (μJy) MIN (μJy)
0521-720	x-ray[423]	05 21 16.5	-72 00 19	283.1	17	44
LMC X-2[488]		80.3188	-72.0053	-32.7		9
4U,2A,1M						

57	Bok	423	Johnston	488	Leong	686	Rapley	802	Tuohy
298	Griffiths	425	Johnston	562	Markert	737	Schnopper		

SOURCE	POSITION TYPE,REF.	RA (1950)	DEC (1950)	LII BII	EQUIVALENT ERROR RADIUS (",90%)	FLUX DENSITY (2-11 keV) MAX (μJy) MIN (μJy)
0531+219	radio[537]	05 31 31.43	21 58 54.40	184.6	0.06	∽100[71]
Crab Pulsar		82.88096	21.98178	-5.8		

5	Apparao	148	Cocke	350	Hewish	786	Taylor
11	Armstrong	171	Crab Nebula	471	Kestenbaum	797	Toor
50	Bennett	240	Fazio	537	Manchester	800	Toor
71	Bradt	261	Fritz	651	Peterson		
125	Cheng	346	Hermsen	761	Staelin		

SOURCE	POSITION TYPE,REF.	RA (1950)	DEC (1950)	LII BII	EQUIVALENT ERROR RADIUS (",90%)	FLUX DENSITY (2-11 keV) MAX (μJy) MIN (μJy)
0532-664	opt[80]	05 32 47.3	-66 24 13	276.3	2	60[223]
LMC X-4[273]		83.1971	-66.4036	-32.5		4
4U,2A						

57	Bok	488	Leong
80	This work	518	Li
128	Chevalier	562	Markert
223	Epstein	686	Rapley
273	2U Cat.	727	Sanduleak
298	Griffiths	737	Schnopper
357	Hiltner	862	White

SOURCE	POSITION TYPE,REF.	RA (1950)	DEC (1950)	LII BII	EQUIVALENT ERROR RADIUS (",90%)	FLUX DENSITY (2-11 keV) MAX (μJy) MIN (μJy)
0532-054	x-ray[81]	05 32 49.2	-05 25 08	209.0	35	5
4U,2A,2S[81]		83.2050	-5.4189	-19.4		(Total flux)
Orion						1.3[81]
Trapezium	opt[732]	05 32 49.0	-05 25 16		1	(Small comp.)
Θ^1 Ori(A-D)	(Θ^1C)	83.2042	-5.4211			

1	Allen	162	2A Cat.	250	4U Cat.	855	White
81	Bradt	199	den Boggende	732	SAO Star Cat.		

SOURCE	POSITION TYPE,REF.	RA (1950)	DEC (1950)	LII BII	EQUIVALENT ERROR RADIUS (",90%)	FLUX DENSITY (2-11 keV) MAX (μJy) MIN (μJy)
0535-668	opt[426]	05 35 46.4	-66 52 10	276.9	3	180[865]
A0538-66[865]		83.9433	-66.8694	-32.2		1[426]
LMC Transient						

57	Bok
80	This work
178	Dachs
426	Johnston
865	White

SOURCE	POSITION TYPE,REF.	RA (1950)	DEC (1950)	LII BII	EQUIVALENT ERROR RADIUS (",90%)	FLUX DENSITY (2-11 keV) MAX (μJy) MIN (μJy)
0535+262	opt[732]	05 35 48.0	26 17 18	181.5	1	2000[718]
A[718],2S,4U?		83.9500	26.2883	-2.6		3
HDE245770						

31	Baratta	247	Fishman	546	Margon	732	SAO Star Cat.
75	Bradt	277	Giangrande	615	Murdin	762	Stier
80	This work	352	Hiltner	693	Rappaport	790	Thomas
116	Carpenter	398	Hutchings	707	Ricker	830	Wade
123	Chartres	517	Li	710	Ricker		
145	Clark	526	Liller	711	Ricketts		
150	Coe	541	Maraschi	718	Rosenberg		

X-RAY	OPTICAL	RADIO	
L(x)=0.8-3.4E38 erg/s[488,562,686,802,298]; Pos.[737,425,423]	F.C.[425]; No candidate, M(v)>-2[425]; D=55 kpc[57]		0521-720 LMC X-2
Pulsar[261]; Other refs. in refs. (171,651,5,797,471,800); γ-ray refs. in refs. (240,346, 50)	Pulsar[148]; Other refs. in refs. (171,5,125)	Pulsar[761]; Other refs. in refs. (350,171,786,11)	0531+219 Crab
Eclipsing (0.2d) binary: P(orb)=1.4d[518,862]; L(x)=<0.2 to 5E38 erg/s[488,686,562,223,298]; ∽40 min high state (60 μJy) with 20s flares[223]; L(x)/L(o)=1 [80]; Pos.[223,737]	Star[223,727,357,128,518,862]; F.C.[223]: Brighter star 4.4 mm due east of inside edge of westernmost point of circle, also see[727]; V=14.0, B-V=-0.1, U-B=-1.1[357]; Type O8 III-V, Ellipsoidal variations, P(orb)=1.4d, λ4686 emission 475 km/s, v(primary)≲35 km/s, A(v)=0.3[128]; D=55 kpc [57]		0532-664 LMC X-4
L(x)≃1E33 erg/s with 2' and 4' collimators [81]; L(x)≃3E33 erg/s with ≳10' coll.[250,162,855,199]; Suggests two components: ≲2' and 6'-12' diameters [81]; Pos. [81]	Trapezium - all four stars in x-ray error region for ≲2' source [81]; F.C.[81]; D=460 pc [1]	Orion nebula complex	0532-054 Orion
Recurrent transient in direction of LMC, four outbursts June-Nov.'77[865,426] with probable 16.7d period[426]; kT=7 keV, June 29, L(x,2-17 keV)∽8E38 at LMC[865]; No persistant component, L(x)< 1E37[426]; L(x)/L(o)=4.8 [80]; Pos.[426], multiple positions eliminated if all four bursts from same source[426].	Star 19[426]; B2Iab[178]; F.C.[426,178] (Fig. 2 of[178]); V=12.8, B-V=-0.1, U-B=-0.9; A(v)≃0.06, probable LMC member, in NGC 2034[178]; D=55 kpc [57]		0535-668
Recurring, pulsating x-ray nova: Apr.'75 (to 2000 μJy[718]), May '77, Dec.'77, and Apr.'78[710,123,145]; P=104s[718,75]; Pulsing to 100 keV, kT=18 keV[707]; Spectra[711,150,707,541]; 50 keV profile[247]; P(orb)≳20d[693,517]; L(x)=2E37 erg/s, L(x)/L(o)=0.08 [80]; Pos.[718,116,517]	Star[718,526,615,546,517]; F.C.[546,517]; HDE245770= SAO077348; V=9.1, B-V=0.5[546], U-B=-0.5[762]; Bpe[352]; O9.7IIe[277]; B0Ve, M(v)∽-4[398]; Variable (yrs.), B=9.4-10.1, like X Per[762]; Search for periodicities, A(v)∽2.6, D∽1-3 kpc[546,830,398]; Obj. prism study [31]	<7 mJy at 15 GHz[790](Apr. '78)	0535+262 HDE245770

SOURCE	POSITION TYPE,REF.	RA (1950)	DEC (1950)	LII BII	EQUIVALENT ERROR RADIUS (",90%)	FLUX DENSITY (2-11 keV) MAX (μJy) MIN (μJy)
0538-641	opt[423]	05 38 39.7	-64 06 34	273.6	3	44[423]
LMC X-3[488]		84.6654	-64.1094	-32.1		<1.7[423]
4U,2A,1M						

57 Bok
80 This work
298 Griffiths
423 Johnston
425 Johnston
488 Leong
562 Markert
686 Rapley
737 Schnopper
802 Tuohy
840 Warren

SOURCE	POSITION TYPE,REF.	RA (1950)	DEC (1950)	LII BII	EQUIVALENT ERROR RADIUS (",90%)	FLUX DENSITY (2-11 keV) MAX (μJy) MIN (μJy)
0540-697	opt[423]	05 40 06.3	-69 46 04	280.2	3	25
LMC X-1[488]		85.0263	-69.7678	-31.5		9[423]
4U,2A,1M						

57 Bok
80 This work
98 Byrne
216 Dufour
225 Epstein
241 Feast
276 3U Cat.
298 Griffiths
402 Isserstedt
423 Johnston
425 Johnston
431 Jones
488 Leong
562 Markert
686 Rapley
737 Schnopper
802 Tuohy
840 Warren

SOURCE	POSITION TYPE,REF.	RA (1950)	DEC (1950)	LII BII	EQUIVALENT ERROR RADIUS (",90%)	FLUX DENSITY (2-11 keV) MAX (μJy) MIN (μJy)
0544-665	x-ray[423]	05 44 11.5	-66 35 24	276.5	60	1.8[423]
H[423],LMC?		86.0479	-66.5900	-31.4		

423 Johnston

SOURCE	POSITION TYPE,REF.	RA (1950)	DEC (1950)	LII BII	EQUIVALENT ERROR RADIUS (",90%)	FLUX DENSITY (2-11 keV) MAX (μJy) MIN (μJy)
0614+091	opt[80]	06 14 22.3	09 09 25	200.9	3	134
4U,2S,1M		93.5930	9.1570	-3.4		<18

80 This work
182 Davidsen
205 Dower
336 HEAO A-3 gp.
613 Murdin
772 Swank
867 Willmore

SOURCE	POSITION TYPE,REF.	RA (1950)	DEC (1950)	LII BII	EQUIVALENT ERROR RADIUS (",90%)	FLUX DENSITY (2-11 keV) MAX (μJy) MIN (μJy)
0620-003	opt[221]	06 20 11.1	-00 19 11	210.0	1	~50,000[207]
A[222],2S		95.0463	-0.3197	-6.5		≲5[250]
Nova Mon '75						
V616 Mon						

53 Bieging
60 Boley
80 This work
115 Carpenter
132 Citterio
151 Coe
188 Davis
207 Doxsey
221 Eachus
222 Elvis
250 4U Cat.
256 French
297 Griffiths
320 Gull
456 Kaluzienski
472 Kestenbaum
477 Kleinmann
532 Little
533 Lloyd
535 Locher
583 Matilsky
635 Oke
636 Oke
638 Owen
712 Ricketts
810 van den Bergh
852 Whelan
877 Wu

X-RAY	OPTICAL	RADIO
L(x)<0.1-3E38 erg/s[488,686,423,802,562,298]; L(x)/L(o)=105[80]; Pos.[686,737,425,423]	Star 8 (=WP)[686,840,425,423]; F.C.[840,425]; BIII-IV[840,425]; V=16.9, B-V=-0.06, U-B=-0.66, variable 0.1 mag.[840]; No emission lines, LMC member[425]; D=55 kpc [57]	0538-641 LMC X-3
L(x)=0.7-2E38 erg/s[488,686,423,802,562,298]; No evidence for finite size[425]; L(x)/L(o)=0.22 [80]; Pos.[737,225,425,423]	Star R148[276,431,686,425,423] =CPD-69°474; F.C.[425]; B5I, A(v)=0.96[241]; V=12.0, B-V=+0.24, U-B=-0.57[431,402,216,840]; M(v)=-7.6 at 55 kpc[425]; Lies in HII region, N159F, and in IR error region, other early type stars in error region (see[425]); D=55 kpc [57]	0540-697 LMC X-1 In region of MC77, possibly an extended HII region [98] - see discussion in ref. (686).
In direction of LMC, L(x)=1.5E37 at LMC[423];	F.C.[423]	0544-665
L(x)=1.3E37 erg/s at 6 kpc, L(x)/L(o)=1300 [80]; Lies in ∿80 sq. deg. error region of a single burst[772]; Pos.[867,205,336]	Star D[867,613,182,205]; F.C.[205]; Variable(?)[613]; Similar to Sco X-1[613,182]; λ4640-50, B=18.8, B-V=0.3, A(v)=1.0, D∿4-8 kpc[182]	0614+091
Bright x-ray nova, Aug.'75, rose ∿8d[222] while kT=30 to 1.5 keV[712]; S(max)∿20-60x Crab[222,712,207,456]; Decay τ∿25d[583,456]; Pulsing <2% 0.2ms-0.25s[207]; Pulsing <0.3% 1-1000s[207,115]; Transient 8d feature[583]; L(x)∿3E38 erg/s at peak [80] and E(total)∿1E45 erg[583] at 1.5 kpc; Nuclear burning outburst untenable[583]; Spectral line[297,472] and polarization[297] limits; Flux >30 keV[151]; L(x)/L(o)=85 [80]; Pos.[207]	Nova Mon 1975[207, 60]=V616 Mon; F.C. [60]; B=20.2[221], to V∿10.4 [60,535]; UBV[256,810]; Decay 0.015 mag/day[533]; Bright state spectra[320,635]; D=1.5-2.5 kpc, L(opt)∿5E35 erg/s[320]; IR Fluxes[477,635,132]; UV Fluxes[877]; A(v)=1.3-3.0[877,320]; Spectra 9/75 to 5/76[852]; Quiet state Nov.'76: K5-7V, V=18.3[636]; D∿1 kpc if star normal[636,852]; Similar outburst in 1917[221]	0620-003 Nova Mon '75 Coincident non-thermal radio flare 0.15-4.6 GHz[638,188, 53,532]; decayed 300 to 55 mJy in 7d at 1.4 GHz[638]

SOURCE	POSITION TYPE,REF.	RA (1950)	DEC (1950)	LII BII	EQUIVALENT ERROR RADIUS (",90%)	FLUX DENSITY (2-11 keV) MAX (μJy) MIN (μJy)
0900-403	opt[732]	09 00 13.2	-40 21 25	263.1	1	280
Vela X-1[130]		135.0550	-40.3569	3.9		<28
4U,1M						
HD77581						

23 Bahcall
25 Bahcall
38 Becker
63 Bolton
80 This work
130 Chodil
248 Forman
250 4U Cat.
273 2U Cat.
356 Hiltner
430 Jones
588 McClintock
611 Morgan
633 Ogelman
692 Rappaport
732 SAO Star Cat.
805 Ulmer
816 van Genderen
843 Watson
880 Zuiderwijk

0918-549	x-ray[205]	09 18 55.6	-54 59 33	275.9	40	13
4U,2S,1M		139.7317	-54.9925	-3.8		6

205 Dower

0921-630	opt[80]	09 21 25.0	-63 04 46	281.8	3	2.3[521]
2S[521]		140.3542	-63.0794	-9.3		

80 This work
521 Li

1052+607	opt[732]	10 52 36.4	60 44 11	145.4	1	3[743]
2A		163.1517	60.7364	51.3		0.8[162]

80 This work
162 2A Cat.
176 Crampton
530 Liller
665 Pounds
732 SAO Star Cat.
743 Schwartz
837 Wallerstein

1118-615	x-ray[227]	11 18 59	-61 35 18	292.5	108	70[227,403]
A[227]		169.7458	-61.5883	-0.8		<5[250]
Transient		(Systematic error: source can be up to 4'to S.W.)				

80 This work
127 Chevalier
192 Davison
227 Eyles
237 Fabian
250 4U Cat.
403 Ives

1119-603	opt[80]	11 19 01.9	-60 20 57	292.1	3	224
Cen X-3[130]		169.7579	-60.3492	0.3		<21
4U,1M						

23 Bahcall
25 Bahcall
47 Bennett
80 This work
95 Brucato
130 Chodil
153 Coe
230 Fabbiano
250 4U Cat.
272 Giacconi
276 3U Cat.
387 Holt
437 Joss
478 Krzeminski
637 Osmer
641 Parkinson
663 Pounds
696 Rappaport
738 Schreier
741 Schreier
803 Tuohy
806 Ulmer

X-RAY	OPTICAL	RADIO	
Pulsating eclipsing Doppler binary; P(orb)=8.97d, eclipse 1.7d[805,248,843]; P=283s[588]; F(x,M)=18.5Mo, a(x)sin(i)=111 lt-s, ecc.= 0.13±0.04[692]; Spin down episode[633]; Cycle to cycle variability[843]; 2-60 keV spectra, 6.8 keV line feature, residual (2%) flux during eclipse [38]; L(x)=1.4E36 erg/s, L(x)/L(o)=3E-3 [80]; Pos.[250]	HD77581[273,805,356,248] =SA0220767; F.C.[248]; B0.5Ib[611]; V~6.9, B-V=+0.5, U-B=-0.5[430]; Spectroscopic[356] and ellipsoidal[430] binary; V and B[816], and 4 color[836] photometry; A(v)=2.2, D~1.4 kpc[356, 23, 63]; Other refs. in refs. (23,25)		0900-403 Vela X-1 [redacted]
Pos.[205]	F.C.[205]		0918-549
L(x)/L(o)=15[80]; Pos.[521]	Star 5[521]; F.C.[521]; $H\alpha$, $H\beta$, λ4686, B~17, similar to Sco X-1[521]		0921-630
Faint high latitude source, listed in 2A preprint only; Confirmed with HEAO-1[743] and Ariel V[665]; L(x)=1E32 erg/s, L(x)/L(o)=7E-3 [80]; Pos.[743]	Star[743,530]=SA0015338; F.C.[743]; Early K with $H\alpha$[530]; Subgiant[837]; Spectrum complex, like RS CVn system[176]; V=8.8[732], D~130 pc[743]		1052+607
Pulsating transient to ~70 μJy, Dec.'74[227,403]; P=405s, hard spectrum[403]; L(x)/L(o)=2 [80]; Pos. (uncertain by 2'-6')[227]	Be star ~4' SW of, but consistant with, x-ray position[127]; No F.C., see[127]; V=12.1, B-V=+0.9, U-B=-0.4, variable, intense UV continuum; $H\beta$, $H\gamma$, HeI, FeII, resembles X Per[127]; RS Cen, a Mira variable, suggested[237] (but see[192])		1118-615
Pulsating Doppler eclipsing binary; P=4.84s[272]; P(orb)=2.087d, F(x,M)=15.5Mo, a(x)sin(i)=39.7 lt-s[738]; ecc.= 0.0008±0.0001[230]; Eclipse 0.45d[738,663]; Pulse shapes[806,803]; Long term behavior[741,47]; Spin down episode[230]; 26-100 keV spectrum[153]; P=43.0d(?)[387]; L(x)=4E37 [80]; L(x)/L(o)=0.05 for B.C.=-3.0 [80]; Pos.[250,641]; Other refs. in refs. (230,696, 437,23,25)	Krzeminski's star[276,738,478]; F.C. [95] (34" S of star 14); O6.5 II-III[637, 23]; V~13.35, B-V=+1.1, U-B=-0.04, A(v)~4.4, D~8 kpc, ellipsoidal binary[478]		1119-603 Cen X-3

SOURCE	POSITION TYPE, REF.	RA (1950)	DEC (1950)	LII BII	EQUIVALENT ERROR RADIUS (",90%)	FLUX DENSITY (2-11 keV) MAX (μJy) MIN (μJy)
1145-619	opt[732]	11 45 33.6	-61 55 44	295.6	1	1000[413]
4U,2S,1M		176.3900	-61.9289	-0.2		7
HEN 715						

78 Bradt
80 This work
205 Dower
242 Feast
276 3U Cat.
413 Jernigan
431 Jones
463 Keenan
540 Maraschi
732 SAO Star Cat.
757 Sofia
863 White

SOURCE	POSITION TYPE, REF.	RA (1950)	DEC (1950)	LII BII	EQUIVALENT ERROR RADIUS (",90%)	FLUX DENSITY (2-11 keV) MAX (μJy) MIN (μJy)
1223-624	opt[80]	12 23 49.7	-62 29 37	300.1	3	100
GX301-2[493]		185.9571	-62.4936	-0.0		9
4U,2S,1M						
WRA 977						

78 Bradt
80 This work
153 Coe
205 Dower
276 3U Cat.
323 Hammerschlag-Hensberge
431 Jones
465 Kelley
493 Lewin
580 Mason
587 McClintock
656 Petro
706 Ricker
708 Ricker
766 Swank
815 van Genderen
823 Vidal
858 White

SOURCE	POSITION TYPE, REF.	RA (1950)	DEC (1950)	LII BII	EQUIVALENT ERROR RADIUS (",90%)	FLUX DENSITY (2-11 keV) MAX (μJy) MIN (μJy)
1239-599	x-ray[205]	12 39 07.5	-59 55 39	301.8	30	10[205]
A[116],2S		189.7813	-59.9275	2.6		

116 Carpenter
205 Dower
388 Huckle

SOURCE	POSITION TYPE, REF.	RA (1950)	DEC (1950)	LII BII	EQUIVALENT ERROR RADIUS (",90%)	FLUX DENSITY (2-11 keV) MAX (μJy) MIN (μJy)
1249-289	opt[335]	12 49 42.9	-28 58 41	303.2	3	5
4U,2A,2AR[846]		192.4288	-28.9781	33.6		
EX Hydrae						

80 This work
162 2A Cat.
250 4U Cat.
335 HEAO A-3 gp.
621 Mushotzky
716 Robinson
846 Watson

SOURCE	POSITION TYPE, REF.	RA (1950)	DEC (1950)	LII BII	EQUIVALENT ERROR RADIUS (",90%)	FLUX DENSITY (2-11 keV) MAX (μJy) MIN (μJy)
1254-690	opt[295]	12 54 20.9	-69 01 07	303.5	4	51
4U,2S,1M		193.5871	-69.0186	-6.4		6[295]

80 This work
153 Coe
205 Dower
295 Griffiths

SOURCE	POSITION TYPE, REF.	RA (1950)	DEC (1950)	LII BII	EQUIVALENT ERROR RADIUS (",90%)	FLUX DENSITY (2-11 keV) MAX (μJy) MIN (μJy)
1258-613	opt[576]	12 58 11.8	-61 19 58	304.1	2	200
GX304-1[587]		194.5492	-61.3328	1.2		6
4U,2S,1M						

1 Allen
78 Bradt
80 This work
205 Dower
276 3U Cat.
388 Huckle
490 Lewin
576 Mason
587 McClintock
591 McClintock
706 Ricker

SOURCE	POSITION TYPE, REF.	RA (1950)	DEC (1950)	LII BII	EQUIVALENT ERROR RADIUS (",90%)	FLUX DENSITY (2-11 keV) MAX (μJy) MIN (μJy)
1417-624	x-ray[7]	14 17 26.2	-62 28 14	313.0	30	22[7]
MX[566],4U,		214.3592	-62.4707	-1.6		8
1M,2S[7]						

7 Apparao
566 Markert

X-RAY	OPTICAL	RADIO
Two(?) periods: 292s and 297s[863]; Brightened to 1000 μJy May '78[413] (S(4U)=80 μJy); L(x)=6E36 erg/s, L(x)/L(o)=0.2 [80]; Pos. [78,205]	HEN 715[276,757,431,540, 78,205] =SA0251595=HD102567; F.C.[78]; D~1.5 kpc [78,863]; V=9.0, B-V=0.2, U-B=-0.9, ΔB~0.02[431]; B1Vne[242]; A(v)~1.6[463, 80]; Variable Hα (~25 min), ~400 km/s (FWHM)[863]	1145-619 HEN 715
Hard and highly variable, (20-40 keV)[493,587,706,708] pulsator (P=699s)[858]; Bright active state to ~1000 μJy and 10s bright flare[465]; L(x-max)~1E37 erg/s at 2 kpc, Strong Fe abs. edge[766]; 26-100 keV spectrum[153]; P=40.8d(?)[580]; L(x)/L(o)=3E-3 [80]; Pos [78,205]	WRA 977[276,823,431, 78,205]; F.C.[823,78]; V=10.8, B-V=1.6, U-B=+0.5, A(v)~5.4, B1.5Ia, D~2.0 kpc[823]; P(orb)=23d (tidal)[323,656], but also see ref. (815)	1223-624 GX301-2 WRA 977
P=191s(?)[388]; Hard spectrum, α=0.8[388]; Pos.[205]	F.C.[205]	1239-599
Dwarf nova system; Variable emission (2-18 keV)[846]; Possible Fe line emission[621] (discussed in[846]); Previous suggested identification: Cluster SC 1251-28[250,162]; L(x)=1.3E32, L(x)/L(o)=2 at opt. min. [80]; Pos.[335]	EX Hydrae[846,335]; F.C.[846]; Dwarf nova, U Gem(?), eclipsing binary (P=99 min.), outbursts: mean period ~465d, duration ~4d, V=13.7-11.5, D(typical)~100pc (see refs. in[716,846])	1249-289 EX Hydrae
26-100 keV spectrum, α=0.6[153]; L(x)/L(o)=2500 [80]; Pos.[205,295]	Star 30[295]; F.C.[205,295]; Blue, λλ4640-50, V=19.1, B=19.2[295]	1254-690
Hard and variable[587,706,490]; Pulsator (P=272s)[591,388]; L(x)=2.1E36 erg/s, L(x)/L(o)=0.3 [80]; Pos. [78,205]	Star 2=MMV[276,576, 78,205]; F.C.[205,576]; Double peaked (300 km/s sep.) Hα, extended atmosphere?, like X Per, B6-B9pe (apparent), B0-B5 V (intrinsic), V=14.7, B-V=1.8, A(v)=6.3, D=2±1 kpc[576]; behind Coalsack (170 pc[1]).	1258-613 GX304-1
Pos.[7]	F.C.[7]	1417-624

SOURCE	POSITION TYPE,REF.	RA (1950)	DEC (1950)	LII BII	EQUIVALENT ERROR RADIUS (",90%)	FLUX DENSITY (2-11 keV) MAX (μJy) MIN (μJy)
1516-569	opt[853]	15 16 48.5	-56 59 13	322.1	1	3000[203,460]
Cir X-1[542]		229.2021	-56.9869	0.0		<5[748]
4U,2S,1M						

55 Bignami
77 Bradt
96 Buff
102 Canizares
133 Clark
152 Coe
203 Dower
250 4U Cat.
255 Forman
285 Glass
290 Goss
322 Gursky
333 Haynes
432 Jones
446 Kaluzienski
460 Kaluzienski
542 Margon
748 Seward
760 Spada
798 Toor
853 Whelan
868 Wilson
870 Wilson

1524-617	opt[80]	15 24 05.8	-61 42 35	320.3	3	225[748]
TrA X-1[662]		231.0242	-61.7097	-4.4		10[748]
A,2S						

77 Bradt
80 This work
406 Jernigan
445 Kaluzienski
541 Maraschi
617 Murdin
662 Pounds
748 Seward

1538-522	opt[80]	15 38 38.6	-52 13 36	327.4	3	30
4U,2S,1M		234.6608	-52.2267	2.1		8[748]

6 Apparao
36 Becker
80 This work
168 Cowley
175 Crampton
195 Davison
197 Davison
276 3U Cat.
640 Parkes
742 Schwartz
748 Seward
811 van den Bergh
820 van Paradijs
872 Wilson

1543-624	opt[80]	15 43 33.7	-62 24 52	321.8	3	64
4U,2S,1M		235.8904	-62.4144	-6.3		21

6 Apparao
80 This work
112 Canizares
228 Eyles
301 Griffiths

1543-475	x-ray[250]	15 43 50.0	-47 33 36	330.9	50	2240
4U,1M		235.9583	-47.5600	5.4		<1.2[748]

45 Belian
250 4U Cat.
513 Li
582 Matilsky
748 Seward

1553-542	x-ray[6]	15 53 55.6	-54 16 15	328.0	35	27[6]
2S		238.4817	-54.2708	-0.9		

6 Apparao
346 Hermsen

1556-605	x-ray[6]	15 56 47.2	-60 35 47	324.1	30	40
4U,2S,1M		239.1967	-60.5964	-5.9		11

6 Apparao
336 HEAO A-3 gp.

X-RAY	OPTICAL	RADIO	
Black hole candidate[255,760,798]; Variable 0.1s to months[542,432,255,760,96,868]; ~8ms burst[798]; Av. spectrum steeper than typical binary[102]; No periods 0.25ms-400s (<1-5%)[203]; Periodic (16.6d) transitions on→off[446,55,203]; Residual flux and fast flares after transition[203]; 25-200 keV spectrum[152]; Brightening 14h after transition[460]; Ephemeris of transition[203]: T=JD 2443291.92+16.62n; L(x)=2E38[203]; L(x)/L(o)≃5[853]; Pos.[250,870,77,322]	Star L[432,133,870,333,853,77]; F.C.[853]; B~22.5, R~16, Strong variable Hα emission ~1000 km/s, A(v)≃11, M(B)≃-4.5 at 10 kpc, OI(?)[853]; IR increases after x-ray transition with 10d decay: 5 Sept.'77 increased 1.5 mag. to K=7.3 in 1h, 0.5d after trans.[285]; Note radio distance	Source near SNR G321.9-0.3[133]; Variable 0.3 to 4.1 Jy in 29h at 5 GHz[333]; Position ≤2" from star, outbursts 12h-2d after transition, non-thermal[853]; D≃8-10 kpc[290]	1516-569 Cir X-1
X-ray nova, Nov.'74[662]; Rise ~19d to ~1000 μJy with precursor, decay τ~52d[445,617]; Soft spectrum[445,541]; ~30d preflare on-state (∿100 μJy)[445]; L(x)/L(o)=2400 [80]; Pos.[617,77]	Coincident optical nova, star N[406,617,77]; F.C.[617]; B>19 to 17.5, Decay τ~117d[617]		1524-617 TrA X-1
Pulsating, eclipsing Doppler binary, P=529s[195,36]; P(orb)=3.7d [36,197]; Ṗ/P<9.7E-4/y, F(x,M)= 13±2.6Mo[197]; L(x)=4E36 erg/s, L(x)/L(o)=0.01 [80]; Pos.[872,6,742]	Star 12[276,872,168,6,742,640,820,175]; F.C. [6,742]; B0I, A(v)~7, Hα, HeI(?)[640,175]; λ4686, K(opt)=33±7 km/s, M(opt)=20±4Mo, M(x)=2.0±0.5Mo, D=7±2 kpc[175]; V~14.5, B-V~1.8[811]		1538-522
L(x)/L(o)=5100 [80]; Pos. [6,301,228]	Star 6 [6,112,301]; F.C. [6]; UV excess, B~20[112]		1543-624
Bright transient; July '71, to ~3000 μJy[582,45]; Detectable until Sept.'72[513]; Pos.[250]	F.C.[582]		1543-475
In 1° gamma ray error circle CG 327-0[346]; Pos. [6]	F.C. [6]		1553-542
Pos [6,336]	F.C. [6]		1556-605

SOURCE	POSITION TYPE,REF.	RA (1950)	DEC (1950)	LII BII	EQUIVALENT ERROR RADIUS (",90%)	FLUX DENSITY (2-11 keV) MAX (μJy) MIN (μJy)
1608-522 MX[566],1M, 4U,2S,XB?	opt[312]	16 08 52.2 242.2175	-52 17 43 -52.2953	330.9 -0.9	1	1100[141] <1.0[748]

6 Apparao
46 Belian
80 This work
141 Clark
217 Duldig
231 Fabbiano
308 Grindlay
312 Grindlay
443 Kaluzienski
447 Kaluzienski
451 Kaluzienski
453 Kaluzienski
566 Markert
748 Seward
781 Tananbaum

SOURCE	POSITION TYPE,REF.	RA (1950)	DEC (1950)	LII BII	EQUIVALENT ERROR RADIUS (",90%)	FLUX DENSITY (2-11 keV) MAX (μJy) MIN (μJy)
1617-155 Sco X-1[267] 4U,2A,1M V818 Sco	opt[756]	16 17 04.49 244.26871	-15 31 14.8 -15.52078	359.1 23.8	0.1	19000 6900

23 Bahcall
68 Bord
73 Bradt
80 This work
103 Canizares
166 Cowley
250 4U Cat.
267 Giacconi
291 Gottlieb
296 Griffiths
321 Gursky
353 Hiltner
355 Hiltner
358 Hjellming
377 Holt
378 Holt
390 Hudson
483 Laros
491 Lewin
602 Miyamoto
608 Mook
626 Neugebauer
723 Sandage
756 Sofia
826 Wade
856 White

SOURCE	POSITION TYPE,REF.	RA (1950)	DEC (1950)	LII BII	EQUIVALENT ERROR RADIUS (",90%)	FLUX DENSITY (2-11 keV) MAX (μJy) MIN (μJy)
1624-490 4U,2S,1M	x-ray[6]	16 24 19.6 246.0817	-49 05 07 -49.0853	334.9 -0.3	35	88 20[748]

6 Apparao
336 HEAO A-3 gp.
726 Sanduleak
748 Seward
871 Wilson

SOURCE	POSITION TYPE,REF.	RA (1950)	DEC (1950)	LII BII	EQUIVALENT ERROR RADIUS (",90%)	FLUX DENSITY (2-11 keV) MAX (μJy) MIN (μJy)
1627-673 4U,2S,1M 2A	opt[590]	16 27 14.2 246.8091	-67 21 16 -67.3544	321.8 -13.1	3	30 10

77 Bradt
80 This work
315 Grindlay
336 HEAO A-3 gp.
401 Ilovaisky
440 Joss
590 McClintock
595 McClintock
695 Rappaport

SOURCE	POSITION TYPE,REF.	RA (1950)	DEC (1950)	LII BII	EQUIVALENT ERROR RADIUS (",90%)	FLUX DENSITY (2-11 keV) MAX (μJy) MIN (μJy)
1630-472 4U,1M	x-ray[871]	16 30 16.5 247.5688	-47 16 51 -47.2808	336.9 0.3	28	1400[455] <2[751]

251 Forman
313 Grindlay
317 Grindlay
435 Jones
436 Jones
448 Kaluzienski
455 Kaluzienski
458 Kaluzienski
751 Share
755 Sims
871 Wilson

X-RAY	OPTICAL	RADIO	
"Steady" source and recurrent nova, to ∽1100 μJy, Nov.'75[443] and July '77[141,447,451]; Variable and flaring[781,451]; Decay 500 to 100 μJy 13 Oct. - 24 Oct. '77[453]; Burster(?)[46,308,781]; L(x)/L(o)=180 for zero intrinsic colors[80]; Pos.[6,231]	Opt. nova[6,312,231]; F.C.[312]; I>20.0 Aug.'76, I=18.2 Aug.'77, A(v)=4.5-7.9[312]	<16 mJy at 15 GHz during post-outburst high state 20 Aug. '77[217]	1608-522
Brightest "steady" source[267,250]; X-ray flares (>20 keV)[491,390]; Highly variable state, to 19,000 μJy when optically bright, relatively steady (∽6,900 μJy) when opt. dim[608,73,103,856]; Search for 0.78d period[377]; Shot noise parameters[378]; High resolution[296,483] and wide band[602] spectra; L(x)=2E37 erg/s at 0.7 kpc, L(x)/L(o)=600[80]; Pos.[321]; other refs. in refs. (355,73)	V818 Sco[321,723]; F.C.[723]; V=12.2 to 13.3, B-V∽0.2, U-B≃-0.8, A(v)∽0.7, variable emission lines, flickering 0.02 mag. in minutes[723]; Bright (B∽12.4) flaring state (∽0.2 mag.; 2-10 min.) and only slower variations (hrs.) at B≥13.0[353,608,73,103]; P=0.787d[291,166,68]; D=0.3 to 1.0 kpc[355]; IR indicates self absorption[626]; Other refs. in refs. (355,23)	Radio flaring to 260 mJy at 2.7 GHz[358,826]	1617-155 Sco X-1
Pos.[871,6,336]	F.C.[6]	"Weak" source[726] (≲1 Jy at 2.7 GHz) G334.9-0.3	1624-490
7.68s pulsator, hard spectrum and variable, τ∽100-300s, P(orb) <0.5d or >35d[695]; Ṗ/P=-1.85E-4/y, transient 1000s oscillations[440]; L(x)/L(o)=600[80]; Pos.[77,336]	Star 4[77,590]; F.C.[77]; UV excess, V=18.5, B-V=0.1, U-B=-1.2[590]; P=7.68s, ≃4% full amplitude[401,595]; Simultaneous x-ray and opt. 5-10 min. flares[595]; Photometry[315]		1627-673
Recurrent transient[251,435]; 4 quasi periodic outbursts up to≲1400μJy(P∽615d)[435]; Late[448] outburst[455,751,458,755]. Decay τ∽45d[458]; α∽4.0[436]; Pos.[871]	No publ. F.C.; Previous candidate[313] lacks emission lines after sky subtraction[317]		1630-472

SOURCE	POSITION TYPE,REF.	RA (1950)	DEC (1950)	LII BII	EQUIVALENT ERROR RADIUS (",90%)	FLUX DENSITY (2-11 keV) MAX (μJy) MIN (μJy)
1636-536	opt[590]	16 36 56.2	-53 39 15	332.9	3	390
4U,MXB,		249.2342	-53.6542	-4.8		140
2S,1M						

80	This work	590	McClintock
111	Canizares	768	Swank
250	4U Cat.	821	van Paradijs
322	Gursky	867	Willmore
373	Hoffman	871	Wilson
410	Jernigan		

SOURCE	POSITION TYPE,REF.	RA (1950)	DEC (1950)	LII BII	EQUIVALENT ERROR RADIUS (",90%)	FLUX DENSITY (2-11 keV) MAX (μJy) MIN (μJy)
1642-455	x-ray[6,322]	16 42 10.1	-45 31 11	339.6	18	640
GX340+0[72]		250.5421	-45.5197	-0.1		170
4U,2S,1M		(Joint SAS-3/HEAO-A3 Position)				

6	Apparao	690	Rappaport
72	Bradt	726	Sanduleak
250	4U Cat.	745	Seaquist
255	Forman	871	Wilson
322	Gursky		

SOURCE	POSITION TYPE,REF.	RA (1950)	DEC (1950)	LII BII	EQUIVALENT ERROR RADIUS (",90%)	FLUX DENSITY (2-11 keV) MAX (μJy) MIN (μJy)
1653-407	opt[732]	16 53 06.9	-40 44 44	344.5	1	42[661]
OAO[661]		253.2788	-40.7456	1.5		$\lesssim$5[250]
V861 Sco						
HD152667						

80	This work
250	4U Cat.
581	Massey
661	Polidan
732	SAO Star Cat.
733	Schild
832	Walker
833	Walker

SOURCE	POSITION TYPE,REF.	RA (1950)	DEC (1950)	LII BII	EQUIVALENT ERROR RADIUS (",90%)	FLUX DENSITY (2-11 keV) MAX (μJy) MIN (μJy)
1656+354	opt[265]	16 56 01.68	35 25 04.8	56.9	0.1	160
Her X-1[778]		245.00700	35.41800	44.8		11
4U,2A,1M						
HZ Her						

16	Avni	262	Fritz
17	Bahcall	265	Gatewood
20	Bahcall	266	Gerend
23	Bahcall	275	Giacconi
25	Bahcall	318	Groth
35	Becker	383	Holt
56	Blumenthal	396	Hutchings
67	Bopp	433	Jones
80	This work	437	Joss
117	Catura	438	Joss
126	Chester	442	Joss
135	Clark	470	Kendziorra
155	Coe	556	Margon
160	Cooke	625	Nelson
172	Crampton	653	Petro
181	Davidsen	658	Petterson
193	Davison	666	Pravdo
206	Doxsey	667	Pravdo
220	Dupree	668	Pravdo
235	Fabian	752	Shulman
243	Fechner	778	Tananbaum
250	4U Cat.	801	Trumper

X-RAY	OPTICAL	RADIO	
Bright bulge source; Burster[768,373]; Bursts: black body spectrum (cooling 28E6-12E6 °K in 50s), intervals 9-12h[373]; ∼10 km radius[373,821]; Steady source variable, kT∼6 keV[373]; L(x)/L(o)=5500 [80]; Pos.[250,867,871,410,322]	Star 3[410,590]; F.C.[410]; λ4686[111]; UV excess, V=17.5, B-V=+0.7, U-B=-0.7[590]		1636-536 MXB
Bright bulge source; Variable τ∼1.5d, factor of 2[255]; Pos.[250,690,871,6,322]	F.C. [6]	G339.6-0.1(?)[726]; Extended source (∼10'), HII region, 3.5 Jy at 5.0 GHz, H109α, D≃2 or 17 kpc[745]	1642-455 GX340+0
Source in 2.5° x 3.5° FWHM FOV, 2 eclipses consistant with P(orb)=7.8d, eclipse duration >1.3d; No (<15%) periods 3-40 min., kT≃11 keV[661]; L(x)=4E35 erg/s, L(x)/L(o)=3E-4 [80]; No precise position.	V861 Sco=HD152667 =SA0227473[661]; B0Iae, P=7.8d spectroscopic binary, F(opt,M)=0.5Mo[832]; UV violet shifted absorption lines[661]; Secondary ≳4Mo[833,661]; Hα variability[832,581]; Member association Sco OB1, D=2.0 kpc[733]; V=6.2, B-V=0.25, U-B=-0.7, A(v)=1.6[733]		1653-407 V861 Sco
Pulsating, eclipsing Doppler binary; P=1.24s, P(orb)=1.7d[778,275]; Eclipse 0.24d[275]; F(x,M)=0.85, a(x)sin(i)=13.2 lt-s[778]; 35d on-off cycle[778,275]; ∼60 keV feature (cyclotron?)[801,155] Turn-ons during extended off state[235,160,433,383,193]; Fe emission[666]; Spectral behavior [35,667]; P=70d(?)[126]; Disk precession(?)[243]; Pulsing >20 keV[438,470,668] and at 0.1-0.4 keV (out of phase)[752,117,438]; Soft flux during OFF state[262]; Pulse profiles[442]; L(x)∼1.0E37 erg/s, L(x)/L(o)∼10 [80]; Pos.[250,206]; Other refs. in refs. (437,243, 193,442,25)	HZ Her[778,135,17]; F.C.[206]; Late A to early F, x-ray heating to B V type spectrum [67,172,181]; B=14.7-13.2 [17,653]; U-B=-0.7 at max.[181]; Optical ∼1.24s pulses[181,318,625]; Spectrum of optical pulses[625]; Disk precession(?)[266,396]; Bowen mechanism[556]; UV spectra[220]; D∼5 kpc [23]; Other refs. in refs. (20,23,56,16,658)		1656+354 Her X-1 HZ Her

SOURCE	POSITION TYPE,REF.	RA (1950)	DEC (1950)	LII BII	EQUIVALENT ERROR RADIUS (",90%)	FLUX DENSITY (2-11 keV) MAX (µJy) MIN (µJy)
1658-298	opt[213]	16 58 55.5	-29 52 26	353.8	3	80[511]
MXB		254.7313	-29.8739	7.3		<5[750]
	80 This work 213 Doxsey 302 Griffiths 502 Lewin 506 Lewin 511 Lewin 750 Share 821 van Paradijs					
1659-487	opt[213]	16 59 01.9	-48 43 06	338.9	3	390
GX339-4[560]		254.7579	-48.7183	-4.3		<14
1M,4U	213 Doxsey 228 Eyles 520 Li 560 Markert 647 Penston 722 Samimi					
1700-377	opt[732]	17 00 32.7	-37 46 29	347.8	1	110
4U,1M		255.1363	-37.7747	2.2		<11
HD153919	23 Bahcall 25 Bahcall 63 Bolton 80 This work 180 Dachs 218 Dupree 238 Fahlman 273 2U Cat. 325 Hammerschlag-Hensberge 326 Hammerschlag-Hensberge		345 Hensberge 427 Jones 428 Jones 585 Matilsky 605 Moffat 646 Penny 703 Rappaport 732 SAO Star Cat. 812 van Genderen 873 Wolff			
1702-429	x-ray[412]	17 02 40.4	-42 57 58	343.9	30	165
4U,2S,1M		255.6683	-42.9661	-1.3		11
MXB?	273 2U Cat. 336 HEAO A-3 gp. 412 Jernigan	512 Lewin 570 Marshall 770 Swank				
1702-363	x-ray[412,322]	17 02 23.4	-36 21 24	349.1	14	1140
GX349+2[72]		255.5975	-36.3567	2.7		420
4U,2S,1M		(Joint SAS-3/HEAO-A3 Position)				
	72 Bradt 250 4U Cat. 255 Forman 283 Glass 287 Glass	322 Gursky 412 Jernigan 586 Mayo 592 McClintock 690 Rappaport	871 Wilson 881 Zuiderwijk 882 Zuiderwijk			
1705-440	x-ray[412]	17 05 18.2	-44 02 10	343.3	20	310
4U,2S,1M		256.3258	-44.0361	-2.3		14
	250 4U Cat. 255 Forman 412 Jernigan 871 Wilson					

X-RAY	OPTICAL	RADIO	
Burster[502]; Intervals (2.0-2.6h) very stable, Oct.'76, June-July '77[506]; Cooling tail, size ∼7 km[821]; Transient episode to 80 μJy Mar.'78[511,750,302] with 40s variability and no observed bursts[511]; L(x)/L(o)=2E3 [80]; Pos.[213]	Star T[213]; F.C.[213]; UV excess, V=18.3, B-V=0.37, U-B=-0.38, λ4686, λ4640-50[213]		1658-298 MXB
Highly variable, factor 60 (1-6 keV)∼100d, High, low, and off states, kT∼1.7 keV in high state, Spectrum harder in low state[560]; Rapid variability, factor of 3 in 40 ms in low state[722]; Steady in high state[520]; L(x)/L(o)∼150[213]; Pos.[213,228]	Star V[647,213]; F.C.[647,213]; Possibly variable 0.5 mag.[647]; V=16.6, B-V=0.84, U-B=-0.24, broad Hα, λ4686, λλ4640-50, A(v)∼2.6 (average in direction)[213]		1659-487 GX339-4
Eclipsing binary P(orb)=3.4d, eclipse 1.1d[427]; L(x)∼3E36 erg/s[427]; L(x)/L(o)=5E-4 [80]; P=97 min(?)[585]; Pos.[703]	HD153919[273,427,703]=SAO208356; F.C.[427]; O6.5f[345,873]; λ4686 etc.[345], Fe XIV[218]; Ellipsoidal[428,646] and spectroscopic[345,873] binary; V=6.6, B-V=0.3, U-B=-0.7[646]; Spectrophotometric[180,238], multicolor[325,812], UV (giving A(V)∼1.5[326]), and λ4686[605] variability; D∼1.7 kpc[873,646,63]; Other refs. in refs.(23, 238,25)		1700-377 HD153919
Steady source[273]; Burster MXB1706-43?[770,570,512]; Pos.[412,336]	F.C.[412]		1702-429
Bright bulge source; Variable, factor of 2 in 1h[255]; Pos.[250,690,871,412,322]	F.C.[412]; Variable late-M giant in error region (8" E of star 2)[881,592,586,882,287]; Not a likely candidate[592]; J=9.06, H=7.72, K=6.87, L=6.06 (Apr. 3, May 21 '78)[283]; A(v)∼6, variable 0.8 mag. in J in 100d, prob. Mira var.[287]		1702-363 GX349+2
Bright bulge source; Variable ∼25% in 2h[255]; Pos.[250,871,412]	F.C.[412]		1705-440

SOURCE	POSITION TYPE,REF.	RA (1950)	DEC (1950)	LII BII	EQUIVALENT ERROR RADIUS (",90%)	FLUX DENSITY (2-11 keV) MAX (μJy) MIN (μJy)
1705-250	opt[300]	17 05 10.4	-25 01 38	358.6	2	3500[842]
H[300]		256.2933	-25.0272	9.1		<1.5[650]
Nova Oph '77						

80 This work
300 Griffiths
450 Kaluzienski
650 Peters
842 Watson

SOURCE	POSITION TYPE,REF.	RA (1950)	DEC (1950)	LII BII	EQUIVALENT ERROR RADIUS (",90%)	FLUX DENSITY (2-11 keV) MAX (μJy) MIN (μJy)
1715-321	x-ray[412]	17 15 31.9	-32 07 15	354.1	60	2000[565]
MX[566],1M		258.8829	-32.1208	3.1		15
2S,4U?						

336 HEAO A-3 gp.
375 Hoffman
412 Jernigan
565 Markert
566 Markert

SOURCE	POSITION TYPE,REF.	RA (1950)	DEC (1950)	LII BII	EQUIVALENT ERROR RADIUS (",90%)	FLUX DENSITY (2-11 keV) MAX (μJy) MIN (μJy)
1728-337	x-ray[410]	17 28 39.6	-33 47 52	354.3	30	210
4U,MXB		262.1650	-33.7978	-0.2		34
2S,1M						

113 Carpenter
260 Friedman
286 Glass
336 HEAO A-3 gp.
371 Hoffman
372 Hoffman
410 Jernigan
476 Kleinmann
498 Lewin
528 Liller
821 van Paradijs
871 Wilson

SOURCE	POSITION TYPE,REF.	RA (1950)	DEC (1950)	LII BII	EQUIVALENT ERROR RADIUS (",90%)	FLUX DENSITY (2-11 keV) MAX (μJy) MIN (μJy)
1728-247	opt[80]	17 28 57.6	-24 42 41	1.9	3	220
GX1+4[492]		262.2400	-24.7114	4.8		34
GX2+5[331]						
4U,2S,1M						

33 Becker
80 This work
185 Davidsen
186 Davidsen
201 Doty
202 Doty
211 Doxsey
282 Glass
331 Hawkins
336 HEAO A-3 gp.
480 Krzeminski
492 Lewin
500 Lewin
578 Mason
706 Ricker
708 Ricker
788 Thomas
858 White

SOURCE	POSITION TYPE,REF.	RA (1950)	DEC (1950)	LII BII	EQUIVALENT ERROR RADIUS (",90%)	FLUX DENSITY (2-11 keV) MAX (μJy) MIN (μJy)
1728-169	opt[80]	17 28 50.1	-16 55 19	8.5	3	330
GX9+9[70]		262.2088	-16.9219	9.0		175
4U,2S,1M						

70 Bradt
80 This work
120 Charles
185 Davidsen
211 Doxsey
228 Eyles
250 4U Cat.
322 Gursky
867 Willmore
871 Wilson

X-RAY	OPTICAL	RADIO
X-ray nova Aug.-Sept. '77[450,300], to 3500 μJy[842]; kT∼3.0 keV (Sept. 18), similar to A0620-00[300]; Light curve[842]; L(x)/L(o)=2500 [80]; Pos.[300]	Nova Oph '77[300]; F.C.[300]; B∼21 to ∼16.5, B-V∼0.75, A(v)∼1.4, λ4686 emission, no absorption lines, D∼3 kpc if L(x)≈1E38 erg/s[300]	1705-250 Nova Oph '77
10 min flare to 2000 μJy[565]; Fast (∼100s) transient(?)[375,412]; Pos.[412,336]	F.C.[412]	1715-321
Bright bulge source; "Slow" burster MXB1728-34[498,371,372]; Bursts: 4-8h intervals, bursting >95% of the time, emission after burst for several minutes[372]; Cooling black body spectra, size ∼10km[372,821]; P=12ms (?) before and during burst[260]; Pos.[113,871,410,336]	F.C.[410]; Several IR sources K<10 within 1' [286,476]; Optical search[528]	1728-337 MXB
Hard, variable[492,706,708,788] pulsator[492,858]; P∼135s (1970)[492,500] to 122.6s (1975)[201] to 116.5s (1978)[202]; $\dot{P}$/P=-0.02/y [33,202]; L(x)∼4E37 erg/s (2-60 keV,10 kpc)[186]; L(x)/L(o)=1.2, L(x)/L(IR)=12.4 [80]; Pos.[578,211,336]	Star GF[331,282,185,186,211]; F.C.[211]; Bright IR object, J=10.4, H=8.8, K=8.1[282]; Hα emission[282,185,186] dominates visible; HeI, FeII, [FeVII] variable x2 (τ=mos.), variable continuum, V=18.7-19.4, A(v)∼5.1, symbiotic (∼M6III+hotter source)[186]; Search for 122s period at Hα, ≲1%[480]	1728-247 GX1+4
Bright bulge source; L(x)/L(o)=400 [80]; Pos.[250,867,871,211,322,228]	Star "DMB"[867,185,120,211]; F.C.[185,211]; UV excess, V=16.6, B-V=0.0, U-B=-0.5, A(v)∼1.0[185]; λλ4640-50, λ4686[120]	1728-169 GX9+9

SOURCE	POSITION TYPE,REF.	RA (1950)	DEC (1950)	LII BII	EQUIVALENT ERROR RADIUS (",90%)	FLUX DENSITY (2-11 keV) MAX (μJy) MIN (μJy)
1730-333	opt[528]	17 30 07.0	-33 21 17	354.8	1	∿800[497]
Rapid Burster[497]		262.5292	-33.3547	-0.2		(Bursts)
MXB,2S						<16[866]
Liller 1						(steady comp.)

8 Ariel-5 Group
212 Doxsey
311 Grindlay
341 Heise
376 Hoffman
411 Jernigan
421 Johnson
475 Kleinmann
497 Lewin
503 Lewin
505 Lewin
506 Lewin
528 Liller
571 Marshall
807 Ulmer
866 White
871 Wilson

SOURCE	POSITION TYPE,REF.	RA (1950)	DEC (1950)	LII BII	EQUIVALENT ERROR RADIUS (",90%)	FLUX DENSITY (2-11 keV) MAX (μJy) MIN (μJy)
1735-444	opt[590]	17 35 19.0	-44 25 19	346.1	3	290
4U,MXB		263.8292	-44.4219	-7.0		110
2S,1M						

80 This work; 305 Grindlay; 322 Gursky; 410 Jernigan; 506 Lewin; 508 Lewin; 590 McClintock; 593 McClintock; 871 Wilson

SOURCE	POSITION TYPE,REF.	RA (1950)	DEC (1950)	LII BII	EQUIVALENT ERROR RADIUS (",90%)	FLUX DENSITY (2-11 keV) MAX (μJy) MIN (μJy)
1742-294	x-ray[412]	17 42 53.6	-29 29 50	359.6	30	120[680]
A[10],2S,		265.7233	-29.4972	-0.4		50[680]
GCX-1[177],						
MXB?						

10 Ariel-5 Group; 177 Cruddace; 228 Eyles; 412 Jernigan; 466 Kellogg; 495 Lewin; 680 Proctor

SOURCE	POSITION TYPE,REF.	RA (1950)	DEC (1950)	LII BII	EQUIVALENT ERROR RADIUS (",90%)	FLUX DENSITY (2-11 keV) MAX (μJy) MIN (μJy)
1742-289	radio[187]	17 42 26.5	-28 59 55	359.9	3	2000[226]
A[226]		265.6104	-28.9986	-0.0		<9[680]
MXB?						

80 This work
90 Branduardi
116 Carpenter
187 Davies
226 Eyles
495 Lewin
680 Proctor
871 Wilson

SOURCE	POSITION TYPE,REF.	RA (1950)	DEC (1950)	LII BII	EQUIVALENT ERROR RADIUS (",90%)	FLUX DENSITY (2-11 keV) MAX (μJy) MIN (μJy)
1743-322	x-ray[322]	17 43 00.5	-32 13 00	357.3	18	770[334]
H[334]		265.7521	-32.2167	-1.8		≲10[250]
		OR				
		17 41 46.1	-32 12 24	357.1	18	
		265.4421	-32.2067	-1.6		

250 4U Cat.; 322 Gursky; 334 HEAO-1 Exp.; 452 Kaluzienski; 875 Wood

X-RAY	OPTICAL	RADIO	
"Rapid burster" MXB1730-335[497]; Up to 4000 bursts/day[506]; ("Type II" bursts[376]); Intervals 6-450s, extended inactive states[503]; 50ms structure[807]; Burst energy proportional to following interval for larger bursts[497,505,866]; In bursting state ~15% of time[311]; Exhibits standard "Type I" bursts at intervals of hours[376] with cooling tail giving ~9 km size[571]; L(av II)/L(av I)≃130[376]; Instantaneous L(x-max)~2E38 erg/s at 10 kpc[497]; No steady emission[497,866]; Pos.[411,871,212]	Globular cluster core[497,8,341,528,475,411,871,212]; F.C.[528]; Highly reddened, compact[528]; Very concentrated (λ=1-5 μm), core diam. 13±1"[475]	≲18 mJy during bursts at 2.7 GHz[421]	1730-333 Rapid Burster
Bright bulge source; X-ray burster[508]; Short bursts 3-7s[508]; Intervals: ~1h-2d[506]; L(x)/Lo=840[80]; Pos.[871,410,322]	Star 5[410,590,593]; F.C.[410]; V=17.5, B-V=+0.2, U-B=-0.8, A(v)~1.0[590], λλ4640-50, λ4686, Remarkably similar to Sco X-1, D=5-10 kpc[593]; Coincident optical and x-ray bursts[305]		1735-444
Variable source[10,680,228] within "extended" 2° GCX region[466]; First reported as transient Mar.'76[10]; Within 20' error region of burster MXB1742-29 (Burst intervals ~0.55d)[495]; Pos.[680,228,412]	F.C.[412]		1742-294 GCX-1
Nova to ~2000 μJy near galactic nucleus, Feb.-Mar. '75[226]; Decay τ=12d for 40d, τ=90d for 140d, α=2 above 5 keV, variable τ~hours[90]; Possible residual flux at 9 μJy[680]; Burster MXB1743-29? (double-triple peaks, 35h intervals)[495]; L(x)=6E38 erg/s[80] at 10 kpc; Pos.[116,871]	No publ. F.C.	Transient Radio[187] to 480 mJy at 0.96 GHz (30 Mar. '75); <55 mJy other times; 53±3" SW of compact source in Sgr A West[187]	1742-289
Transient Aug.'77 to ~770 μJy[452,334]; Long decay, ~50 μJy in Mar.'78[875]; Two precise positions inside NRL region[322]	No publ. F.C.		1743-322

SOURCE	POSITION TYPE,REF.	RA (1950)	DEC (1950)	LII BII	EQUIVALENT ERROR RADIUS (",90%)	FLUX DENSITY (2-11 keV) MAX (μJy) MIN (μJy)
1743-288	x-ray[680]	17 43 52.3	-28 52 01	0.2	60	90[680]
GCX-3[177]		265.9679	-28.8669	-0.2		<7[680]
GX+0.2-0.2[680]		(Joint Ariel-5/Rocket Position)				
A[228],MXB?	177 Cruddace; 228 Eyles; 495 Lewin; 680 Proctor					
1744-361	x-ray[871]	17 44 50.9	-36 06 54	354.1	40	275[191]
A[191]		266.2121	-36.1150	-4.2		<25[116]
	116 Carpenter; 191 Davison; 871 Wilson					
1744-265	x-ray[405]	17 44 49.1	-26 32 50	2.3	4	670
GX3+1[70]		266.2046	-26.5472	0.8		220
4U,2S,1M		(Lunar Occulation)				
	70 Bradt; 211 Doxsey; 250 4U Cat.; 336 HEAO A-3 gp.; 360 Hjellming; 405 Janes; 600 Miley					
1745-203	opt[80]	17 45 54.6	-20 20 36	7.7	5	181
1M,4U		266.4775	-20.3433	3.8		17
NGC 6440		(Center of Globular Cluster)				
	27 Bahcall; 80 This work; 252 Forman; 418 Johnson; 561 Markert					
1746-370	opt[415]	17 46 48.8	-37 02 25	353.5	5	80
4U,2S,1M		266.7033	-37.0403	-5.0		13[415]
MXB?		(Center of Globular Cluster)				
NGC 6441	27 Bahcall; 80 This work; 136 Clark; 276 3U Cat.; 307 Grindlay; 348 Hesser; 400 Illingworth; 415 Jernigan; 515 Li					
1755-338	opt[80]	17 55 21.3	-33 48 15	357.2	3	91
4U,2S,1M		268.8388	-33.8042	-4.9		30
	80 This work; 112 Canizares; 301 Griffiths; 412 Jernigan; 871 Wilson					
1758-250	x-ray[370,412]	17 58 03.5	-25 04 37	5.1	7	1400
GX5-1[70]		269.5146	-25.0769	-1.0		530
4U,2S,1M		(Joint Lunar Occ./SAS-3 Position)				
	70 Bradt; 85 Braes; 190 Davison; 250 4U Cat.; 336 HEAO A-3; 360 Hjellming; 370 Hoffman; 412 Jernigan; 867 Willmore; 871 Wilson					
1758-205	x-ray[198,211]	17 58 33.1	-20 31 58	9.1	9	670
GX9+1[70]		269.6379	-20.5328	1.2		220
4U,2S,1M		(Joint Lunar Occ./SAS-3 Position)				
	70 Bradt; 198 Davison; 211 Doxsey; 250 4U Cat.; 336 HEAO A-3; 360 Hjellming; 867 Willmore; 871 Wilson					
1803-245	x-ray[412]	18 03 45.8	-24 35 38	6.1	30	850[412]
2S		270.9408	-24.5939	-1.9		$\lesssim$10[250]
	250 4U Cat.; 408 Jernigan; 412 Jernigan					
1811-171	x-ray[211]	18 11 37.2	-17 10 16	13.5	20	450
GX13+1[70]		272.9050	-17.1711	0.1		150
4U,2S,1M	70 Bradt; 211 Doxsey; 250 4U Cat.; 336 HEAO A-3; 861 White; 871 Wilson					

X-RAY	OPTICAL	RADIO	
Variable, near galactic center[177,680,228]; Variability with Ariel-5[680]; Lies just outside burster MXB1743-29 error region[495]; Pos.[680,228]	No publ. F.C.		1743-288 GCX-3 GX+0.2-0.2
Transient to ∿275 μJy Feb.-Mar.'76[191]; Pos.[116,871]	No publ. F.C.		1744-361
Bright bulge source; Pos.[250,405,211,336]	F.C.[211,405]	"Weak" source, 11±4 mJy at 1.4 GHz[600]; <5 mJy at 2.7 GHz[360]	1744-265 GX3+1
Transient observed only by OSO-7[561] and Uhuru[252]; Combined position r(e)=8.5'; L(x)=5E37 erg/s [80]	NGC 6440 Globular Cluster[561,252]; F.C.[252]; Cluster parameters, r(core)<6" [27]; D∿10 kpc [27]	Radio survey, weak source at cluster[418]	1745-203 NGC 6440
Highly variable[136]; L(x)=2E37 erg/s [80]; Burster MXB1746-37?[515]; Pos.[415]	NGC 6441 Globular Cluster[276,136,307] core[415]; F.C.[415]; Cluster parameters [27,348,400]; r(core)=9", D=9.3 kpc[400]		1746-370 NGC 6441
L(x)/L(o)=4000 [80]; Pos.[871,412,301]	Star 6[412,112,301]; F.C.[412]; UV excess, B∿19.3, flat featureless continuum[112]		1755-338
Brightest of galactic bulge sources; Pos.[250,370,190,867,871,412,336]	F.C.[370,412]	Source 10±3 mJy at 1.4 GHz (July 2 and Sept 24 '71)[85]; <5 mJy at 2.7 GHz[360]	1758-250 GX5-1
Bright bulge source; Pos.[250,867,871,198,211,336]	F.C.[198,211]	<5 mJy at 2.7 GHz[360]	1758-205 GX9+1
Bright transient to 850 μJy[408,412] May '76; Pos.[412]	F.C.[412]		1803-245
Bright bulge source; No significant variability[861]; Pos.[250,871,211,336]	F.C.[211]	≲10 mJy at 2.7 GHz[861]	1811-171 GX13+1

SOURCE	POSITION TYPE,REF.	RA (1950)	DEC (1950)	LII BII	EQUIVALENT ERROR RADIUS (",90%)	FLUX DENSITY (2-11 keV) MAX (μJy) MIN (μJy)
1813-140	opt[785]	18 13 10.9	-14 03 15	16.4	1	1060
GX17+2[70]		273.2954	-14.0542	1.3		270
4U,2S,1M						

70	Bradt	255	Forman	735	Schnopper
80	This work	336	HEAO A-3 gp.	777	Tananbaum
87	Braes	360	Hjellming	785	Tarenghi
185	Davidsen	367	Hjellming	861	White
211	Doxsey	368	Hoag	871	Wilson
250	4U Cat.	551	Margon		

SOURCE	POSITION TYPE,REF.	RA (1950)	DEC (1950)	LII BII	EQUIVALENT ERROR RADIUS (",90%)	FLUX DENSITY (2-11 keV) MAX (μJy) MIN (μJy)
1814+498	opt[876]	18 14 58.6	49 50 55	77.9	1	7[771]
4U,2A,1M		273.7442	49.8486	25.9		2.5[618]
AM Her						

30	Bailey	340	Hearn
51	Berg	527	Liller
65	Bond	598	Michalsky
80	This work	618	Murray
97	Bunner	671	Priedhorsky
118	Chanmugam	672	Priedhorsky
170	Cowley	675	Priedhorsky
245	Feigelson	763	Stockman
276	3U Cat.	771	Swank
280	Gilmozzi	775	Szkody
330	Harris	782	Tapia
336	HEAO A-3 gp.	876	Wright
338	Hearn		

SOURCE	POSITION TYPE,REF.	RA (1950)	DEC (1950)	LII BII	EQUIVALENT ERROR RADIUS (",90%)	FLUX DENSITY (2-11 keV) MAX (μJy) MIN (μJy)
1820-303	opt[415]	18 20 27.7	-30 23 11	2.8	5	410
4U,1M		275.1154	-30.3864	-7.9		61
MXB,2S		(Center of Globular Cluster)				
NGC 6624						

26	Bahcall	140	Clark	336	HEAO A-3 gp.
80	This work	250	4U Cat.	415	Jernigan
106	Canizares	276	3U Cat.	418	Johnson
110	Canizares	306	Grindlay	523	Liller
138	Clark	310	Grindlay	871	Wilson

SOURCE	POSITION TYPE,REF.	RA (1950)	DEC (1950)	LII BII	EQUIVALENT ERROR RADIUS (",90%)	FLUX DENSITY (2-11 keV) MAX (μJy) MIN (μJy)
1822-371	opt[295]	18 22 22.7	-37 08 03	356.9	4	34
4U,2A,1M		275.5946	-37.1342	-11.3		7

80	This work
295	Griffiths

SOURCE	POSITION TYPE,REF.	RA (1950)	DEC (1950)	LII BII	EQUIVALENT ERROR RADIUS (",90%)	FLUX DENSITY (2-11 keV) MAX (μJy) MIN (μJy)
1822-000	x-ray[210]	18 22 48.5	-00 02 24	29.9	20	62
4U,A[747],		275.7021	-0.0400	5.8		25[747]
2S,1M						

210	Doxsey	336	HEAO A-3 gp.
228	Eyles	747	Seward

SOURCE	POSITION TYPE,REF.	RA (1950)	DEC (1950)	LII BII	EQUIVALENT ERROR RADIUS (",90%)	FLUX DENSITY (2-11 keV) MAX (μJy) MIN (μJy)
1837+049	opt[80]	18 37 29.6	04 59 21	36.1	3	310
Ser X-1[258]		279.3733	4.9892	4.8		140[747]
4U,MXB,		(Southern Component of Close Pair[792])				
2S,1M						

80	This work	336	HEAO A-3 gp.
183	Davidsen	421	Johnson
210	Doxsey	516	Li
215	Doxsey	747	Seward
250	4U Cat.	767	Swank
258	Friedman	792	Thorstensen

X-RAY	OPTICAL	RADIO	
Bright bulge source; Variable thermal spectrum[777,255]; Suggested period (1913s) not confirmed[861]; L(x)=5E36 erg/s, L(x)/L(o)=1500 [80]; Pos.[250,871,211,336]	Star 28 (=TR)[735,360,777,785,211,367]; F.C.[785,211]; Normal G star[185]; Lies in 2" x 5" radio error region[367]; V=17.5, B-V=1.3, U-B=1.1[185,551]; Deep near-IR search[368]; Period search[551]; A(v)$\sim$2.0, D$\sim$1.4 kpc[368,551]	Variable source[360,87,367]; <0.8-3.4 mJy at 5 GHz Mar.-July '77[367]; Flared to 22 mJy May '71[360]	1813-140 GX17+2
Periodic (P=0.129d) ¼ keV (to 700 μJy)[338,340,97] and 2-60 keV (to 7 μJy)[276,771] source; L(x)=8E33 and α=-0.4 (2-60 keV and 300 kpc)[771]; L(x)/L(o)=0.6 [80]; Pos.[336]	AM Her[276,51,338,782,170,340,771]; F.C.[527]; Complex magnetic white dwarf binary system; Emission lines [65]; Periodic (0.129d) in linear and circular polarization[782,598,763], spectroscopy[170,671], and photometry[170,775,672,280,30] with complex phase offsets; V=13.1 to 12.3, B-V=0.7 to 0.4, U-B=-0.8 to -1.0[775]; Flickering and flaring (τ$\sim$min., $\sim$0.3 mag.) [51,775,672,30]; Correlations between intensity and polarization[675]; No periods 2-10,000d[245]; D$\sim$300 pc[118]	<55 mJy at 610 MHz[330]	1814+498 AM Her
Variability down to τ$\sim$3 min.[106]; Burster MXB1820-30[306]; Bursts quasi periodic 2.2-4.3h[138,140]; Bursts ceased when steady source brightened to $\gtrsim$125 μJy[140]; L(x)=1E38 erg/s [80]; Pos.[250,871,415,336]	NGC 6624 Globular Cluster[276] core[415]; F.C.[415,523]; Hα and 1 mag. flare from core[310]; Cluster parameters[523,110]; r(core)=8" [26], D=8.3 kpc[523]	Radio survey, no source in x-ray error region[418]	1820-303 NGC 6624 MXB
L(x)/L(o)$\sim$120 [80]; Pos.[295]	Star S[295]; F.C.[295]; V$\simeq$B$\simeq$16.3, intense λλ4686, 4640-50[295]		1822-371
Pos.[210,336,228]	F.C.[210]; No stars on POSS O		1822-000
Bright bulge source; Burst source[767,516]; Intervals: highly irregular July '76, $\sim$6.3h Aug. '76[516]; L(x)/L(o)=1.2E4 for A(v)=0 [80]; Pos.[250,210,336]	Star D[215,183,210,792]; F.C.[210]; UV excess[183]; Southern component of double (2.2") star, B$\sim$19.2, weak λ4686[792]	$\lesssim$44 mJy during bursts at 2.7 GHz[421]	1837+049 Ser X-1 MXB

SOURCE	POSITION TYPE,REF.	RA (1950)	DEC (1950)	LII BII	EQUIVALENT ERROR RADIUS (",90%)	FLUX DENSITY (2-11 keV) MAX (μJy) MIN (μJy)
1845-024	x-ray[210]	18 45 41.1	-02 28 37	30.4	35	22[747]
A[747],2S		281.4213	-2.4769	-0.4		11[210]
	210 Doxsey 288 Glass 558 Margon 657 Petro 747 Seward					
1850-087	opt[415]	18 50 20.8	-08 46 01	25.4	5	400[156]
A[747],2S,		282.5867	-8.7669	-4.3		4[415]
MXB		(Center of Globular Cluster)				
NGC 6712	80 This work 156 Cominsky 210 Doxsey 336 HEAO A-3 gp. 415 Jernigan 522 Li 652 Peterson 747 Seward 770 Swank					
1905+000	x-ray[210]	19 05 54.9	00 05 37	35.0	35	12[747]
A[747],MXB,		286.4788	0.0936	-3.7		7[747]
4U,2S	210 Doxsey 336 HEAO A-3 gp. 496 Lewin 501 Lewin 747 Seward					
1908+005	opt[791]	19 08 42.9	00 30 05	35.7	2	1000[122]
Aql X-1[258]		287.1788	0.5014	-4.1		<11
4U,A[747], 2S,1M	96 Buff 122 Charles 158 Cominsky 210 Doxsey 258 Friedman 385 Holt 457 Kaluzienski 555 Margon 567 1M Cat. 610 Mook 681 Puetter 747 Seward 783 Tapia 791 Thorstensen 825 Vrba 838 Walter 841 Watson					
1916-053	x-ray[210]	19 16 08.5	-05 19 51	31.4	35	30[747]
4U,A[747],		289.0354	-5.3308	-8.5		9
MXB,2S,1M	37 Becker 80 This work 210 Doxsey 273 2U Cat. 336 HEAO A-3 gp. 510 Lewin 747 Seward 795 Thorstensen					
1956+350	radio[828]	19 56 28.87	35 03 55.0	71.3	0.3	1320
Cyg X-1[258,270]		299.12029	35.06528	3.1		260
4U,1M HDE226868	13 Auriemma 23 Bahcall 25 Bahcall 58 Boldt 61 Bolton 62 Bolton 80 This work 82 Braes 88 Braes 92 Bregman 107 Canizares 200 Dolan 258 Friedman 270 Giacconi 352 Hiltner 360 Hjellming 366 Hjellming 381 Holt 467 Kemp 468 Kemp 538 Mandrou 543 Margon 597 Michalsky 622 Natali 623 Natali 628 Oda 630 Oda 631 Oda 632 Ogawara 673 Priedhorsky 685 Rao 687 Rappaport 688 Rappaport 717 Robinson 720 Rothschild 721 Rothschild 758 Sommer 765 Sutherland 776 Tananbaum 780 Tananbaum 787 Terrell 828 Wade 831 Walborn 835 Walker 836 Walker 847 Webster 850 Weisskopf					

X-RAY	OPTICAL	RADIO
Pos.[210]	F.C.[210]; J-H, H-K colors of star 2 slightly peculiar[288]; Opt. spectrum: normal A dwarf[558,657]	1845-024
Variable sec. to days, flare to 400 μJy in <6 min.[156]; L(x)=5E37 erg/s [80]; Burster MXB1850-08[770,522]; Pos.[210,415,336]	NGC 6712 Globular Cluster[747,156,210] core[415]; F.C.[415]; Cluster parameters, r(core)=49", D=6.8 kpc[652]	1850-087 NGC 6712 MXB
Steady source[747] and burster[496,501]; Pos.[210,336]	F.C.[210]	1905+000 MXB
Recurrent transient, outbursts to ∼1000 μJy (June '75, June '76, June '78), decay τ∼30d [96,457,122]; Flares 1971 and 1973[158,567], Jan.'77[385] to ∼300 μJy; L(x)∼1E38 erg/s at 6 kpc[791]; kT∼5.5 keV (or α∼0.8) [96]; P=1.3d episode[841]; L(x)/L(o)∼1E3[791,555]; Pos.[210]	Star 5[210,791]; F.C.[210,791]; Variable B≳20 to∼16.5, bright during x-ray flares, quiescent state: V=19.2, B-V=1.3, normal ∼K0 spectrum, D>1.7 kpc[791,555]; Bright on Palomar plates, V∼17.1[555]; June '78 outburst[838]: λλ4640-50, λ4686[681,610] and wide band (∼0.15 mag., 1-4 min.) variability[825,783]; V=15.6, B-V=0.9, U-B∼-0.5[825]	1908+005 Aql X-1
Steady source[273] and burster [37,510]; Intervals: 4-5h[510]; Steady source: kT>35 keV, 9.1 keV (Fe XXV abs.?) feature [37]; L(x)/L(o)=1500 [80]; Pos.[210,336]	Star 16[336,795]; F.C.[210]; UV excess, featureless spectrum[795]; B∼19.5 [80]	1916-053 MXB
Black hole candidate[630]; Rapid (to 50ms) aperiodic variability[630,688,632] (and refs. in[107]); Millisecond bursts(?)[720,721,850]; Continual flaring state (1-10s)[107]; X ray transitions (intensity and spectrum)[780,381,200,685,758]; P=5.6d[381]; Shot noise (τ∼0.5s; ∼30%)[787,765], and energy independent[673]; 100 keV-3 MeV (see[538] and refs. therein); L(x)=2E37 erg/s, L(x)/L(o)=2E-2 [80]; Pos.[687]; Other refs. in refs. (58,628, 631,538,107,25)	HDE226868[776,687,360,847,828,61,780,381] =SA0069181; F.C.[687]; O9.7Iab (p-var)[831]; V=8.9, B-V=0.8, U-B=-0.3[352,92]; Spectroscopic binary, P=5.60d[847,61,62]; Periodic linear[468] and circular[597] polarization (5.6d); A(v)∼3.3, D∼2.5 kpc [92,543]; P=39d(?)[467] questioned[835]; V band oscillations (0.1 mag.) during Apr./May '75 x-ray transition[622]; P=83ms? [13] questioned[717]; Long term photometry[836]; P=62 min.(?) during high state[623]; Other refs. in refs. (23,836)	1956+350 Cyg X-1 Radio transition <5 to 15 mJy ∼Apr. '71[360,82]; Coincident with x-ray decrease[780]; Similar radio/ x-ray event Feb. '76 to 23 mJy (1.4 GHz) [88]; Source coincident with star to 0.3" [828]; Other refs. in refs. (366,88)

SOURCE	POSITION TYPE, REF.	RA (1950)	DEC (1950)	LII BII	EQUIVALENT ERROR RADIUS (",90%)	FLUX DENSITY (2-11 keV) MAX (μJy) MIN (μJy)
1957+115 4U,2S,1M	opt[552]	19 57 02.2 299.2592	11 34 16 11.5711	51.3 -9.3	1	57[747] 16
	76 Bradt; 80 This work; 210 Doxsey; 552 Margon; 747 Seward					
2030+407 Cyg X-3[270] 4U,1M	radio[91]	20 30 37.62 307.65675	40 47 12.6 40.78683	79.9 0.7	0.2	430 90
	18 Bahcall; 23 Bahcall; 41 Becklin; 42 Becklin; 43 Becklin; 48 Bennett; 66 Bonnet-Bidaud; 80 This work; 86 Braes; 91 Branson; 101 Canizares; 131 Chu; 250 4U Cat.; 270 Giacconi; 273 2U Cat.; 293 Gregory; 346 Hermsen; 363 Hjellming; 365 Hjellming; 382 Holt; 391 Hughes; 473 Kestenbaum; 481 Lamb; 485 Lauque; 486 Leach; 487 Ledden; 575 Mason; 596 McEllin; 642 Parsignault; 643 Parsignault; 645 Parsignault; 660 Pietsch; 731 Sanford; 746 Serlemitsos					
2127+119 4U,1M, 2A,2S NGC 7078 M15	opt[415]	21 27 33.7 321.8904 (Center of Globular Cluster)	11 57 03 11.9508	65.0 -27.3	5	11 3
	21 Bahcall; 28 Bahcall; 80 This work; 136 Clark; 276 3U Cat.; 310 Grindlay; 336 HEAO A-3 gp.; 415 Jernigan; 418 Johnson; 627 Newell; 652 Peterson; 683 Pye					
2129+470 4U,1M	opt[793]	21 29 36.4 322.4017	47 04 09 47.0692	91.6 -3.0	3	28 9
	80 This work; 793 Thorstensen					
2140+433 SS Cygni	opt[233]	21 40 42.6 325.1775	43 21 51 43.3642	90.6 -7.1	5	20[715] 5[715]
	80 This work; 163 Cordova; 232 Fabbiano; 233 Fabbiano; 281 Glasby; 337 Hearn; 342 Heise; 553 Margon; 691 Rappaport; 715 Ricketts; 716 Robinson; 773 Swank					
2142+380 Cyg X-2[258,270] 4U,2A,1M	opt[268]	21 42 36.9 325.6538	38 05 28 38.0911	87.3 -11.3	0.7	740 220
	80 This work; 250 4U Cat.; 258 Friedman; 268 Giacconi; 270 Giacconi; 360 Hjellming; 380 Holt; 387 Holt; 776 Tananbaum					

X-RAY	OPTICAL	RADIO	
L(x)=7E36 erg/s at 7 kpc, L(x)/L(o)=580 [80]; Pos.[210]	Star M [76,552,210]; F.C.[210]; V=18.7, B-V=0.3, U-B=-0.6, No absorption or emission features detected, A(v)∿1.0, D∿7 kpc[552]		1957+115
P(orb)=4.8h, partial eclipses[642,731, 66,643]; Extended observations[101,486, 645]; Fe line emission[746,473]; P=4.8h at 32-64 keV[660]; P=16.9d(?)[382]; Gamma radiation (P=4.8h)[481] not detected with COS-B[346, 48]; Recent summary in[645]; L(x)=1.2E38 erg/s, L(x)/L(H,K)=95 [80]; Pos.[250]	IR star[273, 86,642,731, 41, 42]; No F.C., see[41]; ∿17 mJy at 2.2μ (H=12.5,K=11.4) [41]; P=4.8h (∿30%, λ=2.2μm), A(K)=1.5 [42]; coordinated observations [43,575]; A(v)∿19 [18]; Note radio distance; Other refs. in ref. (23)	Variable radio source [86,363]; Flared from 10 mJy to 21 Jy (Sept. '72, 10 GHz)[293]; Daily variability[391]; D=10.5±1 kpc[485,131]; Two 1975 outbursts[596, 487]; Other refs. in refs. (365,487)	2030+407 Cyg X-3
Highly variable[136]; L(x)=3E36 erg/s [80]; Pos.[415,336]	M15/NGC 7078 Globular Cluster[276,136,683] core[415]; F.C.[415]; Hα from core[310]; Cluster parameters[627, 21, 28]; r(core)=10" [21]; D=10.0 kpc[652]	Radio survey, no sources in x-ray error region[418]	2127+119 NGC 7078 M15
L(x)/L(o)=160 [80]; Pos.[793]	Star[793]; F.C.[793]; B=16.8-18.0 (P=5.24h), λ4640-50, λ4686[793]		2129+470
¼ keV source during optical flare (variable <7-1600 μJy)[691,337,553] and during opt. low state (15 μJy)[342]; Hard (>1 keV) variable flux in opt. low state[342,715,232] with L(x)=(<1-5)E32; Hard flux suppressed (L(x)<1E32) in flare state[715]; P=8.9s (0.2-0.4 keV) during outburst[163]; Fe line emission[773]; L(x)/L(o)=1.3 [80] (opt. low); Pos.[232] (hard component).	SS Cyg[691,342,715,232]; F.C.[232]; Dwarf nova (U Gem), m(v)=12.1-8.1, outbursts every 33-71d, white dwarf binary, P=6h 38m[716,281], M(v)=7.0-3.0, D∿100 pc (see[715] and refs. therein)		2140+433 SS Cyg
P=11.2d[380,387]; L(x)/L(o)=450 [80]; Pos.[250]	Star 10[270,268,776]; F.C.[268]; V∿15.5, B-V=0.4, U-B=-0.4, A(v)∿1.0, variable[268]	<5 mJy at 2.7 GHz[360]	2142+380 Cyg X-2

REFERENCES

Astr. quants. 1 Allen, C.W., Astrophysical Quantities, 3rd ed., Athlone Press, London (1976).
SMC X-2,3 2 Allen, D., IAU Circ. No. 3143 (1977).
Algol Model 3 Apparao, K.M.V., Nature, 266, 430 (1977).
Crab Nebula 5 Apparao, K.M.V, Astrophys. Sp. Sci., 25, 3 (1973).
2S posits. #6 6 Apparao, K.M.V., Bradt, H.V., Dower, R.G., Doxsey, R.E., et al., Nature, 271, 225 (1978).
1417-624 7 Apparao, K.M.V., Naranan, S., and Kelley, R., in preparation (1978).
Burst posits. 8 Ariel-5 Group, IAU Circ. No. 2929 (1976).
1742-294 10 Ariel-5 Group, University of Birmingham, IAU Circ. No. 2934 (1976).
Crab pulsar 11 Armstrong, J.W., and Coles, W.A., Astrophys. J., 220, 346 (1978).
Glob. clust. 12 Arp, H., in Galactic Structure, eds. Blaauw, A. and Schmidt, M., U. of Chicago Press, p. 401 (1965).
Cyg X-1 13 Auriemma, G., Cardini, D., Costa, E., Giovannelli, F., et al., Nature, 259, 27 (1976).
SMC X-1 15 Avni, Y. and Milgrom, M., Astrophys. J. Lett., 212, L17 (1977).
Her X-1 16 Avni, Y., Lamb, D.Q., Lamb, F.K., and Milgrom, M., Astrophys. J. Lett., 222, L113 (1978).
HZ Her 17 Bahcall, J.N. and Bahcall, N.A., Astrophys. J. Lett., 178, L1 (1972).
Cyg X-3 18 Bahcall, J.N. and Bahcall, N.A., Nature, 247, 446 (1974).
HZ Her 20 Bahcall, J.N., Joss, P.C., and Avni, Y., Astrophys. J., 191, 211 (1974).
NGC 7078 21 Bahcall, J.N., Bahcall, N.A., Weistrop, D., Astrophys. Lett., 16, 159 (1975).
M. blk. holes 22 Bahcall, J.N., and Ostriker, J.P., Nature, 256, 23 (1975).
Binaries 23 Bahcall, J.N., Lectures at Fermi Summer School, Varenna, on the Phys. and Astrophys. of Neut. Stars and Black Holes, N. Holl. Pub. Co., Amster. (1976).
n. s. masses 25 Bahcall, J.N., Ann. Rev. Astr. Astrophys., 16, 241 (1978).
NGC 6624 26 Bahcall, N.A., Astrophys. J. Lett., 204, L83 (1976).
NGC 6440,6441 27 Bahcall, N.A. and Hausman, M.A., Astrophys. J. Lett., 207, L181 (1976).
8 clusters 28 Bahcall, N.A., and Hausman, M.A., Astrophys. J., 213, 93 (1977).
AM Her 30 Bailey, J.A., Mason, K.O., and Parkes, G.E., Mon. Not. R. astr. Soc., 180, 35p (1977).
0535+26 31 Baratta, G.B., Viotti, R., and Altamore, A., Astr. Astrophys. Lett., 65, L21 (1978).
X-Per 32 Barbour, M.S., and Kemp, J.C., IAU Circ. No. 3205 (1978).
GX1+4 33 Becker, R.H., Boldt, E.A., Holt, S.S., et al., Astrophys. J. Lett., 207, L167 (1976).
Her X-1 35 Becker, R.H., Boldt, E.A., Holt, S.S., Pravdo, S.H., et al., Astrophys. J., 214, 879 (1977).
1538-52 36 Becker, R.H., Swank, J.H., Boldt, E.A., Holt, S.S., et al., Astrophys. J. Lett, 216, L11 (1977).
1915-05 37 Becker, R.H., Smith, B.W., Swank, J.H., et al., Astrophys. J. Lett., 216, L101 (1977).
0900-40 38 Becker, R.H., Rothschild, R.E., Boldt, E.A., Holt, S.S., et al., Astrophys. J., 221, 912 (1978).
X Per 40 Becker, R.H., et., al., presented at the HEAO/AAS Mtg., San Diego (1978).
Cyg X-3 41 Becklin, E.E., Kristian, J., Neugebauer, G., et al., Nature Phys. Sci., 239, 130 (1972).
Cyg X-3 42 Becklin, E.E., Neugebauer, G., Hawkins, F.J.,Mason, K.O., et al., Nature, 245, 302 (1973).
Cyg X-3 43 Becklin, E.E., Hawkins, F.J., Mason, K.O., et al., Astrophys. J. Lett., 192, L119 (1974).
X-ray bursts 45 Belian, R.D., Conner, J.P., et al., Proc. Conf. Trans. Sources, Los Alamos (1973); see (513, Li).
Norma bursts 46 Belian, R.D., Conner, J.P., and Evans, W.D., Astrophys. J. Lett., 206, L135 (1976).
Cen X-3 47 Bennett, K., Bignami, G.F., Di Gesu, V., et al., Astr. Astrophys., 51, 475 (1976).
Cyg X-3 48 Bennett, K., Bignami, G.F., Hermsen, W., et al., Astr. Astrophys., 59, 273 (1977).
Crab pulsar 50 Bennett, K., Bignami, G.F., Boella, G., et al., Astr. Astrophys., 61, 279 (1977)
AM Her 51 Berg, R.A., and Duthie, J.G., Astrophys. J., 211, 859 (1977).
X-Per 52 Bernacca, P. and Bianchini, A., IAU Circ. No. 3201 (1978).
0620-00 53 Bieging, J., and Downes, D., Nature, 258, 307 (1975).
Cir X-1 55 Bignami, G.F., Della Ventura, A., Maccagni, D., et al., Astr. Astrophys., 57, 309 (1977).
X-ray stars 56 Blumenthal, G.R. and Tucker, W.H., Ann. Rev. Astr. Astrophys., 12, 23 (1974).
LMC, SMC 57 Bok, B.J., Ann. Rev. Astr. Astrophys., 4, 95 (1966).
Cyg X-1 58 Boldt, E., Holt, S., et al., Proc. of the Calgary Conferences on X-rays in Space, p. 69 (1974).
0620-00 60 Boley, F., Wolfson, R., Bradt, H., Doxsey, R., et al., Astrophys. J. Lett., 203, L13 (1976).
Cyg X-1 61 Bolton, C.T., Nature, 235, 271 (1972).
Cyg X-1 62 Bolton, C.T., Astrophys. J., 200, 269 (1975).
0900, 1700 63 Bolton, C.T., and Herbst, W., Astr. J., 81, 339 (1976).
AM Her 65 Bond, H.E., and Tifft, W.G., Pub. Astr. Soc. Pac., 86, 981 (1974).
Cyg X-3 66 Bonnet-Bidaud, J.M., Ilovaisky, S., Manzo, G., et al., Astr. Astrophys., 62, 275 (1978).
HZ Her 67 Bopp, B.W., Grupsmith, G., and Vanden Bout, P., Astrophys. J. Lett., 178, L5 (1972).
Sco X-1 68 Bord, D.J., Messina, R.J., Mook, D.E., and Hiltner, W.A., Astrophys. J. Lett., 206, L49 (1976).
Rocket pos. 70 Bradt, H.V., Naranan, S., Rappaport, S., and Spada, G., Astrophys. J., 152, 1005 (1968).
Crab pulsar 71 Bradt, H.V., Rappaport, S., Mayer, W., Nather, R.E., et al., Nature, 222, 728 (1969).
GX340+0,349+2 72 Bradt, H.V., Burnett, B., Mayer, W., Rappaport, S., et al., Nature, 229, 96 (1971).
Sco X-1,'71 73 Bradt, H.V., Braes, L.L.E., Forman, W., Hesser, J.E., et al., Astrophys. J., 197, 443 (1975).
0535+26 75 Bradt, H.V., Mayer, W., Buff, J., Clark, G.W., et al., Astrophys. J. Lett., 204, L67 (1976).
0143, 1956 76 Bradt, H.V., Doxsey, R., Jernigan, J.G., and Spada, G., Bull. Am. Phys. Soc., II-21, 544 (1976).
2S posits. #3 77 Bradt, H.V., Apparao, K.M.V., Dower, R., Doxsey, R.E., et al., Nature, 269, 496 (1977).
SAS-3 survey 78 Bradt, H.V., Apparao, K.M.V., Clark, G.W., Dower, R., et al., Nature, 269, 21 (1977).
This work 80 Bradt, H.V., Doxsey, R.E., and Jernigan, J.G., this paper (1978).
Orion 81 Bradt, H.V., and Kelley, R., Astrophys. J. Lett., submitted (1978).
Cyg X-1 82 Braes, L.L.E. and Miley, G.K., Nature, 232, 246 (1971).
X Per 83 Braes, L.L.E. and Miley, G.K., Nature, 235, 273 (1972).
GX5-1 85 Braes, L.L.E., Miley, G.K., and Schoenmaker, A.A., Nature, 236, 392 (1972).

Cyg X-3	86	Braes, L.L.E. and Miley, G.K., Nature, 237, 506 (1972).
GX17+2	87	Braes, L.L.E., and Miley, G.K., IAU Symp. No. 55, X and Gamma Ray Astron., ed: H. Bradt and R. Giacconi, Dordrecht, Reidel (1973).
Cyg X-1	88	Braes, L.L.E., and Miley, G.K., Nature, 264, 731 (1976).
1742-289	90	Branduardi, G., Ives, J.C., Sanford, P.W., et al., Mon. Not. R. astr. Soc., 175, 47p (1976).
Cyg X-3	91	Branson, N.J.B.A., Martin, A.H.M., Pooley, G.G., et al., Nature Phys. Sci., 239, 133 (1972).
Cyg X-1	92	Bregman, J., Butler, D., Kemper, E, Koski, A., et al., Astrophys. J. Lett., 185, L117 (1973).
X-Per	93	Brucato, R.J., and Kristian, J., Astrophys. J. Lett., 173, L105 (1972).
Cen X-3	95	Brucato, R.J., Kristian, J., and Westphal, J.A., Astrophys. J. Lett., 175, L137 (1972).
AqlX-1,CirX-1	96	Buff, J., Jernigan, G., Laufer, B., Bradt, H., et al., Astrophys. J., 212, 768 (1977).
AM Her	97	Bunner, A., in preparation, reported in (771, Swank).
LMC X-1	98	Byrne, P.B. and Butler, C.J., Nature Phys. Sci., 244, 6 (1973).
X Per	100	Campisi, I.E., Treves, A., and Bernacca, P.I., Mon. Not. R. astr. Soc., 176, 225 (1976).
Cyg X-3	101	Canizares, C.R., McClintock, J.E, Clark, G.W., et al., Nature Phys. Sci., 241, 28 (1973).
Cir X-1	102	Canizares, C.R., Li, F.K., and Clark, G.W., Astrophys. J. Lett., 191, L75 (1974).
Sco X-1	103	Canizares, C.R., Clark, G.W., Li, F.K., Murthy, G.T., et al., Astrophys. J., 197, 457 (1975).
Glob-bulge	105	Canizares, C.R., Astrophys. J., 201, 589 (1975).
NGC 6624	106	Canizares, C.R., and Neighbours, J.E., Astrophys. J. Lett., 199, L97 (1975).
Cyg X-1	107	Canizares, C.R., and Oda, M., Astrophys. J. Lett., 214, L119 (1977).
X Per	108	Canizares et al., preprint, withdrawn (1977).
NGC 6624	110	Canizares, C.R., Grindlay, J.E., Hiltner, W.A., Liller, W., et al., Astrophys. J. 224, 39 (1978).
1636-53	111	Canizares, C.R., McClintock, J.E., and Grindlay, J.E., in prep. (1978).
1543, 1755	112	Canizares, C.R., McClintock, J.E., et al., in prep.; see also IAU Circ. No. 3251 (1978).
Bursts	113	Carpenter, G.F., Skinner, G.K., Wilson, A.M., and Willmore, A.P., Nature, 262, 473 (1976).
0620-00	115	Carpenter, G.F., Eyles, C.J., Skinner, G.K., et al., Mon. Not. R. astr. Soc., 176, 397 (1976).
Ariel V pos.	116	Carpenter, G.F., Eyles, C.J., Skinner, G.K., et al., Mon. Not. R. astr. Soc., 179, 27p (1977).
Her X-1	117	Catura, R.C. and Acton, L.W., in X-ray Binaries (NASA SP 389), p. 119 (1976).
AM Her	118	Chanmugam, G., and Wagner, R.L., Astrophys. J. Lett., 213, L13 (1977).
GX9+9	120	Charles, P.A., Thorstensen, J.R., and Bowyer, S., IAU Circ. No. 3096 (1977).
0042+32	121	Charles, P., Thorstensen, J., and Bowyer, S., Mon. Not. R. astr. Soc., 183, 29p (1978).
Aql X-1	122	Charles, P., Holt, S.S., and Sanford, P., IAU Circ. No. 3235 (1978).
0535+26	123	Chartres, M., and Li, F., IAU Circ. No. 3154 (1977).
Crab	125	Cheng, A.F., and Ruderman, M.A., Astrophys. J., 216, 865 (1977).
Her X-1	126	Chester, T.J., Astrophys. J. Lett., 213, L9 (1977).
A1118-61; Be	127	Chevalier, C., and Ilovaisky, S., IAU Circ. No. 2778 (1975).
LMC X-4	128	Chevalier, C. and Ilovaisky, S.A., Astr. Astrophys., 59, L9 (1977).
Cen-Vela	130	Chodil, G., Mark, H., Rodrigues, R., Seward, F., et al., Phys. Rev. Lett., 19, 681 (1967).
Cyg X-3	131	Chu, K.W. and Bieging, J.H., Astrophys. J. Lett., 179, 21 (1973).
0620-00	132	Citterio, O., Conti, G., Di Benedetto, P., et al., Mon. Not. R. astr. Soc., 175, 35p (1976).
Cir X-1	133	Clark, D.H., Parkinson, J.H., and Caswell, J.L., Nature, 254, 674 (1975).
Her X-1	135	Clark, G.W., Bradt, H.V., Lewin, W.H.G., et al., Astrophys. J. Lett., 177, L109 (1972).
3 glob. clus.	136	Clark, G.W., Markert, T.H., and Li, F.K., Astrophys. J. Lett., 199, L93 (1975).
Glob. clus.	137	Clark, G.W., Astrophys. J. Lett., 199, L143 (1975).
NGC6624 brsts.	138	Clark, G.W., Jernigan, J.G., Bradt, H., et al., Astrophys. J. Lett., 207, L105 (1976).
NGC6624 brsts.	140	Clark, G.W., Li, F.K., Canizares, C.R., et al., Mon. Not. R. astr. Soc., 179, 651 (1977).
1608-52	141	Clark, G.W., and Li, F., IAU Circ. No. 3090 (1977).
NGC 1851	142	Clark, G.W., and Li, F., IAU Circ. No. 3092 (1977).
SMC X-2,3	143	Clark, G.W., Doxsey, R., Li, F., Jernigan, J.G., et al., Astrophys. J. Lett., 221, L37 (1978).
0535+26	145	Clark, G.W., and Chartres, M., IAU Circ. No. 3208 (1978).
SMC survey	146	Clark, G.W., Li, F., and van Paradijs, J., Astrophys. J., submitted (1978).
Algol	147	Clark, T.A., Hutton, L.K., Ma, C., Shapiro, I.I., et al., Astrophys. J. Lett., 206, L107 (1976).
Crab pulsar	148	Cocke, W.J., Disney, M.J., and Taylor, D.J., Nature, 221, 525 (1969)
0535+26	150	Coe, M.J., Carpenter, G.F., Engel, A.R., and Quenby, J.J., Nature, 256, 630 (1975).
Cyg X-1,0620	151	Coe, M.J., Engel, A.R., and Quenby, J.J., Nature, 259, 544 (1976).
Cir X-1	152	Coe, M.J., Engel, A.R., and Quenby, J.J., Nature, 262, 563 (1976).
3 sources	153	Coe, M.J., Engel, A.R., and Quenby, J.J., Mon. Not. R. astr. Soc., 177, 31p (1976).
Her X-1	155	Coe, M.J., Engel, A.R., Quenby, J.J., and Dyer, C.S., Nature, 268, 508 (1977).
NGC 6712	156	Cominsky, L., Forman, W., Jones, C., and Tananbaum, H., Astrophys. J. Lett., 211, L9 (1977).
0115+63	157	Cominsky, L., Clark, G.W., Li, F., Mayer, W., and Rappaport, S., Nature, 273, 367 (1978).
Transients	158	Cominsky, L., Jones, C., Forman, W., and Tananbaum, H., Astrophys. J., 224, 46 (1978).
Mass loss	159	Conti, P.S., Ann. Rev. Astr. Astrophys., 16, 371 (1978).
Her X-1	160	Cooke, B.A., and Page, C.G., Nature, 256, 712 (1975).
Stellar winds	161	Cooke, B.A., Fabian, A.C., and Pringle, J.E., Nature, 273, 645 (1978).
2A cat.	162	Cooke, B.A., Ricketts, M.J., Maccacaro, T., et al., Mon. Not. R. astr. Soc., 182, 489 (1978).
SS Cyg	163	Cordova, F., Garmire, G., and Tuohy, I., IAU Circ. No. 3235 (1978).
EX Hydrae	165	Cordova, F., presented at the HEAO/AAS meeting, San Diego (1978).
Sco X-1	166	Cowley, A.P. and Crampton, D., Astrophys. J. Lett., 201, L65 (1975).
Gamma Cas	167	Cowley, A.P., Rogers, L., and Hutchings, J.B., Pub. Astr. Soc. Pac., 88, 911 (1976).
1538-52	168	Cowley, A.P., Crampton, D., Hutchings, J.B., et al., Astrophys. J. Lett., 218, L3 (1977).
AM Her	170	Cowley. A.P., and Crampton, D., Astrophys. J. Lett., 212, L121 (1977).
Crab nebula	171	Crab Nebula, IAU Symp. No. 46, ed: R.D. Davies and F.G. Smith, Dordrecht, Reidel (1971).
HZ Her	172	Crampton, D. and Hutchings, J.B., Astrophys. J. Lett., 178, L65 (1972)
SMC X-2,3	173	Crampton, D., Hutchings, J.B., and Cowley, A.P., Astrophys. J. Lett., 223, L79 (1978).

175-262

1538-52	175	Crampton, D., Hutchings, J.B., and Cowley, A.P., Astrophys. J. Lett., 225, L63 (1978).
1052+607	176	Crampton, D., private communicaton in (743; Schwartz).
Gal. center	177	Cruddace, R.G., Fritz, G., Shulman, S., Friedman, H., et al., Astrophys. J. Lett., 222, L95 (1978).
0535-668	178	Dachs, J., Astr. Astrophys., 18, 271 (1972).
1700-37	180	Dachs, J., Astr. Astrophys., 47, 19 (1976).
Her X-1	181	Davidsen, A., Henry, J.P., Middleditch, J., and Smith, H.E., Astrophys. J. Lett., 177, L97 (1972).
0614+09	182	Davidsen, A., Malina, R., Smith, H., Spinrad, H., et al., Astrophys. J. Lett., 193, L25 (1974).
Ser X-1	183	Davidsen, A., IAU Circ. No. 2824 (1975).
Opt. search	185	Davidsen, A., Malina, R., and Bowyer, S., Astrophys. J., 203, 448 (1976).
GX1+4	186	Davidsen, A., Malina, R., and Bowyer, S., Astrophys. J., 211, 866 (1977).
1742-289	187	Davies, R.D., Walsh, D., Browne, I.W.A., Edwards, M.R., et al., Nature, 261, 476 (1976).
0620-00	188	Davis, R.J., Edwards, M.R., Morrison, I., and Spencer, R.E., Nature, 257, 659 (1975).
GX5-1	190	Davison, P.J.N., Nature Phys. Sci., 246, 90 (1973).
1744-36	191	Davison, P.J.N., Burnell, J., Ives, J., Wilson, A., and Carpenter, G., IAU Circ. No. 2925 (1976).
A1118-61	192	Davison, P.J.N., and Sanford, P.W., Nature, 259, 98 (1976).
Her X-1	193	Davison, P.J.N. and Fabian, A.C., Mon. Not. R. astr. Soc., 178, 1p (1977).
1538-52	195	Davison, P.J.N., Mon. Not. R. astr. Soc., 179, 35p (1977).
SMC X-1	196	Davison, P.J.N., Mon. Not. R. astr. Soc., 179, 15p (1977).
1538-52	197	Davison, P.J.N., Watson, M.G., and Pye, J.P., Mon. Not. R. astr. Soc., 181, 73p-79p (1977).
GX9+1	198	Davison, P.J.N. and Morrison, L.V., Mon. Not. R. astr. Soc., 178, 53p (1977).
Orion	199	den Boggende, A.J.F., Mewe, R., et al., Astron. Astrophys., 62, 1 (1978).
Cyg X-1	200	Dolan, J.F., Crannell, C.J., Dennis, B.R., Frost, K.J., and Orwig, L.E., Nature, 267, 813 (1977).
GX1+4	201	Doty, J., IAU Circ. No. 2910 (1976).
GX1+4	202	Doty, J., et al., presented at the HEAO/AAS Mtg., San Diego (1978).
Cir X-1	203	Dower, R., Bradt, H., and Canizares, C., in prep., see BAAS, 9, 297 and BAAS, 9, 644 (1977); also Dower, R., PhD thesis, MIT (1978).
2S posits. #8	205	Dower, R.G., Apparao, K.M.V, Bradt, H.V., Doxsey, R.E., et al., Nature, 273, 364 (1978).
Her X-1	206	Doxsey, R.E., Bradt, H.V., Levine, A., et al., Astrophys. J. Lett., 182, L25 (1973).
0620-00	207	Doxsey, R.E., Jernigan, G., Hearn, D., Bradt, H., et al., Astrophys. J. Lett., 203, L9 (1976).
2S posits. #2	210	Doxsey, R.E., Apparao, K.M.V., Bradt, H.V., Dower, R.G., et al., Nature, 269, 112 (1977).
2S posits. #5	211	Doxsey, R.E., Apparao, K.M.V., Bradt, H.V., Dower, R.G., et al., Nature, 270, 586 (1977).
1730-335	212	Doxsey, R.E., Bradt, H., Gursky, H., Johnston, M., et al., Astrophys. J. Lett., 221, L53 (1978).
GX339-4,1659	213	Doxsey, R.E., Grindlay, J.E., Griffiths, R.E., et al., Astrophys. J. Lett., submitted (1978).
1S posits.	215	Doxsey, R., IAU Circ. No. 2820 (1975).
R148	216	Dufour, R.J. and Duval, J.E., Pub. Astr. Soc. Pac., 87, 769 (1975).
1608-52	217	Duldig, M., Greenhill, J., Thomas, R., Haynes, R., Simons, L., et al., IAU Circ. 3108 (1977).
1700-37	218	Dupree, A.K., Baliunas, S.L., and Lester, J.B., Astrophys. J. Lett., 218, L71 (1977).
Her X-1	220	Dupree, A.K., ... Boggess, A., ... Macchetto, F., ... Wilson, R., et al., Nature, in press.
0620-00	221	Eachus, L.J., Wright, E.L., and Liller, W., Astrophys. J. Lett., 203, L17 (1976).
0620-00	222	Elvis, M., Page, C.G., Pounds, K.A., Ricketts, M.J.,and Turner, M.J.L., Nature, 257, 656 (1975).
LMC X-4	223	Epstein, A., Delvaille, J., Helmken, H., Murray, S., et al., Astrophys. J., 216, 103 (1977).
LMC X-1	225	Epstein, A., Astrophys. J. Lett., 218, L49 (1977).
1742-28	226	Eyles, C.J., Skinner, G.K., Willmore, A.P., and Rosenberg, F.D., Nature, 257, 291 (1975).
A1118-61	227	Eyles, C.J., Skinner, G.K., Willmore, A.P., and Rosenberg, F.D., Nature, 254, 577 (1975).
Ariel-5 pos.	228	Eyles, C.J., this volume.
Cen X-3	230	Fabbiano, G. and Schreier, E.J., Astrophys. J., 214, 235 (1977).
1608-52	231	Fabbiano, G., Bradt, H.V., Doxsey, R.E., Gursky, H., et al., Astrophys. J. Lett., 221, L49 (1978).
SS Cyg	232	Fabbiano, G,. Gursky, H., Schwartz, D.A., Schwarz, J., et al., Nature, submitted (1978).
SS Cyg	233	Fabbiano, G., pvt. comm. (1978).
Her X-1	235	Fabian, A.C., Pringle, J.E., and Rees, M.J., Nature, 244, 212 (1973).
Tidal capture	236	Fabian, A.C., Pringle, J.E., and Rees, M.J., Mon. Not. R. astr. Soc., 172, 15p(1975).
A1118-61	237	Fabian, A.C., Pringle , J.E., and Webbink, R.F., Nature, 255, 208 (1975).
1700-37	238	Fahlman, G.G., Carlberg, R.G., and Walker, G.A.H., Astrophys. J. Lett., 217, L35 (1977).
Crab pulsar	240	Fazio, G.G., IAU Symp. No. 55, X and Gamma Ray Astr., ed. H. Bradt and R. Giacconi, Dordrecht, Reidel, p. 304 (1973).
LMC stars	241	Feast, M.W., Thackeray, A.D., and Wesselink, A.J., Mon. Not. R. astr. Soc., 121, 337 (1960).
So. B stars	242	Feast, M.W., Stoy, R.H., Thackeray, A.D., et al., Mon. Not. R. astr. Soc., 122, 239 (1961).
Her X-1	243	Fechner, W.B. and Joss, P.C., Astrophys. J. Lett., 213, L57 (1977).
AM Her	245	Feigelson, E., Dexter, L., and Liller, W., Astrophys. J., 222, 263 (1978).
X Per	246	Ferrari-Toniolo, M., Natali, G., Persi, P., and Spada, G., Astr. Astrophys., 61, 47 (1977).
0535+26	247	Fishman, G.J., and Watts, J.W.,Jr., Astrophys. J., 212, 211 (1977).
0900-40	248	Forman, W., Jones, C., Tananbaum, H., et al., Astrophys. J. Lett., 182, L103 (1973).
4U cat	250	Forman, W., Jones, C., Cominsky, L., et al., Astrophys. J. Suppl., in press (1978).
Gal. plane	251	Forman, W., Jones, C., and Tananbaum, H., Astrophys. J. Lett., 206, L29 (1976).
NGC 6440	252	Forman, W., Jones, C., and Tananbaum, H., Astrophys. J. Lett., 207, L25 (1976).
NGC 1851	253	Forman, W., and Jones, C., Astrophys. J. Lett., 207, L177 (1976).
Intens. var.	255	Forman, W., Jones, C., and Tananbaum, H., Astrophys. J., 208, 849 (1976).
0620-00	256	French, H. IAU Circ. No. 2840 (1975).
Algol	257	Frieboes-Conde, H., Herczeg, T., and Hog, E., Astr. Astrophys., 4, 78 (1970).
Early survey	258	Friedman, H., Byram, E.T., and Chubb, T.A., Science, 156, 374 (1967).
1728-33	260	Friedman, H., presented at the HEAO/AAS Meeting, San Diego (1978).
Crab pulsar	261	Fritz, G., Henry, R.C., Meekins, J.F., Chubb, T.A., and Friedman, H., Science, 164, 709 (1969).
Her X-1	262	Fritz, G., Naranan, S., Shulman, S., Yentis, D., et al., Astrophys. J. Lett., 207, L29 (1976).

Low lum. sour. 263 Garmire, G., this volume.

HZ Her posit. 265 Gatewood, G., and Sofia, S., Astrophys. J. Lett., 204, L89 (1976).

Her X-1 266 Gerend, D., and Boynton, P.E., Astrophys. J., 209, 562 (1976).

The beginning 267 Giacconi, R., Gursky, H., Paolini, F.R., and Rossi, B.B., Phys. Rev. Lett., 9, 439 (1962).

Cyg X-1,2 268 Giacconi, R., Gorenstein, P., ... Sandage, A., et al., Astrophys. J. Lett., 148, L129 (1967).

Cyg X-1,2,3 270 Giacconi, R., Gorenstein, P., Gursky, H., and Waters, J.R., Astrophys. J. Lett., 148, L119 (1967).

Uh. gal. scan 271 Giacconi, R., Kellogg, E., Gorenstein, P., et al., Astrophys. J. Lett., 165, L27 (1971).

Cen X-3 272 Giacconi, R., Gursky, H., Kellogg, E., Schreier, E., et al., Astrophys. J. Lett., 167, L67 (1971).

2U cat 273 Giacconi, R., Murray, S., Gursky, H., Kellogg, E., et al., Astrophys. J., 178, 281 (1972).

Her X-1 275 Giacconi, R., Gursky, H., Kellogg, E., Levinson, R., et al., Astrophys. J., 184, 227 (1973).

3U cat. 276 Giacconi, R., Murray, S., Gursky, H., Kellogg, E., et al., Astrophys. J. Suppl., 27, 37 (1974).

0535+26 277 Giangrande, A., Giovannelli, F., Bartolini, C., Guarnieri, A., et al., IAU Circ. No. 3129 (1977)

Algol 278 Gibson, D.M., Viner, M.R., and Peterson, S.D., Astrophys. J. Lett., 200, L143 (1975).

AM Her 280 Gilmozzi, R., Messi, R., and Natali, G., Astr. Astrophys., submitted (1978).

Dwarf novae 281 Glasby, J.S., The Dwarf Novae, American Elsevier, New York (1970).

GX2+5=GX1+4 282 Glass, I.S. and Feast, M.W., Nature Phys. Sci., 245, 39 (1973).

GX349+2 283 Glass, I.S. and Feast, M.W., IAU Circ. No. 3226 (1978).

Cir X-1 285 Glass, I.S., Mon. Not. R. astr. Soc., 183, 335 (1978).

1728-33 286 Glass, I.S., Nature, 273, 35 (1978).

GX349+2 287 Glass, I.S., and Feast, M.W., IAU Circ. No. 3254 (1978).

IR search 288 Glass, I.S., Mon. Not., R. astr. Soc., submitted (1978),

Cir X-1 290 Goss, W.M. and Mebold, U., Mon. Not. R. astr. Soc., 181, 255 (1977).

Sco X-1 291 Gottlieb, E.W., Wright, E.L., and Liller, W., Astrophys. J. Lett., 195, L33 (1975).

X Per 292 Gottlieb, E.W., Wright, E.L., and Liller, W., Astrophys. J. Lett., 202, L13 (1975).

Cyg X-3 293 Gregory, P.C., Kronberg, P.P., Seaquist, E.R., et al., Nature, 239, 440 (1972).

1822, 1254 295 Griffiths, R.E., Gursky, H., Schwartz, D.A., Schwarz, J., et al., Nature, submitted (1978).

Sco X-1 296 Griffiths, R.E., Cooke, B.A., Peacock, A., et al., Mon. Not. R. astr. Soc., 175, 449 (1976).

0620-00 297 Griffiths, R.E., Ricketts, M.J., and Cooke, B.A., Mon. Not. R. astr. Soc., 177, 429 (1976).

LMC 298 Griffiths, R.E., and Seward, F.D., Mon. Not. R. astr. Soc., 180, 75p (1977).

Nova Oph 300 Griffiths, R.E., ... Friedman, H., ... Murdin, P., et al., Astrophys. J. Lett., 221, L63 (1978).

1543, 1755 301 Griffiths, R.E., and the HEAO-A3 group, IAU Circ. No. 3251 (1978).

MXB1659-29 302 Griffiths, R.E., Johnston, M., Bradt, H., Doxsey, R., et al., IAU Circ. No. 3190 (1978).

Transients 303 Griffiths, R.E., Bradt, H.V., Briel, U., Dower, R., et al., this volume (1978).

1735,opt.brst. 305 Grindlay, J.E., McClintock, J.E., Canizares, C.R., et al., Nature, 274, 567 (1978).

NGC6624,brsts. 306 Grindlay, J.E., Gursky, H., Schnopper, H., et al., Astrophys. J. Lett., 205, L127 (1976).

NGC 6441 307 Grindlay, J.E., Schnopper, H., Schreier, E., et al., Astrophys. J. Lett., 206, L23 (1976).

1608-52 308 Grindlay, J.E. and Gursky, H., Astrophys. J. Lett., 209, L61 (1976).

Glob. clust. 310 Grindlay, J.E., and Liller, W., Astrophys. J. Lett., 216, L105 (1977).

MXB1730-335 311 Grindlay, J.E. and Gursky, H., Astrophys. J. Lett., 218, L117 (1977).

1608-52 312 Grindlay, J.E. and Liller, W., Astrophys. J. Lett., 220, L127 (1978).

1630-47 313 Grindlay, J.E., IAU Circ. No. 3104 (1977).

1626-67 315 Grindlay, J.E., preprint (1978).

Gal. center 316 Grindlay, J.E., and Liller, W., this volume.

1630-47 317 Grindlay, J.E., pvt. comm. (1978).

HZ Her 318 Groth, E.J., Astrophys. J., 192, 517 (1974).

0620-00 320 Gull, T.R., York, D.G., Snow, T.P., Jr., and Henize, K.G., Astrophys. J., 206, 260 (1976).

Sco X-1 321 Gursky, H., Giacconi, R., Gorenstein, P., Waters, J.R., et al., Astrophys. J., 146, 310 (1966).

A-3 posits. 322 Gursky, H., Bradt, H., Schwartz, D.A., Schwarz, J., et al., Astrophys J., 223, 973 (1978).

Wra 977 323 Hammerschlag-Hensberge, G., Zuiderwijk, E.J., et al., Astr. Astrophys., 49, 321 (1976).

1700-37 325 Hammerschlag-Hensberge, G. and Zuiderwijk, E.J., Astr. Astrophys., 54, 543 (1977).

1700-37 326 Hammerschlag-Hensberge, G., and Wu, C.C., Astr. Astrophys., 56, 433 (1977).

L.S. cat 327 Hardorp, J., Rohlfs, K., Slettebak, A., and Stock, J., Luminous Stars in the Northern Milky Way, vol. 1, Warner and Swasey Observatory (1959).

Algol 328 Harnden, F.R.,Jr., Fabricant, D., Topka, K., et al., Astrophys. J., 214, 418 (1977).

0042,AM Her 330 Harris, D.E., Bahcall, N.A., and Strom, R.G., Astr. Astrophys., 60, 27 (1977).

GX2+5 331 Hawkins, F.J., Mason, K.O., and Sanford, P.W., Nature Phys. Sci., 241, 109 (1973).

X Per 332 Hawkins, F.J., Mason, K.O., and Sanford, P.W., Astrophys. Lett., 16, 19 (1975).

Cir X-1 333 Haynes, R.F., Caswell, J.L., and Simons, L.W., IAU Circ. No. 2977 (1976).

1743-322 334 HEAO-1 Experimenters, IAU Circ. No. 3113 (1977).

EX Hydrae 335 HEAO A-3 Group, in prep., presented at San Diego HEAO meeting by D. Schwartz (1978).

HEAO posits. 336 HEAO A-3 Group, papers in preparation (1978).

SS Cyg 337 Hearn, D.R., Richardson, J.A., and Li, F.K., Bull. Am. Astr. Soc., 8, 447 (1976).

AM Her 338 Hearn, D.R., Richardson, J.A., and Clark, G.W., Astrophys. J. Lett., 210, L23 (1976).

AM Her 340 Hearn, D.R., and Richardson, J.A., Astrophys. J. Lett., 213, L115 (1977).

MXB1730-335 341 Heise, J. and Grindlay, J., IAU Circ. No. 2929 (1976).

SS Cyg 342 Heise, J., Mewe, R., Brinkman, A.C., et al., Astr. Astrophys. Lett, 63, L1 (1978).

SMC X-1 343 Henry, P. and Schreier, E., Astrophys. J. Lett., 212, L13 (1977).

1700-37 345 Hensberge, G., van den Heuvel, E.P.J., et al., Astr. Astrophys., 29, 69 (1973).

Gamma sources 346 Hermsen, W., Swanenburg, B.N., Bignami, G.F., Boella, G., et al., Nature, 269, 494 (1977).

X-ray trans. 347 Hertz, P., Joss., P.C., and Rappaport, S., Astrophys. J., 224, 614 (1978).

NGC 6441 348 Hesser, J.E. and Hartwick, F.D.A., Astrophys. J., 203, 97 (1976).

Pulsars 350 Hewish, A., Ann. Rev. Astr. Astrophys., 8, 265 (1970).

Algol 351 Hill, G., Barnes, J.V., Hutchings, J.B., and Pearce, J.A., Astrophys. J., 168, 443 (1971).

352-441

O and B stars	352	Hiltner, W.A., Astrophys. J. Suppl., 2, 389 (1956).
Sco X-1	353	Hiltner, W.A. and Mook, D.E., Astrophys. J., 150, 851 (1967).
Opt. review	355	Hiltner, W.A. and Mook, D.E., Ann. Rev. Astr. Astrophys., 8, 139 (1970).
0900-40	356	Hiltner, W.A., Werner, J., Osmer, P., Astrophys. J. Lett., 175, L19 (1972).
LMC X-4	357	Hiltner, W.A., IAU Circ. No. 3039 (1977).
Sco X-1	358	Hjellming, R.M., and Wade, C.M., Astrophys. J. Lett., 164, L1 (1971).
CygX-1,GX17+2	360	Hjellming, R.M. and Wade, C.M., Astrophys. J. Lett., 168, L21 (1971).
Algol	361	Hjellming, R.M., Webster, E., and Balick, B., Astrophys. J. Lett., 178, L139 (1972).
Algol	362	Hjellming, R.M., Wade, C.M., and Webster, E., Nature Phys. Sci., 236, 43 (1972).
Cyg X-3	363	Hjellming, R.M., Hermann, M., and Webster, E., Nature, 237, 507 (1972).
Cyg X-3	365	Hjellming, R.M., Science, 182, 1089 (1973).
Cyg X-1	366	Hjellming, R.M., Gibson, D.M., and Owen, F.N., Nature, 256, 111 (1975).
GX17+2	367	Hjellming, R.M., Astrophys. J., 221, 225 (1978).
GX17+2	368	Hoag, A. and Weisberg, J.M., Astrophys. J., 209, 908 (1976).
GX5-1	370	Hoffman, J.A., Davison, P.J.N., and Morrison, L.V., Nature, 244, 347 (1973).
MXB1728-34	371	Hoffman, J.A., Lewin, W.H.G., Doty, J., et al., Astrophys. J. Lett., 210, L13 (1976).
MXB1728-34	372	Hoffman, J.A., Lewin, W.H.G., and Doty, J., Mon. Not. R. astr. Soc., 179, 57p (1977); erratum in (373, Hoffman, p. L28).
MXB1636-53	373	Hoffman, J.A., Lewin, W.H.G., and Doty, J., Astrophys. J. Lett., 217, L23 (1977).
Fast trans.	375	Hoffman, J.A., Lewin, W.H.G., Doty, J., et al., Astrophys. J. Lett., 221, L57 (1978).
Bursts I,II	376	Hoffman, J.A., Marshall, H.L., and Lewin, W.H.G., Nature, 271, 630 (1978).
Sco X-1	377	Holt, S.S., Boldt, E.A., Serlemitsos, P.J., et al., Astrophys. J. Lett., 205, L27 (1976).
Sco X-1	378	Holt, S.S., Boldt, E.A., Serlemitsos, P.J., et al., Astrophys. J. Lett., 205, L79 (1976).
Cyg X-2	380	Holt, S.S., Boldt, E.A., Serlemitsos, P.J., et al., Astrophys. J. Lett., 205, L143 (1976).
Cyg X-1	381	Holt, S.S., Kaluzienski, L.J., Boldt, E.A., and Serlemitsos, P.J., Nature, 261, 213 (1976).
Cyg X-3	382	Holt, S.S., Boldt, E.A., Serlemitsos, P.J., Kaluzienski, L.J., et al., Nature, 260, 592 (1976).
Her X-1	383	Holt, S.S., Boldt, E.A., Kaluzienski, L.J., Serlemitsos, P.J., et al., Nature, 263, 484 (1976).
Aql X-1	385	Holt, S.S., and Kaluziensky, L.J., IAU Circ. No. 3031 (1977).
0115+63	386	Holt, S.S., and Kaluzienski, L.J., IAU Circ. No. 3161 (1978).
CygX-2,CenX-3	387	Holt, S,S., et al., presented at the HEAO/AAS Mtg., San Diego (1978).
GX304-1,1239	388	Huckle, H.E., Mason, K.O., White, N.E., et al., Mon. Not. R. astr. Soc., 180, 21p (1977).
Sco X-1	390	Hudson, H.S., Peterson, L.E., and Schwartz, D.A., Astrophys. J. Lett., 159, L51 (1970).
Cyg X-3	391	Hughes, V.A., Woodsworth, A., Gregory, P.C., et al., Astrophys. J., 191, 749 (1974).
Gamma Cas	392	Hutchings, J.B., Mon. Not. R. astr. Soc., 150, 55 (1970).
X Per	393	Hutchings, J.B., Cowley, A.P., Crampton, D., et al., Astrophys. J. Lett., 191, L101 (1974).
X Per	395	Hutchings, J.B., Crampton, D., and Redman, R.O., Mon. Not. R. astr. Soc., 170, 313 (1975).
HZ Her	396	Hutchings, J.B. and Crampton, D., Astr. Astrophys., 59, 441 (1977).
SMC X-1	397	Hutchings, J.B., Crampton, D., Cowley, A.P., and Osmer, P.S., Astrophys. J., 217, 186 (1977).
0535+26	398	Hutchings, J.B., Bernard. J.E., Crampton, D., and Cowley, A.P., Astrophys. J., 223, 530 (1978).
NGC 6441	400	Illingworth, G.D., Illingworth, W., Astrophys. J. Suppl., 30, 227 (1976).
1626-673	401	Ilovaisky, S.A., Motch, C., and Chevalier, C., preprint (1978).
LMC	402	Isserstedt, J., Astr. Astrophys. Suppl., 19, 259 (1975).
A1118-61	403	Ives, J.C., Sanford, P.W., and Bell Burnell, S.J., Nature, 254, 578 (1975).
GX3+1	405	Janes, A.F., Pounds, K.A., Ricketts, M.J., Willmore, A.P., et al., Nature, 244, 349 (1973).
TrA X-1=1524	406	Jernigan, J.G., IAU Circ. No. 2817 (1975).
Gamma Cas	407	Jernigan, J.G., IAU Circ. No. 2900 (1976).
1803-24	408	Jernigan, J.G., IAU Circ. No. 2957 (1976).
2S burst pos.	410	Jernigan, J.G., Apparao, K.M.V, Bradt, H.V., Doxsey, R.E., et al., Nature, 270, 321 (1977).
MXB1730-33	411	Jernigan, J.G., (1976), pvt. comm. in (503,Lewin)
2S posits. #7	412	Jernigan, J.G., Apparao, K.M.V., Bradt, H.V., Doxsey, R.E., et al., Nature, 272, 701 (1978).
1145-61	413	Jernigan, J.G., Bradt, H., van Paradijs, J., and Rappaport, S., IAU Circ. No. 3225 (1978).
Glob. clust.	415	Jernigan, J.G., and Clark, G.W., Astrophys. J. Lett., submitted (1978).
0115+63	416	Johns, M., Koski, A., Canizares, C., and McClintock, J., IAU Circ. No. 3171 (1978).
Astron. meas.	417	Johnson, H.J., Ann. Rev. Astr. Astrophys., 4, 193 (1966).
Glob. clust.	418	Johnson, H.M., Astrophys. J., 208, 706 (1976).
NGC 1851	420	Johnson, H.M., Catura, R.C., Charles, P.A., and Sanford, P.W., Astrophys. J., 212, 112 (1977).
Bursts	421	Johnson, H.M., Catura, R.C., Lamb, P.A., White, N.E., et al., Astrophys. J., 222, 664 (1978).
0115+63	422	Johnston, M.D., Bradt, H., Doxsey, R., et al., Astrophys. J. Lett., 223, L71 (1978).
LMC survey	423	Johnston, M.D., et al., in prep., also PhD Thesis, MIT (1978).
LMC X-1,2,3	425	Johnston, M.D., Bradt, H., Doxsey, R., et al., Astrophys. J. Lett, 225, L59 (1978).
0535-668	426	Johnston, M.D., et al., in prep., also PhD thesis, MIT (1978).
1700-37	427	Jones, C., Forman, W., Tananbaum, H., et al., Astrophys. J. Lett., 181, L43 (1973).
1700-37	428	Jones, C., and Liller, W., Astrophys. J. Lett., 184, L65 (1973).
0900-40	430	Jones, C. and Liller, W., Astrophys. J. Lett., 184, L121 (1973).
Opt. candid.	431	Jones, C., Chetin, T., and Liller, W., Astrophys. J. Lett., 190, L1 (1974).
Cir X-1	432	Jones, C., Giacconi, R., Forman, W., and Tananbaum, H., Astrophys. J. Lett., 191, L71 (1974).
Her X-1	433	Jones, C., and Forman, W., Astrophys. J. Lett., 209, L131 (1976).
1630-47	435	Jones, C., Forman, W., Tananbaum, H., and Turner, M.J.L., Astrophys. J. Lett., 210, L9 (1976).
43 spectra	436	Jones, C., Astrophys. J., 214, 856 (1977).
n. s. masses	437	Joss, P.C. and Rappaport, S.A., Nature, 264, 219 (1976).
Her X-1	438	Joss, P.C., Li, F.K., Wang, Y.-M., and Hearn, D.R., Astrophys. J., 214, 874 (1977).
1627-673	440	Joss, P.C., Avni, Y., and Rappaport, S., Astrophys. J., 221, 645 (1978).
Bulge model	441	Joss, P.C., and Rappaport, S., Astr. Astrophys., in press.

Her X-1 442 Joss, P.C., Fechner, W.B., Forman, W., and Jones, C., Astrophys. J., (in press) (1978).
1608-52 443 Kaluzienski, L.J., Holt, S.S., Boldt, E.A., and Serlemitsos, P.J., IAU Circ. No. 2859 (1975).
1524-62 445 Kaluzienski, L.J., Holt, S.S., Boldt, E.A., et al., Astrophys. J. Lett., 201, L121 (1975).
Cir X-1 446 Kaluzienski, L.J., Holt, S.S., Boldt, E.A., et al., Astrophys. J. Lett., 208, L71 (1976).
1608-52 447 Kaluzienski, L.J. and Holt, S.S., IAU Circ. No. 3099 (1977).
1630-47 448 Kaluzienski, L.J. and Holt, S.S., IAU Circ. No. 3104 (1977).
Nova Oph 450 Kaluzienski, L.J. and Holt, S.S., IAU Circ. No. 3104 (1977).
1608-52 451 Kaluzienski, L.J. and Holt, S.S., IAU Circ. No. 3108 (1977).
1743-32 452 Kaluzienski, L.J. and Holt, S.S., IAU Circ. No. 3099 and 3106 (1977).
1608-52 453 Kaluzienski, L.J. and Holt, S.S., IAU Circ. No. 3129 (1977).
1630-47 455 Kaluzienski, L.J. and Holt, S.S., IAU Circ. No. 3144 (1977).
0620-00 456 Kaluzienski, L.J., Holt, S.S., Boldt, E.A., and Serlemitsos, P.J., Astrophys., J., 212, 203 (1977).
Aql X-1 457 Kaluzienski, L.J., Holt, S.S., Boldt, E.A., and Serlemitsos, P.J., Nature, 265, 606 (1977).
1630-47 458 Kaluzienski, L.J., Boldt, E.A., Holt, S.S., Mushotsky, R.F., et al., IAU Circ. No. 3197 (1978).
Cir X-1 460 Kaluzienski, L.J. and Holt, S.S., IAU Circ. No. 3171 (1978).
0449-55 461 Kaluzienski, L.J., Robinson-Saba, J.L., et al., IAU Circ. No. 3174 (1978).
Star collapse 462 Katz, J.I., Nature, 253, 698 (1975).
Spectr. class 463 Keenan, P.C., and Morgan, W.W. in Astrophysics, ed: J. Hynek, McGraw-Hill, NY (1951), p. 23.
GX301-2 465 Kelley, R. and Bradt, H., IAU Circ. No. 3165 (1978).
Gal. center 466 Kellogg, E., Gursky, H., Murray, S., et al., Astrophys. J. Lett., 169, L99 (1971).
Cyg X-1 467 Kemp, J.C., Herman, L.C., Rudy, R.J., and Barbour, M.S., Nature, 270, 227 (1977).
Cyg X-1 468 Kemp, J.C., Barbour, M.S., Herman, L.C., and Rudy, R.J., Astrophys. J. Lett., 220, L123 (1978).
Her X-1 470 Kendziorra, E., Staubert, R., Pietsch, W., et al., Astrophys. J. Lett., 217, L93 (1977).
Crab pulsar 471 Kestenbaum, H.L., Ku, W., Novick, R., and Wolff, R.S., Astrophys. J. Lett., 203, L57 (1976).
0620-00 472 Kestenbaum, H.L., Cohen, G.G., Long, K.S., et al., Astrophys. J. Lett., 208, L27 (1976).
Cyg X-3 473 Kestenbaum, H.L., Long, K.S., Novick, R., et al., Astrophys. J. Lett., 216, L19 (1977).
MXB1730-335 475 Kleinmann, D.E., Kleinmann, S.G., and Wright, E.L., Astrophys. J. Lett., 210, L83 (1976).
1728-337 476 Kleinmann, D.E., and Wright, E.L., in prep. (1978).
0620-00 477 Kleinmann, S.G., Brecher, K., and Ingham, W.H., Astrophys. J., 207, 532 (1976).
Cen X-3 478 Krzeminski, W., Astrophys. J. Lett., 192, L135 (1974).
GX1+4 480 Krzeminski, W., and Priedhorsky, W.C., preprint (1978).
Cyg X-3 481 Lamb, R.C., Fichtel, C.E., Hartman, R.C., et al., Astrophys. J. Lett., 212, L63 (1977).
Sco X-1 483 Laros, J.G., and Singer, S., Astrophys. J., 205, 550 (1976).
Cyg X-3 485 Lauque, R., Lequeux, J., Nguyen-Quang-Rieu, Nature Phys. Sci., 239, 119 (1972).
Cyg X-3 486 Leach, R.W., Murray, S.S., Schreier, E.J., et al., Astrophys. J., 199, 184 (1975).
Cyg X-3 487 Ledden, J.E., Hodge, P.E., and Aller, H.D., Nature, 262, 669 (1976).
LMC, SMC 488 Leong, C., Kellogg, E., Gursky, H., Tananbaum, H., et al., Astrophys. J. Lett., 170, L67 (1971).
GX304-1 490 Lewin, W.H.G., Clark, G.W., and Smith, W.B., Astrophys. J. Lett., 152, L49 (1968).
Sco X-1 flare 491 Lewin, W.H.G., Clark, G.W., and Smith, W.B., Astrophys. J. Lett., 152, L55 (1968).
GX1+4 492 Lewin, W.H.G., Ricker, G.R., and McClintock, J.E., Astrophys. J. Lett., 169, L17 (1971).
GX301-2 493 Lewin, W.H.G., McClintock, J.E., Ryckmann, S.G., et al., Astrophys. J. Lett., 166, L69 (1971).
Gal cen bursts 495 Lewin, W.H.G., Hoffman, J.A., Doty, J., et al., Mon. Not. R. astr. Soc., 177, 83p (1976).
Aql, 1905 496 Lewin, W.H.G., Li, F.K., Hoffman, J.A., et al., Mon. Not. R. astr. Soc., 177, 93p (1976).
MXB1730-335 497 Lewin, W.H.G., Doty, J., Clark, G.W., et al., Astrophys. J. Lett., 207, L95 (1976).
MXB1728-34 498 Lewin, W.H.G., IAU Circ. No. 2922 (1976).
GX1+4 500 Lewin, W.H.G., pvt. comm. in (33, Becker) (1976).
Burst obs. 501 Lewin, W.H.G., Doty, J., Hoffman, J.A., and Li, F.K., IAU Circ. No. 2984 (1976).
MXB1659-29 502 Lewin, W.H.G., Hoffman, J.A., and Doty, J., IAU Circ. No. 2994 (1976).
Burst review 503 Lewin, W.H.G., Mon. Not. R. astr. Soc., 179, 43 (1977).
Burst review 505 Lewin, W.H.G., Proc. of the 8th Tex. Symp. on Rel. Astr., Ann. N.Y. Acad. Sci., 302, 210 (1977).
Burst review 506 Lewin, W.H.G. and Joss, P.C., Nature, 270, 211 (1977).
Burst distr. 507 Lewin, W.H.G., Hoffman, J.A.,..., Swank, J.H., Becker, R.H., et al., Nature, 267, 28 (1977).
MXB1735-44 508 Lewin, W.H.G., Hoffman, J.A., Doty, J., Li, F.K., et al., IAU Circ. No. 3075 (1977).
MXB1916-05 510 Lewin, W.H.G., Hoffman, J.A., and Doty, J., IAU Circ. No. 3087 (1977).
MXB1659-29 511 Lewin, W.H.G., Hoffman, J., Marshall., H., et al., IAU Circ. Nos. 3190 and 3193 (1978).
Burst review 512 Lewin, W.H.G., this volume (1978).
1543-47 513 Li, F.K., Sprott, G.F., and Clark, G.W., Astrophys. J., 203, 187 (1976).
NGC 6441 515 Li, F. and Clark, G., IAU Circ. No. 3095 (1977).
MXB1837+05 516 Li, F.K., Lewin, W.H.G., Clark, G.W., et al., Mon. Not. R. astr. Soc., 179, 21p (1977).
0535+26 517 Li, F.K., Rappaport, S., Clark, G.W., and Jernigan, J.G., Astrophys. J., submitted (1978).
LMC X-4 518 Li, F., Rappaport, S., and Epstein, A., Nature, 271, 37 (1978).
GX339-4 520 Li, F.K., Clark, G., and Rappaport, S., IAU Circ. No. 3238 (1978).
0921-630 521 Li, F.K., van Paradijs, J.A., Clark. G.W., Jernigan, J.G., et al., preprint (1978).
NGC 6712 brst 522 Li, F.K., and Clark, G.W., pvt. comm. (1978).
NGC 6624 523 Liller, M.H. and Carney, B.W., Astrophys. J., 224, 383 (1978).
SMC X-1 525 Liller, W., Astrophys. J. Lett., 184, L37 (1973).
0535+26 526 Liller, W., IAU Circ. No. 2780 (1975).
AM Her 527 Liller, W., Sky and Telescope, 53, 351 (1977).
1728, 1730 528 Liller, W., Astrophys. J. Lett., 213, L21 (1977).
1052+607 530 Liller, W., McCrosky, R.E., Gursky, H., and Griffiths, R., IAU Circ. No. 3176 (1978).
SMC X-3 531 Lindsay, E.M., Astr. J., 66, 169 (1961)
0620-00 532 Little, A.G., Crawford, D.F., and Murdoch, H.S., Nature, 261, 113 (1976).
0620-00 533 Lloyd, C., Noble, R., and Penston, M.V., Mon. Not. R. astr. Soc., 179, 675 (1977).

535-625

0620-00 535 Locher, K., IAU Circ. Nos. 2823, 2840 (1975).
SMC X-1 536 Lucke, R., Yentis, D., Friedman, H., Fritz, G., et al., Astrophys. J. Lett., 206, L25 (1976).
Pulsar param. 537 Manchester, R.N., and Taylor, J.H., Astrophys. Lett., 10, 67 (1972).
Cyg X-1 538 Mandrou, P., Niel, M., Vedrenne, G., Dupont, A., and Hurley, K., Astrophys. J., 219, 288 (1978).
Be stars 540 Maraschi, L., Treves, A., and van den Heuvel, E.P.J., Nature, 259, 292 (1976).
1524, 0535 541 Maraschi, L., Huckle, H.E., Ives, J.C., and Sanford, P.W., Nature, 263, 34 (1976).
Cir X-1 542 Margon, B., Lampton, M., Bowyer, S., and Cruddace, R., Astrophys. J. Lett., 169, L23 (1971).
Cyg X-1 543 Margon, B., Bowyer, S., and Stone, R.P.S., Astrophys. J. Lett., 185, L113 (1973).
X Per 545 Margon, B., Bowyer, S., and Penegor, G., Mon. Not. R. astr. Soc., 176, 217 (1976).
0535+26 546 Margon, B., Nelson, J., Chanan, G., Thorstensen, J.R., et al., Astrophys. J., 216, 811 (1977).
X Per 547 Margon, B., Thorstensen, J.R., Bowyer, S., et al., Astrophys. J., 218, 504 (1977).
0114+650 548 Margon, B., Bradt, H.V., and Dower, R.G. (in preparation); also see IAU Circ. No. 3144 (1977).
0142+614 550 Margon, B., and Bradt, H.V., pvt. comm. (1978).
GX17+2 551 Margon, B., Astrophys. J., 219, 613 (1978).
1956+11 552 Margon, B., Thorstensen, J.R., and Bowyer, S., Astrophys. J., 221, 907 (1978).
Dwarf novae 553 Margon, B., Szkody, P., Bowyer, S., Lampton, M., and Paresce, F., Astrophys. J., 224, 167 (1978).
Aql X-1 555 Margon, B., Katz, J.I., and Petro, L.D., Nature, 271, 633 (1978).
HZ Her 556 Margon, B., and Cohen, J.G., Astrophys. J. Lett., in press (1978).
Opt. review 557 Margon, B., this volume (1978).
1845-024 558 Margon, B.M., pvt. comm. (1978).
GX339-4 560 Markert, T.H., Canizares, C.R., Clark, G.W., et al., Astrophys. J. Lett., 184, L67 (1973).
NGC 6440 561 Markert, T.H., Backman, D.E., Canizares, C.R., Clark, G.W., et al., Nature, 257, 32 (1975).
LMC 562 Markert, T.H., and Clark, G.W., Astrophys. J. Lett., 196, L55 (1975).
0042+327 563 Markert, T.H., Canizares, C.R., Clark, G.W., Li, F.K., et al., Astrophys. J., 206, 265 (1976).
1715-321 565 Markert, T.H., Backman, D.E., and McClintock, J.E., Astrophys. J. Lett., 208, L115 (1976).
OSO-7 sources 566 Markert, T.H., Canizares, C.R., Clark, G.W., et al., Astrophys. J., 218, 801 (1977).
1M cat. 567 Markert, T.H., Winkler, P.F., Laird, F.N., Clark, G.W., et al., Astrophys. J. Suppl., in press.
Gamma Cas 568 Marlborough, J.M., Pub. Astr. Soc. Pac., 89, 122 (1977).
MXB1706-43 570 Marshall, H., Li, F., and Rappaport, S., IAU Circ. No. 3134 (1977).
MXB1730-335 571 Marshall, H.L., Ulmer, M.P., Hoffman, J.A., Doty, J., et al., Astrophys. J., in press (1978).
Gamma Cas 572 Mason, K.O., White, N.E., and Sanford, P.W., Nature, 260, 690 (1976).
X Per 573 Mason, K.O., White, N.E., Sanford, P.W., et al., Mon. Not. R. astr. Soc., 176, 193 (1976).
Cyg X-3 575 Mason, K.O., Becklin, E.E., Blankenship, L., Brown, R.L., et al., Astrophys. J., 207, 78 (1976).
GX304-1 576 Mason, K.O., Murdin, P.G., Parkes, G.E., et al., Mon. Not. R. astr. Soc., 184, 45p (1978).
Accret. torq. 577 Mason, K.O., Mon. Not. R. astr. Soc., 178, 81p (1977).
GX1+4 posit. 578 Mason, K.O., private communication in (185, Davidsen).
GX301-2 580 Mason, K.O., White, N.E., and Sanford, P.W., BAAS, 10, 507 (1978).
1653-40 581 Massey, P., and Conti, P.S., IAU Circ. No. 3247 (1978).
1543-47 582 Matilsky, T.A., Giacconi, R., Gursky, H., et al., Astrophys. J. Lett., 174, L53 (1972).
0620-00 583 Matilsky, T, Bradt, H.V., Buff, J., Clark, G.W., et al., Astrophys. J. Lett., 210, L127 (1976).
1700-37 585 Matilsky, T., La Sala, J., and Jessen, J., Astrophys. J. Lett. 224, L119 (1978); also this volume.
GX349+2 586 Mayo, S.K., Fabian, A.C., Pringle, J.E., Whelan, J.A.J., et al., IAU Circ. No. 3250 (1978).
GX301-2,304-1 587 McClintock, J.E., Ricker, G.R., and Lewin, W.H.G., Astrophys. J. Lett., 166, L73 (1971).
0900-40 588 McClintock, J.E., Rappaport, S., Joss, P.C., et al., Astrophys. J. Lett., 206, L99 (1976).
1636,1735,1626 590 McClintock, J.E., Canizares, C.R., Bradt, H.V., Doxsey, R.E., et al., Nature, 270, 320 (1977).
GX304-1 591 McClintock, J.E., Rappaport, S.A., Nugent, J.J., et al., Astrophys. J. Lett., 216, L15 (1977).
GX349+2 592 McClintock J.E., and Canizares, C.R., IAU Circ. No. 3250 (1978).
1735-44 593 McClintock, J.E., Canizares, C.R., and Backman, D.E., Astrophys. J. Lett., 223, L75 (1978).
1627-673 595 McClintock, J.E., and Canizares, C.R., in prep. (1978).
Cyg X-3 596 McEllin, M., Mon. Not. R. astr. Soc., 175, 5p (1976).
Cyg X-1 597 Michalsky, J.J. and Swedlund, J.B., Astrophys. J., 212, 221 (1977).
AM Her 598 Michalsky, J.J., Stokes, G.M., and Stokes, R.A., Astrophys. J. Lett., 216, L35 (1977).
GX3+1 600 Miley, G.K., pvt. comm. in (405, Janes).
Bulge sources 601 Milgrom, M., Astr. Astrophys., 67, L25 (1978).
Sco X-1 602 Miyamoto, S., Matsuoka, M., Oda, M., and Ogawara, Y., Astr. Astrophys., 65, 329 (1978).
Gam.Cas,X Per 603 Moffat, A.F.J., Haupt, W., and Schmidt-Kaler, Th., Astr. Astrophys., 23. 433 (1973).
1700-37 605 Moffat, A.F.J. and Dachs, J., Astr. Astrophys., 58, L5 (1977).
X Per B 606 Moffett, T.J., and Barnes, T.G., Astrophys. J. Lett., 207, L185 (1976).
X Per 607 Mook, D.E., Boley, F.I., Foltz, C.B., and Westpfahl, D., Pub. Astr. Soc. Pac., 86, 894 (1974).
Sco X-1 608 Mook, D.E., Messina, R.J., Hiltner, W.A., Belian, R., et al., Astrophys. J., 197, 425 (1975).
Aql X-1 610 Mook, D., Kurtz, M., Weed, J., and Johns, M., IAU Circ. No. 3251 (1978).
Blue giant st 611 Morgan, W.W., Code, A.D., and Whitford, A.E., Astrophys. J. Suppl., 2, 41 (1955).
SMC X-2 612 Morton, D.C., and Murdin, P., IAU Circ. No. 3247 (1978).
0614+09 613 Murdin, P., Penston, M.J., Penston, M.V., et al., Mon. Not. R. astr. Soc., 169, 25 (1974).
0535+26 615 Murdin, P., IAU Circ. No. 2784 (1975).
X Per 616 Murdin, P., Penston, M.V., and Penny, A.J., Mon. Not. R. astr. Soc., 176, 233 (1976).
1524-61 617 Murdin, P., Griffiths, R.E., Pounds, K.A., et al., Mon. Not. R. astr. Soc., 178, 27p (1977).
UHGLS 618 Murray, S.S. and Ulmer, M.P., Astrophys. J., 210, 230 (1976).
X Per 620 Mushotzky, R.F., Roberts, D.H., Baity, W.A., et al., Astrophys. J. Lett., 211, L129 (1977).
Clus. spectra 621 Mushotzky, R.F., Serlemitsos, P.J., Smith, B.W., et al., Astrophys. J., 225, 21 (1978).
Cyg X-1 622 Natali, G., Fabrianesi, R., and Messi, R., Astr. Astrophys., 62, L1 (1978).
Cyg X-1 623 Natali, G., and Messi, R., Astr. Astrophys., 67, L33 (1978).
Her X-1 625 Nelson, J.E., Chanan, G.A., and Middleditch, J., Astrophys. J., 212, 215 (1977).

Sco X-1	626	Neugebauer, G., Oke, J.B., Becklin, E., and Garmire, G., Astrophys. J., 155, 1 (1969).
M15/NGC 7078	627	Newell, B., DaCosta, G.S., and Norris, J., Astrophys. J. Lett., 208, L55 (1976).
Cyg X-1	628	Oda, M., Doi, K., Ogawara, Y., Takagishi, K., et al., Astrophys. and Sp. Sci., 42, 223 (1976).
Cyg X-1	630	Oda, M., Gorenstein, P., Gursky, H., Kellogg, E., et al., Astrophys. J. Lett., 166, L1 (1971).
Cyg X-1	631	Oda, M., Sp. Sci. Rev., 20, 757 (1977).
Cyg X-1	632	Ogawara, Y., Doi, K., Matsuoka, M., Miyamoto, S., and Oda, M., Nature, 270, 154 (1977).
0900-40	633	Ogelman, H., Beuermann, K.P., Kanbach, G., et al., Astr. Astrophys., 58, 385 (1977).
0620-00	635	Oke, J.B. and Greenstein, J.L., Astrophys. J., 211, 872 (1977).
0620-00	636	Oke, J.B., Astrophys. J., 217,181 (1977).
Cen X-3	637	Osmer, P.S., Hiltner, W.A., and Whelan, J.A.J., Astrophys. J., 195, 705 (1975).
0620-00	638	Owen, F.N., et al., Astrophys. J. Lett., 203, L15 (1976).
1538-52	640	Parkes, G.E., Murdin, P., and Mason, K.O., IAU Circ. No. 3184 (1978).
Cen X-3	641	Parkinson, J.H., Mason, K.O., and Sanford, P.W., Nature, 249, 746 (1974).
Cyg X-3	642	Parsignault, D.R., Gursky, H., Kellogg, E.M., et al., Nature Phys. Sci., 239, 123 (1972).
Cyg X-3	643	Parsignault, D.R., Schreier, E., Grindlay, J,. and Gursky, H., Astrophys. J. Lett., 209, L73 (1976).
Cyg X-3	645	Parsignault, D.R., Grindlay, J., Gursky, H., and Tucker, W., Astrophys. J., 218, 232 (1977).
1700-37	646	Penny, A.J., Olowin, R.P., Penfold, J.E., et al., Mon. Not. R. astr. Soc., 163, 7p (1973).
Opt. search	647	Penston, M.V., Penston, M.J., Murdin, P., et al., Mon. Not. R. astr. Soc., 172, 313 (1975).
X Per B	648	Penston, M.V., Murdin, P., Penston, M.J., et al., Mon. Not. R. astr. Soc., 176, 237 (1976).
Nova Oph	650	Peters, G., private communication in (300, Griffiths).
Crab pulsar	651	Peterson, L.E., IAU Symp. No. 55, X and Gamma Ray Astr., ed. H. Bradt and R. Giacconi, Dordrect, Reidel, p. 51 (1973).
Glob. clust.	652	Peterson, C.J. and King, I.R., Astron. J., 80, 427 (1975).
HZ Her	653	Petro, L. and Hiltner, W.A., Astrophys. J. Lett., 181, L39 (1973).
Sk 160	655	Petro, L., Feldman, F., and Hiltner, W.A., Astrophys. J. Lett., 184, L123 (1973).
Wra 977	656	Petro, L., pvt. comm. in (78, Bradt).
1845-024	657	Petro, L., pvt. comm. (1978).
Her X-1	658	Petterson, J.A., Astrophys. J., 218, 783 (1977).
Cyg X-3	660	Pietsch, W., Kendziorra, E., Staubert, R., and Trumper, J., Astrophys. J. Lett., 203, L67 (1976).
1653-40	661	Polidan, R.S., Pollard, G.S.G., Sanford, P.W., and Locke, M.C., Nature, 275, 296 (1978).
1524-61	662	Pounds, K.A., IAU Circ. No. 2729 (1974).
Cen X-3	663	Pounds, K.A., Cooke, B.A., Ricketts, M.J., et al., Mon. Not. R. astr. Soc. 172, 473 (1975).
1052+607	665	Pounds, K.A., pvt. comm. in (743, Schwartz).
Her X-1	666	Pravdo, S.H., Becker, R.H., Boldt, E.A., et al., Astrophys. J. Lett., 215, L61 (1977).
Her X-1	667	Pravdo, S.H., Boldt, E.A., Holt, S.S., and Serlemitsos, P.J., Astrophys. J. Lett., 216, L23 (1977).
Her X-1	668	Pravdo, S.H., Bussard, R.W., Becker, R.H., Boldt, E.A., et al., preprint (1978).
SMC, LMC	670	Price, R.E., Groves, D.J., Rodrigues, R.M., et al., Astrophys. J. Lett., 168, L7 (1971).
AM Her	671	Priedhorsky, W.C., Astrophys. J. Lett., 212, L117 (1977).
AM Her	672	Priedhorsky, W.C., and Krzeminski, W., Astrophys. J., 219, 597 (1978).
Cyg X-1	673	Priedhorsky, W.C., Garmire, G.P., Rothschild, R., Boldt, E., et al., preprint.
AM Her	675	Priedhorsky, W.C., Krzeminski, W., and Tapia, S., Astrophys. J., 225, 542 (1978).
Stellar winds	676	Prilutskii, O.F., and Usov, V.V., Sov. Astron., 20, 2 (1976).
SMC X-1	677	Primini, F., Rappaport, S., Joss, P.C., et al., Astrophys. J. Lett., 210, L71 (1976).
SMC X-1	678	Primini, F., Rappaport, S., and Joss, P.C., Astrophys. J., 217, 543 (1977).
Gal. center	680	Proctor, R.J., Skinner, G.K., et al., Mon. Not. R. astr. Soc., submitted (1978); also (228, Eyles).
Aql X-1	681	Puetter, R.C., and Smith, H.E., IAU Circ. No. 3243 (1978).
Gamma Cas	682	Purton, C.R., pvt. comm. in (568, Marlborough).
NGC 1851,7078	683	Pye, J.P., and Cooke, B.A., Nature, 260, 410 (1976).
Cyg X-1	685	Rao, U.R., Kasturirangan, K., Sharma, D.P., and Radha, M.S., Nature, 260, 307 (1976).
LMC	686	Rapley, C.G., and Tuohy, I.R., Astrophys. J. Lett., 191, L113 (1974).
Cyg X-1	687	Rappaport, S.A., Zaumen, W., and Doxsey, R., Astrophys. J. Lett., 168, L17 (1971)
Cyg X-1	688	Rappaport, S.A., Doxsey, R., and Zaumen, W., Astrophys. J. Lett., 168, L43 (1971).
GX349+2,340+0	690	Rappaport, S.A., Zaumen, W., Doxsey, R., Mayer, W., Astrophys. J. Lett., 169, L93 (1971).
SS Cyg	691	Rappaport, S.A., Cash, W., Doxsey, R., et al., Astrophys. J. Lett., 187, L5 (1974).
0900-40	692	Rappaport, S.A., Joss, P.C., and McClintock, J.E., Astrophys. J. Lett., 206, L103 (1976).
0535+26	693	Rappaport, S.A., Joss, P.C., Bradt, H., Clark, G.W., et al., Astrophys. J. Lett., 208, L119 (1976).
1627-67	695	Rappaport, S.A., Markert, T., Li, F.K., Clark, G.W., et al., Astrophys J. Lett, 217, L29 (1977).
X-ray pulsars	696	Rappaport, S.A. and Joss, P.C., Nature, 266, 123 (1977).
Acc. torques	697	Rappaport, S.A. and Joss, P.C., Nature, 266, 683 (1977).
n.s. masses	698	Rappaport, S.A., and Joss, P.C., Ann. N.Y. Acad. Sci., 302, 460 (1977).
0042+32	700	Rappaport, S.A., Clark, G.W., Dower, R., Doxsey, R., et al., Nature, 268, 705 (1977).
0115+63	701	Rappaport, S.A., Clark, G.W., Cominsky, L., and Li, F., IAU Circ. No. 3171 (1978).
0115+63	702	Rappaport, S.A., Clark., G.W., Cominsky, L., et al., Astrophys. J. Lett., 224, L1 (1978).
1700-37	703	Rappaport, S.A., Doxsey, R., and Zaumen, W., pvt. comm. in (427, Jones)
5 sources	706	Ricker, G.R., McClintock, J.E., Gerassimenko, M., et al., Astrophys. J., 184, 237 (1973).
0535+26	707	Ricker G.R., Scheepmaker, A., Ballintine, J.E., et al., Astrophys. J., 204, L73 (1976).
GX1+4,301-2	708	Ricker, G.R., Gerassimenko, M., McClintock, J.E., et al., Astrophys. J., 207, 333 (1976).
0535+26	710	Ricker, G. and Primini, F., IAU Circ. No. 3078 (1977).
0535+26	711	Ricketts, M.J., Turner, M.J.L., Page, C.G., and Pounds, K.A., Nature, 256, 631 (1975).
0620-00	712	Ricketts, M.J., Pounds, K.A., and Turner, M.J.L., Nature, 257, 657 (1975).
0042+32	713	Ricketts, M., and Cooke, B.A., IAU Circ. No. 3039 (1977).
SS Cyg	715	Ricketts, M.J., King, A.R., and Raine, D.J., Mon. Not. R. astr. Soc., submitted (1978).
Opt. novae	716	Robinson, E.L., Ann. Rev. Astr. Astrophys., 14, 119 (1976).

717-806

Cyg X-1	717	Robinson, E.L., Nather, R.E., Africano, J., and Smith, B., Nature, 271, 40 (1978).
0535+26	718	Rosenberg, F.D., Eyles, C.J., Skinner, G.K., and Willmore, A.P., Nature, 256, 628 (1975).
Cyg X-1	720	Rothschild, R.E., Boldt, E.A., Holt, S.S., et al., Astrophys. J. Lett., 189, L13 (1974).
Cyg X-1	721	Rothschild, R.E., Boldt, E.A., Holt, S.S., and Serlemitsos, P.J., Astrophys. J., 213, 818 (1977).
GX339-4	722	Samimi, J., Share, G., Wood, K., Yentis, D., et al., Bull. Am. Phys. Soc., 23, 581 (1978).
Sco X-1	723	Sandage, A.R., Osmer, P., Giacconi, R., Gorenstein, P., et al., Astrophys. J., 146, 316 (1966).
SMC	725	Sanduleak, N. Astr. J., 73, 246 (1968).
Radio search	726	Sanduleak, N. and Dolan, J.F., Astrophys. J. Lett., 187, L73 (1974).
LMC X-4	727	Sanduleak, N. and Philip, A.G.D., IAU Circ. No. 3023 (1976).
SMC X-2	728	Sanduleak, N. and Philip, A.G.D., IAU Circ. No. 3127 (1977).
SMC X-3	730	Sanduleak, N. and Philip, A.G.D., IAU Circ. No. 3134 (1977).
Cyg X-3	731	Sanford, P.W. and Hawkins, F.J., Nature Phys. Sci., 239, 135 (1972).
SAO cat.	732	SAO Star Catalog, Smithsonian Institution, Washington, D.C. (1966)
Sco OB1, 1653	733	Schild, R.E., Hiltner, W.A., et al., Astrophys. J., 156, 609 (1969).
Sagittarius	735	Schnopper, H.W., Bradt, H.V., Rappaport, S., et al., Astrophys. J. Lett., 161, L161 (1970).
Algol	736	Schnopper, H.W., Delvaille, J.P., Epstein, A., et al., Astrophys. J. Lett., 210, L75 (1976).
LMC posits.	737	Schnopper, H. and Delvaille, J., Annals of the N.Y. Academy of Sciences, 302, 300 (1977).
Cen X-3	738	Schreier, E.J., Levinson, R., Gursky, H., et al., Astrophys. J. Lett., 172, L79 (1972).
SMC X-1	740	Schreier, E.J., Giacconi, R., Gursky, H., Kellogg, E., et al., Astrophys. J. Lett., 178, L71 (1972).
Cen X-3	741	Schreier, E.J., Swartz, K., Giacconi, R., Fabbiano, G., et al., Astrophys. J., 204, 539 (1976).
1538-52	742	Schwartz, D.A., Gursky, H., Schwarz, J., Bradt, H., and Doxsey, R., Nature, submitted (1978).
4 weak sources	743	Schwartz, D.A., Bradt, H.V., Briel, U., Dower, R., et al., this volume (1978).
1642-45	745	Seaquist, E.R., Astrophys. J., 211, 547 (1977).
Cyg X-3	746	Serlemitsos, P.J., Boldt, E.A., Holt, S.S., et al., Astrophys. J. Lett., 201, L9 (1975).
Aql-Ser region	747	Seward, F.D., Page, C.G., Turner, M.J.L., et al., Mon. Not. R. astr. Soc., 175, 39p (1976).
S. Milky Way	748	Seward, F.D., Page, C.G., Turner, M.J.L., et al., Mon. Not. R. astr. Soc., 177, 13p (1976).
MXB1659-29	750	Share, G., Wood, K., Yentis, D., Johnson, N., Shulman, S., et al., IAU Circ. No. 3190 (1978).
1630-47	751	Share, G., Wood, K., Byram, E., Meekins, J., Shulman, S., et al., IAU Circ. No. 3197 (1978).
Her X-1	752	Shulman, S., Friedman, H., Fritz, G., Henry, R.C., et al., Astrophys. J. Lett., 199, L101 (1975).
Glob. clust.	753	Silk, J., and Arons, J., Astrophys. J. Lett., 200, L131 (1975).
1630-47	755	Sims, M.R., and Watson, M.G., IAU Circ. No. 3227 (1978).
Sco X-1 pos.	756	Sofia, S., Eichhorn, H., and Gatewood, G., Astron. J., 74, 20 (1969).
Bright candids.	757	Sofia, S., Astrophys. J. Lett., 188, L45 (1974).
Cyg X-1	758	Sommer, M., Maurus, H., and Urbach, R., Nature, 263, 752 (1976).
CirX-1,VelaX-1	760	Spada, G., Bradt. H., Doxsey, R., Levine, A., et al., Astrophys. J. Lett., 190, L113 (1974).
Crab pulsar	761	Staelin, D.H. and Reifenstein, E.C. III, Science, 162, 1481 (1968).
0535+26	762	Stier, M. and Liller, W., Astrophys. J., 206, 257 (1976).
AM Her	763	Stockman, H.S., Schmidt, G.D., Angel, J.R.P., et al., Astrophys. J., 217, 815 (1977).
Cyg X-1	765	Sutherland, P.G., Weisskopf, M.C., and Kahn, S.M., Astrophys. J., 219, 1029 (1978).
GX301-2	766	Swank, J.H., Becker, R.H., Boldt, E.A., Holt, S.S., et al., Astrophys. J. Lett., 209, L57 (1976).
Ser X-1 bursts	767	Swank, J.H., Becker, R.H., Pravdo, S.H., and Serlemitsos, P.J., IAU Circ. No. 2963 (1976).
1636 bursts	768	Swank, J.H., Becker, R.H., Pravdo, S.H., et al., IAU Circ. No. 3000 (1976).
Bursts	770	Swank, J.H., Becker, R.H., Pravdo, S.H., Saba, J.R., et al., IAU Circ. No. 3010 (1976).
AM Her	771	Swank, J.H., Lampton, M., Boldt, E., Holt, S., et al., Astrophys. J. Lett., 216. L71 (1977).
0614+09	772	Swank, J.H., Becker, R.H., Boldt, E.A., et al., Mon. Not. R. astr. Soc., 182, 349 (1978).
SS Cyg	773	Swank, J.H., pvt. comm. in (846, Watson).
AM Her	775	Szkody, P., and Brownlee, D.E., Astrophys. J., 212, L113 (1977).
Cyg X-1,2	776	Tananbaum, H., Kellogg, E., Gursky, H., et al., Astrophys. J. Lett., 165, L37 (1971).
GX17+2	777	Tananbaum, H., Gursky, H., Kellogg, E., and Giacconi, R., Astrophys. J. Lett., 168, L25 (1971).
Her X-1	778	Tananbaum, H., Gursky, H., Kellogg, E.M., et al., Astrophys. J. Lett., 174, L143 (1972).
Cyg X-1	780	Tananbaum, H., Gursky, H., Kellogg, E., et al., Astrophys. J. Lett., 177, L5 (1972).
1608-52	781	Tananbaum, H., Chaisson, L.J., Forman, W., et al., Astrophys. J. Lett., 209, L125 (1976).
AM Her	782	Tapia, S., Astrophys. J, Lett., 212, L125 (1977).
Aql X-1	783	Tapia, S., IAU Circ. No. 3243 (1978).
GX17+2	785	Tarenghi, M. and Reina, C., Nature Phys. Sci., 240, 53 (1972).
Radio pulsars	786	Taylor, J.H. and Manchester, R.N., Ann. Rev. Astr. Astrophys., 15, 19 (1977).
Cyg X-1	787	Terrell, N.J. Jr., Astrophys. J. Lett., 174, L35 (1972).
4 sources	788	Thomas, R.M., Davison, P.J.N., Clancy, M.C., et al., Mon. Not. R. astr. Soc., 170, 569 (1975).
0535+26	790	Thomas, R., Duldig, M., Haynes, R.F., and Murdin, P., IAU Circ. No. 3219 (1978).
Aql X-1	791	Thorstensen, J.R., Charles, P.A., and Bowyer, S., Astrophys. J. Lett., 220, L131 (1978).
Ser X-1	792	Thorstensen, J., Charles, P., and Bowyer, S., IAU Circ. No. 3253 (1978).
2129+470	793	Thorstensen, J., Charles, P., and the HEAO A-3 group, in prep. (1978).
1916-05	795	Thorstensen, J.R., Charles, P.A., Bowyer, S., and the HEAO A-3 group, presented at the HEAO/AAS Mtg. in San Diego (1978).
Algol	796	Tomkin, J,. and Lambert, D.L., Astrophys. J. Lett., 222, L119 (1978).
Crab pulsar	797	Toor, A. and Seward, F.D., Astr. J., 79, 995 (1974).
Cir X-1	798	Toor, A., Astrophys. J. Lett., 215, L57 (1977).
Crab pulsar	800	Toor, A., and Seward, F.D., Astrophys. J., 216, 560 (1977).
Her X-1	801	Trumper, J., Pietsch, W., Reppin, C., Voges, W., et al., Astrophys. J. Lett., 219, L105 (1978).
SMC, LMC	802	Tuohy, I.R. and Rapley, C.G., Astrophys. J. Lett., 198, L69 (1975).
Cen X-3	803	Tuohy, I.R., Mon. Not. R. astr. Soc., 174, 45p (1976).
0900-40	805	Ulmer, M.P., Baity, W.A., Wheaton, W.A., et al., Astrophys. J. Lett., 178, L121 (1972).
Cen X-3	806	Ulmer, M.P., Astrophys. J., 204, 548 (1976).

MXB1730-335	807	Ulmer, M.P., Lewin, W.H.G., Hoffman, J.A., et al., Astrophys. J. Lett., 214, L11 (1977).
X-Per	808	van den Bergh, S., Nature, 235, 273 (1972).
0620-00	810	van den Bergh, S., Astr. J., 81, 104 (1976).
1538-52	811	van den Bergh, S., and Younger, pvt. comm. in (175, Crampton).
1700-37	812	van Genderen, A.M., Astr. Astrophys., 54, 683 (1977).
SMC X-1	813	van Genderen, A.M., Astr. Astrophys., 54, 307 (1977).
Wra 977	815	van Genderen, A.M., Astr. Astrophys., 54, 733 (1977).
0900-40	816	van Genderen, A.M., Astr. Astrophys., 55, 473 (1977).
SMC X-1	817	van Paradijs, J., and Zuiderwijk, E., Astr. Astrophys., 61, L19 (1977).
SMC X-2	818	van Paradijs, J., Schlosser, W., and Tarenghi, M., IAU Circ. No. 3134 (1977).
1538-52	820	van Paradijs, J., IAU Circ. No. 3197 (1978).
Burst prop.	821	van Paradijs, J., Nature, 274, 650 (1978).
NGC 1851	822	Vidal, N.V. and Freeman, K.C., Astrophys. J. Lett., 200, L9 (1975).
GX301-2	823	Vidal, N.V., Astrophys. J. Lett., 186, L81 (1973).
Aql X-1	825	Vrba, F.J., IAU Circ. No. 3243 (1978).
Sco X-1	826	Wade, C.M., and Hjellming, R.M., Astrophys. J., 170, 523 (1971).
Algol	827	Wade, C.M. and Hjellming, R.M., Nature, 235, 270 (1972).
Cyg X-1	828	Wade, C.M., and Hjellming, R.M., Nature, 235, 271 (1972).
0535+26	830	Wade, R.A., and Oke, J.B., Astrophys. J., 215, 568 (1977).
Cyg X-1	831	Walborn, N.R., Astrophys. J. Lett., 179, L123 (1973).
V861 Sco,1653	832	Walker, E.N., Mon. Not. R. astr. Soc., 152, 333 (1971).
V861 Sco,1653	833	Walker, E.N., Mon. Not. R. astr. Soc., 159, 253 (1972).
Cyg X-1	835	Walker, E.N., Watson, M.G., and Holt, S.S., Nature, 270, 229 (1977).
Cyg X-1	836	Walker, E.N. and Quintanilla, A.R., Mon. Not. R. astr. Soc., 182, 315 (1978).
1052+607	837	Wallerstein, G., pvt. comm. in (743, Schwartz).
Aql X-1	838	Walter, F., Charles, P., Thorstensen, J., and Bowyer, S., IAU Circ. No. 3243 (1978).
LMC X-3	840	Warren, P.R. and Penfold, J.E., Mon. Not. R. astr. Soc., 172, 41p (1975).
Aql X-1	841	Watson, M.G., Mon. Not. R. astr. Soc., 176, 19p (1976).
Nova Oph	842	Watson, M.G., Ricketts, M.J., and Griffiths, R.E., Astrophys. J. Lett., submitted (1978).
0900-40	843	Watson, M.G. and Griffiths, R.E., Mon. Not. R. astr. Soc., 178, 513 (1977).
0042+32	845	Watson, M.G. and Ricketts, M.J., Mon. Not. R. astr. Soc., 183, 35p (1978).
EX Hydrae	846	Watson, M.G., Sherrington, M.R., and Jameson, R.F., Mon. Not. R. astr. Soc., submitted (1978).
Cyg X-1	847	Webster, B.L. and Murdin, P., Nature, 235, 37 (1972).
SMC X-1	848	Webster, B.L., Martin, W.L., Feast, M.W., and Andrews, P.J., Nature Phys. Sci., 240, 183 (1972).
Cyg X-1	850	Weisskopf, M.C. and Sutherland, P.G., Astrophys. J., 221, 228 (1978).
0115+63	851	Wheaton, W., and HEAO-A4 group, IAU Circ. No. 3238 (1978).
0620-00	852	Whelan, J.A.J., Ward, M.J., Allen, D.A., et al., Mon. Not. R. astr. Soc., 180, 657 (1977).
Cir X-1	853	Whelan, J.A.J., Mayo, S.K., et al., Mon. Not. R. astr. Soc., 181, 259 (1977).
Orion Neb.	855	White, G.J., and Ricketts, M.J., Astrophys. Lett., 18, 79 (1977).
Sco X-1	856	White, N.E., Mason, K.O., Sanford, P.W., et al., Mon. Not. R. astr. Soc., 176, 91 (1976).
X Per	857	White, N.E., Mason, K.O., Sanford, P.W., et al., Mon. Not. R. astr. Soc., 176, 201 (1976).
GX301-2,1+4	858	White, N.E., Mason, K.O., Huckle, H.E., et al., Astrophys. J. Lett., 209, L119 (1976).
X Per	860	White, N.E., Mason, K.O., and Sanford, P.W., Nature, 267, 229 (1977).
GX17+2,13+1	861	White, N.E., Mason, K.O., Sanford, P.W., Johnson, H.M., et al., Astrophys. J., 220, 600 (1978).
LMC X-4	862	White, N.E., Nature, 271, 38 (1978).
1145-61	863	White, N.E., Parkes, G.E., Sanford, P.W., et al., Nature, 274, 664 (1978).
A0538-66	865	White, N.E., and Carpenter, G.F., Mon. Not. R. astr. Soc., 183, 11p (1978).
Rapid burster	866	White, N.E., Mason, K.O., Carpenter, G.F., et al., Mon. Not. R. astr. Soc., 184, 1p (1978).
Cop. posits.	867	Willmore, A.P., Mason, K.O., Sanford, P.W., et al., Mon. Not. R. astr. Soc., 169, 7 (1974).
Cir X-1	868	Wilson, A.M., and Carpenter, G.F., Nature, 261, 295 (1976).
Cir X-1	870	Wilson, A.M., and Carpenter, G.F., IAU Circ. No. 2965 (1976).
Ariel posits.	871	Wilson, A.M., Carpenter, G.F., Eyles, C.J., et al., Astrophys. J. Lett., 215, L111 (1977).
1538-52	872	Wilson, A.M., and Carpenter, G.F., pvt. comm. in (197, Davison) (1977).
1700-37	873	Wolff, S.C. and Morrison, N.D., Astrophys. J., 187, 69 (1974).
1743-32	875	Wood, K., Share, G., Johnson, N., Yentis, D., and Byram, E., IAU Circ. No. 3203 (1978).
AM Her posit.	876	Wright, E.L., pvt. comm. in (340, Hearn).
0620-00	877	Wu, C.C., Aalders, J.W.G., van Duinen, R.J., et al., Astr. Astrophys., 50, 445 (1976).
SMC X-1	878	Yentis, D., Shulman, S., McKee, J.D., and Rose, W.K., Astrophys. Lett., 19, 53 (1977).
0900-40	880	Zuiderwijk, E.J., Hammerschlag-Hensberge, G., et al., Astr. Astrophys., 54, 167 (1977).
GX349+2	881	Zuiderwijk, E.J., IAU Circ. No. 3221 (1978).
GX349+2	882	Zuiderwijk, E.J., IAU Circ. No. 3250 (1978).

INDEX OF NAMES

Source	X-ray	Optical
0042+327		Star 3
0050-727	SMC X-3	Star 4
0052-739	SMC X-2	Star 5
0053+604		γ Cas
0114+650		Star 1
0115-737	SMC X-1	Sk160
0115+634		Star 1
0142+614		
0304+407		Algol
		β Per
0352+309		X Per
0449-55		
0512-401	MXB	NGC1851
0521-720	LMC X-2	
0531+219	Crab pulsar	Crab pulsar
0532-664	LMC X-4	Star
0532-054		Ori Trapezium
		(θ^1 Ori A-D)
0535-668	A0538-66	Star 19
	(LMC)	
0535+262		HDE 245770
0538-641	LMC X-3	Star 8=WP
0540-697	LMC X-1	Star R148
0544-665	LMC?	
0614+091		Star D
0620-003		Nova Mon'75
		V616 Mon
0900-403	Vela X-1	HD 77581
0918-549		
0921-630		Star 5
1052+607		SAO 015338
1118-615		Be star
1119-603	Cen X-3	Krzeminski's
		star
1145-619		HEN 715
1223-624	GX301-2	WRA 977
1239-599		
1249-289		EX Hydrae
1254-690		Star 30
1258-613	GX304-1	Star 2=MMV
1417-624		
1516-569	Cir X-1	Star L
1524-617	TrA X-1,Nova	Nova N
1538-522		Star 12
1543-624		Star 6
1543-475		
1553-542		
1556-605		
1608-522		Nova
1617-155	Sco X-1	V818 Sco
1624-490		
1627-673		Star 4
1630-472		
1636-536	MXB	Star 3
1642-455	GX340+0	
1653-407	OAO	V861 Sco
		HD 152667
1656+354	Her X-1	HZ Her
1658-298	MXB	Star T
1659-487	GX339-4	Star V
1700-377		HD 153919
1702-429		
1702-363	GX349+2	
1705-440		
1705-250		Nova Oph'77
1715-321		
1728-337	MXB	
1728-247	GX1+4	Star GF
	GX2+5	
1728-169	GX9+9	Star DMB
1730-333	Rapid Burster	Liller 1
1735-444	MXB	Star 5
1742-294	GCX-1	
1742-289	Near Gal.Cen.	Radio trans
1743-322		
1743-288	GCX-3	
	GX+0.2-0.2	
1744-361		
1744-265	GX3+1	
1745-203		NGC 6440
1746-370		NGC 6441
1755-338		Star 6
1758-250	GX5-1	
1758-205	GX9+1	
1803-245		
1811-171	GX13+1	
1813-140	GX17+2	Star 28=TR
1814+498		AM Her
1820-303	MXB	NGC 6624
1822-371		Star S
1822-000		
1837+049	Ser X-1	Star D
	MXB	
1845-024		
1850-087	MXB	NGC 6712
1905+000	MXB	
1908+005	Aql X-1	Star 5,Nova
1916-053	MXB	Star 16
1956+350	Cyg X-1	HDE 226868
1957+115		Star M
2030+407	Cyg X-3	IR star
2127+119		NGC 7078/M15
2129+470		Star
2140+433		SS Cygni
2142+380	Cyg X-2	Star 10

THE OPTICAL COUNTERPARTS OF GALACTIC X-RAY SOURCES

Bruce Margon

Astronomy Department, University of California, Los Angeles, California, USA

ABSTRACT

The current status of optical identifications of compact galactic X-ray sources is reviewed, with particular emphasis on the large group with 2-10 keV luminosities in the 10^{36} -10^{38} erg/sec range. Of the approximately 45 members of this class, a very large majority fall into one of two sub-classes: intrinsically faint, ultraviolet-excess objects similar to Scorpius X-1, and luminous OB stars with weak emission features. Even diverse sources such as bursters, transients, and slow pulsars obey this bifurcation, and it is argued that all of these objects, and also quite possibly the globular cluster sources, are probably extremely similar in their basic nature.

INTRODUCTION

The number of optical identifications of galactic X-ray sources has in the past two years grown almost exponentially with time. This explosive increase, which follows a long period of one or two new identifications per year from Uhuru positions, stems chiefly from positional data returned by satellite-borne modulation collimator experiments aboard SAS-3, Ariel 5, and most recently HEAO-1. These experiments are now routinely providing positions with accuracy of order 30 arcsec, thus finally bringing X-ray astronomy to a point where radio astronomy was 15-20 years ago. At moderate and high galactic latitudes, an error box of this size will generally contain of order five or fewer stars to the magnitude limit of the Palomar Observatory Sky Survey. At low latitude ($b \leq 5^{\circ}$), the boxes are considerably more crowded, but in favorable cases may still contain only a dozen or fewer objects. Thus a "brute force" approach to optical identifications of X-ray sources finally becomes feasible. We may compensate for our ignorance of exactly what to look for in a given source by simply examing every visible object in the error box, with some combination of photometry, spectroscopy, and spectrophotometry. Luckily the vast majority of source counterparts known thus far turn out to be sufficiently unusual in color or spectral features that a single candidate generally emerges from this procedure.

The above technique of course is guaranteed to fail on "normal" appearing counterparts. Thus how many source identifications we are currently overlooking because of our restricted selection criteria must await positions of a few arcseconds accuracy. The wait, of course, will not be long as such data will become available with the launch of HEAO-B at the end of this year. We already know these largely-overlooked "normal stars" are something of a problem. For example, if Capella, a faint soft X-ray source, were moved to the many-kiloparsec distance typical of large numbers of luminous hard X-ray binary stars, its weak K-line emission would be unobservable, and only patient Coude spectroscopy would reveal the binary nature of the system. Sirius, also probably a faint source, would be even harder to optically-identify under these conditions. This confusion is already more than theoretical. The probable optical counterpart of GX17+2 (4U1813-14) was pointed out six years ago [1], but this late K or early G star is spectroscopically [2] and photometrically [3] quite normal. Only a recent, extremely accurate radio position [4] makes the identification reasonably convincing.

In this review I will ignore the distressing possibility that we are thus far overlooking the predominent classes of X-ray source optical counterparts, and instead concentrate on what we do know! This is reasonable on grounds other than optimism: our entire galaxy is quite well-surveyed for steady 2-10 keV sources of luminosity greater than 10^{36} erg/sec. There is a well-known cutoff in the source luminosity function near a few times 10^{38} erg/sec [5], and we obviously already see all of these bright sources. As the current crop of optical identifications is harvested largely from this 10^{36}- 10^{38} erg/sec, 2-10 keV group, we can be reasonably certain that we are discussing a well-defined class of objects which are hopefully intimately related, and do not subdivide into too large a number of subclasses. Thus I will ignore here such "normal" known counterparts as Capella and Sirius, and also the more exotic soft and/or low luminosity objects, e.g. HZ43 [6-8] and SS Cyg [9-11]. It seems quite possible that with these latter objects we are seeing only the tip of the iceberg of the source population, as they are both within about 100 pc; clearly conclusions on these "classes" must await further examples.

With the sample of objects thus restricted, there are perhaps 45 current identifications, depending somewhat on one's standard of proof. This is already obviously far too large to discuss each counterpart individually here, and Professor Bradt has provided an excellent bulk compendium of identifications in his paper at this meeting [12]. Instead I will examine a limited number of the new, recent identifications in an attempt to discern if we understand the trends that are appearing. In short: can we predict now what the next 45 identifications will be like? To reveal my opinion (and thus prejudices!) from the start, I think the answer is yes. With a few glaring exceptions (even in our restricted class under examination), the identifications are breaking down quite nicely into two groups: one with massive ($M \geq 20\ M_\odot$), intrinsically luminous OB primary stars, and the other containing low mass (<2

$M_{\odot}$), less luminous primaries. This bifurcation of the mass function of the primaries fits nicely with theoretical predictions based on mass transfer mechanisms [13], and recognition of these two classes is not a new development (although it is certainly comforting that the classification scheme still works despite the recent very large expansion in the list of identifications). What has emerged just in the past two years, however, is that the optical counterparts of the more exotic X-ray objects which have recently attracted so much attention, namely bursters, transients, globular cluster sources, and slow X-ray pulsars, are usually also well explained within this framework, at least in the cases where sufficient observations are available. This may be taken to imply that all of these diverse X-ray phenomena are in fact manifestations of the same basic underlying machine: a close binary system consisting of one relatively normal star, and one compact, highly evolved object.

IDENTIFICATIONS OF NEW INTRINSICALLY-FAINT COUNTERPARTS

It seems inevitable that the majority of the counterparts of the source group we are discussing are intrinsically faint stars, some interesting claims to the contrary [14, 15] notwithstanding. Despite the tremendous progress in source identification in the past few years, this particular argument could have been made over a decade ago, once the optical counterpart of Sco X-1 was know [16]. This object is the closest luminous (10^{36} erg/sec) extrasolar source, so as long as one accepts a Copernican view that we are not specially-located observers, it seems inevitable that Sco X-1 must be a typical, "normal" source. The absolute visual magnitude of Sco X-1 is not terribly different from the Sun. Without any optical identifications, one can infer strictly from the observed spatial distribution of the bright sources [17] that they are generally many kiloparsecs distant. Thus we expect many sources with distance moduli of 13-14 mag, plus some additional allowance for interstellar extinction. The result is that these "normal" counterparts are V=18 and fainter objects, and thus extraordinarily difficult to identify without our currently-available excellent modulation collimator positions. Little wonder then that the far brighter early supergiant counterparts (Cygnus X-1, Vela X-1, etc.) were found first.

These faint, Sco-like counterparts seem finally to be emerging. Among the identifications in this class are the counterparts of 4U0614+09 [18, 19] and 4U1956+11 [20]. The identifications are difficult to irrevocably confirm, as the counterparts are so faint that they are time-consuming spectroscopic objects even at the largest telescopes. However, both of these stars are extremely ultraviolet and show no underlying stellar absorption spectrum, very similar to Sco X-1. They have also withstood an important test of time: both were suggested when the positional uncertainties of the X-ray sources were of order 2 arcmin, whereas the latest X-ray error circles for these two sources [21, 22] contain a full order-of-magnitude less area and still include the candidates.

If we accept the fact that the Sco-like optical counterparts are a cohesive group (and this now seems inevitable, due to observations described both above and below), then there is some good news and some bad news. The good news is that the considerable effort, both observational and theoretical, that has gone into understanding Sco X-1 in both X-rays and visible light may now be applicable to many other objects. There seem to be direct analogies available, at least observationally: Mason et al. [23] have noted similar X-ray behavior in a group of four sources, and McClintock et al. [24] have optically identified three very important sources (discussed further below) simply by looking for ultraviolet excess stars. The bad news is that all of these objects (0614+09, 1956+11, and the three McClintock et al. identifications, 1626-67,1636-53, and 1735-44) are B=18 and fainter, implying that all spectroscopy on these stars requires telescopes of the 2.5 m class and larger.

Two recent and possibly related identifications also fall into the class of intrinsically-faint counterparts, and may portend a very serious theoretical problem regarding mass transfer. These are the counterparts of the highly variable X-ray sources Aquila X-1 (4U1908+00) and A0620-00 (Nova Monocerotis 1975).

The identification of Aquila X-1 is an interesting case study of rewards for patience. This intensely flaring X-ray source [25, 26] has a much-weaker quiescent component which is reasonably well-located even in the 3U catalog. Thus a variety of groups began acquiring plate material in this field, dating back as long as four years ago. Finally, the plate collection described by Davidsen et al. [27] was used in conjunction with the latest accurate SAS-3 position [22] to achieve the identification [28]. The object proves in the historical data to have sharply correlated optical/X-ray outbursts, which last a few weeks. Unfortunately the object is very faint: the outbursts take this B=20 star to B=17. In Fig. 1 are plates of the field both in quiescence and outburst. The quiescent plate was obtained with the Lick Observatory 0.9 m Crossley reflector, and is described by Margon et al. [29]; the outburst plate is from the Sky Survey, as an eruption was fortuitously underway when those plates were obtained.

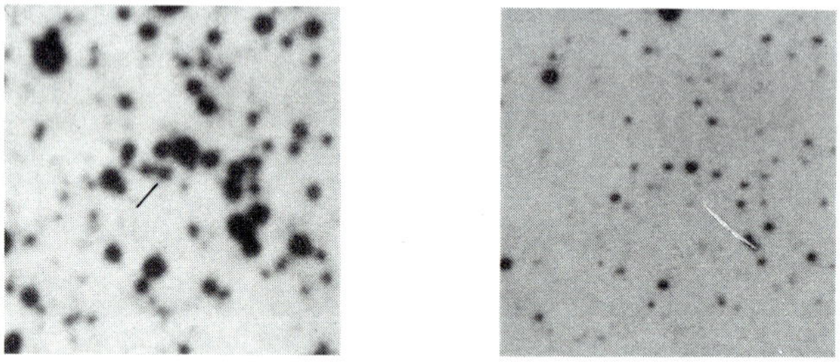

Fig. 1. The Aquila X-1 (4U1908+00) field. Left panel: the Sky Survey E (red) plate (copyright National Geographic Society-Palomar Observatory Sky Survey; reproduced by permission of the Hale Observatories); the counterpart is at the top of the indicator bar. Right panel: a V plate (103aD + GG495) obtained on 1977 August 12 at the Lick Observatory [29]. The faint quiescent counterpart is visible on this original plate at V=19.4, although not prominent in this reproduction. The field in both panels is 2 arcmin on each side. North is to the top and East to the left.

Unfortunately there have been no outbursts since the identification, so we do not yet know the exact timescale or amplitude of the optical eruption, whether there is periodic photometric behavior at maximum light, and what the outburst spectrum looks like. In this connection it is interesting to note that the X-ray data collected before the optical identification [26] seemed to indicate an outburst approximately annually. Now that we know which star to observe, however, there have been no outbursts in last year's or the current observing season! Constant optical monitoring of the object is obviously essential.

The fragmentary data on the quiescent counterpart are already fascinating, however. A minimum-light spectrum (a very difficult achievement for this faint star) indicates a spectral type of late G or early K [28]; the only feature visible is the Mg b band in absorption. This spectrum, obtained with the Lick Observatory 3 m Shane reflector, is reproduced in Fig. 2.

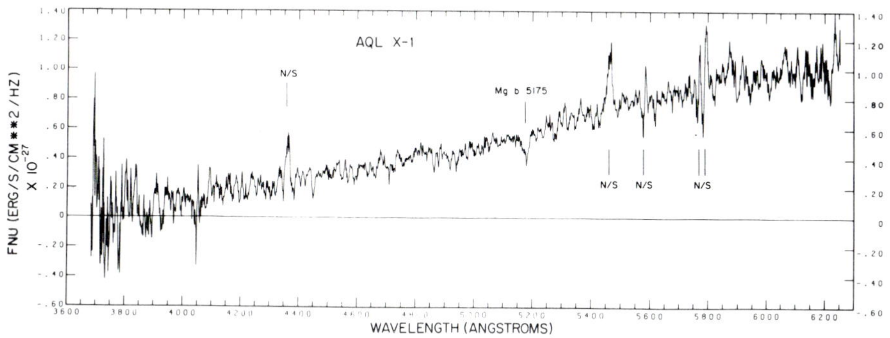

Fig. 2. The spectrum of the optical counterpart of Aquila X-1, obtained with the Robinson-Wampler Image Tube Scanner at the Lick Observatory 3 m Shane telescope [28]. (Copyright by the American Astronomical Society; reproduced with permission.)

An extinction study of the field [29] indicates the counterpart is at or near main-sequence luminosity. This, then, explains the large amplitude of the visible outbursts: the ratio of X-ray to optical luminosity is quite large at these times. For favorable inclination, this will at maximum almost surely cause an observable photometric variation at the binary orbital period (as yet unknown, although a 1.3 day X-ray modulation has been suggested [30]). We can already conclude that the outburst is not Eddington-limited at this 1-2 kpc distance. Furthermore, the observed X-ray to optical luminosity ratio at maximum is a rather direct measure of the solid angle presented by the primary to the secondary. Under the commonly-applied semi-detached Roche model (a KV star seems unlikely to transfer matter via a stellar wind), this then yields a value for the mass ratio. The surprising result [29] is a ratio of 100 - - hardly believable given that we see one of the stars in the system already, and it appears rather normal. Thus perhaps the highly successful Roche model, which seems to so nicely explain other X-ray binaries with low mass primaries, is not applicable to Aquila X-1.

A (hopefully temporary) confusing conclusion like this is perhaps not terribly distressing when it applies to only one system. Unfortunately, we already know of another such X-ray source, whose nature has again recently been elucidated by spectrophotometry: A0620-00, or Nova Monocerotis 1975. Although well-observed by optical astronomers when near maximum light (cf. [31]), the optical counterpart of this source, V616 Mon, is too faint at minimum (V=18, B=20) to reach with all but the largest telescopes. A recent quiescent spectrum [32] also indicates an underlying K dwarf, similar to Aql X-1, although V616 Mon has emission lines when at minimum, unlike the former object. Again the observations may indicate that the Roche lobe is not filled. As neither of the primaries in these two systems would seem sufficiently luminous to support a radiation-driven wind, there obviously remains an intriguing theoretical problem of just how the low-mass transient X-ray sources transfer matter from one component to another.

OPTICAL IDENTIFICATIONS OF BURSTING X-RAY SOURCES

A tremendous amount of excitement has been generated among the X-ray observers during the past few years by X-ray bursters. Several excellent reviews of the X-ray observations have appeared recently [33-35]. Finally the optical observers have been able to join the foray. The accident of history that the first burster [36] was found in a globular cluster, NGC 6624, led many of us to conclude that their useful contributions would have to be made by painfully poking into the centers of the condensed cluster cores. The detection of bursters in one and possibly two further globular clusters (NGC 1851 and perhaps NGC 6441), plus the discovery of a nebulous object near MXB1730-335 [37], only reinforced this conclusion.

Curiously, the subject easily could have developed otherwise. Arthur Davidsen suggested an ultraviolet candidate for Ser X-1 [38] three years ago, and this X-ray source (4U1837+04) later was found to be a burster [39, 40]. Thus a hint appeared that the counterparts of at least some bursters strongly resembled the well-known normal "field" binary sources, but for some reason this did not divert much attention from the concept that bursters and globular clusters were intimately and irrevocably related. At least part of the problem was again faintness: the Davidsen candidate has B=18.5; the first spectrum of the object was reported only recently [41].

What has happened to cool the globular cluster fixation is, of course, the discovery and positioning of many more bursters. The bulk spatial distribution does not particularly suggest a one-to-one association with clusters [33]. More conclusively, at least two and probably three more candidates for optical counterparts of bursters have recently appeared: all are faint stellar objects, definitely not in clusters. Thus although the spatial distribution statistics are intriguing, the recent identification of stellar counterparts would seem to be the irrevocable proof that bursters can and do live outside of globular clusters. This fact, plus the optical properties of the bursters described below, removes (at least in my mind) almost all the motivation for associating massive black holes ($M \gtrsim 100\ M_{\odot}$) with bursting X-ray sources: we know that many, and perhaps all, of the "field" binaries have neutron stars or white dwarfs as secondaries.

The most observations have thus far been reported for the optical counterpart of MXB1735-44. This object, identified by McClintock et al. [24] using a then unpublished SAS-3 position (see [42]), has (B-V) and (U-B) colors very similar to Sco X-1. Fortuitously the galactic latitude is moderate ($b = -7^{\circ}$) so extinction does not destroy the resemblance. What is even more astonishing, however, is the spectrum of the counterpart [43], reproduced in Fig. 3.

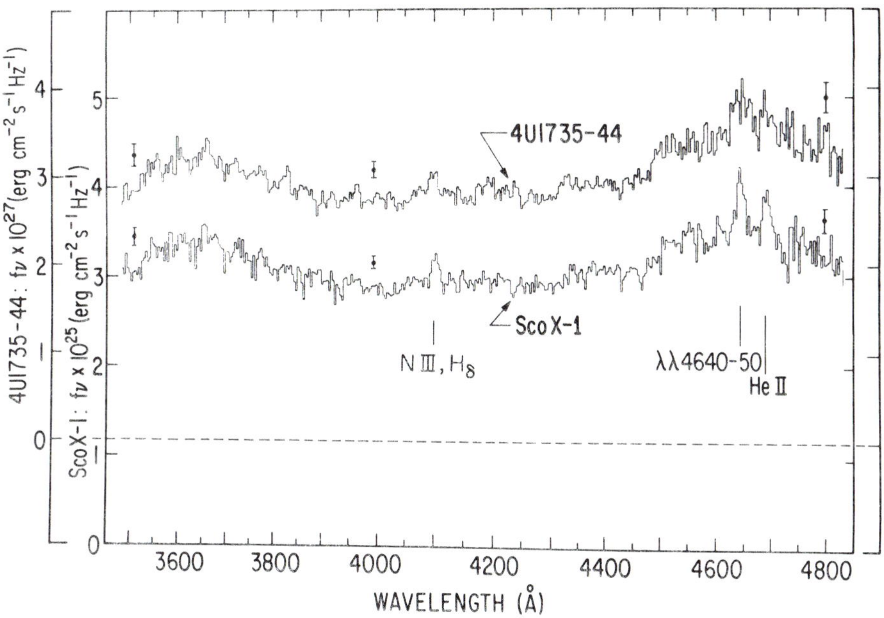

Fig. 3. The spectrum of the optical counterpart of the X-ray burster MXB1735-44, obtained with the SIT spectrograph at the Cerro Tololo 4 m telescope [43]. The spectrum of Sco X-1 is also shown for comparison. (Copyright by American Astronomical Society; reproduced with permission).

The spectral resemblance of MXB1735-44 to Sco X-1, both in emission features and continuum slope, is striking. Here again is an extremely important piece of evidence regarding the nature of X-ray bursters. It obviously heavily favors those burster theories which interpret the sources as only a slight modification on the normal binary models, perhaps inflicted with a "magnetospheric disease" of some sort.

The resemblance of MXB1735-44 and Sco X-1 is not perfect however (even aside from the fact that Sco X-1 does not burst!), so the picture is not yet complete. One of the most striking properties of Sco X-1 is its nearly ubiquitous photometric variability, on all timescales, ranging from minutes to days [44]. I have recently performed the first photoelectric photometry (of which I am aware) designed to detect analogous photometric behavior, if present, in MXB1735-44. These observations were obtained on three nights in 1978 April, with the 1.5 m reflector of the Cerro-Tololo Interamerican Observatory; 1-2 hours of coverage were obtained each night, in both the B band, and also an unfiltered mode where response was determined solely by the S-20 photocathode. The basic time resolution was 30 seconds. Similar technique and duration observations of Sco X-1 would almost unquestionably repeatedly detect the well-known flickering. However, my observations show no statistically-significant photometric variability in MXB1735-44. A sample of these data is shown in Fig. 4.

Inspection of Fig. 4 indicates there is no evidence for fluctuations above the level of about 0.1 mag on this particular night. Further, the nightly mean magnitudes and colors of the star agree on all three nights to about 0.05 mag. These data are also in reasonable agreement with the photometry obtained almost a year earlier by McClintock et al. [24]. This lack of variability thus spoils the perfect analogy to Sco X-1, and the complete picture must again obviously await further observations.

Simultaneous X-ray burst and optical observations would obviously be very useful for all the known counterparts. Although there are no such observations yet reported for the MXB1735-44 system, there have been several successful simultaneous observations of the Ser X-1 counterpart. By "successful" I mean here not only that both X-ray and optical data were obtained, but also that X-ray bursts occured. No optical variability has been seen. At least three such events were observed [35, 45], with current limits on optical-to-X-ray burst energy ratios of order 10^{-4}. Simple X-ray heating models [45] indicate that these limits are at about the level that optical bursts are expected, so continued work in this area is very important.

The remaining two probably stellar optical counterparts of bursters are 4U1636-53 (=MXB1636-53) and 4U1608-52 (="the Norma burster"). The former identification was made [24] again on the grounds of ultraviolet excess. It is probably the least secure at the moment, as its position in the two-color diagram is unusual but not extreme [24], and little other data are currently available. The identification of 4U1608-52 [46] seems quite firm, resting on a star whose I magnitude increased from fainter than 20 to about 18.2, approximately coincident with a brightening of the X-ray source. This region of the X-ray sky is quite complex, containing a steady source, a transient, and a burster, but it seems likely they are all the same object. The X-ray position obtained by combining SAS-3 [47] and HEAO-1 [48] results is very small (about 10x20 arcsec) and contains the faint condidate. Again little other data are available now, but will surely follow shortly.

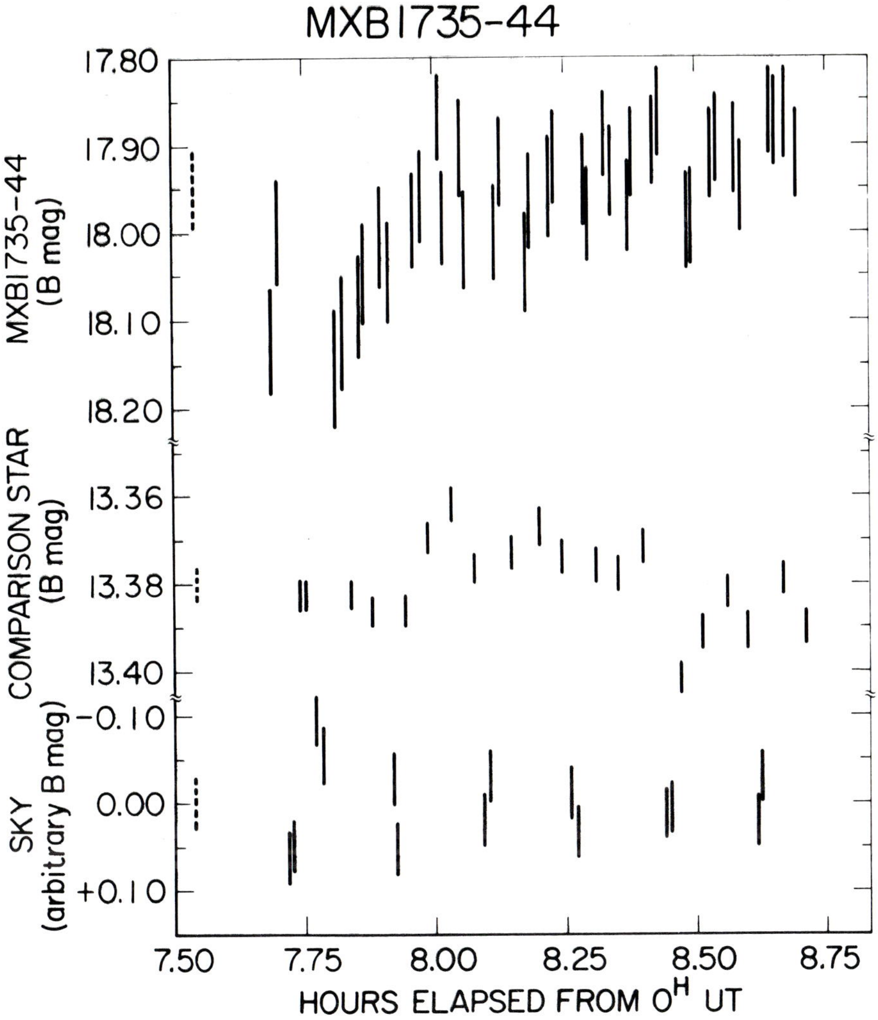

Fig. 4. Photoelectric photometry of the optical counterpart of MXB1735-44, obtained by the author on 1978 April 15 using the 1.5 m reflector of the Cerro Tololo Interamerican Observatory. The broken lines are the mean of the entire data string, while the solid lines are individual 30 second integrations with one standard deviation error bars.

RECENT WORK ON LUMINOUS OPTICAL COUNTERPARTS

In the Introduction I claimed that new counterparts of both groups of primaries discussed by van den Heuvel [13] are appearing, and here I want to at least briefly justify that statement by discussing some new identifications of X-ray systems with luminous, massive early-type primaries. The most thoroughly studied new identification in this class is perhaps 4U1538-52. Here is an interesting example of a relatively obscure, weak galactic source known since the sounding rocket days (probably Norma X-2) which has sprung in a year from a poorly located Uhuru source to an identified system with an almost-complete orbital solution. Stepping stones on this path include the discovery that the X-ray source is a 529 second pulsar [49-51], with Doppler shift yielding an orbital period of 3.73 days and X-ray mass function hinting at an early-type primary; photographic photometry of a large number of stars in the Uhuru and Ariel 5 position boxes [52]; an excellent SAS-3 position [47]; identification of a candidate based on spectroscopy by several groups [53,54], and finally measurement of a spectroscopic radial velocity curve with 3.73 day period [55]. As there are velocities for both components and the eclipse duration is known, we will doubtless rapidly have a mass solution. Ellipsoidal photometric variations may well also be detectable, although unpublished photoelectric observations by the author indicate that they probably do not exceed 0.1 mag in the V band (about the level at which they appear in Cen X-3 [56]). These data, obtained in 1977 April with the Cerro Tololo 1.5 m telescope, yield a 3 night average of V=14.42, (B-V)=2.04, (U-B)=0.90, consistent with expectations for a heavily reddened luminous late O star. It is interesting that we have a solution on hand now for 4U1538-52 and not Cen X-3, despite the fact that the latter system is one magnitude brighter (two magnitudes in B), has (presumably) a similar K velocity, and has been optically identified for 5 years! The necessary optical velocities for Cen X-3 are now in the hands of several observers, and we should expect a mass solution this season.

Another very recent and important identification of an early-type system is the transient [57] 4U0115+63. The 4U position for the object is relatively accurate (about a 4 arcmin square), but insufficient for a confident optical identification. What was obviously needed was an outburst observable by the current generation of modulation collimator experiments, and the source obliged during the first week of 1978. Excellent positions were derived from SAS-3 [58] and HEAO-1 [59], the overlap region restricting the source to a 20 x 30 arcsec area. The brightest visible object in the error box (V=16), a reddened, early type star with Hα emission, has been suggested as the optical counterpart [60].

The X-ray spectrum of 4U0115+63 is quite flat, a property shared by all of the X-ray pulsars [61]. Two previous flat-spectrum transients, A0535+26 and A1118-61, were also X-ray pulsars; the former is quite firmly identified with an early-type emission line star, HDE 245770 [62,63], while the latter has a similar candidate identification [64]. The similarities of 4U0115+63 to these systems thus suggest it may also be a pulsar, and this indeed turns out to be the case [58,59]. The period is 3.6 seconds. The previous pulsing transients showed no detectable Doppler shifts, and thus one could only suspect that they were members of binary systems, by analogy with the steady sources. Information on the system parameters of transients has thus been very scanty, although in A0535+26 the lack of X-ray pulse Doppler shift [65] and reflected optical pulsations [63] hint strongly at long orbital periods.

Observations of 4U0115+63, however, have removed much of this ambiguity, as there an X-ray pulse Doppler shift _is_ detectable, and the orbital period proves to be 24.3 days [66]; further, the orbit is interestingly eccentric, e=0.34. Thus this system provides proof that the X-ray transients are also binary systems. Of particular interest to the optical observer is that at least in principle a mass solution is now possible for this system, as the X-ray velocity curve is now known [66], and the visible component is just barely bright enough that a spectroscopic velocity curve is with some effort probably feasible. As there are no known X-ray eclipses, a formal mass solution will also require observation of ellipsoidal photometric variations [67], but again these are within reach if the mass ratio is not extreme. Thus 4U0115+63 may provide (with patience) the first accurate mass measurement for a transient X-ray source. Given the many similarities in observed properties of the transients and "steady" sources, it now seems quite likely that the two classes are just minor variations on the same theme, and thus not probable that such a mass measurement will lead to any surprises. Nonetheless it would be comforting to actually observationally confirm that the transients and steady sources also share the same range of masses, and 4U0115+63 gives us this opportunity.

Finally, I also want to point out that some quite faint X-ray sources are also yielding identifications with luminous, early type stars. The prototype in this class is X Per, which was suggested as an identification by several workers [68-70] shortly after the appearance of the 2U catalog. It has X-ray luminosity of order 10^{34} erg/sec. Perhaps I am here violating my opening disclaimer about ignoring subluminous sources, but I feel justified as this is a 2-10 keV source, not anomalously soft. The important point is that here observational selection is _surely_ important for sources such as this, which have observed X-ray fluxes near the bottom threshold of the _Uhuru_ survey. Thus one would almost certainly expect new optical identifications in this class as X-ray survey sensitivity improvements occur, and this has now begun. A related object is probably γ Cas, detected in X-rays by both OAO _Copernicus_ [71] and SAS-3 [72]. The X-ray luminosity of this latter object is even less, in the 10^{32} erg/sec range.

It is obviously only a matter of time until we step beyond the 300 pc sphere that contains the above stars, and begin to identify similar objects at the one-to-several kiloparsec distances more typical of the luminous X-ray binaries. In collaboration with H. Bradt and R. Dower of the SAS-3 group, I have recently found such an object. The weak X-ray source 2S0114+650 was discovered by SAS-3 [21], at a flux level equivalent to about 5 _Uhuru_ counts. Thus its absence from earlier surveys does not necessarily imply extreme variability. The accurate modulation collimator position [21] has led to our identification of the source with the reddened, early star LSI +65°010 [73]. Our spectrophotometric observations of this star, obtained with the Lick 3 m and Kitt Peak 2.1 m telescopes, show the spectrum to be relatively normal, with the exception of prominent Hα emission, which is the primary basis for the identification by analogy with the half-dozen other known OBe counterparts. Photoelectric photometry obtained with the Lick 0.9 m Crossley reflector yields V=11.01, (B-V)=1.18, (U-B)=0.07.

A preliminary analysis of the spectrum types the object at about O9Ve, although the photometry (via the "Q" method) yields a slightly later type, perhaps B0. A distance of order 1.4 kpc is indicated, yielding an X-ray luminosity of about 10^{34} erg/sec, i.e., very similar to X Per.

SUMMARY

I have been attempting to suggest that an almost boring regularity is beginning to emerge in new identifications: most are falling into the two categories of Sco-like, or luminous, late O star with weak emission lines. There are glaring exceptions, of course: Cygnus X-3 springs to mind, as well as the bizarre symbiotic star counterpart of GX1+4 [74], an identification that is almost unquestionably correct and received far too little attention thus far. Aside from these occasional renegades, however, the boredom is in a sense a relief: we are beginning to use optical data to systematize the study of the galactic sources.

To support my contention of regularity amongst the identifications, Table 1 shows some linkages amongst sources, i.e., sources which

TABLE 1 Optical Counterparts as Linkages Between Classes of X-ray Sources

Source	Linkage
0900-40	well known field sources -- slow pulsars
0535+26	transients -- slow pulsars
1118-61	OB stars -- slow pulsars
0115+63	
1626-67	Sco-like -- slow pulsars
1735-44	Sco-like -- bursters
1837+04	
1636-53	
1820-30	globular clusters -- bursters
1730-335	
0513-40	
1908+00	well known field sources -- transients
1608-52	transients -- bursters -- field sources

share the properties of many types of X-ray objects at one time discussed as distinct classes. The table indicates that it is almost always possible to find at least one source that provides a bridge, via its optical counterpart, between classes. Perhaps the most tenuously linked class is still the globular cluster sources. On the other hand, even here much of the romance and mystery has recently mitigated. The best studied globular cluster source is probably NGC 6624 (=4U1820-30). The compact central region of this cluster, once considered unusual enough to be given its own name ("Region R" [75]), is now found to be resolved into a few stars [76-78]. An early report of evidence for Hα emission and thus diffuse gas in the core, based on multiaperture photometry [76], has not been confirmed by later, more sensitive slit spectroscopy observations [78-79] and Fabry-Perot interferometry [80]. Figure 5 shows an example of this work, displaying spectra

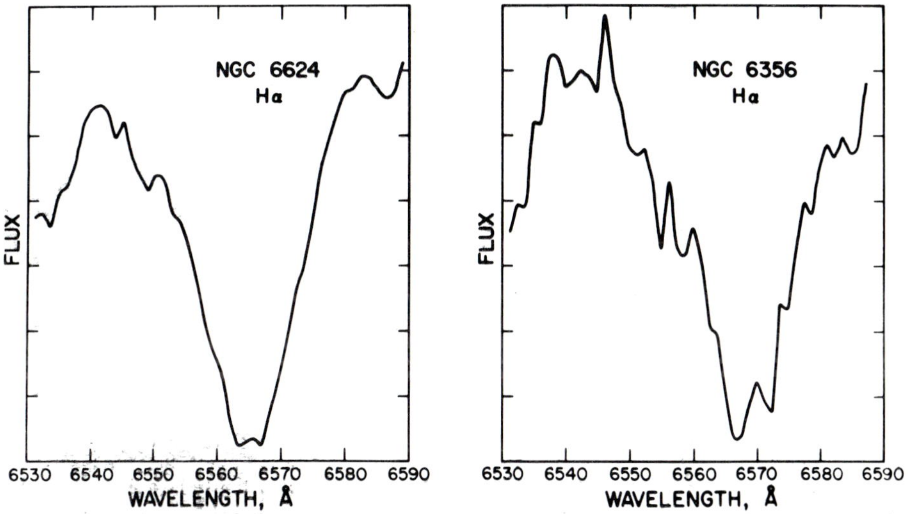

Fig. 5. Spectrophotometry of the region near Hα in the compact core of NGC 6624, an X-ray burster, and NGC 6356, a normal globular cluster, obtained with the Lick 3 m telescope [78].

obtained by the author near Hα of the inner core of both NGC 6624 and NGC 6356. The latter object was picked as a convenient "control" as it is not a known X-ray source, but has magnitude, integrated spectral type, and concentration class similar to NGC 6624. The observed profiles are very similar. While we may not yet know the nature of the globular cluster sources, there seems to be no strong motivation at this time to demand that they are radically different from the "field" objects.

Obviously further optical observations in the near future will help to resolve these uncertainties, and almost certainly also muddy the clear waters I have simplistically tried to portray in this summary!

ACKNOWLEDGEMENTS

I am indebted to Ms. K. B. Kwitter for aid in obtaining and reducing some of the data discussed here. I thank many colleagues for supplying preprints in advance of publication. Some of the author's data presented here were obtained as a Visiting Astronomer at Cerro Tololo Interamerican Observatory and the Kitt Peak National Observatory, which are operated by AURA, Inc., under contract to NSF. I thank Ms. D. Glover for preparing the manuscript, and acknowledge the financial support of the National Science Foundation.

REFERENCES

1. M. Tarenghi and C. Reina, Nature Phys. Sci. 240, 53 (1972).
2. A. Davidsen et al., Astrophys. J. 203, 448 (1976).
3. B. Margon, Astrophys. J. 219, 613 (1978).
4. R. M. Hjellming, Astrophys. J. 221, 225 (1978).
5. B. Margon and J. P. Ostriker, Astrophys. J. 186, 91 (1973).
6. D. R. Hearn et al., Astrophys. J. 203, L21 (1976).
7. B. Margon et al., Astrophys. J. 203, L25 (1976).
8. B. Margon et al., Astrophys. J. 209, 525 (1976).
9. S. Rappaport et al., Astrophys. J. 187, L5 (1974).
10. J. Heise et al., Astr. Astrophys. 63, L1 (1978).
11. B. Margon et al., Astrophys. J. 224, 167 (1978).
12. H. Bradt, this volume.
13. E. P. J. van den Heuvel, Astrophys. J. 198, L109 (1975).
14. H. Gursky, Astrophys. J. 175, L141 (1972).
15. S. Sofia, Astrophys. J. 188, L45 (1974).
16. A. Sandage et al., Astrophys. J. 146, 316 (1966).
17. H. M. Johnson, Astrophys. J. 143, 261 (1966).
18. P. G. Murdin et al., Mon. Not. R. Astr. Soc. 169, 23 (1974).
19. A. Davidsen et al., Astrophys., J. 193, L25 (1974).
20. B. Margon et al., Astrophys. J. 221, 907 (1978).
21. R. Dower et al., Nature, 273, 364 (1978).
22. R. Doxsey et al., Nature 269, 112 (1977).
23. K. O. Mason et al., Mon. Not. R. Astr. Soc. 177, 513 (1976).
24. J. McClintock et al., Nature 270, 320 (1977).
25. J. Buff et al., Astrophys. J. 212, 768 (1977).
26. L. Kaluzienski et al., Nature 265, 606 (1977).
27. A. Davidsen et al., IAU Circular No. 2793 (1975).
28. J. Thorstensen et al., Astrophys. J. 220, L131 (1978).
29. B. Margon et al., Nature 271, 633 (1978).
30. M. G. Watson, Mon. Not. R. Astr. Soc. 176, 19p (1976).
31. S. P. Maran, in X-ray Binaries, NASA SP-389, 293 (1976).
32. J. B. Oke, Astrophys. J. 217, 181 (1977).
33. W. H. G. Lewin and P. C. Joss, Nature 270, 211 (1977).
34. W. H. G. Lewin, Mon. Not. R. Astr. Soc. 179, 43 (1977).
35. W. H. G. Lewin, Ann. N.Y. Acad. Sci. 302, 210 (1977).
36. J. Grindlay et al., Astrophys. J. 205, L127 (1976).
37. W. Liller, Astrophys. J. 213, L21 (1977).
38. A. Davidsen, IAU Circular No. 2824 (1975).
39. J. Swank et al., IAU Circular No. 2963 (1976).
40. F. K. Li et al., Mon. Not. R. Astr. Soc. 179, 21p (1977).

41. J. van Paradijs, IAU Circular No. 3197 (1978).
42. J. G. Jernigan et al., Nature 270, 321 (1977).
43. J. E. McClintock et al., Astrophys. J., 223, L75 (1978).
44. W. A. Hiltner and D. E. Mook, A. Rev. Astron. Astrophys. 8, 139 (1970).
45. A. N. Abramenko et al., Mon. Not. R. Astr. Soc., 184, 27p (1978).
46. J. E. Grindlay and W. Liller, Astrophys. J. 220, L127 (1978).
47. K. M. V. Apparao et al., Nature 271, 225 (1978).
48. G. Fabbiano et al., Astrophys. J. 221, L49 (1978).
49. P. J. N. Davison, Mon. Not. R. Astr. Soc. 179, 35p (1977).
50. R. H. Becker et al., Astrophys. J. 216, L11 (1977).

51. P. J. N. Davison et al., Mon. Not. R. Astr. Soc. 181, 73p (1977).
52. A. P. Cowley et al., Astrophys. J. 218, L3 (1977).
53. G. E. Parkes and K. O. Mason, IAU Circular No. 3184 (1978).
54. J. van Paradijs, IAU Circular No. 3197 (1978).
55. A. P. Cowley et al., IAU Circular No. 3201 (1978).
56. W. Krzeminski, Astrophys. J. 192, L135 (1974).
57. W. Forman et al., Astrophys. J. 206, L29 (1976).
58. L. Cominsky et al., Nature, 273, 367 (1978).
59. M. Johnston et al., Astrophys. J., 223, L71 (1978).
60. M. Johns et al., IAU Circular No. 3171 (1978).
61. S. Rappaport et al., Astrophys. J. 217, L29 (1977).
62. M. Stier and W. Liller, Astrophys. J. 206, 257 (1976).
63. B. Margon et al., Astrophys. J. 216, 811 (1977).
64. C. Chevalier and S. Ilovaisky, IAU Circular No. 2778 (1975).
65. S. Rappaport et al., Astrophys. J. 208, L119 (1976).
66. S. Rappaport et al., Astrophys. J., 224, L1 (1978).
67. J. N. Bahcall, A. Rev. Astron. Astrophys. 18, in press (1978).
68. L. L. E. Braes and G. K. Miley, Nature 235, 273 (1972).
69. S. van den Bergh, Nature 235, 273 (1972).
70. R. J. Brucato and J. Kristian, Astrophys. J. 173, L105 (1972).
71. K. O. Mason et al., Nature 260, 690 (1976).
72. H. V. Bradt et al., Nature 269, 21 (1977).
73. J. Hardorp et al., Luminous Stars in the Northern Milky Way 1 (1959).
74. A. Davidsen et al., Astrophys. J. 211, 866 (1977).
75. N. A. Bahcall, Astrophys. J. 204, L83 (1976).
76. J. E. Grindlay and W. Liller, Astrophys. J. 216, L105 (1977).
77. W. Liller, Ann. N.Y. Acad. Sci. 302, 248 (1977).
78. B. Margon et al., Bull. Am. Astr. Soc. 8, 544 (1976).
79. C. Canizares et al., Astrophys. J., 224, 39 (1978).
80. J. E. Hesser and S. J. Shawl, Astrophys. J., 217, L143 (1977).

A DEEP PHOTOGRAPHIC-INFRARED SEARCH FOR X-RAY SOURCES IN THE GALACTIC CENTER REGION

J. E. Grindlay and Wm. Liller

Center for Astrophysics, 60 Garden St., Cambridge, Massachusetts 02138, U.S.A.

The optical search for the many X-ray sources near the galactic center is badly hampered by the large amount of interstellar absorption between us and the nucleus of the galaxy. Most of the absorbing matter lies in the spiral arm ∿ 3 kpc from the nucleus, and it is most likely that the X-ray sources are inside the arm in the nuclear region. Because X-ray sources are frequently found in globular clusters, we felt it important to take a series of deep photographs in the near infrared in hopes of finding some of these. Also, Oort [1] estimated the number of globular clusters near the galactic nucleus, and the numbers are encouragingly large.

In 1976 and 1977 we took a series of one hour exposures using the 4-meter telescope at CTIO, hypersensitized IV-N plates, and a filter which produced a pass-band of 0.7 to 0.9 μ. The total sky area covered was 4.2 square degrees. (See Fig. 1.) Oort's most conservative model predicts that 3.5 globular clusters should be lurking somewhere within the mosaic shown in Fig. 1 - and the number could be as high as 26 depending on the effects of dynamical friction.

We have carefully scanned these plates microscopically and have not found a single globular cluster - at least not recognizable as such. Maybe this result was to be expected. Becklin, Neugebauer, and Willner [2] estimate $A_V \simeq 28$ mag, corresponding to ∿ 18 mag at the λ_{eff} of our plates. At the distance of the galactic center - 9 kpc - a bright globular cluster seen through the 3 kpc arm would have a photographic infrared magnitude I ∿ 21.5 ... and the estimated limiting magnitude of our plates is ∿ 20. This limit is no fainter because of the strong OH night sky emission.

Let us next consider the nucleus itself. If it is similar to that in M31 [3], it should look much like a super-luminous K star or compact globular cluster. The absolute magnitude of the central 1" of the M31 nucleus is $M_I = -10.7$ (or brighter if there is some interstellar absorption present).

Where should our nucleus be located? Many investigators (Oort, Kellerman, Becklin, Neugebauer, Ekers, Brown and Balick) have identified it with a highly concentrated, probably non-thermal radio source located in the larger thermal radio source known as Sgr A (West) [4]. Lo, *et al.* [5] using 3.8 cm VLBI find the core of this non-thermal source to be ∿ 10 AU in diameter at a radio luminosity, $L_{radio} = 10^{33}$ ergs/s, and brightness temperature $T_B \simeq 10^{10}$ °K.

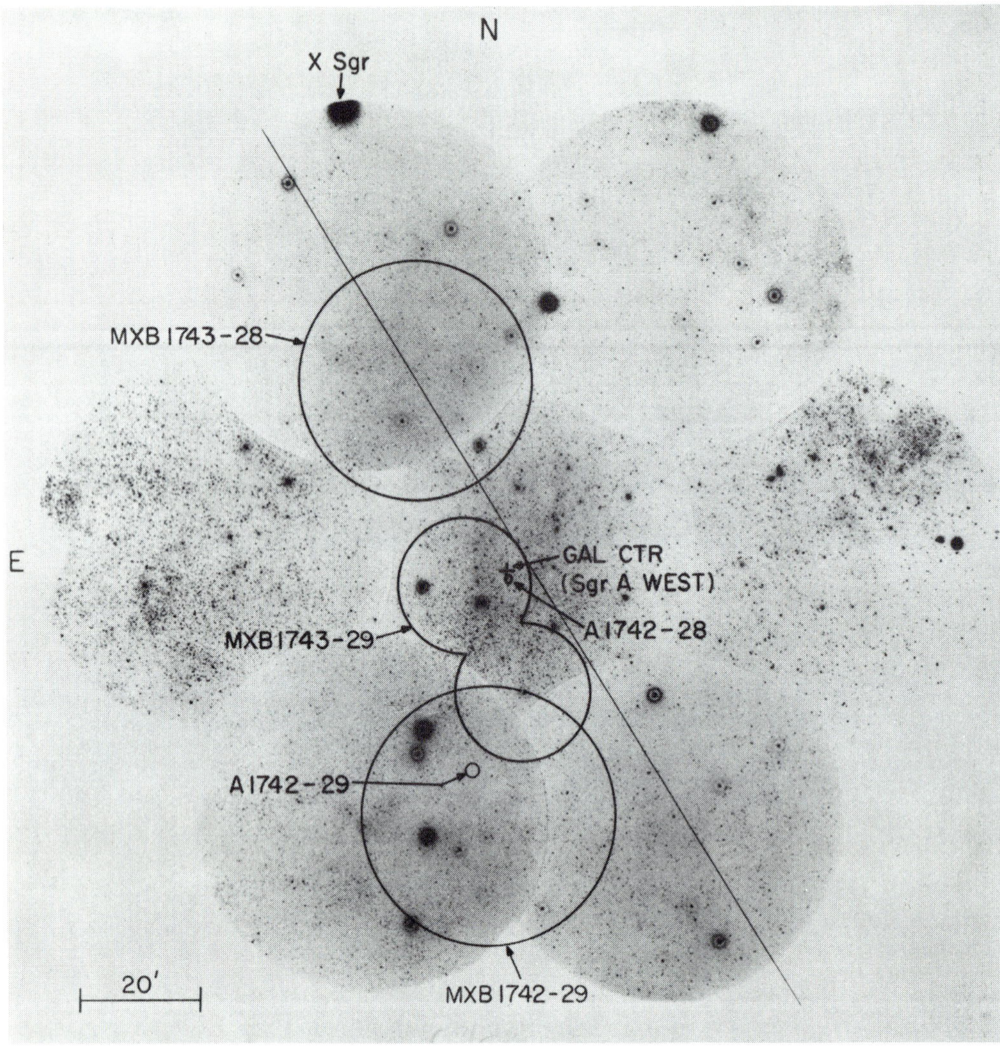

Fig. 1. A mosaic of deep near-infrared photographs of the galactic center regions (4 meter telescope, Cerro Tololo Inter-American Observatory). The approximate pass band is 0.7-0.9 microns. The straight line marks the galactic equator.

Becklin and Neugebauer [6] found a 2.2 μ source there (IRS 16), and on the basis of its spectrum, Willner [2] calls it a cluster - or luminous star.

What do we find at the position of Sgr A (West)? We see what appear to be two stellar-appearing objects: one is 1".4 ± 0".2 N, the other 4".0 ± 0".2 NE of the small diameter non-thermal radio source. (See Fig. 2.) Their location relative to the Becklin-Neugebauer source IRS 16 is difficult to evaluate since the B-N positions are ± 3".

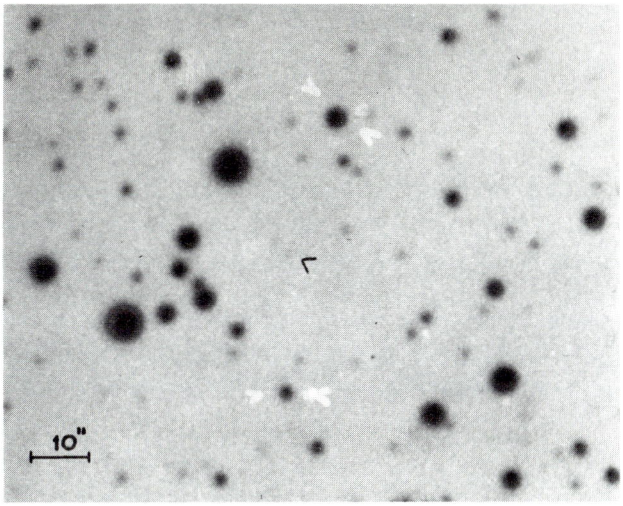

Fig. 2. A composite of 3 deep near-infrared photographs of the galactic center region (see caption, Fig. 1). The star image closest to Sgr A (West) is marked by the black "v". White "v's" mark arbitrary reference stars. North is up, east to the left.

What is the probability of finding a stellar image within 1".4 of the galactic center? From star counts in the vicinity, we calculate that the chances of finding a stellar image within this radius is 0.06. Therefore, it is possible (2 sigma result) that we have recorded an image of the galactic nucleus on our plates.

If we adopt for the stellar image $m_I = 20.0$ and a distance to the galactic center of 9 kpc, then the usual magnitude-distance formula

$$M_I = m_I + 5 - 5 \log r - A_I \quad (1)$$

reduces to

$$M_I = 5.23 - A_I = 5.23 - 0.6 A_V \quad (2)$$

If we accept the estimate that $A_V = 28$, then $M_I = -11.6$. If the spectrum of the M31 nucleus is K0 corresponding to $(V-I)_o = 1.4$, we arrive finally at a visual absolute magnitude, $M_V = -10.2$, in close agreement with the absolute magnitude of the M31 nucleus. The fact that the optical image is displaced from the compact radio source could arise if the optical source were in fact extended (such that it surrounded the radio source) and only a fraction were visible on our plate due to patchy obscuration.

We conclude, therefore, that we possibly have detected the nucleus of our galaxy, the most surprising result being that it appears perfectly stellar on our plates. It is interesting to conjecture what the linear size and mass of such a super star must be.

REFERENCES

1. J. H. Oort, *Astrophys. J. Letters* 218, L97 (1977).
2. S. P. Willner, *Compact Infrared Sources: NGC 7538 and The Galactic Center*, Ph.D. Thesis, Cal. Inst. of Tech. (1976).
3. A. R. Sandage, E. E. Becklin, and G. Neugebauer, *Astrophys. J.* 157, 55 (1969).
4. J. H. Oort, *Ann. Rev. of Astron. and Astrophys.* 15, 355 (1977).
5. K. Y. Lo, M. H. Cohen, R. T. Schilizzi, and H. N. Ross, *Astrophys. J.* 218, 668 (1977).
6. E. E. Becklin and G. Neugebauer, *Astrophys. J. Letters* 200, L71 (1975).

A NEW CLASSIFICATION SCHEME FOR GALACTIC BULGE X-RAY SOURCES

D. R. Parsignault and J. E. Grindlay

Harvard - Smithsonian Center for Astrophysics, Cambridge, Massachusetts 02138, U.S.A.

EXPERIMENT RESULTS

The Hard X-ray Experiment (HXX) on board the Astronomical Netherlands Satellite (ANS), observed 19 strong galactic X-ray sources (listed in Tables 1-4) December 1974 - April 1975, June - October 1975, and March 1976. A description of the HXX, its operation, and its in-flight calibration have been published elsewhere [1].

Using the 15-channel logarithmic PHA of the Large Area Detector (LAD), we studied the X-ray spectra of these sources between 0.93 and 28 keV. Several calibrations of the PHA on the Crab Nebula showed that the LAD was exceptionally stable over the time period of all the ANS observations.

Only data taken in relatively low and constant background were used to calculate minimum χ^2 -fits of the data to model source spectra. The variable parameters of the fits were: 1) a low energy absorption (E_a); 2) a spectral energy index (α) or temperature (kT); and 3) a normalization constant (A). The spectral data were fitted to a power law: $dF/dE = A\,[\exp(-E_a/E)^{8/3}]\,E^{-\alpha}$, and to an exponential model spectrum: $dF/dE = A\,[\exp(-E_a/E)^{8/3}]\,\exp(-E/kT)$, where dF/dE is in units of keV-cm^{-2}-sec^{-1} -keV^{-1}. Fitting the observed spectra to a thermal bremsstrahlung-type spectrum never significantly improved the fit obtained with an exponential-type spectrum: the low energy cut-offs remained essentially unaffected, while temperatures increased by about 20-50%.

An analysis of the spectrum of Cyg X-3 observed by ANS showed a statistically significant excess above the fitted X-ray continuum in two energy channels, between 4.6 and 7.2 keV [2]. These results confirmed the findings by two other groups which identified this same feature as line emission of highly ionized iron [3,4]. We performed a similar search for iron line emission in the 19 source spectra. A best-fit continuum spectrum was thus determined with the 2 PHA channels (4.6 and 7.2 keV) including the iron line omitted. From the differences between the observed number of counts and the fitted number of counts in these two channels, we calculated the equivalent line widths for Fe line emission. The instrument resolution ($\sim$1 keV) at $\sim$6.7 keV did not allow a precise determination of the energy of the observed feature.

SOURCE	E_a (keV)	α [1] or kT (keV) [2]	Eq. Line Width (keV)	I (cts-sec^{-1})
TABLE 1 Exponential-type spectrum preferred (except 4U1811-17) and iron line emission feature.				
4U1617-15	0.69 ± 0.21	2)4.1 - 7.0	0.39 ± 0.06	760 - 860
4U1744-26	1.58 ± 0.19	2)3.1 - 5.4	0.50 ± 0.12	16 - 24
4U1813-14	2.16 ± 0.06	2)3.1 - 5.5	0.34 ± 0.06	29 - 50
4U2124+38	1.16 ± 0.06	2)3.2 - 4.2	{ 0.55 ± 0.12 0.17 ± 0.08	44 - 27
4U1811-17	3.02 ± 0.10 2.60 ± 0.15	1)2.32 ± 0.23 2)2.76 ± 0.57	0.84 ± 0.22 0.60 ± 0.20 }	14
TABLE 2 No spectral-type preference but possible iron line feature.				
4U1642-45	3.52 ± 0.08	1)2.59 ± 0.05	0.62 ± 0.19	19
4U1758-25	2.78 ± 0.03 2.48 ± 0.05	1)1.86 ± 0.07 2)3.18 ± 0.23	0.51 ± 0.05 0.26 ± 0.05	61
4U1820-30	1.40 ± 0.11 0.67 ± 0.25	1)1.12 ± 0.08 2)4.27 ± 0.36	0.67 ± 0.17 0.62 ± 0.17	14
TABLE 3 Exponential-type spectrum preferred and upper limit on iron line emission feature.				
4U1636-53	1.15 ± 0.27	2)2.94 ± 0.48	<0.90	16
4U1702-36	1.43 ± 0.11	2)4.2 - 7.7	<0.32	32 - 58
4U1758-20	1.97 ± 0.19	2)3.83 ± 0.60	<0.60	28
4U1837+04	0.66 ± 0.27	2)3.88 ± 0.41	<0.47	10
TABLE 4 No spectral-type preference and no iron line feature.				
4U0614+09	1.47 ± 0.21 0.67 ± 0.39	1)1.62 ± 0.22 2)3.20 ± 0.48	<1.5 <1.5	2.1
4U1624-49	4.71 ± 0.23 3.94 ± 0.43	1)3.38 ± 0.07 2)2.27 ± 0.51	- -	3.3
4U1728-16	1.74 ± 0.09 0.92 ± 0.21	1)1.43 ± 0.16 2)4.35 ± 0.53	<.57 <.52	13.8
4U1728-24	1.90 ± 1.00 2.20 ± 0.60	1)0.2 - 0.6 2) >30 - 12	- -	5.6 - 2.0
4U1728-33	2.25 ± 0.14 1.95 ± 0.17	1)0.90 ± 0.12 2)7.02 ± 1.24	- -	3.2 - 7.0
4U1735-44	1.78 ± 0.17 1.12 ± 0.42	1)1.26 ± 0.16 2)5.21 ± 0.85	<0.90 <0.78	10.0
4U1755-33	2.0 ± 0.5 1.7 ± 0.5	1)2.1 ± 0.8 2)2.3 ± 0.5	- -	3.6

Table 1 lists the five sources which favored an exponential (or bremsstrahlung) distribution over the range 1 ~20 keV and for which iron line emission was detected. A preference for a given distribution was based on the criterion that the χ^2 value of a particular fit was within 90% confidence level for one spectral type, while the probability of occurrence of the value of χ^2 determined for the other spectral fit was < 5%. (When kT = f(I), the ranges are indicated.) Table 2 gives the results for three sources for which iron line emission may have been detected although with some degree of uncertainty due to either the poor fit of the continua, or apparent Fe line variability (e.g., 4U1820-30). Table 3 includes four sources whose continua were best fitted by an exponential-type spectrum, but for which only upper limits for iron line emission could be inferred. Table 4 lists the remaining sources for which no spectrum preference and no iron line feature could be established. The errors quoted for the parameters correspond to χ^2_{min} + 2.3, which corresponds to $\pm 1\sigma$ [5].

DISCUSSION

We have calculated the line to continuum power ratios for the different sources for which we observed a statistically significant excess above the continuum at ~6.7 keV, and compared them to the calculations of Mewe [6]. This author calculated these ratios as a function of temperature following certain assumptions which should be acceptable for a quiet or moderately active solar corona but, we realize, are probably not satisfied for the objects considered here. The iron abundances in Sco X-1, 4U1744-26, 4U1811-17, and 4U1820-30 are marginally below the cosmic value, N(Fe)/N(H) = 5×10^{-5}, and one-third of that value for 4U1813-14. For 4U2142+38 we found a variable Fe line emission feature which yielded an abundance ratio N(Fe)/N(H) which varied from ~20 - 100% of the cosmic value.

A classification of these galactic X-ray sources can be made using the intensity-temperature correlation (or lack thereof). Figure 1 shows these relationships in sources for which data were available for such an analysis. The bars on each point represent $\pm 2\sigma$ uncertainties. Sources with a direct correlation between kT and I may be similar to Sco X-1 which also shows this relation [7]. We observed the temperature to increase with intensity on a short time scale (~hours) to 4U1702-36, 4U1744-26, and 4U1813-14, and on a long time scale (~six months) for Sco X-1 and 4U1728-24. The conditions of our observations of 4U0614+09 and Sco X-1 (i.e., ~constant intensity) did not allow us to verify this relation on short time scales. In the case of 4U1728-16, the intensity range of our observations corresponds to the ~72-85 cts/62.5 sec measured by Copernicus [8]. Over this range we do not see the intensity-temperature correlation reported by Mason et al. [8], (compare our Fig. 1 to Fig. 2 in Mason et al.). The intensity range over which our observations of 4U1758-25 were made corresponds to ~268-356 cts/62.5 sec measured by Copernicus. Over this range, the temperature is constant, in agreement with the observations of Mason et al. [8].

In addition to being the highest temperature source detected, 4U1728-24 stands out in this survey as almost certainly (given its 122 sec pulsations) being a close binary system.

Over the intensity ranges shown in Fig. 1, no intensity-temperature relations were found for 4U1728-33 and 4U1837+04. In the cases of 4U1642-45 and Cyg X-2, we only obtained limited spectral data and could not establish such a relation.

A long term intensity-temperature relation was found for 4U1642-45 and 4U1820-30. This

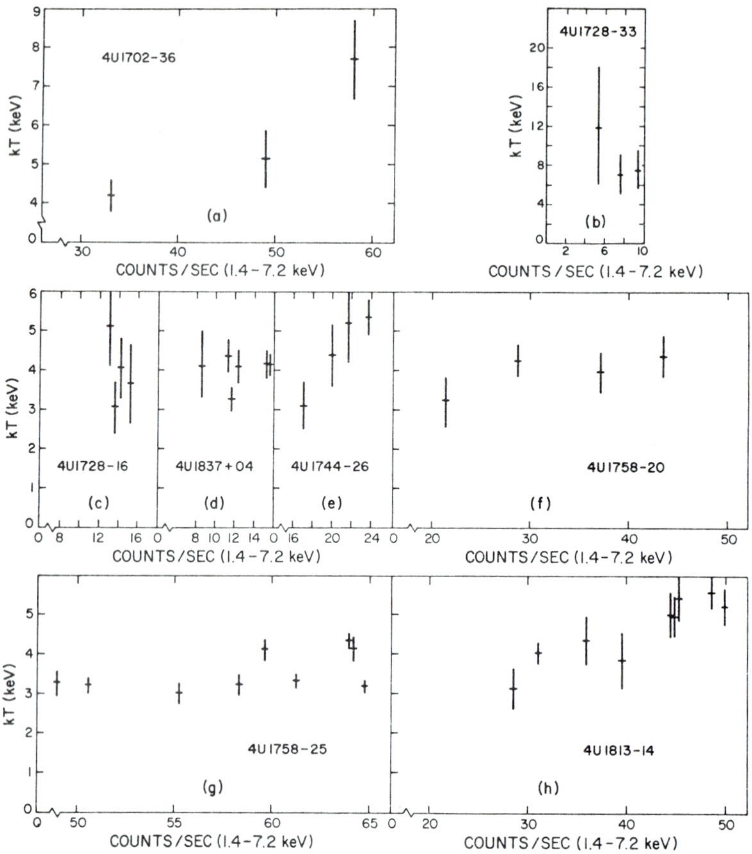

Fig. 1 Temperature (kT) of the best fit exponential spectrum versus source intensity (1.4 - 7.2 keV) observed. The error bars represent $\pm$ 2σ uncertainties.

intensity-temperature relation is characteristic of sources such as Cyg X-3 [2] or Cyg X-1 during its transition phases [9,10]. However, these two sources are differentiated from one another by their variability [11].

4U1837+04 can be singled out by the observation of fast decreases in its intensity which are accompanied by hardening of the X-ray spectrum. These dips in intensity occurred on JD 2,442,503.2 and 689.7. As the intensity decreased from 14.8 to 10 cts-s^{-1}, over a period of $\sim$110 min, the kT increased from 3.70 $\pm$ 0.44 keV to 6.9 $\pm$ 2.0 keV. Such a behavior has previously been observed in Cyg X-1 [12,13,14]. In Cyg X-1 the sharp intensity dips are correlated with the binary phase. Therefore, 4U1837+04 could be a member of a close binary system although no periodicities $\lesssim$ 2 days were observed.

CONCLUSIONS

We summarize our major conclusions as follows:

1. Most GX sources are best fitted in ~1-20 keV range by exponential spectra with kT ≅ 5 keV. However, for some sources (e.g., 4U1642-45 and 4U1758-25) the poor fits to simple exponential, or power law-type distributions indicate a more complex spectrum. Poor fits (i.e., complex spectra) are also sometimes obtained for some sources (4U1820-30) which at other times are well-described by exponential or power law type distributions.

2. Iron line emission at ~6.7 keV is always detected as a persistent spectral feature of 4U1617-15, 4U1744-26, 4U1813-14, 4U2142+38, 4U1642-45, and 4U1811-17, and possibly 4U1758-25. All of these, except 4U1642-45 and 4U1811-17, are observed either in our study or in that of Mason et al. [8] to show a direct correlation between the source temperature kT and the intensity I, in the 1.4 - 7.2 keV band.

3. No burst source counterparts (or candidates) showed a direct correlation between kT and I; at least two showed an inverse correlation. No iron line emission was detected at a 3σ level from burster counterparts during burst active states.

The above conclusions suggest that the strong X-ray sources in the galactic bulge (GX sources), which have previously been grouped together as having similar luminosities ($\geq 10^{37}$ ergs-s^{-1}), similar soft spectra (kT ≃ 5 keV), and a lack of periodicities (i.e., we exclude pulsars like 4U1728-24 with a hard spectrum characteristic of known high mass binaries [15]), may be further divided into at least two classes:

Class 1 GX Sources: Direct kT vs I correlation, usually fitted with an exponential spectrum plus an iron line emission feature, fluctuating primarily on time scales minutes-hours.
Class 2 GX Sources: Inverse (or no) kT vs I correlation, some (or sometimes) fitted by a power law spectra, usually with no iron line emission and generally active on all time scales, but especially on ≳ 3 hours, ~1 day, and 6 month time scales.

We may thus identify the sources in our survey with the classes: 4U0614+09, Sco X-1, and Cyg X-2, all of which have similar optical counterparts, are then Class 1 GX sources. Additional Class 1 GX sources identified with certainty (i.e., all criteria satisfied) are 4U1744-26 and 4U1813-14. The Class 2 GX sources in this study include all the bursters, or at least their candidate steady source counterparts. In addition, 4U1642-45 meets all criteria of a Class 2 source, and should be studied for possible burst activity.

The remaining sources in our study are more difficult to classify on this scheme since all the above criteria may not be satisfied (although "opposite" criteria are not satisfied). Nevertheless, we can probably assign - with decreasing probability - 4U1702-36, 4U1728-16, 4U1811-17, 4U1758-20, and 4U1758-25 all to Class 1 as they show most Class 1 (but no Class 2) characteristics. We are unable to classify 4U1624-49 and 4U1755-33 because of their lower flux and hence poorly determined variability and spectrum parameters.

REFERENCES

1. H. Gursky, H. Schnopper, and D. Parsignault, Astrophys. J. (Letters), 201, L127 (1975).
2. D.R. Parsignault, J. Grindlay, H. Gursky, W. Tucker, Astrophys. J., 218, 232 (1977).
3. P.J. Serlemitsos, private communication (1977).
4. P.W. Sanford, K. Mason, J. Ives, MNRAS, 173, 90 (1975).

5. Y. Avni, Astrophys.J., 210, 642 (1977).
6. R. Mewe, Sol. Phys., 22, 459 (1972).
7. N. White, K. Mason, P. Sanford, S. Ilovaisky, C. Chevalier, MNRAS, 176, 91 (1976).
8. K. Mason, P. Charles, N. White, J. Culhane, P. Sanford, S. Strong, MNRAS, 177, 513 (1976).
9. J. Heise, A. Brinkman, J. Schrijver, R. Mewe, A. den Boggende, E. Gronenschild, D. Parsignault, J. Grindlay, E. Schreier, H. Schnopper, H. Gursky, Nature, 256, 107 (1975).
10. D. Parsignault, J. Grindlay, H. Schnopper, E. Schreier, H. Gursky, Symp. on X-ray Binaries, NASA SP-389, 429 (1975).
11. D.R. Parsignault and J.E. Grindlay, Astrophys.J., (1978)
12. F.K. Li and W.G. Clark, Astrophys.J. (Letters), 191, L27 (1974).
13. K. Mason, F. Hawkins, P. Sanford, P. Murdin, A. Savage, Astrophys. J. (Letters), 192, L65 (1974).
14. D. Parsignault, A. Epstein, J. Grindlay, E. Schreier, H. Schnopper, H. Gursky, Y. Tanaka, A. Brinkman, J. Heise, J. Schrijver, R. Mewe, E. Gronenschild, A. den Boggende, Astr. and Space Sci., 42, 175 (1976).
15. C. Jones, Astrophys. J., 214, 152 (1977).

OPTICAL IDENTIFICATION OF TRANSIENT X-RAY SOURCES POSITIONED WITH THE HEAO-1 MODULATION COLLIMATOR

R. E. Griffiths, H. Bradt, U. Briel, R. Dower, R. Doxsey, G. Fabbiano, H. Gursky, M. Johnston, R. Leach, P. Murdin*, A. Ramsey, D. Schwartz, and J. Schwarz

Center for Astrophysics and Massachusetts Institute of Technology
**Anglo-Australian Observatory*

INTRODUCTION

Between August 1977 and April 1978, six transient X-ray sources were precisely located using the modulation collimator on HEAO-1, and four of these have good optical counterparts. We summarize here the main features of each transient and associated counterpart, and discuss the classification of each (Table 1). The importance of precise positioning and optical identification of transient sources has been highlighted by the optical observations of A0620-00 in a quiescent state [1], revealing a K dwarf. We hypothesize that for four of the transients discussed here (H1705-25, H1743-32, MX1608-52, MXB1659-29) the low mass binary model is applicable. 4U0115+63, a hard, pulsating transient, is identified with a B star/neutron star binary, and for H0449-55 we suggest, from preliminary observations, identification with an M dwarf system.

TABLE 1

Low Mass Binaries "K" + neutron star	"Be" + neutron star
Prototype: A0620-00 = Nova Mon. 1975	Prototype: A0535+26 = HD245770
H1705-25 = Nova Oph '77	4U0115 + 63 (periods $3^s\!.6$, 24^d) = Be
H1743-32 = ?	**M dwarf system**
MX1608-52 = Flaring star	H0449-55 = Me ? (Brief, high gal. lat.)
MXB1659-29 = Blue star	

H1705-25

The X-ray light-curve from Watson et al. [2] is shown in Fig. 1. The X-ray source was positioned using HEAO A3 (scanning modulation collimator) [3] and HEAO A1 (large area sky survey ≡ LASS) and identified with a 16th mag. optical nova found at the Anglo-Australian Telescope in one of the two possible locations (Fig. 2, from [4]). The optical spectrum showed He II λ4686 in emission, on a blue continuum, confirming the identification. The ratio of X-ray to optical luminosities, L_X/L_{opt} was $\sim 10^2$.

This is the third optically-identified X-ray nova of this type, the others being the prototype A0620-00 and A1524-61 [5]. Of the three, A0620-00 is the most well observed and understood, with X-ray emission arising in an accretion disk optically thick to electron scattering [6]. Oke [1] observed the optical counterpart a year after outburst and obtained a spectrum characteristic of a K5-K7 dwarf, together with the remains of an accretion disk, and concluded

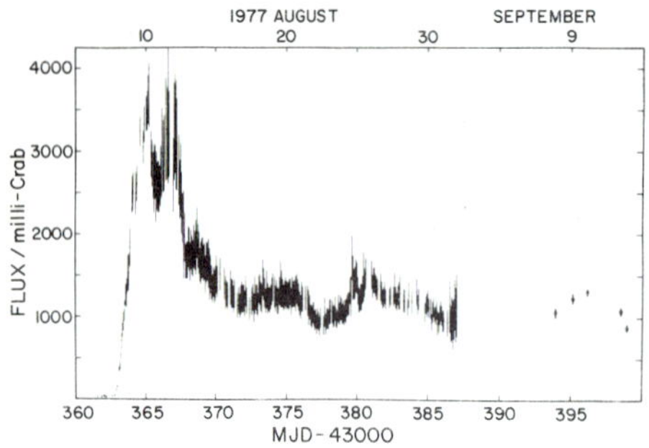

Fig. 1 The X-ray light curve of H1705-25 from [2]. Data mainly from Ariel V SSI.

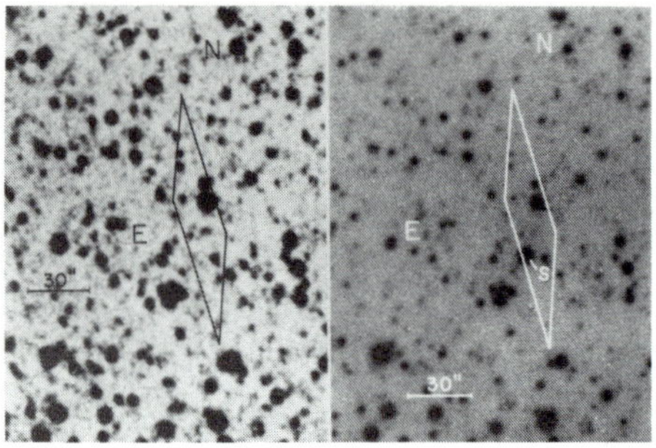

Fig. 2 From [4]. Left-POSS 'E'. Right-blue plate taken at AAT showing the optical nova (S) and the error box for the X-ray transient.

that Roche lobe overflow is unlikely in this system unless the compact star is massive. The alternative model is accretion from a stellar wind [7], with build-up of the disk until instability occurs. This model predicts weak X-ray emission in the quiescent system, which has not been observed either for A0620-00 or H1705-25 [8]. The dynamics of the triggering mechanism for the X-ray/optical novae therefore remains unclear. The link between these transients and their steady X-ray counterparts is provided by Aquila X-1 which shows a K zero dwarf spectrum in quiescence [9].

H1743-32

The second bright X-ray transient observed by HEAO-1, H1743-32, was also positioned using HEAO A1 and A3 [10, 3], resulting in two possible solutions. A UK Schmidt plate of the region was taken on September 4, 1977 (W. Zealey, private communication 1978) near the time of peak X-ray luminosity. Comparison with other plates in a search for the optical candidate is presently under way. Although the source has unfavorable galactic coordinates $\ell^{II} = 357.2$ and $b^{II} = -1.7$, the low-energy X-ray cutoff was not very large, $E_a \sim 1$ keV, and the source may be only ~3 Kpc distant. H1743-32 was unusual in its long decay time as observed by the Ariel V all sky monitor (11). In contrast to H1705-25, the source was still present at ~60 UHURU cts s^{-1} on the second HEAO scan of the galactic hub sources (12). Considering its X-ray spectrum and light-curve, along with lack of pulsations, this transient can be placed in the same class as H1705-25, but further information is pending an optical identification.

MXB1659-29

In the first clear demonstration of the association between a burster and a transient or flaring X-ray source, a position was obtained for a persistent source H1659-29 by HEAO A3. The initial position, from two quick-look orbits (13) has been refined using production data to that overlaid on the Palomar red plate in Fig. 4.

The positions obtained by HEAO A1 (14) and by the SAS-3 slats (15) are also shown in Fig. 3. Of the possible A3 positions, only one falls within the A1 box, and this A3 position is also within the SAS-3 (1976) error ellipse for the burster MXB1659-29 [16]. A comparison of IV N plates taken by Grindlay and by Liller (1978, private communication) has resulted in no detectable variability in any of the optical stars inside the revised MC error box, but Liller (private communication) has pointed out a blue star near the center of the box, labelled S in Fig. 4. Spectra of this star taken at CTIO during this conference confirmed it as the candidate [27].

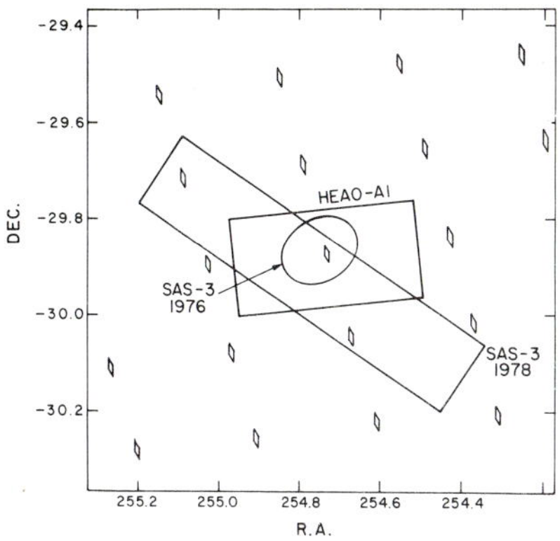

Fig. 3. Positions for the persistent source MXB1659-29. Diamonds are the possible A3 locations.

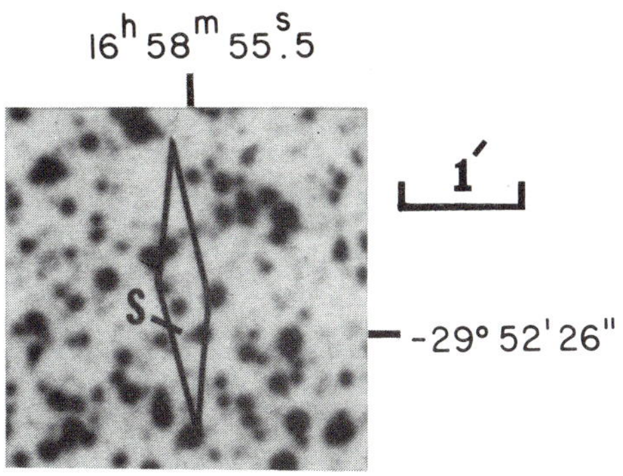

Fig. 4.

MX1608-52

The second possible example of an association between a burster and a flaring X-ray source is that of MX1608-52 [17]. The optical candidate [18] was found on a red-IR plate when the source was flaring, and optical observations in the quiescent state [18] rule out a supergiant or giant star for a source at D < 12 Kpc, as evidenced by the X-ray absorption. Both MXB 1659-29 and MX1608-52 therefore allow classification as low mass binaries involving a neutron star accreting from a dwarf, though the evidence for this is not compelling as yet. Differences in overall X-ray behavior probably reflect differences in accretion rate and/or magnetic field strength of the neutron star.

4U0115+63

The recurrent transient 4U0115+63 was recently studied both by SAS-3 and HEAO-1 [19, 20, 21]. Pulsations with a periodicity of $3.^{s}6$ [19] were observed by both satellites and the spectrum was very hard, with a power law slope $\alpha = -0.4 \pm 0.3$ [20]. This is the first clear evidence for the presence of a neutron star in this kind of system, since the alternative white dwarf models for A1118-61 and A0535+26 are ruled out for 4U0115+63 by the short period. In the first, important demonstration of the binary nature of transient X-ray sources, Rappaport et al. [21] have determined the orbital elements and shown that this system is wider than that of any other known X-ray binary.

Based on the SAS-3 and HEAO A3 positions, an optical identification was made [22] with a star showing strong Hα emission, putting this transient in the same category as A1118-61 and A0535+26 [23, 24] with transient X-ray emission arising at times of enhanced mass loss from the B-type companion to the neutron star.

The hard X-ray spectrum, together with the presence of pulsations, indicates that we are observing material accreting onto the poles of the neutron star rather than the accretion disk as in the first class of transients exemplified by the first two cases above. Steady X-ray counterparts to this class of transients are exemplified by Cen X-3 and SMC X-1.

H0449-55

The first high galactic latitude source observed by HEAO-1, on February 9, 1978 was positioned using experiments A2, A3 and A4, and Fig. 5 shows some of the possible A3 positions inside the A4 box (D. Gruber, private communication, 1978). The source was observed for only 2 orbits and reached a peak intensity of ~0.14 Crab, with a similar spectrum to that of the Crab and no significant low-energy absorption (E < 3 keV) (25).

Of the stars which lie in or near the possible A3 boxes, spectra of some have been taken by one of us (P. M.) at the AAT. The spectrum of the brightest (10^m) star S, indicated in Fig. 8, shows narrow, low excitation emission lines on an M dwarf spectrum with deep Ti O bands. The star closest to this one, labelled T in the figure, shows a similar spectrum. A search through the Harvard plate stacks for variability of the possible candidates for H0449-55 gave null results. A similar, brief high galactic latitude transient was observed by SAS-3 with the suggestion that the emission came from a star of the dM class (26).

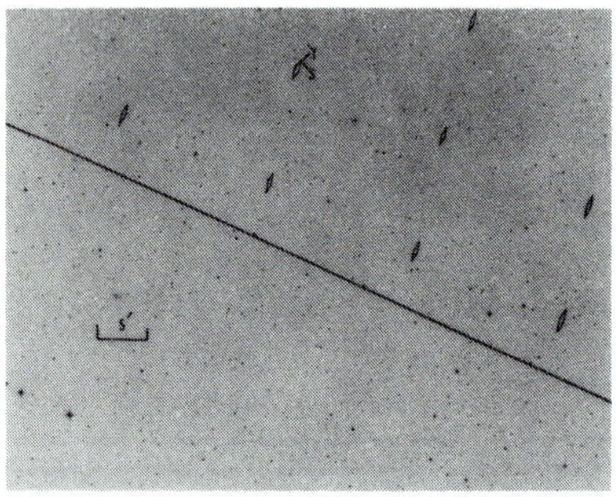

Fig. 5

REFERENCES

1. Oke, J.B. and Greenstein, J.L., 1977, Ap. J., 211, 872.
2. Watson, M.G., Ricketts, M.J., and Griffiths, R.E., 1978, Ap. J., 221, L69.
3. Gursky, H., et al. 1978, Ap. J., 223, 973.
4. Griffiths, R.E., et al. 1978, Ap. J., 221, L63.
5. Murdin, P., et al. 1977, Mon. Not. R. Astr. Soc., 178, 27P.
6. Griffiths, R.E., et al. 1976, Mon. Not. R. Astr. Soc., 177, 429.
7. Amnuel, P.R. and Guseinov, O.Kh., 1976, Sov. Astr. Lett., 2, 152.
8. Share, G., Wood, K., Byram, E., Meekins, J. and Shulman, S., 1978, IAUC 3197.
9. Thorstensen, J., Charles, P., and Bowyer, S., 1978, Ap. J., 220, L131.
10. Doxsey, R. et al., 1977, IAUC 3113.
11. Kaluzienski, L.J. et al. paper presented at the 151st AAS Meeting, Austin, Texas, 1978.
12. Wood, K., Share, G., Johnston, N., Yentis, D., and Byram, E., 1978, IAUC 3203.
13. Griffiths, R. et al. 1978, IAUC 3190.
14. Share, G. et al. 1978 , IAUC 3190.
15. Lewin, W.H.G. et al. 1978, IAUC 3190.
16. Lewin, W.H.G., 1977, Ann. N.Y. Ac. Sci., 302, 210.
17. Fabbiano, G., et al. 1978, Ap. J., 221, L49.
18. Grindlay, J., and Liller, W., 1978, Ap. J., 220, L127.
19. Cominsky, L., et al. 1978, Nature, 273, 367.
20. Johnston, M. et al. 1978, Ap. J. Lett., In press.
21. Rappaport, S., et al. 1978, Ap. J., in press.
22. Johns, M., Koski, A., Canizares, C., and McClintock, J., 1978, IAUC 3171.
23. Chevalier, C. and Ilovaisky, S. 1975, IAUC 2778.
24. Maraschi, L., Treves, A. and van den Heuvel, E.P.J., 1976, Nature, 259, 292.
25. Kaluzienski, L.J. et al. 1978 , IAUC 3174.
26. Rappaport, S. et al. 1976, Ap. J., 206, L139.
27. Grindlay, J.E. and HEAO A3 group, 1978, IAUC 3229.

ACKNOWLEDGEMENTS

It is a pleasure to acknowledge the invaluable assistance of Maureen Conroy in the data analysis. We would like to acknowledge D. Gruber of UCSD for providing the HEAO A4 error box for H0449-55.

HEAO A-1 OBSERVATIONS OF X-RAY EMISSION FROM BL LACERTAE OBJECTS

K. Wood, R. L. Kinzer, J. Samimi, D. Sadeh, D. Yentis, J. Meekins, W. Evans, G. Share, D. McNutt, S. Shulman, E. T. Byram, T. A. Chubb and H. Friedman

Naval Research Laboratory, Washington, D.C. 20375

INTRODUCTION

At least 32 BL Lac objects are known. They resemble quasars, but are distinguished by having flatter radio spectra, steeper optical spectra, a lack of emission lines and greater polarization and variability [1]. Extrapolations of optical spectra lead to low predicted X-ray fluxes, but models of the variable optical and radio emission predict that observable X-ray fluxes will be generated by Compton collisions. The earliest attempts to detect BL Lac objects in X-rays gave only upper limits [2,3] . By combining these flux limits with physical parameters deduced from synchrotron models of the radio emission, upper limits and source distances could be obtained. In some instances such inferred distances were nearly in conflict with distances estimated from redshifts. A sufficiently low X-ray upper limit could produce a contradication that would be resolvable only by either abandoning the cosmological interpretation of redshifts or by revising the emission model. The first BL Lac objects detected in X-rays were Markarian 421 (2A1102+384) [4] and Markarian 501 (4U1651+39) [5]. These two are among those BL Lac objects known to be in elliptical galaxies [6].

High sensitivity observations such as can be made with HEAO-1 can yield new detections and establish X-ray luminosities and variability. Determination of the high energy spectrum also fixes the flux of ionizing radiation near the source, which may be helpful in understanding the absence of emission lines.

OBSERVATIONS

The HEAO A-1 experiment comprises seven proportional counter modules sensitive from 0.5 to 25 keV. Intensity data presented here come from the four scan modules, which have a sensitivity limit of 2×10^{-12} ergs/cm^2 sec. Mechanical collimators on the modules give $1^{o} \times 4^{o}$ (FWHM) fields of view. These data were taken with the spacecraft in scanning mode. A source is seen for 10 seconds each scan and there are 48 scans per day, not all of which are useable.

Faint objects are detected by summing scans of a given region. (Under favorable circumstances a BL Lac object such as MKN 501 (which has a flux of 2 UFU), is detectable in a single scan). Fig. 1 shows successive 1-day scan summations beginning with August 19, 1977. The horizontal coordinate on the figure is the azimuth angle along the scan. The ecliptic plane is at far left and the north

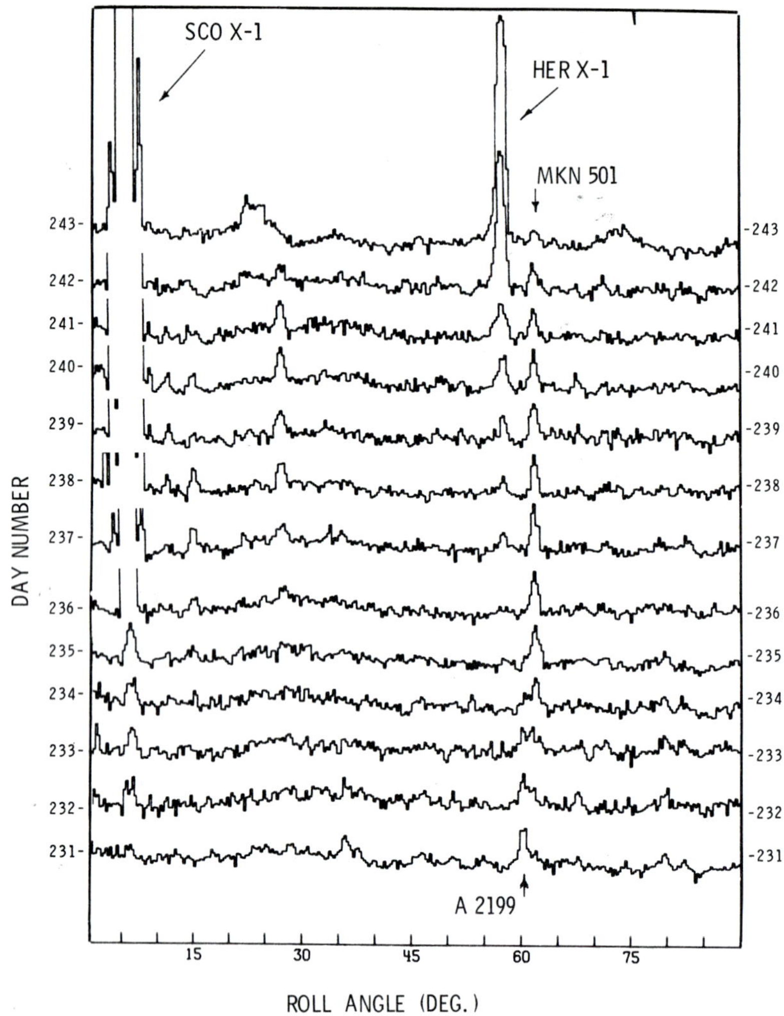

Fig. 1 1 day scan summations near MKN 501.

ecliptic pole is at the far right. Passage of sources through the field of view can be observed directly on this plot. The sensitivity of the instrument is also clear, since the classical sources such as Sco X-1 (6°) and Hercules X-1 (59°) saturate the display, while sources at the limit of sensitivity for previous surveys can be followed in detail from day to day.

MKN 501 is the source near scan azimuth 62°. For the first few days shown on the figure it is confused with the nearby cluster source, Abell 2199. An apparent intensity for a 1-day scan summation is derived by fitting the known collimator response to observed counts. Such intensities have been fit for MKN 501 from each scan summation shown in Fig. 1. It can still be detected on the last scan shown, by which time the apparent intensity is down to 0.2 UFU. By

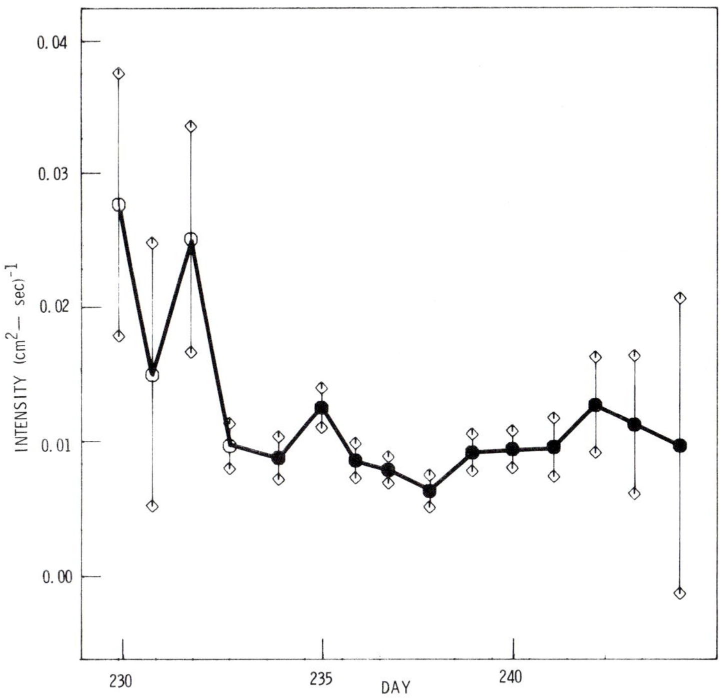

Fig. 2 MKN 501 light curve, days 231-243.

fitting these daily intensities to the collimator response in the 4° direction (perpendicular to the scan), it is possible to derive the source position in ecliptic longitude. The source position can also be established by detecting it in another of the experiment modules which has a smaller field of view, ½ x 1°. Error boxes obtained by either technique include MKN 501, as does a much smaller error box found by the HEAO-A-3 scanning modulation collimators, so that the identification is not in doubt. Using the known optical source position and the instrument collimator response, we correct each measured daily intensity for collimator response, to obtain a light curve for the source. The result is shown in Fig. 2. Since there are fewer counts when the source is near the edge of the field of view, the 1σ error bars are larger there. The four points shown as open circles are from the first four days, when the source is confused with Abell 2199. There is some evidence of variability in the object on day 235 (23 Aug. 1977) at a 2σ level of significance. It is much less dramatic than the variability reported for MKN 421 [4].

In the sky region covered in the first three weeks of HEAO scanning, there occurs only one other known BL Lac object. This is the Zwicky compact galaxy IZw 1727+502, with redshift 0.0554. As with MKN 501, this object was seen distinctly as an X-ray source, although at a level 4 times fainter. Fig. 3 shows this source as it appears in scan summations near its transit on day 248.

The measured intensity of IZw 1727+502 is 0.5 UFU, which is below the sensitivity limits of earlier surveys. Like MKN 421 and MKN 501, it is known to be in a galaxy [7]. The source is detected in independent one-day scan summations and can be followed in the scans for perhaps 11 days, although it is not always possible to get good fits to intensities. Figure 4 shows the error box for the source. The BL Lac object is the only candidate object found by searching catalogs of known sources (excluding normal stars) in this region. The optical spectrum found by Oke [7] falls about a factor of 10 below the observed X-ray flux, if extrapolated to these energies. Thus, the spectrum needs to harden at some intermediate energy. A light curve constructed for IZw 1727+502 in the same manner as that for MKN 501 shows no significant variability.

DISCUSSION

The first two BL Lac objects to enter the field of view of the A-1 experiment detectors (MKN 501 and IZw 1727+502) were both detected over several days, allowing determination of source positions and construction of light curves. Variability was detected for Markarian 501 at marginal significance. The total number of BL Lac objects detected to date in X-ray stands at four, since Mushotzky [8] has reported detection of a further BL Lac object, PKS 0548-322, with the HEAO A-2 experiment. These sources are distinguished within the overall BL Lac sample (summarized in reference [1]) in two respects: all four have known redshifts and all four are shown to be in galaxies. Among the 32 BL Lac sources summarized in [1] , one finds only two others known to be in galaxies but not yet found in X-rays, these being AP Libra and BL Lac itself.

REFERENCES

1. W.A. Stein, et al., Ann. Rev. Astron. Ap., (1976).
2. B. Margon, et al., Ap. J., 207, 359, (1976).
3. M.P. Ulmer, and S.S. Murray, Ap. J., 207, 364, (1976).
4. M.J. Ricketts, et al., Nature, 259, 546, (1976).
5. W. Forman, et al., preprint, (1977).
6. M. Ulrich, et al., Ap. J. 198, 261, (1975).
7. J.B. Oke, Ap. J., 219, L97 (1978).
8. R. Mushotzky, paper presented at Conference on BL Lacertae objects, Pittsburgh (1978).

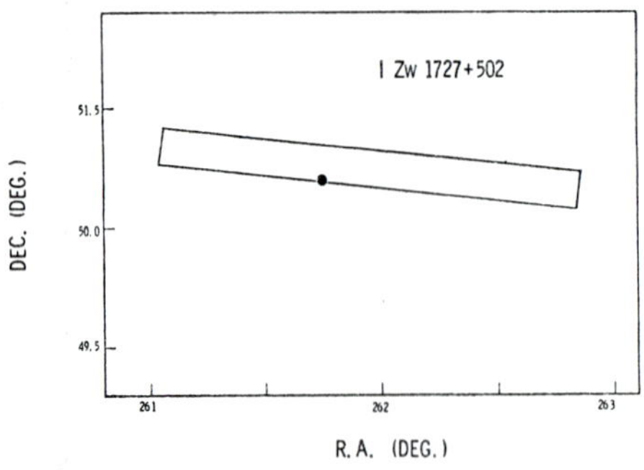

Fig. 4 IZw1727+502 error box, with location of BL Lac object

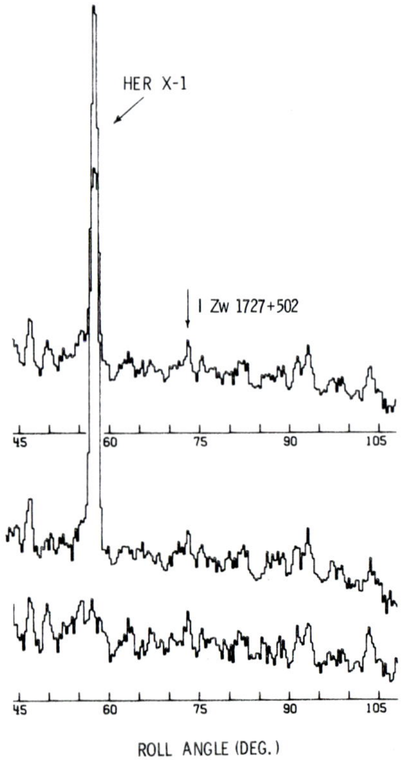

Fig. 3 Scans of IZw1727+502. Below: 4 day sums. Above: 8 day sum. Flux = $(1.4 \pm 0.2) \times 10^{-6}$ Jy.

RECENT RESULTS ON GALACTIC X-RAY SOURCES FROM THE ARIEL V SKY SURVEY INSTRUMENT

M. G. Watson, M. J. Ricketts, G. W. Fraser
and M. R. Sims

X-Ray Astronomy Group, Physics Department, University of Leicester, Leicester LE1 7RH, England

PRELIMINARY RESULTS FROM THE SSI LOW LATITUDE CATALOGUE

Work is presently in progress to produce a complete catalogue of low galactic latitude ($|b| \leqslant 10^o$) x-ray sources from analysis of $\sim$18000 orbits of observations by the Ariel V Sky Survey Instrument (SSI; for details see [1]). The analysis procedures used are essentially the same as those employed in the compilation of the SSI high galactic latitude ($|b| > 10°$; the "2A") catalogue [2] , but the use of the more extensive observations now available gives some improvement in sensitivity over previous work, although this advantage is somewhat offset by confusion problems arising from the higher source density in the galactic plane. The systematic component of source errorboxes (see §5 in [2]) has also been reduced by the use of an attitude refinement technique. Many of the brighter galactic x-ray sources ($I \gtrsim 10$ Uhuru count s^{-1}) have been positioned to an accuracy of < 1 arcminute by a variety of rocket and satellite experiments, notably by observations with the SAS-3 satellite [3]. For the numerous weaker sources in the galactic plane, the SSI observations (with a limiting accuracy $\sim$3 arcminutes) often lead to substantial improvements over earlier surveys (mainly the Uhuru observations [4]). Analysis of the SSI data is also revealing a number of new sources reflecting the improved sensitivity (and, in part, the intrinsic variability of galactic x-ray sources).

Preliminary results for 14 sources in the region $220^o < \ell < 320^o$ are given in Table 1. These sources have the "2A" designation since the eventual low latitude catalogue will supercede previous SSI positions such as those determined by Seward et al [5, 6] from an analysis of only $\sim$30% of the data now available. The list includes 3 previously unreported sources, and 7 sources for which the new errorbox is an improvement (sometimes substantial) over that of the counterpart "4U" [4] and/or "A" source [6] . Two of these sources merit further comment:

2A0655-071 (=MX0656-07). This source was first seen in September 1975 by the SAS-3 satellite [7] and was described as a transient. Subsequent observations by the Ariel V RMC and ASM [8;9] whilst confirming the large variability, indicated that its x-ray light curve did not follow the typical pattern of an x-ray transient. The SSI scanned 2A0655-071 extensively between 1974 November and 1977 December. Over this period the source was only detected during a 45 day interval in 1976 March-April, (between the RMC and ASM observations), with an intensity falling from $\sim$12 to $\sim$4 SSI count s^{-1} (1 SSI count $s^{-1} \approx 2.5$ Uhuru count s^{-1}); at all other times the source was undetectable with typical 3σ upper limits $\leqslant 1$ SSI count s^{-1}.

TABLE 1 Preliminary Catalogue Results

ARIEL NAME	ST *	OTHER NAMES	POSITION RA Dec	l b	ERROR BOX ‡ RA Dec	RA Dec	RA Dec	RA Dec	AREA **	INTENSITY ‡‡ AVE	MAX	MAX/MIN	NOTES
2A0655-071	CON	MX0656-07	103.99 -7.14	220.14 -1.75	104.00 -7.27	103.87 -7.12	103.96 -7.05	104.09 -7.20	0.016	1.0	12.1	>40	highly variable source
2A0709-114	NEW	---	104.32 -11.47	224.14 -3.45	103.80 -11.31	104.77 -11.28	104.78 -11.69	103.81 -11.72	0.310	1.6			
2A0709-221	NEW	---	107.25 -22.18	234.98 -5.86	107.78 -22.30	107.04 -22.66	106.70 -22.05	107.44 -21.69	0.422	1.6			
2A0726-260	IMP	4U0728-25	111.68 -26.08	240.33 -4.11	111.57 -25.99	111.80 -26.02	111.78 -26.14	111.55 -26.11	0.021	2.0			
2A0745-191	IMP	4U0739-19	116.22 -19.12	236.34 2.97	116.35 -19.08	116.22 -19.27	116.04 -19.15	116.17 -18.97	0.038	1.9			
2A0851-469	IMP	A0854-46 4U0842-47	132.86 -46.92	266.98 -1.57	132.93 -46.75	133.01 -47.05	132.70 -47.09	132.62 -46.80	0.051	1.6			
2A0920-628	NEW	---	140.18 -62.83	281.60 -9.22	140.55 -63.01	139.65 -62.98	139.70 -62.63	140.59 -62.65	0.115	2.2			
2A1028-571	CON	A1034-56 4U1036-56	157.12 -57.11	284.79 0.45	156.59 -57.20	157.44 -56.88	157.63 -57.02	156.78 -57.34	0.076	1.6			
2A1134-652	CON	4U1137-65	173.65 -65.28	295.31 -3.79	173.65 -65.44	173.22 -65.29	173.57 -65.12	174.00 -65.27	0.043	5.9	12.8	13	
2A1322-616	CON	4U1323-62	200.65 -61.69	306.97 0.65	200.95 -61.54	200.80 -61.87	200.36 -61.82	200.52 -61.49	0.057	1.9			
2A1343-602	IMP	A1343-60 4U1344-60	205.97 -60.26	309.76 1.62	205.96 -60.05	206.37 -60.31	205.96 -60.47	205.55 -60.21	0.067	1.2			
2A1416-589	IMP	A1415-60	214.05 -58.97	314.04 1.75	213.69 -59.28	213.83 -58.65	214.33 -58.68	214.20 -59.30	0.132	1.4			
2A1439-627	IMP	A1439-61 4U1425-61?	219.76 -62.72	315.24 -2.75	219.51 -63.16	219.43 -62.28	219.95 -62.27	220.05 -63.15	0.176	1.2			Coincident with SNR G315.4-2-3
2A1448-554	IMP	4U1446-554	222.10 -55.45	319.42 3.31	221.84 -55.64	221.87 -55.27	222.29 -55.28	222.27 -55.65	0.071	2.3			

* Status of source: CON = Confirmed; IMP = Improved position
NEW = New source.

** Area of 90% confidence contour (square degrees).

‡ Rectangle just enclosing the 90% confidence contour

‡‡ Intensity in SSI count s^{-1}. Average value only unless clearly variable

These results show that the x-ray light curve of 2A0655-071 is complex with at least two maxima over 1975-76 and a small duty-cycle for x-ray emission.

2A1439-627. The new errorbox for this source is significantly displaced from the Seward et al source A1439-61, and now clearly overlaps the SNR G315.4-2.3 (MSH 14-63) [10] . Since SNRs are not common objects, and at least six are known to emit x-rays >2 keV, this remnant is a reasonable identification on positional grounds alone. Note that the errorbox has been determined assuming a point source - the SSI observations do not preclude extended emission with angular size $\lesssim 1^{\circ}$. X-ray emission from this SNR has been reported previously by Naranan et al [11] (at energies <2 keV) and by Winkler [12] from OSO-7 observations (1-30 keV). In both cases the positional accuracy was slightly worse than that reported here. Winkler's results gave marginal evidence that this source is variable, but this is not supported by the SSI observations.

MSH 14-63 is probably the remnant of the AD185 supernova ([12] and references therein). If this is the case then the detection of 'hard' x-rays supports the general view that significant emission above ~2 keV will originate only from younger SNRs.

OBSERVATIONS OF DWARF NOVAE

Soft x-ray emission ($\lesssim$2 keV) has been detected from a number of dwarf novae, (eg [13]), but to date there is only marginal evidence that any are hard (>2 keV) x-ray emitters (for SS Cygni [14]). SSI observations of two dwarf novae, SS Cygni and Ex Hydrae, now confirm that dwarf novae can indeed be hard x-ray sources.

SS Cygni

X-ray observations of SS Cygni have been made with the SSI over a total of 120 days between 1974 November and 1977 April. The x-ray light curve indicates that the source is highly variable (by a factor >10); of particular interest however is the disappearance of the x-ray source during the peak of the optical outbursts. This remarkable anticorrelation of the x-ray and optical flux is illustrated in Fig. 1 which shows the SSI x-ray observations 'folded' onto one optical outburst according to the outburst phase of the observation (see [15] for more details). During optical outburst the x-ray intensity initially rises and then cuts off very quickly as soon as SS Cyg is brighter than $\sim 9^m$. During optical decline the pattern of events is reversed. Over the optical maximum the SSI observations yield an x-ray intensity of 0.6 $\pm$ 0.6 SSI count s^{-1}.

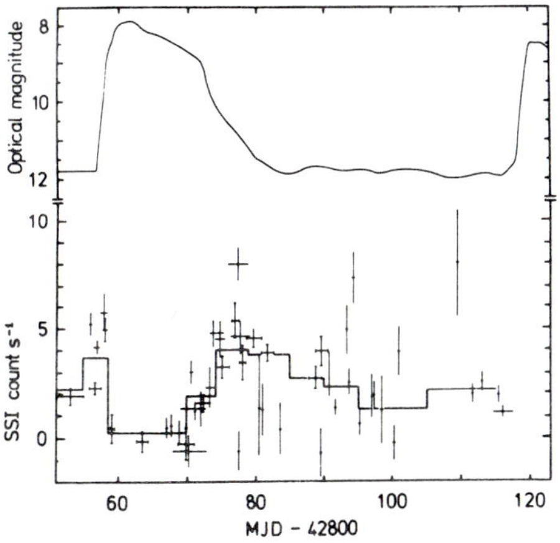

Fig. 1 The SSI observations of SS Cyg 'folded' onto one optical outburst. The histogram shows the weighted mean of the data. From [15].

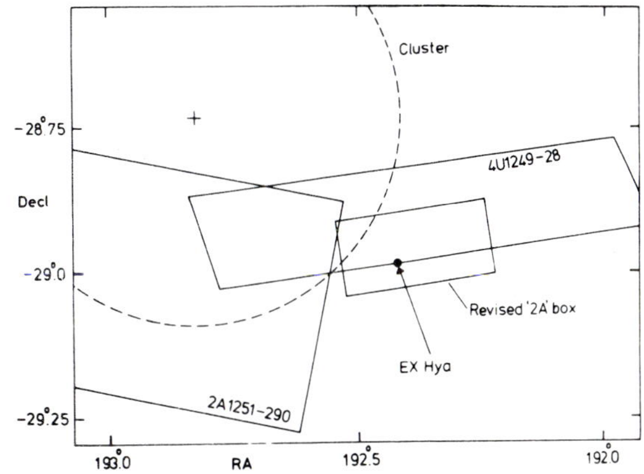

Fig. 2 X-ray errorboxes in the region of Ex Hya together with the cluster SC1251-28.

Ex Hydrae

As a result of the observations of SS Cyg, a sensitive search for x-ray emission from other dwarf novae was carried out using the Point Summation Technique (PST) on the SSI data [16]. This survey gave clear evidence for x-ray emission from Ex Hydrae, a dwarf nova somewhat fainter than SS Cyg with a very long outburst period of ~465 days.

In a follow-up to the PST survey, the SSI observations were used to determine a revised errorbox for 2A1251-290 [2], the SSI source near to Ex Hya, which had previously been tentatively identified with SC1251-28, a cluster of galaxies [17]. The new errorbox, now designated 2AR1249-289, is shown in Fig. 2 with the position of the cluster and previous x-ray errorboxes. Ex Hya lies within 3 arcminutes of the centre of the box, whilst the cluster is well outside; thus on positional grounds alone it must be regarded as a good candidate. Supporting evidence for the identification comes from the SSI x-ray light curve which clearly indicates that the source is variable with a χ^2 probability of less than 0.1% that the source is steady.

Discussion of Dwarf Novae

The generally accepted model for dwarf novae involves a close binary with period <1 day consisting of a white dwarf and late type main sequence star. Mass loss from the main-sequence star via Roche-lobe overflow leads to accretion onto the white dwarf from the accretion disc which is formed (eg [18]). In such a system hard (>2 keV) x-rays can probably be produced only if the accretion flow onto the white dwarf is radial: in the presence of an accretion disc this requires the white dwarf to have a substantial surface field [19]. Detailed investigations of models of this sort [19, 15] have shown that hard x-ray emission is possible for dwarf nova systems but only for a narrow range of accretion rates (F) and magnetic fields (B). In the case of SS Cyg the combined optical/x-ray data allows detailed comparison with theory. The observation that the x-ray emission is effectively switched off when the optical magnitude is brighter than $\sim 9^m$ leads to a determination of the locus of the parameters (F, B) during outburst, giving estimates of $B \approx 10^6$ G with F ranging between $\lesssim 4 \times 10^{16}$ gs^{-1} in quiescence and $\sim 2 \times 10^{19}$ gs^{-1} at outburst [15].

REFERENCES

1 G. Villa et al., M.N.R.A.S., 176, 609 (1976).

2 B.A. Cooke et al., M.N.R.A.S., 182, 455 (1978).

3 H.V. Bradt, this volume.

4 W. Forman et al., Ap. J. Suppl., in press.

5 F.D. Seward et al., M.N.R.A.S., 175, 39P (1976).

6 F.D. Seward et al., M.N.R.A.S., 177, 13P (1976).

7 G.W. Clark, IAU Circ., No. 2843 (1975).

8 G.F. Carpenter et al., IAU Circ., No. 2852 (1975).

9 L.J. Kaluzienski et al., IAU Circ., No. 2935 (1976).

10 D.H. Clark and J.L. Caswell, M.N.R.A.S., 174, 267 (1976).

11 S. Naranan et al., Ap. J. (Lett.), 213, L53 (1977).

12 P.F. Winkler, Ap. J., 221, 220 (1978)

13 S. Rappaport et al., Ap. J. (Lett.), 187, L5.

14 H. Heise et al., Astron. Astrophys., 63, L1 (1978).

15 M.J. Ricketts et al., M.N.R.A.S., in press.

16 M.G. Watson et al., M.N.R.A.S., in press.

17 N.A. Bahcall, Ap. J. (Lett.), 217, L77 (1977).

18 G.T. Bath, in: Structure and evolution of close binary systems, IAU Symp. No.73 Reidel, Dordrecht, 1976.

19 A.C. Fabian et al., M.N.R.A.S., 175, 43.

X-RAY POLARIMETRY

K. S. Long, G. A. Chanan, D. J. Helfand,
W. H.-M. Ku and R. Novick

Columbia Astrophysics Laboratory, Departments of Astronomy and Physics, Columbia University, New York, New York 10027, U.S.A.

INTRODUCTION

As of 1978, two X-ray satellite experiments have been launched whose purposes include the determination of the polarization of cosmic X-ray sources, namely, the Leicester University experiment on *Ariel 5* and the Columbia University instrument on *OSO 8*. Both instruments are Bragg crystal polarimeters. We summarize herein the method by which these instruments operate and some of the results which have been obtained. A precision measurement of the linear polarization in the Crab Nebula at 2.6 and 5.2 keV is included, and evidence is given for polarization in the Cygnus sources and Sco X-1.

BRAGG CRYSTAL POLARIMETRY

A plane crystal illuminated by a parallel beam of X-rays reflects only those X-rays which satisfy the Bragg condition. If the Bragg angle is 45°, Bragg reflection is synonymous with coherent scattering through 90°. Incident photons with electric vectors in the plane of reflection will not be reflected. By rotating the crystal about the axis defined by the beam, polarization can be detected. When a linearly polarized beam is incident on the crystal, the reflected flux will vary with a frequency of twice the rotation frequency of the device, just as it would if one rotated an optical polaroid in a linearly polarized beam of visible light. The simplest design for a Bragg crystal polarimeter consists of a flat crystal panel reflecting X-rays incident at 45° into a proportional counter. The *Ariel 5* polarimeter, which doubles as a spectrometer, is of this type. A serious limitation of this design is that the detector must have the same area as the intercepted photon beam, resulting in a relatively unfavorable signal-to-noise ratio for most sources. This problem can be alleviated by mounting the crystals on a segment of a parabolic surface of revolution so that X-rays are focused onto a smaller detector. In the two focused polarimeters aboard *OSO 8*, the projected area of each crystal panel is 140 cm^2, but the active detector area is ~6 cm^2. For these polarimeters, the modulation as the experiment is rotated about a 100% linearly polarized beam is reduced from 100% to 93%; thus the spectroscopic capabilities are sacrificed, but the background is reduced by more than a factor of 20. The *Ariel 5* and *OSO 8* polarimeters are described more completely by Griffiths *et al.* [1] and Weisskopf *et al.* [2], respectively. Both utilize graphite crystals and are sensitive to X-rays in narrow energy bands near 2.6 keV and 5.2 keV.

OBSERVATIONAL RESULTS

The Columbia polarimeters were used to observe the Crab Nebula for 6 days each in 1976 and 1977. The first unambiguous detection of X-ray polarization in the Crab Nebula was obtained in the 1976 observation [2], which confirmed earlier indications of polarization which had been obtained at Columbia using a sounding rocket [3]. Data contributing to the unpulsed portion of the light curve at 2.6 keV during the two *OSO 8* observations, folded modulo the satellite rotation period, are shown in Fig. 1. Modulation at twice the rotation frequency is clearly apparent.

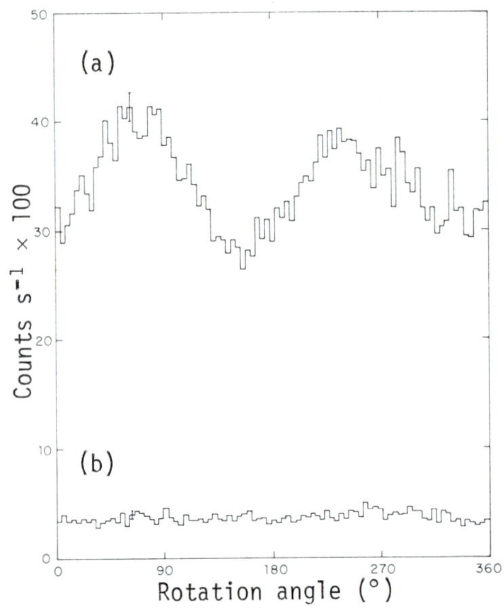

Fig. 1. Average counting rates obtained with the *OSO 8* detectors at 2.6 keV during observations of (a) the Crab Nebula and (b) the earth-occulted instrumental background.

After removing the pulsed emission from the Crab pulsar, the nebula polarization was determined to be 19.2% ± 1.0% at 2.6 keV and 19.5% ± 2.8% at 5.2 keV [4]. These results are consistent with a later, less sensitive polarization measurement using the *Ariel 5* instrument [5]. The position angles determined from the *OSO 8* data are 156.4° ± 1.4° at 2.6 keV and 152.6° ± 4.0° at 5.2 keV. The X-ray measurements are in excellent agreement with the Oort and Walraven [6] determination of the average optical polarization of the nebula of 19% at 162° and firmly establish synchrotron radiation as the X-ray emission mechanism.

The first attempt to measure the polarization of Sco X-1 was made by Angel *et al.* [7] who were able to set a 3 σ upper limit of 20% polarization between 4 and 20 keV using a rocket-borne Thomson-scattering polarimeter. With *Ariel 5*, Gowen *et al.* [8] set a more restrictive constraint of 7.7% (3 σ) polarization at 2.6 keV. More recently, Sco X-1 has been observed with the *OSO 8* polarimeters. Results of a polarization analysis of the time-averaged data, obtained between August 25—30, 1977, are shown in Table 1(a). The apparent polarization of Sco X-1 is 0.6% at 2.6 keV and 1.3% at 5.2 keV. Since polarization is a positive definite quantity, an apparent polarization will always be detected for any finite data set. Therefore, the relevant question is, what is the probability that an unpolarized X-ray source would have produced an apparent polarization larger than what was observed? For Sco X-1, this formal probability is somewhat less than 1% at both energies, strong evidence that a nonstatistical effect has been observed. The modulation was seen throughout the observation and does not appear to have been due to a few transient events. Spurious modulation due to asymmetries in background rates cannot account for the observed result, since the polarized flux would have had to have been ~40% of the total instrumental background. On the other hand, simulations of the

TABLE 1 Polarization of Sco X-1 and Cyg X-1 Obtained with *OSO 8* Polarimeters

Parameter	(a) Scorpius X-1 1977		(b) Cygnus X-1 1975, 1976, 1977	
	2.6 keV	5.2 keV	2.6 keV	5.2 keV†
Source counting rate. .	3.55 s^{-1}	0.40 s^{-1}	0.09 s^{-1}	0.02 s^{-1}
Measured polarization*.	0.56% ± 0.52%	1.28% ± 1.07%	2.5% ± 1.8%	9.8% ± 6.1%
Position angle.	9°	60°	162°	158°
Probability of chance occurrence.	4.1×10^{-3}	1.3×10^{-3}	1.3×10^{-4}	1×10^{-5}

*Uncertainty represents 99% confidence limits. †Polarimeter No. 1 only.

OSO 8 polarimeter response indicate that spurious modulations of the order of 0.5% may be caused by pointing errors or instrumental misalignments. Further analysis of the data for this effect is in progress. At present the *OSO 8* results should be regarded as 99% confidence upper limits to the polarization of Sco X-1 of 1.1% at 2.6 keV and 2.4% at 5.2 keV. These limits can be used to constrain the inclination angle of Sco X-1 if the source geometry is well understood. For example, in the approximation of emission from a semi-infinite plane-scattering atmosphere [9], the 99% confidence limit to the polarization at 2.6 keV restricts the inclination angle to less than 44°, consistent with the recent limit of 30° derived by Perrenod [10] from optical data. In fact, however, Perrenod's best estimate of the inclination angle is 10°, in which case the anticipated polarization would be 0.1% for a simple disk, which is considerably below the sensitivity of either the *Ariel 5* or *OSO 8* experiments.

Cygnus X-1 is distinguished from other binary X-ray sources because there is reasonably good evidence that the mass of the compact object exceeds the mass limit for neutron stars. As a result, three separate observations of Cyg X-1 have been carried out using the polarimeters aboard *OSO 8*. The time-averaged results, indicated in Table 1(b), show evidence of modulation and are consistent with a polarization of 2.5% at 2.6 keV and 9.8% at 5.2 keV. When each year's observations are considered separately, the 2.6 keV modulation appears at almost equal significance in each year's data, while the 5.2 keV results are dominated by the final year's observation. Insofar as possible, the data have been checked for effects which can generate spurious polarization, including the presence of another source in the field of view, effects due to non-Poisson variations in the source strength (by examining the data for modulation at frequencies near twice the rotation period), and modulation due to pointing errors in the satellite. None of these problems is likely to have generated the observed result. A somewhat worrisome problem is that the position angle of the polarization, ~160°, is consistent with the position angle that one would detect if solar X-rays were scattering off the residual atmosphere into the polarimeter. These X-rays would be 100% polarized because they must scatter by 90° to be detected, since the spin axis of the spacecraft is maintained 90° away from the sun. However, solar activity was quite low during the Cyg X-1 observations; no evidence of solar scattering was observed when the instrument was earth-looking; and evidence of modulation was observed from data obtained during satellite night. As a result, it appears likely that these data represent the first detection of X-ray polarization in Cyg X-1. The results are difficult to reconcile with emission from a semi-infinite plane-scattering atmosphere. While a percentage polarization of 2.5% leads to an inclination angle of ~65°, consistent with the lack of X-ray eclipses, a polarization of 9.8% would not lead to the same conclusion. Furthermore, the results at the two energies are only marginally consistent with energy independent polarization, there being an indication of increasing polarization at higher energies. Qualitatively, such an effect is predicted by

Lightman and Shapiro [11] if the Cyg X-1 disk becomes optically thin at an energy of a few kiloelectron volts and if the inner region of the disk is geometrically thin. Alternatively, one might imagine models in which the polarization is due to X-rays scattered off gas streams in the vicinity of the X-ray source. In any event, it is evident that continued and more sensitive measurements of polarization in Cyg X-1 are needed.

Approximately 20 galactic sources have been observed with the polarimeters aboard *Ariel 5* and *OSO 8*. Polarization in at least one source, the Crab Nebula, has been detected. The Sco X-1 and Cyg X-1 results are tantalizing because they begin to constrain emission models for these sources. Other polarization results which have been obtained include an *Ariel 5* upper limit of 2% polarization in the transient A0620-00 [12], an *OSO 8* upper limit of 13.5% to the time-averaged polarization of the pulsating binary Cen X-3 [13], and an apparent detection of ~5% polarization in Cyg X-2 in the 1975 *OSO 8* observations which was not confirmed in the 1976 and 1977 observations, perhaps indicating temporal variability. Unfortunately neither the *Ariel 5* or *OSO 8* polarimeter has sufficient sensitivity to detect polarization in extragalactic sources, such as Cen A. For that, more sensitive but technically feasible experiments are required.

This work was supported by the National Aeronautics and Space Administration under contract NAS5-22408.

REFERENCES

1. R. E. Griffiths, B. A. Cooke, A. Peacock, K. A. Pounds, and M. J. Ricketts, *Monthly Notices. Roy. Astr. Soc.* 175, 449 (1976).

2. M. C. Weisskopf, G. G. Cohen, H. L. Kestenbaum, K. S. Long, R. Novick, and R. S. Wolff, *Astrophys. J. (Letters)* 208, L125 (1976).

3. R. Novick, M. C. Weisskopf, R. Berthelsdorf, R. Linke, and R. S. Wolff, *Astrophys. J. (Letters)* 174, L1 (1972).

4. M. C. Weisskopf, E. H. Silver, H. L. Kestenbaum, K. S. Long, and R. Novick, *Astrophys. J. (Letters)* 220, L117 (1978).

5. K. A. Pounds, private communication, 1978.

6. J. H. Oort and T. Walraven, *Bull. Astr. Inst. Netherlands* 12, 285 (1956).

7. J. R. P. Angel, R. Novick, P. Vanden Bout, and R. Wolff, *Phys. Rev. Letters* 22, 861 (1969).

8. R. A. Gowen, B. A. Cooke, R. E. Griffiths, and M. J. Ricketts, *Monthly Notices Roy. Astr. Soc.* 179, 303 (1977).

9. S. Chandrasekhar, *Radiative Transfer*, Dover Publications, New York, 1960.

10. S. C. Perrenod, *Astrophys. J.* 206, 876 (1976).

11. A. P. Lightman and S. L. Shapiro, *Astrophys. J.* 203, 701 (1976).

12. R. E. Griffiths, M. J. Ricketts, and B. A. Cooke, *Monthly Notices Roy. Astr. Soc.* 177, 429 (1976).

13. E. H. Silver, Ph. D. Thesis, Columbia University, New York, N. Y., 1978.

Low Luminosity Sources and Transients

LOW LUMINOSITY GALACTIC X-RAY SOURCES

G. P. Garmire

Department of Physics, California Institute of Technology, Pasadena, California

ABSTRACT

The number of low luminosity sources found in the Galaxy has increased from a few to more than two dozen in the past three years. The sources range from isolated hot white dwarfs (HZ43) to normal stellar coronae (α Cen), very active coronal stars (RS CVn), flare stars (UV Ceti), interacting winds and stellar surfaces (Algol), and finally to accretion onto white dwarfs (AM Her, steady; U Gem, impulsive). The following review discusses each of these classes in some detail and portrays the wide variety of physical phenomena leading to X-ray emission.

INTRODUCTION

Following the initial discovery of X-ray sources [1] and the subsequent optical identification of a small fraction of the sources found in early surveys, it became clear from the galactic latitude distribution that the brighter galactic X-ray sources are located at distances greater than a kilo-parsec and are more luminous than 10^{35} ergs/sec. Following the UHURU survey Giacconi et al. [2], Margon and Ostriker [3], and Seward et al. [4] made the first attempts at deriving a luminosity function for the galactic X-ray sources. From a statistical point of view, these estimates were important in revealing an upper limit to the luminosity of compact X-ray sources, but without reliable information on the distances to the sources it was not possible to reach any firm conclusions about the population of X-ray sources of low luminosity.

The identification of low luminosity sources is complicated by the fact that they must be relatively nearby and hence will have a very broad latitude distribution, thereby making them difficult to distinguish from the large population of faint extragalactic sources. Nearby bright stars emitting X-rays can be identified even with relatively crude angular resolution, since the probability of a bright star falling in a fraction of a square degree X-ray error box is quite small. The number of stars detected so far in the X-ray band is relatively few (see Table 1) and there are numerous upper limits for stars of all spectral classes [5,6].

In the following sections, I would like to present a review of the currently identified low luminosity X-ray sources. This list is growing daily with the continuation of the HEAO-1 survey. The limited angular resolution of the low energy instrument (1.5°), however, forces us to be a little tentative concerning the identifications of the fainter sources. Future rocket-borne telescopes,

HEAO-B and telescopes on the Space Shuttle will soon make the identifications unique.

TABLE 1 Low Luminosity Galactic X-ray Sources *circa* June 1978

Star	d (pc)	L_X (x 10^{30} ergs/s)	T (10^6 °K)	Reference
Algol Systems				
Algol	30	16	< 30	Schnopper *et al.* [14]
AK Dra (?)	--	--	--	Tuohy *et al.* [16]
RS CVn				
Capella†	14	4	10	Cash *et al.* [10]
λ And†	26	0.6	--	
HK Lac†	150	80	--	Walter *et al.* [11]
σ Gem†	150	100	--	Charles *et al.* [63]
UX Ari	50	21	10	Walter *et al.* [11]
HR 1099	32	30	--	Walter *et al.* [64]
RS CVn	145	60	--	Walter *et al.* [64]
AR Lac	50	10	--	Walter *et al.* [65]
RW U Ma	150	300 (flare)	--	Charles *et al.* [63]
LX Per	145	~ 60	--	Charles *et al.* [63]
AM Her	~ 100	10^3	0.4	Tuohy *et al.* [27]
Cataclysmic Variables				
U Gem	100	200	0.4	Mason *et al.* [37]
			0.2	Rappaport *et al.* [32]
SS Cyg	~ 130	{100 high; 4 low}		Heise *et al.* [35]
White Dwarfs				
HZ 43	65	3 x 10^4	0.11 bb	Margon *et al.* [43]
Fiege 24	90	2.4 x 10^4	0.07 bb	Margon *et al.* [45]
(Sirius B)	2.7	.01	0.2	Lampton *et al.* [54]
Flare Stars				
UV Ceti	2.7	0.06	< 10 peak	Heise *et al.* [46]
YZ CMa	6.6	4	~ 30 peak	Heise *et al.* [46]
Proxima Cen	1.3	~ 1		Haisch *et al.* [48]
Normal Stars				
α Cen	1.3	.003	0.7	Nugent *et al.* [55]
Vega	8.1	~ 0.07	--	Topka *et al.* [56]

TABLE 1 (continued)

η Boo	9.7	~ 0.1	--	Topka *et al.* [56]
12 Peg	1250	10^3	--	Ayres *et al.* [57]
ζ Cyg	58.8	3.9	--	Ayres *et al.* [57]
o^2 Eri	4.9	.03	--	Ayres *et al.* [57]
HR976	53	2 ± 1	--	den Boggende *et al.* [58]
Orion Nebula	450	3.8×10^3	55	den Boggende *et al.* [59]
				Bradt *et al.* [60]

†Long period

THE HEAO-1 A-2 EXPERIMENT

Since much of the data presented below was obtained by the HEAO-1 A-2[1] instrument, a brief description should be included. The low energy portion of the experiment includes two multiwire proportional counters covered by collimators possessing two fields of view for each detector (1.5° x 3°, 3° x 3° and 3° x 3°, 3° x 6°). The latter detector's axis is tilted 6° along the scan path with respect to the first detector's axis. Point source observations are best carried out using the 1.5° x 3° field of view collimator to avoid source confusion. The area of this section of the detector is about 180 cm^2 and it spans the energy range 0.15-3.0 keV. The data from eight scalers (4 for the 1.5° x 3° FOV and 4 for the 3° x 3° FOV) covering the energy bands 0.15-3.0 keV, 0.8-3.0 keV, 0.15-0.5 keV and 0.74-2.0 keV are read out every 1.28 sec, corresponding to about 0.2° of motion along the scan direction. Pulse height data are accumulated for 10.24 sec and read out in 32 pseudo-logarithmic channels. There are numerous other modes of operation available, but these are generally used for pointings and special purpose observations (see Rothschild *et al.* [7] for a more complete description). The spacecraft aspect is determined to an accuracy of about 0.1° for this experiment.

REVIEW OF OBSERVATIONS

RS CVn Stars

The first RS CVn star, Capella, was detected by Catura [8] on a sounding rocket flight. Capella belongs to the class of long period RS CVn systems as noted by Hall [9]. Observations from the ANS satellite [6] reported by Mewe suggested that the emission was variable up to a factor of four on time scales of a day. The HEAO-1 observations reported by Cash [10] determined the temperature to be about 10^7 K and that a strong iron feature (1 keV EW) near 0.8 keV expected from Fe XVII and Fe XVIII was consistent with the 10^7 K temperature and normal iron abundance. The spectrum was found to be predominantly isothermal with very little admixture of other temperatures permitted. No variability has been seen from the HEAO-1 data. The only other long period RS CVn stars detected to this time are λ And, HK Lac and σ Gem but they are so faint that spectra cannot be obtained.

---

[1] The A-2 Experiment is a collaborative effort lead by E. Boldt of GSFC and G. Garmire of CIT, with collaborators at CIT, JPL, UCB, and GSFC.

Turning to the normal RS CVn stars, UX Aries has been studied most extensively by Walter _et al._ [11]. They find a constant X-ray flux to within a factor of 2.5 and a temperature of about 10^7 K, but with an iron abundance that is only 0.03 solar. A low iron abundance has been suggested by Naftilan [12] from optical studies of RS CVn's as a class. Other RS CVn's detected by HEAO-1 are given in Table 1.

The RS CVn stars have been reviewed by Hall [9], who suggests that the strong H and K emission lines of Ca seen outside of eclipse are indicative of chromospheric activity. The multiple periodic nature of the light curve in these binary systems, he explains by a region of large-scale spot activity which darkens one side of the cooler star within 30° of its equator. The darkening produces a "wave" in the long-term light curve which is accounted for by this spotted region and explains the red color of the "wave." If the spotted region is responsible for the X-ray emission, then phase effects should show up in the X-ray light curve of these objects, since for suitable orbital inclinations, the active region should become obscured. A large flare in HR1099 observed by White _et al._ [13] in the X-ray and radio bands supports a flaring model for these stars.

The space density of eclipsing RS CVn systems is estimated to be about $1.0 \pm .2 \times 10^{-6}$ pc^{-3} with a scale height of 110 pc [9]. Walter _et al._ [11] have used these values plus the assumption that the number of noneclipsing systems is ten times more numerous than the eclipsing ones, to obtain an estimate that ten percent of the soft X-ray background might arise from the RS CVn type systems.

Algol Systems

Algol (β Persei) was detected as a 2-6 keV X-ray source by Schnopper _et al._ [14] and confirmed by Harnden _et al._ [15]. The emission is probably variable, ranging between 3×10^6 K and 3×10^7 K for the above observations. Harnden _et al._ [15] constructed a stellar wind model for the source with the K IV subgiant dumping matter onto the B8 V dwarf. They predict a temperature of around 6×10^6 K and a range of X-ray luminosities that extend up to about 10^{33} ergs/s at which point absorption by the streaming matter quenches the X-rays from escaping the system.

The system has been observed by the soft X-ray instruments on HEAO-1. Preliminary reduction of the HEAO-1 data shows that the intensity of the source does not change appreciably over a period of three days and that there is no evidence for an eclipse in the X-ray data within two hours of the optical minimum. This supports the result of Harnden _et al._ [15] that the X-ray emission comes from most of the surface of the B8 V dwarf, or from a region around the dwarf if it is putting out some wind of its own.

Another possible stellar system similar to Algol has been found near AM Her. The source is seen only in the 1/4 keV channel and is visible only when AM Her is in eclipse, since AM Her is so intense when out of eclipse that the new source is lost in the response of the collimator to AM Her. This source position agrees with AK Drac, but the identification is only tentative until a better position is available or coincident temporal behavior is detected [16].

AM Her

The AM Her system has been studied extensively over the past few years following its optical identification by Berg [17], the discovery of soft X-ray emission by

Hearn and Richardson [18], and the optical polarization data which revealed a 3.1 hr period and a 2 x 10^8 Gauss magnetic field [19]. Spectral analysis of the system by Cowley and Crampton [20], Priedhorsky [21], Greenstein et al. [22], Stockman et al. [23], and Priedhorsky et al. [24] determined that one component was a white dwarf and that the companion is probably a dwarf M star. The white dwarf appears to be corotating at its orbital period, since the X-ray period is identical to the optical period to high order. Mass is apparently being transferred from the M star onto the magnetosphere of the white dwarf where the interaction generates the soft X-rays. As the matter pours down onto the surface of the white dwarf and heats up, hard X-rays are generated which extend up to 60 keV and beyond [25, 26]. This system is unique among the low luminosity sources thus far discovered, in that it has a magnetic white dwarf as the source of the gravitational field. The magnetic field complicates the models of the X-ray emission and determines the nature of the optical linear and circular polarization. A detailed study of the X-ray light curve and spectral behavior will be given by Tuohy et al. [27]. An analysis of spherical accretion onto magnetized white dwarf stars has been completed by Masters [28] and Lamb and Masters [29] which shows three spectral features, an ultraviolet optically thick cyclotron peak, a very soft X-ray peak from blackbody emission in the stand-off shock zone and a hard free-free component caused by electron bremsstrahlung from the accreting electrons before they come into equilibrium. Although not all the detected features of the light curve and polarization effects are completely understood, qualitative agreement can be obtained from a model where the magnetic axis of the white dwarf is tilted with respect to the orbital plane and the X-ray eclipses are caused by the emission region being obscured by the body of the white dwarf.

Cataclysmic Variables (CV's)

About twenty cataclysmic variables have been observed by HEAO-1 during optical outburst. Only U Gem was clearly detected during this search [30]. Remarkably the flux from U Gem was nearly 100 times the background noise, so that if the other nineteen CV's were of the same luminosity or spectral shape, a sizeable fraction of them should have been detected. Mason et al. [30] concluded from this study that probably the X-ray emission from the bulk of the CV's was softer than U Gem and hence fell below the threshold for detectability of the HEAO-1 low energy detectors by being too cool rather than by being too low a luminosity. Even the U Gem spectrum is extremely soft, with a fitted thermal bremsstrahlung spectrum of $kT \sim .03$ keV. The physical model for these sources involves accretion beaming onto a disc surrounding one star in a binary pair which forms a bright spot. The star filling its Roche lobe generates the gas stream which originates from the inner Lagrangian point and travels to the accretion disc where the interaction forms the bright/hot spot [31]. An increase in the overflow creates the outbursts. Clearly the size of the accretion disc relative to the mass of the star surrounded by the disc will govern the relative velocities of the stream and disc and hence the temperature of the spot. If the relative velocities are too low, soft X-rays may not be produced. In general this seems to be the case for most CV's based on the HEAO-1 data.

The dwarf nova SS Cyg was observed during outburst by Rappaport et al. [32] as a very soft X-ray source. This was confirmed by Bowyer et al. [33] and by Hearn et al. [34]. During the low state, both hard and soft X-rays were detected by the ANS instruments [35], revealing that the X-rays do not turn off between outbursts. Heise et al. [35] suggest that the hard component requires the existence of a magnetic field to channel the matter flow onto the poles of the white dwarf. The field has not been detected so far, but if it is as low as 10^7 Gauss it would not have been observed [36]. HEAO-1 A-2 data confirm the presence of a hard variable component between outbursts [37] and an enhanced flux near one keV [38].

Isolated White Dwarfs

The source HZ43 is thought to be a white dwarf in a binary system which also contains an M dwarf with weak Balmer emission lines [39]. The source has been observed over a period of some 5 years by Hayakawa et al.[40], Hearn et al. [41], Lampton et al. [42], Margon et al. [43] and by the HEAO-1 A-2 detectors, indicating a fairly steady source of X-ray emission which tends to rule out flare activity in the M dwarf as an explanation for the emission. The more recent data indicate a constant X-ray source to better than 30% [43]. Optical data over a much longer time span support a steady source (Margon op cit). The basic model for this source is a hot DA white dwarf that is still cooling with a possible progenitor of a planetary nebula, or a sd O or sd B star (Margon op cit). HZ43 B is unlikely to play any role in the system because of the wide separation between the two stars. See Table 1 for parameters. A hot coronal model is felt to be implausible by Margon et al. [43], based on the work of Strittmatter et al. [44].

Another hot white dwarf, Fiege 24, was observed by Margon et al. [45]. Again the companion star is an M dwarf which is probably a flare star in this system (Margon op cit). The white dwarf component produces a radiation field adequate to influence the ionization and recombination radiation in the M dwarf's atmosphere. The M dwarf probably does not affect the temperature determined for the white dwarf, but flaring might be expected from this system.

Flare Stars

Two flare stars, UV Ceti and YZ CMi, have been observed with the instruments onboard the ANS satellite [46] during outbursts. The large flare observed from YZ CMi was not covered by any ground-based observations, so the scaling between X-ray energy and any other band is unknown. The spectrum was not analyzed, but the hard X-ray emission is indicative of a nonthermal process. Clearly, better data are needed to draw anything but the most tentative conclusions.

A flare from UV Ceti was covered optically, and showed that a large flux of optical emission accompanied the X-ray emission (Heise et al. op cit). This result has been compared to the solar flare case by Mullan [47], who finds that conductive losses in red flare stars dominate the decay timescale. He finds that the UV Ceti flare observations show a much lower X-ray output compared to optical emission than for the Sun, and this situation leads to the prediction that X-ray flares from flare stars may be very hard to detect in general, since large flares are quite infrequent.

In contrast to UV Ceti, however, Proxima Centauri was observed to be in a bright state in the 44-190 Å band by Haisch et al. [48] on 22^h 38^m UT 1975 July 22, but to lie below the limits of detectability on a scan 24 hours earlier. This behavior was interpreted to imply an X-ray flare was observed. The "flare" gave an estimated ratio $L_x/L_{opt} \approx 2\text{-}80$ which lead them to criticize Mullan's [47] estimate based on the UV Ceti observation and solar data.

Sirius

The ANS focusing telescope first detected the flux from Sirius as reported by Mewe et al. [6]. They found no variation in the X-ray flux on time scales of one and two days, and conclude that the white dwarf is the probable X-ray source.

Upper limits obtained from the Apollo-Soyuz observations of Sirius by Shipman *et al*. [49] were used to derive an upper limit on the helium abundance under the assumption of an optically thick thermal source for the radiation from the white dwarf. The coronal emission models of Mewe *et al*. [50], Hearn and Mewe [51] and Brecher [52] are not supported by the model atmosphere calculations of Shipman *et al*. [49]. The data presented from the Apollo-Soyuz cannot rule out an optically thin coronal emission spectrum. Recent observations by Cash *et al*. [53] rule out a blackbody spectrum for the emission mechanism.

Extensive observations of Sirius have been made from HEAO-1 and will be reported in detail during the conference. The flux appears to be constant over a period of 6 days, but there is evidence for a temperature change during one of twenty-six exposures to the source [54]. The spectrum is extremely soft, being best fit by a thin plasma at a temperature of 2×10^5 °K. No evidence for absorption by interstellar gas is detected, which is consistent with less than about 0.1 atom of hydrogen/cm^3 along the 2.7 pc path to the source. The data neither confirm nor rule out the white dwarf as the emitting object, hence, categorization of this object is ambiguous at this time.

α Cen

The HEAO-1 observations of α Cen reveal a low luminosity soft X-ray source [55] that exhibits a constant flux in the 1/4 keV band over five days of exposure to the source. The temperature is less than 7×10^5 K, which is cooler than the quiet Sun, yet the luminosity of 3×10^{27} ergs/sec is nearly an order of magnitude greater than the solar coronal emission. The low temperature argues against an active corona by analogy with the Sun, yet the high luminosity points toward a larger fraction of the stellar surface contributing to the emission. The emission is probably originating from α Cen A, since it is a G2V close to the Sun, but the dK1 component cannot be ruled out by the observations obtained from HEAO-1. The fact that Mewe *et al*. [6] failed to detect the source at a level corresponding to 1.6×10^{27} ergs/s in the 1/4 keV band suggests the possibility of X-ray variability. The low temperature suggests strong OVI emission which should be detectable by Copernicus spectrometer observations.

Other Stars

There have been observations of several other stars, but the data are sparse at this time. Vega and η Boo have been detected using a rocket-borne focusing telescope [56]. The fluxes are only poorly established because the spectrum is not fixed by the handful of photons detected. Estimated values are given in Table 1. The HEAO-1 low energy instruments failed to detect Vega, but did observe η Boo. Preliminary analysis by Ayres and Linsky [57] of the HEAO-1 low energy data for stellar sources detected emission from 12 Peg, ζ Cyg, and o^2 Eri. These results are still tentative and spectral data have not been examined. The star HR976 has recently been suggested as the identification of a soft X-ray source observed by ANS [58]. One final object of interest is the Orion source reported by the ANS group [59] and recently positioned in the Trapezium region of the Orion Nebula [60]. This source could either be an interacting wind source or a single object in the Trapezium region. The exact nature of the object is not clear since this is a very young region and a compact object would not have had time to evolve, a conclusion reached by den Boggende *et al*. [59].

CONCLUSIONS

The low luminosity X-ray sources in the Galaxy fall into five rough categories thus far: cooling hot condensed stars, coronal emission, flare emission from magnetically active regions, stellar winds (interacting with either accretion discs, stellar surfaces or other winds in a shock zone), and Roche lobe overflow onto an accretion disc around magnetic white dwarfs. The low luminosity is a result of the larger physical size and lower gravity compared to the sources associated with neutron stars and black hole candidates. Except for the accretion onto white dwarfs, the spectra of the low luminosity X-ray sources are softer than the spectra generated in the region surrounding the more condensed objects. The low luminosity objects are by far more numerous than the high luminosity sources in the Galaxy; however, their integrated luminosity is probably less than the total luminosity from neutron stars and black holes. For example, the RS CVn stars number about 10^6 in the Galaxy, but since their luminosity is only of order 10^{31} ergs/s, the total galactic luminosity is approximately 10^{37} ergs/s. The same holds for α Cen-type stars which may number 10^9, but whose luminosity is only of order 3×10^{27} ergs/s.

The theory of corona generation and the production of magnetically active regions has been developed in large part through analogies with the Sun. Now that stellar coronae are being detected, it should be possible to obtain a broader understanding of the phenomena within a variety of astrophysical settings, i.e., differing stellar rotation conditions, surface gravities, chemical composition and temperatures.

The simple models of coronal heating via turbulence from the stellar convective zone may be relevant to stars possessing no magnetic field or differential rotation, but α Aql appears to emit less than 4×10^{27} ergs/s as seen from HEAO-1, which is nearly a factor of two below the predictions of this model [61, 62].

As the analysis of the HEAO-1 sky survey becomes complete, many more examples of the classes of X-ray sources delineated above will become available for study. Probably new classes of objects will be found as well, although the positioning of unknown sources is sufficiently poor that it becomes difficult to make identifications for X-ray sources with faint optical counterparts. HEAO-B will be of tremendous value in providing unambiguous identifications for the many soft sources discovered in the HEAO-1 all sky survey.

This research was supported in part by NASA contract NAS 5-23315.

REFERENCES

1. R. Giacconi et al., Phys. Rev. Lett. 9, 439 (1962).
2. R. Giacconi et al., Astrophys. J. Suppl. 27, No. 237 (1974).
3. B. Margon and J. P. Ostriker, Astrophys. J. 186, 91 (1973).
4. F. D. Seward et al., Astrophys. J. 178, 131 (1972).
5. M. J. Vanderhill et al., Astrophys. J. Lett. 197, L19 (1975).
6. R. Mewe et al., Astrophys. J. Lett. 202, L67 (1975).
7. R. Rothschild et al., to be submitted to Astrophys. J. (1978).
8. R. C. Catura, L. W. Acton, and H. M. Johnson, Astrophys. J. Lett. 196, L47 (1975).
9. D. S. Hall, IAU Colloquium #29, 287 (1976).
10. W. Cash et al., Astrophys. J. Lett., in press (1978).
11. F. Walter et al., Nature, submitted (1978a).
12. S. A Naftilan et al., Proc. Astron. Soc. Pacific 87, 321 (1975).
13. N. E. White et al., Nature, submitted (1978).
14. H. W. Schnopper et al., Astrophys. J. Lett. 210, L75 (1976).
15. F. R. Harnden, Jr. et al., Astrophys. J. 214, 418 (1977).
16. I. R. Tuohy et al., private communication (1978).
17. R. A. Berg and J. G. Duthie, Astrophys. J. 211, 859 (1977).
18. D. R. Hearn, J. A. Richardson, and G. W. Clark, Astrophys. J. Lett. 210, L23 (1976).

19. S. Tapia et al., Astrophys. J. Lett. 212, L125 (1977).
20. A. P. Cowley and D. Crampton, Astrophys. J. Lett. 212, L121 (1977).
21. W. C. Priedhorsky et al., Astrophys. J. Lett. 212, L117 (1977).
22. J. L. Greenstein et al., Astrophys. J. Lett. 213, L121 (1977).
23. H. S. Stockman et al., Astrophys. J. 217, 815 (1977).
24. W. Priedhorsky et al., Astrophys. J., submitted (1978).
25. J. H. Swank et al., Bull. Am. Astron. Soc. 9, 591 (1978).
26. M. J. Coe et al., Nature 272, 37 (1978).
27. I. R. Tuohy et al., COSPAR Conference (1978).
28. A. R. Masters, Ph.D. Thesis, University of Illinois (unpublished) (1978).
29. D. Q. Lamb and A. R. Masters, in preparation (1978).
30. K. Mason et al., Astrophys. J., to be submitted (1978).
31. B. Warner, IAU Symposium #43, 85 (1976).
32. S. Rappaport et al., Astrophys. J. Lett. 187, L5 (1974).
33. S. Bowyer et al., Bull. Am. Astron. Soc. 8, 447 (1976).
34. D. R. Hearn et al., Bull. Am. Astron. Soc. 8, 447 (1976a).
35. J. Heise et al., Astron. and Astrophys., submitted (1977).
36. W. Krzeminski et al., in preparation (1978).
37. K. Mason et al., COSPAR Conference (1978).
38. F. Cordova et al., Bull. Am. Astron. Soc. 9, 592 (1978).
39. B. Margon et al., Astrophys. J. 209, 525 (1976).
40. S. Hayakawa et al., IAU/COSPAR Symposium on Fast Transient X- and Gamma-Rays, Varna, Bulgaria, May (1975).
41. D. R. Hearn et al., Astrophys. J. Lett. 203, L21 (1976).
42. M. Lampton et al., Astrophys. J. Lett. 203, L71 (1976).
43. B. Margon et al., Astrophys. J. 209, 525 (1976).
44. P. Strittmatter et al., Astrophys. J. 174, 91 (1972).
45. B. Margon et al., Astrophys. J. Lett. 210, L79 (1976).
46. J. Heise et al., Astrophys. J. Lett. 202, L73 (1975).
47. D. J. Mullan, Astrophys. J. 207, 289 (1976).
48. B. Haisch et al., Astrophys. J. Lett. 218, L119 (1977).
49. H. L. Shipman et al., Astrophys. J. Lett. 213, L25 (1977).
50. R. Mewe et al., Astrophys. J. Lett. 202, L67 (1975).
51. A. G. Hearn and R. Mewe, Astron. and Astrophys. 50, 319 (1976).
52. K. Brecher, Bull. Am. Astron. Soc. 8, 450 (1976).
53. W. Cash et al., Astrophys. J. Lett., in press (1978).
54. M. Lampton et al., COSPAR Conference (1978).
55. J. Nugent and G. Garmire, Astrophys. J., to be submitted (1978).
56. K. Topka et al., Bull. Am. Astron. Soc. 9, 560 (1978).
57. T. Ayres and J. Linsky, private communication (1978).
58. A.J.F. den Boggende and R. Mewe, COSPAR Conference (1978).
59. A.J.F. den Boggende et al., Astron. and Astrophys. 62, 1-7 (1978).
60. H. Bradt, private communication (1978).
61. C. de Loore, C. de Jager, IAU Symposium No. 37, Non-Solar X- and Gamma-Ray Astronomy, ed. L. Gratton, Dordrecht, Riedel and New York, Springer (1970).
62. P. Gorenstein and W. Tucker, Ann. Rev. Astron. and Astrophys., Vol. 14 (1976).
63. P. Charles et al., COSPAR Conference (1978).
64. F. Walter et al., Astrophys. J., submitted (1978b).
65. F. Walter et al., IAUC 3208 (1978c).

HEAO-1 OBSERVATIONS OF DWARF NOVAE

K. Mason*†, F. Cordova** and J. Swank***

*Astronomy Department and Space Sciences Laboratory, University of California, Berkeley, California 94720, USA
**California Institute of Technology, Pasadena, California 91109, USA
***Goddard Space Flight Center, Code 661, Greenbelt, Maryland 20771, USA

INTRODUCTION

Dwarf novae are generally believed to be close binary systems containing a white dwarf which is accreting material from a non-degenerate companion. The optical characteristics of such stars have been reviewed in [1] and [2]. Their principal distinguishing feature is quasi-periodic outbursts, usually lasting several days and recurring on a timescale of tens of days. The optical outbursts are probably brightenings of an accretion disk surrounding the white dwarf. The discovery that the visually brightest member of this class, SS Cygni, is an X-ray source [3,4,5, 6,7] motivated us to search data from the A-2 experiment‡ on the HEAO-1 satellite for X-ray emission from these stars. The A-2 experiment is making a complete sky survey at energies between 0.2 and 60 keV. About 20 dwarf novae were in optical outburst during the time they were being scanned by HEAO. Of these, only U Geminorum was detected as an X-ray source. X-ray emission was observed from SS Cyg, EX Hydrae and AY Lyrae while they were optically quiescent.

OBSERVATIONS OF SPECIFIC SOURCES

SS Cyg

The HEAO-A2 detectors observed SS Cyg in December 1977 during optical quiescence. Its spectral distribution is not consistent with a power law but could be represented by thermal emission of temperature ∿19 keV. By comparison, the spectral slope measured on two occasions using the OSO-8 detectors gave temperatures of ∿ 10 keV and ∿ 30 keV respectively [Table 1], suggesting that the spectrum is variable. Further evidence in favor of thermal emission is provided by the detection of an iron line feature centered at 6.9 ± 0.1 keV. The line has an equivalent of 550 ± 20 eV, close to that expected for a gas of cosmic abundance at this temperature.

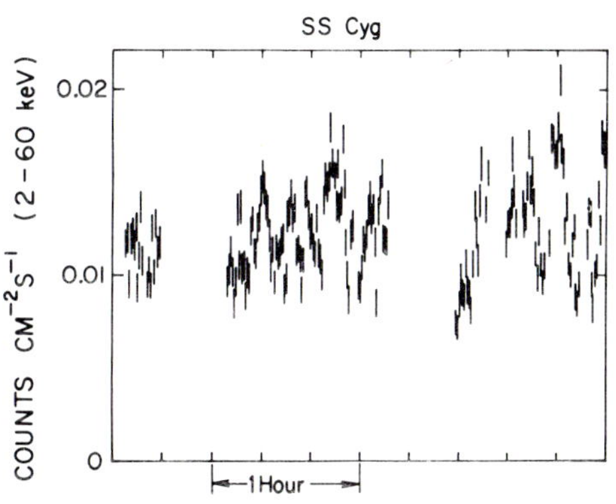

Figure 1 Part of a H.A.E.O. pointed observation of SS Cyg made in the energy range 2-60 kev in December 1977. Each point represents a 40 sec. accumulation.

† Miller Research Fellow
‡A collaborative effort led by E. Boldt (GSFC) and G. Garmire (CIT) with collaborators at CIT, GSFC, UCB and JPL.

The 2-60 keV light curve of SS Cyg measured during a pointed observation with the spacecraft is shown in Fig. 1. The source is clearly variable by a factor of ∿ 2 with a characteristic timescale of a few minutes. However, the data do not favor a single well defined periodicity.

TABLE 1. OBSERVATIONS OF SS CYG AND U GEM

		Outburst		Quiescent		
Group	Energy Band keV	Flux (erg cm^{-2} s^{-1})	Temp. (keV)	Flux (erg cm^{-2} s^{-1})	Temp. (keV)	Ref.
SS Cyg						
MIT Rocket	0.15-0.28	10^{-10}				[3]
SAS 3	0.15-0.28	(3-5) x 10^{11}				[4]
ASTP	0.08-0.25	9 x 10^{-11}				[5]
ANS	0.15-0.28			4 x 10^{-12}		[6]
	1 - 7			7 x 10^{-11}		[6]
OSO-8	1 - 10				10±1	
	1 - 10				∿30	
Ariel V	2-18			3 x 10^{-10+}		[7]
HEAO	0.2-0.5			<5 x 10^{-12}		
	0.5-25			2 x 10^{-10}	19^{+1}_{-5}	
U Gem						
Berkeley rocket	0.15-0.28			<7.5 x 10^{-11}		[16]
Columbia rocket	0.15-0.28			<3 x 10^{-11}		[17]
Ariel V	2 - 18			<1.9 x 10^{-11*}		[14]
HEAO	0.15-0.5	3.2 x 10^{-10+}	0.03±0.01	<10^{-12}		
	0.5-3.0	<2 x 10^{-10}		<2 x 10^{-10}		
	2 - 15	3.8 x 10^{-11}		<5 x 10^{-12}		
EX Hya						
Ariel V	2-18			7.5 x 10^{-11*}		[14]
HEAO	0.15-0.5			<3 x 10^{-12}		
	0.5-10			2 x 10^{-10}		
AY Lyr						
HEAO	0.15-0.5			6 x 10^{-12}	<0.12	

+ at maximum
* time averaged

U Gem

HEAO-1 had begun scanning U Gem a day before the onset of an optical outburst on October 18, 1977. During optical maximum, very strong but variable, 0.2-0.5 keV X-ray emission was discovered from that star. The peak flux was higher than has been reported from SS Cyg during outburst (Table 1). Weak 2-10 keV emission was also detected near the beginning of the optical outburst. These are the first detections of X-rays from U Gem.

The optical, soft X-ray and hard X-ray light curves are shown in Figure 2. The soft emission was not detected strongly until ∿ 1 day after the onset of the optical outburst. However, since the source was extremely variable and we had only a few snapshots of the emission, we cannot be certain that it had not appeared earlier. One explanation of the variability is that the emission is modulated by the 4.25 hr. binary period of U Gem [8]. While all the strong detections do indeed fall within a limited phase interval [Figure 3], we cannot exclude the possibility that this is caused by fortuitous random variability. No orbital modulation can be discerned in the hard X-ray data with an upper limit to the modulated fraction of ∿ 30%. The low energy spectrum of U Gem can be fit by a thermal bremsstrahlung distribution of temperature between 0.02 and 0.04 keV. The slope of the spectrum changes by less than 20% between sightings. The poor statistical quality of the 2-15 keV data preclude the determination of the spectral slope in this band, but the flux observed is clearly inconsistent with an extension of the low energy spectrum.

Other Sources

Analysis of the AY Lyr and EX Hya data (in collaboration with G. Reigler) is in progress. Preliminary work indicates that AY Lyr was detected between 0.2 and 0.5 keV for one day only while the optical star was quiescent.

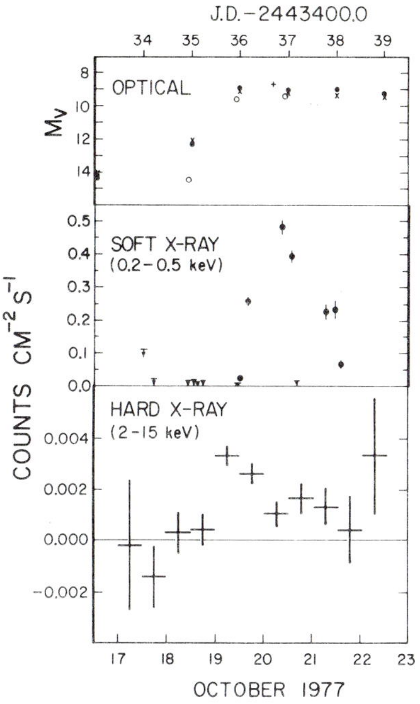

Figure 2 Time history of U Gem in the optical, soft X-ray (15 sec sightings) and hard X-ray (half day averages) bands. Optical data:- (X) - Morgan (IAUC 3125): (●) - A.A.V.S.O. (Mattei, private communication): (+) - British Astronomical Association (Howarth, private communication): (o) - J. Patterson (private communication).

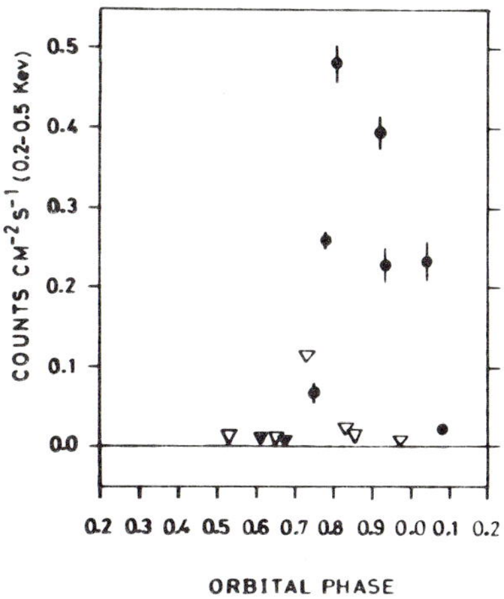

Figure 3 The H.E.A.O. soft X-ray data folded modulo the orbital period of U-Gem. Open symbols - data taken before Oct. 19.0; closed symbols - data taken after Oct. 19.0.

The spectrum of the emission was similar to the low energy spectrum of U Gem during outburst. Hard X-ray emission from EX Hya exhibits similar variability to that observed from SS Cyg [Figure 1]. The spectrum of EX Hya also has an iron emission feature at 6.7 keV [9].

DISCUSSION

A question which immediately arises is why, of the $\sim$ 20 stars scanned by HEAO during optical outburst, only U Gem was detected. The 0.2-0.5 keV upper limits on the remaining stars were typically at least one hundred times less than the maximum flux seen from U Gem. This eliminates a simple distance effect as the cause. Because of the large sample of dwarf novae involved, it is also difficult to invoke selective interstellar absorption, or inclination effects in every case. One possibility is that the temperature of the X-ray emission is too low for significant flux to have been seen. A factor of three reduction in the temperature of U Gem would have rendered it invisible in the HEAO low energy detectors. The hard X-ray survey is not as sensitive to the effects of interstellar absorption or small changes in source temperature, but the degree of hard X-ray emission may be very sensitive to conditions near the emitting star [e.g. 7]. To conclude, we can summarize the major points to have emerged from our investigation of the HEAO data.

1. The soft and hard X-ray emission from U Gem and SS Cyg is linked to the optical state of the system.
2. Both SS Cyg and U Gem have two X-ray spectral components, one soft (kT $\sim$ 30 eV) the other hard (kT $\sim$ 20keV). We note that in this respect, these stars are similar to the peculiar magnetic variable AM Her [10, 11]. The ratio of soft to hard X-ray energy flux from both U Gem and SS Cyg during outburst is $\sim$ 10

(Table 1).

3. The detection of line emission and the shape of the spectrum argue that the hard X-ray flux from SS Cyg and EX Hya originates in a hot, optically thin plasma.

4. The variability of the hard X-ray emission from SS Cyg and EX Hya suggests a cooling timescale of $\lesssim$ 100 sec. This would limit the characteristic size of the emitting region to $\sim 3 \times 10^9$cm, comparable to the dimensions of a white dwarf.

5. X-ray emission of the strength and temperature seen from SS Cyg and U Gem is not common among dwarf novae.

REFERENCES

1. B. Warner, Proc. IAU Symp. 73, Reidel, Dordrecht (1976).
2. E. L. Robinson 1977, Ann. Rev. Ast. Astrophys., p. 119.
3. S. Rappaport et al., Astrophys. J. (Letters) 187, L5 (1974).
4. D. R. Hearn et al., Bull. A. A. S. 8, 447 (1976).
5. B. Margon et al., Astrophys. J. (submitted 1978).
6. J. Heise et al., Astr. Ap. 63 L1 (1978).
7. Ricketts et al., preprint (1978).
8. S. Arnold et al., Astrophys. J. 206, 790 (1976).
9. R. Mushotzky et al., preprint (1978).
10. I. R. Tuohy et al., these proceedings.
11. J. Swank et al., Astrophys. J. 216, L71 (1977).
12. P. Henry et al., Astrophys. J. (Letters), 197, L117 (1975).
13. R. Novick et al., Astrophys. Lett., 16, 67 (1975).
14. M. G. Watson et al., University of Leicester preprint (1978).

HEAO-1 SOFT X-RAY OBSERVATIONS OF SIRIUS

M. Lampton*, I. Tuohy**, G. Garmire** and P. Charles*

**Space Sciences Laboratory, University of California, Berkeley, California 94720, USA*
***Department of Physics, California Institute of Technology, Pasadena, California 91125, USA*

INTRODUCTION

The nearby binary star Sirius has been the subject of a variety of observations from space, which span the soft X-ray, extreme ultraviolet, and UV bands. These observations have been motivated by the possibility of detecting stellar coronae and by model atmosphere calculations which predict appreciable soft X-ray and EUV fluxes from the white dwarf member, Sirius B.

Sirius was found to be a soft X-ray source by the ANS group (1) who derived a luminosity of 10^{28} ergs/sec in the 0.2-0.28 keV band. Other soft X-ray observations have yielded only upper limits to this flux; of particular interest is the OSO-8 determination (2) which lay a factor of four below the ANS detection, and thus indicated that the flux may be variable. Sirius has not been detected at energies above the 0.25 keV band.

Because the angular separation between Sirius A and its white dwarf companion amounts to less than twelve arc seconds, it has not proven possible to distinguish the site of the emission with existing low energy X-ray detectors. Indirect arguments have, however, been put forth based on the expected properties of stellar coronae or photospheres. Shipman (3;4) discussed the possibility of thermal soft X-ray emission originating beneath the photosphere of Sirius B. In this model, the metal-deficient atmosphere yields the observed flux provided that the photospheric temperature exceeds about 32,000 K. However, the model also predicts a strong extreme-ultraviolet (100-1000 Å) flux, which has been ruled out by the recent rocket observations by Cash, Bowyer, and Lampton (5). The possibility that the X-rays are produced by mass transfer from Sirius A onto its companion has been ruled out by Mewe et al. (1) on the basis of the small mass loss rates expected from A Stars. The two presently viable models are coronal emission from a hot envelope surrounding Sirius A (6) or surrounding Sirius B (3;7). We report here the soft X-ray intensity, spectrum and variability of Sirius as determined with the HEAO a-2 experiment†, and comment on the constraints placed on these two coronal models by our data.

†The A-2 experiment is a collaborative effort lead by E. Boldt of GSFC and G. Garmire of CIT, with collaborators at CIT, JPL, UCB, and GSFC.

OBSERVATIONS AND ANALYSIS

Sirius was scanned by the HEAO-1 A-2 experiment Low Energy Detectors (LEDs) from 4 October to 9 October, 1977. These detectors are sensitive to X-rays in the 0.15-2.5 keV band (8). In contrast to most soft X-ray experiments the detectors have sufficient resolution and sensitivity to permit useful spectrum determinations from the pulse height analyzer data within the 0.15-0.28 keV band. Each of the six daily accumulations of the spacecraft scan data showed a count rate peak at the position of Sirius. Twenty-six individual 15 sec scans over Sirius were analyzed to determine the aspect-corrected intensity. These data are plotted in Fig. 1. Most of the scans are consistent with the mean count rate of 18.8±.4 cps. The largest deviation is about -3σ and suggests possible variability, but due to the importance of this conclusion, we treat the evidence with caution.

We have ruled out the possibility of spurious detector response to visible or UV radiation from Sirius by examining scans of the moon and of bright UV stars. The lack of response allows us to set upper limits of 0.01 and 1.0 cps respectively for the Sirius 0.33-0.7 μ and 0.13-0.33 μ flux, i.e. less than 10% of the observed signal. To establish the stability of the detector's gain and resolution, we separately analyzed each day's pulse height spectrum of the Monoceros diffuse soft X-ray enhancement lying 25° to the north of Sirius. These analyses gave values of derived parameters which were consistent within their errors.

The X-ray spectrum of Sirius is extremely soft. In Fig. 2 we show the distribution of the observed counts with respect to pulse height averaged over all days available. In spite of the detector's rising efficiency characteristic over the 0.16 to 0.28 keV interval, the count rates decline rapidly with increasing channel number, which indicates an extremely steep spectrum.

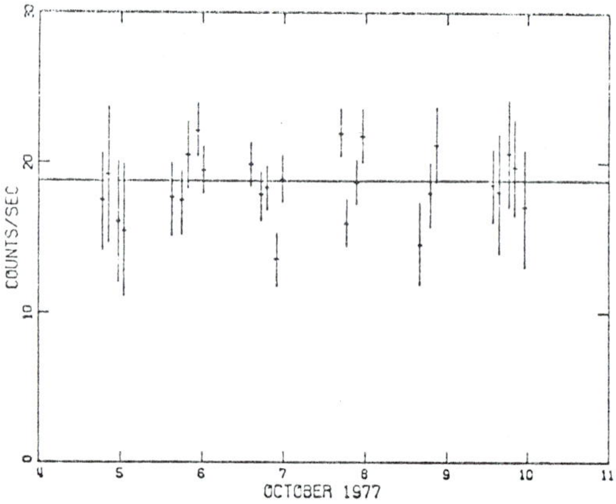

Figure 1. Intensity of Sirius, 0.16-0.28 keV, aspect corrected, between 4-9 October 1977. The solid line represents the weighted mean of the 26 data points.

To quantify our spectrum measurements, we have fitted the weighted mean of the pulse height analyzer data to an exponential spectrum with an energy dependent Gaunt factor; this spectrum models the thermal bremsstrahlung continuum of hydrogen. Although interstellar attenuation is not appreciable for stars as nearby as Sirius, we have included a Brown and Gould (9) absorption cross section with a column density of 10^{18} cm^{-2} (10). The model was convolved with the calibrated detector efficiency and a Poisson resolution function, and the free parameters were adjusted for minimum chi square by means of Marquardt's algorithm (11). Due to the uncertainty in the precise shape of the detector resolution function below 0.1 keV, our data only permit us to set an upper limit on the temperature of the emitting plasma. We find that $T < 2.3 \times 10^5$ K; at this temperature $L_x = 1.2 \times 10^{28}$ erg/sec in the 0.16-0.28 keV band, corresponding to a log emission measure (cm^{-3}) = 55.7. We also find that more elaborate line emission models appropriate to the ten-times-solar abundances of Sirius A (12) fit our data. The temperatures are not much different from the hydrogen case, but the emission measures are smaller by factors of 50-100 due to Si V and O VI line emission.

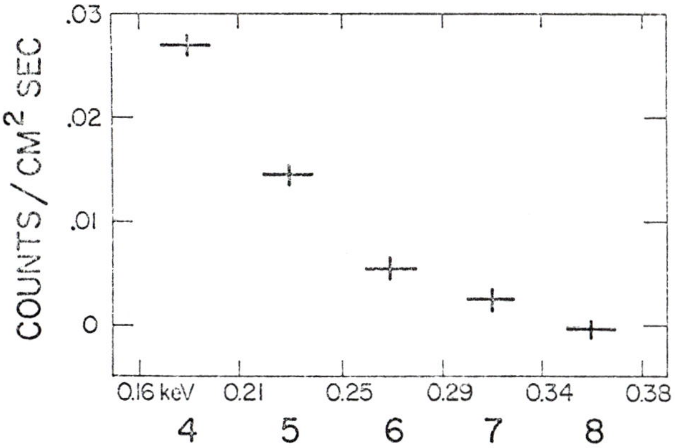

Figure 2. Observed Sirius spectrum after background and exposure corrections. Pulse height analyzer channels and boundaries in keV are shown.

DISCUSSION

The steep spectra reported here are difficult to reconcile with existing models for stellar coronae. Such models for main sequence stars have been computed by Mullan (6) using scaling laws based on Solar properties and by Hearn (13) using the minimum-flux hypothesis. At our observed X-ray luminosity $L_x \simeq 10^{28}$ erg/sec, these models require coronal temperatures of $1\text{-}2 \times 10^6$ K, much higher than the observed spectra allow. Minimum-flux calculations for the white dwarf case (14,15) also require coronal temperatures in this range to produce the observed energy.

Other kinds of observations will prove valuable in understanding Sirius. If the soft X-ray flux originates at Sirius A in a region as cool as the 2×10^5 K derived here, a substantial number of strong UV transitions would be excited. For example, the O VI λ1032 Å line should appear strongly. The absence of this feature would support the Sirius B hypothesis.

A secure detection of X-ray variability would allow a cooling-time test to distinguish the site of the variability. In an impulsively heated plasma, the electron density n_e and volume V can be determined from a measurement of X-ray luminosity and cooling time τ of a gas. For example, $\tau = 10^5$ sec and $L_x = 10^{28}$ erg/sec would require $n_e \simeq 10^{10}$ cm^{-3} and $V \simeq 10^{34} cm^3$ for hydrogen or 10^8 cm^{-3} and 10^{36} cm^3 for an enriched composition. Such figures could suggest densities and dimensions characteristic of main sequence (e.g. Solar) coronae rather than white dwarf envelopes, and could help to decide whether Sirius A or B is the X-ray source.

At radio frequencies, weak variable emission could be caused by either thermal or nonthermal processes in analogy with solar mechanisms or with the much larger scale activity seen in RS CVn X-ray sources (16). In the X-ray band, the discovery of 0.25 keV emission from another early A star, Vega (17), suggests that we may be dealing with a widespread phenomenon unrelated to the compositional peculiarities of Sirius A. Future observations with high angular resolution X-ray telescopes or modulation collimators could directly distinguish the emission site, or high resolution X-ray spectroscopy could be employed to identify the spectrum as a complex of emission lines expected from Sirius A or as a hydrogen continuum characteristic of its white dwarf companion.

ACKNOWLEDGMENTS

We gratefully acknowledge the essential contributions of Elihu Boldt's group at NASA Goddard Space Flight Center in the construction and operation of the A-2 experiment, and thank W. Cash for several useful discussions. This work is supported by CIT44-727600.

REFERENCES

1. R. Mewe et al., Astrophys. J. (Letters) 202, L67 (1975).
2. A. Bunner, private communication (1977).
3. H. Shipman, Astrophys. J. (Letters) 206, L67 (1976).
4. H. Shipman et al., Astrophys. J. (Letters) 213, L25 (1977).
5. W. Cash, S. Bowyer, and M. Lampton, Astrophys. J. 221, L87 (1978).
6. D. J. Mullan, Astrophys. J. 209, 171 (1976).
7. K. Brecher, Bull. Am. Astron. Soc. 8, 450 (1976).
8. R. Rothschild, in preparation (1978).
9. R. L. Brown and R. J. Gould, Phys. Rev. D 1, 2252 (1970).
10. Y. Kondo et al., Astrophys. J. (Letters) 220, L97 (1978).
11. P. R. Bevington, Data Reduction and Error Analysis in the Physical Sciences. McGraw Hill, New York, 1969.
12. S. E. Strom, O. Gingerich, and K. M. Strom, Astrophys. J. 146, 880 (1966).
13. A. G. Hearn, Astron. Astrophys, 40, 355 (1975).
14. A. G. Hearn and R. Mewe, Astron. Astrophys, 50, 319 (1976).
15. M. Lampton and R. Mewe, Astron. Astrophys. (submitted, 1978).
16. P. Charles, F. Walter, and W. Cash, these Proceedings (1978).
17. K. Topka, D. Fabricant, F. R. Harnden, and P. Gorenstein, Bull. Am. Astron. Soc. 9, #4, 560 (1977).

HEAO-1 OBSERVATIONS OF X-RAY EMISSION FROM RS CVn SYSTEMS

Philip Charles, Frederick Walter and Webster Cash

Space Sciences Laboratory, University of California, Berkeley, California 94720, USA

INTRODUCTION

During the last decade several authors have predicted coronal X-ray emission from certain types of stars [1], yet the existence of stellar coronae has never been unambiguously demonstrated, and those observations which do exist [2] conflict with the published theories. The HEAO-1 spacecraft has provided an opportunity to observe stellar sources of soft X-rays in far greater detail than has heretofore been possible. The low energy detectors (LED's) of the HEAO-1 cosmic X-ray experiment† have observed soft X-ray emission from 9 sources which are classified as RS CVn systems [3,4,5,6]. Spectra in the energy range 0.15-2.8 keV have been obtained of 2 of these, α Aur (Capella) and UX Ari (HD 21242). The spectrum of Capella indicates intense line emission at 0.85 keV which is interpreted as emission from an isothermal solar abundance plasma at a temperature of $10.5 \pm 2.8 \times 10^6$ K. The feature is a blend of principally Fe XVII lines at this energy. However, UX Ari does not display emission at 0.85 keV and, interpreted isothermally, indicates an iron abundance at least a factor 10 below the cosmic value. The continuum temperature is again near 10^7 K. The other 7 sources, HR 1099, RS CVn, AR Lac, HK Lac, RW UMa, LX Per and σ Gem, have broadband counting rates consistent with this temperature, and all luminosities are in the range $4 \times 10^{30} - 10^{32}$ erg s^{-1}. We develop a common X-ray emission mechanism for these systems which involves a hot corona maintained by extensive flaring activity. Thus RS CVn systems form a new class of low luminosity, soft X-ray sources. For a full description of the A-2 experiment the reader is referred to Rothschild et al. [7]. For the purpose of this paper we shall consider all the identifications with RS CVn systems as secure. We will now briefly outline the nature of RS CVn systems, then discuss the available X-ray data on each source and finally indicate how a common coronal emission mechanism can account for the observations.

RS CVn SYSTEMS: THE CANONICAL MODEL

The most extensive review of these systems was given by Hall [8]; the canonical model was presented by Hall [9]. Basically, they are slightly evolved [10] binary systems typically consisting of G V and KO IV stars, the latter being responsible

---

†The A-2 experiment of HEAO-1 is a collaborative effort lead by E. Boldt of GSFC and G. Garmire of CIT with collaborators at CIT, GSFC, UCB and JPL.

for strong H and K emission. Those with periods in the range 1-14 days form the standard group while those with periods outside this range form the short- and long-period groups (α Aur, σ Gem and HK Lac are members of the latter). The systems are non-interacting although the evolution is somewhat accelerated by mass loss from an active region on the cooler component. This region is a complex of starspots occupying a large fraction of one hemisphere and can directly account for the observed optical variability. Extensive chromospheric activity, as well as radio and optical flaring, is observed.

X-RAY OBSERVATIONS

The basic observational data on the 9 known X-ray emitting RS CVn stars are summarized in Table 1.

TABLE 1 Observational Properties of RS CVn Systems

	DIST. PC	V_{MAX}	PERIOD DAYS	SP. TYPE	L_X(0.2-2.8 keV) 10^{30} erg s^{-1}
α Aur*	14	0.09	104	G5III/GØIII	4
UX Ari	50	6.5	6.4	G5V/KØIV	21
HR 1099	33	5.9	2.8	G5V/KØV	11
RS CVn	145	8.4	4.8	F4III/KØIV	110
AR Lac	50	6.9	2.0	G2IV/KØIV	15
HK Lac*	150	6.5	24.4	FIV/KØIII	130
RW UMa	150	10.2	7.3	dF9/K1IV	960 (Flare?)
LX Per	145	8.1	8.0	GØV/KØIV	22
σ Gem*	150	4.2	19.6	K1III/?	130

*Members of long-period group.

α Aur

The HEAO-1 0.15-2.8 keV spectrum of α Aur (Capella) is displayed in Fig. 1. The best fitting simple bremsstrahlung spectrum ($\chi_\nu^2 \sim 4.5$) is also shown and there is a clear excess around 0.85 keV. Fitting either a simple continuum plus a line or a full line emission calculation [11] gives equally good fits ($\chi_\nu^2 \sim 1$), the temperature of the latter being 10^7 K. The line emission is principally due to a blend of Fe XVII and XVIII lines and requires the iron abundance to be at its solar value (log Fe = 7.6 with log H = 12.0). In addition the data are inconsistent with the Haisch and Linsky [12] model which explains the α Aur X-ray emission as arising predominantly from a 10^6 K plasma.

UX Ari

The low energy X-ray spectrum of UX Ari is given in Fig. 2. Again a best fitting ($\chi_\nu^2 \sim 1$) temperature of 10^7 K is obtained, but this time using a simple continuum function (exponential with Gaunt factor). Indeed the use of a spectrum which represents the emission of an isothermal solar abundance plasma will only give an acceptable fit if the iron abundance is $\sim$0.03 solar (99% confidence).

As a result of the low observed intensity of the remaining 7 sources we have only been able to calculate source intensities (Table 1) and verify that the counting rates are consistent with temperatures $\sim 10^7$ K. The high intensity of RW UMa only occurred on one day and is probably due to a large flare. Such behavior has already been seen in HR 1099 where an X-ray and radio flare occurred in 1976 [13] and the source is registered as a transient in the 4U catalog [14].

INTERPRETATION

Since these systems are detached [8] with no evidence for the presence of a degenerate component, it is likely that the X-ray emission has a coronal origin.

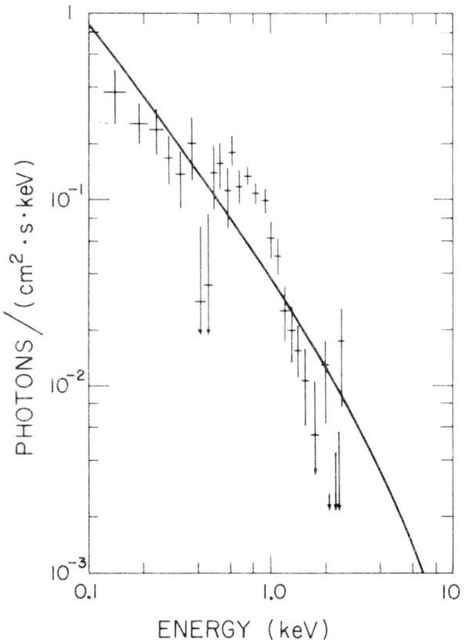

Fig. 1. Soft X-ray spectrum of Capella.
The solid line represents the best fit simple bremsstrahlung spectrum ($T = 5.6 \times 10^7$ K). Line emission is clearly visible at 0.85 keV.

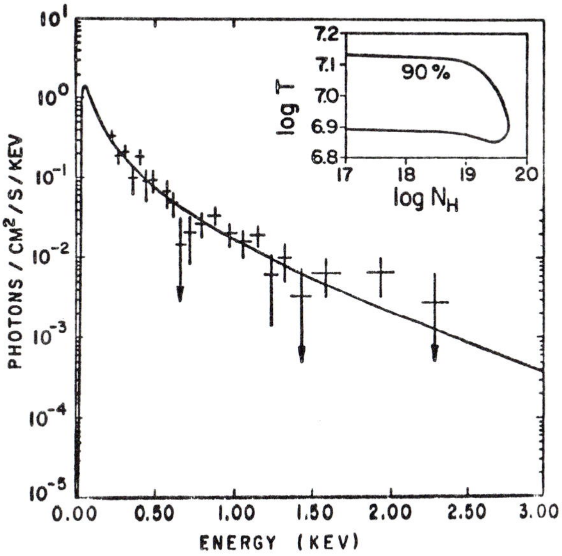

Fig. 2. Soft X-ray spectrum of UX Ari. The insert shows the 90% confidence contour in T - N_H space. The solid line here represents our best fitting spectrum, a bremsstrahlung with $T = 10^7$ K and $N_H = 10^{18}$ cm^{-2}.

Continual flaring activity associated with the extensive starspot systems could provide sufficient energy injection to drive a 10^7 K corona. The observed volume emission measure of $\sim 4 \times 10^{54}$ cm^{-3} in the case of UX Ari can be accounted for with a mass loss rate $\sim 10^{-9} - 10^{-10}$ $M_\odot$ y^{-1}, which is comparable to that required by current theories [10] to explain the orbital period variations.

A more detailed model [15] would magnetically constrain the X-ray emitting region with the starspots and have a cooler stellar wind flowing out of the coronal holes (cf. the Sun). Assuming a typical scale size for the emitting region of 10^{10} cm implies an electron density $\sim 10^{12}$ cm^{-3}, which is optically thin to electron scattering. Hence this model, which applies well to Capella where coronal emission is very likely, given the good fit to an isothermal solar abundance gas, requires a real heavy element underabundance in the case of UX Ari (see also [16] on iron underabundance in RS CVn and RW UMa).

This model is given support by the occasional large flares seen in HR 1099, one of which was coincident with a radio flare, again in close analogy to the Sun. A flare from a large region of sky containing UX Ari has also been reported by Hayakawa et al. [17].

Assuming the space density of all RS CVn systems to be $\sim 10^{-5}$ pc^{-3} with a scale height $\sim$ 100 pc [8] we may estimate the contribution of RS CVn systems to the soft X-ray background. Taking the intrinsic RS CVn X-ray luminosity to be $\sim 2 \times 10^{31}$

erg s^{-1} and the mean interstellar density to be ~ 0.03 cm^{-3} implies an intensity in the LED field of view of $\sim 3.10^{-11}$ erg cm^{-2} s^{-1}. This is approximately 10% of the soft X-ray background at low galactic latitudes.

Future observations should address detailed spectral and temporal studies which investigate spectral changes during flaring events (are they thermal or non-thermal?), and study the X-ray variability in eclipsing systems so as to provide constraints on the location of the emission and on the degree of anisotropy of the corona.

The search for stellar coronal X-ray emission has finally been successful, and it is clear that extensive surface activity is necessary to support such coronae.

REFERENCES

1. P. Gorenstein and W. H. Tucker, A. Rev. Astron. Astrophys. 14, 373 (1976).

2. R. Cruddace, S. Bowyer, R. Malina, B. Margon and M. Lampton, Astrophys. J. (Letters) 202, L47 (1975).

3. W. Cash, S. Bowyer, P. Charles, M. Lampton, G. Garmire and G. Riegler, Astrophys. J. (Letters) 223, L21 (1978).

4. F. Walter, P. Charles and S. Bowyer, Nature, London, in press (1978a).

5. F. Walter, P. Charles and S. Bowyer, Astrophys. J., in press (1978b).

6. F. Walter, P. Charles and S. Bowyer, Astron. J., in press (1978c).

7. R. Rothschild, E. Boldt, S. Holt, P. Serlemitsos, G. Garmire, P. Agrawal, G. Riegler, S. Bowyer and M. Lampton, preprint (1978).

8. D. S. Hall, IAU Colloquium #29, 287 (1976).

9. D. S. Hall, Proc. Astron. Soc. Pac. 84, 323 (1972).

10. D. M. Popper and R. K. Ulrich, Astrophys. J. (Letters) 212, L131 (1977).

11. J. C. Raymond and B. W. Smith, Astrophys. J. (Suppl.) 35, 419 (1977).

12. B. M. Haisch and J. L. Linsky, Astrophys. J. (Letters) 205, L39 (1976).

13. N. E. White, P. W. Sanford and E. J. Weiler, Nature, London, in press (1978).

14. W. Forman C. Jones, L. Cominsky, P. Julien, S. Murray, G. Peters, H. Tananbaum and R. Giacconi, Astrophys. J. (Suppl.), submitted (1978).

15. J. L. Linsky, The Solar Output and Its Variation, Colorado Associated University Press, Boulder, 1977.

16. S. A. Naftilan, Proc. Astron. Soc. Pac. 87, 321 (1975).

17. S. Hayakawa, T. Kato, T. Kohno, K. Nishimura, Y. Tanaka and K. Yamashito, Astrophys. Space Sci. 17, 30 (1972).

WHAT ARE X-RAY BURST SOURCES?

Walter H. G. Lewin

Department of Physics and Center for Space Research, Massachusetts Institute of Technology, Cambridge, MA. 02139, USA

BURST CLASSIFICATION

X-ray bursts have been classified into two distinct groups, Type I and Type II [1]. Type I bursts have been observed from more than 25 sources; they come in intervals from hours to days and generally their spectra soften during burst decay. Type II bursts have been observed from the Rapid Burster and possibly from several binary X-ray sources and pulsars. They occur on time scales of seconds to minutes and do not show the distinct spectral softening as observed in Type I bursts. As I will discuss in this talk:

- Type I burst sources are probably compact objects of solar mass. It may well be that the Type I bursts are due to thermonuclear flashes on the surface of a neutron star.

- Type II bursts are almost certainly the result of instabilities in the accretion flow onto a compact object.

TYPE II BURST SOURCES

Type II Bursts from the Rapid Burster MXB 1730-335

I will first discuss the Type II bursts as observed from the Rapid Burster. Type II bursts may also have been observed from other sources, this will be discussed in the next section.

The rapidly repetitive Type II bursts from the Rapid Burster (MXB 1730-335) are shown in Fig. 1. These bursts do not show any appreciable amount of spectral evolution (other than a change in intensity) [2,3]. Recently, one of my undergraduates, Herman Marshall [4] made a composite of 18 Type II bursts each of about 40 sec duration. The results are shown in Fig. 2. The composite was sliced in nine time bins. The best spectral fit during each of these time intervals is that of a blackbody with an approximately constant temperature of about 18×10^6 K throughout the burst. If indeed the radiation is from a blackbody, then the observed decrease in all energy channels implies that the emitting region is shrinking in time. For spherical symmetry, at a source distance of 10 kpc, the radius is 16±2 km during the first 15 sec, it decreases to 11±2 km between 20 to 30 sec after burst onset and continues decreasing (the size decreases as the square root of the observed burst flux). These sizes suggest that the Rapid Burster is a neutron star or a blackhole of approximately solar mass (see also [5]).

Notice the enhanced emission which is visible in Fig. 2 (right). The emission (most visible in the 3-12 keV range) rises slowly after the large Type II composite burst. It was found [6] that this "steady" emission terminates before the onset of the first of a series of small bursts. The presence of this "steady" emission and its relation to the Type II bursts, strongly suggests that the Type II bursts are instabilities in the accretion flow.

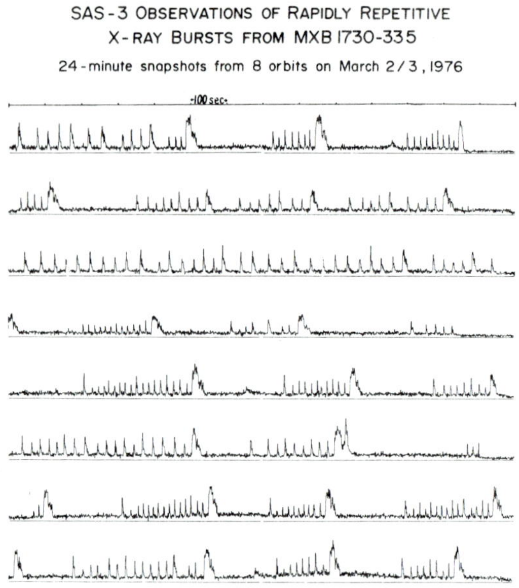

Fig. 1. Rapidly repetitive Type II bursts from the Rapid Burster. Each stretch of data is about 24 min long; occultation by the Earth occurs near the end of each stretch. This figure is from Lewin [9].

In Figs. 3 and 4 I show some recent results from Marshall et al. [4]. Each point in Fig. 3 represents one burst; the integrated burst energy is plotted against the time of Type II burst occurrence. One recognizes immediately two burst patterns. We have designated them Mode I and Mode II. In Mode I, the burst size distribution is double-peaked and the bursts vary by about two orders of magnitude in size (total burst energy). In Mode II, the difference is not more than about a factor of ten. As one can see in Fig. 3, in 1976, the change from Mode I to II occurred in less than a few hours. Based on many observations by SAS-3 and Ariel V [1-3, 7-13], we believe that the Rapid Burster starts its burst active periods in Mode I and dies in Mode II. The Rapid Burster shows active periods of 2-6 weeks at intervals of about six months [1-4, 7-15].

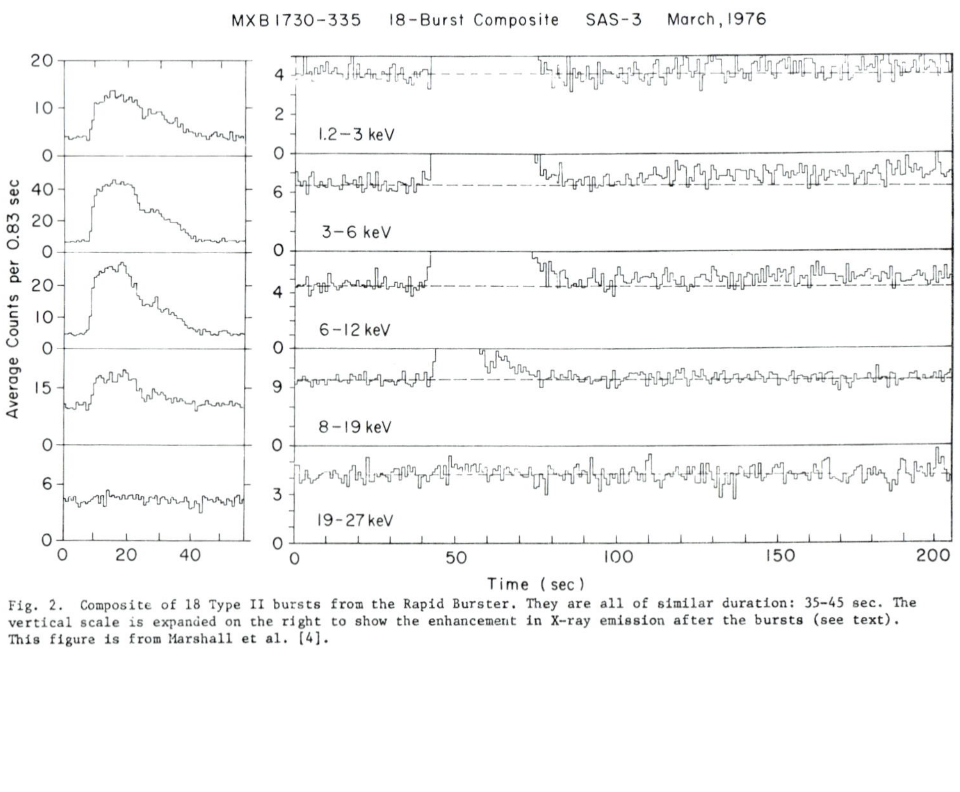

Fig. 2. Composite of 18 Type II bursts from the Rapid Burster. They are all of similar duration: 35-45 sec. The vertical scale is expanded on the right to show the enhancement in X-ray emission after the bursts (see text). This figure is from Marshall et al. [4].

Figure 5 shows the relation between burst energy and the "waiting" time for the next burst to occur. As I earlier pointed out during the Texas symposium in Boston in December 1976 [9], the relation is far from linear for low values of the burst energy (see also [10]).

Before I start to discuss the Type II bursts from other sources, I summarize:

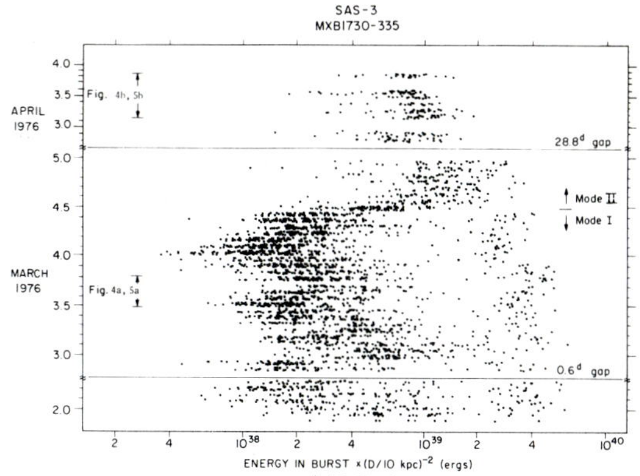

Fig. 3. The total energy in each Type II burst from the Rapid Burster is plotted (abscissa) versus the Universal Time of its occurrence (ordinate). Each dot represents one burst. The two different burst patterns (Mode I and II) are clearly distinguishable. This figure is from Marshall et al. [4].

- The Rapid Burster is probably a neutron star or a black hole of approximately solar mass (more on this later in this talk).

- Type II bursts from the Rapid Burster are almost certainly chopped up "steady" emission due to instabilities in the accretion flow (more on this later in this talk).

- The Rapid Burster is a recurrent 'transient' X-ray source. Periods between turn-ons are about 6 months.

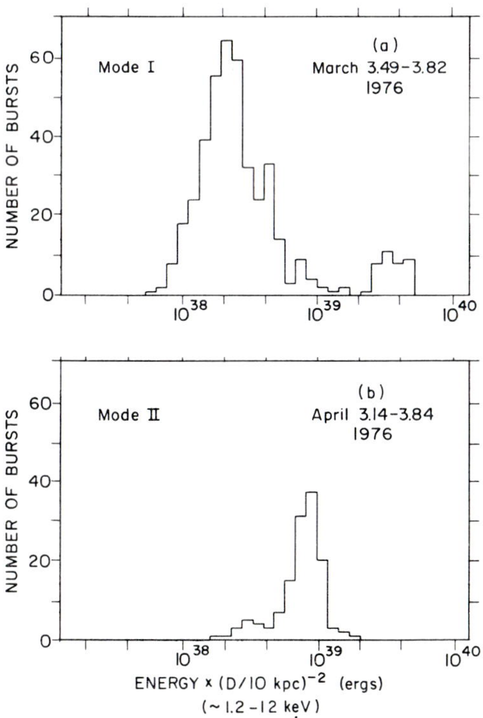

Fig. 4. Distribution of Type II burst energies from the Rapid Burster (a) Mode I data (time interval indicated in Fig. 3). Bursts vary in energy by a factor of about 100. (b) Mode II data. Burst energies range by a factor of ~10 only. This figure is from Marshall et al. [4].

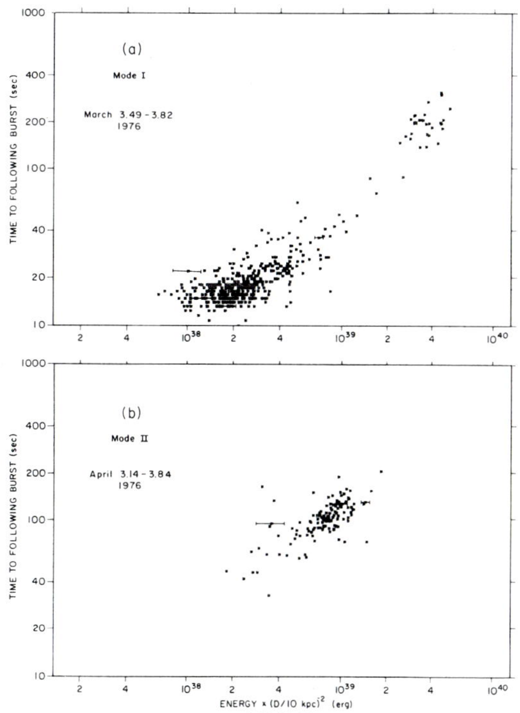

Fig. 5. E-Δt plots for the data shown in Fig. 4. (a) Mode I (same data base as used in Fig. 4a). For low values of E, the corresponding values of Δt are considerably larger than expected from a linear relation. (b) Mode II (data from Fig. 4b). This figure is from Marshall et al. [4].

Type II Bursts from Other Sources

Type II bursts distinguish themselves from Type I bursts in two ways (i) the time scales of recurrence are two to three orders of magnitude shorter and (ii) there is no strong, dominating spectral softening in the burst decay. It was first noted by Hoffman, Marshall and Lewin [1] that the "flare-burst-like" events as observed from the pulsar GX 304-1 [86], and from the binaries Cyg X-1 [87] and LMC X-4 [88] meet the phenomological definition for Type II bursts, and that they may well be caused by the same mechanism as the Type II bursts in the Rapid Burster. In any case, it is very likely that the "burst-like" events from these three sources and the Type II bursts from the Rapid Burster are all due to instabilities in the accretion flow.

Recently, Kelley and Bradt [89] observed a strong burst from the pulsar GX 301-2 (Pop. I, massive binary). The observation was made with the Horizontal Tube system of SAS-3 [49,50] when it was moving over the source (SAS-3 was spinning). The burst was "classic" in that it rose in less than one second and decayed in ∿15 sec. There was no obvious spectral softening in the burst and I suggested the possibility that this was a Type II burst. This, of course, could only be verified by pointing SAS-3 at the source. This was done, and indeed the source was highly active and produced "flare-burst-like" events on time scales of seconds and minutes [90]. The activity continued for several hours.

If the detection of the single burst from GX 301-2 had been found a year or so ago in old SAS-3 data (taken before December 1975 when X-ray bursts were discovered by Grindlay et al. [19]), it would have had a fantastic impact. One would almost certainly have jumped to the conclusion that the nature of burst sources had been unraveled and that most, perhaps all burst sources, and thus the X-ray sources in globular clusters and the galactic bulge sources, are neutron star binary systems. This conclusion may well have been correct, but obviously for the wrong reason. Maybe it was a good thing that the burst from GX 301-2 was observed only recently after the distinction between Type I and II bursts was recognized [1].

Clearly there is no certainty that the GX 301-2 events and the "hick-ups" in GX 304-1, Cyg X-1 and LMC X-4 are Type II bursts in the sense that the mechanism is the same as that in the Rapid Burster. It is an area that should be studied in more detail. In any event, I have not included any of these four sources in the sky map for Type I burst sources (Fig. 6).

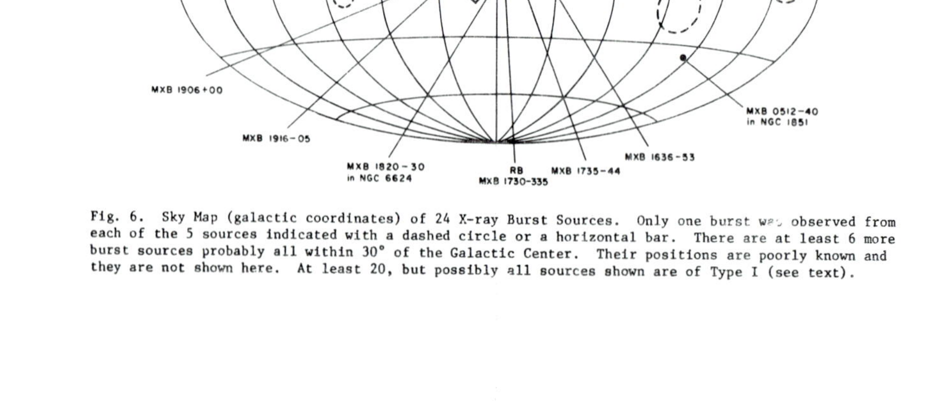

Fig. 6. Sky Map (galactic coordinates) of 24 X-ray Burst Sources. Only one burst was observed from each of the 5 sources indicated with a dashed circle or a horizontal bar. There are at least 6 more burst sources probably all within 30° of the Galactic Center. Their positions are poorly known and they are not shown here. At least 20, but possibly all sources shown are of Type I (see text).

TYPE I BURST SOURCES

Sky Distribution - Globular Clusters

In Fig. 6 I show a sky map of burst sources. The map contains 24 sources. I know of the existence of at least six more but their positions, probably all within $\sim 30°$ of the galactic center, are very uncertain; they are therefore not shown in the map. It is possible that 2 (maybe 3) of the sources shown in Fig. 6 are not of Type I. As an example, we detected a total of only three bursts from MXB 1743-28 (one of the three sources in GCX) and these bursts occurred all three in a 21 min period [42]. Therefore, this source is suspect.

The sources from which only one burst was observed (there are five in Fig. 6) are not necessarily suspect. For instance, the one burst from $\ell^{II} \sim 356.4°$ and $b^{II} \sim 2.3°$ [43] showed the very characteristic softening in the tail ($\sim$100 sec long) and is very likely a Type I burst. The one burst from the area near $\ell^{II} \sim 270°$, $b^{II} \sim -14°$ [91] also showed the softening in the tail. In addition, in a $\sim$20 min period that SAS-3 was continuously observing this area, only one burst was observed. If this burst had been of Type II, several might have been expected in this time. However, the source near $\ell^{II} \sim 80°$, $b^{II} \sim -15°$ is suspect (Jean Swank, private communication).

We earlier reported a source located in a $\sim 2 \times 120$ deg^2 box [18] (only one burst was recorded). I have deleted it from the map in Fig. 6 since the one burst recorded may well have been of Type II.

The distribution shown in Fig. 6 resembles that of the galactic bulge, Pop. II-Type X-ray sources [9,17,18] . At least three, but probably five, are located in globular clusters (see Tables 1 and 2). Certain are MXB 1820-30 in NGC6624 [19-21] and the Rapid Burster (MXB 1730-335) in Liller I; almost certain is MXB 0512-40 in NGC1851 [22,23]; and probable are MXB 1746-37 in NGC6441 [24] and MXB 1850-08 in NGC6712 [25,81].

Identifications

Three burst sources have been identified with faint ($m_B \sim 18$) blue stars: MXB 1735-44 (McClintock's star) [26], MXB 1636-53 [27] and MXB 1837+05 (Davidsen's star) [28,29]. The latter identification is not as rock solid as the first two. The highly variable source MX 1608-52 (4U 1608-52) has been identified optically with a faint star ($m_I \sim 19$) [30]. This may well be a burst source but that is still somewhat uncertain (see Table 2).

These identifications were made possible because of the work of Hale Bradt, Rodger Doxsey, Garrett Jernigan and co-workers who obtained very accurate positions for burst sources using the SAS-3 RMC system (their positions are for the associated "steady" X-ray sources). For more details on the identifications see Bradt, Doxsey and Jernigan (this volume). (See also footnote 2 on next page).

Time Intervals Between Type I Bursts

The time intervals between Type I bursts range from 1-50 hours (and quite possibly much longer). They can be regular as well as erratic. The most regular series was observed from MXB 1659-29 [9,32]. The most erratic series was observed from MXB 1837+05 [33] and from MXB 1735-44 [16,34]. However, at times the bursts from MXB 1837+05 can also be regular [33,83].

MXB 1659-29

MXB 1659-29 is special in several ways. No "steady" source was observed by SAS-3 when bursts from this source were discovered in October 1976 [32]. The upper limit derived for the ratio of energy released in "steady" emission and that in bursts is 25 [16] and this is perhaps a bit low to satisfy simple thermonuclear flash models [35-37] (more on this later in this talk).

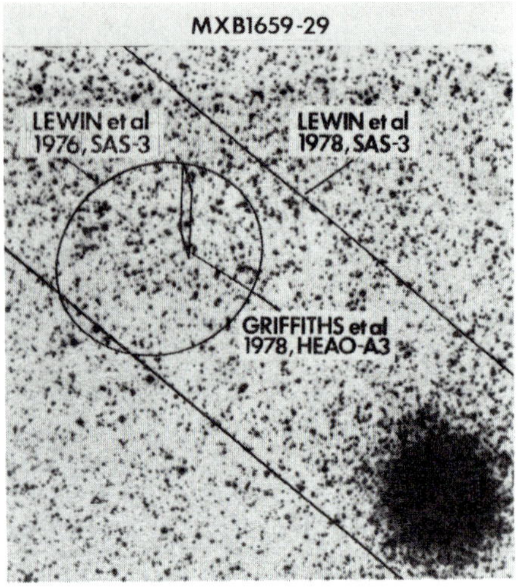

Fig. 7. Location of MXB 1659-29 at a distance of about 20 arc min from the globular cluster NGC6266. The 1976 position [32] was determined from bursts only (there was no observable steady flux at that time). The 1978 positions [39,40] are for the steady source (there were no bursts at that time). This figure was prepared by Oren Grad.

In March 1978, Jeff Hoffman and I attempted to assist HEAO-1 in observing a Type I burst. At that time, the HEAO spacecraft was highly restricted in its pointing capability. It could only be pointed for a few hours continuously and only once or twice a week. The plan was (i) to continuously observe a burst source with SAS-3, (ii) to establish a burst sequence and regularity, (iii) to predict a few hours ahead of time the arrival of a burst, and (iv) to point HEAO-1 at the source in time (footnote 1). Because of its "precise" regularity, MXB 1659-29 was selected first. However, to our great surprise instead of bursts, we observed a strong "steady" source at a level of ∿8% of the Crab [39]. I alerted Rick Dower at MIT and Richard Griffiths at SAO who worked with the HEAO-1 A3 data. Subsequently, the HEAO A3 group determined multiple accurate positions of which one was selected [40] (see Fig. 7). The HEAO-A3 position lies inside the 1976 SAS-3 error box (∿5 arc min radius) which was determined from burst data only [32] and there is little doubt that the source of steady emission in March 1978 and that of bursts in October 1976 is the same (bursts from MXB 1659-29 were also observed during the world-wide burst watch in the summer of 1977 [41]). With the accurate position from HEAO A3, it is expected that an optical identification, probably with a faint blue star, will be made shortly. (footnote 2).

This was the second time that it was found that burst activity is not present when a source is in a "high" state. This correlation was first observed for MXB 1820-30, located in the globular cluster NGC6624 [19-21].

Clark and Li have studied the X-ray source in NGC 1851 extensively. No bursts were ever observed when the steady emission was low (which is the case most of the time). However, they did observe bursts when the steady source was much higher than normal [23]. Thus, based on very limited statistics I tentatively conclude:

- Type I bursts are not observed if the "steady" emission is either very high or very low.

---

[1]Our strategy worked for MXB 1728-34 [38].

[2]A few days after I gave this talk, MXB 1659-29 was identified with an 18 magn. blue star [31]. Thus, the total number of identified burst sources is now (July 25, 1978), four and possibly six (see Tables 1 and 2).

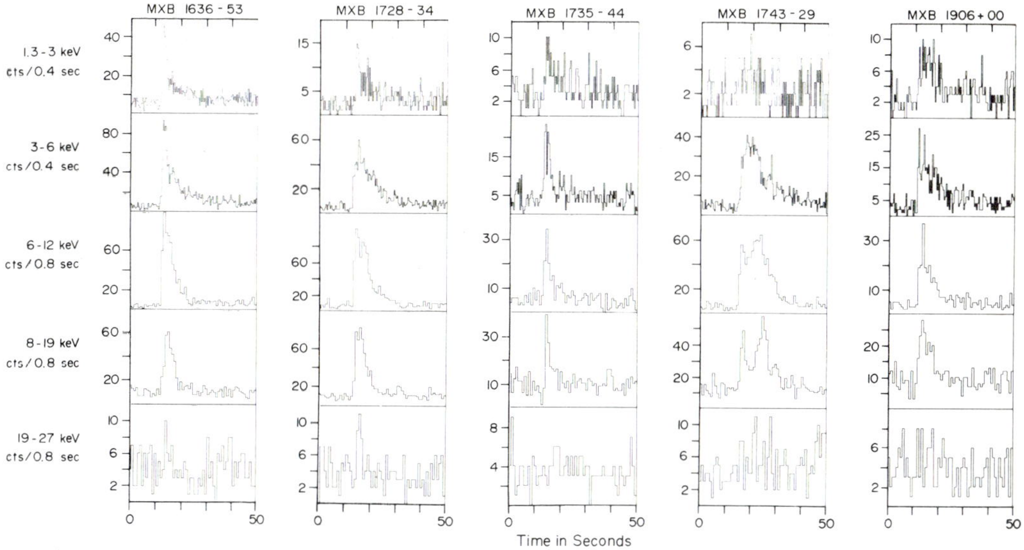

Fig. 8. Profiles (five energy channels) of Type I bursts from five different sources (SAS-3 data). Note that in all cases the gradual decay (tail) persists longer at lower energies than at higher energies. The burst profiles are reasonably distinctive for each particular source. This figure is from Lewin and Joss [16].

The "absence" of Type I bursts for very low steady emission (so far only observed for MXB 0512-40) could well be the result of a selection effect. There is an obvious observational bias against the detection of bursts at very low frequencies. Thus, if the burst frequencies decrease with decreasing steady emission (as expected from the thermonuclear flash models and other models), one would not expect to detect bursts from sources with very low luminosities.

Spectra and Sizes of Type I Burst Sources

Figure 8 shows five bursts from different burst sources. In all five cases, the softening of the spectrum during the burst decay is clearly visible in the burst tails. This distinct spectral softening, first reported for MXB 1743-29 near the galactic center [42], is not present in the Type II bursts from the Rapid Burster. It has been suggested [97] that the observed lengths, energy dependence, and variability of the burst tails can be accounted for by a population of large ($\sim 3\mu m$) interstellar grains. This is perhaps true for some sources but certainly not for the majority. The simple reason is that the tails often vary quite a bit from burst to burst. In fact, we have observed tails from several sources which varied on a time scale of hours from a few seconds to over one hundred seconds. Thus, a large part of the tails must be intrinsic to the bursts.

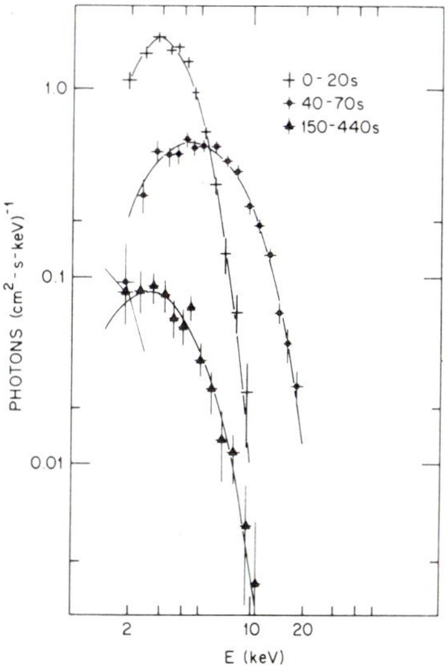

Fig. 9. Incident spectra (three time intervals) of a very long Type I burst (I designated this source XB 1724-30 in Table 2). The solid curves show the best blackbody fits. The observations were made with OSO-8. The figure is from Swank et al. [43].

Jean Swank and co-workers first showed that the best spectral fit to a Type I burst is that of a blackbody [43], (see Fig. 9). Hoffman, Lewin and

Doty [44] showed that the same is true for bursts from MXB 1728-34 and MXB 1636-53. Both groups found peak temperatures of 25 to 30×10^6K (reached at burst maximum) and subsequent cooling. For spherical symmetry and at an assumed source distance of 10 kpc, the calculated radii are comparable to that of a neutron star: Swank et al. [43], observed a decrease in size from $\sim$100 km to $\sim$15 km; Hoffman, Lewin and Doty [44] found sizes of 10-15 km which remained constant throughout the bursts.

Jan van Paradijs [45] carried this idea a bit further. He made an ad hoc assumption that the Type I bursts are standard candles at burst maximum. Thus, $F_{max}\, 4\pi d^2 = L_O$. Here, F_{max} is the maximum observed flux, d is the distance to the source and L_O is the luminosity during burst maximum (the latter was assumed to be a constant for all Type I bursts). Van Paradijs analyzed SAS-3 data from ten burst sources and assumed blackbody radiation for all of them. For spherical symmetry, the radius, R, of an object is:

$$R(t) = T(t)^{-2} \times \left(\frac{F(t)}{F_{max}}\right)^{\frac{1}{2}} \times \left(\frac{L_o}{\sigma 4\pi}\right)^{\frac{1}{2}}$$

Here, T(t) and F(t) are the observed temperature and X-ray flux at time t. Van Paradijs [45] demonstrated that under the above assumptions, the values for R(t) are independent of time and about the same for all ten burst sources. This, of course, greatly strengthens his assumption that the Type I bursts are indeed standard candles at burst maximum. If L_O is chosen to be $\sim 1.8 \times 10^{38}$ erg/sec (Eddington limit for a 1.4 $M_\odot$ object), the radii for all ten burst sources are $\sim$7 km! The conclusion that burst sources are neutron stars (or perhaps blackholes of a few solar masses) becomes almost inescapable.

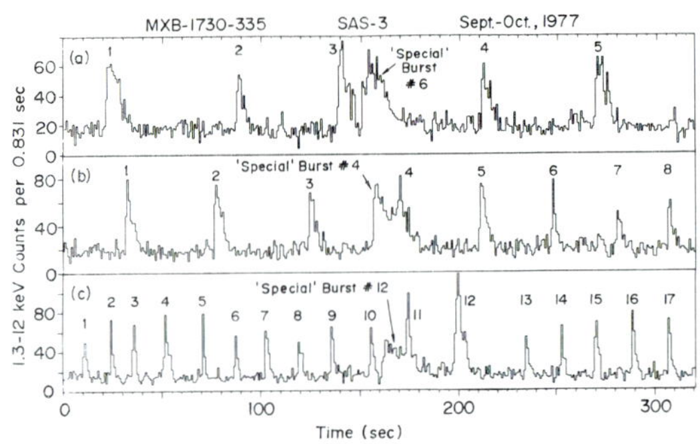

Fig. 10. Discovery of Type I and Type II bursts from the Rapid Burster. The Type I bursts ('special') occur independently of the sequence of the rapidly repetitive Type II bursts (numbered independently). This figure is from Hoffman, Marshall and Lewin [1].

Type I Bursts from the Rapid Burster

The Rapid Burster is the only object that is known to emit both Type I and Type II bursts [1]. It was this discovery that led to the burst classification (Figs. 10 and 11). It was shown recently [4] that the Type I bursts from the Rapid Burster (like those of other sources) are also best fit by blackbody spectra (decreasing temperatures) and that the radius of the object is 8±2 km (assuming spherical symmetry and a source distance of 10 kpc). Thus, both types of bursts from this object indicate sizes comparable to that of a neutron star (or a black hole of a few solar masses).

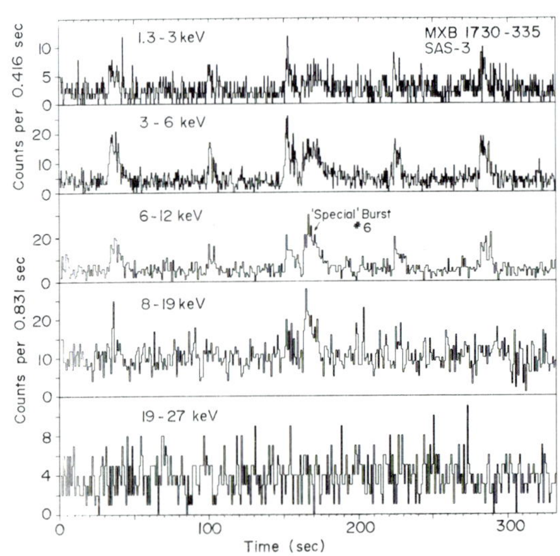

Fig. 11. Spectral information for the data shown in figure 10a. The Type I burst spectrum softens during burst decay; this softening is absent in the rapid (Type II) bursts. This figure is from Hoffman, Marshall and Lewin [1].

If Type I bursts are the result of thermonuclear flashes on the surface of a neutron star, and Type II bursts are chopped up "steady" emission, one expects (assuming isotropic emission) the time averaged ratio of energy emitted in Type II and Type I bursts to be $\gtrsim 10^2$ (see later section on models). The observed value is $\sim$130 [1].

"Steady" X-ray Emission from Burst Sources

There are presently ten Type I burst sources which are known (with a very high degree of certainty) to be associated with a source of "steady" (though variable) X-ray emission [9,16,46]. They are listed in Table 1. Several more have been suggested (Table 2), but in view of possible source confusion (large error boxes for the burst sources in combination with high source density), their association is still somewhat uncertain (in some cases very uncertain). I have added the Rapid Burster to Table 1 since it is fairly certain that the Type II bursts from this object "replace" the "steady" emission as observed from many (not all) Type I burst sources (see above).

TABLE 1

"Steady" X-Ray Sources from which Type I Bursts have been Observed.

Burst Source[+]	Steady Source[+]	Positions of Steady Sources from SAS-3* RA (1950) h	m	s	Dec. °	'	"	Error Circle Radius†	4U Catalogue** Intensity Maximum Ucts	Maximum/Minimum
MXB 0512-40 (in NGC 1851)	2S 0512-400 2A 0512-399 4U 0513-40	05	12	28.7	-40	05	53	20"	18	3
MXB 1636-53 (blue star)	2S 1636-536 4U 1636-53	16	36	57.6	-53	39	21	20"	250	2
	blue star	16	36	56.2	-53	39	15	3"		
MXB 1728-34	2S 1728-337 4U 1728-33	17	28	39.6	-33	47	52	30"	150	5
MXB 1735-44 (blue star)	2S 1735-444 4U 1735-44	17	35	19.5	-44	25	22	20"	210	1.7
	blue star	17	35	19.0	-44	25	19	3"		
MXB 1820-30 (in NGC 6624)	2S 1820-303 4U 1820-30	18	20	28.4	-30	23	14	20"	320	3
MXB 1837+05 (blue star)	2S 1837+049 4U 1837+04	18	37	29.8	04	59	23	20"	280	2
	blue star	18	37	29.4	04	59	29	5"		
MXB 1906+00	2S 1905+000 A 1905+00 4U 1857+01	19	05	54.9	00	05	37	35"	4.05±1.1‡	
MXB 1916-05	2S 1916-053 4U 1915-05	19	16	08.5	-05	19	51	20"	20	2
		from HEAO-A3***							from SAS-3	
MXB 1730-335 Rapid Burster (in Liller I)	H1730-333	17	30	07.4	-33	20	34		$\sim$50 (Type II bursts)	>10
MXB 1659-29 (blue star)	(4U1704-30?) H1658-298	16	58	54.8	-29	49	55		$\sim$80	>15
	blue star	16	58	55.5	-29	52	26	5"		

[+]For references of burst sources and optical identifications, see text.

†Approximately 90% confidence ‡Average Intensity *[71-73] **[82] ***[40,74]

TABLE 2

"Steady" X-ray Sources from which Type I bursts <u>may</u> have been observed

Burst Source		Steady Source						References
Name	Position error degr^2	Name	Position RA(1950) h m s	Position Dec(1950) ° ' "	Position error (arc min)^2	Intensity Maximum ∿Ucts*	Intensity Max/Min	
XB 1608-52	∿3	MX1608-52 4U1608-52 2S1608-523	16 08 51	-52 18 02	0.3	10^3	∿25	[30,75-77,85]
		star	16 08 52.2	-52 17 43	±0.5"			
MXB 1706-43	∿0.3	4U1705-44 2S1705-440	17 05 18.2	-44 02 10	∿0.4	280	3	[25,72,78]
		OR 4U1702-42 2S1702-424	17 02 40.4	-42 57 58	∿0.8	30	3	
XB 1724-30	∿0.5	4U1722-30	17 22 50	-30 31 30	41	7		[43,82]
MXB 1743-29	∿0.2	A1742-289 transient	17 42 26.4	-28 59 56	∿0.6	2×10^3	>200 transient	[42,48,79]
MXB 1742-29	∿0.3	A1742-294 2S1742-294 GCX-1	17 42 53.6	-29 29 50	∿0.8	90	>2	[42,72,79,80]
MXB 1746-37	∿1	4U1746-37 2S1746-370 (in NGC6441)	17 46 48.8	-37 02 25	∿0.8	13		[24,73]
MXB 1850-08	∿1.3	A1850-08 2S1850-087 4U1850-08 (in NGC6712)	18 50 21.9	-8 45 54	∿0.8	4		[25,71,73,81]
XB 06??+??	∿80	4U0614+09				120	5	[82,92-94]
		star	06 14 21.3	09 08 32				

*Approximate values in Uhuru cts/sec are given to allow for an easy comparison with Table 1.

Several large error boxes for burst sources (see Fig. 6) contain more than one "steady" source. There are also some (at least two) that do not contain any known source of "steady" emission. For accurate positions of X-ray sources see also Bradt, Doxsey and Jernigan (this Volume).

How "Steady is "Steady"?

The galactic bulge X-ray sources, of which the burst sources are a member, are often referred to as "mildly" variable. This may be true for some, but it certainly is not true for all. For example, the source 4U1820-30 (in NGC6624) is highly variable, even on a time scale of ten minutes [47]. Long term variability for this source, for MXB 1659-29, and for the Rapid Burster (counting the Type II bursts as steady emission), is in excess of a factor of 10. In fact both MXB 1659-29 and the Rapid Burster could be called "transients" depending on whose criterion one applies. A1742-289, near the galactic center, is a transient [48]. Triple peaked bursts at intervals of ∿35 hours have been observed from MXB 1743-29 (Fig. 8). The transient and burst source may well be the same [42], (see also Table 2).

TABLE 3

Upper limits to periodicities in the "steady" components of Type I burst sources

Source	Upper limits (90% confidence) to pulsed fraction* in %, for periods in the ranges listed		
	2 msec-2 sec	1.6 sec-66 sec	66 sec-10^3 sec
1636-53		1.1	2
1659-29		1.4	4
1728-34		1.4	3
1735-44	8	1.1	2
1820-30	1.5	1.5	1.5†
1837+05		0.7	1.4
1916-05		1.3	15

*fraction of total flux †Period range 66-200 sec.

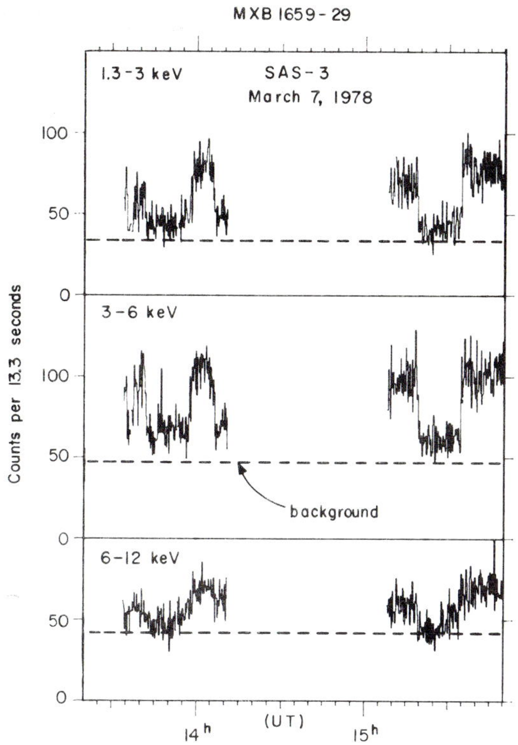

Fig. 12. SAS-3 observations of the "steady" component of MXB 1659-29. The data were taken in March of 1978 when the "steady" emission was first observed [39]. Notice at least five rapid changes (in ~40 sec) in the 3-6 keV data. There are periods of low emission, two of about 15 min. Others were observed (not shown here) which did not last that long. The recurrence of these low states was not periodic. These low states are due to absorption (see text). The source was occulted by the Earth during the gaps in the data. This figure was prepared by Lynn Cominsky and Bill Ossmann.

Rapid variability in MXB 1659-29 has been reported; several intensity changes by factors of four were observed on time scales of tens of seconds [39]. In Fig. 12 I show five such changes. The low states of short duration ($\lesssim$15 min) are due to absorption. A careful comparison of the data has shown that in a transition from "high" to "low", the 1-3 keV counting rate decreases first, then the 3-6 keV and finally the 6-12 keV counting rate. When a transition occurs from "low" to "high", the reverse is the case.

Search for Periodicities and Eclipses

More than 15 burst sources have been studied extensively and for long periods (1-3 weeks) by the Horizontal Tube detectors of SAS-3 [49,50]. The associated "steady" emission of the strongest sources (see Table 1) has been searched for periodicities and the presence of eclipses. None have been found so far. In Table 3 I summarize some of the results as provided by Lynn Cominsky, Herman Marshall and Fuk Li.

The absence of pulsations, periodicities and eclipses in the "steady" emission makes one wonder about proposed binary-neutron star models for burst sources. Yet, I don't believe, on the basis of the presently available information that we have compelling evidence against such models.

- The absence of pulsations may simply mean that the magnetic fields are weak (may be related to their old age).

It has been suggested that magnetic fields of neutron stars decay in time [51-54].

- The absence of eclipses may be due to the geometry of the binary systems.

Ed van den Heuvel has suggested Roche Lobe overflow systems with a very low-mass nuclear burning companion. This idea was worked out in some detail by Joss and Rappaport [55]. In such systems the probability of an eclipse is low (see footnote 3 on next page). Another possibility is perhaps that some of these objects have wide eccentric orbits as observed for the transient pulsar 4U0115+63 [56].

- It is perhaps possible that the reason for the absence of eclipses is simply that the compact objects are single.

Radio-Optical-X-ray Bursts

A world-wide coordinated burst watch was organized in 1977 by the SAS-3 group [41]. Forty four observatories participated from 14 different countries. A total of 120 bursts were observed by SAS-3 in 35 days from ten different Type I burst sources. No simultaneous positive detection was made in either radio, optical, or infrared. We have already published upper limits on radio bursts from MXB 1837+05 [57,58]. The most sensitive measurements were obtained by Ulmer and Hjellming with the VLA [59]. We also published upper limits on optical data from MXB 1837+05 and MXB 1916-05 [60-62]. In Table 4 I summarize the results of the ratios of energy density (accumulated for 10 sec) in the optical and X-ray bursts.

TABLE 4

Results from Simultaneous Optical and SAS-3 X-ray Burst Observations

Burst Source	$(\frac{E_{opt}}{E_x})^*$ burst	$(\frac{E_{opt}}{E_x})$ steady	Number of bursts	aperture telescope (m)	Method	References
MXB 1837+05	$<10^{-4}$	$\sim 4\times10^{-5}$	2	0.5	photography $\sim$B	Abramenko et al. [60] Crimean Obs.
	$<5\times10^{-5}$	$\sim 4\times10^{-5}$	2	1.2	photometry no filter	Bernacca et al. [62] Asiago Observatory
	$<3\times10^{-5}$	$\sim 4\times10^{-5}$	2	1.0	photometry U B	Walker [62] South African Astr. Obs.
	$<6\times10^{-5}$	$\sim 4\times10^{-5}$	2	1.3	photometry V	Backman and Canizares [62] McGraw Hill Obs.
MXB 1916-05 (no optical identification)	$<2\times10^{-3}$		2	0.6	photometry U B V	Takagishi et al. [61] Miyazaki Univ. Japan
MXB 1735-44	$\sim 2\times10^{-5\dagger}$	$\sim 1.3\times10^{-4}$	1	1.5	photometry, 3100-5500 A	Grindlay et al. [63] CTIO

*Observed ratio of energy densities during the first 10 sec of the burst (not corrected for interstellar extinction). Upper limits are 3σ. Optical energies in the range 4000-6000 Å.

†This result was obtained one day after I gave this talk. To date, July 25, 1978, this is still the only positive detection of a simultaneous optical and X-ray burst; we are continuing our efforts to make such observations.

We are pursuing the simultaneous optical observations this summer. Today, (June 1, 1978) and tomorrow, Josh Grindlay of SAO and Jeff McClintock and Claude Canizares of MIT are observing MXB 1735-44 in Cerro Tololo with the 60" telescope (CTIO). Simultaneous SAS-3 observations are being made by Jan van Paradijs, Lynn Cominsky and Fuk Li and we hope to achieve a sensitivity in the optical which is almost an order of magnitude higher than obtained so far (footnote 4 next page).

To conclude the observational part of this talk, I summarize in Table 5 the salient features of Type I and Type II X-ray bursts.

---

[3]About a month after my talk, on July 5, I received a preprint from Milgrom. He suggests a geometry that can explain in a simple way why eclipses are not seen from the strong (best studied) galactic bulge sources. His model predicts that eclipses are expected, however, from the faint (not well studied) sources. I have proposed SAS-3 observations to search for such eclipses. The model also predicts that the steady X-ray emission is strongly anisotropic. Thus, values for α lower than $\sim 10^2$ could be expected (see section on models).

Models

My talk was meant to cover mainly the burst observations and Jim Pringle was asked to cover the theoretical models in some detail. Unfortunately, he could not accept the invitation and I have therefore taken the liberty to use some of his time (not scheduled) to at least mention some recent, exciting work.

TABLE 5

Burst Summary

	Type I	Type II** (Rapid Burster)
Rise time	$\lesssim$1 sec smooth	$\lesssim$1 sec 50 m sec structure
Decay time	3-30 sec	1-30 sec
Tails	$10-10^3$ sec	<10 sec
Best fit spectra T_{max}	Blackbody $\sim 30\times10^6$ K Cooling	Blackbody $\sim 20 \times 10^6$ K No cooling
L_{max}	$\sim 10^{38}$ erg/sec	$\sim 10^{38}$ erg/sec
Energy in burst*	$\sim 10^{39}$ erg	10^{38-40} erg
Radius* (blackbody)	~10 km remains constant during burst decay	~15 km, decreasing during burst decay
Mechanism	Nuclear flash on surface of neutron star?	Instability in accretion flow
Stellar Object	Probably neutron star or black hole of solar mass	Probably neutron star or black hole of solar mass
Single or binary	?	Probably binary

*Assumed distance of 10 kpc.

**Our knowledge on Type II bursts from sources other than the rapid burster is very limited (see text).

Various models have been proposed for X-ray bursts. Most of them fall into two broad categories: (i) instabilities in the accretion of matter onto a compact object and (ii) thermonuclear flashes in matter accreted onto the surface of a neutron star. These two very different type of models have been (and still are) the subject of many heated discussions over the past two years. After the detection by Hoffman, Marshall and Lewin [1] of both Type I and Type II bursts from the same object (the Rapid Burster), it has become clear that perhaps both types of models may apply: (i) accretion instabilities for Type II bursts and (ii) thermonuclear flashes for Type I bursts. Work on accretion instabilities [64-66,84,95,96] was reviewed earlier [16]. In the limited amount of time left, I'd like to talk only about recent, as yet unpublished, work by Joss [37] on thermonuclear flash models.

The idea that bursts might be the result of thermonuclear flashes in matter accreted onto a neutron star was put forward by Maraschi and Cavaliere [67] and by Woosley and Taam [68]. Subsequent work on this idea was done by Joss [35], Lamb and Lamb, [36] and Taam and Picklum [69]. Only recently did Joss [37] succeed in constructing a detailed numerical model for thermonuclear flashes. This model assumes a nonrotating, nonmagnetized spherically accreting neutron star of mass 1.41 $M_\odot$, radius 6.57 km, core temperatures in the range $(2.5-5.7)\times10^8$°K and accretion rates in the range $(0.3-3)\times10^{17}$ g/sec. His model calculations reproduce the observed: (1) burst luminosities, (2) total burst energies, (3) rise times, (4) decay timescales, (5) recurrence times, (6) blackbody radiation, (7) maximum temperature of $(25-30)\times10^6$°K, (8) spectral evolution, and (9) ceasing of burst activity at high steady emission. Figure 13 shows some of Joss' results. As promising as these results are, the model is not without problems as discussed below.

[4]A day after I gave this talk, they succeeded in observing an optical burst simultaneously with an X-ray burst (IAU Circ. 3230; ref 63). The observed 2.8 sec delay in the optical burst relative to the X-ray burst (McClintock, private comm.) suggests that the optical burst is the result of X-ray heating of either a disc, a cloud or a stellar companion. Comparing the steady flux with that at burst maximum, the X-ray signal increased by a factor 6.3 ± 1 and the optical signal by a factor 1.5 ± 0.3. These numbers are consistent with a simple model of X-ray heating of a blackbody (ref. 63).

First, from the model one expects the burst intervals to decrease rapidly with increasing steady emission. Clark and co-workers [21] noticed that during a 3-day observation of MXB 1820-30 the burst intervals decreased by only ∿50% while the steady flux increased by ∿500% at which time the burst activity came to a halt. In the case of MXB 1837+05 [33] and MXB 1735-44 [83], there is no relation at all between the burst intervals and the steady flux. The burst intervals at times are erratic (they vary from ∿1 hour to ∿50 hours) and simultaneous changes in the steady flux are less than a factor of two.

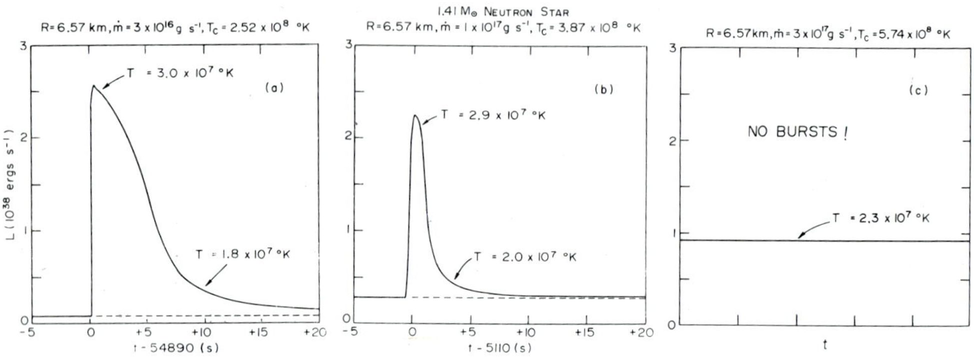

Fig. 13. Thermonuclear flashes on the surface of a neutron star. Temporal evolution of Type I X-ray bursts as calculated by Joss [37] (parameters shown at the top). The dashed line indicates the level of persistent emission due to accretion. a) the burst occurred after ∿15.2 hours of accumulation of accreted material, b) ∿1.4 hours of accumulation (higher accretion rate and higher neutron star core temperature), c) no bursts are observed any time (still higher accretion rate and core temperature; see text). This figure by courtesy of P.C. Joss.

Second, the model predicts that the ratio (α) of the energy emitted in steady emission to that in bursts is $\gtrsim 10^2$ (refs. 35, 36). The observed value of $\alpha < 25$ for MXB 1659-29 is very low [16]. Of course, if the steady emission is not isotropic, α can be much smaller than $\sim 10^2$ and the observed low value of α for MXB 1659-29 does not necessarily impose a problem (suggestion by F.K. Li, private communication; see also footnote 3).

Finally, multiple peaks as observed from various sources (most strikingly from MXB 1743-29; see Fig. 8) are not accounted for by Joss' model calculations.

The above problems can perhaps be overcome if one removes the spherical symmetry (e.g. disc accretion rather than spherical accretion, as assumed in the model).

What Are X-ray Burst Sources?

This question goes to the heart of one of the most intriguing unsolved problems in X-ray astronomy. The right answer would probably tell us what the origin is of most of the galactic bulge sources (including X-ray sources in globular clusters).

Though we do not have conclusive data in hand, the evidence in favor of neutron stars (or perhaps black holes of a few solar masses) is growing, as I have tried to show in this talk. The absence of pulsations is not an embarrassment, it may simply indicate that these objects have no strong magnetic fields. Maybe they were born that way or the field decayed due to old age. It is also possible that the axis of rotation and that of the magnetic dipole gradually coaligned [70]. It remains to be seen whether the burst sources and galactic bulge sources are in general in binary systems. The absence of observed eclipses would argue against this. Yet, the geometry of the binary systems may be such that the chance to observe such eclipses is small [55] (see footnote 3). The observed strong variability in the "steady" emission of burst sources and globular cluster X-ray sources, but particularly the recurrent transient behavior of the Rapid Burster, favors binary systems. The available information is clearly far from conclusive. Yet I'd be very surprised if the majority of the burst sources (members of the galactic bulge sources class) were not in binary systems. The best way to find out about the binary nature of these intriguing objects may well be to make detailed optical studies of the identified stars. Large telescopes and lots of observing time (dark time!) are needed.

Acknowledgements. I thank Roberta Caruso, Lynn Cominsky, Garrett Jernigan, Paul Joss, Fuk Li, Saul Rappaport, Bruno Rossi, Jan van Paradijs and Lorraine Wright for their useful comments and contributions to this manuscript.

Note added on October 11, 1978. I gave this talk on June 1, 1978. Several important results were reported between that time and the date that I finished writing this article (July 25, 1978). They have been added in the form of footnotes. Since July 1978, optical identifications of MXB 1837+05 (IAU Circ 3253) and MXB 1916-05 (Thornstensen) were made with faint blue stars. The first is very faint ($m_B \sim 19.2$) and it lies only $\sim$2 arc sec from the candidate proposed by Davidsen. Thus the total number of optically identified burst sources is five to date (October, 1978).

REFERENCES

1. J. A. Hoffman, H. L. Marshall, and W. H. G. Lewin, Nature 271, 630 (1978).
2. W. H. G. Lewin, IAU Circ. 2922 (1976).
3. W. H. G. Lewin, J. Doty, G. W. Clark et al., Astrophys. J. (Letters), 207, L95 (1976).
4. H. L. Marshall, M. P. Ulmer, J. A. Hoffman et al., Astrophys. J. in press, preprint CSR-P-78-19, (1978)
5. K. Brecher, P. Morrison, and A. Sadun, Astrophys. J. (Letters), 217, L139 (1977).
6. W. H. G. Lewin, J. van Paradijs, H. Marshall et al., priv. comm. manuscript in preparation (1978).
7. K. O. Mason, S. J. Bell-Burnell, and N.E. White, Nature, 262, 474 (1976)
8. M. P. Ulmer, W. H. G. Lewin, J. A. Hoffman et al., Astrophys. J. (Letters), 214, L11 (1977).
9. W. H. G. Lewin, Proc. Eighth Texas Symp. Ann. N.Y. Ac. Sci., 302, 210 (1977).
10. N. E. White, K. O. Mason, G. F. Carpenter et al. in press, Mon. Not. R.A.S. (1978).
11. G. Jernigan, J. E. McClintock, H. Marshall et al., IAU Circ. No. 3204 (1978).
12. H. Marshall and W. H. G. Lewin, IAU Circ. No. 3208 (1978).
13. W. H. G. Lewin, H. Marshall and L. Cominsky, IAU Circ. No. 3211 (1978).
14. W. H. G. Lewin, American Scientist, September-October issue, pg. 605 (1977).
15. J. Grindlay and H. Gursky, Astrophys. J. (Letters), 218, L117 (1977).
16. W. H. G. Lewin and P. C. Joss, Nature, 270, 211 (1977).
17. J. P. Ostriker, Proc. Eighth Texas Symp. Ann. N.Y. Ac. Sci. 302, 229 (1977).
18. W. H. G. Lewin, J. A. Hoffman, J. Doty et al., Nature, 267, 28 (1977).
19. J. Grindlay, H. Gursky, H. Schnopper et al., Astrophys. J. (Letters) 205, L127 (1976).
20. G. W. Clark, G. Jernigan, H. Bradt et al., Astrophys. J. (Letters) 207, L105 (1976).
21. G. W. Clark, F. K. Li, C. R. Canizares, et al. Mon. Not. R.A.S. 179, 651 (1977).

22. W. Forman and C. Jones, Astrophys. J. (Letters), 207, L177 (1976).

23. G. Clark and F. Li, IAU Circ. No. 3092 (1977).

24. F. Li and G. Clark, IAU Circ. No. 3095 (1977).

25. J. H. Swank, R. H. Becker, S. H. Pravdo et al., IAU Circ. No. 3010 (1976).

26. J. E. McClintock et al. Nature 270, 320 (1977) and Astrophys. J. (Letters) 223, L75 (1978).

27. J. E. McClintock, IAU Circ. No. 3088 (1977).

28. Davidsen, A. IAU Circ. No. 2824 (1975).

29. J. van Paradijs, IAU Circ. No. 3197 (1978).

30. J. E. Grindlay and W. Liller, Astrophys. J. (Letters) 220, L127 (1978).

31. J. Grindlay, IAU Circ. No. 3229 (1978).

32. W. H. G. Lewin, J. A. Hoffman and J. Doty, IAU Circ. No. 2994 (1976).

33. F. K. Li, W. H. G. Lewin, G. W. Clark et al., Mon. Not. R.A.S. 179, 21P (1977).

34. W. H. G. Lewin, J. A. Hoffman, J. Doty et al., IAU Circ. No. 3075 (1977).

35. P. C. Joss, Nature 270, 310 (1977).

36. D. Q. Lamb, and F. K. Lamb, Astrophys. J. 220, 291 (1978).

37. P. C. Joss, Astrophys. J. (Letters) in press CSR-P-78-22 (1978).

38. J. Hoffman, W. Lewin, H. Marshall et al., IAU Circ. No. 3190 (1978).

39. W. Lewin et al., IAU Circ. 3190 (1978) and IAU Circ. No. 3193 (1978).

40. R. Griffiths, M. Johnston, H. Bradt et al., IAU Circ. No. 3190 (1978).

41. W. H. G. Lewin IAU Circ. No. 3078 (1977).

42. W. H.G. Lewin, J. A. Hoffman, J. Doty et al., Mon. Not. R.A.S. 177, 83P (1976).

43. J. H. Swank, R. H. Becker, E. A. Boldt et al., Astrophys. J. (Letters), 212, L73 (1977).

44. J. A. Hoffman et al., Mon. Not. R.A.S. 179, 57P (1977) and Astrophys. J. (Letters), 217, L23 (1977).

45. J. van Paradijs, Nature, in press, CSR-P-78-13 (1978).

46. W. H. G. Lewin, Mon. Not. R.A.S. 179, 43 (1977).

47. C. Canizares and E. Neighbours, Astrophys. J. (Letters), 199, L97 (1975).

48. C. J. Eyles, G. K. Skinner, A. P. Willmore et al., Nature, 257, 291 (1975).

49. W. H. G. Lewin, F. K. Li, J. A. Hoffman et al., Mon. Not. R.A.S. 177, 93P (1976).

50. J. Buff, G. Jernigan, B. Laufer et al., Astrophys. J. (Letters), 212, 768 (1977).

51. J. E. Gunn, and J. P. Ostriker, Astrophys. J. 160, 979 (1970).

52. G. Chanmugam and M. Gabriel, Astron. and Astroph. 11, 268 (1970).

53. G. M. Ewart, R. A. Guyer and G. Greenstein, Astrophys. J. 202, 238 (1975).

54. E. Flowers and N. Itoh, Astrophys. J. 206, 281 (1976).

55. P. C. Joss, and S. Rappaport, Astron. and Astrophys. in press, CSR-P-78-02 (1978).

56. S. Rappaport, G. W. Clark, L. Cominsky et al., Ap.J. Lett. 224, L1 (1978).

57. R. M. Thomas, M.L. Duldig, R.F. Haymes et al., Mon. Not. R.A.S. in press, CSR-P-77-37 (1978).

58. H. M. Johnson, R. C. Catura, P. A. Lamb, et al., Astrophys. J. 222, 664 (1978).

59. M. P. Ulmer, R. M. Hjellming, W. H. G. Lewin et al., manuscript in preparation, private comm. (1977).

60. A. N. Abramenko, R. E. Gershberg, E. P. Pavlenko et al., Mon. Not. R.A.S. in press, CSR-P-78-10 (1978).

61. K. Takagishi, K. Nagareda, M. Matsuoka et al., ISAS Res. Note, ISAS RN60, CSR-P-78-31 (1978).

62. P. L. Bernacca, A. Bianchini, A. Walker et al., Mon. Not. R.A.S. in press, CSR-P-78-28 (1978).

63. J. E. Grindlay, J. McClintock, C. R. Cañizares, et al., Nature, in press (1978).

64. F. K. Lamb, A.C. Fabian, J. E. Pringle et al., Astrophys. J. 217, 197 (1977).

65. J. C. Wheeler, Astrophys. J. 214, 560 (1977).

66. J. E. Grindlay, Astrophys. J. 221, 234 (1978).

67. L. Maraschi and A. Cavaliere, in Highlights of Astronomy 4, Part I (1977).

68. S. E. Woosley and R. E. Taam, Nature 263, 101 (1976).

69. R. E. Taam and R. E. Picklum preprint (1978).

70. E. Flowers and M. Ruderman, Astrophys. J. 302, 291 (1977).

71. R. E. Doxsey, K. M. V. Apparao, H. V. Bradt et al., Nature 269, 112 (1977).

72. J. G. Jernigan, K. M. V. Apparao, H. V. Bradt et al., Nature 270, 321 (1977).

73. J. G. Jernigan and G. W. Clark, Astrophys. J., in press, preprint CSR-P-78-26 (1978).

74. R. Doxsey, H. Bradt, H, Gursky et al., Astrophys. J. (Letters) 221, L53 (1978).

75. J. E. Grindlay and H. Gursky, Astrophys. J. 209, L61 (1976).

76. H. Tananbaum, L. J. Chaisson, W. Forman et al., Astrophys. J. (Letters) 209, L125 (1976).

77. G. Fabbiano, H. Bradt, R. Doxsey et al., Astrophys. J. (Letters), 221, L49 (1978).

78. H. Marshall, F. Li and S. Rappaport, IAU Circ. No. 3134 (1977).

79. R. J. Proctor, G. K. Skinner and A. D. Wilmore, Mon. Not. R.A.S. in press, preprint (1978).

80. R. G. Cruddace, G. Fritz, S. Shulman et al., Astrophys. J. (Letters), 222, L95 (1978).

81. F. Li, and G. W. Clark, private communication (1978).

82. W. Forman et al., Astrophys. J. Supplement, in press, 4U Catalogue (1978).

83. W. H. G. Lewin, L. Cominsky and J. van Paradijs, priv. comm. (unpublished SAS-3 results).

84. E. P. T. Liang, Astrophys. J. 211, L67 and 218, 243 (1977).

85. K. M. V. Apparao, H. Bradt, R. Dower et al., Nature, 271, 225 (1978).

86. J. McClintock, S. Rappaport, J. Nugent et al., Astrophys. J. (Letters) 216, L15 (1977).

87. C. Canizares and M. Oda, Astrophys. J. (Letters) 214, L119 (1977).

88. A. Epstein, J. Delvaille, H. Helmken et al., Astrophys. J. 216, 103 (1977).

89. R. Kelley and H. Bradt, IAU Circ. No. 3165 (1978).

90. R. Kelley and H. Bradt, priv. comm. (1978).

91. J. Doty, IAU Circ. No. 2922 (1976).

92. J. Swank, R. Becker, E. Boldt et al., Mon. Not. R.A.S. 182, 349 (1978).

93. P. Murdin, M. J. and M. V. Penston, I. S. Glass et al., Mon. Not. R.A.S. 169, 25 (1974).

94. A. Davidson, R. Malina, H. Smith et al., Astrophys. J. (Letters) 193, L25 (1974).

95. W. Baan, Astrophys. J. 214, 245 (1977).

96. P. Joss and S. Rappaport, Nature 265, 222 (1977).

97. C. Alcock and S. Hatchett, Astrophys. J. 222, 456 (1978).

THE HELIUM ABUNDANCE OF THE WHITE DWARF HZ43 DERIVED FROM ITS SOFT X-RAY SPECTRUM

John Heise* and Henk Huizenga**

**The Astronomical Institute, Space Research Laboratory, Beneluxlaan 21, 3527 HS Utrecht, The Netherlands*
***Cosmic Ray Working Group, Huygens Laboratory, Wassenaarseweg 78, 2300 RA Leiden, The Netherlands*

INTRODUCTION

The hot white dwarf HZ43 is the brightest ultra-soft X-ray source in the sky. The optical, ultraviolet, and extreme ultraviolet emission of this object is compatible with the assumption of a photospheric origin of this radiation [1]. The effective temperature then is 60000 ± 10000 K, derived from ultraviolet fluxes measured with the Astronomical Netherlands Satellite [2]. The extreme ultraviolet (EUV) fluxes [3] constrain the helium abundance relative to hydrogen to be smaller than 10^{-3} and fix the interstellar hydrogen column density towards the source. The total soft X-ray flux measured by SAS-3 [4] is roughly also compatible with the assumption of a photospheric origin, provided that the atmosphere has a helium abundance smaller than 10^{-4}. The observation of hot white dwarf photospheres as soft X-ray sources is very sensitively determined by the helium abundance, [cf. 5] and to a lesser extent by the effective temperature as long as 50000 K $< T_{eff} <$ 150000 K.

MEASUREMENTS

Recently Bleeker et al., [6] measured the ultra soft X-ray spectrum of HZ43 in the range 50 Å - 200 Å with the use of very thin (0.5μ and 1μ polypropylene) window counters. This two-color measurement, together with the pulseheight distribution of the countingrate spectrum indicates the slope of the X-ray spectrum. The blackbody fits to the data given in this paper merely indicate the steepness of the observed spectral shape in the soft X-ray range, as blackbody spectra are not realistic atmosphere models. Indeed, despite the decreasing transmission of the detector window with increasing wavelength (10% at 180 Å for the 0.5 polypropylene window counter) and despite the tendency of flattening because of the poor energy resolution, the resulting pulseheight spectrum still increases with wavelength indicating a very steep incident photon-spectrum.

We fitted these data with a grid of model atmospheres with effective temperatures ranging from 50000 K to 90000 K, the surface gravity ranging from 3.10^7 to 3.10^8, and helium abundance from 0 to 10^{-3}. We folded these models through the two detector responses of the Bleeker et al., [6] experiment, and obtained expected pulseheight distributions, none of which appear to be acceptable fits (reduced $\chi^2 \sim 8$) to the data. This is caused by the fact that all uniform atmospheres predict rather flat soft X-ray spectra owing to the decrease in opacity with decreasing wavelength. The presence of helium suppresses the helium Lyman continuum, shortward of 228 Å, and makes the incident soft X-ray spectra even flatter (cf. dotted line in Fig. 1.)

DISCUSSION

The soft X-ray radiation originates ($\tau_\lambda \sim 2/3$) from layers in the white dwarf atmosphere that range from the top (at 200 Å) down to a depth of $\sim$ 2gr/cm^2 (at 40 Å) at which point the Rosseland mean optical depth is about 40. This comprises about 20 scale heights of approximately a few hundred meters. Since the low helium abundance in DA white dwarfs is thought to be due to gravitational separation, it seems not unreasonable to expect that the helium abundance could change with depth within the range of depths at which the soft X-ray spectrum originates. An increased absorption coefficient at deeper layers due to helium changes the ratio of electron scattering to absorption and therefore the shape of the emerging soft X-ray spectrum. We computed self-consistent models by assuming for simplicity a two layer atmosphere: A pure hydrogen layer on top of a hydrogen-helium atmosphere. The results of some of these models are shown in Fig. 1. Indeed, with a helium abundance of order 0.1, below a Rosseland mean optical depth of order 10, the expected soft X-ray flux is much steeper, whereas the size of the helium Lyman jump $F(228^+)/F(228^-)$ is considerably smaller than in the absence of the hydrogen layer and could even be made to vanish completely. Folding these spectra through the detector responses now gives acceptable (reduced $\chi^2 \sim 1$) fits to the measured pulseheight spectrum.

In conclusion we may say that the measured soft X-ray pulseheight spectrum is only compatible with a photospheric origin in the atmosphere of a hot dwarf if we assume that the helium abundance increases with depth in the atmosphere. More extended calculations will be published elsewhere [7].

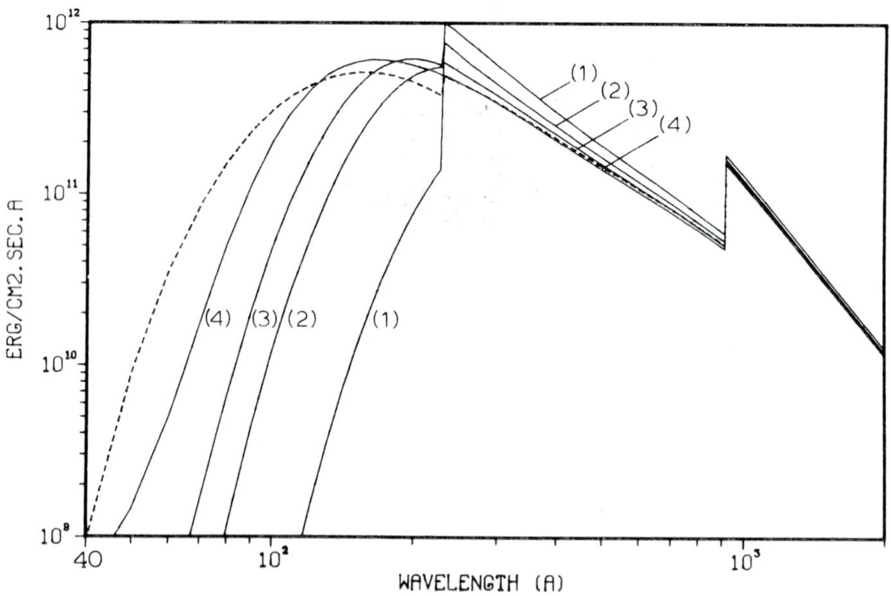

Fig. 1. Emerging flux from modelatmospheres with T_{eff} = 60000 K, log g = 8. Dashed line: a model with a uniform helium abundance of 10^{-5}. Drawn lines: the result of models with hydrogen above a Rosseland mean depth τ and a hydrogen-helium mixture with relative helium abundance 0.1 below that depth. (1) τ = 2, (2) τ = 6, (3) τ = 10, (4) τ = 20.

REFERENCES

1. L.H. Auer and H.L. Shipman, *Ap. J. (Letters)*, 211, L103 (1977).
2. P.R. Wesselius and D. Koester, *Astron. and Astrophysics (in press)* (1978).
3. M. Lampton, B. Margon, F. Paresce, F. Stein and S. Bowyer, *Ap. J. (Letters)*, 203, L71 (1976).
4. D.R. Hearn, J.A. Richardson, H.V.D. Bradt, G.W. Clark, W.H.G. Lewin, W.F. Mayer, J.E. McClintock, F.A. Primini and S.A. Rappaport, *Ap. J. (Letters)*, 203, L21 (1976).
5. H.L. Shipman, *Ap. J. (Letters)*, 206, L67 (1976).
6. J.A.M. Bleeker, J. Davelaar, A.J.M. Deerenberg, H. Huizenga, A.C. Brinkman, J. Heise, Y. Tanaka, S. Hayakawa and K. Yamashita, *Astron. and Astrophysics (in press)* (1978).
7. J. Heise and H. Huizenga, *Astron. and Astrophysics (in press)* (1978).

UPPER LIMITS ON SOFT X-RAY FLUXES FROM SOME CELESTIAL OBJECTS OBSERVED BY "SALYUT 4"

I. L. Beigman*, L. A. Vainshtein*, B. N. Grigorevskaya*, V. M. Lyutyi**, I. P. Tindo*, A. I. Shurygin*

**P. N. Lebedev Physical Institute, Moscow, USSR*
***P. K. Sternberg State Astronomical Inst., Moscow, USSR*

ABSTRACT

The grazing-incidence RT-4 X-ray telescope mounted aboard the "Salyut 4" orbital station was used to search for possible soft X-ray emission in the range 0.16-0.28 Kev from a blue supergiant β Ori (Rigel), the Cepheids W Vir and RR Lyr, and the planetary nebula NGC 6720.

INSTRUMENTATION

The RT-4 telescope has a paraboloidal Ni-coated mirror with a focal length of 62.4 cm and effective area $S \approx 100$ cm^2. The field of view is 3.°4 (FWHM). A two-sectioned ("X-ray" and "anticoincidence shield" sections) proportional counter with 2μ polypropylene window was used as a detector [1] . In the range 0.16-0.28 KeV the efficiency of the telescope is about 2 counts/photon. The instrument is practically insensitive to the X-ray photons with energies $E > E_0$, where $E_0 = 0.284$ KeV is the carbon K-edge. Measurements of the detector background caused predominantly by cosmic rays and UV-radiation were performed using a movable quartz filter periodically placed in front of the counter window. The X-ray intensity was measured during the 38.8 s intervals without the filter. Then the background was measured during the following 8.8 s intervals with the filter. The background counting rate was suppressed by an amplitude discrimination and anticoincidence circuitry. The rejected pulses of the "X-ray" section, together with pulses of the shield section were registered in the "control" channel. The residual background counting rate in the main channel was 0.2-0.3 cps at low latitudes. The RT-4 telescope was mounted on a pointing system; the apparatus is described in more detail elsewhere [2] .

OBSERVATIONS

β Ori. β Ori is a blue supergiant of the class B8 Ia. It is a visually and spectroscopically-double with a period of 21.d9. The distance to β Ori is 930 pc. β Ori was observed on February 4, 1975, at 12^h10^m-12^h36^m (UT). The counting rate averaged over the 38.8 s intervals (without the filter) is shown in the lower part of Fig. 1. Vertical bars in this figure represent the background counting rate (with the filter) averaged over 8.8 s intervals. We shall describe the accepted procedures of the background evaluation. In the first method used the background at the moment of the X-ray observation (without the filter) was determined as a mean of the adjacent counting rates measured with the filter. The result of such a treatment is shown by solid lines in the upper part of Fig. 1. In the second method of background evaluation, the background at the moment of X-ray measurement was derived from the simultaneously measured counting rate in the control channel using a "coefficient of background suppression"(the ratio of the counting rates of the main and control channels at the measurements with the filter). This ratio lies usually between 1/20 and 1/24. The result of application of this method is shown by dashed lines in Fig. 1.

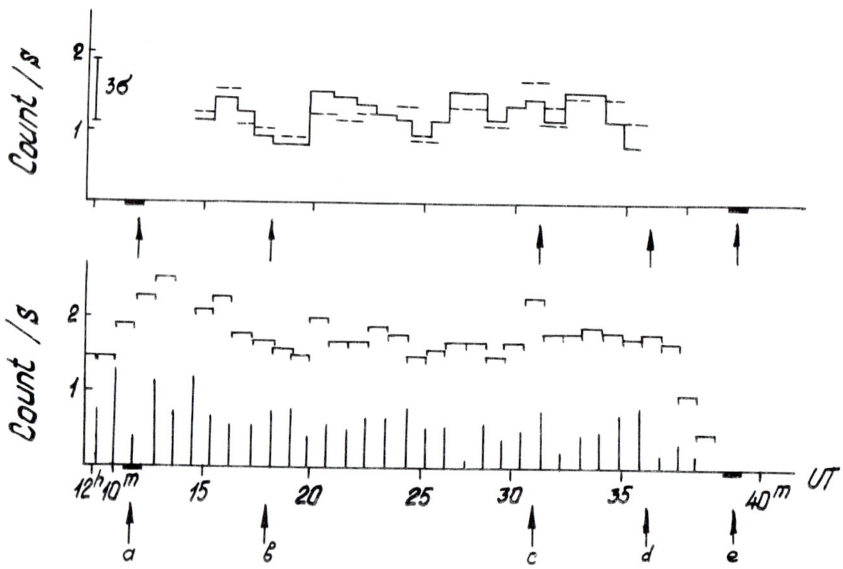

Fig. 1. Observation of β Ori, a - movement of the telescope axis from its basic position to the direction of β Ori. b-d - stabilization, c - Earth equator, e - telescope down.

The counting rate of the main channel measured without the filter averaged over the total observation time ($\approx$1000 s) is 1.90 ± 0.04 cps. The averaged counting rate with the filter is 0.66 ± 0.05 cps. The difference between them is 1.24 ± 0.07 cps. This includes the contributions of possible X-ray flux of β Ori, the diffuse X-ray background and UV-radiation which does not pass through the quartz filter ($\lambda <$ 1900 A). The estimated contribution of the X-ray background is about 1 cps (see, e.g. [3]), thus the soft X-rays from β Ori and the UV-radiation gave together less than $\approx$ 0.3 cps. This gives the upper limit for the β Ori X-ray luminosity of 10^{32} erg/s (0.16-0.28 Kev). This corresponds to an emission measure (at a temperature of $\sim 10^6$ °K) of the order of 5.10^{54} cm^{-3} that agrees well with the upper limit obtained in [4] .

W Vir and RR Lyr. W Vir is a Cepheid star at the distance of 3.3 kpc; it is a supergiant (luminosity class Ib) with the period $P=17.^d27$ and the variable magnitude $m_v=9^m57-10^m.71$. Its spectral type varies between F0-G0. W Vir was observed on June 2, 1975, at $13^h45^m-14^h14^m$ (UT) at phase 0.29. The mean counting rates without and with the filter were 1.2 cps and 0.3 cps. The difference of 0.9 cps between them is attributable to the X-ray background: F_d(0.25 KeV) = 160 phot/cm^2.s.KeV.ster [3] . Thus, the X-ray emission from W Vir does not exceed 0.5 phot/cm^2.s.KeV,giving an upper limit for the X-ray luminosity of 3.10^{34} cm/s, i.e. it is less than $\approx 10^{-3}$ of the total W Vir luminosity. At a temperature of 0.1-0.4 KeV, the upper limit for the emission measure is $\approx 10^{57}$ cm^{-3}.

RR Lyr is a Cepheid with main period $0.^d57$ and secondary period $40.^d8$. Its magnitude varies from $7^m.20$ to $8.^m57$. The spectral type changes between A8 and F7. The distance to RR Lyr is 260 pc. RR Lyr was observed on July 2, 1975 at $03^h24^m-04^h-03^m$ (UT). At that time there was a minimum of the secondary period (phase 0.69) and a maximum of the main period (phase 0.99-0.02). Over almost the whole observation interval the background counting rate was rather high. The difference of the counting rates without and with the filter did no exceed 1 cps, giving an upper limit for the X-ray emission of 0.5 phot/cm^2.s.KeV. The X-ray luminosity is $L_x \approx 2 \times 10^{32}$ erg/s;less than 10^{-3} of the total luminosity of RR Lyr. At a temperature within the range 0.15-0.4 KeV we obtain for the emission measure the upper limit of 10^{55} cm^{-3}.

NGC 6720. NGC 6720 is a planetary nebula of the ring type lying at a distance $\approx$700 pc. Its integral visual magnitude is $9.^m3$ with an extinction $A_V = 1$. The magnitude of the central star - the nucleus of the planetary nebula - is $14.^m7$. The temperature of the nucleus as determined from the optical data is 80 x 10^3K. The observation was made on June 21, 1975 at $11^h49^m-12^h08^m$ (UT). The difference of the averaged counting rates without and with the filter did not exceed 0.2 cps. Taking X-ray radiation of the nucleus being the blackbody one, we can derive the upper limit for its temperature (T). At $E<E_0$ the absorption can be approximately described as $\tau_0 \cdot \exp[-\tau_0(E_0/E)^3]$, where τ_0 is the total optical depth at the photon energy E_0, that is the sum of optical depths of the intrinsic gas in the nebula, the interstellar gas and the window of the counter. Then the counting rate due to X-ray photons is given by E_0 :

$$C_x = N_{pg} \cdot 10^{A_v/2.5} \left[\frac{\lambda_{pg}}{\delta\lambda_{pg}}\right] \left[\frac{E_{pg}}{KT}\right] \cdot S \int_0^{E_0} \left(\frac{E}{E_{pg}}\right)^3 \exp\left[\frac{E}{KT} - \tau_0\left(\frac{E_0}{E}\right)^3\right] \frac{dE}{E} \quad (1)$$

$$N_{pg} = 0.57 \times 10^{7-m_{pg}/2.5}$$

where m_{pg} is the photographic magnitude of the nucleus, A_v is the visual extinction, $\lambda_{pg} = 4800$ A is the effective wavelength of the photographic image of the nebula, $\delta\lambda_{pg}$ is the effective width of the photographic spectral range ($\delta\lambda_{pg}/\lambda_{pg}$ = 0.3), E_{pg} = 2.6 eV is the photon energy corresponding to λ_{pg} , and S is the effective area of the telescope. For evaluation of the integral in (I) we have used the method of the "stationary phase". The hydrogen in the nebula is completely ionized and practically the only absorbing agent there is helium. Taking the intrinsic electron density in the nebula $N_e \sim 10^3 cm^{-3}$ and its size d ~0.2 we obtain τ_{neb} = 0.8. The optical depth of the interstellar gas is τ = $6N_H$. With N_H = 0.5 cm^{-3} and τ_w = 0.7 (optical depth of the counter window [1]) we obtain τ_o = 4.5. Taking into account that $C_x \leqslant 0.3$ cps we find that the temperature of the nucleus is lower than 2.10^5K.

REFERENCES

1. A.D.Ulyanitskaya, I.P.Tindo, D.A.Goganov, V.A.Nazarov, O.D.Lesnyh "Kratkie soobscheniya po fizike" FIAN USSR, No. 6,40 , 1970.
2. I.L.Beigman, L.A.Vainshtein, Yu.P.Voinov, D.A.Goganov, N.I.Komyak, S.L.Mandel'shtam, I.P.Tindo, N.A.Shatsky, A.I.Shuryghin. Trudy FIAN USSR, 77, 3, 1974.
3. A.N.Brunner, P.L.Coleman, W.L.Kraushaar, D.M.Cammon. Astrophys. J.Lett., 192, 67 (1972).
4. R.Mewe, J.Heise, E.H.B.M. Gronenshild, A.C.Brinkman, J.Schrijver, A.J.F.Boggende. Astrophys.Space Sci., 42, No. 1, 217, 1976.

SOFT X-RAY OBSERVATION OF AM HERCULIS

S. Hayakawa, H. Iwanami, H. Kunieda, F. Nagase and K. Yamashita

Department of Physics, Nagoya University, Nagoya, Japan

INTRODUCTION

The binary system AM Herculis has been identified with a soft X-ray source [1]. The binary period of 3.1 hours has been established both for X-ray emission [2, 3,4] and for optical phenomena [5,6,7,8]. The X-ray spectrum has been measured over a wide energy range 0.1 - 60 keV [3,4]. Variabilities of the X-ray flux have been noticed by comparison with various other data at different times [3], and the soft X-ray flux observed by Bunner [4] is several times that measured earlier by Hearn et al [1]. Using large area proportional counters with high efficiency in the low energy range, we have observed the soft X-ray spectrum and the absolute flux of AM Her.

INSTRUMENTATION AND OBSERVATION

Eight proportional counters filled with Ar (90%) + CH_4 (10%) at 1 atm. with three different window thicknesses were employed. The windows consist of polypropylene film coated with Formvar (10 μg cm^{-2}), Lexan (10 μg cm^{-2}) and carbon (34 μg cm^{-2}). Characteristic properties of these three sets of counters are summarized in Table 1.

TABLE 1 Characteristic Properties of Proportional Counters

Window thickness (nominal) polypropylene equivalent		0.9 μm	1.5 μm	4 μm
Transmission probability (%)	C_K	85	74	48
	F_K	63	49	13.5
Effective area (cm^2)		215	423	282
0.15 - 0.5 keV flux of AM Her (counts/cm^{-2} s^{-1})		0.62 ±0.07	0.47 ±0.05	0.23 ±0.04

The field of view of each counter was 10°.7 x 3°.0 (FWHM), the narrow side being parallel to the scan direction. The scan path was determined by early type stars observed with a set of UV detectors.

The instrument on board a sounding rocket was launched at 12^h 22^m UT on 14 September 1977 from Kagoshima Space Center (131° 05' E, 31° 15' N). In the course of scanning along $\ell \simeq 80°$, AM Her was observed for 3.6 seconds, at JD 2443401.01898 and was located within ± 0°.5 of the trajectory of the counter axis at an elevation angle of 60°. The epoch of observation corresponds to the phase $\phi = 0.94$ refering to the linear polarization epoch at JD 2443014.7647 [7] and the binary period of $P = 0^{d}.128926$ [9].

The atmospheric absorption of X-rays is negligible above the rocket altitude of 173 km. UV insensitivity of the counters is confirmed by the insignificant contribution of early type stars to the X-ray counting rate; three UV stars, ι Her (B3, $m_V = 3.8$), αLyr (A0, $m_V = 0.0$) and δCyg (B9.5, $m_V = 2.9$), would have given appreciable peaks in Fig. 1 if the counters were sensitive to UV radiation.

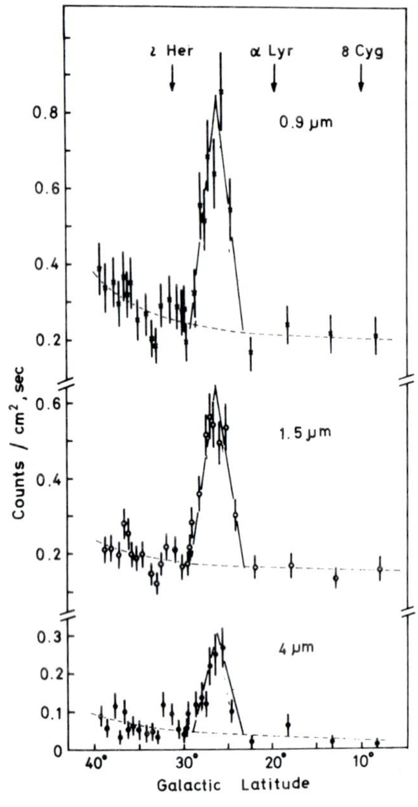

Fig. 1. Counting rates vs. galactic latitude of 0.9 μm, 1.5 μm, and 4 μm counters along the longitude of ℓ=80°. The dashed lines represent the diffuse X-ray background. The collimator responses for soft X-rays from AM Her are shown by triangles. The positions of three UV stars are indicated by arrows.

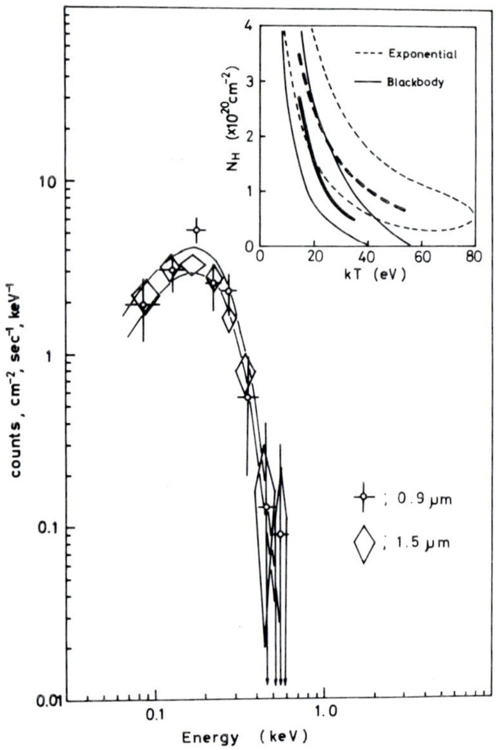

Fig. 2. Pulse height distribution of the soft X-rays from AM Her observed by 0.9 μm and 1.5 μm counters. The superimposed curves are calculated spectra for blackbody model of kT = 25 eV and $N_H = 1 \times 10^{20}$ H atoms cm^{-2}. The inset shows the curves of best-fit sets of kT and N_H (thick lines) and the 90% confidence contours (thin lines) for the exponential and blackbody models.

RESULTS AND DISCUSSION

Figure 1 shows the counting rates versus the galactic latitude in the pulse height range 0.15 - 0.5 keV for three counter sets. The flux values derived for AM Her for the respective counter sets are given in Table 1. No significant counts were observed in the energy range above 0.5 keV. The 2σ upper limit in the energy range 2 ∿ 6 keV was 2.5×10^{-2} counts cm^{-2} sec^{-1}.

The pulse height spectra obtained with 0.9 μm and 1.5 μm counters are shown in Fig. 2. The observed spectra were fitted simultaneously with a simple exponential and a blackbody spectrum with interstellar absorption, each of which is represented by three parameters, the absolute intensity, the apparent temperature kT, and the hydrogen column density N_H. The inset in Fig. 2 shows the curves of best fit sets of N_H and kT together with the 90% confidence contours for these three parameters. In Table 2 several best-fit sets of parameters are shown for these two models. All sets give essentially the same total energy flux in the range 0.1 ∿ 0.5 keV, which is 5×10^{-10} erg cm^{-2} sec^{-1}.

TABLE 2 Sets of Parameters Fitted to the Observed Spectra

Exponential Model			Blackbody Model		
kT	N_H	C	kT	N_H	C
15	3.5	2.6×10^7	15	2.5	1.4×10^9
20	2.5	4.0×10^5	20	1.6	2.4×10^7
30	1.5	5.9×10^3	25	1.0	1.9×10^6
40	1.1	7.8×10^2	30	0.7	3.8×10^5
50	0.75	2.0×10^2			

note; kT in eV, N_H in $\times 10^{20}$ H atoms cm^{-2}, C in photons cm^{-2} s^{-1} keV^{-1}.

The apparent temperature lies in the range of $1.5 \sim 6 \times 10^5$ K, and $1.5 \sim 4 \times 10^5$ K, for the exponential and blackbody spectra, respectively, for N_H greater than 5×10^{19} H atoms cm^{-2}.

The best-fit sets of kT and N_H give nearly equal values of $N_H kT$, which is appreciably greater than that obtained by Bunner [4], who gave kT = 30 eV and $N_H = 5 \times 10^{19}$ H atoms cm^{-2}. This set of parameters would require that the flux ratio of the 4 μm counters to the 0.9 μm counters be greater than 10, in contrast to our observed ratio of 2.7. This fact also supports a higher temperature, or a greater column density. At the phase $\phi = 0.94$ the flux is estimated to be 0.8 times the maximum flux, which is about twice the average flux over the entire light curve [1,2,4]. The average flux in the energy range 0.1 ∿ 0.28 keV is thus found to be 3.1×10^{-10} ergs cm^{-2} sec^{-1}, compared to 0.65×10^{-10} by Hearn et al [1] and 3.9×10^{-10} by Bunner [4].

A typical optical magnitude of AM Herculis is 12.5 mag in the V-band [10,11] and also 12.5 mag in the B-band [12]. The energy flux in the optical range (0.3 - 0.8 μm) corresponding to $V \simeq B = 12.5$ is 1.6×10^{-10} ergs cm^{-2} sec^{-1} [9]. A simple extrapolation of the exponential spectrum gives a much higher optical flux, whereas the blackbody model does not contradict the optical flux only for $kT \gtrsim 20$ eV. This qualitatively restricts the acceptable range of parameters, although information on the spectrum in the XUV region is necessary for quantitative purpose, because the actual spectrum should be complicated by energy dependent opacity.

The total luminosity for the blackbody spectrum estimated from our soft X-ray flux is $(4 \sim 50) \times 10^{33}$ $(d/100\ pc)^2$ erg sec^{-1} for a temperature of $(2 \sim 4) \times 10^5$ K, where d is the distance to AM Her. Thus the radius of the emitting area is $(2 \sim 15) \times 10^7$ cm. Since this is smaller than the radius of a typical white dwarf, the X-rays are probably being emitted from a hot spot on the surface of the white dwarf.

REFERENCES

1. D.R. Hearn, J.A. Richardson and G.W. Clark, Astrophys. J. (Letters) 210, L23 (1976).

2. D.R. Hearn and J.A. Richardson, Astrophys. J. (Letters) 213, L115 (1977).

3. J. Swank, M. Lampton, E. Boldt, S. Hold and P. Serlemitsos, Astrophys. J. (Letters) 216, L71 (1977).

4. A.N. Bunner, Astrophys. J. 220, 216 (1978).

5. W.C. Priedhorsky, Astrophys. J. (Letters) 212, L117 (1977).

6. A.P. Cowley and D. Crampton, Astrophys. J. (Letters) 212, L121 (1977).

7. S. Tapia, Astrophys. J. (Letters) 212, L125 (1977).

8. J.L. Greenstein, W.L.W. Sargent and T.A. Boroson, Astrophys. J. (Letters), 218, L121 (1977).

9. W.C. Priedhorsky and W. Krzeminski, Astrophys. J. 219, 597 (1978).

10. R.A. Berg and J.G. Duthie, Astrophys. J. 211, 859 (1977).

11. P. Szkody and D.E. Brownlee, Astrophys. J. (Letters) 212, L113 (1977).

12. E. Feigelson, L. Dexter, and W. Liller, Astrophys. J. 222, 263 (1978).

COMPACT OBJECTS IN DENSE INTERSTELLAR CLOUDS

A. C. Fabian

Institute of Astronomy, Madingley Road, Cambridge CB3 OHA, England

INTRODUCTION

A lone neutron star or white dwarf may produce detectable emissions at X-ray and other wavelengths if it passes through a dense interstellar cloud. The estimated numbers of lone neutron stars and interstellar clouds are sufficient to make such sources plausible candidates for non-eclipsing sources. The relative velocity with which the compact object traverses the cloud is important in determining the X-ray luminosity and the resultant light curve. In some cases bursts may result [1]. The range of detectable sources of this type and their expected properties are discussed. Globular cluster X-ray sources may be explained in this manner.

Compact galactic X-ray sources are usually associated with binary star systems, an assumption that is often proved observationally correct. It is, however, possible to create a detectable X-ray source by accretion from a dense interstellar cloud onto a compact object. This has been discussed elsewhere [1-6] in various contexts; here I wish to review the situation a little more and emphasise its possible importance in the identification of X-ray sources.

ACCRETION ESTIMATES

The accretion radius of a solar mass object moving with velocity v km s^{-1} in a cloud temperature 100 T_2 K is given by

$$R_A \simeq \frac{10^{16}}{(T_2 + v^2)} \text{ cm.}$$

This generates a luminosity

$$L \simeq 10^{38} n_3 (T_2 + v^2)^{-3/2} \text{ erg s}^{-1}$$

assuming an efficiency of 0.1, appropriate to a neutron star or black hole ($\sim 10^{-3}$ for a white dwarf), and a density of 10^3 n_3 particles cm^{-3}. Talbot & Newman [7] have estimated that the fraction of the Galactic disk occupied by interstellar clouds, exceeding a density n_3, is

$$f (> n_3) \simeq 6 \times 10^{-4} n_3^{-3} \qquad (n_3 > 1)$$

This suggests that the Galaxy contains many dense clouds with compact objects. The appearance of these objects depends upon n_3, T_2, v and other effects discussed below. Also of importance is the opacity of the cloud. X-rays above several kilovolts emerge provided that the radius of the cloud, $r_c \lesssim 3\, n_3^{-1}$ pc.

The hard radiation produced by the compact object may be absorbed by the cloud, affecting the accretion flow [8], [9]. This was pointed out by Mestel [10] as a serious problem in the Hoyle & Lyttleton [11] stellar rejuvenation idea. Recent studies of the radiation feedback in accretion [2], [4], [5], [13] are appropriate to a static object. Its motion may introduce another dimension, radically altering the accretion pattern. One possibility that I have discussed previously [1] involves the build up of a dense shell by the expanding radiation heated gas in the cloud. Passage of the compact object through this shell triggers off a further flash of radiation creating another expanding shell and so on. Radiation bursts with a very short duty cycle may thereby be produced. High luminosities (L up to $\sim 10^{38}$ erg s^{-1}) may arise under certain conditions. It is most unlikely that a steady X-ray source would result.

A very low luminosity may alternatively result if the compact object possesses a large magnetosphere and if n_3 is small. The accreted gas may not penetrate the magnetosphere until cooling has occurred [14], [15]. A stagnant pool of gas may then extend back to the accretion radius (the meaning of which disappears), resulting in most of the gas just flowing past the object and its magnetosphere. This may be relevant to discussions of the Sun and Heliosphere. The accretion rate estimated by Talbot & Newman [7] may be overestimated by their neglect of this effect.

The X-ray events observed by Cooke [16] from the vicinity of the Cometary Globule NGC 5367 may be due to a lone neutron star. The error box of the pulsing X-ray source 3U1626-67 [17] contains a 'fuzzy patch', discernible on the ESO/SRC survey plates and distinctly separate from McClintock's [18] star. 4U1608-52 lies in the vicinity of a remarkable obscuring globule (W. Liller, private communication). Some of these, and other, associations may be coincidental, but observers might consider searching for alternative candidates to blue emission line stars.

One further possible circumstance in which unbound neutron stars accrete gas may be in globular cluster cores. These contain the most massive stars in the clusters, several of which are red giants, possibly undergoing mass loss with $\dot{M} \simeq 10^{-6}\dot{M}_{-6} M_{\odot} yr^{-1}$ The wind velocity is ~ 10 km s^{-1}, leading to $n_3 \simeq 10^{6}\dot{M}_{-6} R_{14}^{2}$ cm^{-3} at a radius of 10^{14} cm. Any neutron stars bound to the cluster but not to the giants has $v \lesssim 10$ leading to

$$L \simeq 10^{37}\, \dot{M}_{-6}\, R_{16}^{2}\ \text{erg s}^{-1}.$$

Thus only 10 red giants plus ~ 100 neutron stars in a core of size $\sim 10^{17}$ cm produces at least one (variable) X-ray source. The heating due to ionization of the gas by accreting neutron stars and white dwarfs may in part be responsible for the expulsion of gas from globular clusters.

I thank the Radcliffe Trust for support.

REFERENCES

1. A. C. Fabian, Nature 268, 607 (1977).

2. J. P. Ostriker, M. J. Rees and J. Silk, Astrophys. Lett. 6, 179 (1970).

3. J. W. Truran et al. Astrophys. J. 211, 539 (1977).

4. F. K. Lamb et al., Astrophys. J. 217, 197 (1977).

5. J. Grindlay, Astrophys. J. 221, 234 (1978).

6. V. A. Hughes and M. R. Viner, Astrophys. J. Lett. 222, L27 (1978).

7. R. J. Talbot and M. J. Newman, Astrophys. J. Suppt. 34, 295 (1977).

8. V. F. Shvartsman, Soviet Astron. A. J. 14, 662 (1971).

9. J. Buff and R. McCray, Astrophys. J. 189, 147 (1974).

10. L. Mestel, Mon.Not.R.astr.Soc. 114, 437 (1954).

11. F. Hoyle and R. A. Lyttleton, Proc. Camb. Phil. Soc. 25, 405 (1939).

12. J. P. Ostriker et al., Astrophys. J. Lett. 208, L61 (1976).

13. L. Cowie, J. P. Ostriker and A. Stark, preprint (1978).

14. J. Arons and S. Lea, Astrophys. J. 207, 914 (1976).

15. R. F. Elsner and F. K. Lamb, Astrophys. J. 215, 897 (1977).

16. B. A. Cooke, Nature, Lond. 261, 564 (1976).

17. S. Rappaport et al., Astrophys. J. Lett. 217, L29 (1977).

18. J. McClintock, I.A.U. Circular 3088 (1977).

Binary X-Ray Emitting Systems — Observations

IRON LINES FROM X-RAY BINARIES

S. H. Pravdo

Laboratory for High Energy Astrophysics, NASA/Goddard Space Flight Center, Greenbelt, Maryland 20771, USA (Also Department of Physics and Astronomy, University of Maryland)

ABSTRACT

In recent years satellite borne X-ray detectors have revealed iron line features in a variety of X-ray sources. Iron line emission from binary X-ray pulsars can be considered unique because the underlying continua can not be characterized by thermal emission from an optically thin high temperature plasma. We will present a summary of the observational evidence from a number of these sources. According to current theories, the continua X-rays from X-ray pulsars originate at or near the surface of neutron stars. These X-rays can then heat or fluoresce iron ions in surrounding physical regions to create the observed emission. Some candidates for the sources of the iron lines include the companion normal star, an accretion disk, or a shell of material at the Alfven surface of the neutron star. Detailed measurements of the line energy, intensity, and width, and the phase dependence of these quantities can help to determine the relative importance of contributions from the candidate sources. We will present evidence that the iron emission from some sources contains more than one component. The spectra of other X-ray binaries including transients and cataclysmic variables also contain iron lines. We will discuss some implications of the observed line emission.

INTRODUCTION

Emission lines reveal interesting physical parameters of X-ray sources. The parameters include elemental abundance, ionization state, temperature, and density in the emitting region. Because of a combination of circumstances, the Kα line from iron is the most prominent feature in the X-ray band. Below $\sim$ 2 keV, interstellar absorption, a multiplicity of line features from several elements, and detector characteristics all contribute to the difficulty of identifying lines. This area of research is now beginning in earnest with the HEAO series of satellite observatories. Above $\sim$ 2 keV, the only lines expected to be readily observable are those due to iron. One reason for this is the relatively high abundance of iron thought to exist in the "cosmic" distribution of high Z elements(1). Table 1 illustrates the prominence of iron line emission in the fluorescence process(2,3). The calculated equivalent width of iron for an E^{-1} photon number spectrum is at least 14 times larger than the equivalent widths of other abundant elements with nearby line energies.

The binary nature of many galactic X-ray sources was discovered with the UHURU satellite(4,5). The X-ray satellite observatories which followed—OSO-7, ANS,

Ariel 5, SAS 3, OSO 8, and HEAO 1--added to this knowledge by identifying diverse types of X-ray binaries, including pulsators, transients, and cataclysmic variables. We have now found examples from each of these classes which contain iron line emission. Clearly many, if not all, X-ray binary systems in our galaxy contain a significant amount of iron as well as, by inference, other high Z elements in their relative, lesser abundances.

Table I
FLUORESCENCE EMISSION

ELEMENT	ABUNDANCE/H (a)	FLUORESCENCE YIELD (b)	Kα LINE ENERGY (keV) (c)	RELATIVE EQUIVALENT WIDTH $dN/dE = E^{-1}$
Si	3.31×10^{-5}	0.043	1.74	0.03
S	1.58×10^{-5}	0.082	2.31	0.03
Ar	7.94×10^{-6}	0.129	2.96	0.03
Ca	2.00×10^{-6}	0.163	3.69	0.01
Fe	3.98×10^{-5}	0.347	6.40	1.00
Ni	2.00×10^{-6}	0.414	7.47	0.07

(a) FROM ALLEN (1973)
(b) FROM BAMBYNEK et al (1972)
(c) FROM HOUSE (1969)

In the following, we will review the observational evidence for iron lines in the spectra of X-ray binaries. The first section contains a brief description of the iron line detectors in current usage and the method of spectral analysis. Other sections discuss classes of binaries as well as the interesting sources Hercules X-1 and Cygnus X-3. Detailed studies of the last two indicate the presence of more than one component in the iron line emission.

THE DETECTORS AND SPECTRAL ANALYSIS

Proportional counters and, in one case, a Bragg spectrometer(6) have detected iron lines from celestial X-ray sources. The former have moderate energy resolution at the iron line energy (1.1 keV full-width-half-maximum resolution at 6.7 keV) while the latter has better resolution (285 eV). Moderate fluxes and broad intrinsic widths (see below) of celestial lines from binaries have favored proportional counters in their detection.

Photoelectric absorption is the physical principle underlying the operation of proportional counters. The quantum efficiency for detecting photons near 6.7 keV is a smooth function of energy for both argon and xenon, the widely used detection gases. This fact aids in the identification of spectral discontinuities due to lines at this energy.

Analysis of spectra obtained with proportional counters is complicated by the singularity of the detector response matrix. This makes it impossible to proceed in one step from the pulse-height-analyzed (PHA) data to the incident spectrum. Instead, we multiply trial models of the incident spectrum by the detector response matrix and compare the resultant to the data using a χ^2 test(7,8). Once an acceptable fit is obtained, it is possible to estimate confidence limits on the model parameters by employing a theorem relating deviations from the minimum best-fit χ^2 value to statistical significance(9,10). The incident spectra presented in this work are determined after an acceptable best-fit model is found. Based on this model and the PHA data, we calculate spectrum-dependent channel-by-channel X-ray efficiencies. The data with accompanying errors are then divided by these efficiencies to give the inferred incident spectrum.

HARD SPECTRUM BINARIES

Many and perhaps all of the class of hard spectrum (power law number index $\lesssim$ 1) binaries are also X-ray pulsars(11,12). They are thought to be rotating magnetised neutron stars. Another characteristic that these sources have in common is broadened iron line emission. Figure 1 illustrates the residuals over the best fit

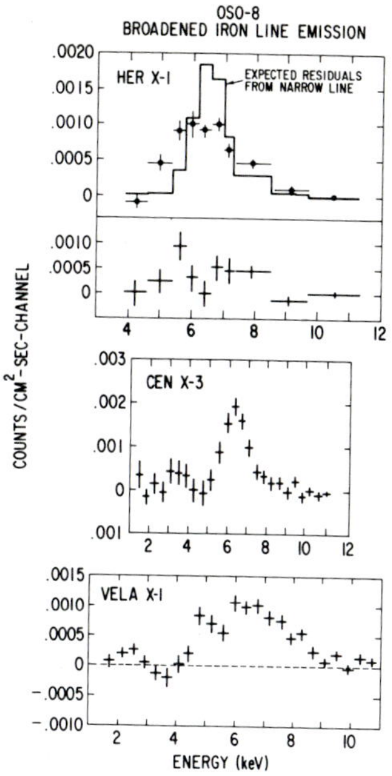

Fig. 1. Residuals near 6.7 keV above the continua of three binary pulsars (13,14,15).

continua near 6.7 keV for three of these binaries(13,14,15). The top plot in Fig. 1 has the predicted residuals for a narrow line superimposed on the data. (These predicted residuals are similar in shape but removed in energy from the residuals observed from in-flight Fe^{55} calibration sources with line energy 5.9 keV). The second part of the top plot demonstrates that, on smaller time scales, there may be structure within the line profile.

The continuum spectra for these sources do not resemble either optically thin bremsstrahlung or blackbody emission except over limited energy ranges. In fact they all can be characterized as a power law continuum with a cutoff at high energies(16,17). The effective X-ray temperatures are > 10^8 K so substantial emission from iron ions in collisional equilibrium at this temperature is not expected. The width of the lines also argues against an equilibrium thermal origin. Iron lines from supernova remnants(18) and clusters of galaxies(19) appear in the same detectors to be intrinsically narrow (FWHM << 1.1 keV). The continua from these last two classes, in contrast to binaries, can be characterized as optically thin thermal bremsstrahlung.

Fluorescence, however, is a non-thermal mechanism which can produce iron line emission with large efficiency(20,21). Fluorescence emission requires a target containing iron nuclei which are not fully ionized, and a source of high energy ($E > E_{THRESHOLD}$) photons. These are present in binary systems. Hatchett and Weaver(21) and Basko(22) conclude that the fluorescence intensity is almost directly proportional to the solid angle, Ω, subtended by the target at the X-ray source. In addition, Basko(22) shows that the fluorescence intensity reflected from a cold atmosphere scales with the integral of the source spectrum above threshold, weighted by E^{-3} near threshold. We can write

$$N_{Line} = \frac{\varepsilon\Omega}{4\pi} \int_{E_{TH}}^{20} \frac{dN}{dE} \left(\frac{E_{TH}}{E}\right)^3 dE$$

for the number of line photons, where ε is an undetermined efficiency factor ($\varepsilon \leq \omega_k = 0.34$, the fluorescence yield of iron), and $E_{TH} = 7.1$ keV. The observed quantities, $\varepsilon\Omega/4\pi$, which are measures of the line production efficiency and solid angle subtended by the target in the fluorescence model, are listed in Table 2 for hard spectrum binaries. The lower limits to the line intensities presented in this and the following Tables are typically 0.8 - 0.9 of the values listed. Upper limits are in some cases proportionally larger.

Figure 2 illustrates an attempt to present a distinction between iron line emission and $\varepsilon\Omega/4\pi$ for hard spectrum binaries on one hand, and the two classes of "thermal" sources mentioned above, on the other. The quantity $\varepsilon\Omega/4\pi$, although it does not have the same meaning as before, is uniquely determined by the temperature (above $\sim$ 3 keV) and the iron abundance in the thermal equilibrium model. We use the model calculations of Raymond and Smith(23). The temperatures of the thermal sources are independently determined by their X-ray continua. This allows

TABLE 2

IRON LINE EMISSION FROM BINARY PULSARS

SOURCE	AVERAGE EQUIVALENT WIDTH (eV)	LINE PHOTONS $CM^{-2}SEC^{-1}$	$\frac{\epsilon\Omega}{4\pi}$
CEN X-3	200	.016	.12
4U1223-62 (GX301-2)	350	.008	.18
4U1626-67	250	.001	.11
HER X-1 (HIGH)	300	.0067	.18
HER X-1 (LOW)	650	.00041	.27
4U0900-40 (VELA X-1)	700	.015	.24

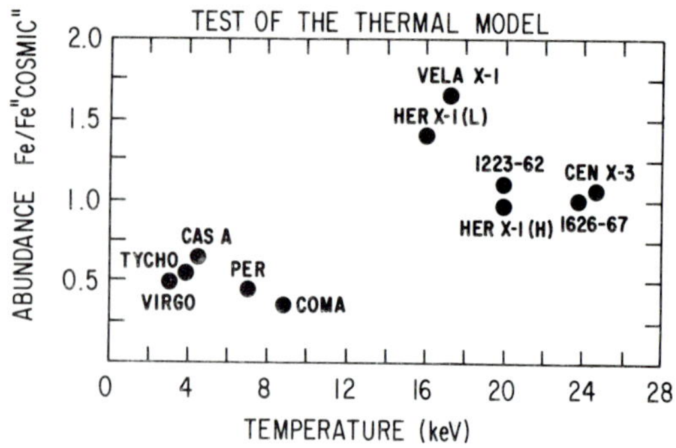

Fig. 2. The implied iron abundance (1) if the source spectra are assumed to be optically thin thermal bremsstrahlung (23) at an effective temperature.

a calculation of the iron abundance relative to that assumed in the model. For the thermal sources, the relative abundance is $\sim\frac{1}{2}$. For the binary sources, the effective temperature is determined from the value of $\epsilon\Omega/4\pi$. In this case the iron abundance is systematically a factor of $\geq$ 2 larger. The iron abundances can be reconciled at the lower value by halving the effective temperatures of the binaries. However these lower temperatures are inconsistent with the hard spectra.

We come to the following conclusions from this analysis. First, the amount of line emission and high energy photons in the spectra of binary sources is not inconsistent with these quantities in the thermal model. The relative iron abundance must, however, be increased from the value determined in the spectra of galactic supernova remnants and clusters of galaxies. There is an alternate explanation for this difference, which is consistent with the non-thermal power law spectra. That is, the line production is not thermal, but fluorescence. The solid angle fractions implied for $\epsilon \leq .34$ are $\Omega/4\pi \gtrsim 0.3$. These fractions generally far exceed those calculated for the binary companion and imply a fluorescence target which is closer to the neutron star. The example of Her X-1 is discussed in the next section.

Hercules X-1

Hercules X-1 has a $\sim$ 35 day high-low cycle as well as a binary and pulsational period(4). Iron line emission has been detected in both the high and low states, but with different characteristics. These observations are discussed in detail by the author(13,24) and will be summarized here.

An iron line feature was discovered in the Her X-1 high state spectrum with OSO-8. The feature is broad (FWHM $\sim$ 2 keV) and appears, at times, to be separated double lines. The second plot in Figure 1 illustrates one example of this double structure, while the top plot shows the time-averaged line profile. Both the line intensity and width can be explained by emission from a shell of material at the Her X-1 Alfven surface. The existence of this shell has been suggested to explain both the intense soft X-ray emission from the source (25,26) and the hard X-ray pulse shape(27).

Figure 3a illustrates spectral data from the Her X-1 high state. A line near 6.7 keV while difficult to see, is statistically very significant. Figure 3b is a demonstration of how Her X-1, itself, reveals the line for visual identification. During intensity dips, low energy continuum photons are photoelectrically absorbed

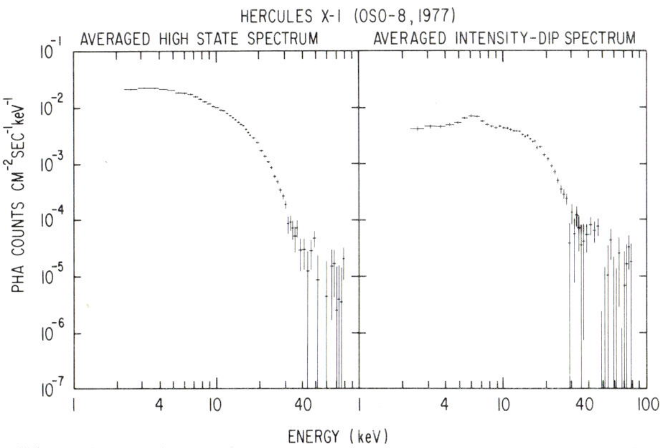

Fig. 3. High state (a) and intensity dip (b) PHA data from Her X-1.

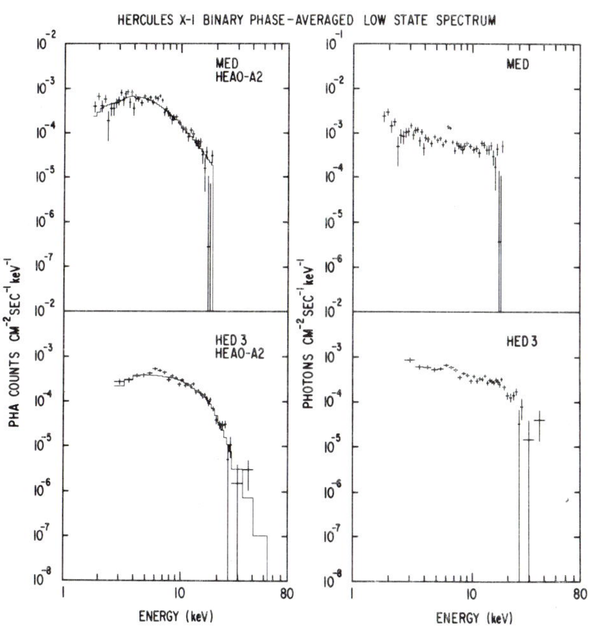

Fig. 4. Her X-1 low state spectra.

and Compton scattered by ambient material (28,29). The reduction in continuum photons near 6.7 keV serves to highlight the line emission, and also indicates that the line is formed outside of the continuum X-ray source, and is not affected as strongly by the intervening matter.

The Alfven shell can provide the necessary line luminosity if it subtends a significant fraction of the solid angle, $\Omega/4\pi \gtrsim 0.5$, around the X-ray source. Two models have been presented to explain the intrinsic width of the line. Ross, Weaver, and McCray (30) showed that a line propagating through a shell with 6 Thomson scattering depths will be broadened to the extent which is observed. Alternatively, in ref. 13 it is suggested that Doppler shifts from a co-rotating subrelativistic ($\beta \sim 0.13$) shell can explain the line width. The latter model can also account for the double line structure, since a uniformly-emitting, rotating ring of material will produce a line profile with peaks at the blue-shifted and red-shifted line energy and a relative deficiency at the unshifted energy.

Her X-1 is a relatively weak X-ray source in the low state with intensity $\lesssim$ 0.3 high state intensity (31,32). The low level or absence of pulsations in this flux(29,33) is interpreted as evidence that much of the low state emission is reflected into the line of sight from ambient material in the system, and not viewed directly from the pulsar. Reference 20 includes the prediction that this reflected flux could be observed and would contain a significant iron K fluorescent line from the surface of the normal star companion, HZ Herculis. The intensity of the reflected line, while too low to be identified in the high state spectrum, can be detected in the low state.

The HEAO 1 A2 experiment pointed at Her X-1 during both the X-ray high and low states(24). The high state observation confirmed the existence, intensity, and intrinsic width of the iron line. Only days later in the low state, the line detected with the same instrument differed. First, the line centroid shifted from 6.8 keV to 6.4 keV. In addition, the low state line has about twice the equivalent width and is intrinsically more narrow (FWHM < 550 eV). Finally, there appears to be a line intensity excess near mid-binary phase relative to early binary phase. Figure 4 illustrates low state spectra.

The ionization state of the fluorescing iron must be less than FeXX so that the line energy is 6.4 ± 0.1 keV(3). This implies an equilibrium temperature $\lesssim 10^7$ K (34,35). Several sources of fluorescence emission in this system have been suggested including the surface of HZ Her and a hot coronal gas either around HZ Her or around the accretion disk(20,22,33). A signature of emission from HZ Her is a marked binary phase dependence, with a peak in intensity at phase 0.5 and essentially no flux at phase < 0.2 (20,21). However, significant line emission has been observed at phase 0.17. This early phase line emission must be attributed to sources other than the HZ Her surface. An excess in the emission near phase 0.5 of ∿ 2 x 10^{-4} photons cm^{-2} sec^{-1} could come from HZ Her. This would correspond to an equivalent width of ∿ 10 eV in the high state spectrum in good agreement with the estimate in ref. 17. We point out, however, that the accompanying continuum component from HZ Her is not detected(20,36). The early phase line emission and the continua at all phases (see Fig. 4) indicate that a hot coronal gas is an important source of emission. The absence of significant low energy absorption or absorption edges in these continua implies an equilibrium temperature $\gtrsim 10^7$ K, which together with the upper limit, above, largely constrains the reflecting plasma temperature to be near 10^7 K.

The low state line emission is only a minor component of the high state emission. It would contribute, on the average, only ∿ 15 eV of the total 300 eV. Since the high state line is absent during the low state, it originates in material near the neutron star. The low state line, however, may be present throughout the high-low cycle. The value of $\varepsilon\Omega/4\pi$ in Table 2 for the low state line is overestimated because most (∿ 97%) of the continuum photons are shielded from the field of view. A corrected value implies that the solid angle subtended is on the order of that subtended by HZ Her, indicating that this fluorescing region is farther removed from Her X-1.

COMPLEX SPECTRA BINARIES

Cygnus X-3 and Circinus X-1 are the well known members of this category. Both sources exhibit spectral variability (7,37) which can be characterized as a change in the relative strengths of two continuum components. Detailed spectroscopy of Cir X-1 including the detection of iron line emission has been performed at present only during a minor X-ray outburst (X-ray intensity ∿ 1/6 Crab intensity--see Fig. 5), and will be discussed below. The following section deals with Cygnus X-3. A general remark about the iron line emission from these sources is that the quantity $\varepsilon\Omega/4\pi$ is a factor of 2-6 larger than the same quantity measured for the hard spectrum binaries (see Table 3); and for $\varepsilon \leq .34$, $\Omega/4\pi$ is as large as 2.

TABLE 3

IRON LINE EMISSION FROM COMPLEX SPECTRA BINARIES

SOURCE	AVERAGE EQUIVALENT WIDTH (eV)	LINE PHOTONS $CM^{-2}SEC^{-1}$	$\frac{\varepsilon\Omega}{4\pi}$
CIR X-1 (HIGH)	160	.0050	.33
CIR X-1 (LOW)	840	.0033	.59
CYG X-3 (HIGH, ϕ=0)	800	.0240	.67
CYG X-3 (LOW, ϕ=0)	1300	.0180	.75

Fluorescence, alone, can not account for the observed line intensity in this case. An additional source of thermal line emission would increase the effective ε if the temperature were ∿ 3 keV since the equivalent width of iron emission reaches a maximum at this temperature (23). We present evidence for multi-component emission.

The two continuum components in the Cir X-1 spectrum can be described as a low temperature (∿ 1 keV) blackbody and a power law extending to high energies. This description is of course

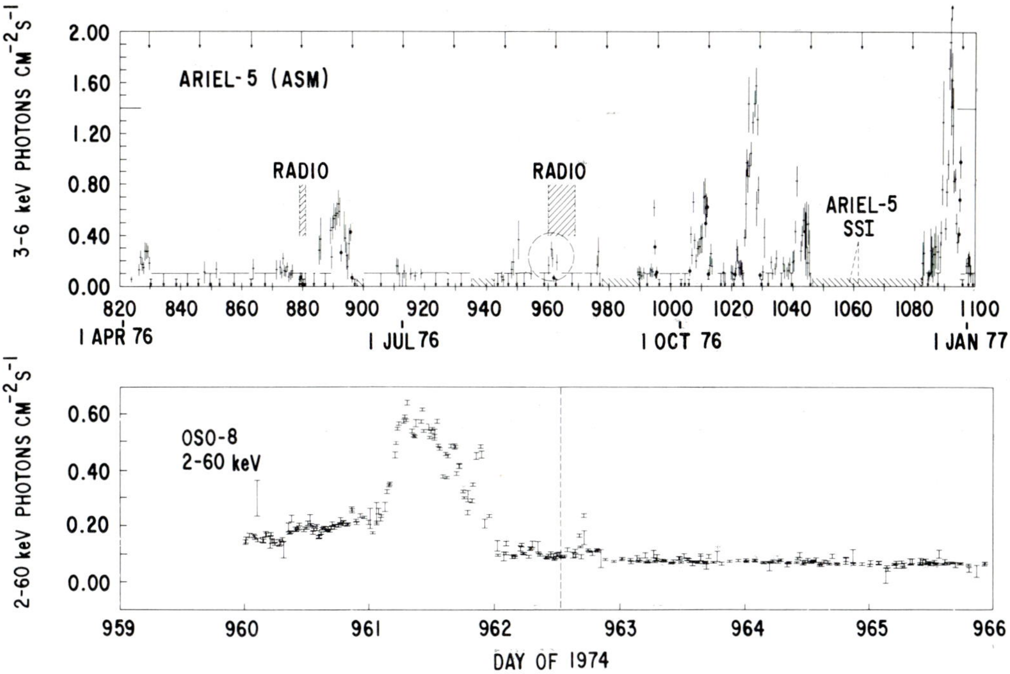

Fig. 5. X-ray light curve of Cir X-1. The text discusses spectroscopy during the outburst circled in the top plot and blown up in the bottom plot.

not unique, as a Wien spectrum or a high temperature thermal continuum can respectively replace the above components over the limited energy ranges. During the X-ray outburst reported in ref. 37, the blackbody source is present and then gradually declines until only the power law source remains. At the same time, the centroid energy of the iron line increases from below 6.4 keV to 6.9 keV. There is evidence that the lowest line energy is less than that expected from fluorescence of neutral iron. It is possible that the presence of the blackbody source is an indicator that the X-rays are probing more deeply toward the compact object. If this is the case, the line emission could arise from a region near enough to the compact object for a gravitational redshift of $\sim$ 0.1. We estimate that for spherically symmetric accretion, it may be possible to obtain the observed line luminosities without so much Thomson scattering that the line feature is destroyed. Thus a line formed in a hot gas ($T \sim 10^8$ K) in this region would be redshifted from an energy of 6.9 keV to an energy below 6.4 keV. Spherically symmetric accretion could occur on a neutron star without a magnetic field strong enough to channel the accreting matter onto 'hot spots'(38). Also, a weak magnetic field ($< 10^{11}$ gauss) insures that the iron line energy is not shifted(39) or radically changed(40) by the Zeeman effect.

CYGNUS X-3

The iron line of Cygnus X-3 is the "Sco X-1" of iron lines. The iron line, itself, is a 20 UHURU count source with a luminosity of 4×10^{36} erg sec^{-1} (distance 10 Kpc). This fact has made possible the detection of the line by two types of X-ray instruments (6,7) as well as a number of experimentors(41,42). Cyg X-3 has a 4.8 hour nearly sinusoidal intensity variation(43) which is intrepreted as a binary period(44,45). Models for this binary system include substantial scattering of X-rays from the compact object to create the observed smooth light curve (46,47).

The intensity and spectrum(7,48) are known to vary substantially over long time scales (∿ months). Figure 6 shows the Cyg X-3 light curve for over three years of data obtained with the Ariel 5 All Sky Monitor. Observations with HEAO 1 occurred during the relatively high intensity state (2-6 keV) shown in Fig. 6. Line emission was present at about the same intensity as in lower states. A re-analysis of the spectrum reported in ref. 7, using improved knowledge of the detector response, shows line emission is present with an equivalent width ∿ 1/5 of that in lower states but with about the same intensity. The 2-6 keV intensity at that time was twice as large as the intensity during the HEAO 1 observation. This demonstrates that iron line emission is present from this source at nearly the same intensity (within a factor of 2) regardless of the intensity-spectral state. Figure 7 illustrates the spectrum (PHA data) during two intensity states.

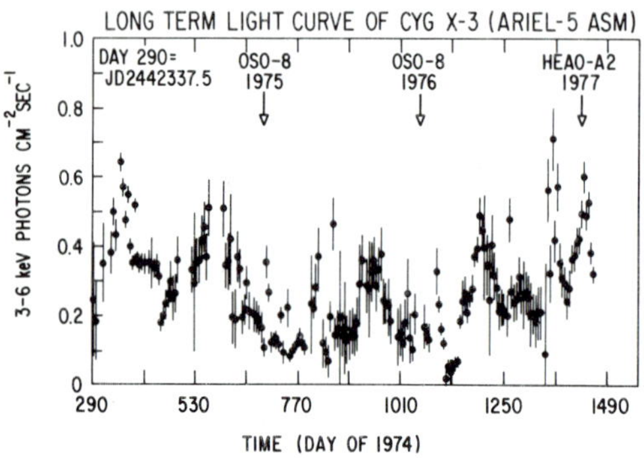

Fig. 6. X-ray light curve of Cyg X-3.

Recent analysis of OSO-8 data by Becker et al(49) has detected 4.8 hour periodic changes in the continuum spectrum and in the centroid energy of the line emission. These low state observations now support models (46) which predict continuum changes. Figure 8 illustrates that the line energy shifts toward 6.4 keV at phase 0.5 and toward 6.7 keV at phase 0. The line feature however has nearly constant equivalent width. It can be characterized as either a single broad feature or two lines with variable relative intensities. During a similar intensity state, the Bragg spectrometer(6) on OSO-8 observed a line at 6.8 keV with an equivalent width ∿ ½ that measured by the proportional counter. This is further evidence that the line feature is complex--i.e. the spectrometer detected only that fraction of the emission.

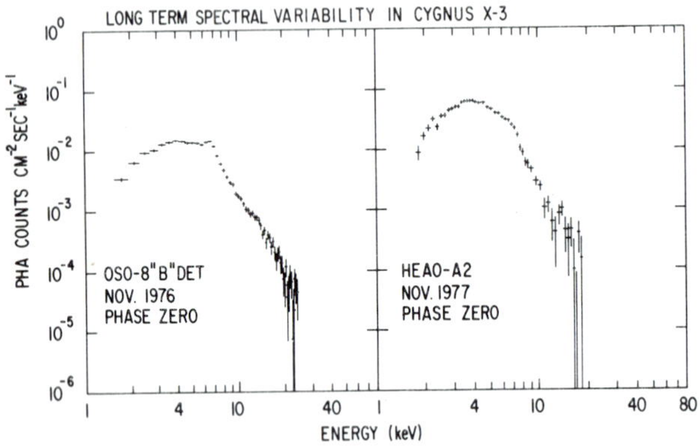

Fig. 7. Cyg X-3 high and low state PHA data.

In ref. 49 a two-component model is discussed. A 6.7 keV line with nearly constant 4.8 hour intensity originates in a volume on the order of the system size from an ∿ 3 keV plasma. There is, in addition, a 6.4 keV fluorescence line, with an intensity which is proportional to the number of high energy photons. At phase 0.5 the centroid energy is thus shifted toward 6.4 keV.

The fluorescing iron must be contained in a plasma which is cooler than 3 keV so that the line energy is 6.4 keV(3,34,35). The normal star companion is an unlikely candidate for the fluorescent target since models predict either an equivalent width which is too low(22,47) or a constant line intensity(20).

Alternatively, if a self-consistent model involving hot and cool plasma is not devised, the possibility of a gravitationally redshifted line from hot material should be considered (see above). A redshifted line would vary in intensity if

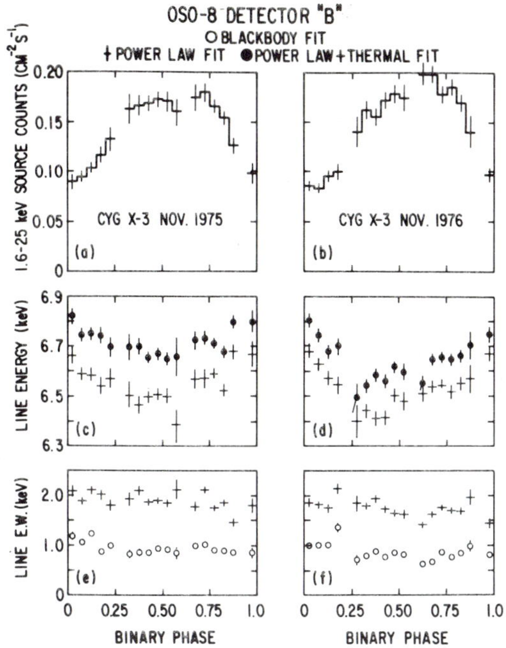

Fig. 8. The parameters of the Cyg X-3 iron line as a function of 4.8 hour phase.

the compact object is obscured during the binary orbit. We emphasize that the high value of $\epsilon\Omega/4\pi$ implies multi-component emission.

TRANSIENTS

HEAO 1 has detected iron line emission in the spectra of the two transient sources listed in Table 4. The Norma transient(50) has a continuum spectrum similar to that of Sco X-1 and to that of galactic bulge sources from which iron emission has also been reported(51, 52) (see a following section). If this source is identified with the Norma burster(53), this demonstrates a similarity in iron abundance between a burst source and other binaries. The line feature is again quite broad (FWHM $\gtrsim$ 1.5 keV).

The transient, 4U0115+63, was detected (54,55) in an X-ray high state during early 1978. At that time it was found to be an X-ray pulsar, and periodic Doppler shifts in the pulse period revealed its binary nature(56). The X-ray spectrum(57) is flat below 10 keV and falls off exponentially above this energy. In addition there is broad iron line emission. Thus, qualitatively, the spectrum of this transient is identical to that of the binary pulsars discussed earlier, and in this respect the transient behavior seems incidental. We expect the X-ray source emitting region to be near the surface of the neutron star and the iron line emitting region to be cooler material, perhaps in an Alfven shell. The value of $\epsilon\Omega/4\pi$ for this source is similar to that of the other pulsars. A possible fluorescence target region could then subtend about the same solid angle in this source as in the others.

The spectra of other transients has not revealed line emission. Both A0620-00(58) and Nova Ophiuchi(59) have upper limits to an iron line equivalent width below the detections listed in Table 4. This has been interpreted as an indication that a large scattering depth is present in the source which broadens any line emission beyond detectability(58). We note the interesting correlation thus far between iron line emission in transients and recurrence(60,61) of X-ray outbursts(59). This could indicate a correspondence between the Thomson scattering optical depth during an X-ray high state and the time scale between high states.

TABLE 4

IRON LINE EMISSION FROM TRANSIENTS

SOURCE	AVERAGE EQUIVALENT WIDTH (eV)	LINE PHOTONS $CM^{-2}SEC^{-1}$	$\frac{\epsilon\Omega}{4\pi}$
NORMA TRANSIENT (4U1608-52)	275	.044	.21
4U0115+63	880	.054	.35

CATACLYSMIC VARIABLES

The white dwarf binary systems, Am Her and SS Cygni, have been identified as soft X-ray(62) and hard X-ray(63,64) sources. Recently Kα iron line emission has been detected in the spectrum of

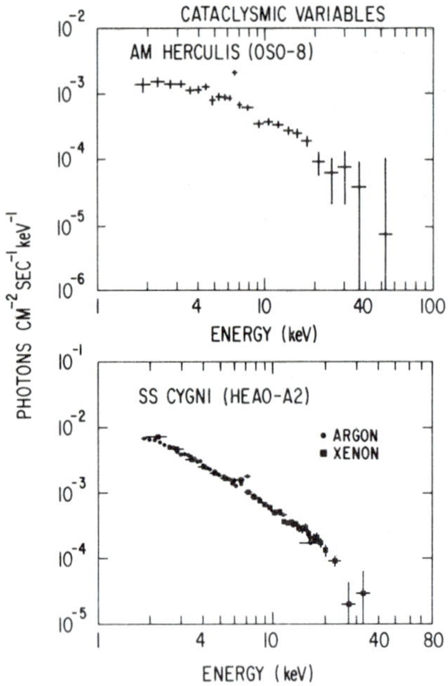

Fig. 9. Hard X-ray spectra of Am Her and SS Cygni

both(65) (see Fig. 9). In AM Her the very high temperature (> 65 keV) and the large value of $\varepsilon\Omega/4\pi$ (Table 5) suggest strongly that fluorescence is a line production mechanism. For $\varepsilon \lesssim \omega_k$, then $\Omega/4\pi \geq 1.4$. An additional thermal component of line emission, an overabundance of iron, or an underestimation of the number of high-energy continuum photons could explain this high value. The most likely explanations (i.e. without extreme contributions from afore-mentioned effects) imply that a large fraction of the solid angle is subtended by a fluorescence target. Candidates for the fluorescence target include a hot corona around the white dwarf or the white dwarf surface. Future observations of the binary phase dependence of the line emission (especially near eclipse) will help determine the distance between the source of the hard X-rays and the source of the fluorescence emission.

The continuum spectrum of SS Cygni, when line emission was observed, could be characterized as optically thin thermal bremsstrahlung with temperature $\lesssim$ 20 keV. At this time there is no evidence that the iron line is not of thermal origin although the implied iron abundance is a factor of $\sim$ 2 larger than that in other thermal sources (see a preceeding section). The SS Cygni X-ray temperature may significantly change on long time scales ($\sim$ months) (65). A test of the thermal model would be to search for correlations between the continuum temperature and the line equivalent width.

BINARIES WITH OPTICALLY THICK SPECTRA

Iron line emission has been reported in both Sco X-1 type sources(51) and in galactic bulge sources(52). We feel that the large width of these features makes it difficult to obtain accurate line parameter estimates without a detailed knowledge of the continuum. For this reason, we shall return to a discussion of these sources in the future, upon completion of a program of modelling the effects of many Thomson scatterings on both the continuum spectrum and the line profile.

SUMMARY AND THE FUTURE

We have attempted to present an informative, if not exhaustive, review of the observations of X-ray line emission. The capabilities of present day detectors in establishing physically interesting facts about X-ray sources is demonstrated. High resolution observations of iron lines in the post-HEAO age will initiate another quantum leap forward in understanding. In the near future, HEAO B will perform high resolution observations of lower energy (< 4 keV) lines. If, however, fluorescence is the dominant source of emission in binary sources, then the lines of lower Z elements may be difficult to detect in binaries (see Table 1).

TABLE 5

IRON LINE EMISSION FROM CATACLYSMIC VARIABLES

SOURCE	AVERAGE EQUIVALENT WIDTH (eV)	LINE PHOTONS CM^{-2}SEC^{-1}	$\frac{\epsilon\Omega}{4\pi}$
AM HER	940	.00072	.46
SS CYGNI	550	.00044	.22

At the present time, observations with the HEAO-1 A2 experiment are extending the

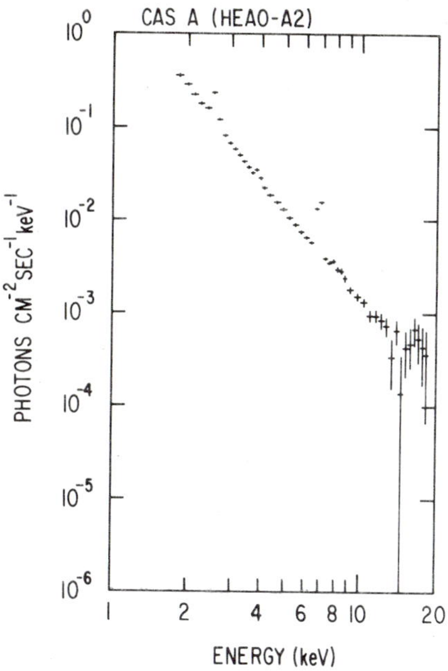

Fig. 10. The Cas A X-ray spectrum.

knowledge of X-ray line emission with the detections of lines due to other atomic transitions of iron(66), as well as lines due to other elements. Figure 10 shows the incident spectrum of Cas A. The first iron line from a supernova remnant(18,67) and now the first sulphur line (∿ 0.026 photons cm^{-2} sec^{-1}) have been detected from this source. These observations hold promise that, at least for X-ray sources with thermal continua, HEAO will greatly expand our knowledge of X-ray line emission.

ACKNOWLEDGMENTS

We thank the scientists in the X-ray astronomy group at the GSFC Laboratory for High Energy Astrophysics for interesting discussions and for their assistance in this review. We particularly acknowledge J. Robinson-Saba, Drs. R. Becker, J. Swank, L. Kaluzienski, L. A. Rose, D. Worrall, and S. Holt for permission to discuss their results prior to publication. We also thank S. Shrader, F. Shaffer, and L. White for extraordinary technical assistance. Finally we thank Dr. B. Smith and Dr. R. Weaver for useful discussions.

REFERENCES

1. Allen, C. W. 1973, Astrophysical Quantities (3rd ed. London: Athlone).

2. Bambynek, W., Craseman, B., Fink, R. W., Freund, H. V., Mark, H., Swift, C. D., Price, R. E., and Rao, P. V. 1972, Rev Mod Phys 44, 716.

3. House, L. L. 1969, Ap. J. (Supple) 18, 21.

4. Tananbaum, H., Gursky, H., Kellog, E. M., Levinson, R., Schreier, E., and Giacconi, R. 1972, Ap. J. (Lett) 174, L143.

5. Schreier, E., Levinson, R., Gursky, H., Kellogg, E., Tananbaum, H., and Giacconi, R. 1972, Ap. J. (Lett) 172, L79.

6. Kestenbaum, H. L., Long, K. S., Novick, R., Weisskopf, M. C., and Wolff, R. S. 1977, Ap. J. (Lett) 216, L19.

7. Serlemitsos, P. J., Boldt, E. A., Holt, S. S. Rothschild, R. E., and Saba, J.L.R. 1975, Ap. J. (Lett) 201, L9.

8. Pravdo, S. H., Becker, R. H., Boldt, E. A., Holt, S. S., Serlemitsos, P. J., and Swank, J. H. 1978, Ap. J 223, in press.

9. Lampton, M., Margon, B. and Bowyer, S. 1976, Ap. J. 208, 177.

10. Avni, Y. 1976, Ap. J 209, 16.

11. Ulmer, M. P. 1975, Ap. J. 196, 827.

12. Rappaport, S., Markert, T., Li, F. K., Clark, G. W., Jernigan, J. G. and McClintock, J. E. 1977, Ap. J. (Lett) 217, L29.

13. Pravdo, S. H., Becker, R. H., Boldt, E. A., Holt, S. S., Serlemitsos, P. J., and Swank, J. H. 1977, Ap. J. 215, L61.

14. Swank, J. H., private communication.

15. Becker, R. H., Rothschild, R. E., Boldt, E. A., Holt, S. S., Pravdo, S. H., Serlemitsos, P. J. and Swank, J. H. 1978, Ap. J. 221, 912.

16. Holt, S. S., Boldt, E. A., Rothschild, R. E., Saba, J. R., and Serlemitsos, P. J. 1974, Ap. J. 190, L109.

17. Becker, R. H., Boldt, E. A., Holt, S. S., Pravdo, S. H., Rothschild, R. E., Serlemitsos, P. J., Smith, B. W., and Swank, J. H. 1977, Ap. J. 214, 879.

18. Pravdo, S. H., Becker, R. H., Boldt, E. A. Holt, S. S., Rothschild, R. E., Serlemitsos, P. J., and Swank, J. H. 1976, Ap. J. (Lett) 206, L41.

19. Serlemitsos, P. J., Smith, B. W., Boldt, E. A., Holt, S. S., and Swank, J. H. 1976, Ap. J. (Lett) 211, L63.

20. Basko, M. M., Sunyaev, R. A., and Titarchuk, L. L. 1974, Astr. and Ap. 31, 249.

21. Hatchett, S. and Weaver, R. 1977, Ap. J. 215, 285.

22. Basko, M. M. 1978, preprint.

23. Raymond, J. and Smith, B. W. 1977, Ap. J. Supple 35, 419.

24. Pravdo, S. H., Boldt, E. A., Holt, S. S., Rothschild, R. E., and Serlemitsos, P. J. 1978, submitted to Ap. J. (Lett).

25. McCray, R. and Lamb, F. K. 1976, Ap. J. (Lett) 204, L115.

26. Basko, M. M. and Sunyaev, R. A. 1976, M.N.R.A.S. 175, 395.

27. Basko, M. M. and Sunyaev, R. A. 1976, Astr Zh 53, 950 (transl. Sov Astr 20, 537).

28. Giacconi, R., Gursky, H., Kellogg, E., Levinson, R., Schreier, E. and Tananbaum, H. 1973, Ap. J. 184, 227.

29. Becker, R. H., Boldt, E. A., Holt, S. S., Pravdo, S. H., Rothschild, R. E., Serlemitsos, P. J., Smith, B. W. and Swank, J. H. 1977, Ap. J. 214, 879.

30. Ross, R. R., Weaver, R., and McCray, R. 1978, Ap. J. 219, 292.

31. Fabian, A. C., Pringle, J. E., and Rees, M. J. 1973, Nature 244, 212.

32. Cooke, B. A. and Page, C. G. 1975, Nature 256, 712.

33. Jones, C., and Forman, W. 1976, Ap. J. (Lett) 209, L131.

34. Jordan, C. 1970, M.N.R.A.S. 148, 17.

35. Raymond, J. 1978, private communication.

36. Felsteiner, J. and Opher, R. 1976, Astr and Ap 26, 189.

37. Saba, J. R., Boldt, E. A., Holt, S. S., Serlemitsos, P. J. and Swank, J. H. 1977, Bull. of AAS 9, 297.

38. Lamb, F. K., Pethick, C. J., and Pines, D. 1973, Ap. J. 184, 271.

39. Sarazin, C. L. and Bahcall, J. N. 1977, Ap. J. (Lett) 216, L67.

40. Rau, A.R.P., Mueller, R. O. and Spruch, L. 1975, Phys Rev A 11, 1865.

41. Sanford, P. W., Mason, K. O., and Ives, J. 1975, M.N.R.A.S. 173, 9.

42. Parsignault, D. R., Grindlay, J., Gursky, H. and Tucker, W. 1977, Ap. J. 218, 232.

43. Parsignault, D. R., Gursky, H., Kellogg, E. M., Matilsky, T., Murray, S., Schreier, E., Tananbaum, H. and Giacconi, R. 1972, Nat Phys Sci 239, 123.

44. Pringle, J. E. 1974, Nature 247, 21.

45. Parsignault, D. R., Schreier, E., Grindlay, J. and Gursky, H. 1976, Ap J. (Lett) 209, L73.

46. Davidson, A. and Ostriker, J. P. 1974, Ap. J. 189, 331.

47. Milgrom, M. 1976, Astr and Ap 51, 215.

48. Leach, R. W., Murray, S. S., Schreier, E. J., Tananbaum, H. D., Ulmer, M.P., and Parsignault, D. R. 1975, Ap. J. 199, 184.

49. Becker, R. H., Robinson-Saba, J. L., Boldt, E. A., Holt, S. S., Pravdo, S. H., Serlemitsos, P. J., and Swank, J. H. 1978, submitted to Ap. J. (Lett).

50. Kaluzienski, L. J., Holt, S. S., Boldt, E. A., Rothschild, R. E., and Serlemitsos, P. J. 1978, Bull. of the APS 23, 581.

51. Swank, J. H. 1977, Bull. of the APS 22, 588.

52. Parsignault, D. R. and Grindlay, J. 1978, Ap. J., in press.

53. Grindlay, J. and Gursky, H. 1976, Ap. J. (Lett) 209, L61.

54. Holt, S. and Kaluzienski, L. 1978, IAU Circ No. 3161.

55. Cominsky, L., Clark, G. W., Li, F., Mayer, W., and Rappaport, S. 1978, IAU Circ No. 3161.

56. Rappaport, S., Clark, G., Cominsky, L. and Li, F. 1978, IAU Circ No. 3171.

57. Rose, L. A., Pravdo, S. H., Kaluzienski, L. J., Marshall, F. E., Holt, S. S., Boldt, E. A., Rothschild, R. E. and Serlemitsos, P. J. 1978, preprint.

58. Griffiths, R. E., Ricketts, M. J., and Cooke, B. A. 1977, M.N.R.A.S. 177, 429.

59. Kaluzienski, L. J., and Swank, J. H. 1978, private communication.

60. Fabbiano, G., Bradt, H. V., Doxsey, R. E., Gursky, H., Schwartz, D. A., and Schwartz, J. 1978, Ap. J. (Lett) 221, L49.

61. Forman, W., Jones, C., and Tananbaum, H. 1976, Ap. J. (Lett) 206, L29.

62. Hearn, D. R., Richardson, J. A., and Clark, G. W. 1976a, Ap. J. (Lett) 210, L23.

63. Berg, R. A. and Duthie, J. G. 1977, Ap. J. 211, 859.

64. Heise, J., Mewe, R., Brinkman, A. C., Gronenschild, E.H.B.M., den Boggende, A.J.F., Schrijver, J., Parsignault, D. R., and Grindlay, J. E. 1978, Astr and Ap. 63, L1.

65. Swank, J. H., Boldt, E. A., Holt, S. S., Pravdo, S. H., Rothschild, R. E., and Serlemitsos, P. J. 1978, Bull. of the APS 23, 581.

66. Lea, S., Mason, K., Reichert, G., and Riegler, G. 1978, this conference.

67. Davison, P.J.N., Culhane, J. L., and Mitchell, R. F. 1976, Ap. J. (Lett) 206, L37.

BRAGG CRYSTAL SPECTROSCOPY ON THE OSO 8 SATELLITE

K. S. Long, G. A. Chanan, D. J. Helfand, W. H.-M. Ku and R. Novick

Columbia Astrophysics Laboratory, Departments of Astronomy and Physics, Columbia University, New York, New York 10027, U.S.A.

INTRODUCTION

The Columbia University Bragg crystal spectrometer aboard *OSO 8* has been used to obtain high resolution spectra from a variety of cosmic X-ray sources since orbit was achieved in June 1975. The *OSO 8* instrument has been described by Kestenbaum *et al.* [1]. It is a graphite spectrometer having a resolution of 10 eV at 2 keV and 285 eV at 6.7 keV. It can be used to obtain the spectra of X-ray sources between 2 and 10 keV, although the typical exposure for an observation of one day's duration is rather modest, ~400 cm^2 s. A typical source is observed from 1 to 15 days. For the strongest sources, the spectrometer has sufficient sensitivity to obtain detailed continuum spectra, while for a much larger group of sources, it permits one to place meaningful upper limits, typically 0.02 photons $(cm^2\ s\ keV)^{-1}$ on the strength of narrow line emission. Positive evidence for iron line emission near 6.7 keV has been obtained for several sources including Cyg X-3. This evidence, significant at the 2.5 σ level, represents the first detection of iron line emission with a dispersive spectrometer.

OBSERVATIONAL RESULTS

An example of the continuum spectra which can be obtained with *OSO 8*, a spectrum of Sco X-1 obtained in 1976, is shown in Fig. 1. The systematic errors in the inversion from observed counts to flux are estimated to be <10%. The spectrum is relatively featureless. The best fit bremsstrahlung spectrum (a) has a temperature of 7.2 keV but is a poor fit; χ^2 = 560 for 271 degrees of freedom. An improved fit (b), χ^2 = 398, is obtained by modeling the Compton evolution of a bremsstrahlung spectrum in a hot isothermal plasma, as discussed by Long and Kestenbaum [2]. The characteristic temperature is ~1.8 keV, lower

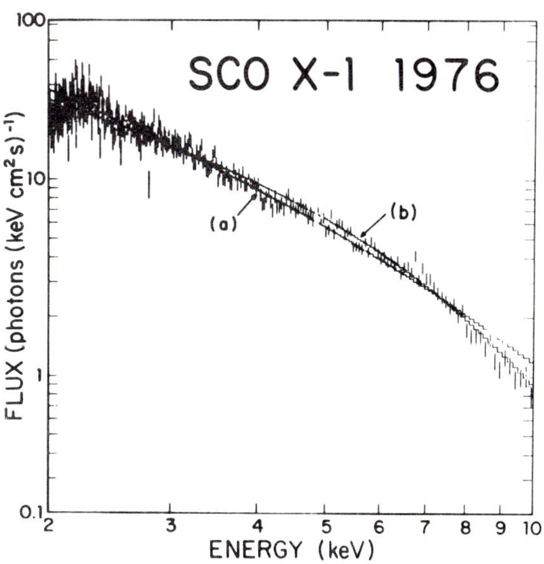

Fig. 1. The spectrum of Scorpius X-1.

than for a simple bremsstrahlung fit, and the number of electron scatters required is of order 100. Similar results were obtained in an analysis of another observation in 1975 where χ^2 with 267 degrees of freedom improved from 397 to 354. Since photons must diffuse out of the X-ray plasma which constitutes Sco X-1, 100 electron scatters are roughly equivalent to an optical depth of 10 to the center of the source. In an optically thin plasma with a temperature of 5–9 keV, line emission would be anticipated from the helium-like triplet of Fe XXV with an equivalent width of 400–500 eV. The apparent strength of the iron line feature is 22.0 ± 9.3 eV when data from both observations of the *OSO 8* spectrometer are combined, consistent with an electron scattering depth of 10.

The data collected from observations of more than 30 galactic sources including Sco X-1 have been searched for evidence of narrow line emission. Results of the search for narrow line emission due to iron near 6.7 keV are tabulated in Table 1. No line has been detected at the 3 σ level or greater, although at lesser statistical significance, features have been detected in the spectra of Cyg X-3 and Cen X-3, as well as Sco X-1. Typical 3 σ upper limits for iron line emission are of order 0.02 photons $(cm^2\ s)^{-1}$ for the better studied sources. Line limits for emission from other ions, including Si XIII, Si XIV, S XV, and S XVI, have also been obtained and will be presented elsewhere.

The source for which we have obtained the strongest evidence for iron line emission with the *OSO 8* Bragg crystal spectrometer is Cyg X-3 [3]. In observations in September 1975 and January 1976, the observed line flux was 0.020 ± 0.008 photons $(cm^2\ s)^{-1}$; in January 1977, it was 0.0097 ± 0.0043 photons $(cm^2\ s)^{-1}$. The line width is consistent with the instrumental resolution. The energy of the apparent flux enhancement is consistent with emission from moderately ionized iron but excludes both Fe XXV and Fe XXVI. The equivalent width of the feature in 1977 was 220 ± 100 eV, somewhat smaller than the 420 ± 170 eV width of the earlier observations.

Using proportional counters aboard *OSO 8* and, more recently, *HEAO 1*, the X-ray group at the Goddard Space Flight Center (GSFC) has obtained definitive evidence for iron line emission in a wide variety of astrophysical objects [4,5]. Some of the Columbia spectrometer and GSFC proportional counter results are compared in Table 2(a). The crystal spectrometer results for Cyg X-3 that we have reported here are consistent with the proportional counter data which indicate an average flux of ~0.02 photons $(cm^2\ s)^{-1}$. We at Columbia have also obtained evidence in Cen X-3 for an iron line feature with a measured intensity of 0.011 ± 0.004 photons $(cm^2\ s)^{-1}$. In general, the GSFC proportional counter measurements are more sensitive than the Columbia Bragg crystal spectrometer results. The spectrometer upper limits are consistent with the GSFC detections, with the possible exception of the spectrometer line limit for the Norma transient (about 40% below the GSFC detection). However, there are several large discrepancies in Table 2(b) between the Columbia upper limits on the iron line equivalent widths for a group of galactic bulge sources and the detection of iron lines in these sources reported by Parsignault and Grindlay [6,7], using the proportional counters aboard ANS. We note that these discrepancies are reduced if the emission lines are substantially broader than the spectrometer resolution of 285 eV.

This work was supported by the National Aeronautics and Space Administration under contract NAS5-22408.

TABLE 1 Three-Sigma Sensitivities to Narrow Line Emission from Iron (6.7 keV)

Source Name	Epoch (yr/days)	Exposure (s/1°)	3 σ Sensitivity [photons $(cm^2\ s)^{-1}$]
0531+21	5/286-288, 5/349-353	260	0.033
"	6/348-357	507	0.024
0535+26	6/347-355	468	0.026
0620-00	5/290-291	42	0.060
"	5/299	15	0.110
"	6/001-009	272	0.019
0900-40	7/138-154	642	0.013
1118-60	6/092-100	366	0.020
"	7/087-099	666	0.013
1516-56	6/111-114, 7/213	60	0.075
1543-47	6/115-116	123	0.025
1608-52	7/201-206	325	0.027
1617-15	5/190-195	170	0.134
"	6/150-155, 7/138-142	159	0.125
1636-53	6/117-131, 6/164-183, 7/247, 7/252-253	905	0.012
1642-45	6/132-133, 6/166-169	161	0.046
1656+35	5/214-215, 6/039-040, 6/161-162, 6/218	182	0.030
1658-48	6/132-133, 6/160-164, 6/170-171, 6/179-182, 7/191-211	320	0.033
1700-37	6/135-141, 6/154-169, 7/193-194	931	0.017
1702-36	6/136-141, 6/153-169, 7/192-193	863	0.015
1705-44	6/133-134, 6/159-170, 7/195-209	251	0.027
1728-16	5/192,196, 6/166-168, 7/159-161	115	0.036
1735-28	6/170-171, 7/169-189	394	0.028
1735-44	6/135-136, 6/152-154, 6/172-173, 6/182-184, 7/197-198, 7/206,210	199	0.028
1744-26	5/184-187, 6/171-172, 7/167-186	689	0.022
1746-20	5/190-195, 6/170-171	158	0.023
1758-20	5/182-195, 6/172-173, 7/164-166	166	0.028
1758-25	5/176-179, 5/184-188, 6/172-173, 7/167-184	634	0.022
1811-17	6/173-174, 7/164	65	0.067
1813-14	6/173-174, 7/162-163	70	0.055
1820-30	6/175-179, 6/183-184	236	0.023
1837+04	6/104, 6/174-177, 7/120, 7/129-130, 7/156	191	0.025
1908+00	5/208-211, 6/181-185, 6/223-224, 7/122-127	250	0.026
1956+35	6/191-197	281	0.023
"	7/111-112, 7/137-143	286	0.026
2030+40	5/220-221, 6/183-189	379	0.024
"	7/018-031	1054	0.013
2142+38	5/251-259, 6/265-267, 7/227-228	136	0.031
2321+58	5/223-224, 5/268-270, 6/281-289	665	0.017

TABLE 2 Iron Line Measurements

Source Name	Epoch (yr/days)	Columbia Experiment: 3 σ Sensitivity: Line Flux [photons (cm² s)⁻¹]	Columbia Experiment: 3 σ Sensitivity: Equiv Width (eV)	Columbia Experiment: Measured Line Intensity: Line Flux [photons (cm² s)⁻¹]	Columbia Experiment: Measured Line Intensity: Equiv Width (eV)	Columbia Experiment: Measured Line Intensity: Line Energy (keV)	GSFC [4,5] Measured Line Flux [photons (cm² s)⁻¹]	ANS [6] Measured Equiv Width (eV)
(a) Galactic X-Ray Sources								
Vela X-1 (0900-40)	7/138-154	0.013	...	...	...	...	0.015	...
Cen X-3 (1118-60)	6/092-100	0.020	...	...	...	...		...
	7/087-099	0.013	...	0.011	...	6.59	0.016	
Norma (1608-52)	7/201-206	0.027	150	...	...	...	0.044	...
Cyg X-3 (2030+40)	5/220-221, 6/183-189	0.024	500	0.020	420	6.84	0.024 (high)	...
	7/018-031	0.013	300	0.010	220	6.52	0.018 (low)	
Cas A (2321+58)	5/223-224, 5/268-270	0.017	5600	...	...	...	0.008	...
(b) Galactic Bulge Sources								
Sco X-1 (1617-15)	5/190-195	0.134	44	0.086	28.1	6.6	...	340–640
	6/150-155	0.125	36	0.058	16.9	6.6		
1642-45	1976	0.046	170	...	...	...	...	620
1702-36	1976	0.015	75	...	...	...	...	390
1744-26	1975,1976	0.022	55	...	...	...	...	570
1758-25	1975-1977	0.022	110	...	...	...	...	230–540
1811-17	1976,1977	0.067	250	...	...	...	...	540–860
1813-14	1976,1977	0.055	250	...	...	...	...	320–620
1820-30	1976	0.023	300	...	...	...	...	620–670
1837+04	1976	0.025	380	...	...	...	...	630
2142+38	1975,1976	0.031	400	...	...	...	...	170–760

REFERENCES

1. H. L. Kestenbaum, G. G. Cohen, K. S. Long, R. Novick, E. H. Silver, M. C. Weisskopf, and R. S. Wolff, *Astrophys. J.* 210, 805 (1976).

2. K. S. Long and H. L. Kestenbaum, *Astrophys. J.* 226 (Nov. 15, 1978), in press.

3. H. L. Kestenbaum, W. H.-M. Ku, K. S. Long, E. H. Silver, and R. Novick, *Astrophys. J.* 226 (Nov. 15, 1978), in press.

4. S. H. Pravdo, this volume.

5. S. H. Pravdo, R. H. Becker, E. A. Boldt, S. S. Holt, R. E. Rothschild, P. J. Serlemitsos, and J. H. Swank, *Astrophys. J. (Letters)* 206, L41 (1976).

6. D. R. Parsignault and J. E. Grindlay, *Astrophys. J.* (1978), in press.

7. D. R. Parsignault and J. E. Grindlay, this volume.

SEARCH FOR SI XIII AND SI XIV LINE EMISSION FROM COSMIC SOURCES

D. R. Parsignault and J. E. Grindlay

Harvard - Smithsonian Center for Astrophysics, Cambridge, Massachusetts 02138

THE BRAGG SPECTROMETER

The Astronomical Netherlands Satellite (ANS) was launched into polar orbit on 30 August 1974. One of the experiments, the Hard X-ray Experiment (HXX), contained a Bragg spectrometer which used two sets of PET crystals tuned to detect Si XIII and Si XIV line emission at ~1.85 and 2.0 keV. A detailed description of this instrument and of its operation, together with some preliminary results, can be found in references [1] and [2]. Table 1 gives parameters of the Bragg spectrometer relevant to the data analysis.

CALIBRATION OF THE SPECTROMETER

The in-orbit calibration of the complete Bragg spectrometer system, including the crystals, could only be performed using a narrow emission feature from a cosmic source, or using the continuum emission from a strong X-ray source. As discussed below, no discrete features were found in any of the observed sources. The only X-ray sources whose continua were strong enough to be used in calibration were Sco X-1 and the transient source A0620-00.

Table 2 lists the count rates detected by the two Bragg detectors during the Bragg scans of Sco X-1 and A0620-00, while Fig. 1 shows the plotted data. The horizontal lines correspond to the average backgrounds. The small bars perpendicular to these lines and those associated with each point represent ± 1σ uncertainties.

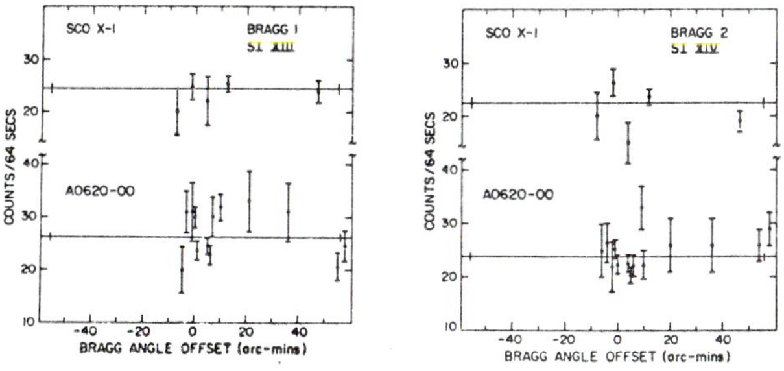

Fig. 1 Calibration data of the Bragg Crystal Spectrometer obtained from the X-ray continua of Sco X-1 and A0620-00.

These counting rates were then compared to the counting rates observed for similar satellite positions in orbit, while monitoring other weaker sources whose continua could not significantly contribute to the observed counting rates as seen by the Bragg detectors (listed as AVBKGD in Table 2). The differences, N_C, are the net contributions of the X-ray continua. We also monitored the background radiation environment, using the LAD (energy range 1.4 - 7.2 keV) at offsets $\geq$ 15 arcmin, and the so-called background register which records the number of events in the LAD and in the Bragg detectors which are rejected by the PSD, and the overscale events (E > 28 keV). The numbers in parentheses correspond to $\pm$ 1σ uncertainties.

Calculations of the expected number of counts due to the continua of Sco X-1 and A0620-00 were performed with the parameters listed in Table 2, and the calculated $F(E)_C$, the continuum counting rate per unit area per unit Bragg angle. The results are listed in Table 3, and we can see that there is good agreement between the expected and observed intensities, indicating that the Bragg spectrometer was operating properly.

EXPERIMENT RESULTS AND DISCUSSION

Our program of observations consisted of measuring the line emission of highly ionized silicon, the helium-like Si XIII, and the hydrogen-like Si XIV, from several different types of X-ray emitting objects. We observed three known binary systems (Cyg X-1, X-2, X-3), three strong galactic sources (GX5-1, GX17+2, 4U1957+11), three supernova remnants (Crab Nebula, Cas A, Tycho), and three clusters of galaxies (Perseus, Virgo, Coma). We also observed the X-ray Nova A0620-00, and Sco X-1 as mentioned above. The data on the latter object were not usable at the Bragg angles of interest because these particular observations were made in a high background radiation environment.

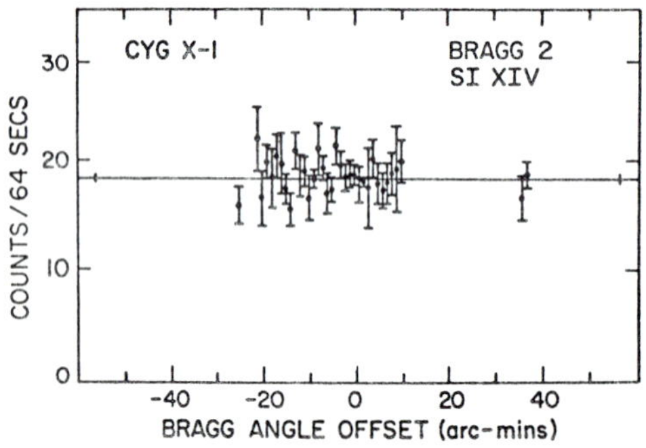

Fig. 2 Sample data of a scan of Cyg X-1 made with Bragg 2 crystal (Si XIV). A narrow line emission feature, if present, would appear at Bragg angle offset 0 $\pm$ 1 arcmin.

Figure 2 shows a typical sample of data obtained with the Bragg 2 crystal spectrometer (Si XIV) on the X-ray source Cyg X-1. As for all the rest of our data, no line feature is apparent at a confidence level > 99% (> 3σ). We therefore proceeded to calculate 3σ upper limits for a narrow line emission of Si XIII at λ = 6.649 A (1.865 keV), and λ = 6.684 A (1.855 keV), and Si XIV at λ = 6.184 A (2.00 keV). These upper limits are obtained by $N_{L,m} = 3\sqrt{\underline{n}\,\sigma_B^2 + \sigma_T^2}$ where σ_B is the uncertainty in the background over one arcmin interval, σ_T is the uncertainty in the line plus background, and $\underline{n}$ is the number of arcmin-channels over which a discrete line should extend. The resolution of the PET crystal is 1.2 eV (FWHM) which corresponds to 2.6 arcmin, and the pointing accuracy of the spacecraft is $\pm$ 0.5 arcmin. Thus, in order to maximize the sensitivity of the measurements for a point source, we summed the data in 3 arcmin bins around the critical Bragg angles (n = 3 for a point source). In the case of an extended object, n is equal to the size of the object in arcmin plus 3.

Table 4 summarizes the results obtained on the different X-ray sources at the observation periods indicated (column 2). Column 3 indicates the number of continuum photons at the given energy $F(E)_c$, while column 4 shows the total observing times at the given Bragg angles. Column 5 gives the 3σ upper limits, $N_{L,m}$, in units of photons per 64 seconds since the spectrometer PHA was read out once every 64 seconds. Finally, columns 6 and 7 list the calculated equivalent line widths in eV, and count-cm^{-2}-sec^{-1} corresponding to the $N_{L,m}$ values.

CONCLUSIONS

Our data have shown no positive detection, at a level of confidence > 99%, of Si XIII and Si XIV discrete line emission in the different types of cosmic objects observed. In the case of A0620-00, the very stringent upper limit of 1.0 eV seems to rule out any narrow line emission altogether. This upper limit is one of order of magnitude smaller than that reported by OSO-8 [3] and ~factor 4 smaller than the value obtained by Ariel 5 [4].

The most stringent upper limits for discrete line emission from Cyg X-1 were obtained from our May observations: 2.4 eV and 5.0 eV for the Si XIII ^1P and ^3P transitions respectively, and 3.8 eV for Si XIV. The absence of line emission would tend to confirm the accepted picture for this close binary system in which the X-ray emission at E $\lesssim$ 10 keV emanates from an optically thick disk [5].

In the case of Cyg X-2, calculations based on the Tucker-Koren model [6] for Si XIV line emission from a thin plasma at ~40 x 10^6 °K yield an equivalent line width of 14 eV, which is consistent with our 3σ upper limit of 26 eV for our two sets of observations.

Our results for the 3σ upper limits for Si XIII and Si XIV line emission from Cyg X-3 are somewhat more stringent than the OSO-8 upper limits [7].

Our present 3σ upper limits of 0.014 ph-cm^{-2}-sec^{-1} (Si XIV) and 0.026 ph-cm^{-2}-sec^{-1} (Si XIII) line emission are a factor ~2 lower than the values of ~0.025 and 0.040 ph-cm^{-2}-sec^{-1}, respectively, reported by OSO-8 for the case of unbroadened line emission [8]. One should note that the sensitivity of our measurements must be substantially reduced by the fact that the lines are expected to be broadened by the Doppler effect to mass motion of the plasma which has been estimated to be expanding at ~1400 km-sec^{-1} [9].

In the case of Cas A, we found a 3σ upper limit for Si XIV line emission of 0.016 ph-cm^{-2}-sec^{-1}. Our result is to be compared to a line intensity, I_L < 0.030 ph-cm^{-2}-sec^{-1} inferred by Ariel 5 Bragg spectrometer measurements [10] and is also a factor ~2.5 below the proportional counter results of Hill [11].

Our upper limit of 0.009 ph-cm^{-2}-sec^{-1} in the case of Tycho is a factor of 3 smaller than the results obtained by Ariel 5 for the entire SNR, and is consistent with the positive detection made by the LRL group [11]. Again, the sensitivity of our measurements could be reduced significantly by the Doppler shift effect on the discrete line emission if the X-ray emitting region in Cas A was expanding with the same velocity as the radio and optical shell (~7000 km-sec^{-1}) [12,13]. However, if the X-ray emitting cloud velocity is no more than a few hundred km-sec^{-1} [14], it would imply that the silicon overabundance is less than the factor 8 suggested by Hill's measurements. Nevertheless, our data do not exclude significant contributions from broadened line shapes, but for broadening $\gtrsim$ 10 eV it is not possible to put meaningful upper limits on the equivalent line widths. Even larger line broadening may be expected from electron scattering processes and by the dynamical effects of accretion onto a disk. In the case of SNR, the bulk motion of the gas, at velocities of the order of 1000 km-sec^{-1} or more, would contribute substantially to the broadening emission lines. In any event, it is obvious from our results that a greater sensitivity than that of our instrument is necessary to detect the unbroadened line cores emanating from either compact or diffuse cosmic sources.

REFERENCES

1. H.W. Schnopper, J.P. Delvaille, A. Epstein, K. Kalata, and R. Sohval, Space Sci. Instr., 2, 243 (1976).

2. H. Gursky, H. Schnopper, and D. Parsignault, Astrophys. J. (Letters), 201, L127 (1975).

3. H.L. Kestenbaum, G.G. Cohen, K.S. Long, R. Novick, E.H. Silver, M.C. Weisskopf, and R.S. Wolff, Astrophys. J. (Letters), 208, L27 (1976).

4. R.E. Griffiths, M.J. Ricketts, and B.A. Cooke, MNRAS, 177, 429 (1976).

5. K.S. Thor and R.M. Price, Astrophys. J. (Letters), 195, L101 (1975).

6. W.H. Tucker and M. Koren, Astrophys. J., 168, 283 (1971).

7. H.L. Kestenbaum, K.S. Long, R. Novick, M.C. Weisskopf, R.S. Wolff, Astrophys. J. (Letters), 216, L19 (1977).

8. H.L. Kestenbaum, private communication (1978).

9. L. Woltjer, A. Rev. Astron. Astrophys., 10, 129 (1972).

10. R.E. Griffiths, B.A. Cooke, A. Peacock, K.A. Pounds, M.J. Ricketts, MNRAS, 175, 449 (1976).

11. R.W. Hill, G.A. Burginyon, F.D. Seward, Astrophys. J., 200, 158 (1975).

12. K.V. Bychkov, Sov. Astr. As., 17, 577 (1974).

13. C.F. McKee, Astrophys. J., 188, 335 (1974).

14. A.G. Sgro, Astrophys. J., 197, 621 (1975).

TABLE 1 Bragg Spectrometer Parameters

	Bragg 1, Si XIII	Bragg 2, Si XIV
Crystal	PET (2d = 8.7 A)	
Effective Collecting Area, A	43.3 cm^2	40.6 cm^2
Net Detecting Efficiency	0.44	0.47
Peak Efficiency	0.17	0.17
Resolution ΔE (FWHM)	1.2 eV	1.2 eV
$\Delta \theta$ (FWHM)	2.6 arcmin	2.6 arcmin

TABLE 2 Calibration Data of the Bragg Crystal Spectrometer

	Bragg 1 (cts/64 sec)	Bragg 2 (cts/64 sec)	Bkgd Reg (cts/64 sec)	LAD 1 (cts/64 sec)	LAD 2 (cts/64 sec)
Sco X-1	24.38 (1.08)	22.38 (1.03)	8730 (578)	421.7 (4.8)	291.3 (4.0)
A0620-00	26.26 (0.70)	23.87 (0.67)	8544 (226)	425.2 (7.3)	291.1 (6.0)
AVBKGD	21.68 (0.41)	19.19 (0.39)	8330 (290)	418.2 (4.0)	280.0 (3.0)

TABLE 3 Expected and Observed Continuum Fluxes Detected by the Bragg Crystal Spectrometer

	Expected Continuum (cts/64 sec)		Observed Continuum (cts/64 sec)	
	Bragg 1	Bragg 2	Bragg 1	Bragg 2
Sco X-1	4.2 (0.8)	3.8 (0.8)	2.7 (1.2)	3.2 (1.1)
A0620-00	6.2 (1.2)	5.2 (1.0)	4.6 (0.8)	4.7 (0.8)

TABLE 4: 3σ Upper Limits for Narrow Line Emission of Si XIII and Si XIV.

BRAGG 2: Si XIV, $1s^2S_{1/2}-2p^2P_{1/2,\,3/2}$, 6.184 Å

Source	Date	$F(E)_c$ (cts/cm^2-sec-keV)	T (sec)	$N_{L,m}$ (cts/64 sec)	ΔE (eV)	$I_{L,m}$ (cts/cm^2-sec)
Crab	March 75	1.62	3392	3.00	9	0.014
	Sept. 75	1.59	3904	3.40	10	0.016
A0620-00	Sept. 75	22	960	4.74	1.0	0.023
4U1957+11	April 75	0.06	512	4.57	367	0.022
GX 5-1	March 75	0.52	768	4.69	43	0.023
GX 17+2	March 75	0.40	576	4.64	56	0.022
Cyg X-1	Nov. 74	0.27	1920	2.70	48	0.013
	May 75	3.06	2176	2.43	3.8	0.012
	Nov. 75	1.39	512	5.00	17	0.024
Cyg X-2	Dec. 74	0.59	2816	2.90	24	0.014
	June 75	0.97	576	5.55	28	0.027
Cyg X-3	Nov. 74	0.10	640	4.41	213	0.021
	May 75	0.06	1920	2.93	236	0.014
Cas A	Jan. 75	0.20	1088	4.13	100	0.020
	July 75	0.20	822	3.31	80	0.016
Tycho	Jan. 75	0.07	960	5.18	357	0.025
	July 75	0.07	7104	1.91	132	0.009
Perseus	Feb. 75	0.05	2368	5.19	500	0.025
	Aug. 75	0.05	2112	3.36	324	0.016
Virgo	Dec. 74	0.05	4416	3.32	320	0.016
Coma	Dec. 74	0.03	3520	2.95	474	0.014

BRAGG 1: Si XIII, $1s^2\,{}^1S_0-1s2p\,{}^3P_{2,1}$, 6.684 Å

Source	Date	$F(E)_c$ (cts/cm^2-sec-keV)	T (sec)	$N_{L,m}$ (cts/64 sec)	ΔE (ev)	$I_{L,m}$ (cts/cm^2-sec)
Crab	March 75	1.70	448	5.42	15	0.026
	Sept. 75	1.70	576	5.30	15	0.026
4U1957+11	April 75	0.06	320	6.37	512	0.031
	Oct. 75	0.06	1856	3.16	.254	0.015
GX 17+2	March 75	0.29	320	6.70	111	0.032
Cyg X-1	May 75	3.50	1152	3.63	5.0	0.018
	Nov. 75	1.41	192	8.00	27	0.039
Cyg X-2	June 75	1.07	192	8.89	40	0.043
	Dec. 75	0.53	768	4.73	43	0.023
Cyg X-3	Nov. 74	0.14	704	4.15	143	0.020
Tycho	July 75	0.08	448	5.00	302	0.024
Perseus	Aug. 75	0.06	3008	3.79	305	0.013
Coma	Dec. 74	0.03	3776	4.26	685	0.021

BRAGG 1: Si XIII $1s^2\,{}^1S_0-1s2p\,{}^1P_1$, 6.649 A

Source	Date	$F(E)_c$ (cts/cm^2-sec-keV)	T (sec)	$N_{L,m}$ (cts/64 sec)	ΔE (eV)	$I_{L,m}$ (cts/cm^2-sec)
Cyg X-1	Nov. 74	0.27	1088	3.63	65	0.018
	May 75	6.13	1344	3.00	2.4	0.014
	Nov. 75	1.55	128	11.5	36	0.055
Cyg X-3	Nov. 74	0.10	704	4.55	220	0.022
	May 75	0.04	448	5.44	655	0.026

ANS: EVIDENCE FOR SOFT X-RAY EMISSION FROM THE BINARY HR976

A. J. F. den Boggende and R. Mewe

The Astronomical Institute, Space Research Laboratory, Beneluxlaan 21, 3527 HS Utrecht, The Netherlands

OBSERVATIONS

With the soft (0.16-0.284 keV) and medium energy (1-8 keV) X-ray detectors aboard the Astronomical Netherlands Satellite (ANS) (cf., Brinkman et al., [1], 1974; den Boggende and Lafleur, [2], 1975) a region around the position of the binary HR976 (=HD20210) was observed during Febr. 14 - 15, 1975. The instruments have fields of view (FOV) of 34' (FWHM, circular) and 38' x 75' (FWHM), respectively.

Assuming a flat X-ray spectrum, the measured (3.8σ) signal in the soft X-ray interval corresponds to 5.5(±1.4) x 10^{-12} erg/cm^2 s at Earth, while in the 1-3.5 keV interval a 3σ upperlimit corresponding to 3.8 x 10^{-11} erg/cm^2 s was obtained (den Boggende et al., [3], 1978).

ANALYSIS

An inspection of the sky within the circular FOV (34' FWHM) around HR976 reveals no visual objects other than the $8^m_.3$ G5 star DM34 606 at the edge of the FOV and two fainter (m_V > 10 mag) stars (Whipple, [4], 1971). After considering various alternative X-ray source candidates, we tentatively identified the X-ray source with HR976. This is a single-line spectroscopic binary with a primary Am star (Abt and Bidelmann, [5], 1969), classified as A2/A8/F2 according to the K/H/Metallic line strength (Osawa, [6], 1959). From various observational quantities its mass and radius are derived to be 1.7 $M_\odot$ and 1.56 $R_\odot$, respectively. With the adopted distance of 53 pc, an X-ray luminosity L_X(0.16-0.284 keV) of 2(± 1) x 10^{30} erg/s is obtained and a 3σ upper limit L_X(1-3.5 keV) $\leqslant$ 1.3 x 10^{31} erg/s.

Den Boggende et al., [3] (1978) have argued that accretion onto the secondary of matter from a stellar wind of the primary cannot provide sufficient X-ray energy to explain the observations. The same conclusion holds for a possible corona around the secondary. We are left with the possibility of emission from a corona around the primary Am star, which will be investigated in the next section.

A CORONA AROUND THE PRIMARY OF HR976

The presence of a hot (few million degrees), tenuous gas (so-called corona) around a star is well-known for the Sun. This gas seems to be heated by mechanical energy, coming from underlying layers, and transported by waves. In late type main sequence stars like the Sun (G2 V), the mechanical flux F_a is believed to be generated in the convection zone below the surface of the photosphere. Calculations of De Loore, [1] (1970) showed that F_a reaches a maximum for early F to late A type stars. It seems therefore obvious, that a corona will envelop the primary Am (A8 V from the H line spectrum) star in HR976.

In order to derive some physical parameters of a corona, which would be able to emit the amount of X-rays as observed, we have used some simplified coronal models. The basic difficulty in such models is the lack of knowledge of the details of the loss of mechanical energy by dissipation and radiation mechanisms in the chromosphere and transition region. This denotes that one will anticipate only an order of magnitude result for the calculated parameters from such kinds of models.

A specific coronal model has been developed by Hearn, [8] (1975). His method is based on determining the temperature of a corona, *which minimizes the total energy flux* necessary to maintain the corona against energy losses due to thermal conduction towards the photosphere, radiation, and mass loss by a stellar wind. The resulting *minimum flux coronae* (MFC) have temperatures T and base pressures p_o, which are single-valued monotonic functions of the total acoustic energy input flux F_a. In his first paper, Hearn, [8] (1975) incorrectly assumed that the total radiative energy coefficient j_o was constant in the transition region, but in a later paper (Hearn, [9], 1977) he made a better estimate by taking into account the temperature dependence of j_o both in the transition region and the corona.

We have applied the MFC model calculations in which the total radiative emissivity, as reported by Raymond et al., [10] (1976) for a plasma with solar abundances, is approximated by:

$$j_o = 2 \times 10^{-19} \, T^{-0.55} \quad (\mathrm{erg/cm^3\,s})$$

for the temperature range 3×10^5 K < T < 5×10^7 K. The soft (0.16-0.284 keV) X-ray luminosity L_X is given by:

$$L_X = EM \, j_X$$

where j_X is the soft X-ray emissivity, taken to be equal to the average of the results of Tucker and Koren, [11] (1971), and those of Mewe, [12,13] (1972; 1975). Further, the emission measure EM is calculated to be:

$$EM = 2\pi R^2 N_e^2 (H/2)$$

where R is the radius of the star. The electron density N_e is equal to:

$$N_e = 0.52 \, p_o/kT$$

and the scale height H of the corona equals:

$$H = f_R \, kt/\mu mg.$$

For a corona with solar abundances, the mean molecular weight $\mu = 0.6$, while m is the proton mass, g is the surface gravity of the star and f_R is a numerical correcfactor, derived from Hearn, [8] (1975, Eq. (18)):

$$f_R \approx 0.832 + \frac{4.1284}{x} - \frac{5.072}{x^2} \quad (1.05 \leqslant x < 15)$$

where (cf. Hearn, [8], 1975, Eq. (13)):

$$x = \mu mg\ R/2kT.$$

Note that $f_R = 1$ for a pure hydrostatic corona ($x \gg 10$). Detailed analysis shows that $L_X \propto j_X/j_o$ and that for typical coronal temperatures ($T \approx 10^6$ K) j_X and j_o depend approximately in the same way on the heavy-element (e.g., Si, F_e) abundances. Therefore, the calculations done for solar abundances are probably equally well applicable for stars with (apparently) enhanced metal abundances, as in Am stars.

The resulting values for L_X(0.16-0.284 keV), p_o, and F_a are plotted as a function of the coronal temperature T in Fig. 1. As can be seen, the observations reconcile a MF corona with the following properties:
$T = (2\text{-}3) \times 10^6$ K, p_o = 5-20 dyne/cm^2 and $F_a = (1\text{-}10) \times 10^8$ erg/cm^2 s.

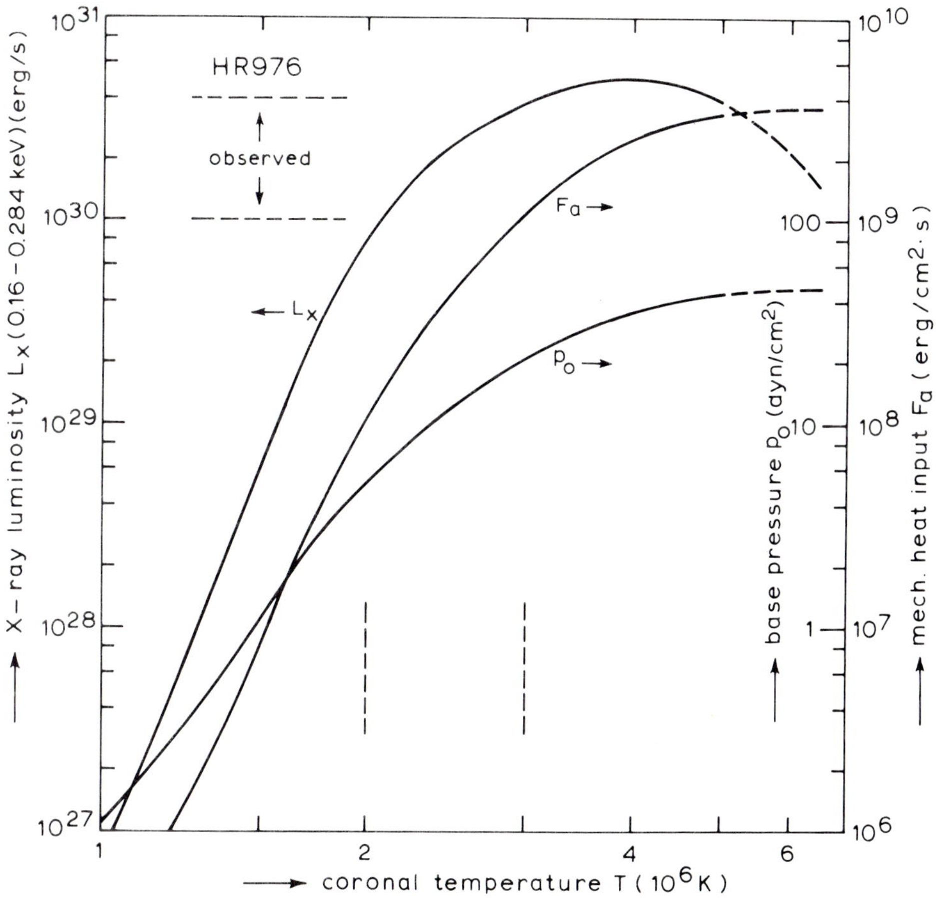

Fig. 1. Results of MFC model calculations for X-ray luminosity L_X, base pressure p_o, and heat input F_a as a function of temperature T. The range of observed L_X is indicated, as well as the corresponding temperature range. Above $T \gtrsim 5.10^6$ K the calculations are less reliable.

The question arises: can such a flux F_a be produced in a star like the primary in HR976? De Loore's, [7] (1970) calculations give for the mechanical flux emerging from the photosphere a value of about 1.5 x 10^8 erg cm^2/s in the case of an A8V star, well in the range of the obtained F_a values. However, the derived F_a is the flux which actually enters the corona after suffering losses in the underlying chromosphere and transition region. In the case of the Sun, only a small fraction ($\lesssim$ 0.1) of the photospheric flux survives to reach the corona! On the other hand, recently developed convection theory (Toomre et al., [14], 1976) seems to predict a much larger acoustic flux than that following from De Loore's, [7] (1970) calculations.

Although large uncertainties exist about the energy dissipation processes in the layers below the corona and about the energy production in the convection region, the energy requirements to produce a corona around the primary in HR976 appear not to be excessive, at least on an order of magnitude basis.

REFERENCES

1. A.C. Brinkman, J. Heise, C. de Jager, *Philips Techn. Rev.*, 34, 43 (1974).
2. A.J.F. den Boggende, H.Th.J.A. Lafleur, *IEEE Trans. Nucl. Sci.*, NS-22, 555 (1975).
3. A.J.F. den Boggende, R. Mewe, J. Heise, A.C. Brinkman, E.H.B.M. Gronenschild, J. Schrijver, *Astron. Astrophys.*, 67, L29 (1978).
4. F.L. Whipple, SAO Star Catalog, part 1, Smithsonian Institute, Washington, D.C. (1971).
5. H.A. Abt, W.P. Bidelmann, *Astrophys. J.*, 158, 1091 (1969).
6. K. Osawa, *Astrophys. J.*, 130, 159 (1959).
7. C. de Loore, *Astrophys. Space Sci.*, 6, 60 (1970).
8. A.G. Hearn, *Astron. Astrophys.* , 40, 355 (1975).
9. A.G. Hearn, *Solar Phys.*, 51, 159 (1977).
10. J.C. Raymond, D.P. Cox, B.W. Smith, *Astrophys. J.* 204, 290 (1976).
11. W.H. Tucker, M. Koren, *Astrophys. J.* 168, 283; 170, 621 (1971).
12. R. Mewe, *Solar Phys.*, 22, 459 (1972).
13. R. Mewe, *Solar Phys.*, 44, 383 (1975).
14. J. Toomre, J.-P. Zahn, J. Latour, E.A. Spiegel, *Astrophys. J.* 207, 545 (1976).

HEAO-1 OBSERVATIONS OF AM HERCULIS BETWEEN 0.1 AND 3 keV

I. R. Tuohy*, F. K. Lamb*,**, G. P. Garmire*
and K. O. Mason***

**Department of Physics, California Institute of Technology, Pasadena, California 91125*
***University of Illinois at Urbana-Champaign, Urbana, Illinois 61801*
****Space Sciences Laboratory, University of California, Berkeley, California 94720*

OBSERVATIONS

In this paper we report the results of HEAO-A2 scanning observations of AM Her in 1977 September-October; a more complete report will be given elsewhere [1]. The HEAO-A2 instrument is described in detail by Rothschild et al. [2]. Briefly, the data reported here were obtained by LED1 which is sensitive to X-rays in the interval 0.1-3 keV and has two coaligned fields of view measuring 1.5° and 2.8° FWHM in the scan direction. The angular response of the detector perpendicular to the scan path is 3° FWHM for each field of view. Owing to its high ecliptic latitude, AM Herculis was within 3° of the HEAO scan plane for an extended period of ~ 20 days between September 22 and October 12, 1977. Approximately 40 source transits occurred each day, but only an average of 4 scans per day were usable due to earth occultations and various LED operating constraints.

A light curve of AM Her was produced in the following manner. Each source sighting (duration ~ 20 sec) was fitted with the detector angular response and the resulting flux was then corrected for source aspect. Only data acquired when the source was within 2.2° of the scan plane were accepted in order to avoid large aspect correction factors. The corrected source intensities were then folded with a period of 0.128927 day using the epoch JD 2443014$^{d}_{.}$7647 for phase zero. This epoch corresponds to the center of the linear polarization event observed by Tapia [3], while the period is that derived from circular polarization and optical observations during 1976 and 1977 [4]. There is at present no evidence for period change or for any difference between the photometric and polarimetric periods.

Figure 1a shows the light curve obtained for the 1.5° field of view of LED1. The flux observed in the energy range 0.15-0.5 keV was typically 8 x 10^{-10} ergs cm^{-2} s^{-1} at the maximum of the X-ray light curve. The mean light curve is qualitatively similar to that found by SAS-3 in May 1976 [5]. Note however that there is evidence of a residual flux throughout the X-ray minimum (phase 0.00-0.15). There is also pronounced scatter in the light curve over the phase interval 0.2-0.8 corresponding to variations in the flux by factors ~ 2-3 from sighting to sighting. However, there is typically only one sighting during each orbit of the binary system, so that the present data do not show whether this variability extends to shorter time scales.

Spectral data for each source transit were obtained in the energy range 0.18-3 keV, although the source was not detected above ~ 0.6 keV. In order to search for spectral variability, we have computed a hardness ratio, defined as the ratio of the (0.30-0.60) to (0.18-0.30) keV count rates, for each sighting. Figure 1b shows the results of folding the hardness ratio values with the 3.1 hour period. No pronounced spectral variation with phase is evident, although the fluctuations in the data are larger than those expected from statistics (reduced $\chi^2 \sim 2$), suggesting the possibility of small variations in the source spectrum.

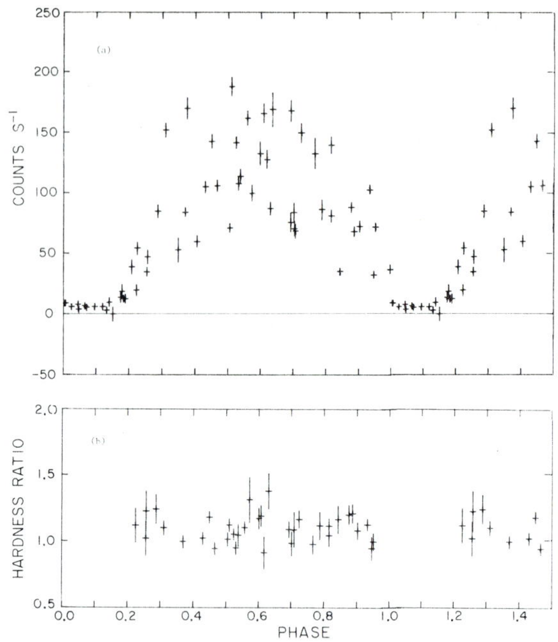

Fig. 1 (a) Light curve of AM Her in the energy range 0.18-3 keV. The data are plotted for 1.5 cycles of the 0.128927 day period.
(b) Hardness ratio (see text) of AM Her as a function of source phase. Error bars are ± 1σ.

In order to obtain an improved measurement of the AM Her spectrum, the high voltage to LED1 was briefly increased during a single ground station pass, thus decreasing the low energy electronic threshold to 0.11 keV. At this time the source was at phase 0.39. The resulting data were fitted in turn with a blackbody, and an exponential spectrum with an energy dependent Gaunt factor [6], allowing for absorption in the interstellar medium [7]. Figure 2 shows the best fit blackbody spectrum corrected for the detector response, together with the 90% confidence contours [8] for both spectral types. Either model provides an acceptable

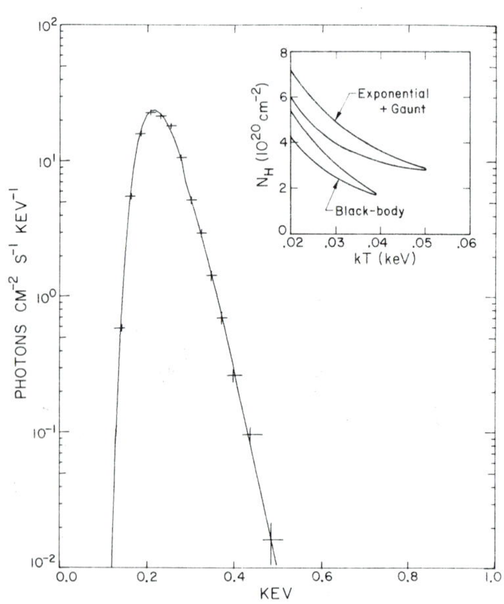

Fig. 2 Inferred incident spectrum of AM Her for a blackbody model with kT = 0.025 keV and $N_H = 3.75 \times 10^{20}$ H-atoms cm^{-2}. The inset shows the 90% confidence contours for blackbody and exponential models.

fit to the data, but note that the measured values of kT and N_H are highly correlated.

It is clear from Figure 2 that there is a definite turnover in the observed spectrum at ~ 0.2 keV. From the chi-squared contours, we can set lower limits to the column density of 1.7 x 10^{20} (blackbody) and 2.8 x 10^{20} H-atoms cm^{-2} (exponential) at the 90% confidence level. From Figure 2 we can also set 90% confidence upper limits to the source temperature of 4.5 x 10^5 K and 5.8 x 10^5 K for blackbody and exponential spectra respectively.

IMPLICATIONS

AM Her is believed to be a binary system containing a strongly magnetic (B ~ 10^8 gauss), relatively massive (M ~ 1-1.4 $M_\odot$) white dwarf rotating in synchronism with the orbital motion of the system and accreting matter from a low mass (M ~ 0.25-0.4 $M_\odot$) companion (see, e.g., Chanmugam and Wagner [9]; Stockman et al. [10]; Priedhorsky and Krzeminski [4]. The qualitative features of X-ray production by accretion onto such a star have been discussed by several authors (see Lamb and Masters [11] and references therein).

One of the principal conclusions of this work is that the spectrum of the radiation from the plasma at the surface of such a star consists of three components: a high-temperature (T ~ 10^8 - 10^9 K) component produced by electron-ion bremsstrahlung in the hot post-shock plasma, a much lower temperature (T ~ 10^5 - 10^6 K) blackbody component produced by the heated surface of the white dwarf, and a still softer component produced by optically thick electron cyclotron emission in the post-shock region. Lamb and Masters [11] find that typically $L_{bb} \sim L_{cyc} + L_{brems}$, where L_{bb}, L_{cyc}, and L_{brems} denote the luminosities in the blackbody, cyclotron, and bremsstrahlung components.

If we assume a homogeneous emission region and interpret the observed soft X-rays as the tail of the blackbody component, the temperature, spectral flux, and total luminosity associated with this component can have the values listed in Table 1 and be consistent with the observations reported here. The blackbody

TABLE 1 Blackbody Parameters Consistent with the Observed Spectrum of AM Her

T_{bb}(eV)	$N_H(10^{20}\ cm^{-2})$	B_1 (ergs $cm^{-2}\ s^{-1}\ keV^{-1}$)[a]	L_{bb} (ergs s^{-1})[b]
40	1.70	4.2 x 10^{-4}	8.4 x 10^{33}
35	2.15	1.2 x 10^{-3}	1.4 x 10^{34}
30	2.75	5.1 x 10^{-3}	3.2 x 10^{34}
25	3.75	4.2 x 10^{-2}	1.3 x 10^{35}
20	5.00	8.0 x 10^{-1}	1.0 x 10^{36}
16	6.50	3.2 x 10	1.6 x 10^{37}

[a]The spectral flux f_ν^{bb} is given by $B_1(E/1\ keV)^3\ (e^{E/kT} - 1)^{-1}$.

[b]Blackbody luminosity for a distance of 100 pc.

interpretation of the soft X-rays implies that the total luminosity of the system is completely dominated by soft X-ray and UV emission (the high inferred luminosity of the blackbody component requires a comparable UV luminosity due to cyclotron emission).

Assuming that the blackbody interpretation is correct, the flux observed at optical wavelengths can be used to bound T_{bb} from below. The most severe constraint is imposed by the requirement that the V-band cyclotron flux not exceed the observed V-band flux. For the model developed by Lamb and Masters [11] applied to AM Her, the cyclotron component has a luminosity comparable to that of the blackbody component but shows a Rayleigh-Jeans spectrum with a brightness temperature $\sim T_e$, the electron temperature in the post-shock region. Equating the cyclotron luminosity and the blackbody luminosity inferred from the present observations, the spectral flux due to cyclotron emission in the optical region is $\sim 10^{-26}$ $(\nu/2\ \mathrm{eV})^2$ $(T_e/10\ \mathrm{keV})$ $(B_1/10^{-4})$ ergs cm^{-2} s^{-1} Hz^{-1}, where B_1 is given in Table 1. This spectral flux is just consistent with the observed V-band flux if $T_{bb} \sim 40$ eV and $T_e \leqslant 30$ keV. Taken together with the results in Table 1, these constraints suggest (1) that there is a blackbody component with a temperature $T_{bb} \sim 40$ eV and a luminosity $\sim 8 \times 10^{33}$ $(D/100\ \mathrm{pc})^2$ ergs s^{-1}, (2) that there is a cyclotron component in the UV with a comparable luminosity, and (3) that the total luminosity of AM Her is ~ 20 times that in hard X-rays. The existence of such a copious soft X-ray and UV flux may account in part for the presence of strong emission lines [12, 13, 10, 14].

ACKNOWLEDGEMENTS

We are grateful to E. Boldt, P. Agrawal and G. Riegler for their essential contributions in the development of the LED detectors. It is a pleasure to thank D. Q. Lamb, A. R. Masters and W. C. Priedhorsky for helpful discussions and for communicating their results in advance of publication. F. K. Lamb acknowledges the support of an Alfred P. Sloan Foundation Research Fellowship, and K. O. Mason, that of a Miller Fellowship. This work was supported by NASA (NAS 5-23315) and by NSF (PHY 78-04404).

REFERENCES

1. I. R. Tuohy, et al., submitted to Astrophys. J. Lett. (1978).
2. R. E. Rothschild, et al., in preparation (1978).
3. S. Tapia, Astrophys. J. Lett. 212, L125 (1977).
4. W. C. Priedhorsky, and W. Krzeminski, Astrophys. J. 219, 597 (1978).
5. D. R. Hearn, and J. A. Richardson, Astrophys. J. Lett. 213, L115 (1977).
6. E. Kellogg, J. R. Baldwin, and D. Koch, Astrophys. J. 199, 299 (1975).
7. R. L. Brown, and R. J. Gould, Phys. Rev. D1, 2252 (1970).
8. M. Lampton, B. Margon, and S. Bowyer, Astrophys. J. 208, 177 (1976).
9. G. Chanmugam, and R. L. Wagner, Astrophys. J. Lett. 213, L13 (1977).
10. H. S. Stockman, et al., Astrophys. J. 217, 815 (1977).
11. D. Q. Lamb, and A. R. Masters, in preparation (1978).
12. W. C. Priedhorsky, Astrophys. J. Lett. 212, L117 (1977).
13. A. P. Cowley, and D. Crampton, Astrophys. J. Lett. 212, L121 (1977).
14. J. L. Greenstein, et al., Astrophys. J. Lett. 218, L121 (1977).

AM HERCULIS TYPE BINARIES

G. Chanmugam and R. L. Wagner

Department of Physics and Astronomy, Louisiana State University, Baton Rouge, Louisiana 70803, USA

INTRODUCTION

AM Herculis, VV Puppis and AN Ursae Majoris are members of a newly discovered class of remarkable variable stars with strong and variable (linearly and circularly) polarized light [1-3]. In the prototype star AM Her, soft [4] and hard [5] X-ray variations, synchronous with the 3.1 hr spectroscopic binary period [6] have been reported. A model of a synchronously rotating, strongly magnetic white dwarf accreting matter from a lower main-sequence companion has been presented [7,8]. Because of the high magnetic moment of the white dwarf, the formation of an accretion disk similar to those in cataclysmic variables is unlikely [7,8] unless the white dwarf is rotating with a period ~1 min [9]. However, no periodic light variations on time scales ~1 min [10] nor any optical evidence of a disk has been observed. Estimates of the magnetic field strength $B \sim 10^8$ gauss in the AM Her type binaries are based on the assumption that the polarized light is due to cyclotron emission at the fundamental frequency $\omega_c = eB/mc$, where e and m are the charge and mass of the electron [1,7]. However, an intense source must, by Kirchhoff's law, (self-) absorb the radiation. Cyclotron self-absorption in thermonuclear reactors has been studied by a number of authors [11-13]. For frequencies $\omega \gg \omega_p$ the plasma frequency, the radiation is the sum of contributions from individual particles [13]. Computation of the intensity is in general complicated but the results may be more easily obtained for a two-dimensional Maxwellian distribution of electrons [14]. In this paper, the angle-integrated emissivity for such a distribution is calculated for accretion columns in the AM Her type binaries. The maximum frequency ω^* for which the column is opaque is calculated for parameters appropriate to such systems. In AM Her, cyclotron self-absorption can explain the coincidence of the primary optical minimum and the standstill in the fractional circular polarization with the X-ray maximum.

CYCLOTRON SELF-ABSORPTION

Consider a plasma slab of thickness ℓ with a uniform magnetic field B parallel to the slab. Let $\boldsymbol{\beta} = \mathbf{V}/c$ be the normalized electron velocity and let $\beta_{\parallel}$, $\beta_{\perp}$ be the components parallel and perpendicular to the field. For simplicity, choose $\beta_{\parallel} \ll \beta$ so that the distribution function is the relativistic two-dimensional Maxwellian:

$$f(\beta) = N\theta/(1+1/\theta)\gamma^4 e^{-\theta(\gamma-1)} , \qquad (1)$$

where $\gamma = (1-\beta^2)^{-1/2}$, $\theta = mc^2/kT$, T is the temperature and N the electron number density. The factor $1/\theta$ was neglected in [12]. This is valid for $\theta >> 1$.

Let $I_n(\omega)$ be the angle-integrated emissivity for the n'th harmonic and let $B(\omega)$ be the Rayleigh-Jeans flux ($\hbar\omega << kT$) for two polarizations and both faces of the slab. Then if collisions are neglected it follows from [12] and Eq. (1) that

$$\frac{I_n(\omega)}{B(\omega)} = \frac{\pi\Lambda n^{-2} x^{-3} \theta^2}{(1+1/\theta)} e^{-\theta(1/x-1)} \{J_{2n}[2n(1-x^2)^{1/2}]$$

$$- (1-x^2)^{-1/2} J_{2n+1}[2n(1-x^2)^{1/2}] - x^2(1-x^2)^{-1/2} \sum_{j=1}^{\infty} J_{2n+2j+1}[2n(1-x^2)^{1/2}]\} \qquad (2)$$

Here $x = \omega/n\omega_c$, $\Lambda = \omega_p^2 \ell/\omega_c c$ and J_m is the m'th Bessel function of the first kind. If $I(\omega) = \sum_{n=1}^{\infty} I_n(\omega)$ then $I(\omega)/B(\omega)$ is approximately the average optical depth τ. Figure 1 illustrates an example of $I(\omega)/B(\omega)$. For a given value of Λ, $I(\omega)/B(\omega) \leq 1$ for frequencies ω greater than a critical value ω^*, so that radiation with frequencies $\omega \leq \omega^*$ is self-absorbed and appears black-body like. Figure 2 gives $m^*(=\omega^*/\omega_c)$ as a function of Λ for various temperatures. For $\theta = 10$ ($kT \approx 50$keV), the individual lines are substantially broadened so that a number of harmonics will contribute to the emission at a given frequency. We include many more harmonics than Ref. [12] which accounts for the higher values of m* obtained. For $\theta =$ 50 ($kT \approx 10$keV) the lines are sharper. On the other hand the correction to the distribution function reduces $I_n(\omega)$ so that the values of m* are slightly lower than those in [12].

APPLICATIONS

We now apply the above results to AM Her with the warning that an accretion column has not the same geometry as the ideal plasma slab discussed above. At the base of the accretion column $N \approx 10^{16}$ cm^{-3} [5] while the thickness of the column $\ell \sim 10^8$ cm [7,15]. Thus for $B \sim 10^8$ gauss we have $\Lambda \sim 10^7$. From Fig. 2 we note that for $kT \sim 1 - 10$ keV, $m^* \simeq 5$-10 with m* increasing with both Λ and T. For high values of m*, the optically thin emission would be perpendicular to the field, linearly polarized [16], and could give rise to the observed sharpness of the linear polarization pulses [17]. Alternatively, the latter could be explained as a result of Faraday rotation and depolarization in the column [17,18].

Let us consider the possibility that the circularly polarized light arises from cyclotron emission. Since the observed maximum fractional circular polarization $|q_{max}|$ is ~10 to 35%, and since the strongly polarized cyclotron radiation may be diluted by other (unpolarized) light, the cyclotron luminosity $L_{cyc} \sim \alpha L_{opt}$, where the optical luminosity $L_{opt} \sim 10^{33}$ ergs sec^{-1} and $0.1 \lesssim \alpha \leq 1$. If the cyclotron emission comes from the surface of a cylinder of radius ℓ and height h,

$$L_{cyc} \approx 2\pi \ \ell h(kT)(m^*\omega_c)^3/12\pi^2 c^2 , \qquad (3)$$

[19,20]. Since the source of the circularly polarized light is not eclipsed, it

cannot be at the magnetic pole but must be at a height $\gtrsim 0.2$ R above the pole, where R is the radius of white dwarf. Furthermore, the broadband nature of the polarized radiation is unlikely to be due to thermal broadening [21] ($\Delta\omega \sim \omega_c (2kT/mc^2)^{1/2}$) as the required temperature $\approx$250 keV would be too high. Thus the broadening is probably due to the inhomogeneity in the dipole field over the emitting region. This requires h~R/2 (say). Setting $R \sim 10^9$ cm, $m^*\omega_c \simeq 2\times10^{15}$ rad sec^{-1} [8] and $\ell \sim 10^8$ cm we find that for $L_{cyc} \sim 10^{32}$ rad sec^{-1} one requires kT~1 keV. For $B \sim 10^8$ gauss, kT~1 keV, $\Lambda \sim 10^7$, $m^* \approx 5$ which is reasonable considering the uncertainty in the luminosity and other parameters. Note that B cannot be much larger than 10^8 gauss in the emitting region as no increased emission in the ultraviolet is observed in AM Her. For lower magnetic fields ($\lesssim 10^7$ gauss), in order to keep $m^*\omega_c$ fixed one would require much larger values of m* and hence much higher temperatures, contrary to observation. Furthermore, for m*>5 it would be difficult to obtain significant circular polarization by cyclotron processes. Circular polarization may, however, be obtained from an optically thick source if there is a large temperature gradient [21] but it is not clear whether one can obtain the high values of fractional circular polarization q observed.

Even for small values of m*, more radiation would be emitted in directions perpendicular to the field rather than along it. This could account for the standstill in q which occurs at X-ray maximum and possibly the concurrent primary optical minimum.

This research was supported under NSF grants AST 76-06807 and AST 75-01890-A02. This work is LSU Observatory Contribution No. 136.

REFERENCES

1. S. Tapia, *Astrophys. J. (Letters)* 212, L125 (1977).

2. W. Krzeminski and K. Serkowski, *Astrophys. J. (Letters)* 216, L45 (1977).

3. S. Tapia, *IAU Circ.*, No. 3054 (1977).

4. D. R. Hearn and J. A. Richardson, *Astrophys. J. (Letters)* 213, L115 (1977).

5. J. Swank, M. Lampton, E. Boldt, S. Holt and P. Serlemitsos, *Astrophys. J. (Letters)* 216, L71 (1977).

6. W. C. Priedhorsky, *Astrophys. J. (Letters)* 212, L117 (1977).

7. G. Chanmugam and R. L. Wagner, *Astrophys. J. (Letters)* 213, L13 (1977).

8. H. S. Stockman, G. D. Schmidt, J. R. P. Angel, J. Liebert, S. Tapia and E. A. Beaver, *Astrophys. J.* 217, 815 (1977).

9. A. C. Fabian, J. E. Pringle, M. J. Rees and J. A. J. Whelan, *M.N.R.A.S.* 179, 9P (1977).

10. R. P. Berg and J. G. Duthie, *Astrophys. J.* 211, 859 (1977).

11. B. A. Trubnikov and A. E. Bazhanova, *Physics of Plasma and Problems of Controlled Thermonuclear Reactions*, Vol. 3, 141 (1959) (London: Pergamon).

12. J. L. Hirshfield, D. E. Baldwin and S. C. Brown, *Physics Fluids* 4, 198 (1961).

13. W. E. Drummond and M. N. Rosenbluth, *Physics Fluids* 6, 276 (1963).

14. S. Tamor, *Nucl. Fusion* 18, 229 (1978).

15. F. K. Lamb, C. J. Pethick and D. Pines, *Astrophys. J.* 184, 271 (1973).

16. B. A. Trubnikov, *Physics Fluids* 4, 195 (1961).

17. H. S. Stockman, *Astrophys. J. (Letters)* 218, L57 (1977).

18. G. Chanmugam and R. L. Wagner, *Astrophys. J.* 222, 641 (1978).

19. A. C. Fabian, J. E. Pringle, and M. J. Rees, *M.N.R.A.S.*, 175, 43 (1976).

20. A. R. Masters, J. E. Pringle, A. C. Fabian and M. J. Rees, *M.N.R.A.S.* 178, 501 (1977).

21. Yu. N. Gnedin and R. A. Sunyaev, *Astron. Astrophys.* 36, 379 (1974).

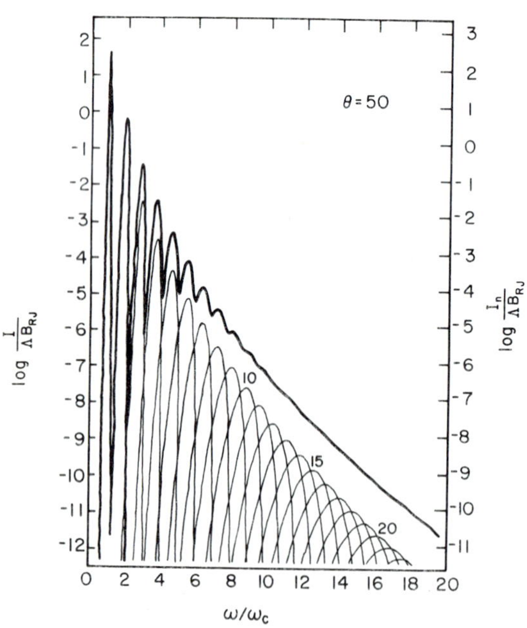

Fig. 1. $I_n/\Lambda B_{RJ}$ and $I/\Lambda B_{RJ}$ (thick line) versus ω/ω_c at θ=50.

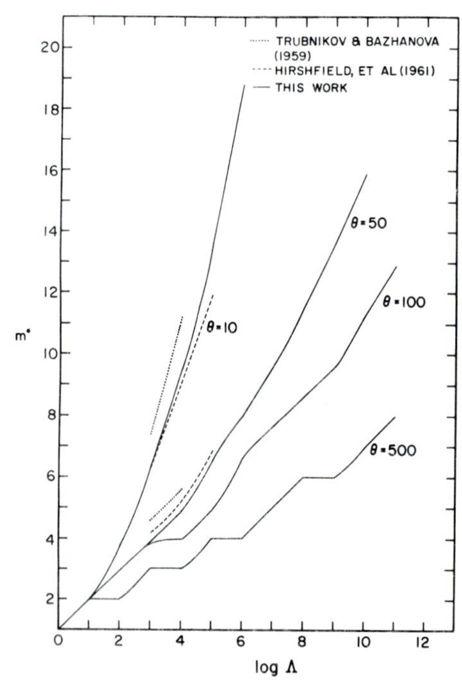

Fig. 2. m* versus log Λ for various values of θ.

VARIABILITY OF CYGNUS X-1 ON MILLISECOND TIMESCALES

J. F. Meekins, E. T. Byram, R. Hedler, W. Evans, K. Wood, D. Yentis, T. Chubb and H. Friedman

E.O. Hulburt Center for Space Research, Naval Research Laboratory, Washington, D.C. 20375

During the first two weeks of November 1977, the detectors of the Naval Research Laboratory instrument aboard the HEAO-1 spacecraft scanned over Cygnus X-1 while it was in the low state. While over a ground station, the full 128 Kbps capability of the spacecraft telemetry system can be used to transmit our data in real time, if so commanded. We were fortunate in that during two of the many scans over the Cygnus X-1 source, the spacecraft was both over a ground station and so commanded. One ground station pass occurred at 1002 UT on 6 November 1977; the other occurred at 1218 UT on 7 November 1977.

Data from our instrument were transmitted in our bit mode telemetry format, i.e., the 1.024 MHz spacecraft clock is counted down to 128 Kbps; if an X-ray event occurs, the square wave pulse (of the counted down clock) immediately following is inhibited from the data stream. In this way, we are able to determine the time of X-ray events to within 10 microseconds. The number of 128 Kbps clock pulses occurring between each X-ray event comprised the data on which this analysis is based.

In Fig. 1a and 1b, we show the data obtained on the 6 November and the 7 November ground station passes, respectively, after sorting the X-ray events into 40 msec bins. The apparent source width in each of the scans is produced by our one degree FWHM collimators scanning over the source. The large fluctuations in these data are similar in character to those reported by earlier observers.

In Table 1, we list the major features of the data obtained during each ground station pass. The background count rates were determined from data taken adjacent to each source; the total source counts include subtraction of background; the mean source rate is determined from the total width; and the peak count rate is determined from the total source counts assuming a constant source.

TABLE 1 Major Features

	6 November	7 November
Background count rate (counts $msec^{-1}$)	0.12	0.16
Total source counts (counts)	16100	10700
Mean source count rate (counts $msec^{-1}$)	1.3	0.85
Peak count rate (counts $msec^{-1}$)	2.8	1.8

The data obtained on each of these passes were sorted into groups of 1 msec, 3 msec, 10 msec, 30 msec, and 100 msec bins. Chi-squared was determined for all the non-overlapping 10 bin sets contained in the 9 seconds of data centered on the source for each of these bin groups. The mean of each 10 bin set was determined, i.e. not assumed, so in each set there are 9 degrees of freedom. The collimator response was included in each calculation.

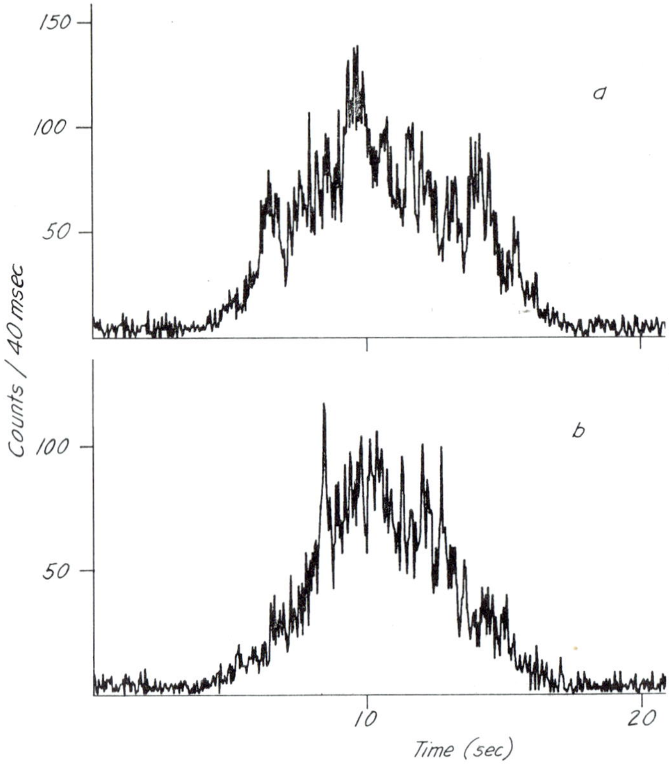

Fig. 1 X-ray scans through Cygnus X-1 obtained on 6 November 1977 (a) and 7 November 1977 (b).

In Fig. 2 we show the integrals of the chi-squared distributions vs. chi-squared obtained from the data of the 6 November and 7 November passes, respectively. The horizontal axis is chi-squared per degree of freedom; the vertical scale is logarithmic. The vertical intercept should be one for each pair of curves - we have plotted the pairs one above the other for clarity. The number of ten-bin sets used to produce these curves range from 8 (100 msec bins) to 900 (1 msec bins). The likelihood of the top two curves occurring by chance is extremely small (each contains one chi-squared value greater than 90). The bottom two curves however, are essentially those we would expect for a constant source. Significant departures from the expected (ideal) integral of the chi-squared distribution occur only for time scales longer than 10 msec (i.e., chi-squared from ten 10 msec bins). The shortest timescales for which Cygnus X-1 variability has been detected here are most correctly described as shorter than 100 msec and longer than 10 msec.

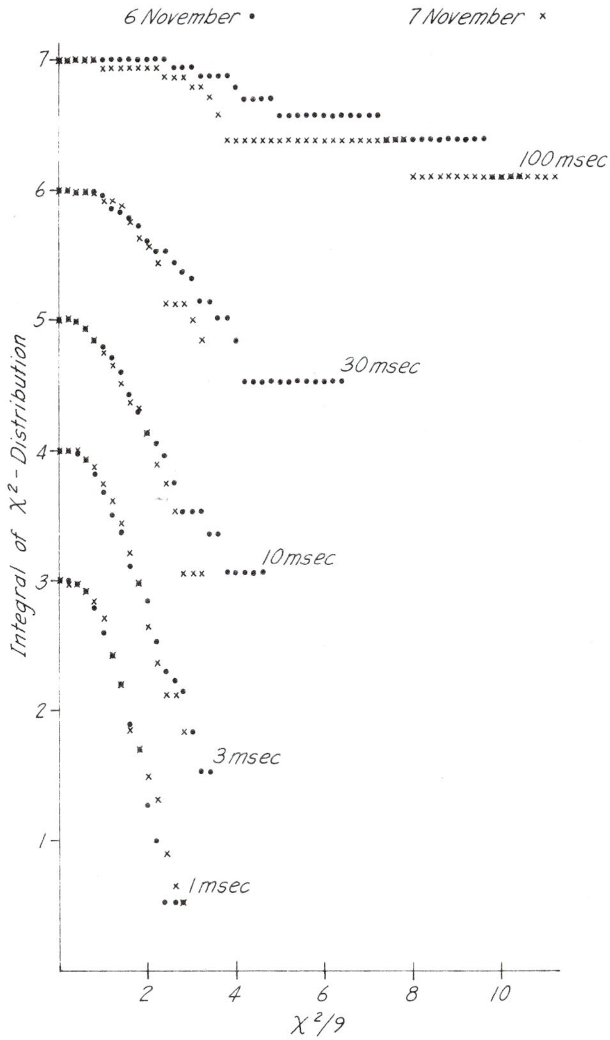

Fig. 2 Integral of the distributions of chi-squared. The vertical scale is logarithmic (see text).

A possible objection to this result is that the ideal chi-squared distribution may not apply when the expected count in 1 msec is small. To test this, integrals of chi-squared distributions were also obtained for 3 counts per 10 bins and 10 counts per 10 bins by Monte-Carlo simulations. The integral distribution of the latter was indistinguishable from the ideal (many-count, 9 degrees of freedom) integral distribution; the integral distribution of the former was distinguishable from the ideal only because of the steps formed by the three possible configurations of 3 counts distributed in 10 bins (1110000000, 1200000000, 3000000000).

We conclude, therefore, that we may indeed compare the integrals of the chi-squared distribution obtained from these data with the ideal. Further, this comparison shows evidence for Cygnus X-1 fluctuations on timescales of 10 msec to

100 msec and longer, but no evidence for fluctuations on timescales of 3 msec to 30 msec and shorter.

The submillisecond variations reported by Rothschild et al. [1 and 2] were not observed, but the two analyses differ in two respects. First, the thirteen events they reported stem from an analysis of approximately ten times the data tested here, so we may have simply missed such submillisecond source variations. Second, while the shortest times over which data was binned (1 msec) are similar, the times over which the means were determined differ. Rothschild et al. used a time of ~400 msec to determine the mean (against which the number of counts in each 1 msec bin are tested) whereas we used 10 msec.

To explore the consequences of using long timescales for the determination of the mean, we tried the extreme available to us, i.e., the full scan duration of 12.6 sec. The count in each of the central 9107 1 msec bins obtained from the 7 November pass was tested against that expected from a steady source (of the same mean intensity as Cygnus X-1) passing through the field of view. Fifty-two bins were found with an expected occurrence probability $<10^{-3}$ (versus the 9.1 expected); five of these were found with an expected occurrence probability $<1.5 \times 10^{-4}$ (versus 1.4 expected). Chi-squared for the frequency distribution of count values was >200 for eleven degrees of freedom. While this result implies that the source is not steady over the 12.6 seconds used to determine the mean, it does not imply millisecond variability. If we isolate these 52 bins and test them by our earlier chi-squared technique (mean determined from ten 1 msec bins) we find them to be unremarkable (the highest chi-squared obtained was 22.2). Thus, the 52 bins occur in regions of enhanced local means, and, as the earlier analysis showed, can be interpreted as expected fluctuations about that enhanced local mean.

Testing by either method can establish that there is variability on timescales shorter than that used to determine the mean; but, the shortest duration of intrinsic variability is determined by the shortest timescale over which means are taken and variability is evident.

References

1. R.E. Rothschild et al., Astrophys. J. 189, L13 (1974).

2. R.E. Rothschild et al., Astrophys. J. 213, 818 (1977).

CIRCINUS X-1 OBSERVATIONS WITH HEAO A-1

D. Sadeh*,**, H. Smathers*, K. Wood*, J. Meekins*, D. Yentis*, W. Evans*, E. T. Byram*, T. A. Chubb* and H. Friedman*

**E.O. Hulburt Center for Space Research, Naval Research Laboratory, Washington, D.C. 20375*
***University of Tel-Aviv, Tel-Aviv, Israel*

Circinus X-1 (4U1516-56) was first detected in 1969 [1] and has since been studied in X-ray, radio, and optical wavelengths. Particular interest in this source has resulted from its resemblance to Cygnus X-1, which is suspected to be a black hole. Both objects have X-ray luminositites in range 10^{37} - 10^{38} erg s^{-1} [2,3]. Both exhibit rapid intensity variations with timescales as low as 1 ms reported by [4,5], and longer timescales of seconds to days [6,7,8]. Both X-ray sources are in binary systems; the orbital period reported for Cir X-1 is 16.585 days [9]. Circinus X-1 differs from Cygnus X-1 in that the total mass of the binary pair is unknown. While an optical candidate has been found [10], there is no estimate of its mass, so that the case for black hole is not as strong for Cir X-1. Variable infrared emission 11 has been reported from the companion. There have been conflicting reports of X-ray periods on the order of seconds: Margon et al. [1] suspected a period of 0.685 sec and Toor [4] reported a possible 2.2 second period. Circinus X-1 happened to be in an active state when the HEAO-1 Observatory was scanning its vicinity in August 1977.

OBSERVATIONS

Prior to 0200 UT on August 28, 1977 the observed intensity of Circinus X-1 was low. At that time there was a dramatic increase and the source continued in a high state for two more days, after which it left the field of view. Throughout this time the average intensity varied by as much as a factor of 2 during the 15-30 minute intervals between successive scans.

The data base accumulated during the high state consists of 77 scans with modules 1-4 (1° x 4° collimator) and 8 good scans with module 7 (8° x 2°). Five of the scans with modules 1-4 utilized 5 ms timing resolution, but all of the module 7 scans had 320 ms timing resolution. The total observing time from all scans is 20 min.

Erratic intensity variations were observed on all scans, and on many scans no pattern in the variability was evident. A notable exception was the pass shown in Fig. 1, where a train of pulses lasting for 40 seconds was seen by module 7. The time between successive pulses was 2.5 seconds. On several scans, modules 1-4 observed similar pulsing. Figure 2 shows two scans, one with the pulsations and one without.

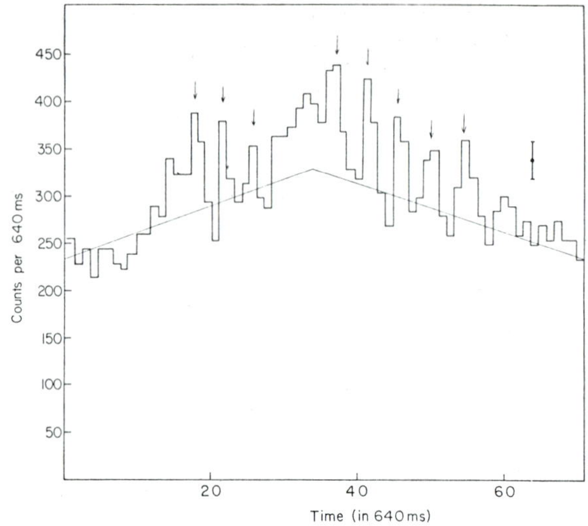

Fig. 1. A scan through Cir X1 with module 7 using 640 ms sums. The triangular response of the collimators is represented by the two straight lines. A steady source output should follow these lines, but instead regularly spaced pulses appeared. The spacing appeared to be 2.5 sec (there are 10 pulses in 39 bins i.e., the period is 2.50±0.06 sec).

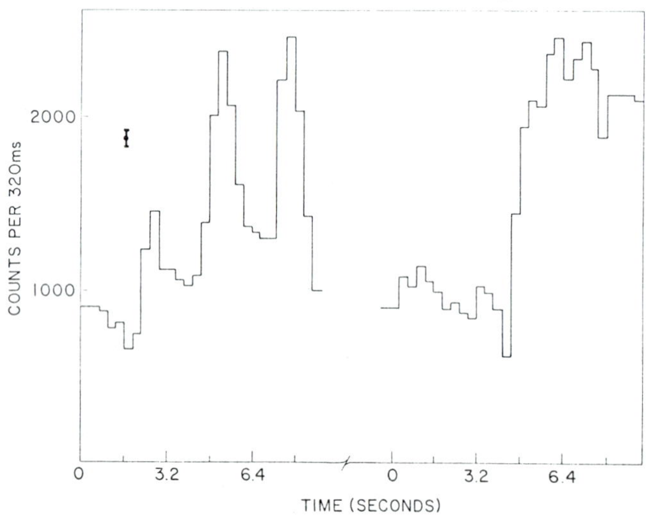

Fig. 2. Two consecutive scans from module 1-4 (corrected for collimator response). On the left one can see a train of three 2.5 sec pulses; on the right, the normal erratic behavior of Cir X1.

Despite the fact that not all scans showed periodic pulsations, the pattern on the scan shown in Fig. 1 was sufficiently distinct (with individual pulses standing out at 5σ) that a more systematic search for periodicity was undertaken. In this search, four hypotheses were considered:

1) the pulsing in Fig. 1 was of instrumental origin,
2) Cir X-1 varies periodically with a 2.5 sec period,
3) the source is not strictly periodic, but has some quasi periodicity,

such a characteristic recovery interval following each burst or perhaps a periodic component which is sometimes clearly present and other times absent or masked by aperiodic variability, and

4) the intensity variations are aperiodic, i.e., burst times are completely random (shot noise).

To test the third conjecture required high sensitivity, so that individual bursts could be isolated and their times could be determined.

Instrumental Effects

Module 7 has intermittently exhibited a high, irregular count rate, thought to be due to electrical noise. As we do not fully understand the nature of the malfunction, there remains some possiblity that the periodicity in Fig. 1 is an instrumental effect. It should be kept in mind that scans from modules 1-4, which are free of this problem, show the same quasiperiodic pulses (Fig. 2).

Periodic Pulsations

The search for periodic pulsations was conducted using three methods. (i) Fourier analysis. Each of the scans from module 7 was Fourier analyzed. Only one showed a significant peak at 2.5 seconds. The probability for the peak was 1/220. Scans from modules 1-4 were not Fourier analyzed, there being only 32 points per scan, versus 320 in module 7. (ii) Epoch folding. The epoch folding technique is more sensitive than Fourier analysis and can be used when the period is approximately known. Again, only the single pass from module 7 showed a significant peak. (iii) Epoch foldimg of multiple passes. It is possible to epoch fold data from several scans of the source by inserting strings of zeroes to fill the half-hour intervals between scans. A search for periods near 2.5 seconds with this method failed to find a significant periodicity.

Quasiperiodicity vs. Random Bursting

A comparative test of the hypotheses of quasiperiodicity and random bursting was made using the distribution of intervals between successive bursts. A rise in count rate was called a burst if and only if it stood out by at least 2σ above immediately adjacent data. The time of the burst was identified with the local maximum, and the inter-burst interval was the difference in time between two successive bursts. Figure 3 shows the distribution of such intervals.

If the bursts are randomly distributed in time, the distribution of intervals will be exponential, with a characteristic time which is the inverse of the mean rate. The circles show the distribution expected if we fit the characteristic interval to the observed rate of bursting. There is a large excess in the number of intervals falling near 8 bins, or 2.5 seconds. The chi-square value for the observed distribution is 57, giving a probability of 10^{-5}. The poor fit is due primarily to the contribution from intervals near 2.5 seconds.

CONCLUSIONS

A random distribution of burst times fails to provide a good fit to the distribution of intervals between bursts, because there are many intervals near 2.5 seconds. Although this suggests a periodicity, and despite the fact that highly periodic behavior is seen on one scan, strict periodicity does not fit the observed intensity variations. An alternative interpretation is that there is a characteristic delay following one burst before another can occur, and that in short data samples this sometimes gives the appearance of periodicity. This would mean that times of burst events are not completely independent, as in the shot noise picture.

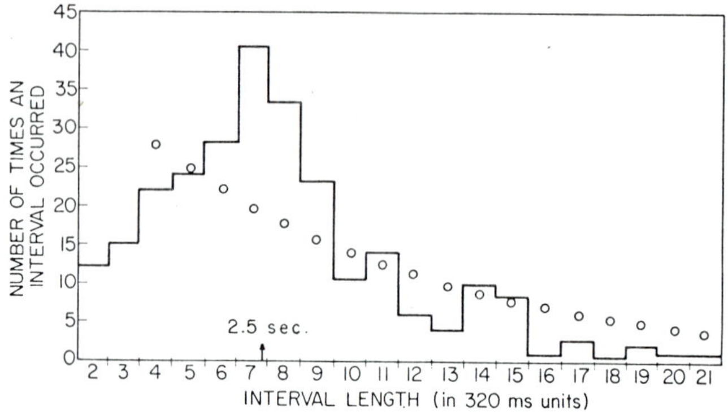

Fig. 3 The histogram shows the distribution of intervals between successive bursts seen in modules 1-4 and module 7. Circles represent the values that would be expected for a shot noise model.

REFERENCES

1. B. Margon, M. Lampton, S. Bowyer, and R. Cruddace, Ap. J. (Letters) 169, L23 (1971).

2. M. Oda, Space Science Review 20, 756, (1977).

3. W.M. Goss and U. Mebold, Mon. Not. R. Astr. Soc. 181, 255 (1977).

4. A. Toor, Ap. J. (Letters) 215, L57 (1977).

5. R.E. Rothschild, E.A. Boldt, S.S. Holt and P.J. Serlemitsos, Ap. J. 213, 818 (1977).

6. C.R. Canizares, F.K. Li, and G.W. Clark, Ap. J. (Letters) 191, L75 (1974).

7. G. Spada, H. Bradt, R. Doxsey, A. Levine and S. Rappaport, Ap. J. (Letters) 190, L113 (1974).

8. P.S.N. Davison and I.R. Tushy, Mon. Not. R. Astr. Soc. 173, 33 (1975).

9. L.J. Kluziensky, S.S. Holdt, E.A. Boldt and P.J. Serlemitsos, Ap. J. (Letters), 208, L71 (1976).

10. S.K. Mayo, J.A.J. Whelan and D.T. Wichramasinghe, I.A.U. Telegram No. 2957, (1976).

11. I.S. Glass, I.A.U. Telegram No. 3095, (1977).

12. M.C. Weisskopf, S.M. Kahn and P.G. Sutherland, Ap. J. (Letters) 199, L146 (1975).

13. P.G. Sutherland, M.C. Weisskopf and S.M. Kahn, Ap. J. 219, 1029 (1978).

RECENT RESULTS FROM OBSERVATIONS OF 4U1700-37 USING SAS-3

Terry Matilsky

Department of Physics, Rutgers University, New Brunswick, N.J. 08903, USA

Introduction

The X-ray source 4U 1700-37 was first detected by the UHURU satellite in 1970 December. (Jones et.al. 1973).[1] The object is in a binary orbit of $3\overset{d}{.}4$, and the extreme 07 f star HD153919 is the confirmed optical counterpart (Penny et.al. 1973, Hutchings, et.al. 1973).[2,3] The low energy cutoff is the strongest of all X-ray binaries, and the source is variable on the timescale of minutes and seconds. The spectrum is extremely hard. This fact has led to several short term ($<1^{d}$) observations to search for a rotation period, with no success until now. We report here the first detection of a 97^{m} periodicity in the 1.0-10.0 KeV X-ray flux of 4U1700-37.

The Data

From 1977 March 27 to 1977 March 31, the Y-axis detectors of the SAS-3 X-ray observatory were pointed at 4U1700-37. Only data from the horizontal tube collimators (80 cm^2 effective area, $1\overset{o}{.}7$ FWHM) were used, since we could avoid contamination with all other nearby sources using this system. This was achieved by offsetting the detectors to totally avoid GX 349+2. The spacecraft was stable to within $\pm0\overset{o}{.}2$ from the desired position, and variation in source intensity due to satellite drift was held to $\lesssim 10\%$. Aspect was monitored throughout satellite night-time and trim commands were issued periodically to assure a uniform pointing direction.

The data from 1-10 keV are shown in Fig. 1. The times when the satellite was earth-blocked or in the South Atlantic Anomaly are eliminated. Several orbits remained unanalyzed due to data dropouts and other data transmission defects.

The sampling rate of the data shown is once every $0\overset{s}{.}831$ (= one minor frame or MF), but is displayed in Fig. 1 compressed by a factor of 64 for convenience. Sensitive spectral resolution between 1-10 keV is available, but will be reported elsewhere. The data used, unless otherwise indicated, sums all the energy channels (A, B and C) from

1-10 keV.

Of crucial importance in this analysis are the inclined tube data. The inclined tube detectors are identical to the horizontal ones, but are displaced at $+6^{o}$ to the Y axis. Thus, these detectors view the X-ray background while maintaining similar sampling and energy characteristics. This enables us to use these data to isolate any periodicities that may occur due to satellite orbital effects or particle events.

As a further check against anomalous particle events, the satellite was instructed to "dither" about the source position $\pm 5^{o}$ to allow the horizontal tubes to sample the X-ray background for approximately two minutes each orbit. This occurred just prior to entrance into earth blockage, and of course, these data are also eliminated in Fig. 1. No anomalous particle intensity was seen and the average orbital background count rate and spectrum remained within count statistics throughout the observation. Altogether, about 50^{m} of each orbit of $\sim 94^{m}$ contain usable data.

The Analysis

Fourier transforms of the horizontal tube data (1-10 keV) yielded a normalized Power Density Spectrum amplitude of 7200 at a period of $96\overset{m}{.}8$. The only sidebands of significance (PDS = 3500) in the transform occur at 105^{m} and 92^{m} corresponding to the 97^{m} amplitude beating against a low frequency component of the window function. The inclined tube data yields maximum normalized PDS amplitudes of 63 and 49 at 101^{m} and 94^{m}. These are quite naturally explained as the synodic and sidereal periods of the satellite. It should be noted that all references to the inclined tube data refer only to orbits from 532 on, since a ground controlled satellite command affecting these detectors rendered earlier inclined tube data unusable. Thus, strictly speaking, all comparisons with the horizontal tubes should use only the latter half of the data.

For this reason, as well as the desire to check whether the periodicity at 97^{m} is due to only several isolated bursts of activity from the source, fortuitously commensurate with 96.8^{m}, the data from various subsets of the observation were folded at $94\overset{m}{.}5$ (the satellite period), 96.8^{m}, and 98.6^{m} and compared for consistency with respect to phase, and percentage of modulation.

In Fig. 2, we display the entire 1-10 keV data set, folded modulo 5808^{s}, corresponding to the maximum amplitude PDS frequency. Phase 0.0 on all plots corresponds to J.D. 2442500.0. In Fig. 3, we compare identical data intervals from the inclined tubes (sampling only background) and the horizontal tubes. We use only those horizontal tube data where usable inclined tube data also exist. Note that if the 5808^{s} periodicity were merely an artifact of the satellite orbital period (or a complex function thereof), we would expect this to show up as a modulation in the inclined tube data. None is observed.

When the data are folded modulo $94\overset{m}{.}5$ (the satellite period) and $98\overset{m}{.}6$, the coherent modulation in the horizontal tube data vanishes, yielding an upper limit of $\sim 8\%$ in both cases.

These analyses, while useful in establishing the reality of the $96^{m}.8$ periodicity, do not answer the question of repeatability, i.e. is this result merely the superposition of a few quasi-periodic flares from this quite variable source, or is it a persistent phenomenon detectable at all times?

To answer the above query, the following analyses were made: the 7 most intensely variable orbits were removed from the data set, i.e. a fold was done with all data except those from orbits 524, 531, 532, 533, 538, 550 and 551. Then, these orbits alone were folded modulo 5808^{s}. The results are shown in Fig. 4. Note the persistence of the modulation even when the strongest sources of the variation are removed, as well as the phase consistency between the two data sets. Finally, we mention the data from the xenon tube collimator (XCTA: 8-18 keV) also displays a 50% modulation when folded modulo the best fit period.

Results and Interpretation

All tests on the $\sim 3^{d}.5$ data train seem to indicate a persistent, phase-coherent modulation of 4U1700-37 at a period of $97^{m} \pm 2^{m}$. Note that this error does not mean that the variation is within 3σ of the satellite period. It merely reflects the difference in frequency of adjacent Fourier components. In other words, the modulation damps out (within statistics) within 2 minutes on either side of 97^{m}. Thus our accuracy excludes (using these data) determination of Doppler effects due to the source's $3^{d}.4$ period of revolution.

However, 97^{m} is the longest (rotational) period observed in a compact X-ray source to date. According to Rappaport and Joss (1977),[4] such a long period would predict a $|\dot{p}/p|$ of about 1 $year^{-1}$. Therefore, this time would be considerably shorter than its known lifetime. There are several intriguing possibilities to explain this anomaly:
(1) The source's periodicity may not be rotational in origin.
(2) The source may be a white dwarf, where the increased moment of inertia lends to much slower spin-up (or spin-down) timescales.
(3) The source may not have an accretion disc at all. Clearly, further studies of possible secular changes in this periodicity are needed.

Since the source is rotating (?) so slowly, for the first time it will be possible to follow changes across the pulse continuously. Hints of what might be happening may be gleaned from Figs. 2 and 3. Note that the onset and suspension of the prominent X-ray pulse intensity appear at approximately phases 0.85 and 0.1, respectively. This may correspond to the sudden entrance and exit of the beamed flux to our line of sight. But within that phase fraction, much variability is present, cf. Figs. 3 and 4. This is similar to the pulse structure of other pulsars, (for example CP1919). Perhaps we are seeing a variable hot spot near the magnetic pole of the X-ray source.

Finally, we point out the high pulsed fraction of the source. If we use the inclined tube data of Fig. 3 as a representative background level, we find that all subsets of data have a pulsed fraction in 1-10 keV X-rays within 15% of the 60% we find for all the data shown in Fig. 2.

As this was going to press, we received a preprint (Kruszewski, 1978)

[5] in which it is stated that optical pulsations consistent with 97^{m} are seen in the source when the X-ray star is in front of the Of companion. This occurs at the level of $\Delta m_V = \pm 0.02$.

We acknowledge extremely helpful and inciteful discussions with W. Lewin, F. Primini, G. Jernigan, J. Ostriker, and R. Stellingwerf. J. Jessen and G. LaSala helped immeasurably with the data analysis. This work was supported in part by NASA under contracts NAS5-11450 and NG 5195.

References

1. C. Jones, W. Foreman, H. Tananbaum, E. Schreier, H. Gursky, E. Kellogg, and R. Giacconi, Ap. J., 181, L43 (1973).

2. A. J. Penny, R. Olowin, J. Penfold, and P. Warren, M.N.R.A.S., 163 (1973), 7 P.

3. J. B. Hutchings, A. D. Thackeray, B. L. Webster, and P. J. Andrews, M.N.R.A.S., 163 (1973), 13 P.

4. S. Rappaport, and P. C. Joss, Nature, 266, 683 (1977).

5. A. Kruszewski, preprint (1978).

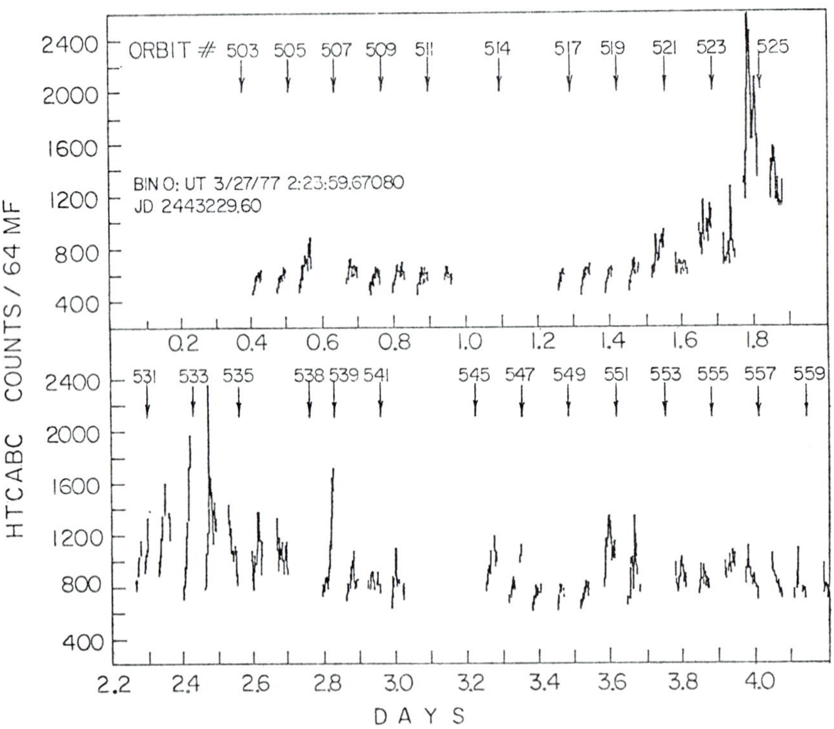

Fig. 1: Counting rate data for 4U1700-37 in the energy interval 1-10 keV. Data taken during times when the satellite was in earth occultation or in the South Atlantic Anomaly are not shown. Other gaps in the data result from telemetry dropouts.

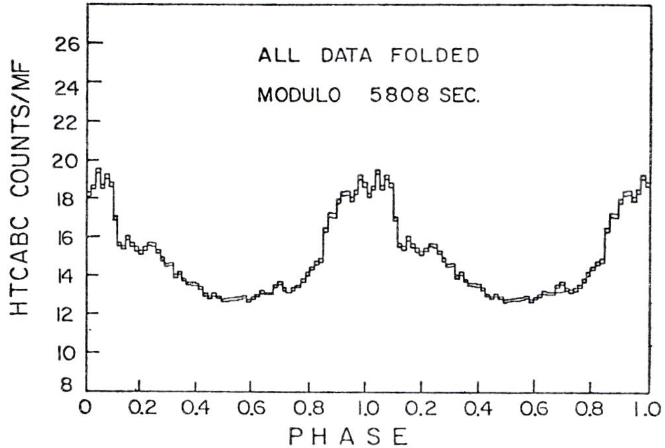

Fig. 2: The average pulse profile derived from the entire observation. The data are folded into 64 bins modulo the 5808^s period and are repeated in a second set of bins. The 1-10 keV horizontal tube data intensity is plotted per minor frame (= $0^s.831$). Count statistics are shown by the double lines drawn for each bin.

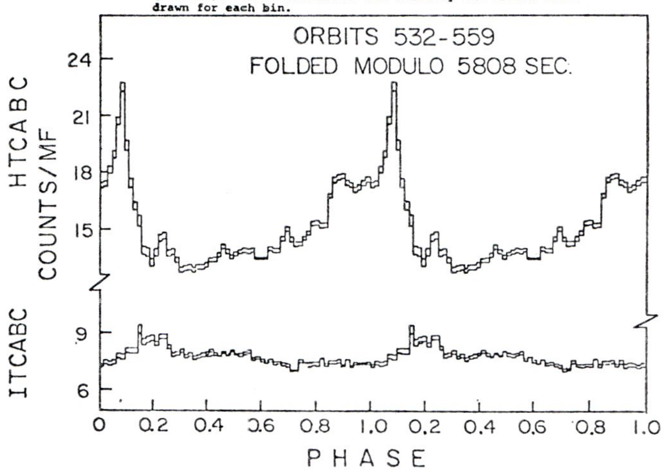

Fig. 3: Identical data intervals from the horizontal tubes (HTCABC = 1-10 keV) looking at 4U1700-37, and the inclined tubes (ITCABC: 1-10 keV) looking at background are folded into 64 bins modulo the source period of 5808^s. These data are repeated in a second set of bins.

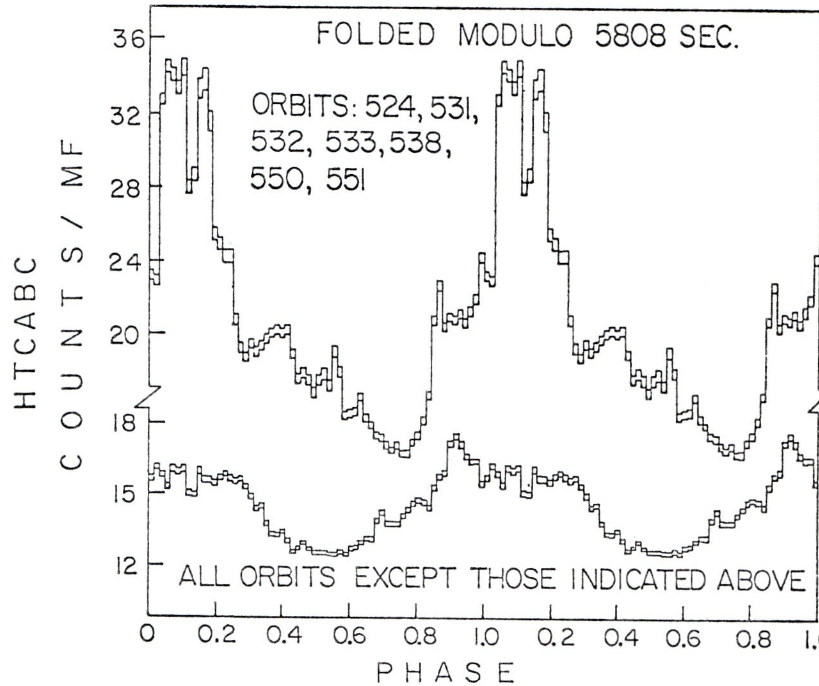

Fig. 4: Two subsets of the horizontal tube data are shown folded in 64 bins modulo the source period of 5808^s. The upper curve shows the 1-10 keV flux from the seven most variable orbits in the data. The lower curve represents the entire data interval from the horizontal tubes excluding these seven orbits shown on the upper curve.

RECENT OBSERVATIONS OF HERCULES X-1 WITH HEAO-1 AND OSO-8

S. H. Pravdo*, R. H. Becker*, R. W. Bussard*,
E. A. Boldt, S. S. Holt, R. E. Rothschild**,
P. J. Serlemitsos and J. H. Swank

Laboratory for High Energy Astrophysics, NASA/Goddard Space Flight Center, Greenbelt, Maryland 20771, USA
**Also Dept. of Physics & Astronomy, University of Maryland*
***Now with University of California (San Diego)*

INTRODUCTION

The Goddard Space Flight Center X-ray spectroscopy experiments on OSO-8 and HEAO-1 have monitored the X-ray emission from Hercules X-1 in overlapping observations from August 24 - September 13, 1977. The X-ray light curve determined from HEAO-1 indicated that for ∿ 0.5 day before the catastrophic X-ray intensity turn-on, there was 1.2 second pulsed emission from Her X-1 at a level intermediate between the high and low state intensities. We discuss this in relation to models for the 35 day cycle. We also present results of a search for ∿ 58 keV cyclotron line emission in the spectrum observed with HEAO-1 in February 1978.

Hercules X-1 was discovered to be an X-ray pulsar in a binary system by Tananbaum et al.(1). Since that time it has become the subject of intense study aided by the remarkable stability (time scale ∿ years) in its X-ray properties. In August - September 1977 the Goddard Space Flight Center experiments on the earth-orbiting satellites HEAO-1 and OSO-8 performed overlapping observations of Her X-1. Figure 1 shows the composite light curve obtained. Three general features shown here are now familiar:

1) An X-ray high state lasts for ∿ 11 days out of each ∿ 35 day cycle(2). The onset is rapid (∿ hours) while the decline is gradual.

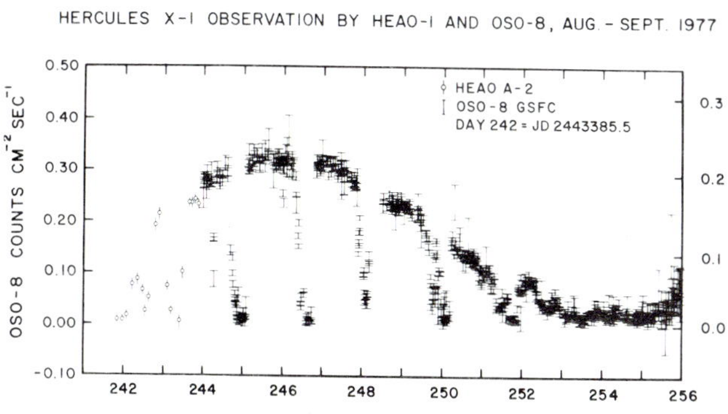

Fig. 1. Light curve of Hercules X-1. The horizontal error bars do not represent the temporal integration time which is 600 sec.

2) Her X-1 is eclipsed by its binary companion HZ Herculis(3) every 1.7 days.
3) Dips in the X-ray intensity occur at progressively earlier phase during the high state.

In this paper we will discuss the HEAO-1 observations near the onset of the high state. The aim is to help distinguish between current theories for the origin of 35-day-related phenomona.

The HEAO-A2 experiment pointed at Her X-1 three times between February 24-27 1978 during an X-ray high state. We also present results of a search for ∿ 58 keV cyclotron line emission(4).

OBSERVATIONS NEAR ONSET OF THE HIGH STATE

The X-ray light curve near high state onset is shown in Fig. 2a. Low state emission is evident for at least 4 days before turn-on. It has been reported that the spectrum at this 35 day phase is similar to the high state spectrum(5) and that 1.2 second pulsations are present at some phases(6) and not at others(7).

Figure 3 shows the light curve folded over the pulse period for this data. We find no pulsations and derive a 2σ pulsed fraction upper limit of 0.03 (see also Pravdo et al.(8)).

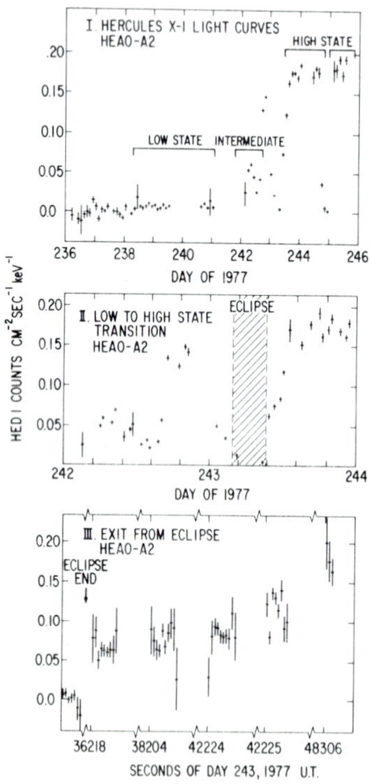

Fig. 2. Detailed light curves of Her X-1.

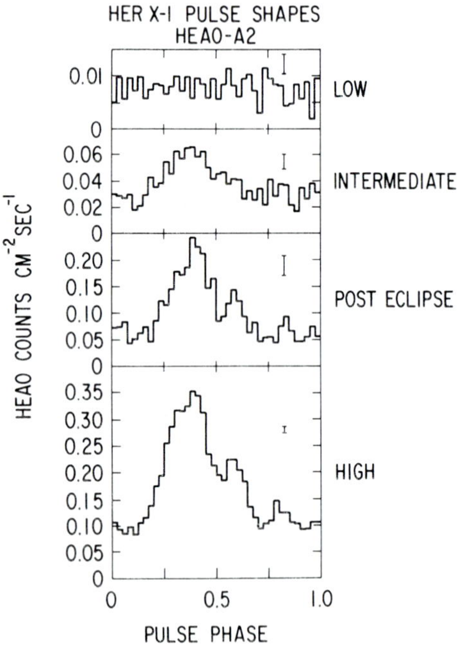

Fig. 3. Pulse of Her X-1 during different phases of the 35 day cycle.

For ∿ 0.5 day before X-ray turnon, Her X-1 was in an intermediate state, at ∿ 0.2 the high state intensity (see Fig. 2b). The X-ray spectrum (Fig. 4a) shows substantial low energy absorption corresponding to (3.0 ± 1.1) x 10^{23} H atoms

cm^{-2} of cool material in the line of sight[9,10]. This is in contrast to the high and low state spectra both of which indicate a much lower amount of low energy absorption[5,11,12]. It is consistent with the absorption seen in spectra during high state intensity dips(7). Figure 3 demonstrates that this emission is pulsed.

Figure 2c illustrates the exit from the first X-ray eclipse after the high state began. Each data point is 5.12 seconds of integration and the large time gaps are due to the ∿ 30 minute period of the great circle scans of the detectors. The figure shows that the X-ray intensity did not return to its high state value until ∿ 10000 seconds after the end of a normal eclipse. This is marked contrast to the typical egress from eclipse which can take only ∿ 100 seconds(13). The spectrum during this time also indicates a large amount of low energy absorption with N_H $(1.8 \pm 0.5) \times 10^{23}$ H atom cm^{-2} (Fig. 4b). Pulsations are present (Fig. 3).

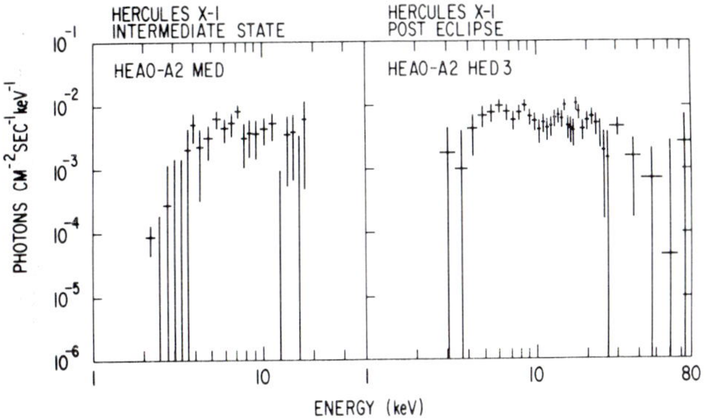

Fig. 4. Her X-1 intermediate state and post-eclipse incident spectra.

Similar features in the X-ray light curve were observed by Davison and Fabian(14). In addition, an "anamolous" dip (i.e. not in the pre-eclipse dip series) also occurred in the suceeding binary orbit in both observations. It still seems difficult, even on the basis of these detailed repetitions in the X-ray light curve, to determine their cause. A free precession of the neutron star(15), precession of the tilted, twisted accretion disk(16,17), asynchronous rotation of Hz Her(18), and a third member of the stellar system(19) have been suggested as the underlying 35 day clock mechanism. Difficulties exist with many of these proposals(20,21).

The HEAO-1 results demonstrate the stability in the 35 day cycle when compared to previous observations. The discovery of pulsations from the Her X-1 for ∿ 0.5 day before the abrupt X-ray turn on suggests that "turn on" can instead be defined as the time when 1.2 second pulsations begin. With this definition the 35 day period may be more exactly defined. The two pulsation turn on times presently available (the present result and the 1975 OSO-8 result(7)) yield a high-low period value consistent with the long term period(22) determined with the Ariel 5 ASM of $34.875 \pm .025$ days.

This possible exact periodicity in pulsation turn ons could relate to a stable, opaque, occulting body. The jitter in catastrophic intensity turn ons would then be due to a variable scale height of cool semi-transparent material located above the opaque body. This material is likely to be associated with a discontinuous mass transfer from HZ Her into the accretion disk which is also responsible for the pre-eclipse dips(23). It is also probably responsible for the gradual, absorbed egress from the following eclipse. In reference 14 it is suggested that this post-eclipse dip is an extension of the intensity dip which began before eclipse. In this case the dip duration of ∿ 10 hours is ∿ 3 times longer than typical intensity dips.

The 35 day cycle may be a precise periodicity. HEAO-1 will attempt to observe pulsation turn ons in the future to decide this question. The X-ray spectrum during the gradual ∿ 0.5 day turn on is different from the spectrum during the decline of the high state (ref. 2, this work). In particular low energy absorption (cool material) is present in the former but not in the latter. These results should further constrain models for the occulting material.

SEARCH FOR 58 keV LINE EMISSION

Two high energy detectors (sensitive to energies 2-80 keV) of HEAO-A2 pointed at Her X-1 during the high state. Figure 5 shows the incident spectrum from one of these detectors. There is no apparent emission feature near 58 keV. We have calculated a quantitative upper limit to line emission with data from the other detector which has about twice the detection efficiency at high energies. The exposure to the source in the high energy channels is 7.4×10^4 cm^2-sec, at 58 keV.

To do this, we add together data from three observations. The result is the observed-binary and pulse-phase-averaged intensity upper limit. Background is the diffuse background obtained at celestial positions from which the diffuse background has a negative fluctuation from its average value (see Boldt et al. (24) this conference). In addition this background is chosen with a more strict ambient particle contamination criterion, so that the resulting net counts are, perhaps, overestimated. We then assume that there is no continuum emission near 58 keV--i.e. only line photons are present. This introduces an additional overestimation of the observed line counts. The energy resolution of this detector at 59 keV is determined to be 12% from an inflight Am^{241} calibration source.

Figure 6 shows the net counts observed from this high energy detector. Also plotted are the predicted number of counts in each channel due to a narrow line (1 keV FWHM) at 58 keV with 1×10^{-3} photons cm^{-2} sec^{-1} and a broad (10 keV FWHM) line at 58 keV with 1.5×10^{-3} photons cm^{-2} sec^{-1}. These are our 2σ upper limits to line emission at this energy. They are a factor of three below the intensity detected two years earlier(4). They are just consistent, however, with a subsequent detection at a lower level(25). As can be seen from Fig. 6 our sensitivity increases and upper limit decreases as the line is assumed to be more narrow. If the feature is determined to be broad this could support a cyclotron absorption(26) rather than an emission origin. The pulse phase-dependent cutoff in the > 20 keV spectrum of Her X-1 has already been interpreted as a consequence of absorption(27).

Phase-averaged line emission from Her X-1 at ∿ 58 keV must be variable in intensity by at least a factor of two if all the measurements are to be reconciled. This is in contrast to the < 20 keV continuum emission which does not show this large factor of variability from high state to high state. We note that our third point at Her X-1 occurred at a binary and 35 day phase chosen to duplicate those in ref. 4. In this observation and in previous observations of Her X-1 at this phase(13) an intensity dip is in progress. These data, taken alone, also do not show a 58 keV line feature, although statistics are worse. Thus, a binary and ∿ 35 day phase correlation in the line intensity is also not supported. Clearly further observations are needed to understand the nature of this emission.

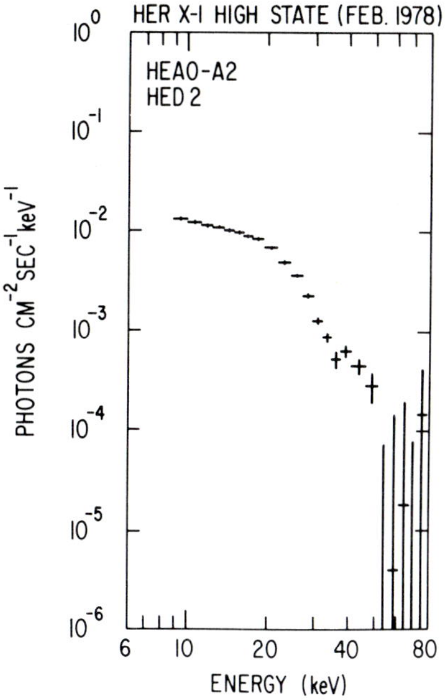

Fig. 5. High state spectrum of Her X-1.

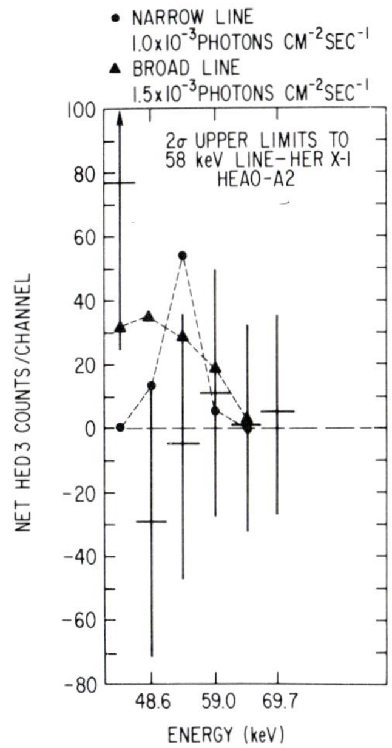

Fig. 6. Upper limit to 58 keV line emission.

REFERENCES

1. Tananbaum, H., Gursky, H., Kellogg, E. M., Levinson, R., Schreier, E., and Giacconi, R. 1972, Ap. J. (Letters) 174, L143.

2. Giacconi, R., Gursky, H., Kellogg, E., Levinson, R., Schreier, E., and Tananbaum, H. 1973, Ap. J. 184, 227.

3. Liller, W. 1972, IAU Circ., No. 242.

4. Trumper, J., Pietsch, W., Reppin, C., Voges, W., Staubert, R. and Kendziorra, E. 1978, Ap. J. (Letters) 219, L105.

5. Jones, C. and Forman, W. 1976, Ap. J. (Letters) 209, L131.

6. Voges, W., Pietsch, W., Reppin, C. and Trumper, J. 1978, IAU Circ., No. 3184.

7. Becker, R. H., Boldt, E. A., Holt, S. S., Pravdo, S. H., Rothschild, R. E., Serlemitsos, P. J., Smith, B. W. and Swank, J. H. 1977, Ap. J 214, 879.

8. Pravdo, S. H., Boldt, E. A., Holt, S. S., Rothschild, R. E., and Serlemitsos, P. J. 1978, submitted to Ap. J. (Letters).

9. Brown, R. L. and Gould, R. J. 1970, Phys. Rev D 1, 2252.

10. Fireman, E. L. 1974, Ap. J. 187, 57.

11. Clark, G. W., Bradt, H. V., Lewin, W.H.G., Markert, T. H., Schnopper, H. W., and Sprott, G. F. 1972, Ap. J. (Letters) 177, L109.

12. Holt, S. S., Boldt, E. A., Rothschild, R. E., Saba, J. R. and Serlemitsos, P. J. 1974, Ap. J. 190, L109.

13. Pravdo, S. H. 1976, Ph.D. Thesis, University of Maryland.

14. Davison, P.J.N. and Fabian, A. C. 1977, MNRAS 178, 1P.

15. Brecher, K. 1972, Nature 239, 325.

16. Roberts, W. J. 1974, Ap. J. 187, 575.

17. Petterson, J. A. 1975, Ap. J. (Letters) 201, L61.

18. Bisnovatyi-Kogan, G. S. and Komberg, B. V. 1975, Astr Zn 52, 457, (transl. Sov. Astr. 19, 279, 1975).

19. Mazeh, T. and Shaham, J. 1977, Ap. J. (Letters) 213, L17.

20. Chevalier, R. A. 1976, Ap. Letters 18, 35.

21. Shaham, J. 1978, Ap. J. (Letters) 213, L17.

22. Holt, S. S., Boldt, E. A., Kaluzienski, L. J. and Serlemitsos, P. J. 1978, preprint.

23. Gerend, D. and Boynton, P. E. 1976, Ap. J. 209, 95.

24. Boldt, E. A., Stottlemyer,, A., Shafer, R., Marshall, F., Mushotzky, R., Holt, S., Rothschild, R., and Serlemitsos, P. J. 1978, this conference.

25. Trumper, J. et al. 1978, Proceedings of the Symposium on Gamma-Ray Spectroscopy, NASA/GSFC, in preparation.

26. Daugherty, J. K. and Ventura, J. 1978, Phys. Rev D, in press.

27. Pravdo, S. H., Bussard, R. H., Boldt, E. A., Holt, S. S., Serlemitsos, P. J., and Swank, J. H. 1978, Ap. J, in press.

DISTRIBUTION OF LARGE SCALE GALACTIC X-RAY SOURCES

R. Rothenflug, R. Rocchia and M. Cassé

Division de la Physique, Section d'Astrophysique, Centre d'Etudes Nucléaires de Saclay, B.P. 2 - F 91190 Gif-sur-Yvette, France

A statistical association of bright X-ray sources with extreme population I objects is expected from observations and models invoking mass transfer in binary systems (1). Thus the overall distribution of X-ray sources may reflect the distribution of young galactic objects, which seem to be linked to the gas distribution (2). We are interested in testing the idea that the overall distribution of X-ray sources is reminiscent of the gas distribution. This is certainly an oversimplified view, in particular because identified X-ray sources are not all associated with early type stars (3).

Any model of spatial distribution of GXRS is constrained by observations : (i) the number-intensity distribution, (ii) their distribution versus galactic longitude, (iii) their distribution versus galactic latitude. We use data from the 4th Uhuru catalog for which the sky coverage is considered as complete for intensity $S > 5$ Uhuru-U-c. s^{-1} (4)

BASIC INGREDIENTS OF COMPUTED DISTRIBUTIONS

They are the following : (i) The volume density of GXRS at a given position in the galactic disk is governed by $dN_x/dV \propto \rho_g^n$ where n is a free parameter and ρ_g the volume gas density, i.e. the formation rate of GXRS is given by the Schmidt law (5), (6). For neutral hydrogen, we use the density distribution of Burton and Gordon (7). Two distributions of molecular hydrogen (Gordon and Burton, hereafter GB (8) and Cesarsky et al., hereafter CCP (9)) have been introduced in our calculations as extreme cases (10). (ii) We parametrize the luminosity function by a power law $dN/dL_x \propto L_x^{-\alpha}$ where α is a free parameter and L_x the X-ray luminosity, law compatible with observations (11), (12). Adopted lower and upper limits are $L_{min} = 10^{34}$ ergs s^{-1} (not critical) and $L_{max} = 2 \times 10^{38}$ ergs s^{-1}.

NUMBER INTENSITY AND LONGITUDE DISTRIBUTIONS

The shape of the computed number intensity distribution is quite insensitive to the spatial distribution of GXRS, but depends on the index α of the luminosity function, since the excursion in luminosity is much larger than that of the gas density. At the opposite, the computed distribution in longitude is sensitive to the spatial distribution of GXRS, i. e. to the index n. Figure 1 shows the fit obtained on the symmetrized longitude distribution. The fit made on the differential intensity distributions (fig. 2a) gives $\chi^2 = 5.1$ for 5 degrees of freedom. The central region (20° both ways for the Galactic Center) deserves a special discussion. Figure 2b shows the difference between observed and computed (n = 1.35, α = 1.1, GB distribution) differential intensity distributions for this region. An excess in the observed number of sources with intensity S > 200 U. c. s^{-1} relative to the expected one could be due to a different class of sources (13).

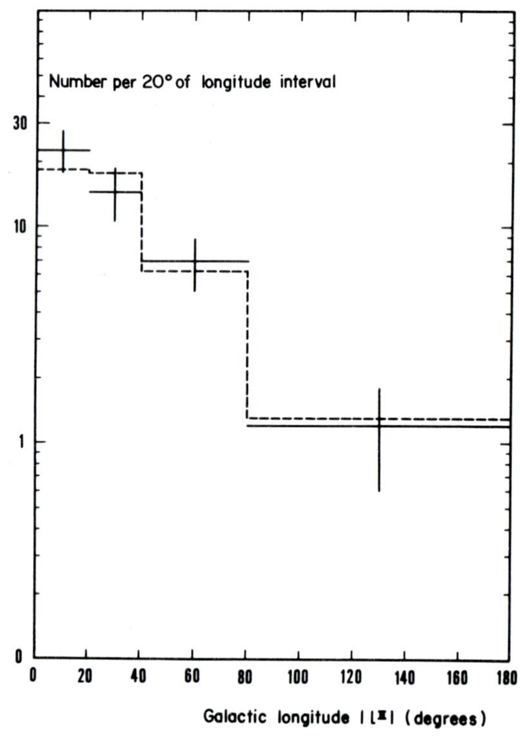

Fig. 1 - Symmetrized longitude distribution. Sources with S > 5 U. c. s^{-1} have been grouped in increasing longitude ranges (for statistical significance) for which the number of sources per 20° interval is plotted with the corresponding error bar. Conventional criteria were used to select potential GXRS (14). Dotted line : computed longitude distribution for n = 1.35 , α = 1.1 and the Gordon and Burton gas distribution (see text). For CCP distribution, n = 1.9.

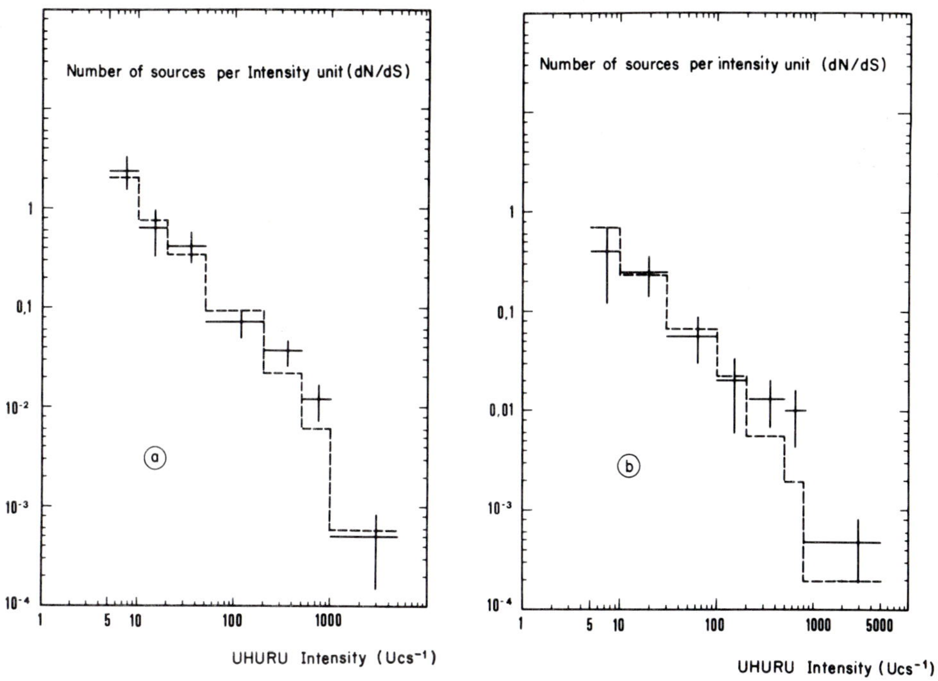

Fig. 2 - Number-Intensity (logN - log S) diagrams. a) for all galactic sources $0° \leq |b^{II}| \leq 20°$; b) for the central region $0° \leq |\ell^{II}| \leq 20°$, $0° \leq |b^{II}| \leq 20°$. Dotted line : computed number-intensity distributions for $n = 1.35$, $\alpha = 1.1$ and the Gordon and Burton gas distribution (see text). For CCP distribution, $n = 1.9$.

LATITUDE DISTRIBUTION

GXRS extend to greater heights above the galactic plane than the interstellar gas (15). This observation suggests that most of the GXRS are indeed migrating systems, in analogy with pulsars (16), (17) as already predicted and discussed (18), (19), (20). Note that our principal assumption for binary X-ray sources remains valid as long as the observed GXRS have not moved away from their birth place by more than 500 pc in the radial direction, since the radial distribution of the gas is averaged in rings with $\Delta R = 500$ pc. At a given position, the height of GXRS depends on the initial velocity perpendicular to the galactic plane, on the adopted gravitational field, and on the runaway time (time elapsed between the beginning of the runaway and the epoch of observation). Available evolution scenari of massive binary X-ray sources give a runaway time between 3 to 5×10^6 yrs (21). A scale height of 650 pc is required to fit the latitude distribution of outer regions ($60° < |\ell^{II}| < 180°$). With 5×10^6 for the runaway time, we get $v_z \simeq 130$ km s^{-1} for the perpendicular velocity. Figure 3 shows the observed and predicted latitude distributions for $0° < |\ell^{II}| < 60°$ (inner regions) and $60° < |\ell^{II}| < 180°$ (outer regions).

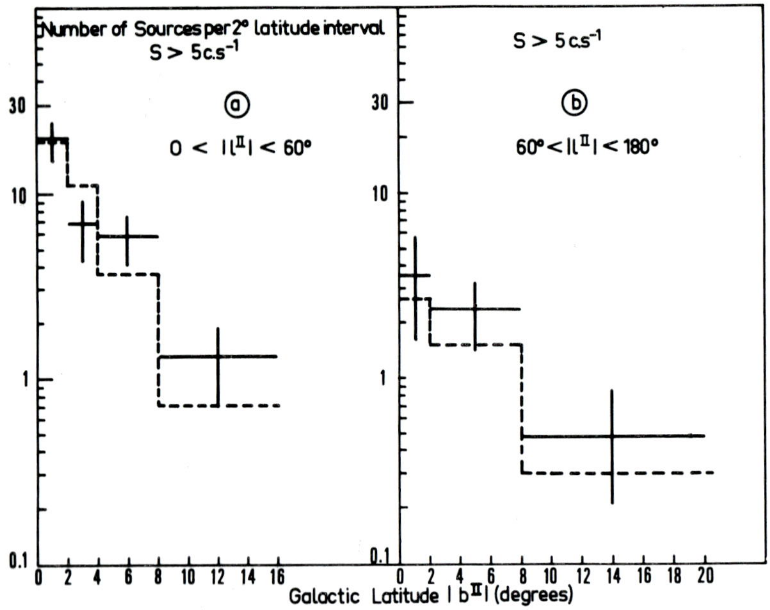

Fig. 3 - Latitude distribution of all GXRS. a) Inner regions ($0° < |\ell^{II}| < 60°$; b) Outer regions ($60° < |\ell^{II}| < 180°$. Dotted line : computed distribution with $\tau_R = 5 \times 10^6$ yr and $v_z = 130$ km s^{-1}.

CONCLUSION. The conclusions can be summarized as follows : a) most of GXRS contain an extreme population I object and their birth rate is related to the gas density prevaling on their birth place ; b) most of the GXRS are runaway systems with perpendicular velocity $v_z \simeq 100$ km s^{-1}. A more detailed discussion will be given in a forthcoming paper.

REFERENCES

1. R. MacRay, 1977, in Highlights of Astronomy, vol.4, part 1.
2. W.B. Burton, 1976, Ann. Rev. Astron. and Astrophys., 14, 275.
3. H. Gursky and E. Schreier, 1975, Galactic X-ray sources in Neutron Stars, Black Holes and Binary X-ray sources (Reidel).
4. W. Forman et al., 1978, 4th Uhuru Catalog of X-ray sources, submitted to Ap. J. Suppl.
5. M. Schmidt, 1959, Ap. J., 129, 243.
6. M. Schmidt, 1959, Ap. J., 137, 758.
7. M.A. Gordon and W.B. Burton, 1976, Ap. J., 208, 346.
8. W.B. Burton and M.A. Gordon, 1976, Ap. J. (Letters), 207, L 189.
9. C.J. Cesarsky et al., 1977, Astron. and Astrophys., 60, 139.
10. J. Guibert et al., 1978, Astr. and Ap., 68, 1.
11. B. Margon and J.P. Ostriker, 1973, Ap. J., 186, 91.
12. T.H. Markert et al., 1977, Ap. J., 218, 801.
13. C.R. Canizares, 1975, Ap. J., 201, 589.
14. T. Matilsky et al., 1973, Ap. J., 181, 753.
15. T. Matilsky, 1977, Ap. J. (Letters), 217, L 83.
16. J.E. Gun and J.P. Ostriker, 1970, Ap. J., 160, 979.
17. J.H. Taylor and R.N. Manchester, 1977, Ap. J., 215, 885.
18. C. De Loore and J.P. De Grève, 1975, Ap. Space Sci., 35, 241.
19. W. Sutantyo, 1975, Astron. and Astrophys., 41, 47.
20. D.J. Helfand and E. Tademaru, 1977, Nature, 267, 130.
21. E.P.J. Van den Heuvel, 1977, Annals of the N.Y. Acad. Sci., 302, 14.
22. K.A. Innanen, 1965, Ap. J. 143, 153.

ARIEL V RMC OBSERVATIONS OF THE GALACTIC PLANE

A. M. Wilson

Department of Space Research, University of Birmingham, Edgbaston, Birmingham B15 2TT, England

INTRODUCTION

The Ariel V Rotation Modulation Collimator is one of six experiments on board the British satellite, Ariel V, which was launched in 1974 October. The satellite is spin stabilised, with four of the experiments, including the RMC, aligned along the spin axis.

The field of view of the RMC is limited by a honeycomb collimator, giving a FWHM response of 17°. As the satellite rotates, X-ray flux from within the field of view is modulated by two parallel sets of grids located in front of the proportional counters. A point source will therefore produce a characteristic ''signature'' in the data. When the time-dependent modulation is deconvolved, the response of the instrument to a point source can be shown to be a two-dimensional Bessel Function [1]. Thus the image consists of a peak surrounded by quite strong side lobes. The resolution of the RMC is of the order of the FWHM of the Bessel function peak and is $\sim 0.5^{\circ}$ for the Ariel V RMC.

The important feature of the RMC is that the positional resolution is two-dimensional, in comparison with the essentially one-dimensional resolution obtained by a scanning slat collimator. It is therefore possible for the RMC to resolve individual point sources without difficulty even in a region such as the galactic centre where there are many strong sources within a few degrees of each other.

ANALYSIS OF SINGLE ORBIT DATA

Ariel V data is integrated over one orbit ($\sim$100 m) and then dumped to a ground station. It can usually be assumed that the drift in attitude of the spin-axis is negligible in this time. Thus, analysis of the data collected in one orbit gives the coordinates, relative to the spin-axis, of each significant source within the field of view. Unfortunately, due to the failure of the star-tracker, the position of the spin-axis is not known precisely, and it is necessary to use one or more of the observed X-ray sources as positional "benchmarks" in order to obtain absolute positions for the other sources. By combining results from the analysis of several hundred orbits it is possible to build up a "chain" of observations which is fixed on the sky by means of one or more benchmark sources. Clearly, benchmark sources must be strong and have known independently determined positions.

A relaxation algorithm has been developed which optimises the positions of all the non-benchmark sources. It is also possible to estimate the error in position of each source using this algorithm. However, in a chain of observations, the positions of the sources are highly interdependent, and it is necessary to use some care in estimating the errors. The procedure used is to fix a source at a set of positions away from its optimum and then employ the algorithm to optimise all the source positions except that for the source under consideration. This gives a χ^2 contour for the position of the source which can then be used to obtain the 90% confidence region. The shape of this contour usually differs little from a circle.

Using the method outlined above, positions of 21 sources near the galactic centre have been published [2]. In most cases the precision of the results is better than 1 arc min and for some sources 20 arc sec precision has been achieved.

COMBINATION OF SINGLE ORBIT DATA

In the analysis of a single orbit the sensitivity achieved is such that a source at $\sim$40 Uhuru cts s^{-1} near the centre of the field of view would just be significant in the data. Clearly by considering many orbits together, the integration time is increased and thus a better sensitivity can be obtained. In performing the summation it is necessary to allow for any changes in the spin-axis direction between orbits.

The first technique developed for this analysis is essentially a χ^2 fitting algorithm. A small grid ($0.1^\circ \times 0.1^\circ$) of possible positions on the sky is considered, and best fit mean intensity of a source at each grid position is calculated. A χ^2 value, indicating the goodness of fit, is obtained for each of these postulated sources. The position of minimum χ^2 is then found by an iterative minimisation algorithm. Note that the positions are fixed in R.A. and Dec. which means that the R.A. and Dec. of the spin-axis must be known for each orbit under consideration. However, this is not a problem in regions of sky where there are strong sources, since the single orbit analysis provides this information.

The minimisation algorithm provides an intensity, R.A. and Dec.for each source under consideration. Additionally a convenient estimate of the significance of the source is also obtained, since this is numerically equal to the square root of the decrease in χ^2 obtained by fitting this source. Empirical tests show that a significance of 3.8σ would have a 1% probability of being obtained by chance in a pure noise dataset at a random position on the sky. Thus a 3.8σ source may be regarded as significant at the 99% confidence level.

The advantage of this algorithm is that it uses the original time-dependent RMC data as its input and gives a position and intensity for each source directly. The principal disadvantage is that its radius of convergence in position is rather small, about 0.3°. Therefore it is unsuitable for searching for previously unreported sources, or for locating sources with a large uncertainty in position. For these situations a mapping algorithm has been developed. This deconvolves the time-dependent data and produces a position dependent map of a small region of sky. Each orbit's data is deconvolved in turn and added into a summed map, allowing for any drift in spin-axis direction. A point source should give the characteristic Bessel function peak and side-lobes described earlier. Thus a search for significant peaks on the map can be made in order to find any new sources or sources with very imprecise positions which may be within the region of the map.

THE GALACTIC PLANE FROM $l^{II} \sim 20^{\circ}$ TO $l^{II} \sim 50^{\circ}$

Ariel V observations of the region defined by $20^{\circ} \lesssim l^{II} \lesssim 50^{\circ}$ and $|b^{II}| \lesssim 15^{\circ}$ have totaled $\sim$60 days in four epochs - June 1975, May 1976, October 1976 and March-April 1977. The data summation techniques have been used to search for previously unreported sources in the region, typical sensitivities (at a 5σ level) ranging from 0.4 RMC counts s^{-1} for $|b^{II}| \simeq 0^{\circ}$ to 1.0 counts s^{-1} for $|b^{II}| \simeq 15^{\circ}$ (1 RMC count s^{-1} = 13.5 μJy). A 5σ level is regarded as significant for the detection of a previously unreported source because of the large number of independent picture elements in a map (there are typically $\sim$ 2-3 peaks above 3σ on a map). Table 1 lists the previously unreported sources in the region which were detected in this way.

TABLE 1 New Sources detected in the $20^{\circ} \lesssim l^{II} \lesssim 50^{\circ}$ region by Ariel V RMC

Source	Position				Error Circle Radius (90%)	Intensity (Counts s^{-1})
	α(1950)	δ(1950)	l^{II}(deg)	b^{II}(deg)		
A1835-08	$18^h35^m45^s$	$-8^{\circ}21' 15''$	24.060	-0.927	$4'.5$	0.6
A1850+01	$18^h50^m17^s$	$1^{\circ}15' 35''$	34.267	0.293	$5'$	0.4
A1856+08	$18^h56^m40^s$	$8^{\circ}11' 00''$	41.146	2.068	$4'$	0.5

SOURCE POSITION MEASUREMENTS

Analysis of Ariel V data obtained during 1974-76 gave positions for 21 galactic plane sources [2]. We are currently analysing the more recent data and also applying the data summation techniques to obtain positions for the weaker sources which are not detected at a significant level in single orbits. Some preliminary results are shown in Table 2.

TABLE 2 New Ariel V RMC Positions

Source	Position				Error Circle Radius (90%)
	α(1950)	δ(1950)	l^{II}(deg)	b^{II}(deg)	
4U1538-52	$15^h38^m39^s.0$	$-52^{\circ}12' 05''$	327.436	2.184	$2'.3$
4U1543-62	$15^h43^m34^s.3$	$-62^{\circ}25' 10''$	321.754	-6.340	$56''$
4U1658-48	$16^h59^m01^s.6$	$-48^{\circ}42' 26''$	338.938	-4.319	$48''$
4U1728-16 (GX9+9)	$17^h28^m50^s.6$	$-16^{\circ}55' 49''$	8.510	9.034	$16''$
4U1822-00	$18^h22^m49^s.5$	$-00^{\circ}02' 55''$	29.934	5.785	$60''$

Figure 1 shows the new Ariel V position for GX9+9 (4U1728-16) together with previous measurements of its position. The SAS-3 [3], HEAO-1 [4] and Ariel V results are all in good agreement and support the optical candidate proposed by Davidsen et al [5]. The earlier Uhuru [6] and Copernicus [7] results are clearly inconsistent.

Figure 2 shows the Ariel V position for 4U1658-48. To date the only other position measurement for this source is the Uhuru result (this may be because of its highly variable nature). The overlap of the error regions is only marginal and a search of the Ariel V error circle for an optical counterpart is highly desirable.

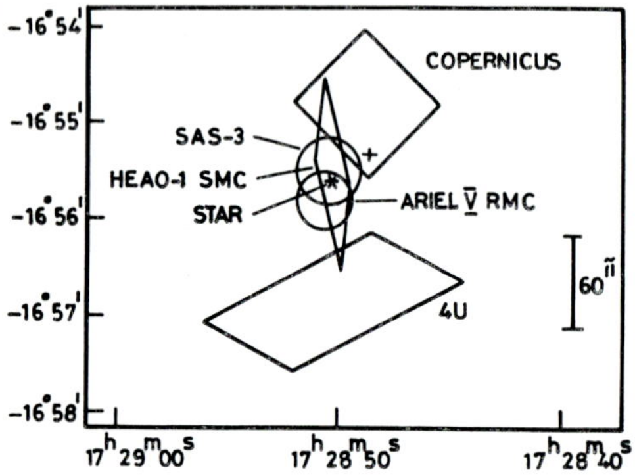

Fig.1. The Ariel V RMC position for GX9+9 (4U1728-15) together with other published positions

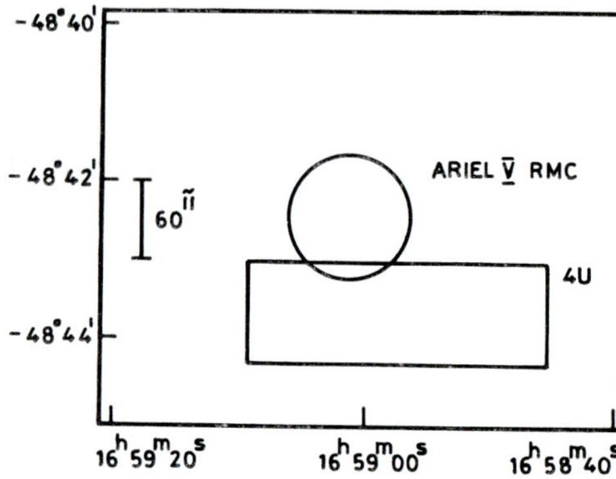

Fig.2. The Ariel V RMC position for 4U1658-48 together with the Uhuru position

CONCLUSIONS

The Ariel V RMC has high positional resolution and a fairly large field of view, and is ideally suited to making observations of the galactic centre region, with its high density of strong sources. Although the sensitivity is fairly modest, most of the sources can easily detected in a single orbit of data integration, and combinations of many orbits can give an increase of at least an order of magnitude in sensitivity, making observations of weaker sources also possible.

ACKNOWLEDGEMENTS

I am grateful to Professor A.P.Willmore and the other members of the Ariel V team at Birmingham for helpful discussions in connection with the present work. I also thank the staff at the UK Control Centre at the Appleton Laboratory, Slough, for their unflagging efforts in satellite data processing and distribution. I acknowledge the financial support of the Science Research Council.

REFERENCES

1. A.P.Willmore, Mon.Not.R.astr.Soc., 147, 387 (1970)
2. A.M.Wilson, G.F.Carpenter, C.J.Eyles et al., Astrophys.J. (Letters) 215, L111 (1977)
3. R.E.Doxsey, K.M.V.Apparao, H.V.Bradt et al.,Nature, 270, 586 (1977)
4. D.A.Schwartz, J.Schwartz, H.Gursky et al., Centre for Astrophysics Preprint CFA/HEA-77-190 (1978)
5. A.Davidsen, R.Malina and S.Bowyer, Astrophys.J., 203, 448 (1976)
6. W.Forman, C.Jones, L.Cominsky et al.,Astrophys.J.(Suppl.) in press (1978)
7. A.P.Willmore, K.O.Mason, P.W.Sanford et al., Mon.Not.R.astr.Soc., 169, 7 (1974)

ARIEL V RMC OBSERVATIONS OF THE GALACTIC CENTRE

C. J. Eyles

Department of Space Research, University of Birmingham, Edgbaston, Birmingham B15 2TT, England

INTRODUCTION

The Ariel V RMC (Rotation Modulation Collimator) was conceived as an instrument for measuring the positions of the brighter X-ray sources to 1 ' and in practice accuracies $\sim 20''$ have been achieved [1]. Additionally, because of the large (17° FWHM) field of view of the instrument, together with the way the Ariel V observing program has been organised, the RMC has performed valuable work by monitoring fairly large areas of sky over long time intervals (in some cases for many tens of days) and this has led to the discovery of a number of transient sources and results for time variations of variable X-ray sources, e.g. [2,3].

As described in the preceding paper we have recently developed techniques for summing data collected over many orbits of the spacecraft by using transformations to allow for changes of the spacecraft attitude. Potentially this increases the sensitivity of the instrument from ~ 40 Uhuru counts s^{-1} for a single orbit to ~ 4 Uhuru counts s^{-1} for 100 orbits (~ 7 days). Two techniques have been developed - firstly there is a "hill-climbing" technique which optimises the position and intensity of a known source to give the best fit to data accumulated over many orbits. This is used for determining positions for the weaker sources (i.e. sources not detected at a significant level in single orbits). A second technique is used to form a "map" of a region of sky, typically $8^{\circ} \times 8^{\circ}$, representing a summation of the data over many orbits. By searching such a map for significant peaks a region of sky may be surveyed for new sources - in particular regions of the sky having a high density of sources (e.g. the galactic bulge) may be surveyed in this way without encountering the problems of source confusion suffered by scanning survey instruments such as Uhuru and the Ariel V Sky Survey Instrument (SSI). Source confusion only becomes a problem for the RMC when the separation of two sources is of the order of the FWHM of the response of the instrument to a point source (separations $\lesssim 0.5^{\circ}$). We present here results obtained using these summation techniques for the region in the immediate vicinity of the galactic centre.

THE GALACTIC CENTRE REGION

The region in the immediate vicinity (i.e. within about 2°) of the galactic centre has been within the effective field of view of the Ariel V RMC for a total of about 90 days over a period of 3 years. We have also made observations of the region using a coded mask X-ray telescope flown on Skylark rocket SL1501 in June

1976 [4].

Observations of this region by the Uhuru satellite [5] revealed the presence of a weak source, designated 4U1743-29, which was about 2° in extent and centred on the galactic nucleus. The data were also consistent with the source being a complex of at least 3 unresolved point sources rather than an extended source. A bright transient source, 4U1735-28, located about 2° from the galactic centre was also observed. It reached a peak intensity of 565 Uhuru counts s^{-1} in March 1971. A similar transient, A1742-289, has been observed by Ariel V [2], with a peak intensity of 160 RMC counts s^{-1} (1 RMC count s^{-1} = 12.14 Uhuru counts s^{-1}) in February 1975. This source was located within 1 ' of the galactic nucleus. However Ariel V RMC data shows that the X-ray emission from the region is usually dominated by the highly variable source A1742-294. Figure 1 shows a light curve of this source. It was first observed in March 1976 [6] but might well have been present in our February 1975 data at a similar level since it is separated by only $\sim 0.5°$ from the transient A1742-289 which was bright during our February-March 1975 observations of the region and would not have been resolved from the latter. Recent data shows that A1742-294 was still present in May 1978 with an intensity of $\sim$6 RMC counts s^{-1} ($\sim$80 μJy).

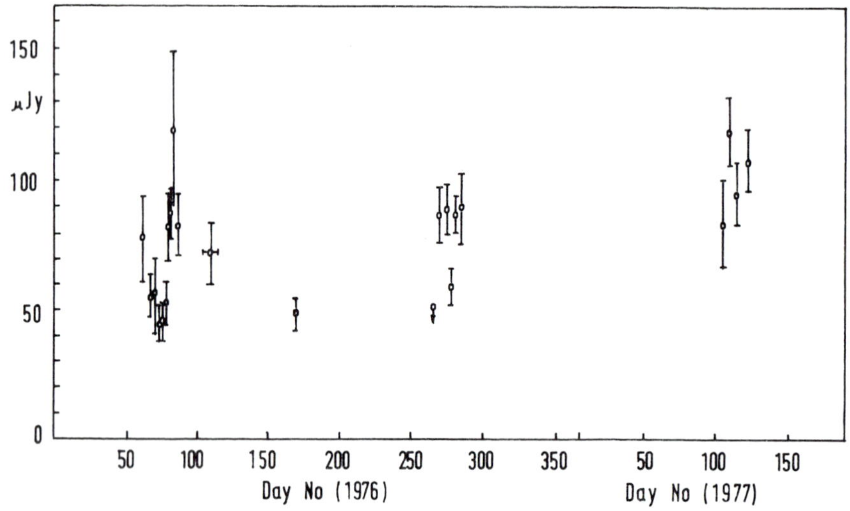

Fig.1. Light curve of A1742-294 (13.5 μJy = 1 RMC count s^{-1})

Two other sources in this region have been observed in the RMC data at a significant level by applying the data summation techniques. One of these, GX+0.2,-0.2, was first detected in the SL1501 rocket data and was found to be present in the RMC data some 100 days later at a rather higher intensity (see Fig.2). This behaviour could be reconciled with a characteristic transient-type light curve but it seems much more likely that the source is another highly variable source. The other, A1741-302, was detected only by summing large blocks of RMC data and was not detected in the SL1501 data. Little can be deduced about the variability of this source. Two sources, GX+1.1,-1.0 and GX+0.2,-1.2, which were detected in the rocket data have not been observed in the RMC data. Table 1 summarises the results for the four sources in the region which were observed in the RMC data. An upper limit of 0.6 RMC counts s^{-1} may be placed on the average intensity of the galactic centre sources which were not detected by the RMC.

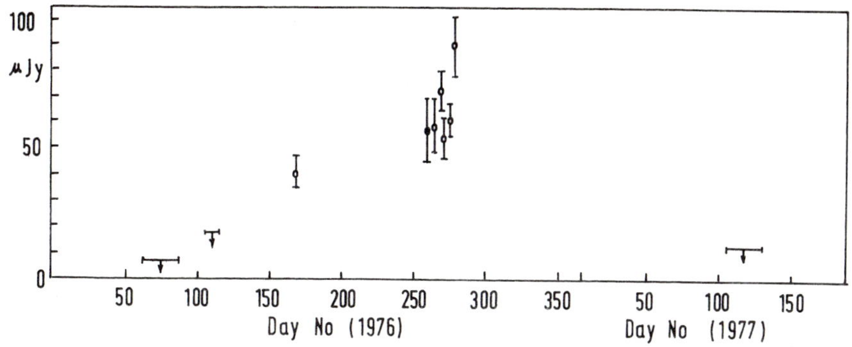

Fig. 2. Light curve of GX+0.2,-0.2

TABLE 1 Sources within 2° of Galactic Centre observed by Ariel V RMC

Source	Position				Error Circle Radius	Intensity
	α(1950)	δ(1950)	l^{II}(deg)	b^{II}(deg)	(90%)	(counts s^{-1})
A1741-302	$17^h41^m54^s.7$	-30°15' 36"	358.797	-0.608	4'.8	0.3 - 0.9
A1742-294	$17^h42^m52^s.2$	-29°29' 03"	359.566	-0.377	1'	3.3 - 8.8
A1742-289	$17^h42^m26^s.4$	-28°59' 56"	359.929	-0.043	27"	160*
GX+0.2,-0.2 (A1743-288)	$17^h43^m52^s$	-28°52' 16"	0.202	-0.240	1'.4	<0.7 - 6.6

*Intensity at peak in Feb.1975; $\lesssim$0.6 counts s^{-1} in 1976, 1977 data

Although the RMC is not capable of imaging extended sources, by subtracted from the total count rate in the detector the contributions due to the charged particle background and known point sources in the field of view, it has been found that there is a considerable enhancement of the diffuse background flux in the direction of the galactic centre. The enhancement can be represented by a line source parallel to the galactic plane of length 9.5 ± 2.5 degrees centred on the galactic centre. The shape of the enhancement is also consistent with the distribution of total matter in the galactic centre region. It is of course not possible to say whether this is a genuinely diffuse source or a population of weak, unresolved point sources. If, for example, the emission is due to globular clusters the intensity would imply an average luminosity of 10^{35}erg s^{-1} per cluster [7].

CONCLUSIONS

It is clear that the galactic centre region shows considerable X-ray activity - a number of transient and other variable sources having been observed in a very small region of the sky. However it can be shown that the density of sources in the region is not incompatible with the distribution of mass within the galaxy, particularly if Population II/disc Population matter contributes to the X-ray emission [4]. The RMC data (excluding the February 1975 data) has been examined for evidence of emission from the galactic nucleus and a marginally significant flux of 0.6 ± 0.2 RMC counts s^{-1} has been found, although this could be due to residual emission from A1742-289. An upper limit of 2 x 10^{36} erg s-1 (2-10 keV) is placed on emission from the nucleus.

ACKNOWLEDGEMENTS

The author acknowledges the other members of the Ariel V team at Birmingham - namely Professor A.P.Willmore, Drs. G.F.Carpenter and G.K.Skinner and Mr. A.M.Wilson - all of whom have contributed to the results presented here. The result on the background enhancement towards the galactic centre is due to Dr. B. Ramsey. Credit is also due to the staff of the Ariel V Control Centre at the Appleton Laboratory for their successful operation of the satellite since its launch in October 1974.

REFERENCES

1. A.M.Wilson, G.F.Carpenter, C.J.Eyles *et al.*, *Astrophys.J. (Letters)*, 215 L111 (1977)
2. C.J.Eyles, G.K.Skinner, A.P.Willmore and F.D.Rosenberg, *Nature*, 257 291 (1975a)
3. C.J.Eyles, G.K.Skinner and A.P.Willmore, *Mon.Not.R.astr.Soc.*, 173, 63P (1975b)
4. R.J.Proctor, G.K.Skinner and A.P.Willmore, *Mon.Not.R.astr.Soc.* In press
5. E.Kellogg, H. Gursky, H.Murray *et al.*, *Astrophys.J. (Letters)*, 169, L99 (1971)
6. G.F.Carpenter, C.J.Eyles, G.K.Skinner *et al.*, *Mon.NotR.astr.Soc.*, 179, 27P (1977)
7. B.Ramsey, private communication

Theories on Accretion in Binary Systems

HER X-1 CYCLOTRON LINES

J. Trümper

Max-Planck-Institut für Physik und Astrophysik,
Institut für extraterrestrische Physik, 8046 Garching, West Germany

ABSTRACT

The hard X-ray spectrum of Hercules X-1 contains spectral features at 58 keV, and probably at 110 keV, which have been interpreted as first and second harmonic cyclotron lines generated in the hot polar plasma of the magnetic neutron star. The corresponding magnetic field strength is 5.3×10^{12} Gauss (Trümper et al., 1978). Following this discovery, several balloon and satellite observations of Hercules X-1 have been made in hard X-rays and we summarize the present observational evidence regarding the cyclotron lines.

The measurement of cyclotron lines represents the first, and possibly the only, direct method of determining neutron star magnetic field strengths. Furthermore, the knowledge of cyclotron line widths and intensities in different harmonics should be of key importance for an understanding of the physical conditions in polar hot spots and accreting columns. In this context we briefly summarize theoretical work which has been done in attempts to interprete the observed spectral features.

INTRODUCTION

Cyclotron line spectroscopy rests on the well-known and simple fact that the transverse energy of electrons moving in a homogeneous magnetic field is quantized in multiples of $\hbar\omega_H$ where ω_H = eH/mc is the Larmor frequency. The cross sections and probabilities of the corresponding electron transitions are very large and cyclotron features both in emission or absorption are dominant in the spectra of magnetized plasmas under a wide range of physical conditions.

In fusion plasmas at kilogauss magnetic fields the resonance occurs in the microwave range. Cyclotron lines up to rather high orders ($s \geq 30$) have been observed in laboratory plasmas and cyclotron resonant heating and losses play a major role in this branch of science and technology. In high magnetic field facilities ($\sim$ 300 KG) the resonance is in the infrared range and cyclotron line spectroscopy has become a powerful tool of exploring the band structure of semiconductors. For white dwarfs having even somewhat larger magnetic fields (10^5 to 10^8 Gauss) cyclotron radiation plays an important role for the optical ultraviolet emission and I am sure that we will hear more about that in the next talk [1].

Finally, in the case of neutron stars, we know from radio pulsar physics, binary neutron star spin-up and general considerations about flux compression during gravitational collapse that their surface magnetic field strengths must be of the order of 10^{12} Gauss. Consequently, the fundamental cyclotron frequencies

$$\hbar\omega_H = 11 \text{ keV} \times H_{12}$$

lie in the X-ray domain where H_{12} is measured in 10^{12} Gauss.

The influence which cyclotron resonance absorption and scattering may have on the spectral time variability of rotating accreting neutron stars has been discussed by Elsner et al. [2], with particular emphasis on slow rotators like Vela X-1 which exhibit a very complex energy dependence of pulse shapes.

Another important aspect is that the X-ray emitting hot plasma of an accreting neutron star must be confined to a rather small region which is located at the magnetic poles very close to the stellar surface. Therefore, the magnetic field is expected to be quite homogeneous in the emission region leading to the possibility of rather narrow cyclotron resonances. This had been first stressed by Gnedin et al. [3] and Basko et al. [4] several years ago.

PRESENT OBSERVATIONAL EVIDENCE

The first evidence for a rather strong and narrow cyclotron feature was found in spectral data of Her X-1 which were obtained with a Phoswich scintillation counter during a balloon flight from Palestine, Texas, by a collaboration of the Astronomical Institute of the University of Tübingen and MPI Garching [5, 6].

Since its discovery by Uhuru [7] the complex spectral variability of Her X-1, connected with the 1.24 sec rotational period, the 1.7 day orbital period and the 35 day cycle, had been studied by a large number of investigators. A comprehensive review of this work is given in [8]. As far as the high energy flux ($\geq$ 20 keV) is concerned it was known in 1976 that the source is pulsing up to 30 keV [9] and that at $\sim$ 25 keV a spectral break occurs beyond which the spectrum steepens dramatically [10], thus hampering observations at higher energies. The May 3, 1976 observation established for the first time that the 1.24 sec pulsations continue to be present up to $\sim$ 80 keV [11]. It also showed that beyond the spectral break the 1.24 sec main pulse is characterized by a narrow single peak as compared with a double peak normally observed below $\sim$ 25 keV [11].

At $\sim$ 58 keV a strong and rather narrow line feature was found in the pulsed spectrum, containing a few percent of the total X-ray luminosity [5]. It was immediately clear that such a feature could not be produced by atomic or nuclear processes for abundance reasons. On the other hand, in view of the large magnetic fields expected for neutron stars and the large abundance of electrons in the emitting plasma, a cyclotron origin seemed entirely plausible. Interpreted as first harmonic cyclotron emission, the corresponding magnetic field strength is 5.3×10^{12} Gauss, uncorrected for the gravitational redshift ($\sim$ 10 to 30 %).

Of course it appears highly desirable to look for signatures which unequivocally would prove the cyclotron origin of the lines. One possibility is to measure the characteristic polarization of the cyclotron radiation. However, apart from physical difficulties [12] such an undertaking seems prohibitive for present instruments in view of the low line fluxes. The other, much more promising, method is to search for second and even higher harmonic cyclotron features which can be expected to occur at their characteristic energies, provided the physical conditions (high temperature, nonthermal processes) allow their excitation.

Now there was a chance to search for the second harmonic with enhanced sensitivity, since during the 4.5 hours of the May 3, 1976 observation Her X-1 underwent a flux decrease by a factor of ~ 3 which may be related to the onset of a pre-eclipse dip [11]. Taking the data from the beginning of the flight where the signal-to-noise ratio was highest, spectra could be obtained which clearly show the 58 keV peak, but in addition exhibit a significant spectral excess around 110 keV (Fig. 1a). This is just the energy where, including relativistic corrections, the second harmonic is expected to occur. Figure 1b shows one of the deconvoluted spectra derived from these data [6].[1]

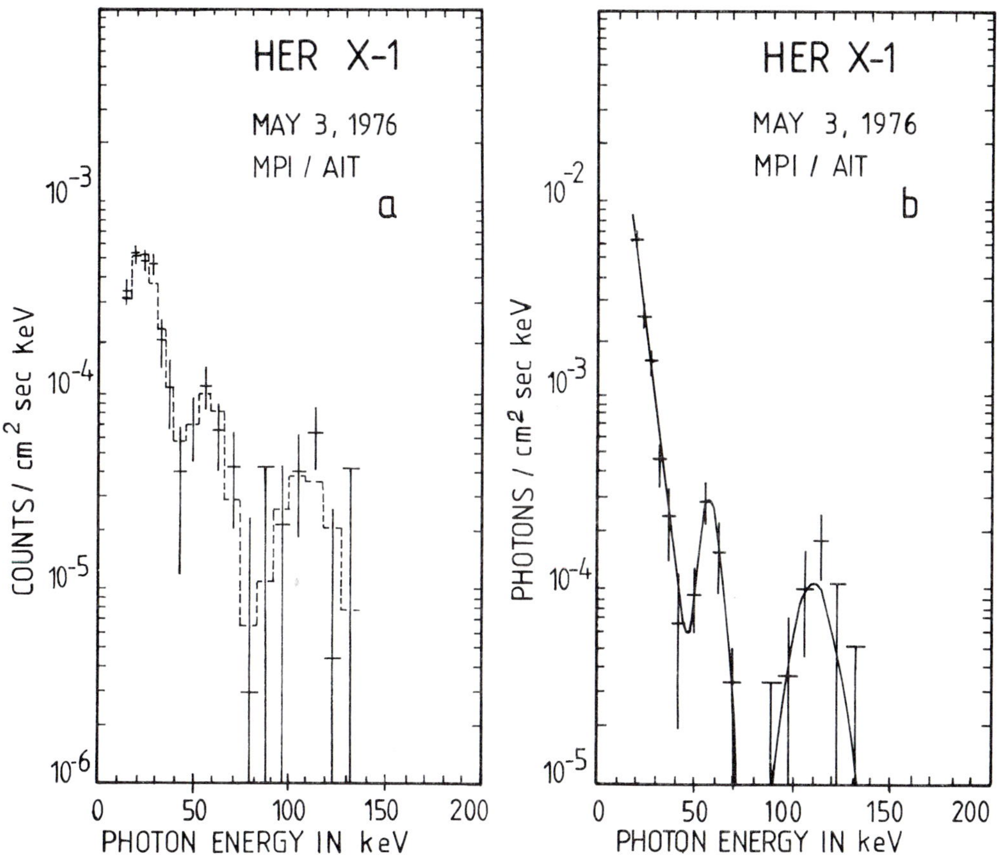

Fig. 1 Count rate (a) and deconvoluted photon (b) spectra of Her X-1. The spectra refer to the pulse peak of the 1.24 sec pulsation.

[1] The spectrum shown in Fig. 1a is the so-called Pulse minus Off-Source (POS) spectrum and identical to that published in [6] except for a linear instead of a logarithmic scale on the abscissa.

In an effort to check these findings, a much larger (factor 6.5) Phoswich scintillation detector was built by the MPI/AIT collaboration and flown three times from Palestine, Texas in fall 1977 at different phases of the Her X-1 35 day cycle [13, 14, 15].

During the flight of September 3, 1977 Her X-1 was in the "on state", at a phase 0.12, (with respect to turn-on) close to that of the May 3, 1976 observation. The new spectral data obtained, which will be presented by W. Voges et al. [16] at this meeting, clearly confirm the existence of a pulsed spectral feature. The excess over the extrapolated continuum has a statistical significance of $\sim 11\,\sigma$, and the best fitted line energy is 55.4 ± 2.7 keV consistent with the earlier result, within the error limits. However, there are also significant differences as compared with the May 3, 1976 data: 1) Both continuum and line flux were lower by a factor of ~ 2 to 2.5 on September 3, 1977. 2) The second harmonic could not be detected. It seems possible that the absence of the second harmonic peak is related to the lower X-ray luminosity, e.g. if there is a correlation between source luminosity and temperature.

Considerable efforts have been undertaken during the last 18 months by several groups to search for Her X-1 cyclotron lines. Table 1 contains a summary of results reported so far. An Ariel-5 measurement made with the Imperial College hard X-ray detector during the entire February 1977 "on state" of the 35 day cycle indicated a line flux at 64 ± keV which is roughly one order of magnitude larger than the other quoted line fluxes. However, this result has a rather marginal statistical significance (at most 2 - 3 σ) and is contradicted by the upper limits quoted in Table 1. These show that a line feature at ~ 60 keV is not significantly stronger when averaged over a full "on state" as compared to the fluxes measured during the relatively short balloon flight [19].

The upper limit reported by Pravdo et al. [23] is of particular significance. It has been obtained during the February 1978 pointing of HEAO-1 using the A-2 proportional counter. This detector has little efficiency at ~ 58 keV but a considerably better energy resolution than scintillation counter spectrometers. The quoted upper limit corresponds to an assumed line width of 10 keV. It is consistent with the 1977 balloon result, but inconsistent with the 1976 data, underlining that Her X-1 was particularly bright at the beginning of the 1976 balloon flight.

Most recently, measurements with the high-energy X-ray detector on HEAO-1 have confirmed the existence of a pulsed spectral excess in the Her X-1 spectrum at ~ 58 keV. The data were taken during three satellite orbits on February 24, 27 and 28, 1978 and were treated independently by the UCSD and MIT groups collaborating on the A-4 experiment [20, 21a, 21b]. The observed excesses over the extrapolated lower energy continua have a high statistical significance. If a continuum plus emission line is fitted to the data, the resulting line energies are 61 ± 4, 60 ± 3 and 59 ± 4 keV, which agree with the MPI/AIT results within the error limits. The line fluxes observed during the first two orbits when Her X-1 was in its high state are close to the MPI/AIT results from 1977. The third observation took place during an intensity dip where both continuum and line flux were lower by a factor ~ 3. At the position of the second harmonic the HEAO-1 data yield 3 σ upper limit of $\sim 10^{-3}/\mathrm{cm}^2$ sec, which is consistent with the September 3, 1977 MPI/AIT balloon result.

An important question is whether other compact X-ray sources can be found which exhibit similar spectral features. Prime candidates for such a search are the other ~ 14 pulsating X-ray sources. The NRL group has investigated four of them (Cen X-3, GX 1 + 4, GX 301-2, GX 304-1) during a balloon flight late in 1977 in Australia [22].

TABLE 1. Reported Line Fluxes from Her X-1 around 58 keV

Observation	Ref.	Instr. Area (cm^2)	Energy Resol. at 60 keV FWHM %	Line Flux ($cm^{-2}sec^{-1}$)
MPI/AIT Balloon '76	5,6	120	22	(~5σ)
MPI/AIT Balloon '77	16	800	21	(~11σ)
Ariel-5 '77	17	8	42	(~2-3σ)
Saclay Balloon ('75)	18	380	25	
OSO-8 Proportional Counter '77	19	~20 (60 keV)	12	
OSO-8 Scintillation Counter '77	19	28	40	
HEAO-1 A-2 '78	23			
HEAO-1 A-4 '78	20 21a,21b	~200	~29	

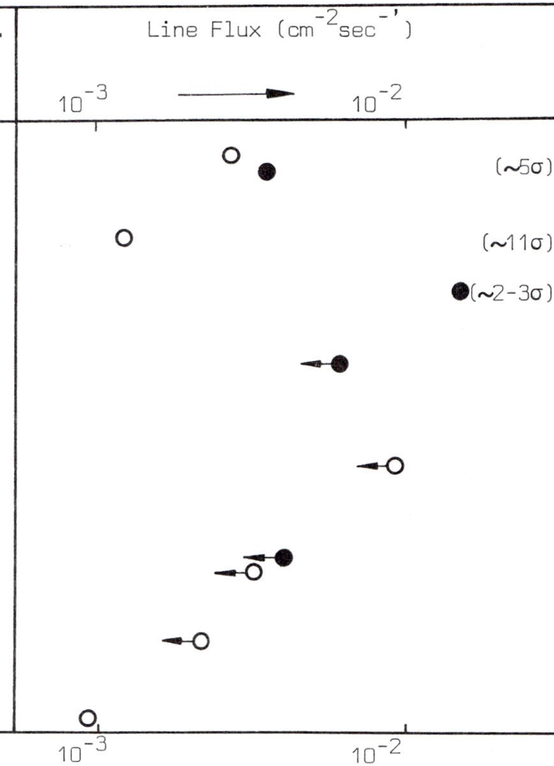

○ pulsed line flux

● time integrated line flux

←○ 3σ upper limit

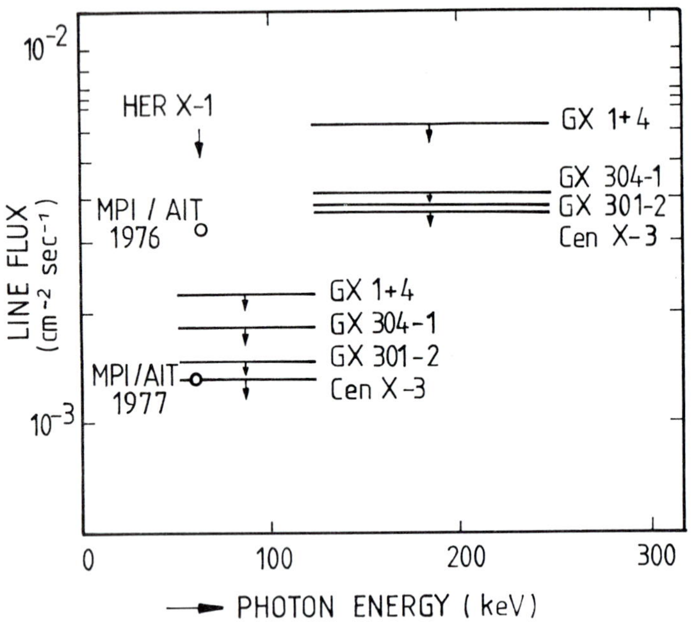

Figure 2 depicts the preliminary results of the analysis which so far has been limited to an inspection of the time integrated spectra and to the energy range > 50 keV, corresponding to magnetic field strengths > 5 x 10^{12} Gauss. Only upper limits could be obtained which are comparable to the reported Her X-1 line fluxes. A line search in the pulsed spectra which has not been done yet, will have a factor of 3 higher sensitivity [22].[2]

Fig. 2 Summary of 3 σ upper limits on spectral lines in the time integrated flux of pulsating X-ray sources [22].

INTERPRETATION OF THE LINE FEATURES

The Her X-1 data obtained by the MPI/AIT and UCSD/MIT groups on five different sightings clearly show a tremendous spectral excess in the 50 - 65 keV region over the extrapolated lower energy (≥ keV) continuum. This excess can hardly be produced by a second continuous spectral component like a power law or exponential tail or a black-body spectrum. A notch appearing in some of the spectra at ~ 45 keV and a rather steep spectral slope beyond ~ 65 keV are indicative of a line feature [16].

It has been pointed out repeatedly that a line in that energy region cannot be produced by atomic or nuclear processes [11], while a cyclotron origin appears possible and plausible. However, in this case there are two possibilities [6], c.f. also Fig. 3.

- There may be a cyclotron emission line at ~ 58 keV, and a possible second one at ~ 110 keV

- There may be a cyclotron absorption feature at ~ 40 keV, and a

[2] See also note at the end of the paper

possible second one at ~ 80 keV. In particular, there may be a rather broad absorption feature between the spectral break at 25 keV and ~ 45 keV.

It appears difficult to distinguish between either alternative on the basis of the available experimental data. This had been stressed already in [6] since spectral fits with exponential continuum plus emission and absorption line respectively gave equally good χ^2 fits.

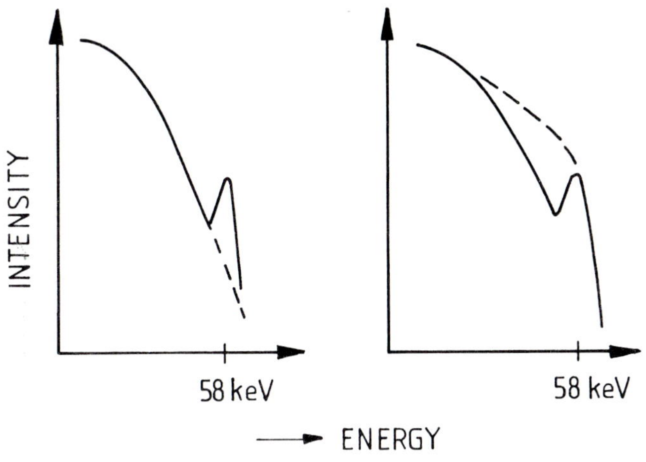

Fig. 3 Interpretation of the spectral feature at 58 keV as emission or absorption line.

Regarding the determination of magnetic field strengths, there is little difference between the two alternatives; the corresponding values are ~ 5.3×10^{12} Gauss and ~ 4×10^{12} Gauss, which both have to be corrected for the gravitational redshift.

If, however, one wants to go one step further and use cyclotron line spectroscopy (line widths and shapes, equivalent line widths) as a means of plasma diagnostics, it is necessary to know whether one has the case of emission or absorption. It seems possible that this problem can be solved with a minimum of theoretical assumptions on the basis of experimental data alone, if signatures can be detected which are typical either for absorption or emission. Asymmetrical spectral shapes or characteristic dependence of spectral shape on pulse phase will be important in this context.

THEORETICAL CONSIDERATIONS

Can theoretical arguments and model considerations help us to come to a conclusion? The occurrence of cyclotron emission lines in the spectra of accreting neutron stars had been suggested by Gnedin et al. [3] and Basko et al. [4] as early as 1974. More recently models of the radiating hot polar plasma have been discussed by Sunyaev et al. [24] and Meszaros [25] which support an emission line explanation.

On the other hand, an interpretation of the Her X-1 spectral feature in terms of cyclotron absorption was put forward by Daugherty et al. [26] early in 1977. Very recently this view has been supported by Lamb [27], Fabian [28] and Pravdo et al. [23].

The answer to the question, whether the cyclotron resonance manifests itself in emission or absorption depends on the distribution of density and temperature in the outer layers of the emitting plasma. In the case of Her X-1 we know from the overall spectrum that the temperature of the "visible" surface is of the order of $kT \sim 8$ to 20 keV, and the magnetic field strength is $\sim 5 \times 10^{12}$ Gauss. The narrowness of the spectral feature limits magnetic line broadening (5,6), and in the case of a dipolar field the radial extent of the emission region cannot be larger than $\sim$ 10 % of the stellar radius. For a width of the accretion column of $a = 10^5$ cm, a mass accretion rate $\dot{M} = 10^{17}$ g/sec ($L = 10^{37}$ erg/sec) leads to a plasma density in the free fall region of $n \approx 10^{21} cm^{-3}$. The actual densities in the emission region will be higher since the material is decelerated there. However, it is very likely that they are well below $10^{28} cm^{-3}$. It then holds

$$\text{free free opacity } \varkappa_{ff} \ll \text{Thomson opacity } \varkappa_{Th}$$

Under such conditions the continuum intensity must be below the black body limit by a factor $\varkappa_{ff}^{1/2} \varkappa_{Th}^{-1/2}$ (e.g. Felten et al. 1972 [35]).

In the case of a dense plasma ($\varkappa_{ff}^{1/2} \varkappa_{Th}^{1/2} a \gg 1$) one has the following situation: Since the cyclotron opacity is very high near the resonance ($\varkappa_c \gg \varkappa_{Th}$) the "resonance photons" come from a very thin outer layer as opposed to the "continuum photons" which escape from greater depths. If there is an outward temperature decrease in the surface layers one expects cyclotron absorption to occur ("Fraunhofer cyclotron lines").

On the other hand, cyclotron line emission can take place if the dense and hot polar plasma responsible for the continuum is surrounded by an even hotter plasma which is optically thin for Thomson scattering but has a substantial optical thickness in the cyclotron resonance. In this case "corona type cyclotron lines" would be emitted. Even if there is no outward temperature increase, cyclotron emission could appear in certain angular intervals due to the strong dependence of the cross sections on the polar angle (Sunyaev et al., 1978, [24]).

Unfortunately, so far we have only crude models but no self-consistent picture of the three-dimensional density and temperature distributions in the emission region. This would require the simultaneous solution of the following problems:

- how is the kinetic energy of the infalling gas converted into thermal and turbulent energy and other modes like plasma waves etc.?
- how does the radiation field influence the plasma flow?
- what does the radiative heat transfer look like and what role is played by heat conduction?

Basic ingredients for such a theory are the photon cross sections in a highly magnetized plasma, which strongly depend on frequency and direction. Until recently the Thomson cross sections under such conditions were explicitly known only in the cold plasma limit and for the limiting cases $\omega/\omega_H \to 0$ and polar angles $\theta \sim 0$ and $\theta \sim \pi/2$ [29]. A considerable extension to other frequencies and angles has recently been worked out by Börner et al. [30] and Ventura [31]. The influence of relativistic effects on cross sections has been studied by Herold [32]. The effects of vacuum polarization on the Thomson cross sections which become important in strong magnetic fields have been treated by Mészáros et al. [33].

Thus, the cyclotron line observations have revitalized theoretical work in this field and one may hope that ultimately we will know more not only about neutron star magnetic field strengths but also about the interesting and exotic physical processes which determine the appearence of these objects.

NOTE ADDED AFTER THE CONFERENCE

Recently the hard X-ray experiment on HEAO-1 has detected a spectral feature near 20 keV in the pulsed spectrum of the transient X-ray binary 4U0115 + 63 which may be due to cyclotron absorption [34].

REFERENCES

1. D. Q. Lamb, this volume
2. R.F. Elsner et al., Nature 262, 356 (1976)
3. Yu. N. Gnedin et al., Astron. Astrophys. 36, 379 (1974)
4. M. M. Basko et al., Astron. Astrophys. 42, 311 (1975)
5. J. Trümper et al., Annals New York Academy of Sciences 302, 538 (1977)
6. J. Trümper et al., Astroph. J. 219, L105 (1978)
7. H. Tananbaum et al., Astroph. J. 174, L143 (1972)
8. S. H. Pravdo, Ph. D. Thesis, Univ. of Maryland (1976)
9. P. C. Joss et al., 1976, Astroph. J. 214, 874 (1977)
10. R. H. Becker et al., Astroph. J. 214, 879 (1977)
11. E. Kendziorra et al., Astroph. J. 217, L93 (1977)
12. R. N. Novick et al., Astroph. J. 215, L117 (1977)
13. W. Voges et al., IAU Circular 3184 (1978)
14. R. Staubert et al., this volume
15. J. Trümper, Proc. GSFC Symposium on Gamma Ray Spectroscopy in Astrophysics (1978)
16. W. Voges et al., this volume
17. M. J. Coe et al., Nature 268, 508 (1977)
18. P. Durouchoux et al., Proc. GSFC Symposium on Gamma Ray Spectroscopy in Astrophysics (1978)
19. B. R. Dennis et al., Proc. GSFC Symposium on Gamma Ray Spectroscopy in Astrophysics (1978)
20. J. L. Matteson et al., this volume

21a R. E. Rothschild, Workshop on Cyclotron Lines from Accreting Neutron Stars, Munich, 1978

21b W. H. G. Lewin, Workshop on Cyclotron Lines from Accreting Neutron Stars, Munich, 1978

22. J. D. Kurfess et al., APS Spring Meeting 1978 and private communication

23. S. H. Pravdo et al., this volume

24. R. A. Sunyaev et al., to be published

25. P. Meszaros, Proc. GSFC Symposium on Gamma Ray Spectroscopy in Astrophysics (1978)

26. J. K. Daugherty et al., Phys. Rev. D., in press

27. F. K. Lamb, Workshop on Cyclotron Lines from Accreting Neutron Stars, Munich, 1978

28. A. C. Fabian, Workshop on Cyclotron Lines from Accreting Neutron Stars, Munich, 1978

29. J. Lodenquai et al., Astrophys. J. 190, 141 (1974)

30. G. Börner et al., 1978 to be published

31. J. Ventura, 1978 to be published

32. H. Herold, 1978 to be published

33. P. Meszaros et al., 1978 to be published

34. W. Wheaton et al., IAU Circular 3238 (1978)

35. J. E. Felten & M. J. Rees, 1972. Astron. Astrophys. 17, 226

Binary X-Ray Emitting Systems — Theory

EVOLUTION OF BINARY X-RAY SYSTEMS

B. Paczyński

N. Copernicus Astronomical Center, Polish Academy of Sciences, 00-716 Warszawa, ul. Bartycka 18, Poland

Among the galactic X-ray sources there is one subgroup: massive binaries, which has been very well studied observationally and theoretically. The optical components are O or B type giants, nearly filling their Roche lobes, with masses in the range of 15 to 30 solar masses. The X-ray components are either neutron stars (X-ray pulsars) like Cen X-3, Vela X-1, SMC X-1, 3U 1700-37, 3U 1223-62, 3U 1538-52, or possibly black holes like Cyg X-1 and Cir X-1. LMC X-4 and 4U 1145-61 may be similar systems. The evolution of these systems has been discussed by van den Heuvel and Heise [1], van den Heuvel and De Loore [2], Massevitch et al. [3], Ziółkowski [4], and most recently it was reviewed by van den Heuvel [5]. There is a major unsettled problem here: what is the duration of the bright X-ray phase in the evolution of a massive binary? There are attempts to provide the answer by analyzing the statistics of massive X-ray binaries and massive non-X-ray stars. The original estimates by van den Heuvel and others indicated a very short duration of the X-ray phase of evolution, about ten thousand years, essentially consistent with the thermal, i.e. Kelvin-Helmholtz time scale of the massive OB type giant. More recently, Ziółkowski [6] came to the conclusion that the X-ray phase should last for up to a milion years, which is close to the nuclear time scale. If that is true, then the OB giants in X-ray binaries should be still burning hydrogen in their cores. The observed spin-up time scales for X-ray pulsars are in the range 100 to 10,000 years, much less than the 1,000,000 years suggested for the lifetime. This leads to the conclusion [6] that the X-ray phase is intermittant: the spin-ups are associated with accretion and the "on" phase, while there are also spin-downs associated with the "off" phases. Ziółkowski [6] requires also a substantial mass loss from those systems to explain the overluminosity of the OB type components.

Future evolution of massive X-ray binaries must lead to the formation of a contact system because of the evolutionary expansion of the OB giant. According to Flannery and Ulrich [7] subsequent expansion of a common envelope should lead to mass ejection from the binary system through the outer Lagrangian point. According to Taam, Bodenheimer and Ostriker [8] subsequent expansion of a common envelope should lead to the formation of a giant or supergiant star with two compact cores spiraling towards each other. This phase may end in two ways. Either the giant star loses its envelope and the two cores are left as a very short period binary like the binary radio pulsar [9], or the two cores merge and a single red supergiant is formed [10].

There is plenty of observational evidence that all massive X-ray binaries are interactive systems, i.e. the matter is presently transferred from one component to another, most likely by means of a stellar wind. Her X-1/HZ Her is the only low mass system for which we have convincing evidence of interaction, in this case through the Roche lobe overflow. This system shares many properties with well known cataclysmic binaries: novae and dwarf novae. One component is close to the main sequence and fills up its Roche lobe. Matter flows through the inner Lagrangian point towards the compact component and forms an accretion disk around it. The compact component is at the final stage of stellar evolution, but the total mass and the angular momentum of the binary are small. A very considerable loss of mass and angular momentum is required during the previous evolution of such a binary. Her X-1/HZ Her, just like cataclysmic variables must have been initially a long period binary. As a result of evolutionary expansion of the initially more massive component a contact system, and later a common envelope binary was formed. The core of the evolved star and the main sequence companion spiralled towards each other. At a certain time the common envelope was lost and a short period binary was left [11], [12], [13], [8]. Later the core of the initial primary exploded leaving a neutron star remnent. Finally, the secondary evolved off the main sequence and filled up its Roche lobe, giving rise to mass transfer onto the compact component. At this phase of evolution the system became an X-ray binary.

It is generally believed that the majority of strong galactic X-ray sources are interacting close binaries. Those sources which do not have OB type giant optical companions are believed to be low mass binaries, possibly the descendants of cataclysmic variables [14], [5], [15], [16]. However, it should be emphasized that even though there is marginal evidence for the binary nature of Sco X-1 and Cyg X-2 [17], there is certainly no direct evidence for a binary nature or for mass transfer between components in the galactic bulge sources, the globular cluster sources, or in the bursters, i.e. in the type 2 sources of Joss and Rappaport [15]. A standard model for a type 2 source requires a neutron star or a black hole as the accreting component, and a low mass dwarf overflowing its Roche lobe as the mass losing component. This model can explain the very high ratio of X-ray to optical luminosity for the type 2 sources; typically a ratio of 1,000:1.

It is striking that none of the type 2 sources is eclipsing, and none of their dwarf companions shows the reflection effect modulated with the orbital period, as seen in the Her X-1/HZ Her system. Joss and Rappaport realized this difficulty, but considered it to be not devastating to the model. The difficulty can be reduced if the accretion disk around the component is geometrically and optically thick, so that it can shield the dwarf companion from the X-ray flux. In this case an observer close enough to the orbital plane to see the eclipses would not see any X-ray emission at all, because the compact source would be permanently obscured by the disk [16]. The disks in cataclysmic variables may be thick [18], and the same may be true for the type 2 X-ray binaries.

There is one highly uncomfortable property of the proposed model for the type 2 X-ray sources. It leaves no observable evidence for the existence of the dwarf companion to the X-ray source,The optical light from the type 2 sources may be understood either as the accretion luminosity of the disk itself, or as due to reprocessing of X-rays by the disk [19], [20]. Therefore, one may ask why do we need the unobservable dwarf companion at all? It is generally considered to be needed as a supply of matter to be accreted by the compact component of the binary. However, this need exists only if accretion disks cannot store matter for an appreciable length of time, a common but unjustified assumption. In fact there is hardly any observational or theoretical reason to believe that disks cannot store matter. Let us suppose they can. In that case a massive disk can provide the supply of matter needed to power the accreting compact object and produce strong X-ray emission [21], [22]. There are various ways in which a massive disk can form. It can be left over from the collapse of a rotating stellar core. The inner parts of the core could form a neutron star or a black hole, while the outer parts could form a disk. A massive disk can be also formed as a result of mass transfer in a close binary. Nothing is known about the lifetimes of the disks. Perhaps some of them can survive for more than the Hubble time, and we may have a disk in a detached binary left over from the previous evolutionary phases.

It seems entirely possible that type 2 sources are compact objects with massive accreting disks. They may be either single objects or detached binaries at the present time. I believe that the main unsolved problem for the type 2 sources is their present nature. Are they low mass interacting binaries, or are they just compact objects with massive disks around them? In the second case the next question is: are these single objects or are they detached binaries Is it possible that some of the sources, in particular those in globular clusters, are massive black holes accreting from the interstellar medium [23], [24]? Since the nature of the type 2 sources is so poorly known, it is premature to discuss their evolutionary status.

There is a number of transient X-ray sources which seem to be detached binaries. The best studied are the 104 second pulsar A 0535+26 [25], the 3.6 second pulsar 4U 0115+63 [26], and the nonpulsing transients: A 0620-00 [27], and Aquila X-1 [28]. The first two have OB type optical companions and are probably related to massive X-ray binaries. However, the transient pulsars are almost certainly members of well detached binaries. Therefore any mass transfer through the stellar wind is very inefficient. It seems possible that because of the low accretion rate, matter accumulates near the neutron star during the "off" state and gets accreted during the "on" state. It is interesting that the period of the X-ray pulsar A 0535+26 was always decreasing during the X-ray "on" state, while there was very little change in the pulsar period between the eruptions. It is possible that this transition source demonstrates what Ziółkowski [6] claims the massive X-ray binaries are doing on a somewhat longer time scale. In the case of A 0620-00 and Aquila X-1 the optical companion is well-detached; it is a dwarf star and any stellar wind is almost certainly very weak. It is hard to understand how such a hypothetical wind could possibly feed the X-ray source [27], [28]. Very likely the source of matter for accretion is a massive disk around the compact component.

Cyg X-2, one of the best-studied type 2 sources, and an old candidate for low mass binary status, has unknown orbital period. There are claims that the period is close to 11 days [29], [17]. If this period is correct, then the binary with a dwarf optical companion must be detached, and Cyg X-2 would be the prime candidate for a compact source with a massive accretion disk.

It is a great pleasure to acknowledge many stimulating and very informative discussions with Dr. Janusz Ziółkowski.

REFERENCES

1. E. P. J. van den Heuvel and J. Heise, Nature, 239, 67 (1972).
2. E. P. J. van den Heuvel and C. De Loore, Astr. Astrophys., 25, 387 (1973).
3. A. G. Massevitch, A. V. Tutukov and L. R. Yungelson, Astrophys. Space Sci., 40, 115 (1976).
4. J. Ziółkowski, Annals N. Y. Acad. Sci., 302, 47 1977 .
5. E. P. J. van den Heuvel, Annals N. Y. Acad. Sci., 302, 14 (1977).
6. J. Ziółkowski, in preparation (1978).
7. B. Flannery and R. Ulrich, Astrophys. J., 212, 533 (1977).
8. R. E. Taam, P. Bodenheimer and J. P. Ostriker, Astrophys. J., 222, 269 (1978).
9. J. H. Taylor, R. A. Hulse, L. A. Fowler, G. E. Gullahorn and J. M. Rankin, Astrophys. J., 206, L53 (1976).
10. K. S. Thorne and A. Żytkow, Astrophys. J., 212, 832 (1977).
11. B. Paczyński, in: IAU Symposium 73, p. 75 (Structure and Evolution of Close Binary Systems, eds. P. Eggleton, S. Mitton and J. A. Whelan, D. Reidel, Dordrecht, Holland, 1976).
12. H. Ritter, in: IAU Symposium 73, p. 205 (Structure and Evolution of Close Binary Systems, eds. P. Eggleton, S. Mitton and J. A. Whelan, D. Reidel, Dordrecht, Holland, 1976).
13. R. F. Webbink, PhD thesis, Cambridge University 1975 .
14. H. Gursky, in: IAU Symposium 73, p. 19 (Structure and Evolution of Close Binary Systems, eds. P. Eggleton, S. Mitton and J. A. Whelan, D. Reidel, Dordrecht, Holland, 1976).
15. P. C. Joss and S. Rappaport, Nature, in press (1978).
16. M. Milgrom, Astr. Astrophys., 67, L25 (1978).
17. A. P. Cowley, Annals N. Y. Acad. Sci., 302, 1 (1977).
18. J. Madej and B. Paczyński, in: IAU Colloquium 42, p. 313 (The Interaction of Variable Stars with their Enviroment, eds. R. Kippenhahn, J. Rahe and W. Strohmeir, Bamberg, 1977).
19. N. I. Shakura and R. A. Sunyayev, Astr. Astrophys., 24, 337 (1973).
20. V. M. Lyuty and R. A. Sunyaev, Astr. Zurnal, 53, 511 (1976).
21. B. Paczyński and M. Jaroszyński, Acta Astr., 28, 111 (1978).
22. Ch. Alcock and B. Paczyński, in preparation (1978).

23. J. Bahcall and J. P. Ostriker, Nature, 256, 23 (1975).
24. J. Silk and J. Arons, Astrophys. J., 200, L131 (1975).
25. S. Rappaport, P. C. Joss, H. Bradt, G. W. Clark and J. G. Jernigan, Astrophys. J., 208, L119 (1976).
26. S. Rappaport, G. W. Clark, L. Cominsky, P. C. Joss and F. Li, Astrophys. J., 224, L1 (1978).
27. J. B. Oke, Astrophys. J., 217, 181 (1977).
28. B. Margon, J. I. Katz and L. D. Petro, Nature, 271, 633 (1978).
29. S. S. Holt, E. A. Boldt and P. J. Serlemitsos, Astrophys. J., 205, L143 (1976).

SPIN DOWN OF NEUTRON STARS IN X-RAY BINARY SYSTEMS

R. E. Davies and J. E. Pringle

Institute of Astronomy, Madingley Road, Cambridge CB3 0HA, England

INTRODUCTION

More than half of the X-ray sources that have been observed to display regular pulses have a period, P, longer than a hundred seconds. We adopt the usual model in which the periodic X-ray source is taken to be a magnetised neutron star lying in a close binary system so that the neutron star accretes material from the wind of its companion, and the regular pulses arise from the rotation of the neutron star. By analogy with pulsars the neutron star is assumed to be born spinning rapidly ($P < 1s$). It then undergoes a phase in which its rotation is slowed down, initially by conventional pulsar energy loss mechanisms, and subsequently by the interaction between its magnetosphere and the stellar wind. We concentrate here on the latter process.

SPIN DOWN MODELS

Several authors have suggested order-of-magnitude estimates for the torque and rate of loss of rotational energy experienced by the neutron star based on various assumptions about this interaction (A: Illarionov and Sunyaev, [1]; Fabian, [2]; B: Davidson and Ostriker, [3]; Kundt, [4]; C: Shakura, [5]; Lipunov and Shakura, [6]). Denoting by μ ('Mach number of rotation') the ratio of the rotational velocity of the boundary of the magnetosphere (at radius r_m) to the escape velocity there (v_m) (i.e. $\mu = (2\pi/P).(r_m/v_m)$, torque (B) exceeds torque (A) by a factor μ and torque (C) exceeds torque (A) by a factor μ^2. We will present more detailed arguments elsewhere to show that (A) is the only consistent expression for the torque, and note here that (B) and (C) arise from making arbitrary assumptions about the angle of inclination of the field lines at the boundary.

It has usually been taken for granted that accretion commences once the neutron star has spun down to its so-called equilibrium period, P_{eq}, at which the magnetosphere equals the co-rotation radius (i.e. $\mu \sim 1$), leading to $P_{eq} \propto \dot{M}^{-3/7}$, where $\dot{M}$ is the accretion rate. The observed sources with long periods consequently require a low $\dot{M}$, implying slow spindown on a timescale longer than the main sequence lifetime of its companion ([2]; Savonije and van den Heuvel, [7]). However, a further necessary condition for accretion is that the gas near the magnetosphere must cool faster than it is heated by the dissipation of rotational energy occurring there. If, as is the case for low $\dot{M}$, cooling is not sufficiently fast, the neutron star will not accrete even though $P > P_{eq}$ ($\mu < 1$).

There is thus a further phase of spindown - the subsonic propeller - in which turbulence near the magnetosphere continues to slow down the rotation until conditions are such that cooling is sufficiently fast to allow accretion (and spin-up) to begin. The important point is that $\dot{M}$ is not limited by the value derived from P_{eq}, so that shorter spindown timescales than were previously calculated are possible. Details of the relevant timescales and of the expected periods will be presented elsewhere [8].

REFERENCES

1. A. F. Illarionov and R. A. Sunyaev, Astr. Astrophys., 39, 185 (1975).

2. A. C. Fabian, Mon. Not. R. astr. Soc., 173, 161 (1975).

3. K. Davidson and J. P. Ostriker, Astrophys. J., 179, 585 (1973).

4. W. Kundt, Physics Letters, 57A, 195 (1975).

5. N. I. Shakura, Soviet Astr. Lett., 1, 223 (1975).

6. V. M. Lipunov and N. I. Shakura, Soviet Astr. Lett., 2, 133 (1976).

7. G. J. Savonije and E. P. J. van den Heuvel, Astrophys. J., 214, L19 (1977).

8. R. E. Davies, A. C. Fabian and J. E. Pringle, preprint (1978).

PERIOD CHANGES IN NEUTRON STAR X-RAY SOURCES ACCRETING FROM A DISK

P. Ghosh* and F. K. Lamb*,**

*Department of Physics, University of Illinois at Urbana-Champaign
**California Institute of Technology

INTRODUCTION

The qualitative features of disk accretion by rotating magnetic neutron stars were first described by Pringle and Rees [1] and Lamb, Pethick, and Pines [2]. These authors argued that a slowly rotating neutron star accreting matter from a Keplerian disk should spin up as a consequence of the torque exerted on the star by the accreting matter. Recently, Ghosh, Lamb, and Pethick [3] have given detailed solutions for plasma flow within the magnetosphere and have shown that these solutions can be used to derive bounds on the accretion torque which are quite general and do not require a detailed knowledge of the flow in the transition zone between the disk and the magnetosphere. Although the actual value of the accretion torque cannot be calculated without knowing the structure of the flow in the transition zone, these bounds show that the earlier estimate of the torque on slow rotators is correct if the zone is thin, and that in this case the same estimate is approximately correct even for fast rotators.

There is now strong observational evidence that many pulsating X-ray sources are rotating magnetic neutron stars which are accreting matter from a binary companion. Measurements of secular spin-up time scales (see [4] and [5]) are generally consistent with the above theoretical estimates. However, the observed spin-up time scale in Her X-1 is substantially longer than the slow rotator estimate, while the sources Her X-1, Cen X-3, and 4U 0900-40 all show episodes of spin-down in addition to their secular trend toward spin-up [6-8].

Progress in understanding these observations has been frustrated in part by the absence of a detailed, quantitative model of disk accretion by magnetic neutron stars. The purpose of this paper is to describe such a model, which makes definite predictions regarding the inner radius of the disk, the radial and vertical structure of the transition zone, and the accretion torque. One of our main conclusions is that in disk accretion the transition zone is not thin, and recognition of this is an important step in understanding the behavior of the periods of Her X-1, Cen X-3, and 4U 0900-40. A more detailed account of these calculations will be given elsewhere [9,10]. Here we restrict ourselves to a description of the basic model and a brief summary of the observational implications.

THE MODEL

In roughly spherical accretion by a magnetic neutron star, currents circulating in the transition region between the magnetosphere and the outside plasma completely screen the stellar magnetic field from the plasma, confining the field to the interior of the magnetospheric cavity. In accretion from a thin disk, this picture is not self-consistent.

To see this, suppose that current sheets on the surfaces of the disk were initially to screen the disk completely. The magnetic field above and below the disk would then be oppositely directed and of the same magnitude as the unscreened stellar field. Such a strong field will rapidly invade the disk as a result of turbulent diffusion, development of the Kelvin-Helmholtz instability, and reconnection to small-scale fields within the disk [10]. Thus the star's magnetic field cannot be completely screened from the disk, and some stellar field lines connect through it.

The stellar field lines that connect through the disk are distorted by the azimuthal and radial motion of the disk plasma. Consider first the azimuthal motion. It generates a toroidal field B_ϕ comparable to the poloidal field B_p in a time $\sim |\Omega_K - \Omega_s|^{-1}$, due to the different angular velocities of the disk plasma and the star. Here $\Omega_K = (GM/r^3)^{1/2}$ and Ω_s is the angular velocity of the star. Were the magnetic field amplification limited only by ordinary or Bohm diffusion, B_ϕ would become extremely large. Before this can happen, the toroidal magnetic field, which is of opposite sign above and below the midplane of the disk, reconnects. Thus, in a steady state the azimuthal pitch, $\gamma_\phi = B_\phi/B_z$, is limited to the value given by the condition that the rates of reconnection and amplification of the toroidal field balance. A similar analysis can be carried through for the distortion caused by the radial motion of the disk plasma. The final result is that those stellar field lines that thread the disk are pinched inward at the midplane and stretched in the direction of the disk orbital motion, but on average the rate of slippage of field lines through the disk plasma balances the rate at which the field lines are transported by the plasma.

The picture which emerges from this analysis is as follows. When it is far away, the motion of the accreting matter is determined by the effective viscosity in the disk and is unaffected by the magnetic field of the star. However, as it approaches the magnetosphere it enters a transition region where its motion is determined more and more by magnetic stresses associated with the stellar field and less and less by viscous stresses. The transition region ends and magnetospheric flow begins when the viscous stresses are negligible compared with the magnetic stresses and the flow is field-aligned. The structure of the accretion flow is shown schematically in Fig. 1.

APPLICATION TO PULSATING X-RAY SOURCES

Among the most significant properties of the pulsating X-ray sources is the rate of change of spin, which reflects both the sign and the magnitude of the accretion torque and the moment of inertia of the X-ray star. For example, their secular spin-up rates have been used to identify these sources as neutron stars.

For a fixed mass-accretion rate, our solutions show that the accretion torque can be <u>either</u> positive (causing spin-up) <u>or</u> negative (causing spin-down), depending on the angular velocity of the star, Ω_s. The spin-up torque on slow rotators is $\sim N_0 = (GMr_A)^{1/2}$, whereas that on fast rotators is much smaller; this provides a natural explanation of why Her X-1 has an average spin-up rate much below the slow rotator

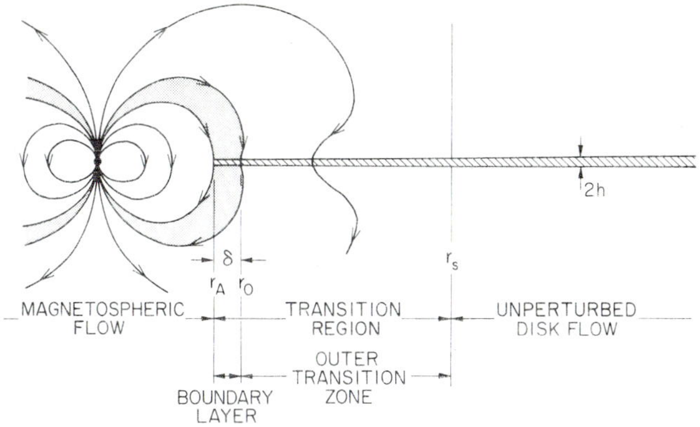

Fig. 1. Schematic picture of disk accretion by magnetic neutron stars. Beyond the radius r_s at which the stellar magnetic field is completely screened, the disk flow, of vertical thickness 2h, is unperturbed by the magnetosphere. In the transition region between r_s and r_A, the disk flow changes into magnetospheric flow. The transition region divides into two parts, a broad outer transition zone where viscous stresses dominate magnetic stresses, and a narrow boundary layer where the magnetic stresses dominate.

estimate. For sufficiently fast rotators, the total torque is negative. Finally, for Ω_s larger than a certain critical value, there are no stationary flow solutions, indicating that steady accretion is not possible for such fast rotators. Alternatively, one can consider the torque on a star of given spin period, P, as a function of the mass-accretion rate, as shown in Fig. 2. At high accretion rates the torque is $\sim N_0$ whereas at lower rates the torque is reduced. At sufficiently low rates the total torque is negative. Finally, at very low rates there are no stationary solutions, indicating that steady accretion is not possible.

Given a spin period P and luminosity L, and a particular neutron-star model, the present model of disk accretion gives a secular spin-up time scale T_s which depends only on the single parameter μ, the stellar magnetic moment. We have adjusted μ to give the best possible agreement with the value of T_s observed in the eight pulsating X-ray sources for which data are available. In general the model gives satisfactory agreement with the observations, including those of Her X-1, for moments $\sim 10^{30}$ gauss cm^3. An exception is the source 4U 0900-40, for which no agreement is possible unless μ is $\sim 10\text{-}10^2$ times larger.

In addition to their secular spin-up, the most carefully studied sources Her X-1, Cen X-3, and 4U 0900-40 show substantial period fluctuations on time scales of weeks to months, with occasional episodes of spin-down [6-8]. Fluctuations in the accretion torque are one possible explanation for this phenomenon [11]. Our calculations show that torque fluctuations of the required size can be caused by fluctuations in the mass-accretion rate [10].

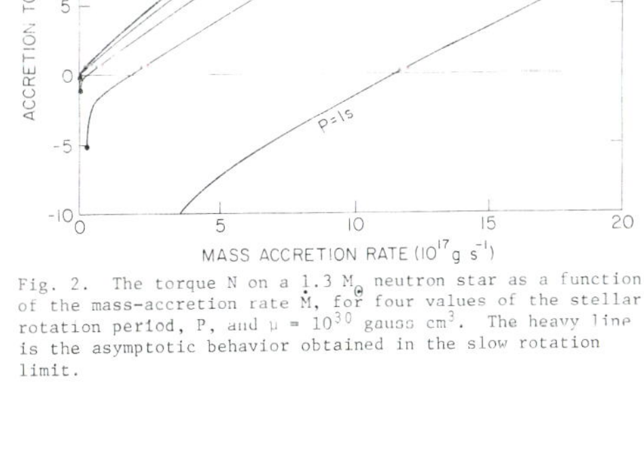

Fig. 2. The torque N on a 1.3 $M_\odot$ neutron star as a function of the mass-accretion rate $\dot{M}$, for four values of the stellar rotation period, P, and $\mu = 10^{30}$ gauss cm^3. The heavy line is the asymptotic behavior obtained in the slow rotation limit.

Finally, we remark that a large fraction of the known pulsating sources have long periods, even though they are, paradoxically, observed to be spinning up on very short time scales, $\sim$ 50-100 years. It is important to understand how these sources maintain their slow spin rates over long time scales. The braking torque found by the present calculation for very low accretion rates may be one way, if the long period sources have low states during which the accretion rate is much reduced.

This work was supported in part by NSF grant PHY78-04404 at Illinois and NASA contract NAS 5-23315 at Caltech. One of us (F.K.L.) acknowledges the support of an Alfred P. Sloan Foundation Research Fellowship.

REFERENCES

1. J. Pringle and M. J. Rees, Astr. Ap. 21, 1 (1972).

2. F. K. Lamb, C. J. Pethick, and D. Pines, Astrophys. J. 184, 271 (1973).

3. P. Ghosh, F. K. Lamb, and C. J. Pethick, Astrophys. J. 217, 578 (1977).

4. S. Rappaport and P. C. Joss, Nature 266, 683 (1977).

5. K. O. Mason, M.N.R.A.S. 178, 81P (1977).

6. R. Giacconi, in: Astrophysics and Gravitation, Proceedings of the 16th International Solvay Conference, Brussels: Université de Bruxelles, 1974, p. 27.

7. G. Fabbiano and E. J. Schreier, Astrophys. J. 214, 235 (1977).

8. R. H. Becker et al., Astrophys. J. 221, 912 (1978).

9. P. Ghosh and F. K. Lamb, Astrophys. J. Lett. 223, L83 (1978).

10. P. Ghosh and F. K. Lamb, submitted to Astrophys. J. (1978).

11. F. K. Lamb, D. Pines, and J. Shaham, Astrophys. J. 224, 969 (1978).

FORMATION OF STRUCTURED CONFINED CORONAE ON ACCRETION DISKS

A. A. Galeev †, R. Rosner and G. S. Vaiana ‡

Harvard-Smithsonian Center for Astrophysics, Cambridge, MA, USA

INTRODUCTION

Our work focusses upon the interaction of magnetic fields and flows within accretion disks, previously discussed by Eardley and Lightman [1], Shields and Wheeler [2], Bisnovatyi-Kogan and Blinnikov [3], Ichimaru [4], and others. The ultimate aim is to understand the physics of compact binary X-ray sources exhibiting strongly fluctuating hard and soft X-ray emission, as well as possible anti-correlation between the hard and soft emission components. The prototype in mind is Cyg X-1 (cf. review by Oda [5]).

We are going to assume that we understand part of the problem, namely that the ultimate energy source is mass accretion, that an accretion disk has formed about the massive compact binary component, and that the soft X-ray emission largely derives from this moderately hot (10^6 K), optically-thick disk. What we shall focus on is:

(a) The evolution of magnetic fields within a convective disk; to anticipate, we conclude that within the inner portion of the disk, reconnection cannot limit field growth and that approximate equipartition is attained. Field amplification will therefore lead to flux eruption from the disk.

(b) The consequences of dissipating these erupted fields above the disk, where reconnection processes have substantially shorter time scales. The key results are that the closed field topology above the disk, together with plasma heating resulting from reconnection, leads to the formation of an ensemble of very hot confined plasma structures whose emission time scales are dictated by the heating process; further, that the cooling of these plasma structures is strongly influenced by the soft X-ray luminosity of the underlying accretion disk.

We emphasize that the structured accretion disk corona under discussion is substantially different from coronae recently discussed by others (cf. Liang and Price [6]) -- both qualitatively and quantitatively -- because first, the plasma confinement obtained by the erupted magnetic field allows substantially higher coronal temperatures to be attained and, second, the spatially-inhomogeneous nature of the resulting accretion disk corona is an essential aspect for the

---

† On leave from the Space Research Institute, Moscow (USSR)
‡ And Osservatorio Astronomico di Palermo, Italy

temporal fluctuations in its emission. Similar considerations apply to stellar coronae and have received detailed observational support in the case of the solar corona (cf. Vaiana and Rosner [7]).

MAGNETIC FIELD EVOLUTION

We turn first to the problem of field evolution within the accretion disk. Field amplification results from the joint action of differential rotation and convection within the disk; Fig. 1 indicates schematically the flow pattern for a typical convective cell. The positive feedback provided by convection can be shown to lead to exponential growth, with a characteristic time scale given by the

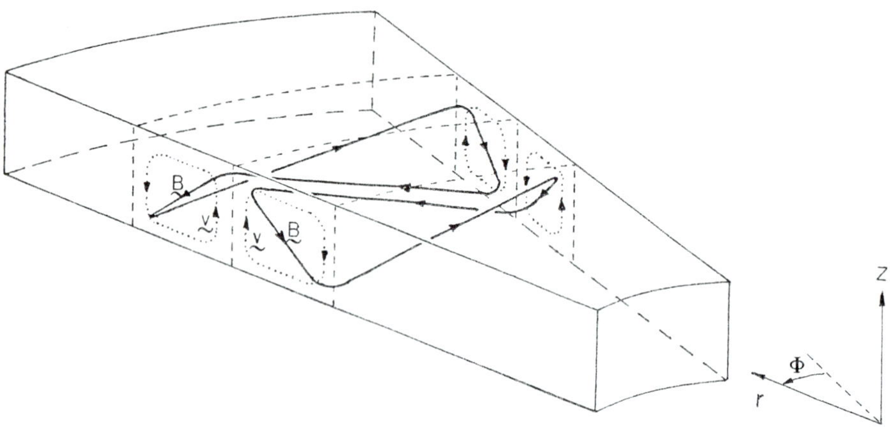

Fig. 1

expression $\tau_G \simeq r/3v$ sec, where v is the convective speed and r the radial distance from the central compact object. Note that convection is essential, as without it, only linear field growth is obtained.

What ultimately limits field growth? The obvious candidate is reconnection within the disk. If we are conservative and choose the most efficient such process (Petschek-type reconnection; cf. Priest and Soward [8]), we obtain an upper bound for the dissipation rate, $\tau_R^{-1} \simeq 2.1 \times 10^{-2}\ v/z_o\ \mathrm{sec}^{-1}$, where z_o is the thickness of the disk. Reconnection is unimportant if the growth rate τ_G^{-1} exceeds the dissipation rate τ_R^{-1}; this condition obtains in the inner portion of the disk, within approximately 10 r_o, where r_o is the last (innermost) stable orbit of the disk. This calculation is based upon the disk structure calculated by Bisnovatyi-Kogan and Blinnikov [3], but is a result whose essentials are not very sensitive to the details of such calculations. Field amplification is hence limited by non-linear effects which come into play when Lorentz forces begin to be effective at suppressing convection. One can then show that the field energy density becomes comparable to the thermal plasma energy density, a result one could have gotten simply by invoking equipartition. However, note that the latter argument carries no weight by itself, as one has to show explicitly -- as we have done above -- that reconnection does not act to limit field strengths before approximate equipartition is attained. Once this stage has been reached, density contrasts between magnetized and relatively unmagnetized plasma lead to buoyancy effects and the eruption of flux from the disk; a schematic view of the result of this process is shown in Fig. 2.

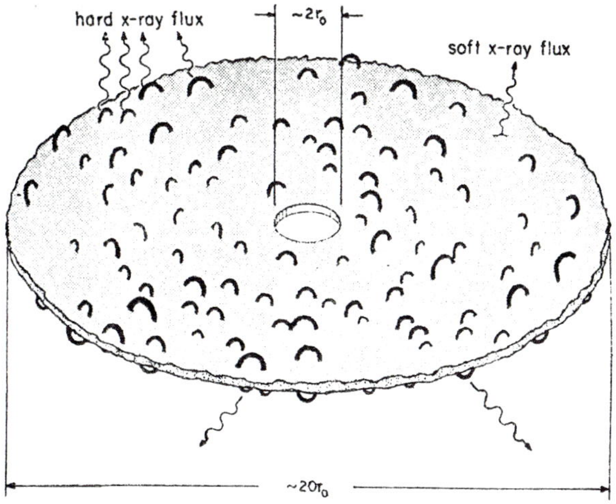

Fig. 2. Schematic drawing of inner accretion disk geometry

FORMATION OF DISK CORONA

What happens once the flux has erupted? The key elements in our analysis are that the emerging fields are sheared by disk fluid motions, that the consequent free energy is rapidly thermalized by tearing mode instabilities (cf. Galeev [9]), and that the confinement of the heated plasma by the magnetic field leads to enhancements both in temperature and density within the erupted flux loops. The detailed energy balance analysis within these confined plasma structures then yields the results:

1. Compton-cooling by soft photons from the disk can be an important loss mechanism for the hot coronal loops;
2. The energy release process is stochastic and hence so is the emission from individual loop structures;
3. The high temperatures attained ($\sim 10^9$ K) result from plasma confinement, and are bounded above by heat conduction to the cooler disk.

The detailed arguments are as follows: for each loop structure we solve the energy balance problem, including heating by reconnection (E_H), thermal conductive loss to the disk (F_c), bremsstrahlung losses (E_R), and Compton-cooling losses (E_{cc}; these depend upon the disk soft X-ray luminosity), but excluding synchrotron cooling, which can be shown to be relatively ineffective under the assumed conditions; similarly, cooling by e^--e^+ pair production has been neglected. Thus we solve

$$E_H - \frac{\partial}{\partial s} F_c - E_R - E_{cc} = 0 \qquad (1)$$

with

$$E_H \simeq 1.4 \times 10^{-2}\, v_A \cdot (B^2/4\pi) \cdot (2/R) \text{ erg s}^{-1} \text{ cm}^{-3}$$

$$F_c \simeq -\kappa_{QL}\, (kT_e/m_p)^{1/2}\, nkT_e \text{ sgn } dT/ds \text{ erg s}^{-1} \text{ cm}^{-2}$$

$$E_R \simeq 1.3 \times 10^5\, p^2\, T^{-3/2} \text{ erg s}^{-1} \text{ cm}^{-3}, \text{ and}$$

$$E_{cc} \simeq 4\sigma_T n(\varepsilon c) kT_e/m_e c^2 \text{ erg s}^{-1} \text{ cm}^{-3}$$

Here v_A is the Alfven speed associated with the sheared field B, R the loop major radius, κ_{QL} the turbulent (quasi-linear) heat conduction coefficient, σ_T the Thompson cross-section, and εc the soft X-ray flux from the disk; n, T and p are the loop density, temperature, and pressure, respectively.

For a low disk luminosity ($\sim$ 0.02 x Eddington limit), Compton-cooling is insignificant. The loop temperature attained is then $\sim 3 \times 10^9$ K, and one can check for self-consistency by calculating the Comptonization parameter y $[\equiv (4kT_e/m_e c^2)(n\sigma_T R)^2]$. We find that it is less than unity, so that Compton-cooling is indeed unimportant. The energy released in the reconnection process, which has a typical time scale of 1 sec, is of the order of a few times 10^{35} ergs s^{-1}. A source such as Cyg X-1 would therefore require a corona consisting of an ensemble of roughly 100 such loop structures.

In the high disk luminosity limit, Compton cooling losses (unsaturated Comptonization of soft X-ray photons from the disk) can dominate all other losses; in this limit, above roughly 0.06 L_c, the loops are somewhat cooler $[T_e \simeq 2 \times 10^9\ K \propto (L/L_c)^{-1/5}$, L_c the Eddington luminosity limit] and -- calculating the comptonization parameter y again -- we see that our calculation is indeed self-consistent as $y \gtrsim 1$. Note that the loop temperature in this limit depends weakly upon the inverse of the disk luminosity, so that in this case the soft and hard emission components are indeed inversely correlated.

A more detailed presentation will be published elsewhere. The authors thank Drs. A. Lightman and S. Shapiro for enlightening comments. This work was supported by the Langley-Abbot Program of the Smithsonian Institution.

REFERENCES

1. D.M. Eardley and A.P. Lightman 1975, *Ap. J.* **200**, 187.
2. G.A. Shields and J.C. Wheeler 1976, *Ap. J. Lett.* **17**, 69.
3. G.S. Bisnovatyi-Kogan and S.I. Blinnikov 1977, *Astron. Ap.* **59**, 111.
4. S. Ichimaru 1977, *Ap. J.* **214**, 840.
5. M. Oda 1977, *Space Sci. Rev.* **20**, 757.
6. E.P. Liang and R.H. Price 1977, *Ap. J.* **218**, 247.
7. G.S. Vaiana and R. Rosner 1978, *Rev. Astron. Astrophys.* **16** (in press).
8. E.R. Priest and A.M. Soward 1976, IAU Symp. **71**, 353.
9. A.A. Galeev 1978, *Phys. Fluids* (in press).

RADIATION-DRIVEN WINDS IN BINARY X-RAY SOURCE SYSTEMS

Keith MacGregor and Peter A. J. Vitello

Harvard-Smithsonian Center for Astrophysics, Cambridge, Massachusetts 02138, USA

ABSTRACT

The nature of a radiatively-accelerated wind in a binary X-ray source system is investigated. It is found that the presence of even a weak X-ray source gives rise to a wind which is significantly different from that of an isolated primary.

INTRODUCTION

The optical components of several of the known binary X-ray source systems have been identified as luminous O and B stars [1]. Observations indicate [2] that stars of these spectral types generally possess strong stellar winds, which are thought to be driven by the force resulting from the absorption of radiation by the ultraviolet resonance lines of ions in the flow [3]. As was originally suggested by Davidson and Ostriker [4], accretion from such a wind onto a compact companion is a likely source for the observed X-ray emission. These authors pointed out that for systems in which the accreting object is a white dwarf or a neutron star, the X-ray luminosity $L_x \propto \dot{M}_p/v^4_{rel}$, where $\dot{M}_p$ is the mass loss rate from the primary, and v_{rel} is the wind velocity relative to the secondary. Since L_x depends sensitively upon the magnitude of v_{rel}, a consistent theoretical description of such systems requires a detailed knowledge of the wind velocity profile, which may differ significantly from that of a wind emitted by an isolated primary [5]. In this paper we present the results of an investigation into the hydrodynamics of line-driven winds in binary systems containing a weak X-ray source. Our treatment includes the gravitational, ionization, and heating effects due to the secondary on the acceleration of the wind. We find that the changes in the acceleration which result from the presence of even a weak X-ray source can profoundly alter the wind velocity profile.

LINE-DRIVEN WINDS IN BINARY X-RAY SOURCE SYSTEMS

In this section we outline the methods and approximations used in describing a radiatively-accelerated stellar wind in the presence of an X-ray source. A more detailed discussion of these points will be left to a forthcoming paper.

We consider a binary system in which a compact, X-ray emitting object of mass M_x and radius R_x is located a distance d from the center of an early-type star of mass M_p and radius R_p. For simplicity, we neglect the orbital motion of the secondary and treat both components of the system as non-rotating. We assume that the wind can be represented as a steady, radial flow and confine our attention to the line joining the centers of the two objects, with the distance r measured from the center of the primary being our independent variable. The flow is accelerated by the thermal and radiation pressure forces and is acted upon by the gravitational forces of both components of the system. In calculating the radiation pressure force, we take into account the forces arising from (1) the scattering of photons from the primary by free electrons in the wind, and (2) the absorption of photons from the primary by the ultraviolet resonance lines of ions of C, N, and O. In determining the force exerted on the wind material due to resonance lines, we use the approximate results of Castor [6] for the force associated with lines formed in a rapidly expanding stellar atmosphere. In order to determine the thermal and ionization structure of the wind, thermal and ionization equilibrium is assumed. We have taken the wind as well to be optically thin to the radiation fields of both components of the system. The flow is followed outward from the surface of the primary to the position of the secondary, ignoring the formation of a bow shock, with the treatment of the hydrodynamics being similar to that of Castor, Abbot, and Klein [7].

RESULTS

We present here the results of a calculation for a model with the following physical parameters: $M_P = 15M_\odot$, $R_P = 20R_\odot$, $M_x = 1M_\odot$, $d = 2R_p$. The radiation field of the primary was represented by a blackbody spectrum with brightness temperature $T_r = 3 \times 10^4$°K for $\lambda \geq 912\text{Å}$, and $T_r = 2.1 \times 10^4$°K for $\lambda < 912\text{Å}$. For the X-ray source, a bremmstrahlung-type spectrum of the form

$$L_\varepsilon = (L_x/\varepsilon_{max}) \exp\left[-(\varepsilon - \varepsilon_{min})/\varepsilon_{max}\right], \quad \varepsilon \geq \varepsilon_{min}$$

$$= 0 \qquad \varepsilon < \varepsilon_{min}$$

was assumed, with ε_{min} = lkeV and ε_{max} = 20keV. Several values of L_x were considered.

We first consider a model in which the efficiency for converting accreted matter into X-rays is assumed to be negligible (i.e. $L_x = 0$). Our results for this case are given by the dashed curves in Fig. 1-4. For this solution, $\dot{M}_p \simeq 4.6 \times 10^{-8}\ M_\odot$/yr. Interior to the sonic point ($r \simeq 1.013R_p$), the outward acceleration of the gas is provided mainly by the force due to electron scattering. Beyond the sonic point, the value of Γ (which gives the ratio of the total radiation pressure force to the gravitational force of the primary) is dominated by the forces arising from the resonance lines, and the flow rapidly attains a velocity $\simeq$ 1000 km/s with Γ becoming approximately constant. Throughout the wind, T $\simeq$ constant and is determined primarily by the balance between photoionization heating and radiative recombinative cooling. For distances larger than the distance to the Lagrange point ($r = 1.46\ R_p$), the net gravitational force on the flow is directed outward. As the wind approaches the vicinity of the compact object, the velocity gradient increases rapidly, which in turn gives rise to a significant increase in Γ [6]. For $r \leq 1.97\ R_p$, the wind is primarily accelerated by line forces, while at larger distances, the gravitational force due to the secondary is dominant.

Gravitational capture of the wind will presumably occur if it encounters the compact object with an impact parameter less than the accretion radius R_A. We take R_A to be the distance (measured from the center of the compact object) at which the gravitational potential energy of the flow, GM_x/R_A, becomes comparable with its kinetic energy, v^2_{rel}. For the present model this gives $R_A \simeq 10^9$ cm. We note, however, that the enhanced acceleration of the wind in the neighborhood of $r \simeq d$ is likely to be a function of impact parameter. An upper bound for R_A can be obtained by considering accretion from the wind of an isolated primary. This is roughly equivalent to extrapolating the flow from the region of constant Γ to the position of the compact object ignoring the enhanced acceleration and yields a value for $R_A \simeq 10^{10}$ cm.

To investigate the effects of a weak X-ray source on the wind, we next consider a model with the same physical parameters as before, but with $L_x = 7.5 \times 10^{33}$ erg/s. Our results for this wind are given by the solid curves in Figs. 1-4. For this flow, $\dot{M}_p \simeq 5.5 \times 10^{-8}$ $M_\odot$/yr, representing an increase of roughly 20% over the value obtained for $L_x = 0$. At intermediate distances, the thermal structure of the wind is essentially unchanged, but the X-rays from the secondary markedly increase the degree of ionization of the gas. For $r \lesssim 1.5\ R_p$, the presence of additional ions with strong resonance lines which can couple effectively to the primary radiation field results in a radiation pressure force significantly larger than that obtained in the previous solution for this region of the flow. At larger distances, the wind is heated by X-ray photoionization, and the remaining L-shell electrons are removed from the ions by collisional and radiative ionization. As a result, the force due to resonance lines decreases, and Γ ultimately becomes equal to its electron scattering value. The velocity of the wind in this region is substantially higher than that for the previous solution ($v \simeq 3000$ km/s as opposed to $v \simeq 1500$ km/s for $L_x = 0$), and the flow coasts to the orbit of the secondary with little subsequent acceleration until gravitational effects become important. Again we find $R_A \simeq 10^9$ cm. In the immediate vicinity of the X-ray source, all ions are completely stripped of their electrons, and the gas temperature rises rapidly to a value $T \simeq \varepsilon_{max}/4k \simeq 5.8 \times 10^7$ °K determined by the balance between Compton heating and cooling.

DISCUSSION

In this paper we have considered the nature of a line-driven wind in a binary system with a weak X-ray source. We find that the gravitational and ionization effects due to the presence of the compact object lead to an enhanced acceleration in the wind. This suggests that the use of wind velocity profiles characteristic of isolated early-type stars in the modeling of binary X-ray sources will lead to an overestimate of the rate at which matter is accreted from the wind, and hence to an overestimate of the X-ray luminosity as well.

We note, however, that as one increases the value of L_x, the region of high temperature and high degree of ionization is expected to extend to greater distances from the X-ray source. This may result ultimately in a net reduction in the relative velocity of the wind at the orbit of the compact object, and thus to an increased accretion rate. Therefore, it is not clear at this time whether a self-consistent solution for the system consists of a low value of L_x with a high velocity wind, or a high value of L_x with a low velocity wind. In fact, one cannot rule out the possibility that both solutions may exist.

ACKNOWLEDGEMENTS

The authors would like to acknowledge useful discussions with Dr. J. Raymond and Dr. G. Rybicki.

REFERENCES

1. P. S. Conti, Astr. and Astrophys. 63, 225 (1978).

2. T. P. Snow and D. C. Morton, Astrophys. J. Suppl. 32, 429 (1976).

3. L. B. Lucy and P. Solomon, Astrophys. J. 159, 879 (1970).

4. K. Davidson and J. P. Ostriker, Astrophys. J. 179, 585 (1973).

5. S. Hatchett and R. McCray, Astrophys. J. 211, 552 (1977).

6. J. I. Castor, M.N.R.A.S. 169, 279 (1974).

7. J. I. Castor, D. C. Abbot and R. I. Klein, Astrophys. J. 195, 157 (1975).

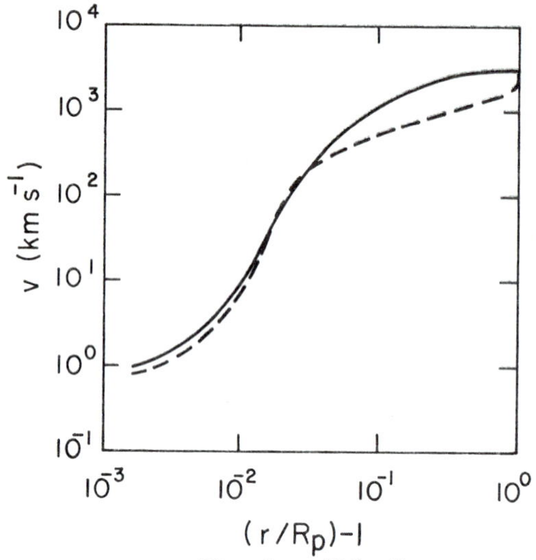

Fig. 1. Velocity

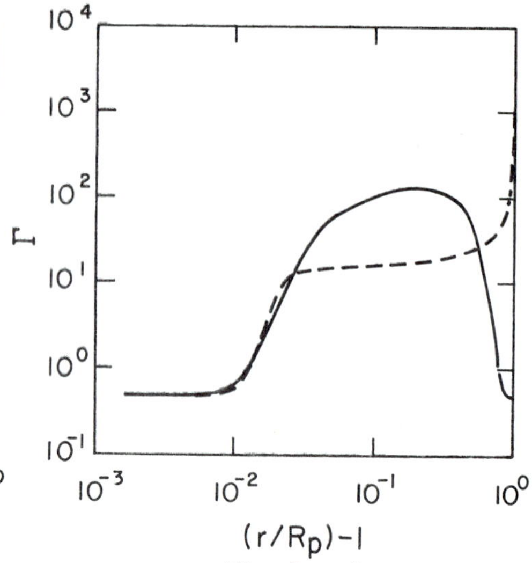

Fig. 2. Γ

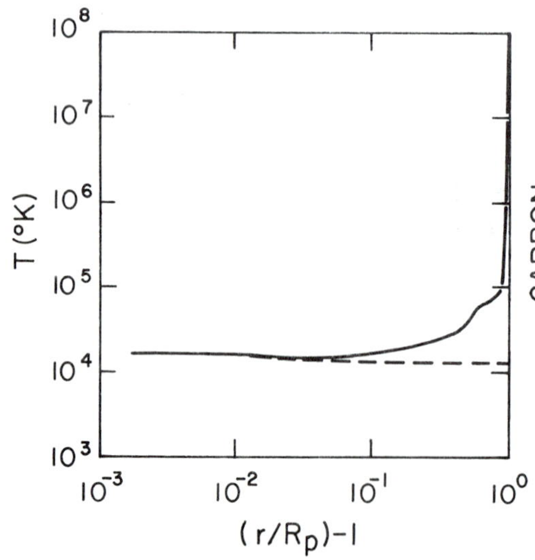

Fig. 3. Gas temperature

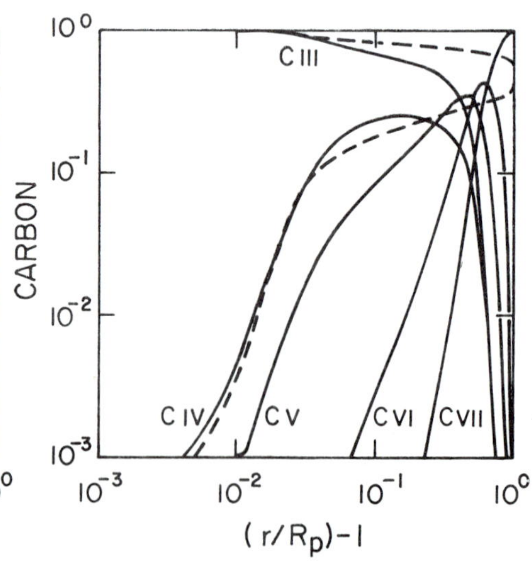

Fig. 4. Carbon ionization fractions

TURBULENT ACCRETION ONTO MASSIVE BLACK HOLES

L. Maraschi*,**, G. C. Perola*, C. Reina* and A. Treves*

*Istituto di Fisica dell'Università, Via Celoria 16, I 20133 Milano
**Laboratorio di Fisica Cosmica del CNR, Via Bassini 15, I 20133 Milano

Observations of the X-ray emission of type I Seyfert galaxies [1,2] have shown that the nuclei of some of these objects have luminosities above 2 keV which are comparable or larger than in the optical band. For these objects the standard picture of accretion disks is inadequate, since for the large central masses required in galactic nuclei (10^6-$10^{10} M_\odot$) the typical temperatures are in the UV band [3,4] .

In this paper a spherical model involving turbulence and magnetic dissipation is examined. We have determined the temperature profile of the accreting gas for a rather wide range of hole masses (from 10^6 to 10^9 $M_\odot$) and find that the inner regions attain values within 10^9 and 10^{10} °K. The energetically dominant emission mechanism from these regions is not bremsstrahlung, but rather cyclotron-synchrotron radiation from the thermal plasma. The typical frequencies of this radiation are between 10^{13} and 10^{14} Hz. However, when the optical depth to Thomson scattering becomes of order unity or larger, the comptonization process is important, and the spectrum of the emergent radiation extends to $h\nu \simeq 3kT$. We therefore attempt to explain the continuum emission of the Seyfert nuclei from the optical to the medium energy X-rays with this process.

BASIC FORMULAE

The luminosity corresponding to an accretion rate $\dot{M}$ is

$$L = \frac{\varepsilon \dot{M} GM}{\bar{r}} = \varepsilon 1.5\times 10^{45}\ \dot{M}_{25}\ \text{erg/sec} \tag{1}$$

where $\bar{r}=6\ GM/c^2=9\times 10^{13}\ M_8$ cm, M_8 is the mass of the hole in units of $10^8 M_\odot$, and ε is the efficiency of conversion of the gravitational power into radiation. Assuming free fall, the density is given by

$$n = 3.5\times 10^9\ x^{-3/2} \dot{M}_{25}\ M_8^{-2}\ \text{cm}^{-3} \tag{2}$$

where x is the radial distance in units of $\bar{r}$. The equipartition magnetic field is

$$B = 2.5\times 10^3\ \dot{M}_{25}^{\frac{1}{2}}\ M_8\ \text{Gauss} \tag{3}$$

and the optical depth for Thomson scattering is

$$\tau_T = 0.42\ \dot{M}_{25}\ M_8^{-1}\ x^{-\frac{1}{2}} \tag{4}$$

The dissipative heating is given by

$$\Gamma = 16.4\ \dot{M}_{25}\ M_8^{-3}\ x^{-4}\ \text{erg cm}^{-3}\text{sec}^{-1} \tag{5}$$

where it is supposed that 25% of the gravitational power is transformed in the heating of the gas i.e. $\varepsilon \approx 0.25$ (see [5]).

Particle acceleration in the electric field produced by magnetized turbulent cells is shown in [6] to be an efficient mechanism to produce γ-rays peaked at

$$h\nu = 0.1\sqrt{\frac{1}{3}}\ x^{-\frac{1}{2}}\ \frac{2\pi mc^2}{\alpha} \approx 25\ x^{-\frac{1}{2}}\ \text{MeV} \tag{6}$$

where α is the fine structure constant.

TEMPERATURE PROFILE

The temperature variation of the infalling gas is given by the equation

$$\frac{dT}{dx} + \frac{T}{x} = \frac{4\pi m_p}{3\dot{M}}\ r^{-3}\ x^2\ (\Lambda - \Gamma) \tag{7}$$

where T is the temperature and Λ is the energy loss rate per unit volume. When the cooling is efficient, as for accretion rates and masses relevant to the case under consideration, an outer quasi-isothermal zone is formed where free-free and line emission dominate the energy loss. From the condition

$$\Gamma = \Lambda = n^2\ \tilde{\Lambda}\ (T) \tag{8}$$

the inner radius of this zone is estimated to be

$$x_{is} \simeq 1.4\times10^{-18}\dot{M}_{25}^{-1}\ M_8\ \tilde{\Lambda}^{-1} \tag{9}$$

where $\tilde{\Lambda}$ is the cooling factor for line emission and depends both on the temperature and on the ionization process.

Within x_{is} the most important cooling mechanism are cyclo-synchrotron and Compton scattering. One has to take into account that the plasma is opaque to the synchrotron photons, and therefore the quantity of interest, as shown in [5], is the frequency ν^+ at which the plasma becomes transparent. The form given in [7] for the synchrotron absorption coefficient was adopted here. The cooling rate due to the single scattering of thermal synchrotron photons by the electrons is given by

$$\Lambda_{es} = \frac{32}{3}\ \frac{\pi}{c^2}\ \left(\frac{kT}{m_e c^2}\right)^2\ n_e\ \sigma_T\ \nu^{+3} \tag{10}$$

We have neglected in first approximation the effects of multiple Compton scattering, which on the other hand begins to be important for $\tau \gtrsim 1$.

We have obtained the temperature profiles for various values of the hole mass ($10^6 < M < 10^9\ M_\odot$) and of the accretion rate. All the profiles are characterized by a rather steep rise starting at x_{is} up to a maximum value between 10^9-10^{10}°K, after which the temperature declines very slowly as x decreases (see Tab. 1).

SPECTRUM OF THE OUTGOING RADIATION

The shape of the spectrum depends critically on the optical depth to Thomson scattering and therefore we shall consider separately the case $\tau<<1$ and $\tau\geq 1$.

i) $\tau<<1$

In this case the energetically dominant emission mechanism is synchrotron radiation from the 10^9-10^{10}°K electrons, which yields a flat spectrum between 10^{13}-10^{14} Hz. The contribution of Compton scattering is minor and does not exceed a frequency of 10^{14} Hz. The bremsstrahlung photons reach frequencies of the order of 10^{21}Hz, but the energy they carry is only a small fraction of the total emission power.

ii) $\tau \gtrsim 1$

When the optical depth to Thomson scattering is of order one or greater, two effects come into play. The first is that convection of radiation into the hole becomes important and decreases the efficiency of energy extraction. The second is that comptonization of the synchrotron photons can induce a drastic change in the form of the spectrum. In order to estimate the importance of the convection process, we compare the free fall time, $t_{ff} \approx r/v$, with the diffusion time

$$t_d \sim \frac{n\sigma_T r^2}{c} = \frac{\sigma_T}{c} \frac{\dot{M}}{4\pi v m_p} \qquad (11)$$

where v is the infall velocity. The two time scales are equal at a radius $r^+=\dot{M}\bar{r}/\dot{M}_c\varepsilon$ which is obviously greater than $\bar{r}$ when $\dot{M}>\varepsilon\dot{M}_c$ ($\dot{M}_c \equiv L_E\bar{r}/\varepsilon GM$; $L_E \equiv 4\pi GM\, m_p c/\sigma_T = 1.3\times 10^{46} M_8$ ergsec^{-1}). Therefore, assuming that only the regions with $r>r^+$ contribute to the outgoing luminosity, one obtains

$$L_{out} = \frac{\varepsilon\dot{M}GM}{r^+} = \varepsilon^2 \dot{M}_c \frac{GM}{\bar{r}} = \varepsilon L_E \qquad (12)$$

Note that in this regime L_{out} is independent of the accretion rate. The maximum optical depth for the outgoing radiation in the two regimes ($\bar{r} \gtrless r^+$) depends on the accretion rate as follows,

$$\tau\,(\bar{r}) = 3.6 \frac{\dot{M}}{\varepsilon\dot{M}_c} \quad \text{for } \dot{M}<\varepsilon\, \dot{M}_c \qquad (13a)$$

$$\tau\,(r^+) = 3.6\left(\frac{\dot{M}}{\varepsilon\dot{M}_c}\right)^{\frac{1}{2}} \quad \text{for } \dot{M}>\varepsilon\, \dot{M}_c \qquad (13b)$$

For a detailed discussion of comptonization we refer to [8] . In this paper, Kompaneets equation, with the addition of a diffusive term is studied for a monochromatic photon source at frequency ν_o surrounded by a static, homogeneous plasma sphere. It is shown that the resulting spectrum depends on the ratio of the radius R of the sphere to a critical radius $R_c=\lambda\pi(m_e c^2/4kT)^{\frac{1}{2}}$ where $\lambda=(\sigma_T n)^{-1}$. For $R/R_c>>1$ the resulting spectrum approaches a Wien distribution, while for $R/R_c<<1$ the emission is still peaked at ν_o but with a power law tail of spectral index $\alpha>1$. In particular, for $R/R_c=0.8$, the spectrum has an index $\alpha=1$ from ν_o to $\nu=3kT$, where T is the electron temperature. We adapt the results given in [8] to our case by approximating the infalling gas with a homogeneous sphere of radius $\bar{r}$ (or r^+ if $r^+>\bar{r}$), density $n(\bar{r})$ or $n(r^+)$, and temperature $T=\xi\ 10^9$°K, where ξ, taken as T($\bar{r}$ or r^+) + $T_{Max}/2\times 10^9$, is of the order of 1-10, provided that r^+ is not much larger than the radius at which the temperature reaches $T=T_{Max}$. In this approximation we have

$$R_c = 9.4\times 10^{14}\, M_8^2\, \dot{M}_{25}^{-1}\, \xi^{-\frac{1}{2}} \quad \text{for } \dot{M}<\varepsilon\, \dot{M}_c \qquad (14a)$$

$$R_c = 9.4\times 10^{14}\, M_8^2\, \dot{M}_{25}^{-1}\, \xi^{-\frac{1}{2}} \left(\frac{\dot{M}}{\varepsilon\dot{M}_c}\right)^{3/2} \quad \text{for } \dot{M}>\varepsilon\, \dot{M}_c \qquad (14b)$$

and correspondingly

$$\bar{r}/R_c = 9.6\times 10^{-2} M_8^{-1}\dot{M}_{25}\ \xi^{\frac{1}{2}} = 0.23\tau_T(\bar{r})\xi^{\frac{1}{2}} = 0.83\ \dot{M}/\varepsilon\dot{M}_c\ \xi^{\frac{1}{2}} \qquad (15a)$$

$$r^+/R_c = 0.83 \left(\frac{\dot{M}}{\varepsilon \dot{M}_c}\right)^{\frac{1}{2}} \xi^{\frac{1}{2}} \qquad (15b)$$

From this it appears that when $\dot{M}$ approaches $\dot{M}_c$, irrespective of the mass of the hole (except through ξ) comptonization becomes the dominant effect in shaping the spectrum. It would seem that effective values of $R/R_c>1$ can be attained even before $\dot{M}>\varepsilon\dot{M}_c$ (see Table 1). Correspondingly we expect spectra which can be power laws from the optical to the hard X-rays, with spectral index which can be as small as 0.5 or even spectra which peak at 3kT with a rather minor energy flux (and also photon flux) in the optical - UV region.

DISCUSSION

We have shown that the temperature attained by the infalling gas are of the order of 10^9-10^{10}°K for a wide range of masses of the hole and of accretion rates. However the dominant emission process is not bremsstrahlung and the photons emitted in the synchrotron process are confined to the infrared portion of the spectrum. Only a scattering process can lead to a radically different frequency distribution in the emitted photons, and the efficiency of this process depends very critically on the accretion rate. When this rate increases, the scattering process yields power law spectra extending up to 3kT and eventually, when it approaches the critical value corresponding to the Eddington luminosity, most of the emitted energy is concentrated in hard X-rays. If the model applies to galactic nuclei, one would therefore expect a great variety in the frequency distribution of the central source. It is tempting to consider the "normal" galaxies with a nuclear IR excess and the Seyfert galaxies with a strong X-ray emission as two rather extreme manifestations of the same process. With regard to "normal" galaxies, measurements of the 10μ flux of a complete sample [9] show that the IR excess lead to a very high values of the ratio L_{10}/M where L_{10} is the 10μ luminosity of their nuclei, and M is the nuclear mass as deduced from the velocity dispersion in their star content. In our model the excess IR luminosity could be explained by accretion on a central hole, when the accretion rate is small enough for the accreting gas to be essentially transparent to Thomson scattering. At the other extreme Seyfert nuclei are known to contain an "excess" of UV ionizing radiation, and are also very powerful X-ray emitters. From Table 1 in [2] , the ratio L_X/L_U (nuclear) is of the order of unity, corresponding to a spectral index α between the two "frequencies" of about 1. In the present model, following the results of [8] , it appears that when comptonization is efficient enough to produce a "hard" optical spectrum up to the UV, then the spectrum extends up to 3kT with a constant spectral index $\alpha \gtrsim 0.8$. Therefore the UV and the X-ray photons would have in this model the same origin, and the correlation between UV and X-ray powers would find a straightforward explanation. On the other hand it is important to realize that small changes in the ratio R/R_c imply large variations in the spectral shape: while with R/R_c=0.8 the spectrum is a power law with $\alpha \simeq 1$, with R/R_c=3 most of the emergent flux is concentrated in the Wien peak at 3kT, with a relatively minor contribution in the optical and UV. Therefore we would expect from this model a rather wide range in the ratio L_X/L_U, and in particular values much greater than 1. The fact that such a spread in L_X/L_U has not been observed so far, may represent a problem for this model, but it is also conceivable that the available data are affected by severe selection effects. Recent measurements of one of the Seyferts, NGC 4151, beyond the medium energy X-rays, have led to the surprising result that its spectrum flattens above E=20 keV and continues with $\alpha \simeq 0$ up to 100 keV [10] [11 and references therein]. In the present model one can attribute the high energy X-rays to the non thermal process proposed by [6]. The major problem, however, is in the energetics, because we should require that perhaps as much as 90% of the available energy goes into the relativistic particle channel. However we do not know at the moment of a physical argument strongly prohibiting this unbalance.

TABLE 1.

M_8	$\dot{M}_{25}$	$L_{tot} = \frac{\varepsilon G M \dot{M}}{\bar{r}}$ (erg/sec)	$\xi = \left(\frac{T(\bar{r})+T_{max}}{2\times10^9\,^oK}\right)$	$\nu^+(r=\bar{r})$ (Hz)	$\frac{\bar{r}}{R_c}$	α
0.01	1.7×10^{-2}	2.4×10^{42}	8	1.5×10^{14}	0.44	2.7
	4.1×10^{-2}	6×10^{42}	6.7	1.4×10^{14}	1	0.8
0.1	1.4×10^{-1}	2×10^{43}	9.9	5.7×10^{13}	0.41	2.8
	2.9×10^{-1}	4.2×10^{43}	8.2	8.3×10^{13}	0.70	1.15
1	1.4	2×10^{44}	8.7	4.4×10^{13}	0.4	2.8
	8.2	1.2×10^{45}	6.1	5.2×10^{13}	1.9	0.12
10	13.9	2×10^{45}	13.2	1.16×10^{14}	0.47	2.4
	41.7	6×10^{45}	9.4	1.8×10^{14}	1.2	0.6

REFERENCES

1. H. Elvis et al., Mon.Not. Roy. Astr. S. 183, 129 (1978)
2. H. Tananbaum et al., Astrophys. J. 223, 74 (1978)
3. J. Pringle and M. J. Rees, Astron. Astrophys. 21, 1 (1972)
4. N. J. Shakura and R. A. Sunyaev, Astron. Astrophys. 24, 337 (1973)
5. P. Meszaros, Astron. Astrophys. 44, 59 (1975)
6. L. Maraschi and A. Treves, Astrophys. J. 218, L 113 (1977)
7. A. G. Pacholczyk, Radio Astrophysics, Freeman, San Francisco, 1970
8. J. I. Katz, Astrophys. J. 206, 910 (1976)
9. G. H. Riecke and M. J. Lebofsky, Astrophys. J. 220, L 37 (1978)
10. G. Di Cocco et al., Nature 270, 319 (1977)
11. G. Auriemma et al., Astrophys. J. 221, L 7 (1978)

COS-B OBSERVATIONS OF COMPACT X-RAY SOURCES

Lydie Koch-Miramond

Division de la Physique, Section d'Astrophysique, Centre d'Etudes Nucléaires de Saclay, B.P. 2 —91190 Gif-sur-Yvette, France

The ESA satellite COS-B carries a collimated proportional counter sensitive in the energy range 2 to 12 keV, aligned with the γ-ray instrument, as described in detail by Boella (1). In the normal mode X-ray counts are accumulated during 25.4 s and transmitted every 102.4 s. In the pulsar mode the arrival time of X and γ-ray photons can be related to Universal Time to 0.25 ms. The minimum continuous observation period for a pointing is one month. Apart from the two γ-ray pulsars in Crab (PSR 0531+21) and Vela (PSR 0833-45) regions, five strong X-ray sources were observed. Two, Cen X3 and Vela X1 are known to be pulsating members of binary star systems, two others Cyg X1 and Cir X1 to be erratic variables and Cyg X3 a modulating variable. In addition 25 radio pulsars were in the field of view continuously during one month with an efficiency of detection greater than 50% between August 1975 and June 1977.

X-RAY LIGHT CURVE STRUCTURE OF THE CRAB PULSAR

During the first period of observation between 1975 August 17 and September 17 more than 5.5×10^6 events were recorded from PSR 0531 in the 2-12 keV range. The 2 peaks are separated by 13.5 ± 1.0 ms (phase separation 0.41 ± 0.03). The main pulse width is 1.5 ms and the secondary pulse width is 3.5 ms at half height. A complex interpulse structure appears clearly and the pulse fraction derived from these measurements was $8.5 \pm 0.8\%$, Kanbach (2). Possible variations in intensity, period and light curve structure of PSR 0531+21 during the observation periods remain to be studied.

SEARCH FOR X-RAY EMISSION FROM RADIO PULSARS

Among the 25 radio pulsars in the field of view of COS-B, 11 have well known period P and period derivative $\dot{P}$. Due to the high temporal resolution and long observation time, the sensitivity limit of COS-B is 2.5×10^{-12} erg $cm^{-2}\ s^{-1}$, i.e. 2×10^{-3} Crab. This limit is 100 times better than the limit attained by Ricker (3) on ten pulsars and 8 times better than the limit reported by Pravdo (4) on PSR 0833-45. In order to predict the X-ray emission of pulsars, we must know the efficiency of conversion η_x of the rotational energy loss $\dot{E}$ in X-ray energy. For the Crab pulsar $\eta_x \approx 10^{-3}$. Assuming with Bonnet-Bidaud (5) a variation of η_x with "age" similar to the variation η_γ

with "age" proposed by Buccheri (6), one predicts the X-ray emission of pulsars shown in fig. 1 where $\Phi_x = \eta_x \frac{\dot{E}}{4D^2}$, D being the pulsar distance. Among the 7 candidates detectable by COS-B, PSR 0740-28 is particularly interesting because P and $\dot{P}$ are well known and the error box of the X-ray source 4 U 0742-28 lies less than 30 seconds of arc away. From the light curve obtained by Bonnet-Bidaud (8) using 834278 events and P = 166.75244751 ms, $\dot{P} = 0.16871 \times 10^{-13}$, $\ddot{P} = 0$, T = 2442554.8 J.D. One deduces a preliminary upper limit of the X-ray flux of PSR 0740-28 as $\sim 2.5 \times 10^{-12}$ erg cm^{-2} s^{-1}. Full analysis is in progress following the method of (4) and using contemporary P and $\dot{P}$ deduced from γ-ray analysis. The same preliminary upper limit can be set to X-ray emission of the Vela pulsar observed during the period 1975 October 20 to November 28. As shown by Bonnet-Bidaud (9) no pulsation is detectable using an average of 334552 events per bin and the following parameters : T_o = 2442703 J.D., P = 89.234746642 ms, $\dot{P} = 0.125 \times 10^{-12}$, $\ddot{P} = 0$. This limit is 8 times better than the limit reported by Pravdo (4) on OSO 8, but far below the predicted flux showing no simple correlation between X and γ-ray emissions of radiopulsars.

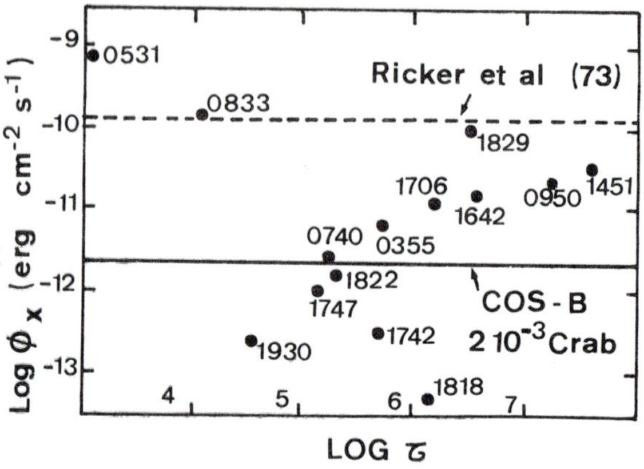

Fig. 1 - Predicted X-ray flux (2-12 keV) of radio pulsars versus "age" using the Crab pulsar spectrum as a standard, as compared to (3) and COS-B sensitivity limits.

RESULTS OBTAINED ON VELA X1

This source has been observed by COS-B between 1975 November 8 and November 28. Ögelman (10) detected the previously observed 283 s period of pulsations and used it to determine the orbital parameters of the binary system. They found that the pulse period for November 1975 was : P = 282.919 ± 0.002 s ; that is 0.0274 ± 0.0020 s longer than that determined from the earlier SAS-3 observations. This implies an average time variation of the period between the 2 observations of $\dot{P}/P = +(2.6 \pm 0.2)10^{-4}$ yr^{-1}. But new observations by SAS-3 in May 1976 show a further decrease of the period in good agreement with previous Copernicus and SAS-3 data (Sanford (11). For a full discussion of fluctuations on spin velocity of the object it is necessary to measure P and $\dot{P}$ during the other periods of COS-B observations.

CIR X1 RESULTS

COS-B observed this source from 1976 February 24 to March 24. Bignami (12) observed two peaks of enhanced X-ray emission and confirmed the periodic behaviour of this source as reported by Kaluzienski (13). The separation between 2 adjacent maxima is 16^d34 with a duty cycle of the periodic phenomenon confined to ~ 0.3 of the period. Bignami suggested that a weak emission from the source is present outside the peaks.

CYGNUS X1 AND CYGNUS X3 RESULTS

Cyg X1, Cyg X3 and 3U 1953+31 were in the field of view of COS-B between 75 Nov. 28 and Dec. 24 and were seen with a geometrical efficiency of 0.59, 0.41 and 0.37 respectively. The dominant contribution (more than 80%) came from Cyg X1 and Manzo (14) reported the X-ray light curve for this period, during which the source appears to be in the high state as confirmed by the contemporary ARIEL V result (15). The chaotic behaviour of the emission was observed at hours and minutes time scales taken during the "flare" observed by COS-B between Dec. 20.87 and Dec. 21.25, 1975. On Cyg X3 Bonnet-Bidaud et al. (16) were able to detect the 4.8 h modulation period at least during the last 3 weeks of observation when the source recovered from a low intensity state : both the mean flux and the modulation amplitude have varied during the month of observation.

CEN X3 RESULTS

This source was observed by COS-B between 1975, Dec. 24 and 1976, Jan. 23. Preliminary results obtained on 10% of the data were published by Bennett (17). Bonnet-Bidaud (18) used the complete data set to study the shape of the light curve during 14 subsequent binary cycles, which show good agreement with the zero-phase point and eclipse duration derived from Tuohy (19) and Schreier (20). During the 2 first cycles observed, a transition from a low to a high state similar to the turn-on reported by Schreier (21) was observed. An asymmetry is present in these first two cycles, the intensity being lower in the second part of the light curve after the mid-phase spike ; this is interpreted as due to a variation of absorption versus phase. In all subsequent cycles, the uneclipsed part is quite flat, while absorption affects only 15% of the period at most, nearest to the eclipse. Figure 2 shows the total number of counts received during the uneclipsed part of each binary cycle : the source is seen at the highest intensity level during cycles 7, 8 and 9 ; during the earlier cycles (3 to 6) and the later ones (10 to 14) the intensity remains fairly constant but at lower levels. The transitions between the four intensity levels all occur within one day. During the 3 high intensity cycles the average counting rate is 31 counts/second. Using the exponential spectral shape with kT = 5.6 keV fitted by Swank (22), and assuming a distance of 8 kpc (Krzeminski (23), one obtains an X-ray luminosity in the 2-12 keV range of 6.3×10^{37} ergs/second. Bonnet-Bidaud (18) concluded that COS-B observed Cen X3 in a high state at a higher luminosity than previously reported. They argue that such a luminosity probably rules out the stellar wind accretion as the only source of X-rays and that critical lobe overflow is present at least during the high states of the source. They also propose the time delay between the transitions in luminosity of cycles 2 to 3 and of cycles 9 to 10 to be a measure of the radial flowtime of

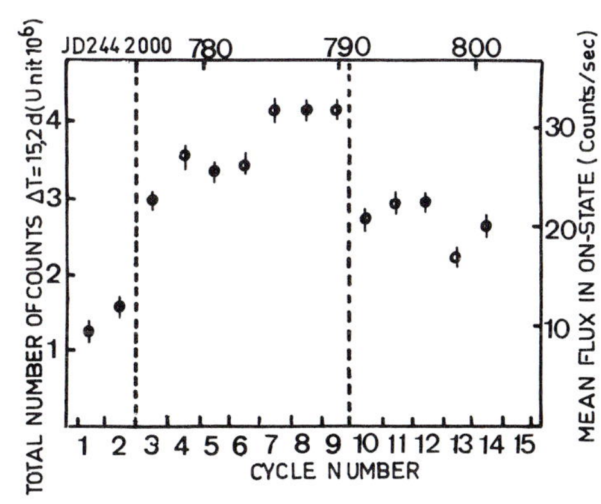

Fig. 2 - Total number of counts received during the uneclipsed parts of each of the 14 binary cycles of Cen X3 versus cycle number.

matter through the disk. In this scenario the low intensity of cycles 1 and 2 is interpreted as due to absorption of X rays by the large quantity of gas expected to be ejected from the primary star to the surroundings of the disk. The clear transition between cycles 2 and 3 is apparently caused by a decrease in the feeding of matter to the circumstellar environment. The subsequent decrease of feeding of the disk will appear later as a decrease of X-ray luminosity seen as the intensity transition between cycles 9 and 10 after the matter has passed through the disk. If this interpretation is correct the radial flowtime of matter through the disk is 15 ± 1 days.

Acknowledgements. I thank my colleagues J. M. Bonnet-Bidaud, N. D'Amico, N.R. Robba, and M. Van der Klis for results on Cen X3, PSR 0833-45 and PSR 0740-28 prior to publication and thank all the people whose efforts have made the COS-B mission successful.

REFERENCES

1. Boella, G. et al., 1974, Proceedings of Ninth Eslab Symposium, ESRO SP 106, p. 353.
2. Kanbach, G. et al., 1977, Proceedings of 12th Eslab Symposium ESA SP 124, p. 21.
3. Ricker, G.R. et al., 1973, Ap. J. (Letters) 186, L 111.
4. Pravdo, S.H. et al., 1976, Ap. J. (Letters) 208, L 67.
5. Bonnet-Bidaud, J.M., 1978, private communication.
6. Buccheri, R. et al., 1978, submitted to Nature.
7. Buccheri, R. et al. 1977, Contribution to International School of Relativistic Astrophysics, ERICE.
8. Bonnet-Bidaud, J.M., D'Amico, N., 1978, private communication.
9. Bonnet-Bidaud, J.M., D'Amico, N., Robba, N.R., 1978, private communication.
10. Ögelman, H. et al., 1977, Astr. and Astroph., 58, 385.
11. Sanford, P.W., 1977, 8th Texas Symposium on Relativistic Astrophysics, Annals of the New York Academy of Science, 302, 386.
12. Bignami, G.F. et al., 1977, Astron. and Astrophys., 57, 309.
13. Kaluzienski, L.J. et al., 1976, Ap. J. (Letters), 208, L 71.
14. Manzo, G. et al., 1978, Astron. and Astrophys., 62, 265.
15. Holt, S.S. et al., 1976, Nature, 261, 213.
16. Bonnet-Bidaud, J.M. et al., 1978, Astron. and Astrophys., 62, 275.
17. Bennett, K. et al., 1976, Astron. and Astrophys., 51, 475.
18. Bonnet-Bidaud, J.M. et al., 1978, Astron. and Astrophys., to be published.
19. Tuohy, I.R., 1976, M.N.R.A.S., 174, 45.
20. Schreier, E. et al., 1972, Ap. J. (Letters), 172, L 79.
21. Schreier, E. et al., 1976, Ap. J., 204, 539.
22. Swank, J. et al., 1975, Symposium on X-Ray Binaries, NASA X-660-75-285.
23. Krzeminski, W., 1974, Ap. J. (Letters), 192, L 135.

OBSERVATIONS OF AN X-RAY QSO

G. R. Ricker*, G. W. Clark*, R. E. Doxsey*,
R. G. Dower*, J. G. Jernigan*, C. R. Canizares*,
J. P. Delvaille**, G. M. MacAlpine***,
R. M. Hjellming +

**Department of Physics and Center for Space Research,*
Massachusetts Institute of Technology, Cambridge, MA 02139
***Smithsonian Astrophysical Observatory, Cambridge, MA 02138*
****University of Michigan, Ann Arbor, MICH 48109*
+National Radio Astronomy Observatory, Socorro, NM 87801

INTRODUCTION

In this paper we report X-ray, optical, IR and radio measurements of a new QSO. This QSO, designated MR2251-178, is the first one to be initially discovered by its X-ray emission [1]. It is only the second X-ray emitting QSO discovered, 3C273 being the previous one [2]. In the following, we present the positional information for MR2251-178, then discuss its intensity in various energy ranges, and give some details of its optical spectrum. Finally, we report on its optical behavior during the past ninety years of astronomical photography.

DISCOVERY

In Fig. 1, the rectangle designated "A" is an X-ray error box which had previously been established by the Leicester University group using the Ariel V satellite [3]. There are of the order of a thousand optical objects visible inside this box down to the plate limit. Last August, we examined this error box with the rotating modulation collimator system on the SAS-3 X-Ray Observatory. When we analyzed our data, we found a 1.5 Uhuru count X-ray source inside the small circle labeled "S". The circle within the inset is 40 arc sec in radius, and is the preliminary SAS-3 error region for the X-ray source. Of the two objects designated inside the circle, Object 1 appeared to be a galaxy, based on PDS measurements of the Palomar glass plates at Kitt Peak. Object 2, on the other hand, had a hard, star-like image. S. Kleinmann measured the positions for both 1 and 2 on the Grant Engine at Kitt Peak. G. MacAlpine then examined these objects on the Tololo-Michigan Objective Prism Survey plates and found that Object 2 showed bright emission lines, from which he estimated $\alpha \simeq 0.068$. Meanwhile, R. Hjellming observed the SAS-3 error circle with the VLA. He found that there was a dominant point radio source at Object 2, and an extended one near Object 1.

From the positional coincidence and the strong optical emission line spectrum reminiscent of a Type 1 Seyfert galaxy, we concluded that Object 2 was very likely the X-ray source, and in fact that it was a low z QSO - based on its hard, star-like image on the Palomar Sky Survey Plate, and its luminosity.

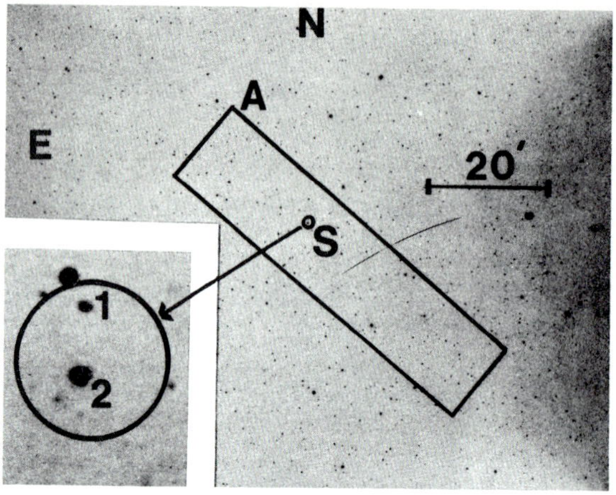

Fig. 1. The QSO MR2251-178 and the X-ray source 2A2251-179. "A" and "S" denote the Ariel V error box and the SAS-3 error circle, respectively. The inset shows the preliminary SAS-3 error circle enlarged by a factor of 20. Object 2 is MR2251-178. Photographs are taken from the Palomar Sky Survey Plates (© National Geographic Society).

IMPROVED POSITION AND OPTICAL STRUCTURE

We continued to analyze our X-ray data, and finally came up with the error circle shown in Fig. 2, which is a 4 meter plate taken at our request by J. Graham at Cerro Tololo. The error circle (90% confidence) is the final result of the SAS-3 data analysis. It shows that Object 1 is excluded as the X-ray source. On this plate, unlike the Palomar plate, Object 2 seems to show a fuzzy patch to the southwest. PDS scans on this plate are presently in progress.

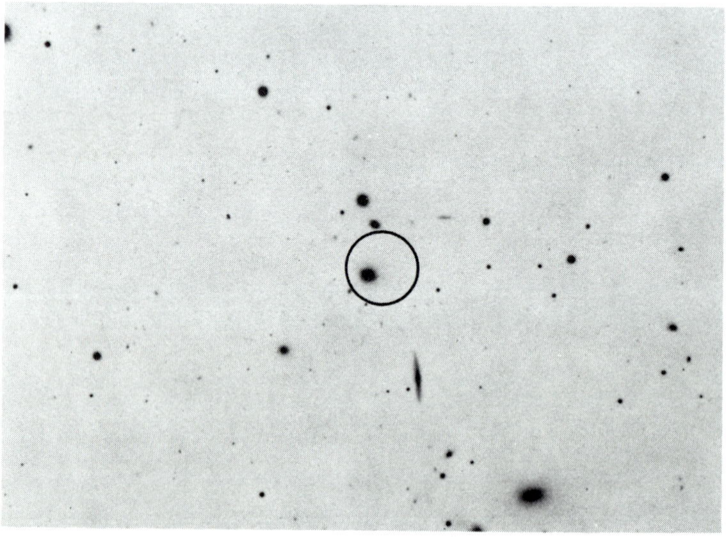

Fig. 2. The QSO MR2251-178. The circle is the final SAS-3 error circle (90% confidence). (CTIO 4m plate).

RELATED OBSERVATIONS

Table 1 summarizes the information we have gathered on MR2251-178 during the past year. The X-ray position was determined from the SAS-3 data, while the optical coordinates are from the glass plate measurements at Kitt Peak. The radio measurements are from the VLA. The redshift value was recently determined from scanner spectra taken at the McGraw Hill Observatory [4]. This value is within the error of our earlier 0.068 measurement [1]. The optical colors were measured at McGraw Hill by D. Backman of M.I.T. We originally estimated the object to have a B magnitude of ~ 16 based on Palomar Plate measurements, but it turns out to be ~ 4 times brighter in 1977, twenty-five years later.

TABLE 1 Observations Summary

MR 2251-178

	R.A. (1950)	Dec. (1950)	2σ Error Radius
X-ray	$22^{H}51^{M}25^{S}.4$	-17°50'40"	30"
Optical	22 51 25.89	-17 50 54.2	1
Radio	22 51 25.90	-17 50 53.5	2

Optical Colors:
V = + 14.11
B-V = + 0.52
U-B = - 0.86

Redshift: Z = 0.065

Absolute Magnitude:
M_V = - 24.2

X-ray Luminosity (2 - 10 keV):
1977.6 (SAS-3): 5×10^{44} ergs s^{-1}
1975 (2A) : 1.6×10^{45} ergs s^{-1}

LUMINOSITY AND SPECTRUM

The implied absolute V magnitude of -24.2 is in line with those of other low z QSOs, such as the 14 discussed by Baldwin [5]. (The object was actually brighter than -25 around 1970; Fig. 5, discussed below.) For the X-ray luminosity, the 1977 value is from SAS-3, while the 1975 value is from Ariel V. At its peak X-ray brightness, the X-ray:optical luminosity ratio appears to be about ~ 3:1, which is why we refer to it as an X-ray QSO.

In Fig. 3, we show an optical spectrum taken with the Mark I scanner on the 52 inch telescope at McGraw Hill in November 1977. The spectrum extends from 3600 Å to just beyond Hα. All the lines we have identified are labeled. The most notable feature of the spectrum is the extreme width of the Hα emission line. The FWZI is between 30,000 and 40,000 km/s. Otherwise, the spectrum is quite normal for a QSO, and it fits in rather well with the collection of 14 low z QSOs studied by Baldwin [5]. This spectroscopic normalcy may possibly imply that all QSOs are powerful X-ray emitters, or alternatively, that strong X-ray emission does not produce a characteristic optical signature.

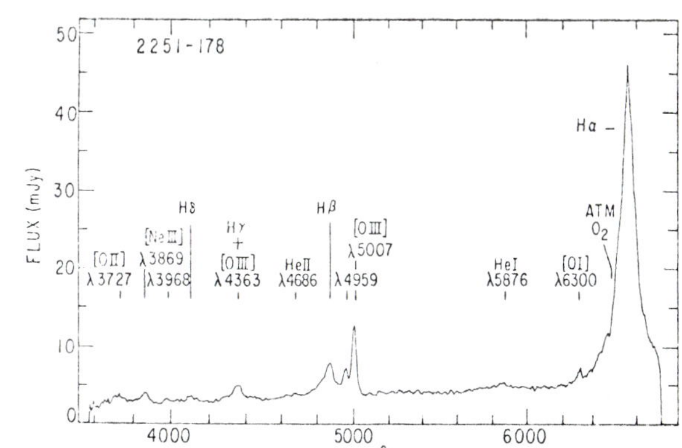

Fig. 3. Optical spectrum of MR2251-178.

Figure 4 gives the overall spectrum of the QSO, from the radio to the X-ray region. The 3 short bars marked J, K and L are IR color measurements made by S. Kleinmann, D. Sargent and R. Joyce. The other measurements plotted are 1977 observations discussed earlier. The dashed line is a fit to the IR, optical and X-ray points.

PHOTOMETRIC HISTORY

Because of the brightness of MR2251-178, and its location near the ecliptic, we had reasonable hope that there would be good coverage of it in the Harvard Plate Stack collection. As it turned out, we were able to find usable plates extending all the way back to the year 1889. Figure 5 shows a portion of the magnitude measurements we made on the QSO, covering the period up to 1977. During this time, the object went from fainter than 16th magnitude (in 1935) up to brighter than 14th (in 1970). The filled triangle near the right edge is the photometry we performed in November 1977 at McGraw Hill. Superimposed on the gradual trends are numerous bumps and wiggles on timescales of months to about a year, particularly evident in the mid 1940s. The method we used for analyzing this data is the standard one developed by the Lillers and their associates [6]. With this method, the uncertainties in magnitude estimates are about a tenth of a magnitude, and within this uncertainty, we did not detect fluctuations on timescales shorter than about one month.

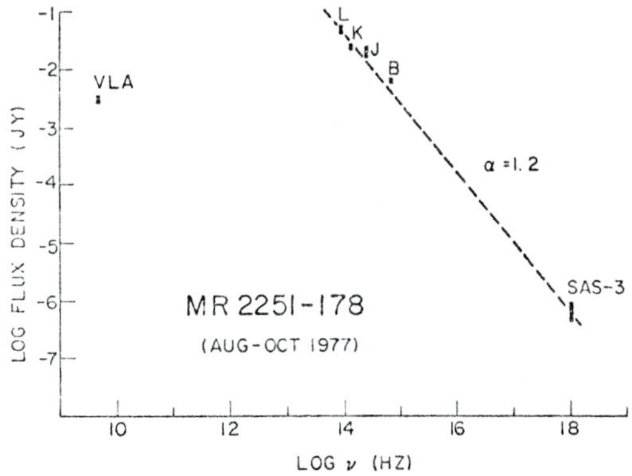

Fig. 4. Spectrum of MR2251-178.

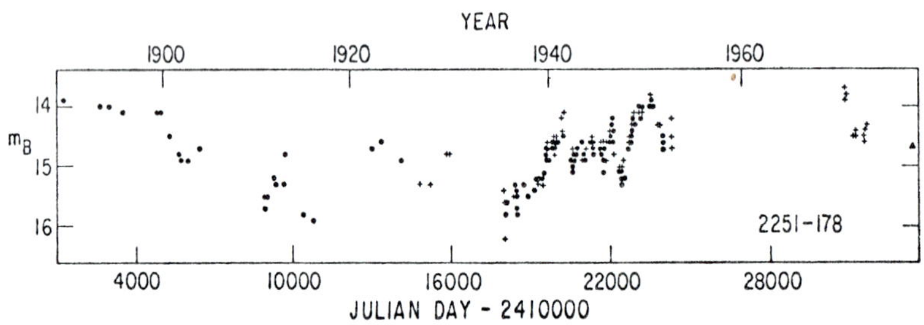

Fig. 5. Optical variability of MR2251-178.

Acknowledgements We acknowledge support from the staffs of the Center for Space Research at M.I.T. and the Goddard Space Flight Center of NASA. We thank H. Bradt, B. Cooke, M. Elvis, R. Kelley, W. Lewin, M. Liller, J. McClintock, J. Pye, R. Remillard, and B. Rossi for helpful discussions and assistance. The National Radio Astronomy Observatory is operated by Associated Universities Inc., under contract with the National Science Foundation. This work was supported in part by the National Aeronautics and Space Administration under contract NAS5-11450.

REFERENCES

1. G. Ricker, G. Clark, R. Doxsey et al, Nature, 271, 35 (1978).

2. S. Bowyer, M. Lampton, J. Mack et al, Astrophys. J. (Letters), 161, L1 (1970).

3. B. Cooke, M. Ricketts, T. Maccacaro et al, Mon. Not. R.A.S., 182, 489 (1978).

4. C. Canizares, J. McClintock, and G. Ricker, Astrophys. J. (Letters), in the press.

5. J. Baldwin, Astrophys. J., 201, 26 (1978).

6. M. Liller and W. Liller, Astrophys. J. (Letters), 199, L133 (1975).

TWO DIVERSE EXTRAGALACTIC X-RAY OBJECTS QSO 0241+622 AND THE ELLIPTICAL GALAXY NGC2110

H. Bradt*, B. Burke*, C. Canizares*, R. Dower*, J. McClintock*, B. Margon**, R. Hjellming***, H. Helmken+, K. Apparao++, G. Bignami+++ L. Maraschi+++, M. Johns§ and A. Koski§

*Massachusetts Institute of Technology, Cambridge, MA, USA
**University of California, Los Angeles, CA, USA
***National Radio Astronomy Observatory, Socorro, NM, USA
+Center for Astrophysics, Harvard University, Cambridge, MA, USA
++Tata Institute, Bombay, India
+++Lab. Fis. Cos. Tech. Rel., CNR, Milan, Italy
§University of Michigan, Ann Arbor, MI, USA

ABSTRACT

A precise position of 4U0241+61 obtained with SAS-3 has led to its identification as the third X-ray QSO (0241+622). This QSO has a red shift of $z = 0.0438$, an absolute magnitude of $M_V \sim -26$, an X-ray flux of 1.8 μJy (2-11 keV), and exhibits quite intense radio emission (0.2-0.5 Jy). It is the closest known QSO and would be the brightest in the sky (∿11 mag) if it did not lie in the galactic plane (b^{II}=+2.4). As a class, the three known QSOs (3C273, MR2251-178, and 0241+622) are the brightest (apparent magnitude) of all known quasars and are among the closest to our Galaxy. They are associated with X-ray sources which are among the faintest known. This indicates that the space density of QSOs is greater than hitherto believed, and that faint X-ray sources are good tracers of QSOs.

In another observation, the same detectors recorded the existence of a hitherto unreported X-ray source which is coincident with NGC2110. This galaxy is unusual as an X-ray source in that it is classified as elliptical and exhibits no evidence for a bright Seyfert-like nucleus. It does, however, exhibit strong narrow emission lines.

Optical, X-ray, and radio data for both sources will be presented.

FIRST SPECTRUM OF AN EXTRA-SOLAR OBJECT IN THE EXTREME ULTRAVIOLET: THE WHITE DWARF HZ 43

R. F. Malina, S. Bowyer and F. Paresce

Astronomy Department and Space Sciences Laboratory, University of California, Berkeley, California 94720, U.S.A.

The discovery of stellar extreme-ultraviolet (EUV) sources has opened up a new band of the spectrum, from 100-1000 Å, to astronomers. The EUV survey carried out during the Apollo-Soyuz mission led to the discovery of five EUV sources [1] of which the brightest was identified as the hot white dwarf HZ 43 [2]. The Apollo-Soyuz telescope consisted of four nested grazing incidence paraboloids focused on a channel electron multiplier. The instrument operated in a broad-band photometric mode with a field of view of 2°. Two lines of instrumental development have been followed in succeeding instruments at Berkeley [3]. In order to increase the instrumental sensitivity, optics and detectors with imaging capability have been built and flown to carry out searches for new sources [4]. This paper describes a second instrument[3] which has been used to carry out a spectroscopic observation of the hot white dwarf HZ 43.

INSTRUMENTATION

A schematic of the instruments is shown in Fig. 1. The instrument consists of an EUV telescope, and an EUV spectrometer housed in a sounding rocket shell 44 cm in diameter and 176 cm in length. The total instrument weight is 110 kg. The telescope is an f/10 Wolter-Schwarzschild Type II system [5]; the maximum diameter of the primary of 38 cm. The distance from the front of the primary to the focal plane is 120 cm. The telescope was fabricated from forged aluminum blanks electroplated with copper. The figure was machined at the Lawrence Livermore Laboratories using the diamond-turning technique [6] The surface was then coated with nickel and polished; the final axis image had a diameter of better than 10 arc seconds in visible light, and on the order of 30 arc seconds in the extreme ultraviolet.

The spectrometer is shown schematically in Fig. 2. The spectrometer is housed in vacuum box held below 10^{-5} torr by a small ion pump. In flight the entrance to the vacuum box is covered by a 1500 Å thick aluminum filter; the filter is used to reduce potential geocoronal Ly-α contamination and shield against outgassing products from the instrument and sounding rocket shell.

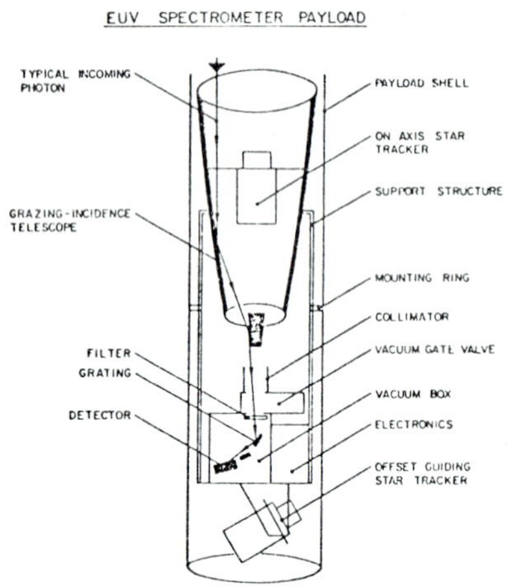

Fig. 1. Schematic of the EUV telescope and spectrometer.

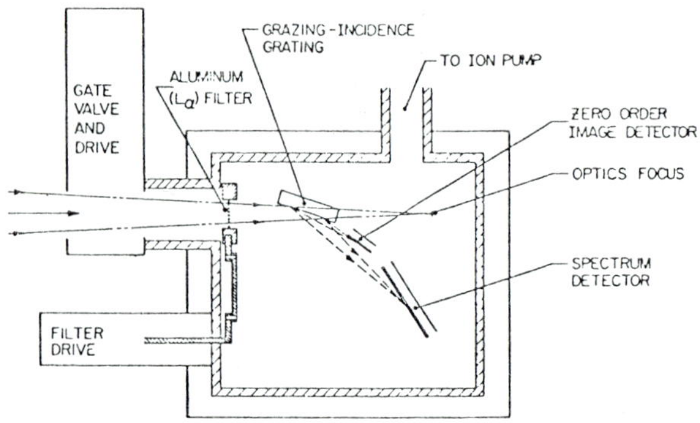

Fig. 2. Schematic of EUV Spectrometer

The spectrometer was designed to provide a resolution of 15 Å throughout the EUV. The adopted solution is based on a Type III or aberration corrected grazing incidence configuration [7]. The concave grating is located in the converging beam from the optics allowing a very compact layout. The grating parameters are given elsewhere [3].

The zero order image from the grating is focused on a 25 mm diameter microchannel plate. The first order spectrum image is focused on a 70 mm x 20 mm microchannel plate. In order to increase the quantum efficiency of this detector for the spectrometer configuration, the front plate is cut at a bias angle of 40°. A resistive anode was located behind each microchannel plate detector and read out to provide two dimensional imaging as described by Lampton and Paresce [8].

OBSERVING PROGRAM

The instrument was launched from the White Sands Missile Range on a Black Brant sounding rocket on April 10, 1978. The prime target for the flight was the hot white dwarf HZ 43 which was successfully observed for 200 seconds. The image recorded by the spectrum detector is shown in Fig. 3. The spectrum signal is seen as a dark band in the lower half of the detector plane. Wavelength increases from left to right. The curvature in the spectrum is due to residual distortion in the resistive anode, and the downward displacement of the spectrum from the center of the detector is due to a pointing offset. The spectrum is strongly detected from 170 Å to 400 Å with second order image being detected beyond that wavelength. The spectrum is not detected shortward of 170 Å due to the cutoff of the aluminum filter. There is a decrement at 228 Å, which corresponds to the photoelectric absorption of HeII. No other features appear significantly above the continuous spectrum.

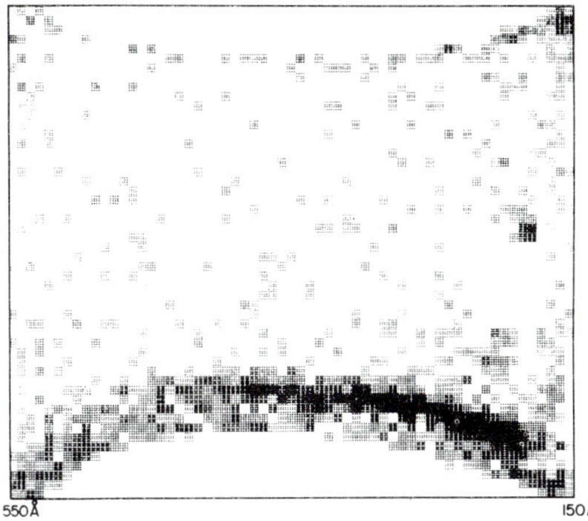

Fig. 3. Two dimensional image of the spectrum of HZ 43 as recorded by the detector. Dark band in lower portion of figure is the spectrum; curvature is due to residual distortion of detector. Wavelength decreases from left to right.

DISCUSSION

The observation of the spectrum firmly identifies HZ 43 as the source of the EUV radiation, improving the error box to the ±5 arc min entrance aperture of the spectrometer. The spectrum obtained of HZ 43 provides valuable new information on both HZ 43 itself and the intervening interstellar medium. Extensive optical, soft X-ray and EUV studies [9] provide an overall description of the source. HZ 43 is a detached binary system consisting of a DAwk hot white dwarf and a dM3·5e companion; the distance to the system derived from spectroscopic and trigonometric parallax is 65 ±15 pc [9]. The surface temperature and emission process is however very poorly determined and depends strongly on the assumed surface composition and stellar parameters. Margon et al. [9] found that for a pure black body, the available data was consistent with a temperature of 110,000°K ±10,000°K. Durisen et al. [10] however, found that for a solar composition atmosphere, a temperature higher than 125,000°K was required. Finally a hot corona around the white dwarf, although less likely on theoretical grounds, could still explain the broad band EUV and X-ray data [9]. The observed spectrum is, however, not consistent with a coronal model.

The observation of the HeII absorption edge can be used to put a tight constraint on the allowable surface helium number density. Preliminary photospheric models [11] indicate that in order to explain the observed decrement, the number density of helium relative to hydrogen must be in the range of 10^{-4} to 10^{-5}. Thus the models proposed by Durisen et al. [10] can be ruled out.

Detailed modelling of both the continuum and the helium 228 Å decrement will allow a self-consistent determination of the temperature and the surface helium fraction [12]. The observed decrement can also be used to set an upper limit of 4×10^{17} cm^{-2} on the column density of ionized helium in the intervening interstellar medium [12].

FUTURE PROSPECTS

The successful spectroscopic observation of HZ 43 is an encouraging confirmation of the potential of EUV astronomy. The instrument described in this paper is capable of obtaining moderate resolution spectra of all the known stellar EUV sources. Such spectroscopy will of course be essential to understanding the emission processes in the coronas of nearby stars, such as Proxima Centauri, which can be expected to have rich emission line spectra. It has already been shown to provide information unobtainable by other means on the properties of hot white dwarfs. New instrumentation extending the wavelength range and increasing the wavelength resolution is feasible and currently under development at Berkeley.

ACKNOWLEDGEMENT

We acknowledge the considerable help of many colleagues at Berkeley in the development of this instrument, in particular M. Lampton, G. Penegor, R. Kimble and W. Cash. The fabrication of the optics was carried out at the Lawrence Livermore Laboratories by J. Bryan and P. Baker. We also acknowledge the support of the Goddard Sounding Rocket Division. This work was supported by NASA grant NGR 05-003-450.

REFERENCES

1. F. Paresce, Earth Extraterr. Sci. 3, 55 (1977).

2. M. Lampton et al., Astrophys. J. (Letters), 203, L71 (1976).

3. M. Lampton et al., Proc. Soc. Photo-Opt. Instrum. Eng. 106, 93 (1977).

4. W. Cash, S. Bowyer and M. Lampton, Astrophys. J. (Letters), 221 L87 (1978).

5. H. Wolter, Annls Phys., 10, 286 (1952).

6. T.T. Saito, Appl. Opt., 14, 1773 (1975).

7. S.A. Jobin Yvon, "Handbook: Diffraction Gratings Ruled and Holographic," 19 (1975).

8. M. Lampton and F. Paresce, Rev. Sci. Instrum., 45, 1098 (1974).

9. B. Margon et al., Astrophys. J., 209, 525 (1976).

10. R.H. Durisen et al., Astrophys. J. (Letters), 206, L149 (1976).

11. L. Auer, private communication (1978). H. Shipman, private communication (1978). J. Heise, private communication (1978).

12. R. Malina, in preparation (1979).

Diffuse Galactic Phenomena, Interstellar Medium and Supernovae

DIFFUSE GALACTIC EMISSION

W. L. Kraushaar

Department of Physics, University of Wisconsin,
1150 University Avenue, Madison, Wisconsin 53706, U.S.A.

The purpose of this review is to summarize the observations of the soft X-ray background radiation in the energy region below 1 keV. In addition, since the purpose of the observations is ultimately to understand the origin of the radiation and how it relates to other astrophysical phenomena, I shall attempt to indicate conclusions regarding the origin that can be drawn from the observations independent of specific models. Models of one kind or another have accompanied almost every new set of observations and while they are essential to progress, it seems well to pause at this point and emphasize what we know rather than what might be.

The data we have are of two sorts--spatial distributions or maps in broad energy bands and energy spectra in the form of proportional counter pulse height distributions. Both the spatial distributions and the energy spectra are severely limited in resolution by the state of technology and observing time.

Representative maps [1,2,3,4] for the energy band just below the carbon K-edge (0.284 keV) are shown in Figs. 1, 2, 3, 4, and for the energy band just below the boron K-edge (0.188 keV) in Fig. 5 [1]. The most prominent feature in these maps, aside from the Cygnus Loop and Vela supernova remnants which are too intense for the graphical display techniques used, is the enhancement along ℓ=25° in the northern Galactic hemisphere. This feature is also evident in the higher energy soft X-ray bands and has been associated with the North Polar Spur radio feature. Borken and Iwan [5] have published higher spatial resolution maps of this region. In other regions of the sky there is a general tendency of the X-ray intensity to anti-correlate with the column density of neutral interstellar gas, as is discussed later.

Some spatial distribution data are also available in the energy region above the carbon K-edge in the general region of 1 keV [3,6,7]. These data will largely be omitted from this discussion because separation of the three components--diffuse galactic emission, emission from galactic or extragalactic discrete sources, and diffuse extragalactic emission--is so difficult. For example, the apparently diffuse extragalactic power law spectrum is only known above 2 keV and so must be extrapolated to energies below 1 keV and then corrected for interstellar absorption. The energy region near 1 keV is important, however, and deserves more attention than it has received.

The conclusions of the discussion that follows may be summarized as follows:

1. No observational evidence supports the hypothesis that a significant fraction

of the diffuse soft X-ray intensity below 0.284 keV is of extragalactic origin. (We assume hereafter that all is of galactic origin.)

2. The radiation is really diffuse, not the superposition of unresolved stellar sources.

3. The radiation is of thermal, not non-thermal, origin.

4. Most of the E<0.284 keV radiation we detect at the earth is of relatively local origin. By this we mean that there is less than $\sim 10^{20}$ atoms cm^{-2} of intervening absorbing interstellar gas.

5. The temperature of the emitting regions is in the range $T=10^{6\pm0.5}$K.

6. The pressure in the hot emitting regions exceeds 10^4 cm^{-3} K and is therefore appreciably greater than the nominal interstellar pressure, 2000 cm^{-3} K.

Galactic vs. Extragalactic Origin

Following the first observations of the soft diffuse background radiation by Bowyer *et al.*[8], there was widespread speculation that the emission was from a hot intergalactic medium of density sufficient to close the universe. Qualitative support for an extragalactic origin was provided by the observation that the intensity fell off with increasing column density of interstellar X-ray absorbing gas. The intensity fall-off data, however, were severely scattered and not nearly as steep as would be predicted from photoelectric absorption. Further, the intensity in directions near the galactic plane, where the opacity to outside the galaxy is complete, was 20-25% as large as the maximum intensity seen anywhere. These observations have been verified in more recent and more complete surveys.

Enthusiasm for some appreciable fraction of the soft X-ray background being of extragalactic origin persisted, however. Shadowing experiments, where one attempts to measure the intensity decrease that results from an intervening extragalactic object, could in principle separate galactic and extragalactic contributions. A summary of the results of these attempted shadowing observations is shown in Table 1. No shadowing by the gas of the Small Magellanic Cloud (SMC), the Large Magellanic Cloud (LMC), or M-31 has been detected. Nor has absorption shadowing by detailed features in the structure of gas in our own galaxy been demonstrated. There is always the possibility, of course, that soft X-ray emission by the chosen shadowing object just fills in to compensate for the absorption of the extragalactic intensity. With measurements sensitive to gas in our own as well as in three external galaxies, this possibility seems contrived. In any event, there is no observational evidence requiring that a significant fraction of the soft X-ray intensity below 0.284 keV be extragalactic.

TABLE 1 Shadowing Observations to Test Extragalactic Origin Hypotheses

	Fraction of observed flux that must be of local (galactic) origin
SMC + Local Gas [9,10]	>75%
M31 + Local Gas [11]	>48%
LMC + Local Gas [12]	>35%
LMC + Local Gas [13]	90-94%

Diffuse vs. Point Source Origin

We know of no observational evidence that supports the hypothesis that an appreciable fraction of the intensity below the carbon K-edge comes from unresolved point sources. Gorenstein and Tucker [14] and Levine *et al.* [15] have used measurements of the apparent granularity of the background radiation to set lower limits on the space density of stellar objects that could be responsible for the observed emission. Their values are 0.01 and 0.2 pc^{-3}, respectively. The 0.01 value is larger than the space density of any stellar type, such as white dwarves, that could reasonably be expected to be X-ray emitters. The 0.2 value is larger than the space density of all stars of whatever type.

In another approach, Vanderhill *et al.* [16] set upper limits to the soft X-ray luminosity of the nearest approximately 50 stars over what amounted to a randomly selected area of the sky. The known space densities of these stars with the measured luminosity upper limits were not consistent with a point source region.

Stellar or point sources of very soft X-ray emission are known and are discussed by Gordon Garmire in this volume [17]. The time-average luminosities and space densities of these objects, in agreement with the granularity arguments above, fall short of accounting for the measured diffuse intensity.

Thermal vs. Non-Thermal Emission

The best current evidence for thermal emission is the incompatibility of known non-thermal processes. These have been discussed in some detail by Hayakawa [18] and by Williamson *et al.* [6]. In Table 2 is given a summary of possible non-thermal mechanisms, the particle energy required, and a short statement of the reasons for rejection.

TABLE 2

Mechanism	Particle Energy	Remark
1. Bremsstrahlung	$E_e > 300$ eV	$\zeta_H >> 10^{-15} s^{-1}$
2. Synchrotron	$E_3 \sim 3 \times 10^{13}$ eV	$T_{1/2} \simeq 5000$ years
3. Inverse Compton on 3K	$E_e \simeq 250$ MeV	Too many 100 MeV Gamma Rays
4. Inverse Compton on Starlight	$E_e \simeq 6$ MeV	ISM overpressured by 1000
5. Transition radiation	$E_e \simeq 50$-500 MeV	Too many 100 MeV Gamma Rays
6. Sub-relativistic cosmic ray ion line emission	$E_n \simeq 1$-10 MeV nucleon^{-1}	$\zeta_H >> 10^{-15} s^{-1}$

Electrons of energy 300 eV or greater can certainly produce bremsstrahlung soft X-rays in collisions with the gas atoms of the interstellar medium. But the process is notably inefficient in that most of the electron's energy goes into ionization and excitation, not X-ray emission. If an electron flux large enough to supply the soft X-ray emission is postulated, the accompanying ionization rate is far greater than that permitted by the *Copernicus* measurements of the relative abundances of HD, OH, and H_2 [19].

Electrons of $3x10^{13}$ eV can radiate soft X-rays via the synchrotron process in the galactic magnetic field. But such energetic electrons lose half their energy with a characteristic time scale of 5000 years. The storage time of cosmic ray electrons is about 10^7 years [20]. It follows that the replenishment time must be of this order also--certainly not as small as 5000 years.

Electrons of 250 MeV are required to make soft X-rays via inverse Compton scattering on the 3 K radiation. But 250 MeV electrons would also make gamma rays via bremsstrahlung on the interstellar gas, and the required flux would make too many gamma rays.

For inverse Compton production on starlight, only 6 MeV electrons are required. But the required flux of these electrons would overpressure the interstellar medium by a factor of 1000 or so. The resulting shocks would make hot gas, thus emitting X-rays in any case.

Transition radiation from electrons of hundreds of MeV colliding with dust can be in the soft X-ray region, but the process is less efficient than inverse Compton processes on the 3 K radiation, and that is already in deep trouble.

When low energy cosmic ray ions slow down through ionization losses in the interstellar medium, they can pick up electrons in excited states which then emit, for example, Lyman-α type X-rays as they fall into the ground state. This has been considered in detail by Boldt and Pravdo [21] and by Watson [22]. The difficulty is that again the required flux of low energy cosmic rays would produce an ionization rate in the interstellar gas in excess of $10^{-15} s^{-1}$.

Conceivably, failure to identify a suitable non-thermal emission process could indicate our lack of imagination rather than support for a thermal origin. The *Copernicus* measurements of O VI [23,24] in the absorption spectrum of early-type stars show that gas near $3x10^5$ K exists in the interstellar medium, and this at least provides observational plausibility that other and possibly related regions exist at somewhat higher temperatures to supply the soft X-ray emission. A definitive proof of thermal emission would be provided by detection of the emission lines expected from an astrophysical plasma near 10^6 K. Hayakawa *et al.* [25] have interpreted enhancements near 0.65 keV in proportional counter pulse height spectra taken inside the radio feature Loop I as emission from O VII and O VIII. The pulse height spectra are shown in Fig. 6.

If this important suggestion of line emission is confirmed it will lend considerable support to the idea that the North Polar Spur or Loop I is, as has often been supposed, an ancient supernova remnant. A definitive demonstration that the diffuse soft X-ray background radiation is thermal in other than this special direction awaits the more widespread detection of emission lines.

Local vs. Remote Emission Regions

There are a number of observations that indicate that the bulk of the soft X-ray diffuse emission below 0.284 keV arrives at the solar system having suffered little or no energy-dependent absorption.

The observations of Yentis *et al.* [26], Cash *et al.* [27], and Levine *et al.* [15,28] all used proportional counters equipped with very thin polypropylene windows. The data have been interpreted by the authors as indicating the presence of very low energy X-rays ($E \lesssim 0.1$ keV) with an intensity that is increasing rapidly with decreasing X-ray energy. At 0.1 keV the Brown and Gould [29] effective absorption cross sections correspond to unit optical depth at a hydrogen column density of less than $2x10^{19}$ cm^{-2}. That these observations imply lack of significant absorption is a reasonable qualitative conclusion. But the emission spectrum is unknown and could conceivably have a very steeply rising low energy spectrum and so overcome the absorption which is steeply falling with X-ray energy.

A stronger case for the same conclusion can be made from the data of Sanders *et al.* [30]. In these observations two quite distinct energy bands were defined by two types of counter windows. As mentioned earlier, one transmits below the boron K-edge at 0.188 keV, and one transmits below the carbon K-edge at 0.284 keV. Data in these two energy bands, B and C respectively, are now available over the southern galactic hemisphere. They are shown in the maps of Figs. 1 and 5. When the data are binned in ranges of galactic latitude and the intensity in a given direction are plotted against N_H, the column density of atomic hydrogen, the plots shown in Fig. 7 result. At mid-latitudes we see a general tendency for the intensity in both energy bands to decrease with increasing N_H. If the intensity decrease were due to an energy-dependent absorption effect, both the C- and B-band data should fall much more rapidly with N_H, and the decrease in B-band should be steeper by a factor of two or more than that in C-band.

It follows from the observed lack of energy-dependent absorption in B- and C-bands that either there is very little absorption or that there is absorption but the spatial distribution of the gas is such that the energy dependence is suppressed. We examine below the logical consequences of these possibilities.

Under the first hypothesis the emission region must be within a distance corresponding to a column density of neutral gas less than 10^{20} cm^{-2}. Then, as discussed by Sanders *et al.* [30], the scattered but nonetheless real tendency for the X-ray intensity to decrease with increasing N_H can be taken as a displacement effect. A larger than average X-ray intensity indicates an emission region of larger than average extent along the line of sight, and the increased extent of the X-ray emitting region leaves less room for the cold neutral gas. A necessary consequence of this hypothesis is that the X-ray emitting and cold neutral gas components of the interstellar medium occupy comparable and large fractions of the interstellar volume.

Under the second hypothesis, all but a small fraction of the neutral interstellar gas would be clumped in structures optically thick in both B- and C-bands. Optically thick clouds, the only type structure discussed here, were discussed some time ago in the X-ray absorbing context [8,31] and are an integral part of the recent interstellar medium model of McKee and Ostriker [32]. Variations in the number density of clouds, n_c, would result in variations in N_H that would therefore anti-correlate with variations in X-ray intensity. We show below that available X-ray data do not favor this second hypothesis.

The high resolution 21 cm studies of Griesen [33] and Dickey *et al.* [34] show that cloud structures do not have characteristic dimensions less than $r_c \simeq 0.5$ pc. The clouds must have column densities not much less than $N_{HC} \simeq 4x10^{20}$ cm^{-2} in order to be optically thick in C-band and the smoothed density of neutral gas is $\bar{n}_H \simeq 1$ cm^{-2}. The number density of clouds must therefore be $n_c \simeq 10^{-2}$ pc^{-3} with a mean free path $\lambda_c \simeq 130$ pc. (The corresponding values adopted by McKee and Ostriker [32] are

$n_c = 5.9 \times 10^{-3} pc^{-3}$, $\lambda_c \simeq 88$ pc, and $\sqrt{\overline{r^2}} \simeq 0.811$ pc.) These parameters appear to be in conflict with the X-ray background granularity measurements of Levine *et al.* [15]. These authors set a 3σ upper limit of 9% to the granularity, δI/I, for a 0.01 steradian solid angle. They then used a Monte Carlo simulation to set a lower limit on the space density, n, of uniform luminosity point source emitters embedded in an absorbing medium of mean free path λ. Their result was:

$$n > 0.2 \left(\frac{100\ pc}{\lambda}\right)^3 pc^{-3}.$$

P. M. Fried of this laboratory has considered the granularity that results when thick absorbing clouds of uniform size and density are either embedded in a uniform emitting medium or located in front of a uniform emitting region and has shown that it is approximately the same as the granularity in the complementary situation considered by Levine *et al.*. The measured absence of granularity therefore implies $n_c > 0.1\ pc^{-3}$. Consistency with this second hypothesis, however, requires $n_c < 0.01 pc^{-3}$. The observational means exist for improved granularity measurements, and gas distributions other than spherical should be investigated. On the basis of the information available, however, the X-ray data seem to favor the hypothesis that there is little or no interstellar absorption between us and the sources of the very soft diffuse X-ray background.

Temperatures of the Emission Regions

The various observational groups have analyzed their proportional counter pulse height or X-ray filter data in terms of temperature, T, of an emitting hot gas and N_H, a measure of the amount of interstellar absorption. As pointed out before, all measurements to date have poor spectral resolution. Nevertheless, all observers agree that the gas responsible for the emission must be near 10^6 K. Williamson *et al.* [6], de Korte *et al.* [3], Burstein *et al.* [7], and Levine *et al.* [15] suggest that more than one temperature is required. Hayakawa *et al.* [35] are able to fit their data with a single temperature, although they too require more than one temperature for regions inside Loop I.

As is discussed by Dr. Hayakawa in this volume [36], these differing conclusions may be due, in part at least, to differences in the assumed emission models. Unfortunately, the calculations of thermal emission models are complicated and dependent upon elemental abundances and atomic physics parameters that are not well known. It is hence not surprising that significant differences exist, for example, in the detailed emission spectra predictions of Kato [37] and Raymond and Smith [38]. In addition, while one model may be closer to the truth than others for an equilibrium plasma, transient effects which result in ionic abundances that are not in equilibrium with the electron temperature may make no equilibrium model appropriate. Shapiro and Moore [39] have considered some aspects of the time-dependent problem.

These difficult questions simply indicate the need for more work. On the one hand observations capable of detecting and measuring the relative strengths of spectral lines are required. On the other hand, we need a better understanding of the emission spectrum of hot astrophysical plasmas, particularly under non-equilibrium conditions. In the meantime it is difficult to see how, even with the present observational and theoretical uncertainties, the responsible emitting gas can have temperatures far outside the range $T = 10^{6\pm0.5}$ K.

Pressure of the Emitting Gas

Observational evidence indicates that the regions of thermal X-ray emission are significantly overpressured relative to the nominal interstellar pressure of 2000 cm^{-3} K. This fact has been emphasized before [7, 40, 41].

The soft X-ray intensity that originates in a region of unabsorbed thermal emission $\overline{n_e^2}d$. If the electron density, n_e, is not uniform through the contributing distance, d, a filling factor f may be introduced such that $\overline{n_e^2}d = fn_e{}^2d = fp^2d/4T^2$ where p and T are the pressure and temperature respectively, in a region of electron density n_e. The pressure is then

$$p = 2T(\overline{n_e^2}d/fd)^{1/2}$$

$$p > 10^4 \ cm^{-3} \ K.$$

The inequality reflects the facts that a.) temperatures different than the 10^6 K we have used require larger values of $\overline{n_e^2}d = 2.5 \times 10^{-3} \ cm^{-6}$ pc to account for the observed intensity data over most of the sky, b.) in evaluating $\overline{n_e^2}d$ we have assumed no absorption of the emitted radiation and have used the calculations of Raymond and Smith which require less emission measure than required by the Kato calculations, c.) normal, not depleted interstellar abundances have been assumed, and d.) filling factors must of course be one or less. For d we have used 100 pc since values greatly in excess of this value would conflict with both of the hypotheses discussed that can account for the observed lack of energy dependent absorption.

This work has been supported in part by NASA grant NGL 50-002-044 and NASA contract NAS5-22410.

REFERENCES

1. P. M. Fried and J. A. Nousek, Unpublished theses, University of Wisconsin, Madison, June 1978.

2. A. Davidson, S. Shulman, G. Fritz, J. F. Meekins, R. C. Henry, and H. Friedman, *Astrophys. J.* 177, 629 (1972).

3. P. A. de Korte, J. A. Bleeker, A. J. Deerenberg, Y. Tanaka, and K. Yamashita, *Astrophys. J. (Letters)* 190, L5 (1974).

4. A. N. Bunner, P. L. Coleman, W. L. Kraushaar, and D. McCammon, *Astrophys. J. (Letters)* 172, L67 (1972).

5. R. J. Borken and D. C. Iwan, *Astrophys. J.* 218, 511 (1977).

6. F. O. Williamson, W. T. Sanders, W. L. Kraushaar, D. McCammon, R. J. Borken, and A. N. Bunner, *Astrophys. J. (Letters)* 193, L133 (1974).

7. P. Burstein, R. J. Borken, W. L. Kraushaar, and W. T. Sanders, *Astrophys. J.* 213, 405 (1977).

8. C. S. Bowyer, G. B. Field, and J. E. Mack, *Nature* 217, 32 (1968).

9. D. McCammon, A. N. Bunner, P. L. Coleman, and W. L. Kraushaar, *Astrophys. J. (Letters)* 168, L33 (1971).

10. D. McCammon, S. S. Meyer, W. T. Sanders, and F. O. Williamson, *Astrophys. J.* 209, 46 (1976).

11. B. Margon, S. Bowyer, R. Cruddace, C. Heiles, M. Lampton, and T. Troland, *Astrophys. J. (Letters)* 191, 117 (1974).

12. S. Rappaport, A. Levine, R. Doxsey, and H. V. Bradt, *Astrophys. J. (Letters)* 196, L15 (1975).

13. K. S. Long, P. C. Agrawal, and G. P. Garmire, *Astrophys. J.* 206, 411 (1976).

14. P. Gorenstein and W. H. Tucker, *Astrophys. J.* 176, 333 (1972).

15. A. Levine, S. Rappaport, J. Halpern, and F. Walter, *Astrophys. J.* 211,215 (1977)

16. M. J. Vanderhill, R. J. Borken, A. N. Bunner, P. H. Burstein, and W. L. Kraushaar, *Astrophys. J. (Letters)* 197, L19 (1975).

17. G. Garmire, this volume.

18. S. Hayakawa, *I.A.U. Symposium No. 55, X- and Gamma-Ray Astronomy*, H. Bradt and R. Giacconi, eds., Madrid 1972, Reidel, p. 235.

19. L. Spitzer and E. B. Jenkins, *A. Rev. Astron. Astrophys.* 13, 133 (1975).

20. G. Hartmann, D. Müller, and T. Prince, *Phys. Rev. Letters* 38, 1368 (1977).

21. E. A. Boldt and S. H. Pravdo, *Astrophys. J.* 200, 727 (1975).

22. W. D. Watson, *Astrophys. J.* 206, 842 (1976).

23. E. B. Jenkins, *Astrophys. J.* 219, 845 (1978).

24. E. B. Jenkins, *Astrophys. J.* 220, 107 (1978).

25. S. Hayakawa, T. Kato, F. Nagase, K. Yamashita, T. Murakami, and Y. Tanaka, *Astrophys. J. (Letters)* 213, L109 (1977).

26. D. J. Yentis, R. Novick, and P. Vanden Bout, *Astrophys. J.* 177, 365 (1972).

27. W. Cash, R. Malina, and S. Stern, *Astrophys. J. (Letters)* 204, L7 (1976).

28. A. Levine, S. Rappaport, R. Doxsey, and G. Jernigan, *Astrophys. J.* 205, 226 (1976).

29. R. L. Brown and R. J. Gould, *Phys. Rev. D.* 1, 2252 (1970).

30. W. T. Sanders, W. L. Kraushaar, J. A. Nousek, and P. M. Fried, *Astrophys. J. (Letters)* 217, L87 (1977).

31. A. N. Bunner, P. L. Coleman, W. L. Kraushaar, D. McCammon, T. M. Palmieri, A. Shilepsky, and M. Ulmer, *Nature* 223, 1222 (1969).

32. C. F. McKee and J. P. Ostriker, *Astrophys. J.* 218, 148 (1977).

33. E. W. Greisen, *Astrophys. J.* 203, 371 (1976).

34. J. M. Dickey, E. E. Salpeter, and Y. Terzian, *Astrophys. J. (Letters)* 211, L77 (1977).

35. S. Hayakawa, T. Kato, F. Nagase, K. Yamashita, and Y. Tanaka, *Astron. Astrophys.* 62, 21 (1978).

36. S. Hayakawa, this volume.

37. T. Kato, *Astrophys. J. Suppl.* 30, 397 (1976).

38. J. C. Raymond and B. W. Smith, *Astrophys. J. Suppl.* 35, 419 (1977).

39. P. R. Shapiro and R. T. Moore, *Astrophys. J.* 207, 460 (1976).

40. A. N. Bunner, P. L. Coleman, W. L. Kraushaar, D. McCammon, and F. O. Williamson, *Astrophys. J.* 179, 781 (1973).

41. P. R. Shapiro and G. B. Field, *Astrophys. J.* 205, 762 (1976).

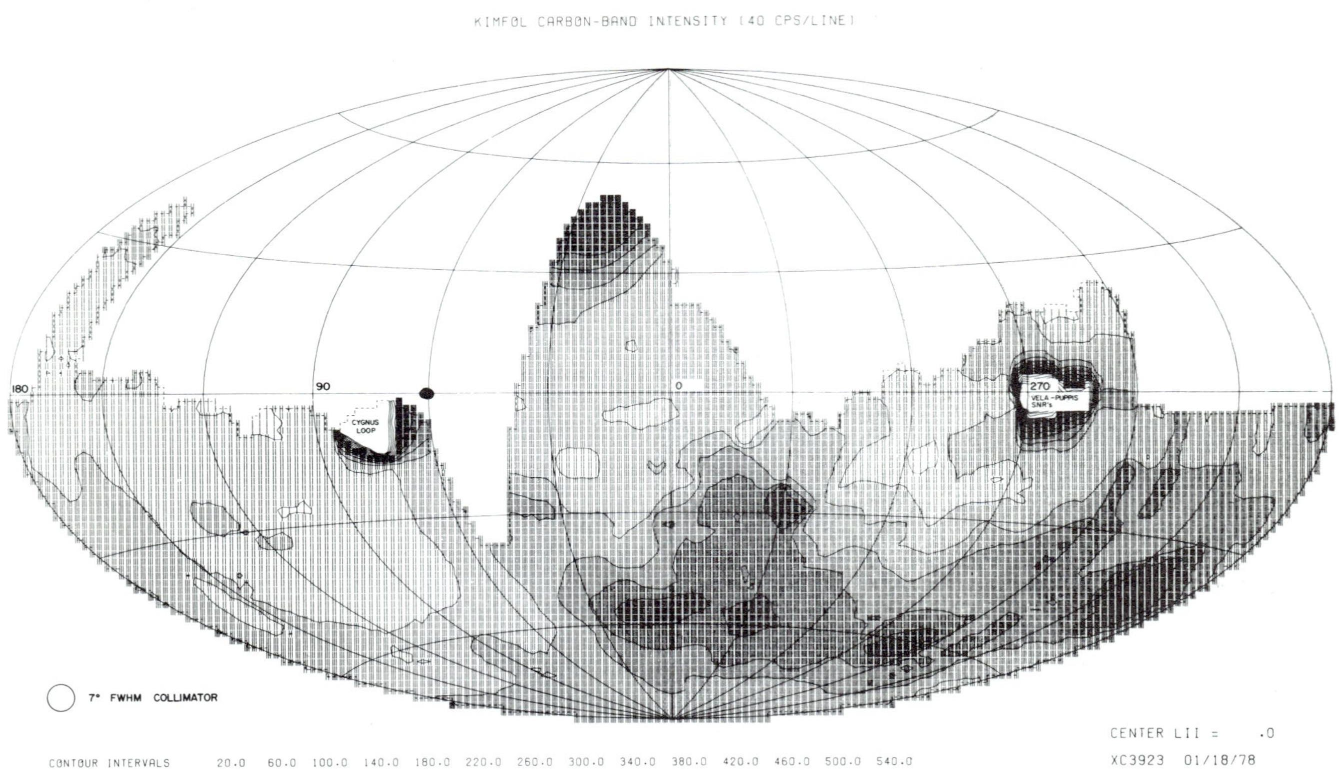

Fig. 1. Spatial distribution of X-ray intensity in the C-band (E<0.284 keV) in galactic coordinates. The data were acquired during five controlled aspect rocket flights by the Wisconsin group [1].

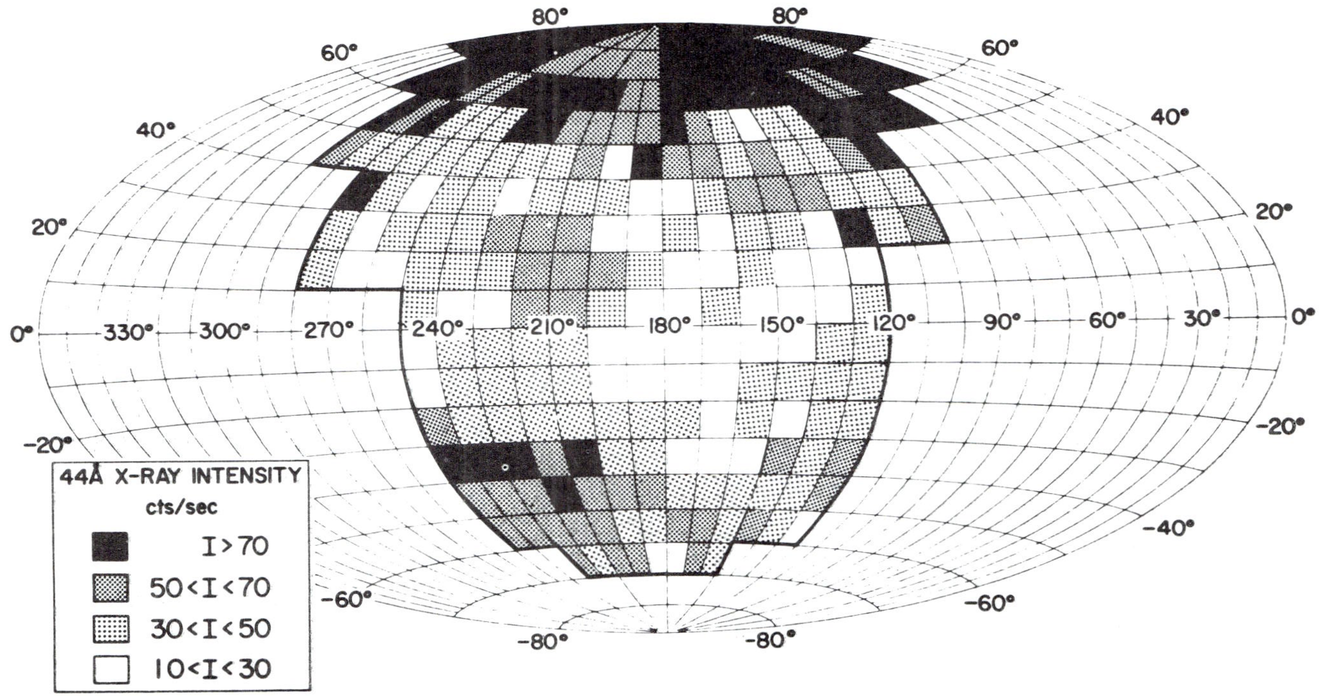

Fig. 2. Spatial distribution of X-ray intensity in the C-band. The data were acquired during a free-spinning rocket flight by the NRL group [2]. (Figure courtesy of Davidsen *et al.* and *The Astrophysical Journal*, University of Chicago Press, copyright 1972.)

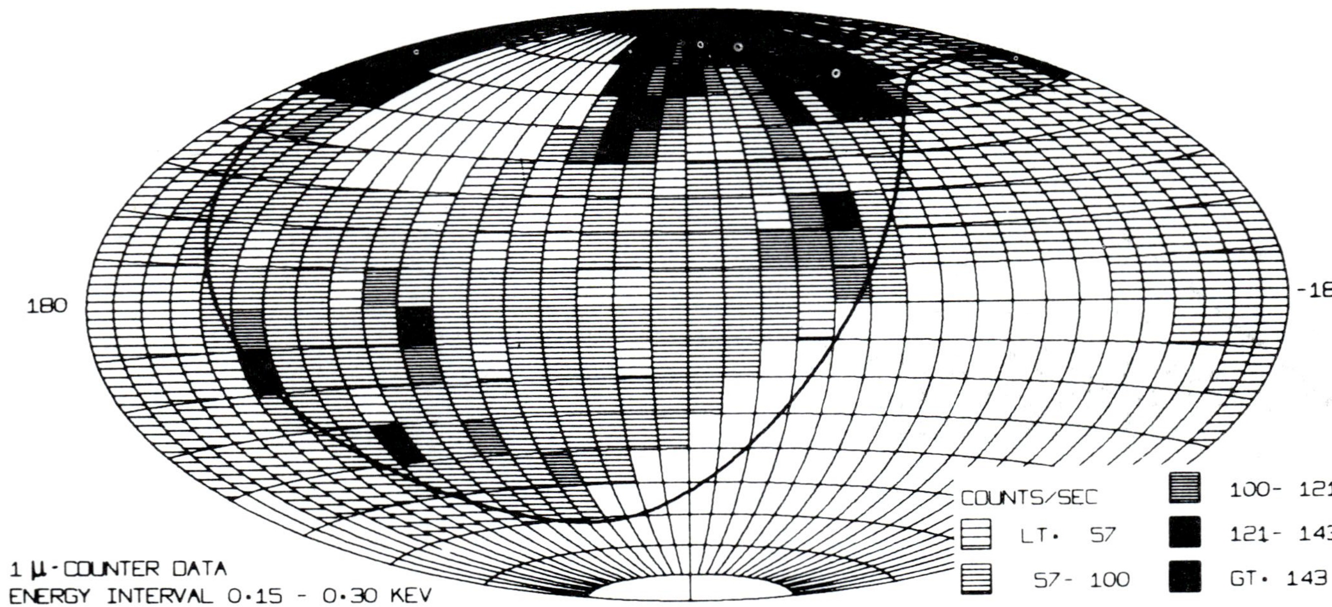

Fig. 3. Spatial distribution of X-ray intensity in the C-band. The data were acquired during a free-spinning rocket flight by the Leiden-Nagoya group [3]. (Figure courtesy of de Korte *et al.* and *The Astrophysical Journal*, University of Chicago Press, copyright 1974.)

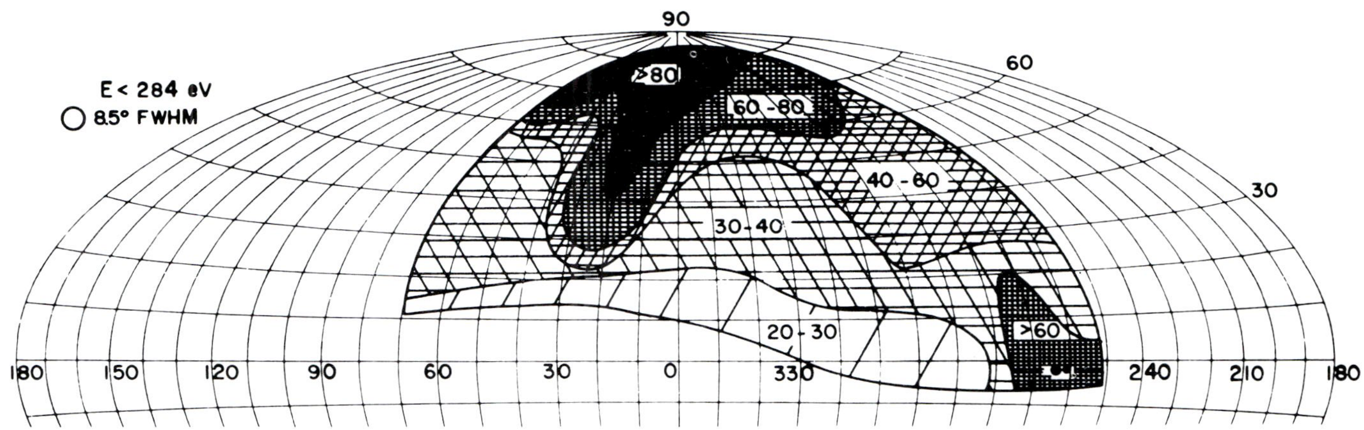

Fig. 4. Spatial distribution of X-ray intensity in the C-band. The data were acquired during a free-spinning rocket flight by the Wisconsin group [4]. (Figure courtesy Bunner *et al.* and *The Astrophysical Journal*, University of Chicago Press, copyright 1972.)

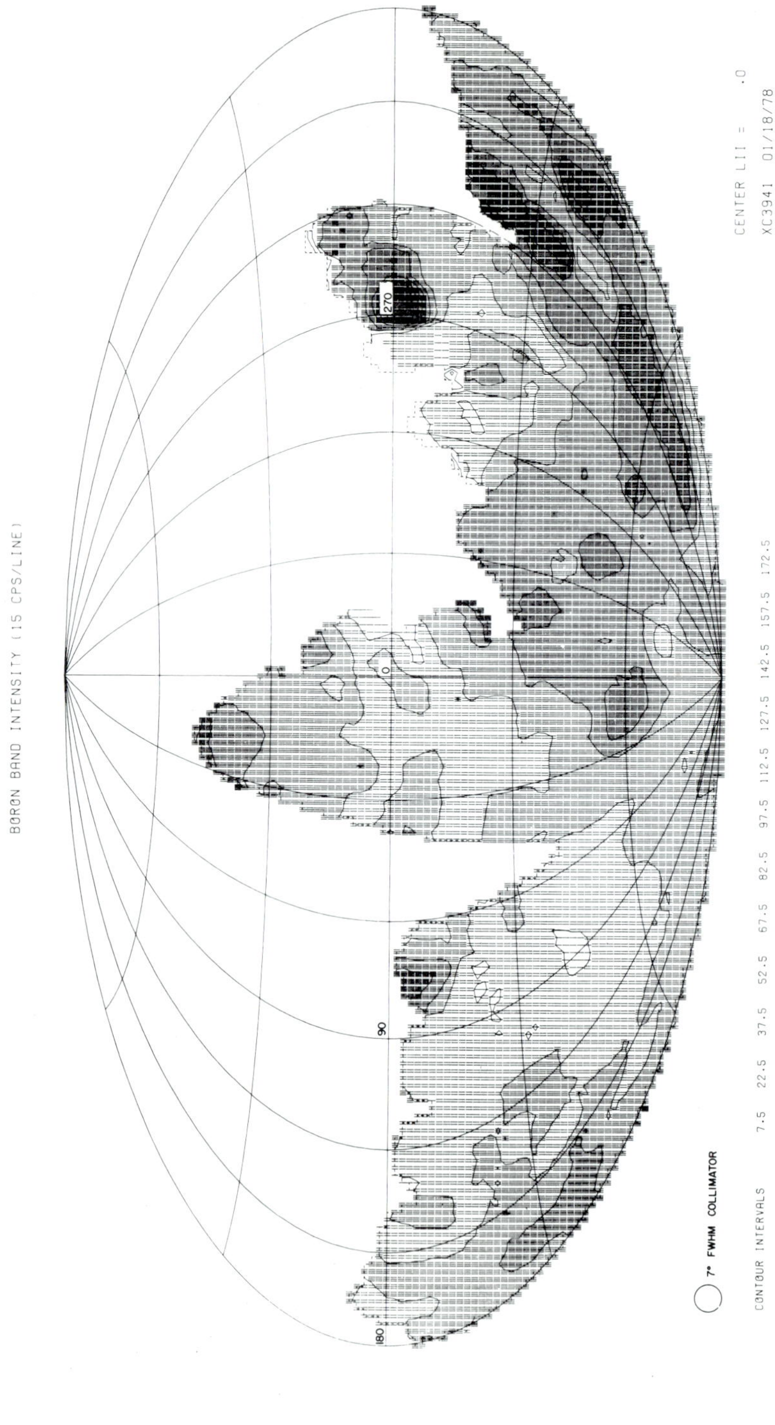

Fig. 5. Spatial distribution of X-ray intensity in B-band ($E < 0.188$ keV). The data were acquired during the same rocket flights as those shown in Fig. 1 [1].

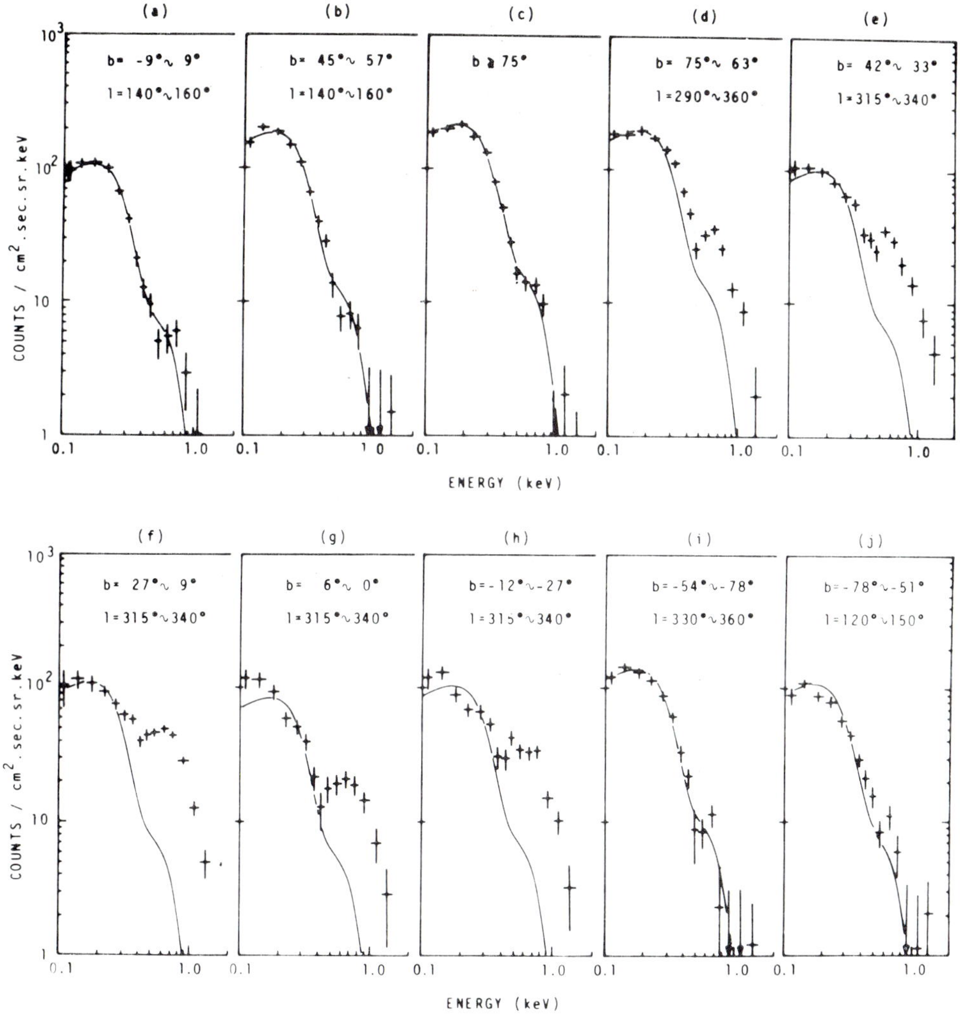

Fig. 6. Proportional counter pulse-height spectra taken inside (d,e, f,g,h) and outside (a,b,c,i,j) Loop I. The data are from Hayakawa *et al.* [25]. (Figure courtesy of Hayakawa *et al.* and *The Astrophysical Journal*, University of Chicago Press, copyright 1977.)

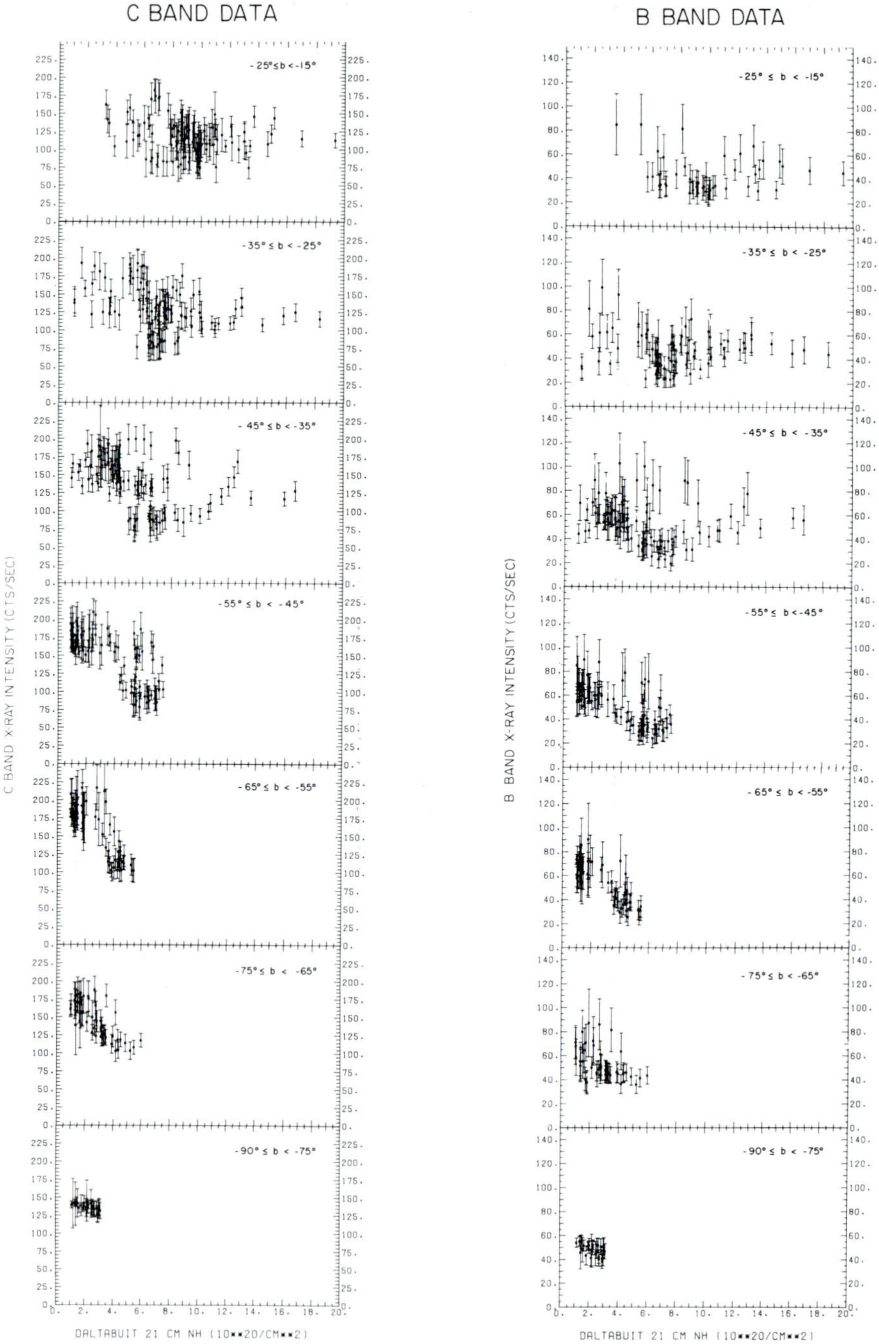

Fig. 7. C-band and B-band intensity data plotted versus N_H in 10° ranges of galactic latitude over the southern galactic hemisphere. The data are from Sanders *et al.* and maps of intensity data are shown in Figs. 1 and 5. (Figure courtesy of Sanders *et al.* and *The Astrophysical Journal*, University of Chicago Press, copyright 1977.)

OBSERVATION OF SOFT X-RAY DIFFUSE COMPONENT AND THE CYGNUS LOOP WITH GAS SCINTILLATION PROPORTIONAL COUNTER

H. Inoue*, K. Koyama*, M. Matsuoka*, T. Ohashi*,
Y. Tanaka* and H. Tsunemi**

**Institute of Space and Aeronautical Science, University of Tokyo, Tokyo, Japan*
***Department of Physics, Osaka University, Osaka, Japan*

Current investigations of soft X-ray diffuse component have strengthened the view that the observed radiation below 1 keV is mostly thermal emission originating from hot plasma of temperature $\sim 10^6$K within our galaxy [1]. Observational evidence of the thermal origin, however, is yet to be obtained, though various considerations hardly admit other mechanisms. The direct proof would be the detection of emission lines which are characteristic of the thermal emission for the expected temperature. A similar argument holds for the soft X-ray emission from supernova remnants. Detailed study of the energy spectrum is essential for elucidating the evolution of the supernova remnants and thereby the heating of the interstellar matter.

We conducted a rocket observation of soft X-rays above 0.1 keV with gas scintillation proportional counters. This type of the counter provides twice as good an energy resolution as that of conventional proportional counters. In this paper, we present the preliminary results on the energy spectrum of the diffuse component of the N_H-minimum region in Hercules and on the spectrum of the Cygnus Loop.

The X-ray detector comprises four identical sets of gas scintillation proportional counter filled with 1.1 atm. of xenon, the total effective area being 53 cm^2. The window consists of polypropylene film of 120 $\mu g\ cm^{-2}$ thickness. The field of view is 10° x 20°. The properties of the gas scintillation proportional counter were extensively examined in the laboratory from 109 eV(Be-K) through 5.9 keV(Mn-K) [2]. The rocket was launched on Sept. 14, 1977, at 12:22 UT., from Kagoshima Space Center, University of Tokyo. The counters were sealed with ejectable covers until the rocket reached 70 km altitude. Inflight calibrations were provided by ^{55}Fe(5.9 keV) and Al-K(1.5 keV). All four counters functioned properly throughout the flight, and data from all four counters were added together for the analysis. The energy resolution inflight was determined to be 11 % FWHM at 5.9 keV and 48 % FWHM at 0.28 keV, respectively.

N_H-MINIMUM REGION IN HERCULES

An extended region of N_H-deficiency in Hercules was cross-scanned. The region is roughly centered at $\ell \simeq 80°$ and $b \simeq 40°$, and the N_H-value goes down to 1.7×10^{20} H atoms cm^{-2}. A distinct enhancement in the L-band (< 0.28 keV) intensity is observed in correlation with the diminishing N_H-value. The observed pulse-height spectrum of the diffuse background in this region is shown in Fig. 1(a), after subtraction of the power-law component. The spectrum reveals a well-defined peak at about 0.57 keV. The shape of this peak is in good agreement with that expected for a single emission line convolved with the resolution of the present counter. From the peak position, we may identify this to be OVII line (0.57 keV). This line will not be resolved with the resolution of ordinary proportional counters, however.

The observed spectrum is then compared with model calculations of the thermal spect[ra] of thin hot plasma either by Kato [3] or by Raymond and Smith [4] for cosmical abundances of element. These model spectra are found to satisfactorily fit the observational result for a temperature $T \simeq 1.5 \times 10^6$K. Fig. 1(b) shows Kato's spectrum which is fitted to the observed data in Fig. 1(c). As a matter of fact, the OVII line at 0.57 keV is the single prominent line in the M-band (> 0.28 keV) for this temperature.

The present result is probably the first clear evidence for the presence of the emission line from highly ionized atoms in soft X-ray diffuse component, and we now consider that the thermal origin of this component, if not entire, is established.

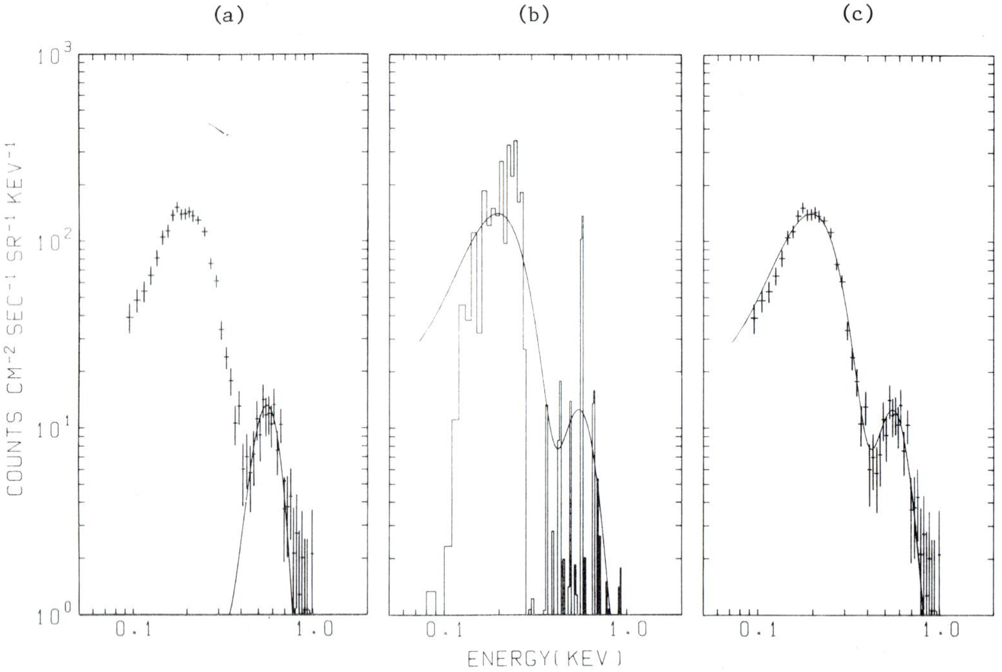

Fig. 1 (a) The pulse-height spectrum for the N_H-minimum region in Hercules. The curve indicates the spectrum expected for a single line at 0.57 keV. (b) The thermal emission spectrum by Kato [3] for $T = 1.5 \times 10^6$ K and an intervening interstellar matter $N_H = 4 \times 10^{19}$ H atoms cm^{-2}. The solid curve represents the convolution with the counter resolution function. (c) The convolved spectrum in (b) fitted to the observed data.

THE CYGNUS LOOP

The Cygnus Loop, an intense soft X-ray source, has been observed by several groups previously [5][6][7]. However, considerable differences exist between the publishe[d] spectra. In the present experiment, the Cygnus Loop was observed for about twenty seconds. The observed spectrum of the Cygnus Loop is shown in Fig. 2(a), correcte[d] for the background and the effective exposure. As compared with the spectrum of t[he] Hercules N_H-minimum in Fig. 1(a), the Cygnus Loop exhibits a much more pronounced hump in the M-band. Neither a thermal Bremsstrahlung (free-free) spectrum, nor one with a single OVIII line (0.65 keV) in addition, fits the observed spectrum. The M-band peak is much broader than that for a single line.

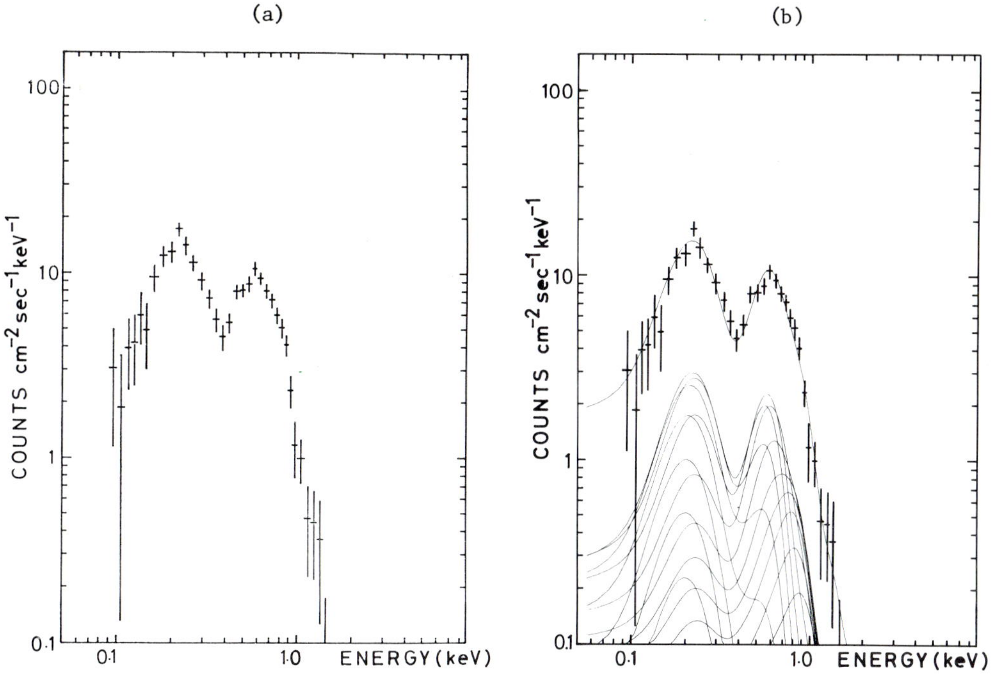

Fig. 2 (a) The observed spectrum of the Cygnus Loop.
(b) The best-fit thermal spectrum composed of multiple components of different temperatures, for the dependence of the emission measure $\propto T^{-2.5}$. Fifteen components in the range of $Log_{10}T$ 5.6 − 7.0 with a step of 0.1 and their sum fitted to the observed spectrum are shown.

We then tested the thermal spectra [3][4] for a single temperature. No tolerable fit was obtained for any set of T and N_H with the assumption of cosmical abundances of element. One might consider the case of anomalous element abundances, but there is no compelling evidence for this at present.

We instead attempt fitting with multiple components of different temperatures, imposing a restriction that all components undergo the same interstellar absorption. It turns out that satisfactory fits are obtained this way. For the fitting with two components, the following set gives the best fit to the observed spectrum with χ^2-min. = 18.8 for 27 degree of freedom: T_h = 4 x 10^6K, T_ℓ = 1.3 x 10^6K, the ratio of emission measures $n_h^2V_h / n_\ell^2V_\ell$ = 0.21, and N_H = 4.75 x 10^{20}H atoms cm^{-2}, where subscripts h and ℓ specify the higher- and the lower-temperature components, respectively. The derived value of N_H is in agreement with that estimated from interstellar reddening [8], for the assumed 770 pc distance for the Cygnus Loop.

In the frame of the adiabatic blast wave theory, the two components are interpreted to arise from two zones with different ambient gas densities, $n_{oh} \simeq 0.17\ cm^{-3}$ and $n_{o\ell} \simeq 0.52\ cm^{-3}$ with subscript o to represent the state prior to the compression by the shock wave. The theory, however, does not explain the association of the filaments with the L-band X-ray emission, as observed in the published maps [7][9]. It, therefore, appears that the two-component model is somewhat artificial.

For further elaboration, we assume that the emission measure $n(T)^2V(T)$ is distributed according to a power law of temperature, i.e. $n(T)^2dV(T) \propto T^{-\gamma}dT$. Employing Kato's spectrum over the range of $\mathrm{Log}_{10}T$ from 5.6 to 7.0 with a step of 0.1, we do find a good fit for N_H = 4.5 (3.9 — 5.0) x 10^{20}H atoms cm^{-2} and γ = 2.5 (2.2 — 2.7) with χ^2-min. = 26.5 for 29 degree of freedom, where the values in parentheses indicate 90 % confidence limits. Fig. 2 (b) shows the components and their sum for the best fit case.

Sedov's self-similar solution for the adiabatic phase of supernova remnants in a uniform interstellar medium gives a temperature dependence of the emission measure which can be approximated by a power law with γ = 3.3, largely exceeding the acceptable range. This result indicates that more emission measure is required at higher temperature range than that expected from Sedov's solution.

This feature can be qualitatively understood in terms of the evaporation of HI clouds inside the supernova remnant. Following the picture of McKee and Ostriker [10], let us consider the interstellar space that is mostly filled with tenuous plasma ($n \lesssim 10^{-2} cm^{-3}$) with small-scale HI clouds ($n \gtrsim 10\ cm^{-3}$) embedded therein. In such a circumstance, the hot plasma responsible for observed soft X-rays is expected to derive from the evaporation of these HI clouds. After the shock wave has passed, HI clouds will be evaporated and further heated by the very hot environment, thereby enhancing the emission measure at high temperatures as required. Those clouds where the shock front is now passing would be heated to a temperature over 10^4K and appear as the filaments. Hot gas evaporated from the surface of these clouds accounts for the associated X-ray emission. This process would operate generally through the adiabatic phase of supernova remnants in a similar interstellar condition.

We note that the spectra of Vela SNR as well as the Loop I (the North Polar Spur) look very similar, if interstellar absorption is taken into account, to that of the Cygnus Loop, despite their different physical appearances. This fact supports the hypothesis that the process described above is commonly operative in each of these supernova remnants.

REFERENCES

1. Y.Tanaka and J.A.M.Bleeker, *Space Science Reviews* 20, 815 (1977); references therein.

2. H.Inoue, K.Koyama, M.Matsuoka, T.Ohashi, Y.Tanaka and H.Tsunemi, *ISAS preprint* RN 59 (1978), submitted to *Nuclear Instr. and Meth.*

3. T.Kato, *Astrophys.J. Suppl.* 30, 397 (1976).

4. J.C.Raymond and B.W.Smith, *Astrophys.J. Suppl.* 35, 419 (1977).

5. P.Gorenstein and W.H.Tucker, *Annual Review of Astron.& Astrophys.* 14, 373 (1976); references therein.

6. P.A.Charles, J.L.Culhane, J.C.Zarnecki, *Astrophys.J.(Letters)* 196, L19 (1975).

7. E.H.B.M.Gronenschild, R.Mewe, J.Heise, A.C.Brinkman, A.J.F.den Boggende and J.Schrijver, *Astron.& Astrophys.* 49, 153 (1976).

8. C.Ryter, C.J.Cesarsky and J.Audouse, *Astrophys.J.* 198, 103 (1975).

9. S.Rappaport, W.Cash, R.Doxey, G.Moore and R.Borken, *Astrophys.J.(Letters)* 186, L115 (1973).

10. C.F.McKee and J.P.Ostriker, *Astrophys.J.* 218, 148 (1977).

THE CYGNUS LOOP AS OBSERVED BY ANS

E. H. B. M. Gronenschild

The Astronomical Institute, Space Research Laboratory, Beneluxlaan 21, 3527 HS Utrecht, The Netherlands

With the ANS satellite we have observed the Cygnus Loop twice during 1975 in a set of 150 pointings uniformly distributed over the entire optical extent (cf., [1]). With the Algebraic Reconstruction Technique (ART) (cf., [2]) we have obtained detailed X-ray brightness maps in the energy ranges 0.16-0.284 keV, 0.32-0.46 keV, and 1-2.5 keV. The maps are binned in cells of 5'x6.'7 and from the reconstruction of a point source the actual resolution has been derived. In the soft X-ray band the resolution is found to be better than 10'x12', in the other two mentioned energy ranges it is about 25'x50' and 35'x75', respectively.

The soft X-ray brightness distribution shown in Fig. 1 is in many respects similar to the optical and radio (cf., [3]) distributions, especially at the north-east filaments NGC 6992-5. We observe two distinct maxima coinciding with the radio maxima [3] and a minimum in between where also the optical filaments are very narrow (see also [4]). The outer X-ray edge reaches to at least 10' farther out than the optical edge and follows very closely the outer radio contours. This is also the case west of the western filament NGC6960. In the north the X-ray emission is not correlated with the optical filaments. Just east of NGC6974 we see intense X-ray emission with almost no optical emission. The X-ray distribution extends to about a declination of 32°.5, coincident with a faint Hα filament [5].

The X-ray maps in the higher energy ranges can not easily be compared with the optical or radio maps due to the worse resolution. However, they can be described with emission features in the north and near NGC6992-5. In general our maps show similar structure as previous X-ray maps ([2], [6], [7]), although detailed differences are present. In particular, we do not observe enhanced emission from the central region.

Binning the ART maps in cells measuring 30'x40' we have made spectral fits to the pulse-height data in each such cell. The standard errors to the derived counting rates in each cell have been obtained by using Monte Carlo simulations. The result is that the temperature varies over the region investigated between 10^6 K and 5×10^6 K (see Fig. 2). The derived hydrogen column density varies in the range $(1-12)\times10^{20}$ atoms cm^{-2} with an upper limit of 10^{22} atoms cm^{-2}. From the patchy temperature distribution it is clear that the interstellar medium is far from homogeneous. Modeling the Cygnus Loop in the form of emission emanating from a thin shell with a thickness of $^1/12$ of the radius yields density inhomogeneities in the range 0.1-2 cm^{-3} (see Fig. 3). The density structure is suggestive of a large cloud in front of and across the source with a depth of 20-200 pc. The cloud is also visible in radio

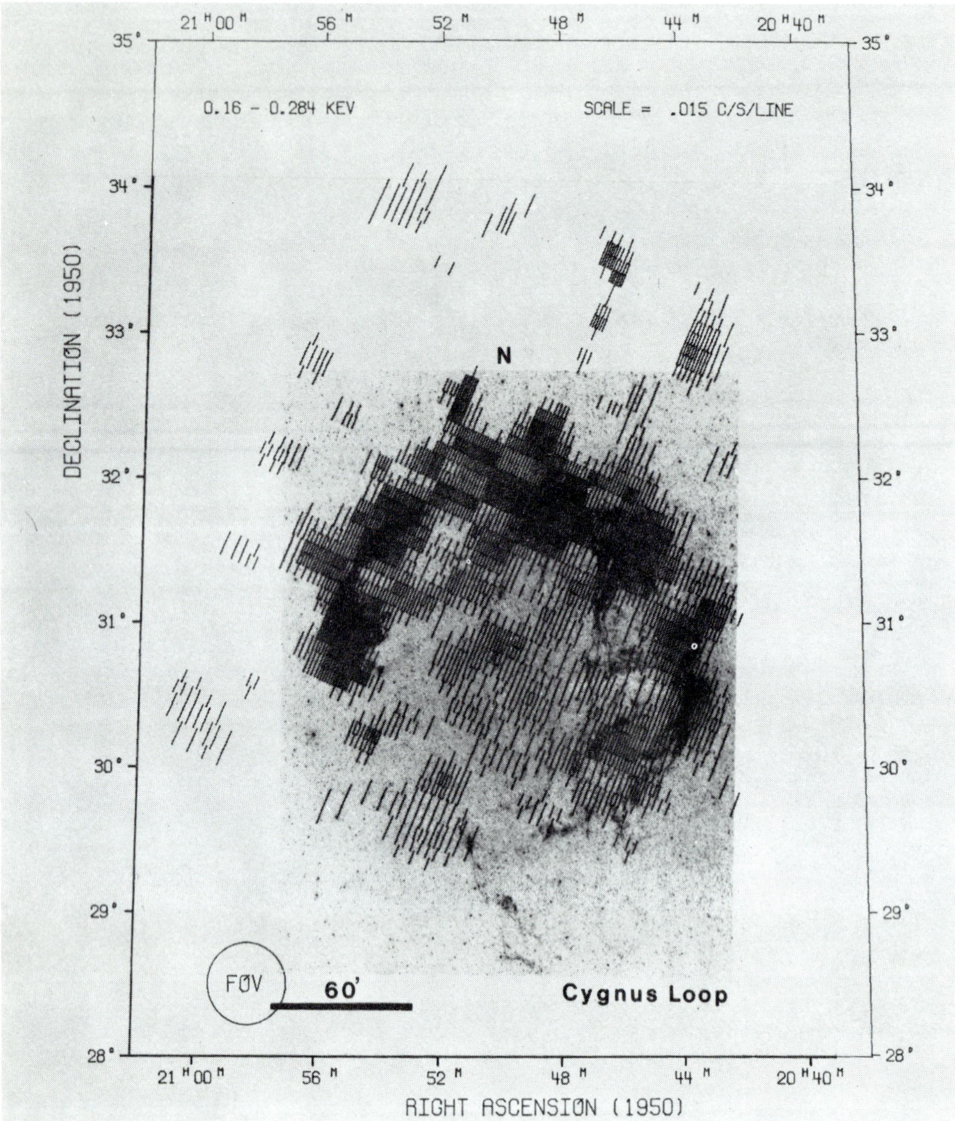

Fig. 1. Soft X-ray brightness distribution superposed on an Hα photograph of the Cygnus Loop (Van den Bergh *et al.* [14]. The distribution is displayed in cells measuring 5'x6'.7 in which the density of lines is linearly proportional to the observed counting rate. The actual resolution is about 10', the field of view (FOV) of the detector is shown in the lower left-hand corner. The isolated 'sources' along the edges may not be real but artifacts produced by the reconstruction technique.

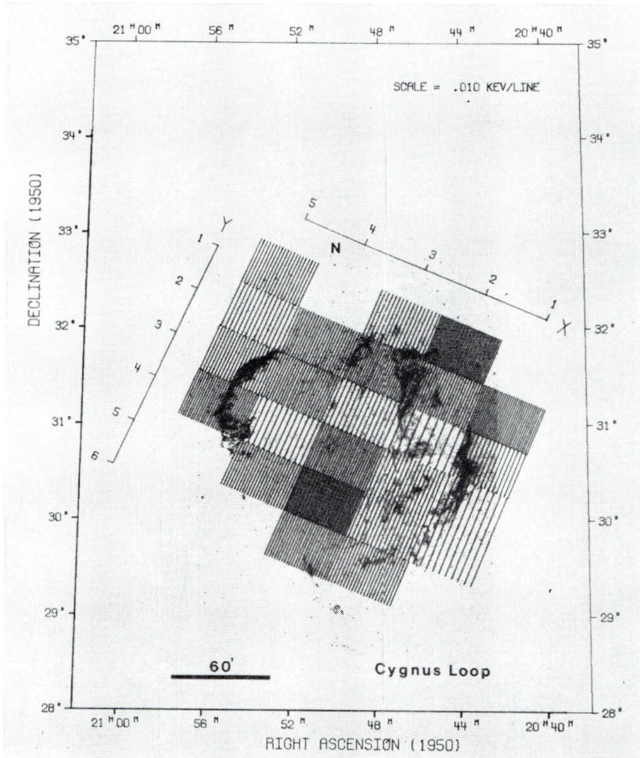

Fig. 2. Distribution of the X-ray temperature in bins of 30'x40'. Every line in each bin corresponds to a temperature of 0.01 keV or about $1.2x10^5$ K. The result has been obtained by fitting the reconstructed pulse-height data with a continuum and line spectrum. The lowest temperature found is about 10^6 K, the highest about $5x10^6$ K, just south of the optical centre. Note the low temperatures along the western filament.

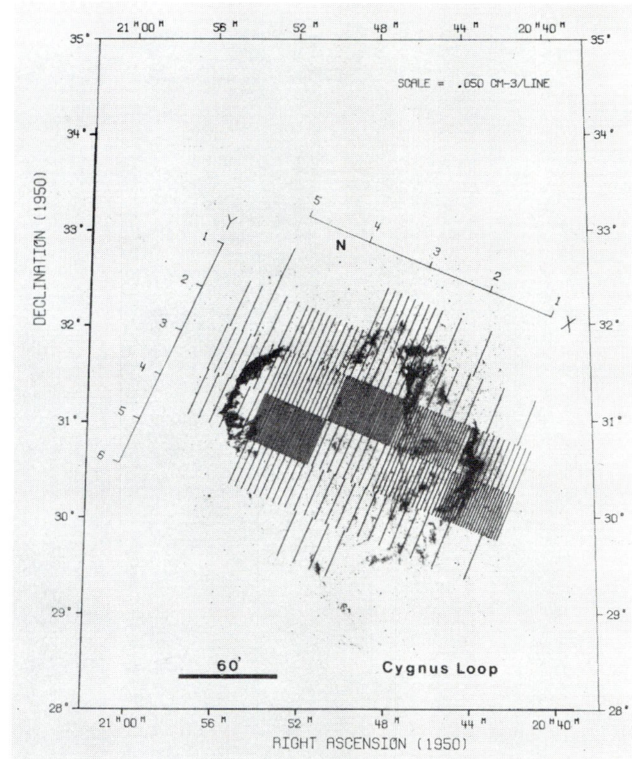

Fig. 3. Distribution of the ambient interstellar density. The result is obtained by assuming that the X-ray emitting plasma is confined in a thin shell with a thickness of 1/12 of the radius. Each line in a bin corresponds to 0.05 cm^{-3}. Note the high densities across the source between a declination of 30° and 31°.

data [8] and in the west the cloud contains CO molecules [9].

The recent negative search for O VII line emission near 22 Å [10] may be well explained with the derived temperature structure of the source.

Comparison of the observed X-ray fluxes and the coronal [FeXIV] 5303 Å line fluxes ([11],[12]) results in an underabundance of iron compared with standard cosmical values of at least a factor 3 in the north and a factor 7 near the centre. The reported discovery of the forbidden [FeX] 6374 Å line in the western filament NGC6960 [13] is consistent with the derived temperature and X-ray flux if a normal iron abundance is assumed.

The overall behaviour of the supernova remnant may be best approximated by the following set of parameters: X-ray temperature $T_X = (2.26\pm0.10)\times10^6$ K; explosion energy $E_o = (1.1\pm0.3)\times10^{50}$ erg; mean ambient interstellar density n_o = (0.15±0.03 cm^{-3}, and age t = (21000±4000) years. The interstellar hydrogen column density for the integrated source is $N_H = (4.04\pm0.14)\times10^{20}$ atoms cm^{-2}.

REFERENCES

1. E.H.B.M. Gronenschild, R. Mewe, J. Heise, A.C. Brinkman, A.J.F. den Boggende and J. Schrijver, *Astron. & Astroph. Letters*, 49, 153 (1976).
2. J.C. Stevens and G.P. Garmire, *Astroph. J. Letters*, 180, L19 (1973).
3. N.J. Keen, W.E. Wilson, C.G.T. Haslam, D.A. Graham and P. Thomasson, *Astron. & Astroph.*, 28, 197 (1973).
4. P.A. Charles, J.L. Culhane and J.C. Zarnecki, *Astroph. J. Letters*, 196, L19 (1975).
5. T.R. Gull, R.A.R. Parker and R.P. Kirshner, *Supernovae*, p71, Ed. D.N. Schramm, Reidel, Dordrecht (1977).
6. S. Rappaport, W. Cash, R. Doxsey, G. Moore and R. Borken, *Astroph. J. Letters*, 186, L115 (1973).
7. S. Rappaport, R. Doxsey and A. Solinger, *Astroph. J.*, 194, 329 (1974).
8. L.K. de Noyer, *Astroph. J.*, 196, 479 (1975).
9. N.Z. Scoville, W.M. Irvine, P.G. Wanner and C.R. Predmore, *Astroph. J.* 216, 320 (1977).
10. J.P.W. Stark and J.L. Culhane, *Mon. Not. R. Astr. Soc.*, 184, 509 (1978).
11. B.E. Woodgate, H.S. Stockman Jr., J.R.P. Angel and R.P. Kirshner, *Astroph. J. Letters*, 188, L79 (1974).
12. B.E. Woodgate, R.P. Kirshner and R.J. Balon, *Astroph. J. Letters*, 218, L129 (1977).
13. P.V. Shcheglov, *Sov. Astron. A.J.*, 11, 567 (1968).
14. S. van den Bergh, A.P. Marscher and Y. Terzian, *Astroph. J. Suppl.*, 26, 19 (1973).

SOFT X-RAY STRUCTURE OF THE NORTH POLAR SPUR RIDGE AND THE LUPUS REGION

J.A.M. Bleeker*, J. Davelaar*, A.J.M. Deerenberg*, S. Hayakawa**, K. Yamashita** and Y. Tanaka***

**Cosmic Ray Working Group, Huygens Laboratory, Leiden, The Netherlands*
***Department of Physics, Nagoya University, Nagoya, Japan*
****Space Science Division, ISAS, University of Tokyo, Tokyo, Japan*

ABSTRACT

During a rocket borne soft X-ray experiment, a one dimensional X-ray focussing collector with an angular resolution of 15 arc minutes, performed a crossing over the northern ridge of the North Polar Spur near $b^{II} = 750$, $\ell^{II} \approx 350°$. The width and position of the ridge are energy dependent. The structure is complicated and may contain more than one component. The width of the X-ray ridge at this latitude is about 4 degrees in the range 0.5 - 1 keV and appreciably smaller than at $b^{II} = 30°$. The relation between the positions of the X-ray and radio ridges will be discussed.

The Lupus region, an extended soft X-ray enhancement, has been observed in the energy range from 60 to 3000 eV during a rocket borne experiment. The instrumentation consisted of a one dimensional X-ray focussing collector and eight collimated large area detectors with angular resolution of 0.25° and 3°, respectively. The enhancement is, in one dimension, resolved into at least four sources: SN 1006, an extended soft source, coincident with the Lupus Radio Loop, and two as yet unidentified bright spots. The diameters of SN 1006 and the Lupus Loop agree with their radio counterparts. Spectral analysis of the Lupus source reveals a temperature of three million degrees, the evolutionary phase of the loop is propably comparable to the Cygnus Loop and Vela. A lower limit of 340 parsec for the distance to the source is derived.

SOFT X-RAY OBSERVATION OF THE NORTH POLAR SPUR AND AN HI HOLE IN HERCULES

S. Hayakawa, H. Iwanami, H. Kunieda, F. Nagase and K. Yamashita

Department of Physics, Nagoya University, Nagoya, Japan

INTRODUCTION

The energy spectrum of diffuse soft X-rays is obtained from the pulse height distributions measured with three sets of proportional counters with different spectral sensitivities. In the present paper we report on the results for two interesting features, the North Polar Spur (NPS) in the mid-latitude region and an HI hole in Hercules.

Three sets of proportional counters on board a sounding rocket were equipped with plastic windows, as described in a separate paper [1]. The window thicknesses were 0.9, 1.5 and 4.0 μm polypropylene equivalent so as to provide three colour photometry. The field of view of each counter was 10°.7 x 3°.0 (FWHM), the long side being perpendicular to the scan direction. The counter axis crossed NPS three times in the region ℓ = 0° - 45° and b = 20° - 50°, passed the Hercules hole ($\ell \simeq$ 80°, b $\simeq$ 40°) in two mutually perpendicular directions, and drifted along $\ell \simeq$ 80° through AM Her to the Cygnus Loop.

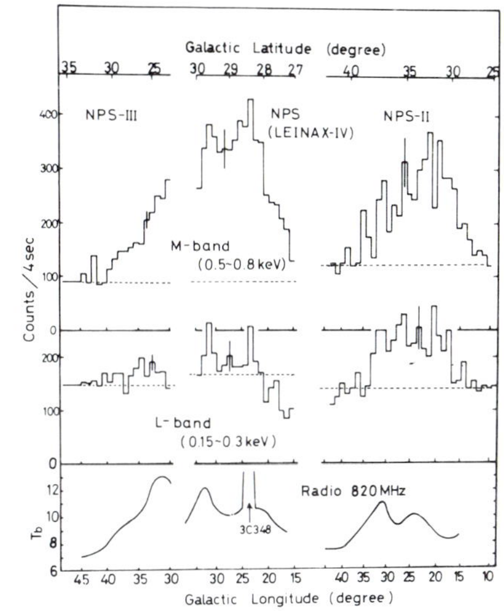

Fig.1. X-ray profiles of the North Polar Spur in comparison with radio profiles. The upper two diagrams represent the X-ray counts per 4 seconds in the M(0.5-0.8 keV)- and L(0.15-0.3 keV)-bands. The results for NPS-II and NPS-III are obtained by the present expriment, while the result in the middle is due to LEINAX-IV [2]. The dashed lines indicate the counting rates outside NPS. The brightness temperature profiles at 820 MHz [3] in the same regions are shown for comparison.

NORTH POLAR SPUR

Three segments crossing the NPS limb at b = 47°, 38° and 25° are called NPS-I, II and III, respectively. The observation of NPS-I was restricted to the inner region of NPS since an inflight calibration started about 8° before reaching the limb, while only the outer region was observed for NPS-III since slow scanning started at the top of the limb. A complete profile was available for NPS-II at a scan speed of 1°.25 s^{-1}.

The X-ray profiles of NPS-II and NPS-III are shown in Fig.1 separately for the L-band (0.15 - 0.3 keV) and the M-band (0.5 - 0.8 keV) of 1.5 μm counters along with those at b ≃ 30° obtained by a previous rocket observation (LEINAX-IV) [2]. The brightness temperature of 820 MHz radio continuum [3] along these scan paths are shown in the figure for comparison. The flux in the M-band shows a prominent feature similar to the profile of radio continuum for NPS III and NPS (LEINAX-IV), whereas the peak position for NPS II seems to be shifted inward by about 10°. The L-band flux is only weakly enhanced also in these regions of NPS, as already noted by earlier observations [4 - 6].

The pulse height spectra observed with three sets of counters for NPS-II (ℓ=15° - 30°) are shown in Fig.2. These spectra were obtained by subtracting those outside of NPS in the region (ℓ = 37° - 44°), assuming that they represent the spectra for the foreground and background regions of NPS.

These three spectra are fitted simultaneously with a single thin, hot plasma model. Using Kato's second compilation [7] with cosmic abundances [8], we obtain the electron temperature and the column density of absorbing matter in the ranges kT = 0.22 - 0.27 keV and N_H = (1.5 - 0.5) x 10^{20} H atoms cm^{-2}, respectively. The best fit values are kT = 0.25 keV and N_H = 1 x 10^{20} H atoms cm^{-2} with χ^2_{min} = 32.7 for the degree of freedom 44. The calculated spectra for the best fit parameters are shown by solid curves in Fig.2. The assumptions of abundances and ionization equilibrium imply that X-rays are emitted mainly from shock-heated clouds which are rather densely populated in the mid-latitude region, at variance with X-ray emission from the northern limb region [4].

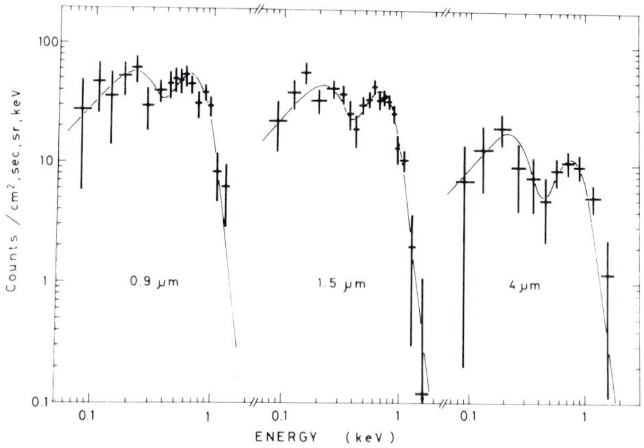

Fig.2. The X-ray spectra from NPS-II measured with three sets of counters with window thicknesses 0.9, 1.5 and 4.0 μm. The experimental points are obtained by subtracting the flux values for the outer region of NPS. The solid curves represent model spectra for kT = 0.25 keV, N_H = 1 x 10^{20} H atoms cm^{-2}, and the cosmic abundances.

HERCULES HOLE

A low N_H region in Hercules was scanned along b ≃ 43° from ℓ ≃ 67° to ≃ 93° and along ℓ ≃ 80° from b ≃ 50° to ≃ 37°. This region, called the Hercules hole, is characterized by a high L-band flux and a low M-band flux. No apprecaiable spatial variation of the flux over the hole was observed.

The X-ray spectra for the hole measured with three sets of counters are shown in Fig.3. These spectra can be fitted with a thin hot plasma model by Kato [7] with kT = 0.134 keV for cosmical abundance or kT = 0.109 keV for depleted abundance and both with $N_H = 1 \times 10^{-19}$ H atoms cm^{-2}. The spectrum with the 1.5 μ counters is identical with the spectrum over the belt along ℓ = 150° and 330° outside radio loops [9]. The fact that a single spectrum holds everywhere except for the regions invaded by SNR's is interpreted in terms of a hot plasma cloud surrounding the solar system. The spectrum for the Hercules hole demonstrates that the hole may also belong to the plasma cloud. The spectrum observed with the 0.9 μ counters indicates finite interstellar absorption with $N_H = (0.5 - 2) \times 10^{19}$ H atoms cm^{-2}, although it depends slightly on the model calculation for thin hot plasma.

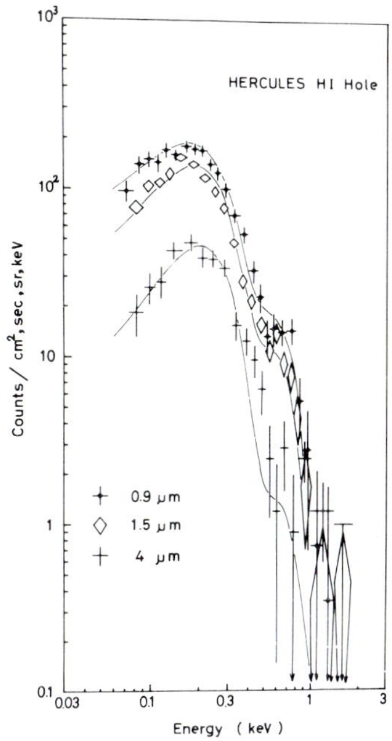

Fig.3. The X-ray spectra for the Hercules HI hole measured with three sets of counters. The solid curves represent model spectra for KT = 0.11 keV, $N_H = 1 \times 10^{19}$ H atoms cm^{-2}, and the depleted abundances with depletion factors 3, 10, 10, and 10 for C, Mg, Si and Fe, respectively.

DISTRIBUTION OF HOT PLASMA CLOUDS

In fig. 4 the ratios of the M- to L-band fluxes (M/L ratio) for the Hercules HI hole regions are given by solid circles against the total hydrogen column density obtained from 21 cm observations [10]. The M/L ratios for other regions outside NPS are also shown in the figure by open circles. Our previous results [9] for outside Loop I are shown by crosses.

For the simplified model that the interstellar matter along the line of sight consists of n hot clouds of the same temperature T_o and of cool absorbing clouds of optical depth τ(E) for each in between, the M/L ratio R is given by

$$R = \frac{\Lambda_M(T_o, E_M)\ \{1-e^{-n\tau(E_M)}\}\ \{1-e^{-\tau(E_L)}\}}{\Lambda_L(T_o, E_L)\ \{1-e^{-n\tau(E_L)}\}\ \{1-e^{-\tau(E_M)}\}}, \tag{1}$$

where Λ_M and Λ_L are the volume emissivities for the M- and L-bands, respectively. If only a hot plasma cloud sorrounding the solar system contributes to observed X-rays, the above expression is reduced to a constant value

$$R_o = \frac{\Lambda_M}{\Lambda_L}, \tag{2}$$

where $\sigma(E)$ is the absorption cross section. In the limit $\tau(E) << 1$, which means that a single cool cloud is transparent, the expression (1) is reduced to

$$R_1 = \frac{\Lambda_M\{1 - e^{-\sigma(E_M)N_H}\}\sigma(E_L)}{\Lambda_L\{1 - e^{-\sigma(E_L)N_H}\}\sigma(E_M)} . \qquad (3)$$

In Fig.4, R_o and R_1 are shown by the dashed line and the dott-dashed curve, respectively, for $E_L = 0.2$ keV and $E_M = 0.6$ keV. Almost all data points lie between these curves. This fact implies that the soft diffuse X-rays in the direction of $\ell \simeq 150°$ can be attributed to the hot plasma cloud surrounding the solar system, whereas hot plasma clouds located beyond absorbing regions contribute to observed X-rays in other directions.

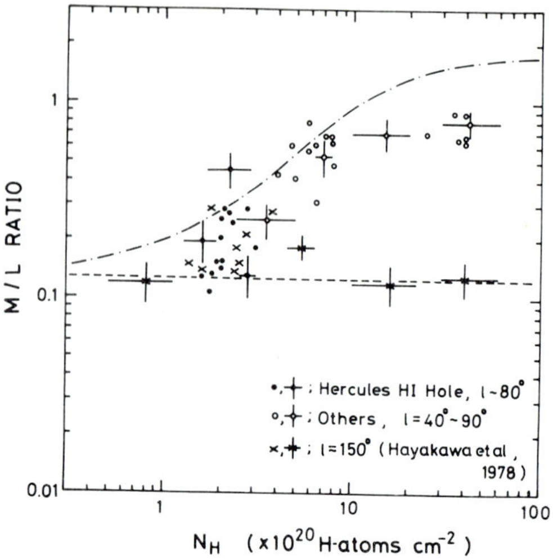

Fig.4. The ratio of the M-band flux to the L-band flux versus the hydrogen column density for the Hercules hole (solid circles) and mid-latitude regions of $\ell = 40° - 90°$ (open circles) obtained by the present experiment as well as for the regions along $\ell \simeq 150°$ (crosses) [9]. The dashed and the dott-dashed lines represent the M/L ratios predicted by equations (2) and (3), respectively.

REFERENCES

1. S. Hayakawa, H. Iwanami, H. Kunieda, F. Nagase, and K. Yamashita, Proceedings of COSPAR/IAU Symposium on X-Ray Astronomy, 1978.

2. S. Hayakawa, K. Yamashita, Y. Tanaka, J.A.M. Bleeker, J. Devalaar, and A.J.M. Deerenberg, COSPAR, 1977.

3. E.M. Berkhuijsen, Astron. Astrophys. 14, 359 (1971).

4. S. Hayakawa, T. Kato, F. Nagase, K. Yamashita, T. Murakami, and Y. Tanaka, Astrophys. J. Lett. 213, L109 (1977).

5. R.J. Borken and D.C. Iwan, Astrophys. J. 218, 511 (1977).

6. S. Hayakawa, T. Kato, F. Nagase, K. Yamashita, and Y. Tanaka, submitted to Publ. Astron. Soc. Japan.

7. T. Kato, IPPJ-AM-4, Inst. Plasma Physics, Nagoya University (1978).

8. C.W. Allen, Astrophysical Quantities, The Athlone Press (1973).

9. S. Hayakawa, T. Kato, F. Nagase, K. Yamashita, and Y. Tanaka, Astron. Astrophys. 62, 21 (1978).

10. C. Heiles, Astron. Astrophys. Suppl. 20, 37 (1975).

HOT INTERSTELLAR MEDIUM

Satio Hayakawa

Department of Physics, Nagoya University, Nagoya, Japan

INTRODUCTION

In recent years two new components of the interstellar medium (ISM) have been found important; they are the molecular clouds in which hydrogen mostly forms molecules and the hot interstellar medium (HISM) in which heavy elements are highly ionized. Nearly a half of the mass of ISM is in the form of hydrogen molecules, whereas nearly a half of the volume of ISM is occupied by HISM.

The HISM has been observed by ultraviolet absorption lines of highly ionized ions such as OVI, as reviewed by Spitzer and Jenkins [1] and by Jenkins [2,3], and by the diffuse component of soft X-rays, as reviewed by Tanaka and Bleeker [4], by Tanaka [5] and by Kraushaar [6]. The origin of HISM has been suggested by Cox and Smith [7] as due to repeated heating by supernova remnants (SNR). A number of SNRs are observed to be extended soft X-ray sources, and their X-ray features are interpreted in terms of SNR models [8,9]. The most extended SNR source is found for a radio loop with which the North Polar Spur (Loop I) is associated [10 and references therein]. The reason why Loop I is of a large size can be attributed to its proximity and a low density of ISM through which the SNR expands. Moreover, the ISM surrounding Loop I is hot and contributes to a major part of soft X-rays observed. In other words, the HISM surrounds the solar system and is likely to be an old SNR [11], thus supporting the suggestion by Cox and Smith [7]. The ISM may be populated by such hot components, as demonstrated by a numerical calculation [12].

Since SNRs are considered to be a main source of HISM, we begin with a review of SNR models. As a number of reviews of SNR models are available, we concentrate ourselves to recent works which are not fully discussed in the latest review by Chevalier [13].

In the interpretation of X-ray features of SNRs it is important to remark that the ionization equilibrium does not always hold in the SNR. In an early stage of SNR, moreover, the equipartition of energy between ions and electrons hardly holds. These facts require a revised theory in correlating observed X-ray spectra with SNR parameters. Even if the ionization equilibrium is reached, some divergent conclusions are drawn from the same observational result, because of differences in the ion abundances and the line emission powers which are based on different theoretical results of electron-ion collisions. We compare several theories of X-ray emission and present a theory which should be referred to.

SUPERNOVA REMNANT MODELS

The SNR expands according to the following sequences. The free expansion stage persists until the mass of ISM swept becomes comparable to the mass of supernova ejecta. This is followed by the adiabatic expansion stage in which a blast wave propagates through the ISM, the energy radiated being small compared with the blast wave energy. Then the matter at the shock front is cooled by radiation but continues to expand without losing its momentum. In this non-adiabatic stage a cold shell is formed at the front, but the matter behind the front is hot enough to emit radiation. As the expansion ceases because of momentum dissipation, a hot medium is left and is still capable to emit radiation.

Theoretical models of SNR expanding in a cold, homogeneous medium have been discussed by a number of authors [13 and references therein]. Several coments on the canonical model may be useful for detailed comparison of SNR models with X-ray features observed.

The adiabatic expansion as represented by the Sedov solution may not hold because of efficient heat conduction by hot electrons, but the expansion is likely to be isothermal. As demonstrated by Sollinger et al [14], however, X-ray features predicted by the adiabatic and the isothermal models are of minor difference because of a high concentration of matter into the SNR shell. Hence no appreciable difference arises if comparison is made with adiabatic models, provided that heat conduction across the shock front is suppressed by magnetic fields. There are two reasons to suspect that heat conduction may not be so efficient as one might think; the electron temperature is lower in the early stage of expansion, and the temperature gradient is lower for the case of weak shock than expected from the canonical model.

Electron Temperature

The energy of a blast wave first dissipates to heat up ions, and the ion energy is gradually transferred to electrons mainly by Coulomb collision. The temperatures of ions and electrons, T_i and T_e, are generally different from one another and are related to the average temperature T employed in current SNR models as

$$(n_i + n_e)\,T = n_i T_i + n_e T_e, \qquad (2.1)$$

where n_i and n_e are the number densities of ions and electrons, respectively. If T_e is significantly smaller than T, heat conduction is not efficient since conductivity is proportional to $T_e^{5/2}$. The radial profile of electron temperature has been calculated by Itoh [15] for the adiabatic blast wave model, as shown in Fig.1.

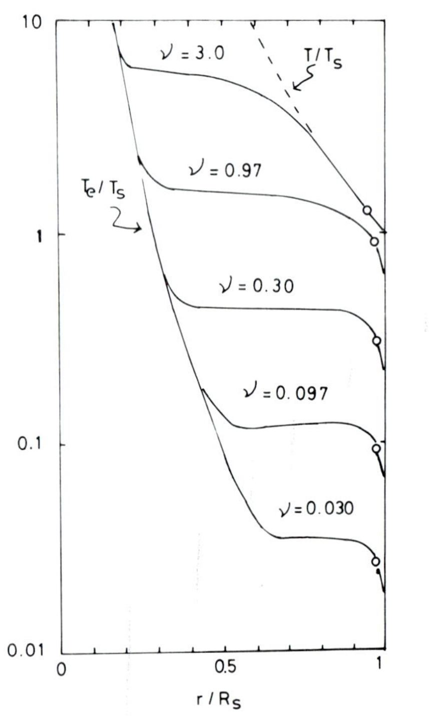

Fig.1. The radial profiles of T_e/T_s and T/T_s for the Sedov SNR. The Parameter ν represents the reduced age, $\nu = 1.5(n_o^{8}/E_{50}^{3})^{1/14} t_4$. The circles represent the effective electron temperatures for X-ray emission. T_s is the temperature at the shock front.

The value of T_e/T in the SNR shell is approximately expressed by

$$T_e/T = 1.5\ (n_o^8/E_{50}^3)^{1/14}\ t_4 \equiv \nu, \tag{2.2}$$

where E_{50} is the total energy in units of 10^{50} erg, n_o is the IMS density in cm^{-3} and t_4 is the age in 10^4 yr.

Equation (2.2) demonstrates that the energy equipartition between ions and electrons does not hold for young SNRs such as Cas A. Even if the age is large, the equipartition is not reached for SNR expanding in a low density ISM. This is the case for Loop I, since n_o is smaller than $10^{-2}\ cm^{-3}$ despite $t_4 \simeq 3$.

Ion Abundances

The ionization equilibrium is difficult to hold since either the recombination times or the ionization times or both for some ions are longer than the time scale of expansion if the age and/or the density are small.

For ions with one or two orbital electrons, recombination takes place mainly by radiative processes. The time for radiative recombination is

$$t_r \simeq 1 \times 10^6\ z^{-2}\ n_e^{-1}\ T_{e6}^{1/2}\ \text{yr}, \tag{2.3}$$

where z is the ion charge. This is longer than the ages of most SNRs observed if $n_e z^2 < 30$. For ions with outer shell electrons, dielectronic recombination is so fast that the ionization equilibrium is reached in most of practical cases. The ionization times of closed shell ions are relatively long compared with those with electrons outside the closed shell. If, for example, the SNR expands in a hot gas of $T_o \simeq 10^6$ K, the most abundant oxygen ions, OVII, are partly converted to OVIII but not appreciably to OIX, although OIX would be more abundant than OVIII under the ionization equilibrium at an electron temperature of about 3×10^6 K.

The time variations of ion abundances are shown for two constant values of electron temperature in Fig.2. Ionization starts with the equilibrium abundances at 10^6 K, at which oxygen is mostly in the stage of OVII and the ionization stages of iron are distributed over ions with a few M-shell electrons. As the temperature is increased to T_e at $t = 0$, the degree of ionization increases, and the ion abundances reach constant values when ionization and recombination are balanced. For iron ionization stops at FeXVII, a closed shell ion, for $T_e = 300$ eV, but spills over a little to FeXVIII for $T_e = 600$ eV. Since binary collisions dominate both for ionization and recombination, the variation of ion abundances is expressed as a function of nt, the electron density times time. Hence the evolution of ionization depends on the density of medium in which the blast wave expands.

The general interstellar medium consists of a tenuous gas and clouds embedded therein, as depicted by McKee and Ostriker [16]. In the solar neighbourhood the gas is hot ($T_e \sim 10^6$ K) and tenuous ($n \sim 10^{-2}\ cm^{-3}$), and Loop I is a SNR expanding in such a medium. Neither equipartition nor ionization equilibrium holds in the SNR shell formed in the tenuous gas, since the value of nt is estimated to be only slightly higher than $10^{10}\ cm^{-3}$ s. In the hot shock front formed in clouds, however, the value of nt is considered to be appreciably greater than $10^{11}\ cm^{-3}$ s, and both equipartition and ionization equilibrium hold. Hence the X-ray spectrum has to be interpreted in terms of a superposition of these two components.

In the energy range 0.5 - 1 keV the emission lines of OVII (570 eV), OVIII (650 eV) and FeXVII (830 eV) contribute to X-ray emission. In the tenuous plasma in the SNR shell the OVII and OVIII lines give a major contribution, whereas in the hot envelope of clouds the FeXVII line is expected to be strong. A mixture of these lines can be found necessary from multi-colour observations of the North Polar Spur, as shown in Fig. 3 [17, 18].

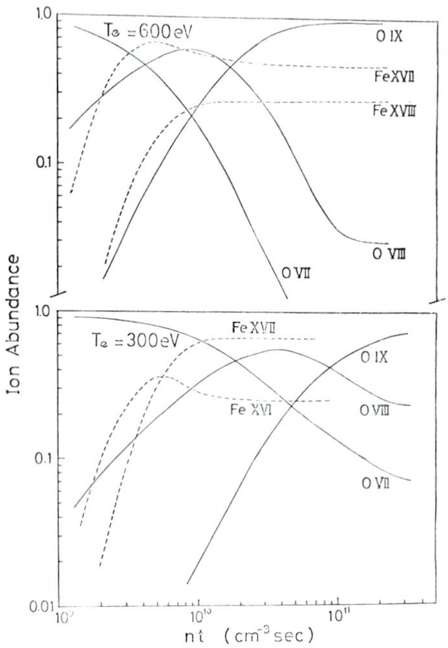

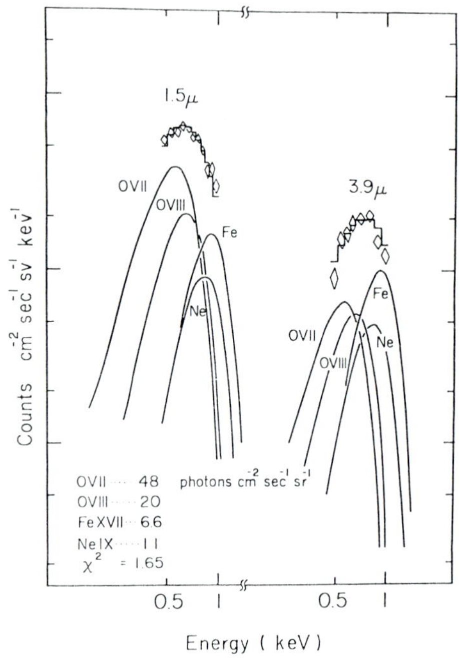

Fig.2. Time variations of ion abundances in SNR. The equilibrium abundances at 10^6 K at t = 0 are assumed. Abscissa: the electron density times the time after the electron temperature is raised to T_e. Ordinate: the fractional ion abundances of each element.

Fig.3. X-ray spectrum of the North Polar Spur. The spectrum in the range 0.5-1 keV is decomposed into the contributions of three lines, OVII (570 eV), OVIII (650 eV), FeXVII (830 eV) and NeIX (920 eV). Their contributions to the pulse height spectra for two proportional counters with window thicknesses 1.5 μm and 3.9 μm are given by solid curves. Their sums shown by solid lines are compared with experimental results.

Shock Structure

The smallness of the interstellar density assumed above is based on the emission measure and unappreciable absorption of soft X-rays outside the northern limb of Loop I [11]. This is also supported by the fact that Loop I is of large size compared with other SNRs. In the Sedov model the shock temperature T_s is related to the SNR radius R_s in pc as

$$T_s = 63(E_{50}/n_o R_s^3) \text{ keV}. \tag{2.4}$$

Since $T_s \simeq 0.5$ keV and $2R_s = 230 \pm 135$ pc, we have $E_{50}/n_o \simeq 10^{4\pm1}$. This requires a small value of n_o since the SNR energy hardly exceeds 10^{52} ergs. The small density is also obtained from the absolute flux of X-rays associated with the SNR shell. The peak emission measure at the northern limb, derived taking into account the X-ray emissivity as discussed above, is very small,

$$EM_I = \langle \eta^2 n_o^2 \rangle K R_s = 4 \times 10^{-3} \text{ cm}^{-3} \text{ pc}, \quad (2.5)$$

where η is the density jump and K is a numerical factor which slowly decreases as η increases. For $\eta = 4$, in the case of strong shock, $K = 0.35$ and $\langle n_o^2 \rangle^{1/2} \lesssim 4 \times 10^{-3}$ cm^{-3}.

Even if the density is low, the pressure of the HISM is not always negligible because of a high temperature, which is of the order of 10% of the shock pressure of Loop I. The pressure gives a significant effect on the SNR structure unless the shock radius is much greater than R_o [19]. The critical radius R_o is defined by

$$R_o = (E/4\pi n_o kT_o)^{1/3} = 12.6(E_{50}/n_o T_{o6})^{1/3} \text{pc}, \quad (2.6)$$

where T_{o6} is the temperature of IMS in units of 10^6 K. The pressure of ISM affects the SNR structure in the following manner; the density jump, the density gradient, and the temperature gradient decrease. Assuming the density jump of $\eta = 3$, we can reproduce the X-ray profile of Loop I, as shown in Fig.4. The X-ray enhancement is prominent in the energy range 0.5 - 0.8 keV, whereas only a weak enhancement occur in the low energy range 0.16 - 0.3 keV. This is because the emission of OVII and OVIII lines is strong under the non-equilibrium condition of ionization, whereas the contribution from the surrounding HISM is large in the low energy range, as will be discussed latter.

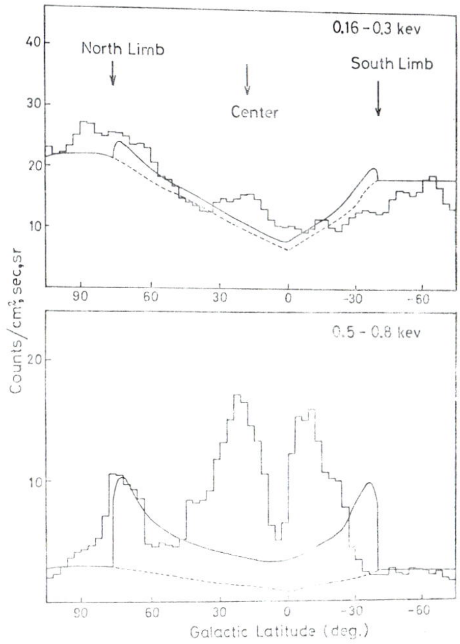

Fig.4. X-ray profile of Loop I. The histograms represent the observed flux after subtracting the counter background and the extragalactic component. The dashed lines represent the contribution from the foreground and background media. The solid lines are obtained by adding the contribution of SNR which is assumed to be spherical. The shape in the southern hemisphere should be modified by taking into account the inhomogencity of ISM.

LINE EMISSION POWERS

The interpretation of observed results depends on the theory of line emission. The emission power of a line is given by the product of the ion abundance and the collisional excitation rate, provided that the X-ray source under consideration is a thin plasma. Several compilations of line emission powers do not always give the same results, and observed results analysed with different compilations lead one to conclusions inconsistent with each other. The following differences significantly affect the conclusions.

The Number of Lines

The ion lines are so densely populated that a large number of weak lines give a significant contribution to the emission rate. Kato [20] has compared the spectra calculated with the compilations by Landini and Fossi [21], 40 lines in the range 1 - 100 Å, by Tucker and Koren [22], 459 lines in the range 1 - 70 Å, and by Kato [23], 795 lines in the range 1 - 250 Å. Fig.5 shows that the power calculated by Landini and Fossi is appreciably weaker at gaps between strong lines than those by two others.

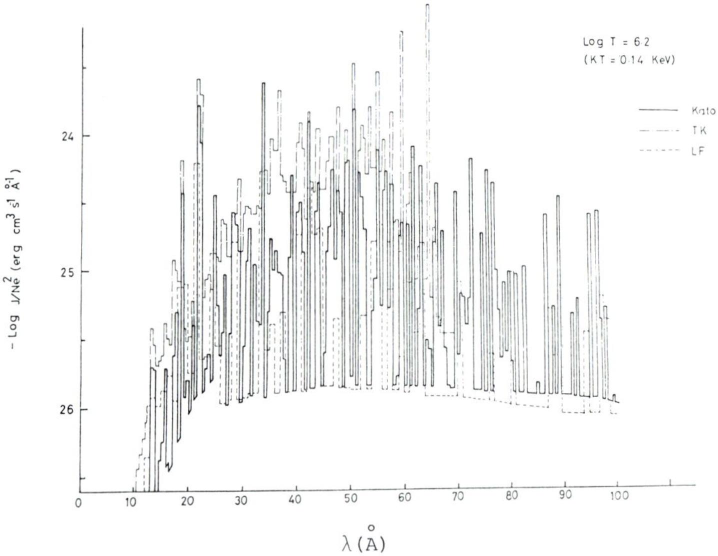

Fig.5. X-ray spectra at $T = 1.6 \times 10^6$ K for cosmic abundances calculated by Landini and Fossi (LF) [21], by Tucker and Koren (TK) [22] and by Kato [23]. The line intensity is equal to the differential emission rate time 1 Å.

Ion Abundances

Under the ionization equilibrium the ion abundances are obtained by the balance between collisional ionization and electron recombination. Differences in the rate coefficients of these processes give rise to different ion abundances.

The equilibrium abundances calculated with ionization cross sections given by Lotz [24,25] and by Summers [26] are compared in Fig.6. The maximum abundance is attained at a lower temperature in the former case, simply because the ionization cross section is larger. Lotz's cross section for NIII and OIII are in better agreement with experimental ones than Summers'. Seaton's cross section [27], which are essentially the same as Lotz's, were employed by Jordan [28,29] to calculate ion abundances which were adopted by Tucker and Koren [22], and by Kato [23], whereas Summers' cross sections were adopted by Raymond amd Smith [30]. Since the temperatures for the maximum abundance are different in these two cases, the rate coefficients of, for example, Si VII (82 Å) and Si VIII (61 Å) for the maximum abundances would be about twice larger in the case of Raymond and Smith [30] than in Kato [23], even if the same formula for collisional excitation were used.

In Jordans' ion abundances the dielectronic recombination rates are based on Burgess' theory [31]. In this theory, however, autoionization into excited states is not taken into account, as correctly pointed out by Jacobs et al [32]. If this is taken into account, the dielectronic recombination rate is reduced, and the temperature for the maximum abundance is lowered as shown in Fig.7. Accordingly the line emission rate is reduced. Since the corrected recombination rates have been calculated only for iron ions, the revised values of ion abundances are not yet available for other ions of interest.

Excitation Rates

In Fig.5 we see that Tucker and Koren [22] gave generally higher emission rates than Kato [23]. This is due to larger values of the Gaunt factor. For H- and He-like ions there are unappreciable differences among different compilations. For more complicated ions the Gaunt factor given by Mewe [33] is adopted by Kato [23]. Its values for allowed transitions with $\Delta n \neq 0$ are appreciably lower than those adopted by Tucker and Koren [22], whereas those for $\Delta n = 0$ are about twice larger than those by Pottasch [35]. Mewe's values are in good agreement with those calculated with the distorted wave approximation [35]. Raymond and Smith [30] have adopted the Gaunt factors for strong lines from Tucker and Koren [22] and from Pottasch [34], which are significantly greater than Mewe's [33].

The difference in the Gaunt factor affects the emission rate more strongly than that in the temperature dependence of ion abundances, as shown

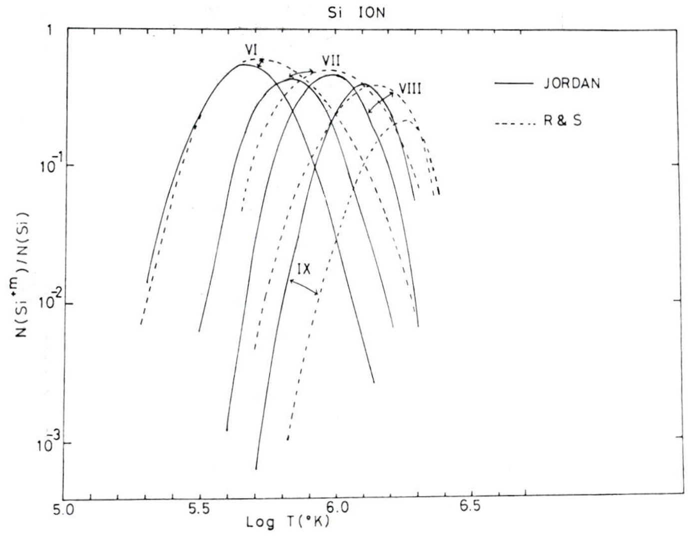

Fig.6. The equilibrium abundances of Si VII and Si VIII calculated by Jordan [28,29] who employed the ionization cross sections essentially equal to those given by Lotz [24,25] - solid lines - and by Raymond and Smith (R&S) [30] with Summers [26] - dashed lines -.

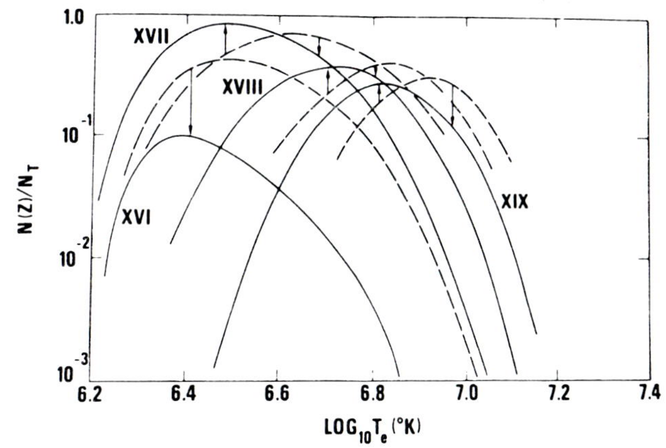

Fig.7. The equilibrium abundances of iron ions with the recombination rates by Burgess [31] - dashed lines - and by Jacobs et al [32] - solid lines-.

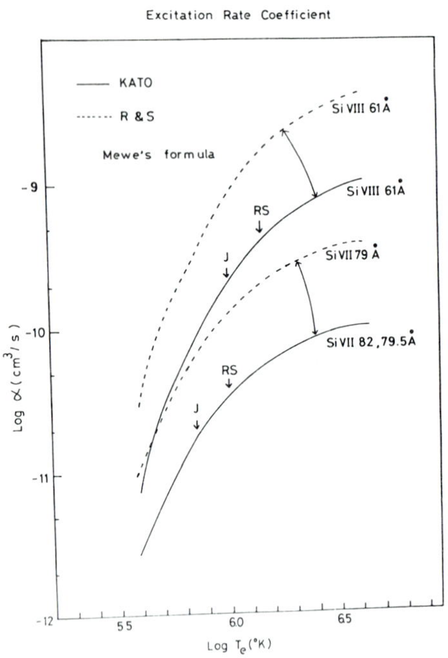

Fig.8. The rate coefficients of excitation of Si VII (82 Å) and Si VIII (61 Å); the solid lines are calculated by Kato with use of Mewe's Gaunt factors, and the dashed lines represent those by Raymond and Smith. The temperatures for the maximum abundances calculated by Jordan and by Raymond and Smith are indicated by arrows, J and RS, respectively.

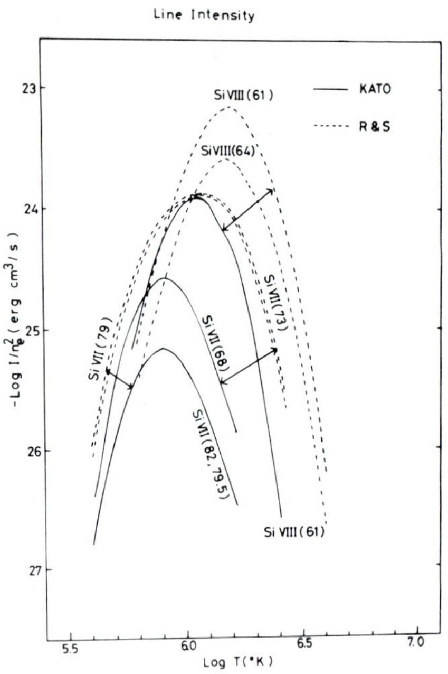

Fig.9. The emission rates of strong lines of silicon ions calculated on the basis of Kato [23] and of Raymond amd Smith [30]. Different wavelengths have been adopted by these authors as the representative wavelengths of a multiplet system.

in Fig.8. The emission rates of strong lines of silicon ions given by Kato [23] and by Raymond and Smith [30] are compared in Fig.9. The values of Raymond and Smith are in general greater by a factor of about five than those by Kato. Similar differences are also found for lines of Mg, S and Fe ions at energies around 200 - 250 eV.

Example of Different Interpretations of the Same Experimental Result

The Difference in the emission rate results in different interpretations of an experimental result. Kraushaar [6] observed the diffuse component of soft X-rays with three spectral bands, the B-band of a transmission window peaked at the boron K-edge (190 eV), the C-band of window at the carbon K-edge (280 eV) and the M-band effective in the range above 500 eV. He plotted the observed results in the f_B - f_M diagram, where f_B and f_M are the fractions of counting rates in the B- and M-bands, respectively. Using Kato [23] and Raymond and Smith [30], Nousek [36] obtained f_B - f_M trajectories for varying temperature as shown in Fig. 10.

The observed points fall inside a closed curve which does not intersect the trajectory based on Raymond and Smith but does intersect that based on Kato. From the former Kraushaar [6] concluded the two-component model of diffuse soft X-rays. However, the trajectory calculated by using Kato's emission rates is significantly shifted rightward and crosses the observed points, as shown in Fig.10. The difference between two trajectories can be explained as follows. The M-band fluxes in both cases are in approximate agreement with one another, since OVII and OVIII lines (H- and He-like) are mainly responsible for the M-band. On the other hand, the C-band flux by Raymond and Smith is much greater than that by Kato, since Si lines are mainly responsible for the C-band. As a result, both f_B and f_M are significantly reduced for the case of Raymond and Smith, if f_C is much larger than f_B anf f_M. If Kato's emission rates are adopted, one cannot claim the two-component model but may favour the single component model as advocated by Hayakawa et al [11].

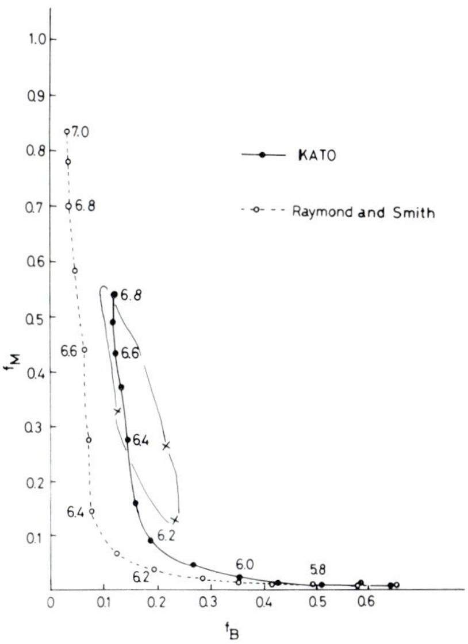

Fig.10. f_B - f_M diagram. The diffuse component of soft X-rays observed by the Wisconsin group [6] gives the fractional counting rates f_B (boron window) and f_M (medium energy band) inside the closed curve. The crosses give results of high statistical accuracies. The solid and the dashed lines represent the trajectories for varying temperature (the numbers attached representing log T) calculated for cosmic abundances without interstellar absorption with use of Kato [23] and of Raymond and Smith [30], respectively.

Preference of one over the other will be concluded by comparison with experimental excitation rates. At the present, however, no experimental data to be compared are yet available. Nevertheless, Kato's compilation is preferred because the Gaunt factors she adopted from Mewe [33] are in good agreement with theoretical results for Fe X, XI and XIV with use of the distorted wave approximation [37]. Hence we recommend to use Kato's compilation together with that by Stern et al [38] for a long wavelength range, since these two give consistent results in the overlapping region.

DISTRIBUTION OF HISM

Since the absorption coefficient of X-rays depends strongly on energy, absorption features of the soft X-ray spectrum would indicate how cool media are interwoven with hot X-ray emitting plasmas.

Let us consider a simplified model that along the line of sight there are ν hot regions of temperature T, electron density n and linear size r, and between two adjacent hot regions there is a cool cloud of optical depth $\tau(E)$. The X-ray flux of energy E is then given by

$$j(E) = \frac{\lambda(T, E)}{4\pi} n^2 r \frac{1 - e^{-\nu\tau(E)}}{1 - e^{-\tau(E)}} , \tag{4.1}$$

where $\lambda(T, E)$ is the emission rate of photons with energy E in units of $cm^3\ s^{-1}$. Since $\tau(E)$ is a rapidly decreasing function of energy E, the value of $j(E)/\lambda(E)$ increases with ν which is proportional to the hydrogen colum density N_H. If we select two energy bands with average energies E_L and E_M ($>E_L$), the flux ratio j_M/j_L is an increasing function of N_H.

The flux ratio for the pulse height ranges 0.5 - 0.8 keV and 0.16 - 0.3 keV observed for different sky regions without SNR features is plotted against N_H in Fig. 11. In the direction of $\ell = 150°$ no appreciable N_H-dependence is observed [11], whereas the relation (4.1) approximately holds in the direction of $\ell = 40° - 90°$ [39] and also in some other directions [40]. This implies that the hot region in the direction of $\ell = 150°$ is extended without containing an appreciable amount of absorbing matter ($N_H < 5 \times 10^{19}$ cm^{-2}). This has provided a basis of proposing a model that solar system is surrounded by a hot plasma bubble. In other directions, however, absorbing clouds intervene hot plasmas, and the hot regions beyond the absorbing clouds contribute to observed X-rays. Hence the hot plasma bubble in the solar neighbourhood is more or less disk-like with the long side parallel to the circle along $\ell = 150°$. The distribution of interstellar dust as obtained from extinction by FitzGerald [41] shows a similar feature as seen from Fig. 12.

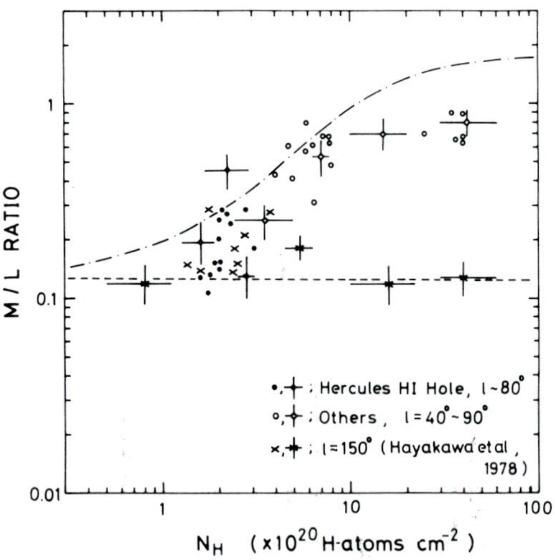

Fig. 11. The flux ratio F_M/F_L vs N_H. The filled and open circles represent data for the Hercules HI hole and the northern region of $\ell = 40° - 90°$, respectively, as given in [39], and the crosses represent data for $\ell \simeq 150°$ as given in [11]. The dashed line indicates the M/L ratio expected for a single hot plasma model proposed in [11], whereas the dott-dashed line indicates the relation calculated by equation (4.1).

In the direction near the north galactic pole outside the northern limb of the North Polar Spur, properties of the hot plasma bubble are obtained as follows [11,17]. The emission measure in this direction is

$$EM_{ISM} = \langle n_o^2 \rangle R_{ISM} \simeq 3 \times 10^{-2}\ cm^{-6}\ pc. \tag{4.2}$$

The X-ray profile for the northern limb of Loop I as shown in Fig.4 can be fitted with $\eta = 3$ and $K = 0.23$, and equation (2.5) thus gives

$$\langle n_o^2 \rangle R_s \simeq 2 \times 10^{-3}\ cm^{-6}\ pc. \tag{4.3}$$

For $R_S = 50$ pc we obtain

$$R_{ISM} \simeq 750 \text{ pc}, \ \langle n_o^2 \rangle^{1/2} \simeq 6 \times 10^{-3} \text{cm}^{-3}. \quad (4.4)$$

A larger value of R_S results in a smaller density and a greater bubble size. Such a low density given in equation (4.4) may be attainable in a medium swept by a SNR blast wave. This means that the hot plasma bubble is also an old SNR whose age is estimated from instability of a disk-like bubble of overpressure to be several million years. Loop I is a SNR expanding in a low density medium formed by the old SNR, thus supporting the idea by Cox and Smith [7].

ACKNOWLEDGEMENT

The author expresses his thanks to Dr. T. Kato who gave useful advices on X-ray line emission and supplied numerical results. He also thanks B. Gaffet, H. Itoh, H. Iwanami, T. Kato, W.L. Kraushaar, H. Kunieda, F. Nagase, J. Nousek, Y. Tanaka and K. Yamashita for making their results available prior to publication.

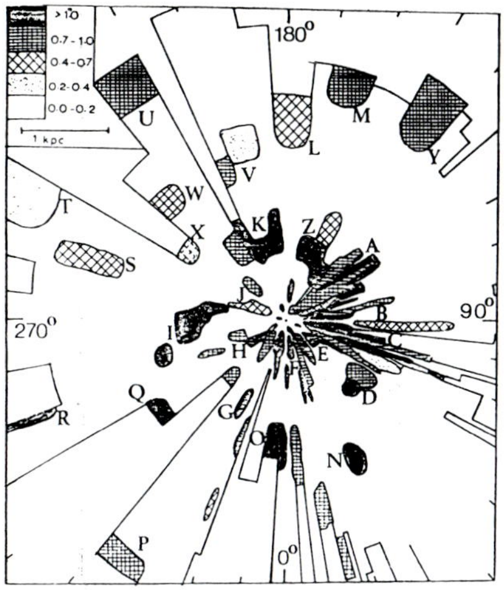

Fig. 12. The distribution of interstellar dust in the galactic plane given by FitzGerald [41]. The galactic longitude is represented by ℓ^I. The gap in the direction of $\ell^I \sim 180°$ indicates a HISM extended in the direction of $\ell^{II} \sim 150°$.

REFERENCES

1. L. Spitzer, and E.B. Jenkins, Ann. Rev. Astron. Astrophys. 13,133. (1975).

2. E.B. Jenkins, Paper at 1977 COSPAR.

3. E.B. Jenkins, Astrophys. J. 219, 845; 220, 107 (1978).

4. Y. Tanaka, and J.A.M. Bleeker, Space Sci. Rev. 20, 815 (1977).

5. Y. Tanaka, Paper at 1977 COSPAR.

6. W.L. Kraushaar, Paper at Honolulu A.A.S. Meeting (1977).

7. D.P. Cox, and B.W. Smith, Astrophys. J. Lett. 189, L105 (1974).

8. P. Gorenstein, F.R. Harnden, Jr., and W.H. Tucker, Astrophys. J. 192, 661 (1974).

9. D.H. Clark, and J.L. Culhane, Mon. Not. Roy. Astr. Soc. 175, 573 (1976).

10. S. Hayakawa, T. Kato, F. Nagase, K. Yamashita, T. Murakami, and Y. Tanaka, Astrophys. J. Lett. 213, L109 (1977).

11. S. Hayakawa, T. Kato, F. Nagase, K. Yamashita, and Y. Tanaka, Astron. Astrophys. 62, 21 (1978).

12. B.W. Smith, Astrophys. J. 211, 404 (1977).

13. R.A. Chevalier, Ann. Rev. Astron. Astrophys. 15, 175 (1977).

14. A. Sollinger, S. Rappaport, and J. Buff, Astrophys. J. 201, 381 (1975).

15. H. Itoh, Publ. Astron. Soc. Japan 29, 813 (1977).

16. C.F. McKee, and J.P. Ostriker, Astrophys. J. 218, 148 (1977).

17. S. Hayakawa, T. Kato, F. Nagase, K. Yamashita, and Y. Tanaka, submitted to Publ. Astron. Soc. Japan.

18. S. Hayakawa, K. Yamashita, Y. Tanaka, J.A.M. Bleeker, J. Davelaar, and A.J.M. Deerenberg (1978) Paper at 1977 COSPAR.

19. B. Gaffet, Astrophys. J., in press.

20. T. Kato, IPPJ-AM-4, Institute of Plasma Physics, Nagoya University (1978).

21. M. Landini and B.C. Monsignori Fossi, Solar Physics 20, 322 (1971).

22. W.H. Tucker and M. Koren, Astrophys. J. 168, 283 (1971).

23. T. Kato, Astrophys. J. Suppl. 30, 397 (1976).

24. W. Lotz, Report, Institut für Plasma Physik IPP 1/62, (1967).

25. W. Lotz, Report, Institut für Plasma Physik, IPP 1/76 (1978).

26. H.P. Summers, Mon. Not. Roy. Astr. Soc. 169, 663 (1974).

27. M.J. Seaton, Planet. Space Sci. 12, 55 (1964).

28. C. Jordan, Mon. Not. Roy. Astr. Soc. 142, 501 (1969).

29. C. Jordan, Mon. Not. Roy. Astr. Soc. 148, 17 (1970).

30. J.C. Raymond, and B.W. Smith, Astrophys. J. Suppl. 35, 419 (1977).

31. A. Burgess, Astrophys. J. 141, 1588 (1965).

32. V.L. Jacobs, J. Davis, P.C. Lepple, and M. Blaha, Astrophys. J. 211, 605 (1977); Astrophys. J. 215, 690 (1977).

33. R. Mewe, Astron. Astrophys. 20, 215 (1972).

34. S.R. Pottasch, Bull. Astr. Inst. Netherlands 19, 113 (1967).

35. J. Davis, P.C. Kepple, and M. Blaha, J. Quart. Spectrosc. Radiat. Transfer. 15. 1145 (1974).

36. J. Nousek, private communication.

37. H.E. Mason, Mon. Not. Roy. Astr. Soc. 170, 651 (1975).

38. R. Stern, E. Wang, and S. Bowyer, Astrophys. J. Suppl. 37, 195 (1978).

39. S. Hayakawa, H. Iwanami, H. Kunieda, F. Nagase, and K. Yamashita, (1978b) in this volume.

40. H. Iwanami, H. Kunieda, F. Nagase, K. Yamashita, and T. Murakami, (1978) to be published.

41. M.P. FitzGerald,Astr. J. 73, 983 (1978).

A SURVEY OF SUPERNOVA REMNANTS WITH THE HEAO A-1 EXPERIMENT

S. Shulman, D. Yentis, J. Meekins, K. Wood, R. L. Kinzer, G. Share, W. Evans, D. McNutt, E. T. Byram, T. A. Chubb and H. Friedman

Naval Research Laboratory, Washington, D.C. 20375

INTRODUCTION

The A-1 experiment on board the HEAO-1 Observatory is carrying out an all-sky survey of X-ray sources. As part of this larger effort we are conducting a systematic search for X-ray sources at positions of known optical and radio supernova remnants (SNR). Catalogs of such positions are found in references [1] and [2]. X-ray fluxes detectable with the A-1 instrument are expected for many of these sources.

The A-1 Experiment is sensitive in the energy range from 0.5 to 25 keV. Four of the seven modules (the "scan modules") are used to make the majority of new source detections. These modules are mounted on the -Y side of the Observatory and have the narrow direction of their $1^{o} \times 4^{o}$ collimators aligned with the scan plane of the spacecraft. The spin axis is kept toward the sun at all times, causing the scan plane to advance roughly one degree per day.

Data from all useable scans of a given sky region are aligned and summed. These sums are then used to search for sources. The instrument can detect sources in the 0.6-8 keV energy band with fluxes as low as 2×10^{-12} ergs cm^{-2} s^{-1}, or a factor of 10^{3} fainter than the Cygnus Loop. It is not always possible to detect all sources down to this limit because of source confusion. The confusion problem is somewhat worse in the plane than at high galactic latitudes.

OBSERVATIONS

During the first 23 days of experiment operation, the scan plane intersected the galactic plane in two regions, at galactic longitudes 130° - 160° and 310° - 340°. We report here on the supernova remnants identified in the first of these regions. Confusion problems are worse in the second region, and results for it will be given at a later date.

No X-ray SNR are known between longitudes 130° and 160°. Known optical and radio remnants here are 3C58, HB3, HB9 and CTB13. Figure 1 (top) shows a superposition of 4 days of data, from days 230-235 of 1977. A weak emission from the galactic plane is visible. The strong source at the right of the figure is the Perseus Cluster (intensity 47 UFU). New source locations are obtained by fitting the known collimator response to observed counts. One fit gives a line of position

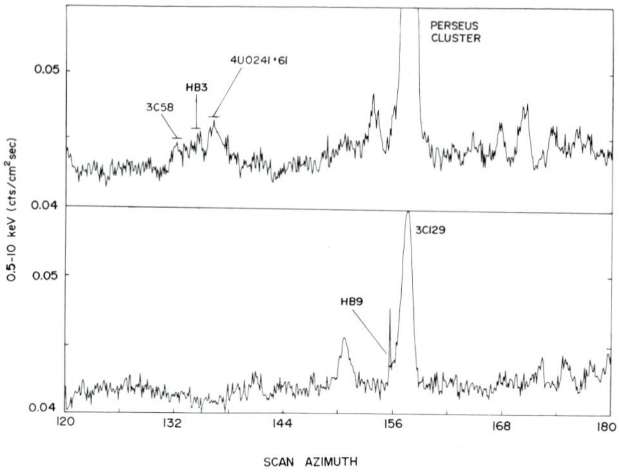

Fig. 1 Superposition of 4 days of scan data from days 230-235 of 1977 (top) and days 250-253 of 1977 (bottom).

perpendicular to the scan plane. A series of fits (from successive summations) shows the rise and fall of the source as it passes through the long direction of the collimator, and can be used to get the position in the direction perpendicular to the scan. In the region shown here it is assumed there are three sources to be fit, as shown in the figure. One of these is the Uhuru source, 4U0241+61. The feature labelled 3C58 has an error box that includes that remnant and no other known objects, excluding normal galactic stars. (The possibility that a star is the correct identification cannot be excluded, and this same caution applies to the other new source discussed below.) The central feature has a line of position which includes another SNR, HB3, but the fit to successive daily intensities yields a 90% confidence error box that excludes HB3. Figure 2 shows error boxes for both sources.

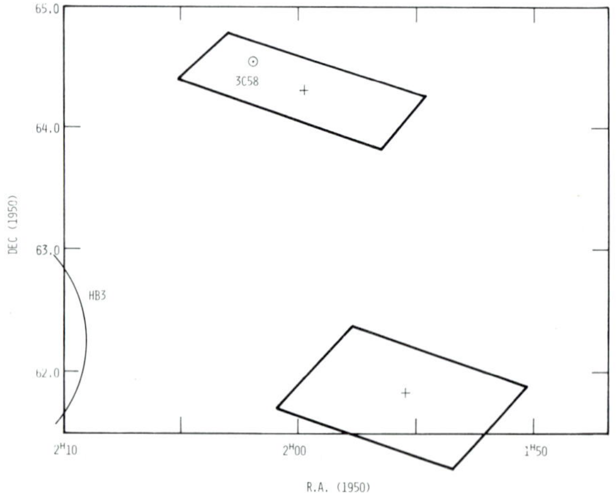

Fig. 2 Location of 2 new sources detected by the A1 experiment. One error box includes the SNR 3C58, while the other excludes HB3.

Figure 1 (bottom) shows the last four days of the 23-day period for which scan superpostions are available. The response to the previously-known source identified with the 3C129 cluster gives an asymmetrical collimator response, indicating a second source is present, but confused with the cluster. The error box derived for this source places it inside the 2° diameter of the radio SNR, HB9. It is not possible to determine the extent of the X-ray source because of the confusion. (This may eventually become possible with the fine-mapping module of the experiment, which has a field of view ½° x 1). Fig. 3 gives positions for the radio and X-ray sources.

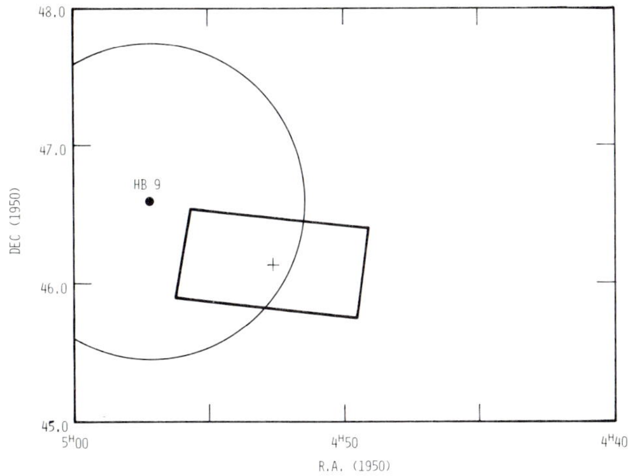

Fig. 3 Location of a new source which coincides with the SNR HB9.

TABLE 1.

SNR	DIST (kpc)	DIAM (pc)	N_H ($\times 10^{21}$ cm^{-2})	F_x(0.5-10 keV) ($\times 10^{-10}$ ergs/cm^{2}s)	L_x(0.5-10 keV) ($\times 10^{35}$ ergs/s)
3C58	10.9	12	3	0.1 - 0.2	3 - 8
HB9	1.1	43	1	0.4 - 1	0.06 - 0.2
HB3	1.9	41	2	<0.3 - 1	<0.2 - 0.8
CTB13	1.1	66	1.5	<0.07 - 0.2	<0.01 - 0.04

Table 1 summarizes fluxes and 2 upper limits for remnants in the surveyed region. The detection of 3C58 is the more noteworthy of the two new detections. This is a young remnant, since it has a small diameter. It has an optical counterpart which has been identified with the historical supernova, SN1183 [3]. In both the radio and optical it appears similar to the Crab. The X-ray luminosity at 0.5 - 10 keV is also comparable to the Crab (10^{36} erg s^{-1}). This suggests that the observed X-ray emission may be synchrotron radiation and not bremsstrahlung from a shock-heated plasma. The energetics would indicate that a pulsar is required to provide relativistic electrons. No radio pulsar is known at this location. A search for an X-ray pulsar is planned when HEAO-1 is next able to point at the region, in August, 1978.

HB9 has an X-ray luminosity of 10^{34} erg s^{-1} which makes it similar to other older remnants in the adiabatic expansion phase. The identification of the X-ray source with the SNR is supported by these physical considerations. In the same way, it is somewhat surprising that HB3 is not seen, since its diameter suggests it is the same age as HB9. The upper limit for CTB13 is consistent with its large diameter, indicative of an age 10^5 yr, long enough to have cooled below X-ray emitting temperatures.

REFERENCES

1. D. Downes, Astron. J., 76, 305 (1971).
2. D.H. Clark, and J.K. Caswell, MNRAS., 174, 267 (1976).
3. S. van den Bergh, Ap. J., 220, L9 (1978).

SOME RECENT CORONAL LINE OBSERVATIONS OF SUPERNOVA REMNANTS

J. C. Zarnecki*, R. Lucke**, B. E. Woodgate**,
J. L. Culhane* and D. Socker**

**Mullard Space Science Lab., University College London,
Holmbury St. Mary, Dorking, Surrey, England*
***Goddard Space Flight Center, Greenbelt, Maryland, U.S.A.*

INTRODUCTION

Some recent optical coronal line observations of supernova remnants (SNR's), carried out at Cerro Tololo (CTIO) and the Anglo-Australian Observatory (AAO), are described here. These have, in general, been motivated by the detection of soft X-ray emission from several SNR's. If such emission is thermal in nature, then the plasma responsible for its production is likely to be at the appropriate temperature (a few million K) to be producing, in addition, coronal line emission, and in particular [Fe X] and [Fe XIV]. Therefore, the possibility of studying such X-ray emitting regions at optical wavelengths suggests itself.

PUPPIS-A

The most recent high resolution spectrum of this source [1] tends to indicate that two different temperature components are required to fit the data adequately. The two temperatures are approximately 2.5 and 12.5 x 10^6 K with a common column density in the range 3 - 20 x 10^{20} atoms/cm^2. There also exists some indication as to the possible spatial locations of these two components. Data from the soft X-ray telescopes on the Copernicus satellite, with their small fields of view, have been analyzed [2] to show that for a series of pointing positions in the eastern part of the source, a significantly softer spectrum is required to fit the data than for the remainder of the source. In addition, data from a Lockheed sounding rocket [3] has also shown good evidence for a localized region of softer emission within the source. Furthermore, the detection of O VIII line emission from Puppis-A [4] is indicative of plasma with a temperature of approximately 2.5 x 10^6 K. The spatial relationship of these features is shown in Fig. 1. The uncertainty in the location of the O VIII emission ($\pm$ 6 arc min.) allows it to be consistent with both the Copernicus and Lockheed data, if indeed all three are observations of the same region.

In order to confirm the existence of this lower temperature component, the eastern part of the source was searched for [Fe X] and [Fe XIV] emission using a differential filter photometer with a 5 arc min. field of view. No [Fe X] was observed but strong [Fe XIV] emission was detected from one region. This is also shown in Fig. 1 and is seen to be consistent with the position of the soft component implied by the X-ray data. Since the filter passband was ∿ 8 Å, definite confirmation of the line identification was obtained from an image tube spectrogram taken on the 4m. telescope at CTIO and a spectrum taken by Murdin and Clark at the AAO.

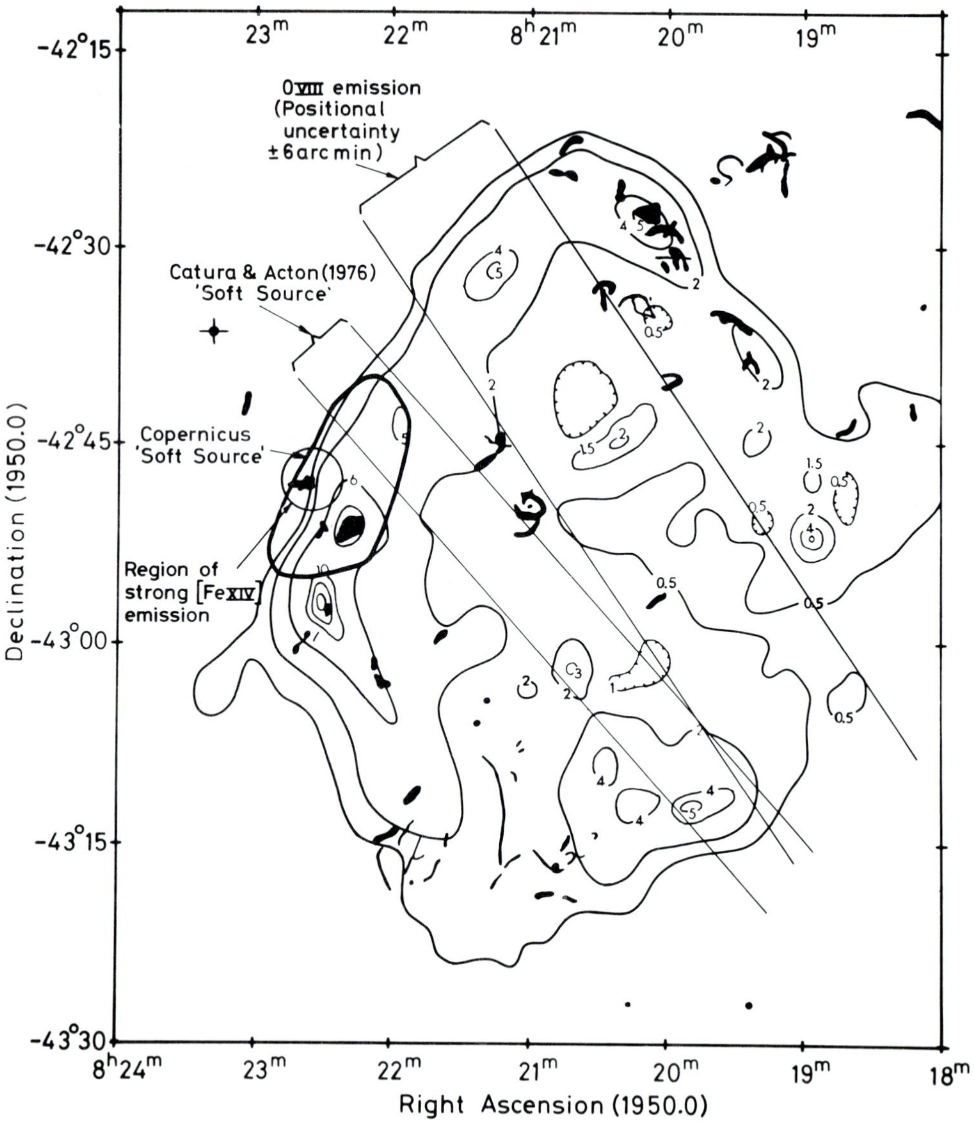

Fig. 1. Spatial relationship of Puppis-A emission features. The main radio contours and optical filaments are shown as well as the soft X-ray features.

The X-ray and optical observations taken together seem to imply the existence of a cool localized component superimposed on a more extended hotter region. A possible explanation is that the expanding shock front has interacted with a relatively dense interstellar cloud. The details of such a situation have been considered in some detail by Sgro [5]. He has shown that if the pre-shocked cloud density is less than a certain critical density, the result will be a 'hot' cloud radiating in soft X-rays and possibly coronal lines. In order to fit the observations, the cloud must have the following properties:-

$$T \sim 2 - 3 \times 10^6 \text{ K}$$

$$N_e \sim 3 - 10 \text{ cm}^{-3}$$

$$d \sim 2.5 - 6 \text{ pc}$$

In addition, in order to explain the observed intensity of [Fe XIV], it is possible that iron is underabundant with respect to the cosmic value. The cooling time for such a cloud would be of the order of $t_c \sim 10^4$ yr., which is comparable with the age of the remnant ($\sim 3 - 10 \times 10^3$ yr.).

These predicted parameters may be usefully compared with those of the so-called standard interstellar cloud as described by Spitzer [6]. They are as follows:-

$$\left.\begin{array}{ll} N_e & \sim 10\text{cm}^{-3} \\ d & \sim 7 \text{ pc} \end{array}\right\} \text{mean values but with a large scatter}$$

These are not too dissimilar from the parameters required to fit the observations of the Puppis-A low temperature component. The higher temperature component ($\sim 12.5 \times 10^6$ K), on the other hand, can most likely be identified with the interaction of the expanding shock front with the ambient interstellar medium.

THE LMC SNR N49

The SNR, N49, in the LMC was observed at the AAO by Murdin, Clark and Culhane [7]. It was chosen because its radio properties imply it may be at the appropriate evolutionary stage to be emitting in soft X-rays and also coronal lines. The spectrum shows a broad feature centred on 5303 Å, and its identification with [Fe XIV] seems likely. This feature is ~ 50% broader than other lines in the spectrum and suggests velocities of the order of several hundred kilometres/sec.

Based on the observed coronal line intensity, it is possible to make some predictions as to the likely X-ray flux from this remnant. Assuming a temperature of 2×10^6 K and a density behind the shock front of 4 cm^{-3}, we obtain an emission measure at this temperature of $\sim 10^{59}$ cm^{-3}. Then, using the calculations of Tucker and Koren [8] for the emissivity of a hot coronal-like plasma, we may predict the likely X-ray flux from this source. These are shown in Table 1 for $N_H = 0$ and 10^{21} atoms/cm^2. Based on spectral observations of known LMC X-ray sources,one does not expect the absorption to be too great.

The predicted fluxes in the soft X-ray bands, though small, are at the sort of intensities that would make observations of this source seem very worthwhile. If successful, this would represent the first SNR outside the galaxy to be studied in X-rays.

TABLE 1 N49 Predicted Parameters

Flux in [Fe XIV]	6.5×10^{-15} erg cm^{-2} sec^{-1}
Volume of 'hot' plasma ($T = 2 \times 10^6$ K)	6×10^{57} cm^3
Emission measure (assuming $N_e = 4$ cm^{-3})	10^{59} cm^{-3}

Energy Band (keV)	Predicted X-ray flux (photons cm^{-2} sec^{-1})	
	$N_H = 0$	$N_H = 10^{21}$ cm^{-2}
0.28 - 0.50	$\sim 3 \times 10^{-3}$	$\sim 6 \times 10^{-4}$
0.50 - 1.55	$\sim 7 \times 10^{-4}$	$\sim 4 \times 10^{-4}$
1.55 - 8.3	$< 10^{-5}$	$< 10^{-5}$

REFERENCES

1. J. C. Zarnecki *et al.*, *Ap. J.* 219, L17 (1978).

2. J. C. Zarnecki, Ph.D. Thesis, University of London, 1977.

3. R. C. Catura and L. W. Acton, *Ap. J.* 207, L163 (1976)

4. J. C. Zarnecki and J. L. Culhane, *M.N.R.A.S.* 178, 57P (1977).

5. A. G. Sgro, *Ap. J.* 197, 621 (1975).

6. L. Spitzer, *Diffuse Matter in Space*, Interscience, New York, 1968.

7. P. Murdin, D. H. Clark and J. L. Culhane, *M.N.R.A.S.* in press.

8. W. H. Tucker and M. Koren, *Ap. J.* 168, 283 (1971).

HEAO A-1 SEARCH FOR X-RAY PULSATIONS FROM THE PULSAR PSR 0833-45 IN VELA

D. J. Yentis*, K. Wood*, D. Sadeh**, W. Evans*, W. N. Johnson*, J. F. Meekins*, S. Shulman*, D. McNutt*, E. Byram*, T. A. Chubb* and H. Friedman*

*Naval Research Laboratory, Washington, D.C., U.S.A.
**University of Tel-Aviv, Israel

In the ten years since their discovery, some 100 radio pulsars have been identified and studied. The fastest of these is the 33-millisecond period pulsar NP0532 in the Crab Nebula. Pulsations from the Crab have now been observed over 15 decades in energy: from the radio, infrared and optical, through X- and γ-ray [1]. The Vela pulsar, PSR 0833-45, is the third fastest radio pulsar with a period of about 89 milliseconds. Pulsed emission from this object is now well established in γ-rays [2,3]. Recently, optical pulsations have been reported [4] with a magnitude of $\sim$24 [5]. In fact, Vela is the only other radio pulsar, besides the Crab, to have been observed beyond radio wavelengths. However, unlike the Crab, the existence of X-ray pulsations at the radio period have not been established. Published results reporting positive three-sigma detections [6,7] are, unless the X-ray pulsations are time dependent, inconsistent with other results establishing lower upper limits [8,9, 10].

HEAO Observations

On November 27 and December 1, 1977, the HEAO A-1 experiment was pointed at the position of the Vela radio pulsar for approximately three hours. The mode of operation in both pointed observations was to accumulate 5-millisecond integrations in a single fixed energy bandwidth. The first observation used two of the four "scan" detector modules (designated modules 3 and 4) configured to integrate events between 2 and 15 keV. With a total collecting area of 3200 cm^2 and a $1^{\circ} \times 4^{\circ}$ (FWHM) field of view, approximately 1.77 million events were recorded in 6100 seconds. The second observation, eight days later, used the single wide-field "timing" detector module (module 7) configured to integrate events in the 10 to 20 keV energy range. In this observation, with 1800 cm^2 of collecting area and an $8^{\circ} \times 2^{\circ}$ (FWHM) field of view, approximately 905,000 events were recorded in 5100 seconds.

The Vela pulsar is located in a complex part of the X-ray sky defined by two extended regions of soft (<2 keV) X-ray emission associated with the two supernova remnants, Vela X ($\sim5^{\circ}$) and Puppis A ($\sim0\overset{\circ}{.}5$). The pulsar is located about $1\overset{\circ}{.}5$ from the center of Vela X and about 2° from Puppis A. Uhuru has reported [11, 12] a 9.1 ufu non-pulsed "dc" source less than $1\overset{\circ}{.}2$ in extent and coincident with PSR 0833-45. The Uhuru source dominates the Vela X nebula above 2 keV with an energy power-law spectral index of 1.1 from 2-10 keV. Ariel-V has recently reported [13] a 2° extended source centered on the pulsar.

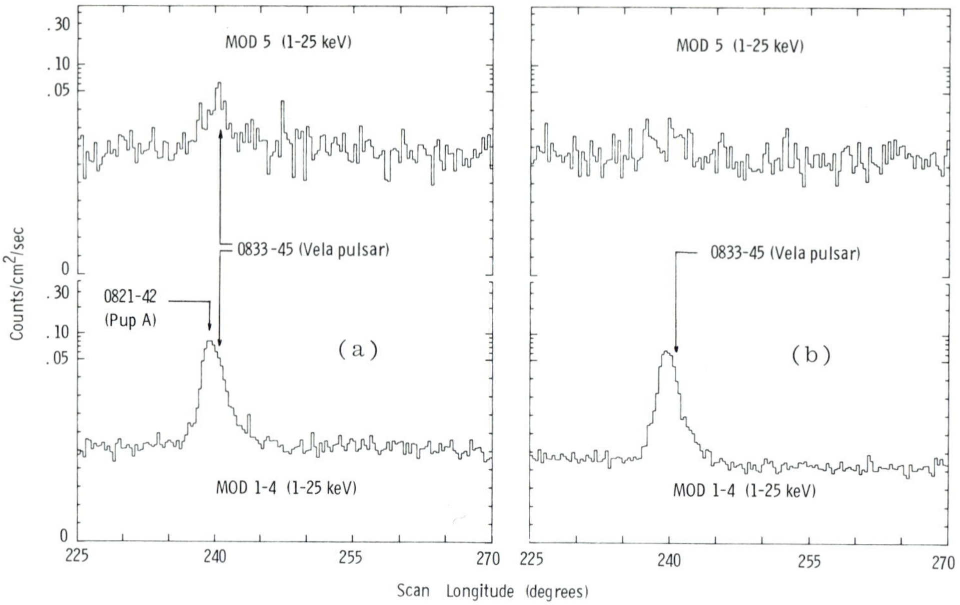

Fig. 1 X-ray scans through the Vela-Puppis region on (a) 25 November 1977 and (b) 26 November 1977. The integration step is 0°.3 of scan longitude or about 1.5 seconds. (Note: above 0.050 counts/cm^2/sec, the scale becomes logarithmic)

Figure 1 shows two HEAO A-1 scans through the Vela-Puppis region which occurred one and two days prior to the first pointed observation. In the earlier of these scans, the coordinates of PSR 0833-45 lie within (but not in the center of) the 1° x $1/2^{o}$ (FWHM) field of view of one of the two narrow-field "mapping" detectors (module 5). The look direction of this detector precedes that of the "scan" modules by $1/3^{o}$ so that it will observe a discrete source approximately one satellite spin-axis step or 12 hours before a scan module. We note in Fig. 1a that the peak intensity in module 5 is at about one-half the level (no collimator correction has been applied) we would expect in our detector from the Uhuru "dc" source. Also, the X-ray flux appears to be extended by more than 1° in the scan direction of module 5 which is consistent with the Ariel-V observation. In Fig. 1b, the Vela "pulsar" has disappeared from module 5 and is becoming more central in the field of view of the scan modules. However, there remains a large contribution from Puppis A to the total observed X-ray flux.

Pulsar Analysis

Dr. George Downs of the Jet Propulsion Laboratory provided NRL with a radio-period model for Vela to cover the time of our observation [14]. A search for pulsations at the period predicted by the model, but corrected to the geocenter, was performed using light-curve epoch folding. Figure 2 is a plot of χ^2 versus period for data obtained in the December 1 observation. The maximum in the distribution is within the search resolution (10 nanoseconds) of the radio period. With 18 5-millisecond bins in the light curve, the probability for the peak χ^2 (26.2) to occur at random is 5%. This result, while intriguing, is for now a two-sigma upper limit. The observation of November 27 was similarly analyzed. No significant peak was found near the radio period. This is believed to be due to a significantly reduced signal-to-noise in these data because of increased particle background and increased X-ray contribution from Puppis A in the lower-energy data channels used for this measurement.

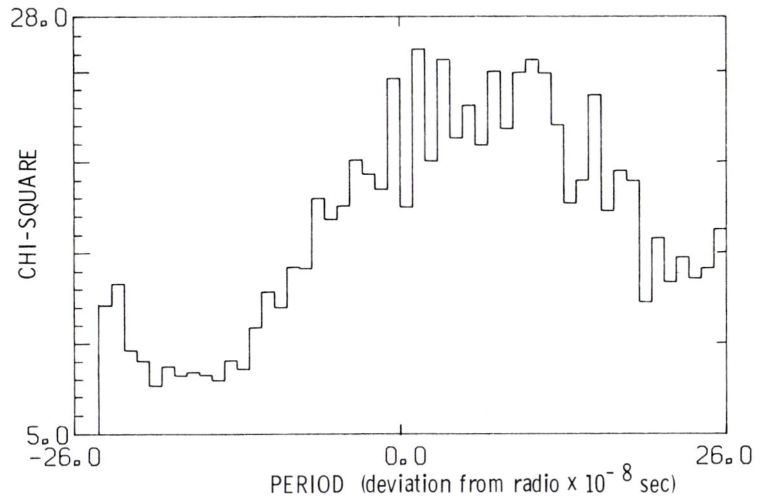

Fig. 2 χ^2-distribution versus pulse period (see text).

Discussion

Figure 3 is a broadband plot of the spectral flux intensity of the Vela pulsar and nebula. The HEAO upper limit is better than an order of magnitude below the OSO-8 upper limit [10]. If we assume a $\lesssim$5-millisecond wide radio-like pulse profile [15], the HEAO upper limit corresponds to a time-averaged intensity of $\lesssim 2 \times 10^{-12}$ ergs/cm^2/sec in the range 10-20 keV. If we assume a 50% pulse duty cycle, which is inclusive of optical [4] and γ-ray [2,3] pulse profiles, our upper limit intensity is $\lesssim 7 \times 10^{-12}$ ergs/cm^2/sec. For a distance to the source of 500 pc, the corresponding source luminosities are $\lesssim 7 \times 10^{-31}$ ergs/sec and $< 2 \times 10^{-31}$ ergs/sec, respectively. The upper limit on the pulsed fraction is 8 - 24% of the Uhuru source intensity.

The upper-limit pulsed flux is a factor of $2\text{-}7 \times 10^{-4}$ of the Crab nebula or 0.001 - 0.004 of the Crab pulsed flux (assuming a 20% pulsed fraction) in the 10-20 keV energy range. For comparison, the Vela-to-Crab pulsed-flux ratio is approximately 0.01 in the radio, 0.0015 in the optical, and $\gtrsim 1$ in the γ-ray. Interpolation between the Vela pulsar optical and γ-ray fluxes leads to an estimated X-ray pulsed flux about an order of magnitude below the present upper limit. This is possibly an overly cautious estimate: the same procedure applied to the Crab pulsar would underestimate the X-ray flux by an order of magnitude.

We are continuing to observe the Vela pulsar in an attempt to detect X-ray pulsations with longer duration high-sensitivity measurements. These data will be presented in future papers.

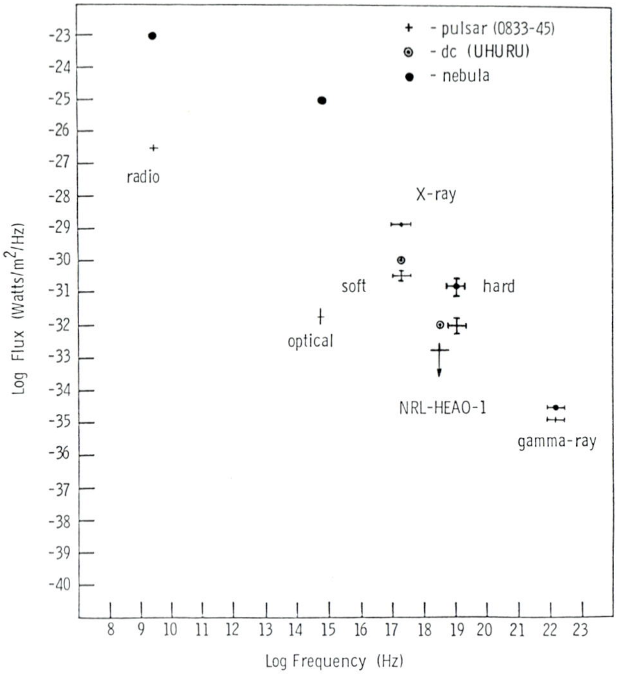

Fig. 3 Vela pulsar and nebula spectral flux density: radio [16, 17], optical [4,5,17], soft X-ray [7], present experiment, hard X-ray [6], and γ-ray [2].

REFERENCES

1. A. Hewish, in: *A. Rev. Astron. Astrophys.*, Annual Reviews Inc., California, 1970, pp. 265-296.
2. D.J. Thompson *et al.*, *Astrophys. J.* 214, L17 (1977).
3. K. Bennett *et al.*, *Astr. Astrophys.* 61, 279, (1977).
4. P.T. Wallace *et al.*, *Nature* 266, 692 (1977).
5. I.J. Danziger, private communication (1978).
6. F.R. Harnden Jr. *et al.*, *Astrophys J.* 172, L91 (1972).
7. F.R. Harnden Jr. *et al.*, *Nature* 241, 107 (1973).
8. W.E. Moore *et al.*, *Astrophys. J.* 189, L117, (1974).
9. S. Rappaport *et al.*, *Nature* 251, 471, (1974).
10. S.H. Pravdo *et al.*, *Astrophys. J.* 208, 471 (1974).
11. E. Kellog *et al.*, *Astrophys. J.* 183, 935 (1973).
12. R. Giacconi *et al.*, *Astrophys J.* Supp. Ser. No. 237 27, 37 (1974).
13. A. Smith, *Mon. Not. R. Astr. Soc.* 182, 39P (1978).
14. G. Downs, private communication (1977).
15. M.M. Komesaroff *et al.*, *Nature* 252, 210 (1974).
16. G.S. Downs *et al.*, *Nature* 222, 1257 (1969).
17. A.N. Bunner, in: *The Gum Nebula and Related Problems*, ed. S.P. Maran, J.C. Brandt, T.P. Stecher, Goddard Space Flight Center Report No. X-683-71-375, 1971, pp. 169-179.

Extragalactic Compact and Extended Sources

OBSERVATIONAL PROPERTIES OF EXTRAGALACTIC X-RAY SOURCES

K. A. Pounds

X-ray Astronomy Group, Physics Department, University of Leicester LE1 7RH, England

INTRODUCTION

At the last IAU Symposium on X-ray Astronomy, held in 1972 in Madrid, data from the Uhuru satellite survey was beginning to make a major impact. Later that year the 2U catalogue was published [1] containing 125 x-ray sources, some 14 of which had extragalactic identifications which appeared reasonable on positional grounds. In 1974 the expanded 3U catalogue came out [2], with 161 sources and ~20 extragalactic counterparts. This number comprised 5 (galactic-type) sources in the Magellanic Clouds, a small number of outstanding active galaxies (3C 273, NGC 4151, NGC 5128, M87) and 11 associations with rich clusters of galaxies. The establishment of powerful, spatially extended, x-ray emission from rich clusters was the outstanding result of extragalactic x-ray astronomy at that time.

Over the subsequent four year period, other satellite (and rocket-borne) experiments have been flown and the available data have increased substantially. The importance of x-radiation in the study of active galaxies has been established, while the first significant spectral information and crude x-ray maps of nearby clusters have appeared. The purpose of the present review is to briefly summarise the current status of these observations and prepare some of the ground for the subsequent discussion [3] by Professor Rees on the nature of the extragalactic x-ray sources.

THE UHURU AND ARIEL V SKY SURVEYS

The great majority of new extragalactic sources (post-1974) have resulted from a refinement of 3U catalogue positions (with data from Copernicus, Ariel V, SAS-3, ANS, and by a revision and addition of the Uhuru data base, resulting in the recent 4U catalogue [4]), and from the new and independent Ariel V (2A) catalogue [5].

The Ariel V Sky Survey Instrument (SSI) is described elsewhere [6,5] and consists essentially of a scanning, fan-beam proportional counter system, quite similar to that used on Uhuru. Both surveys were sensitive over the energy band 2-20 keV and both reached a limiting source detection of ~1.5 Uhuru counts sec^{-1}(~0.5 Ariel count sec^{-1}). Two important differences exist between the 3U and 2A catalogues, however, which are especially relevant to the present review. First, the 2A catalogue is based on a more uniform sky coverage, whilst it also has a smaller range of (and particularly many fewer large) error boxes. These differences appear to be largely responsible for the higher yield of source identifications in the 2A catalogue. A comparison of the 2A and 4U catalogues, the latter including new data from the period 1972-74 (after the spacecraft star sensors had failed, but with aspect recovered from a solution of theon-board magnetometers) that 4U includes many of the new 2A sources, and others not detected b . A greater spread of positional uncertainty remains, however, in 4U. Figure 1 compares the distribution of error box areas (the area of the rectangle circumscribing the 90% confidence contour of source position) in the 2A and 4U catalogues, for the section of sky mainly relevant to the present review, viz. at galactic latitude $|b|>10^o$. Sources identified with galactic objects or the Local Group are excluded, to give a clearer comparison of the large scale extragalactic field. Also shown for the two catalogues are the corresponding numbers of sources now identified with an extragalactic object. Clearly, most of the unidentified 4U sources remain so because of their large (>1 deg^2) error boxes. A further comparison shows a mean error box of 0.15 deg^2, with 8 of 77 sources of area >0.5 deg^2, compared with a mean of 0.7 deg^2 in 4U, with 105 of 187 sources of >0.5 deg^2.

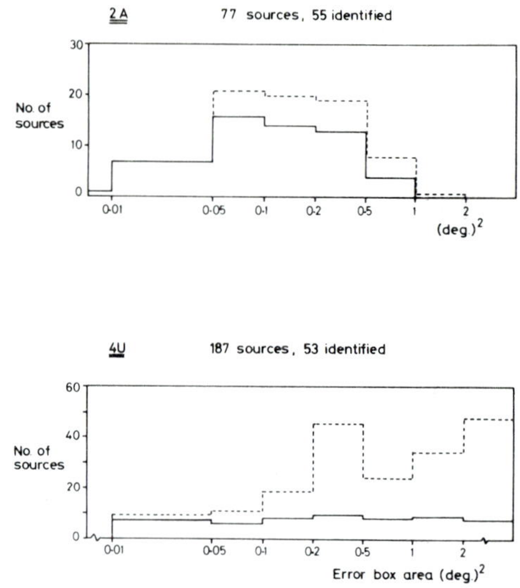

Fig. 1

X-RAY EMISSION FROM CLUSTERS OF GALAXIES

Identification of X-ray Clusters

Table 1 lists 41 rich clutters of galaxies which are (in the author's view) reliably identified with 2A or 4U sources. They are all of Abell distance class 4, or the equivalent for the southern clusters, with the exception of A1146 (D=5, R=4, BM I, 2A error box 0.37 deg^2), A2204 (D=5, R=3, BM II, revised 2A error box 0.6 deg^2), and SC 0316-44 (D=5, R=1, BM I, 2A error box 0.2 deg^2). The richness and centrally condensed nature of these three exceptions provides strong support for their inclusion in the light of the subsequent discussion of the properties of x-ray clusters The random occurrence of a cluster of D≤4 in any x-ray error box in the 2A catalogue (total area for 107 sources ~15 deg^2) is only 0.3, indicating few, if any, chance coincidences. A second requirement for a source to be listed in Table 1 is that the 2A x-ray flux should be compatible with a non-varying source. This criterion has ruled out two rich southern clusters, CA 0340-538 (D=3, R=2), and Sersic 40/6 (D=3, R=3), both of which appear good on positional grounds alone. Possibly a field source or an active galaxy in each of these clusters is contributing a substantial fraction of the recorded x-ray flux.

The distribution of clusters in Table 1 shows 16 clusters now identified with 3U sources, including several confirmed by later, improved positions, and 17 x-ray clusters first reported in the 2A catalogue. More recent additions to the list have come from the 4U catalogue [7] and from a search of a wide list of clusters [8] using the whole Ariel V data base and a point-summation technique (PST), in which some additional sensitivity is achieved by summing all data regardless of scan position angle about a chosen point in the sky. In total 41 x-ray clusters are listed 38 of which are in common with the list (also based on 2A and 4U), recently published by Jones and Forman [7].

Cluster Morphology

Several authors have discussed the correlation of detected x-ray emission with the morphology of a cluster of galaxies. The most recent and careful analyses have been by Bahcall [9,10] and McHardy [11] using 2A catalogue sources, and by Jones and Forman [7] using 2A and 4U identifications. Bahcall and McHardy both find a correlation of x-ray luminosity with cluster richness, enhanced if a dominant central galaxy is present, and with a scarcity of spirals in the cluster. McHardy finds a positive correlation of x-ray emission with richness and with Bautz-Morgan type, while Jones and Forman believe cluster richness is the dominating factor. Bahcall argues persuasively that the 'central galaxy density' is a better measure than cluster richness, being more closely related to the core region of a cluster, where most of the x-radiation arises.

Figure 2, after Bahcall [9], shows the x-ray luminosities L_x, of the x-ray clusters listed in Table 1 for which the redshifts are known, plotted against the central galaxy density $\overline{N}_0$ within 0.5 Mpc of the cluster centre. For greater consistency the x-ray luminosities are are all derived from 2A catalogue fluxes. Flux values in the 4U catalogue are averages of positive detections only and thus overstate fluxes, particularly of the fainter sources. The correlation of L_x with $\overline{N}_0$ is good over the range of two orders in luminosity and the scatter small, particularly since all Bautz-Morgan (and Rood-Sastry) types of cluster have been plotted together. Conversely the small scatter demonstrates the major dependence of L_x is on $\overline{N}_0$. This is, of course, entirely consistent with the now favoured model for cluster x-ray emission, as thermal bremsstrahlung from a hot intracluster gas whose density is proportional to the density of galaxies.

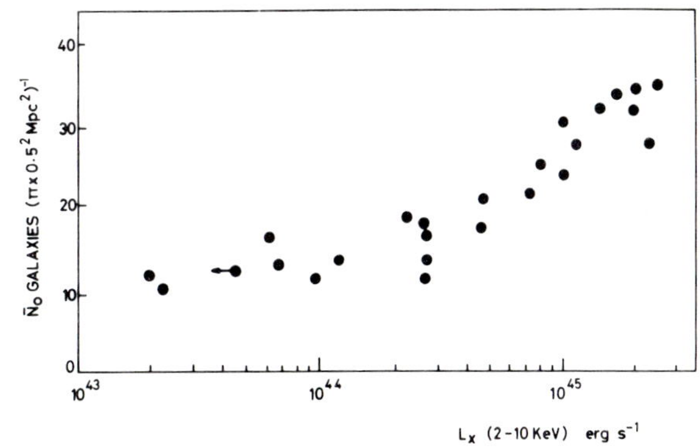

Fig. 2

Figure 3 shows a second impressive correlation, of cluster x-ray luminosity with the fraction of galaxies F(Sp) which are spirals. Again the values of L_x are from 2A fluxes, while measurements of F(Sp) are taken from Bahcall [10] McHardy [11] and - for several rich southern clusters - from Tytler and Vidal [12]. The latter authors pointed out the exceptionally high F(Sp) of SC 1252-28, for its reported x-ray intensity, and it is interesting to note that the corresponding 2A source has now been re-identified [13] with the dwarf nova EX Hydrae. The clear relation between spiral fraction and x-ray luminosity is interesting in

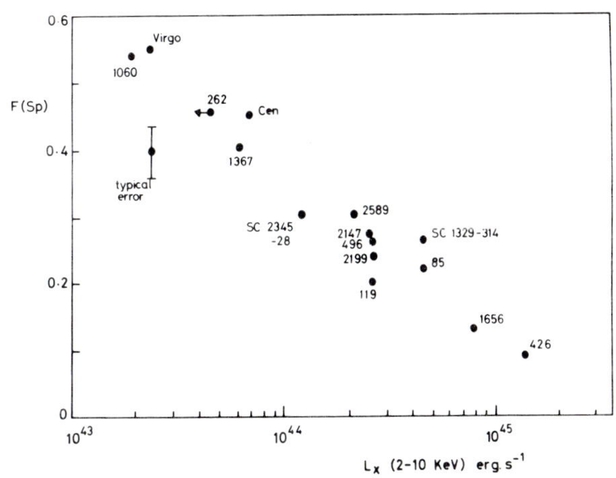

Fig. 3

that it is perhaps the clearest evidence yet that the evolution of galaxies is affected by their membership of a rich cluster and, also, is again fully consistent with the thermal bremsstrahlung model for the x-radiation. In a later paper in this volume, Dr Culhane uses data on the luminosity and spectral temperature, kT, of a sample of x-ray clusters to express the ram pressure experienced by galaxies moving through the postulated, dense intracluster medium and finds F(Sp) to be inversely related to this quantity, implying effective stripping of the interstellar gas in spirals as they move through the inner regions of a rich cluster.

The X-ray Luminosity Function of Clusters

Using a statistically complete sample of x-ray clusters (Abell + Virgo), McHardy [11] has calculated the x-ray luminosity function of clusters, by the standard maximum volume method and the known 2A catalogue sky coverage correction. The resulting luminosity function is shown in Fig. 4. The upper limit at the high luminosity end of the function corresponds to the fact that no sources of power greater than $10^{45.75}$ ergs s^{-1} are seen with a flux greater than 1 Ariel count s^{-1}; the survey is complete to this limit over 90% of the sky. This is therefore a very strong upper limit and essentially truncates the luminosity function. It can be seen that the total density of x-ray clusters approaches that of all Abell clusters, ρ_A, at a luminosity of $\sim 10^{43}$ ergs s^{-1}. The slight excess above ρ_A is caused by the fact that the maximum accessible volume of space in which sources of low luminosity can be seen is small compared with the inverse of the space density of such sources. It is therefore predicted that all clusters will emit x-rays at a luminosity greater than 10^{43} ergs s^{-1} in the 2-10 keV range, so that no distinction should be made between x-ray and non x-ray clusters in future. The only allowable distinction is between powerful and weak x-ray sources.

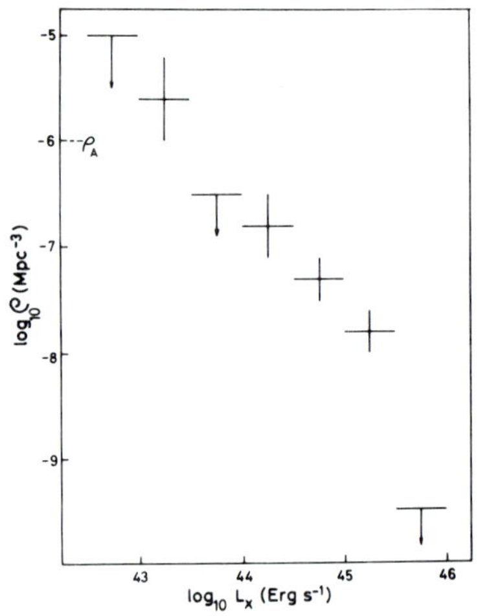

Fig. 4

The contribution of clusters to the x-ray diffuse background, calculated from the luminosity function of Fig. 4 (ie neglecting evolutionary effects) is ~7% at 2-10 keV. [41]

X-ray Spectra of Clusters

Remarkably, considering the low spectral resolving power of current proportional counter detectors, the improved statistical precision provided by the non-scanning collimated detectors on Ariel V and OSO-8, has yielded x-ray spectra of considerable significance in the study of x-ray clusters. The breakthrough came with the MSSL Ariel V measurement of the Perseus cluster in 1975, showing clear evidence for a monoenergetic feature near 7 keV [14]. The interpretation of this feature (Fig. 5) as thermal line emission from highly stripped iron nuclei (Fe XXIV-VI) provided, at once, the strongest evidence for hot plasma (T $\sim 10^7$ - 10^8 K) in the cluster, together with a unique determination of the abundance of iron in the intracluster medium. Rapidly, other x-ray clusters revealed a similar feature and a recent summary of the GSFC OSO-8 data [15] on 20 x-ray clusters showed 8 to have positive iron-line emission (though one of these is now probably EX Hydrae - and dwarf novae also appear to show iron lines!) with the others having upper limits all consistent with an iron abundance ~0.5 of the solar value.

In addition to the detection of iron lines, the MSSL and GSFC spectra provide a measure of the mean temperature of intracluster gas. In the majority of cases a thermal bremsstrahlung fit gives best agreement with the measured 2-20 keV spectra. Values of kT range from ~2-12 keV, with a tendency for higher values for clusters of greater x-ray luminosity, [15, 16].

The gross properties of x-ray clusters are summarised in Table 2, where it is noted that the various correlations of L_X and kT with cluster morphology, etc, are consistent with the view that the x-ray emission arises from hot gas held in the gravitational potential of the constituent galaxies (and gas).

Fig. 5

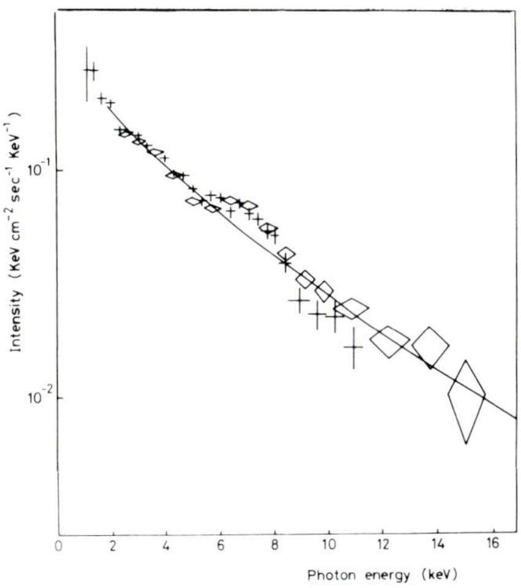

TABLE 1 X-ray Sources Identified with Clusters of Galaxies

3U catalogue identifications	(11)	Virgo, Coma, Perseus, Centaurus, Cyg A 3C 129, A 262, 401, 1367, 2256, 2319
3U identifications confirmed by 2A position	(5)	A 478, 754, 1060$^{(a)}$, 2147, 2199
Kellogg list confirmed by 2A position	(1)	A 496
New identifications in 2A catalogue (b)	(17)	A 85, 119, 133, 644, 1146, 1795, 1904, 2029, 2142, 2312, 2440, 2589, SC 0316-44 0626-54, 1344-32, 2009-57, KE 44
Post 2A catalogue identifications		
(i) SSI observations (c)	(3)	A 999, 1644, 2204
(ii) 4U catalogue (d)	(4)	A 539, 2065, 2657, 2666
Total =	41	X-ray clusters identified with SSI and Uhuru sources.

Notes

(a) First confirmed by Ariel V CPC expt. (Ives and Sanford) [48]
(b) Exclude SC 1251-29 and SC 1326-31, now re-identified with EX Hya Watson [13] and MCG-6-30-15 (SAS group, private comm.)
(c) McHardy [11] and Ricketts [8]
(d) Forman et al [4]

TABLE 2 Summary of X-ray Properties of Rich Clusters

Luminosity function shows (at 2-10 keV)
- all clusters have $L_X \geq 10^{43}$ and $< 10^{45.75}$ erg s^{-1}
- contribution of ~7% to x-ray background

Thermal spectra with KT ~3-12 keV
- plus iron line → Fe/H ~0.5 solar

Various correlations
- Lx v richness or central galaxy density
- Lx v BM type (or RS cD etc)
- Lx v scarcity of spirals
- kT v velocity dispersion
- kT v central galaxy density
- ram pressure v scarcity of spirals

support model in which x-rays arise from hot gas held in gravitational well of cluster.

As yet the data do not allow firm statements on the source of the gas or of the mechanism of heating, although the presence of large amounts of iron argues that a substantial fraction of the emitting gas is not primordial. It will be important to discover whether the iron line emission is widely dispersed through the cluster, is concentrated in higher density regions in the cluster core, or lies nearby individual galaxies [17].

Structure of Nearby Rich Clusters

Uhuru found evidence for spatial extension,[1] typically of 30-50 arc minutes, in 6 x-ray clusters (Virgo, Perseus, Coma, Centaurus, A262 and A2256) [18]. This result needs confirmation for the last two clusters, since A262 is very faint (the Ariel V sky survey 3σ upper limit is only 0.4 of the reported 3U intensity) whilst the linear size of x-ray emission from A2256, at D=3, is ~4Mpc. However, for the nearby rich clusters in Virgo, Perseus and Coma, the extension is in no doubt and recently crude x-ray maps have been obtained. Data on these clusters is now briefly reviewed.

[1]Being twice the 'core radius' of the best-fit isothermal gas sphere.

Perseus. Observations with the small field, non-imaging telescope on Copernicus first resolved a separate component of x-radiation from the massive central galaxy NGC1275, [19] together with the core of the cluster emission. The Berkeley and Columbia groups made several scans across Perseus with narrow fan-beam detectors and confirmed discrete emission from NGC1275 (25 ± 10% of the total Perseus flux at 0.5-3 keV) together with an extended source lying along the well-known line of galaxies, from NGC1275 to IC 310 [20]. The most recent and complete map to date is reproduced in Fig. 6. This is from a 120 sec exposure with a Kirkpatrick-Baez imaging telescope, flown in December 1975 on an Aerobee 350 rocket by the CFA group [21] The energy range is 0.5-1.5 keV and the pixels of 1.4 x 1.4 arc minutes (although the telescope resolution was ~4 arc min). A bright, unresolved source accounting for 25% of the total flux is seen to be coincident with NGC1265 (within the rocket aspect uncertainty of 3 arc min), superimposed on a diffuse emission extending over the whole field (1° in diameter). No discrete emission from NGC1265 is seen and no evidence is found for the enhanced emission along the line of galaxies. The most straightforward interpretation of Fig. 6 is that the Perseus cluster emission is symmetric about the cluster centre with a measured 'core radius' of the hot gas ~17 arc min. From observations with the SAS-3 instruments, Ulmer and Jernigan [22] find a similar value at 1.5-5.5 keV, increasing to 40^{+6}_{-16} arc min at 5.5-15.5 keV.

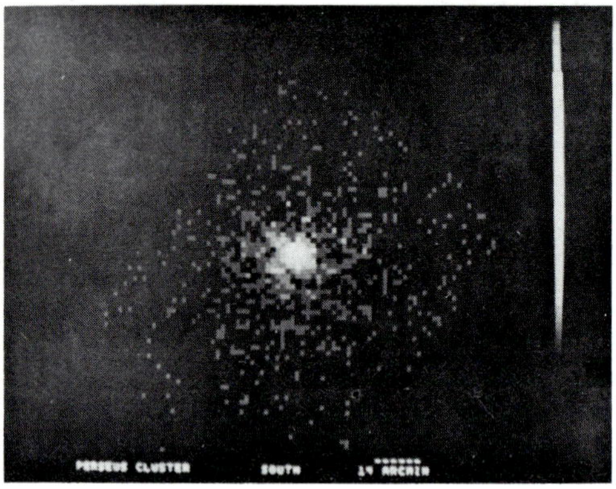

Fig. 6

Virgo. It was never clear from the Uhuru data how much, if any, of the extended x-ray emission was associated with the cluster rather than with the giant elliptical galaxy M87. This question has recently been clarified by Lawrence [23] using Ariel V SSI data. He finds evidence for two components at 2-20 keV, one being centred on M87 and the other ~$0\overset{o}{.}5$ to the NW, close to the cluster centre as defined by the distribution of elliptical galaxies [24]. Figure 7 shows the resulting SSI map of the Virgo region, with a source ($\lesssim 1^{o}$ in diameter) centred on M87 and an extended source associated with the cluster as a whole. Independent support for this conclusion is provided by spectral studies of Virgo with the Ariel V CPC. Davison [25] finds a best fit spectrum with two temperature components, of 1 keV and 7 keV, with the harder emission arising from a region to the NW of M87.

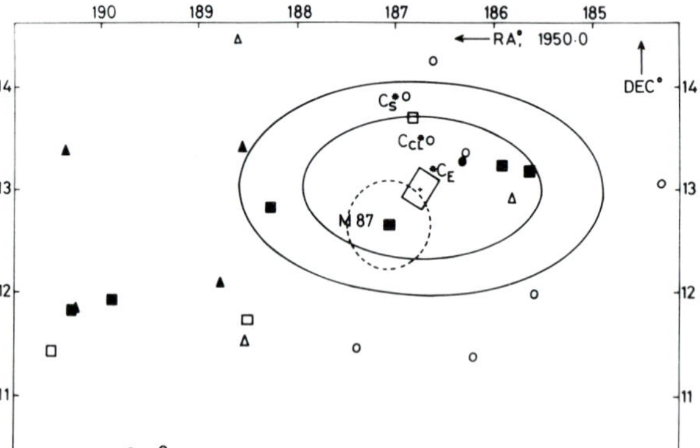

Fig. 7

An x-ray image of Virgo, taken with the CFA Kirkpatrick-Baez telescope, shows [26] an extended, diffuse emission at 0.15-1.5 keV, centred on M87 and with a 'core radius' of ~15 arc min. Thus it seems that the Virgo region contains a hard, extended source associated with the cluster as a whole, together with a softer source, also extended, centred on M 87.

Coma. This cluster, the archetypal evolved cluster, differs from Perseus and Virgo in having no active central galaxy. Copernicus observations [27] indicated a 'core radius' at 0.5-1.15 keV of 40 arc min, slightly larger than the Uhuru 2-6 keV value. A recent map obtained with the CFA rocket-borne telescope shows [28] a featureless emission, well-fitted by an isothermal gas sphere model with no evident emission from the prominent galaxies NGC 4874 and NGC 4869.

SEYFERT GALAXIES

X-ray Seyferts as a Class

Prior to the appearance of the 2A catalogue, data on the x-ray emission from active galaxies was scanty. This is now no longer so, with the number of identified sources of this type comparable to that of clusters. The most notable increase in a particular astronomical class is for Type 1 Seyfert galaxies. Table 3 summarises the present position, with 25 Seyferts now reliably identified with 2A, Uhuru or SAS-3 x-ray sources. As seen from the table, the majority of the Seyfert identifications are based on 2A catalogue sources and the establishment of this second main class of extragalactic x-ray emitter is one of the most important achievements of the Ariel V survey. Now, as with clusters, there are sufficient identified Seyferts to establish their x-ray luminosity function and study the relation of x-ray emission with other Seyfert properties. Spectral data on the x-ray Seyferts is still patchy, but NGC 4151 is notable in having the hardest spectrum, now detected beyond 1 MeV [29], of any cosmic x-ray source. This bright x-ray Seyfert also shows a substantial low energy cut-off, [30] and evidence of time variability on a time-scale of several days, [30,31] with one report of flare lasting only 730 secs. [32].

The x-ray luminosities of the Seyferts in Table 3 extend from $10^{42.1}$ erg s^{-1} (NGC 3227) to $10^{44.8}$ ergs s^{-1} (3C 390.3), a range closely comparable with that of clusters. Figure 8 shows the distribution of L_X for detected Seyferts against red shift z, and it is interesting to note the range of luminosities covered by the Seyferts dxtends more than two thirds of the way from NGC 3227 to the quasar 3C 273. A similar demonstration of H_β and $L\alpha$ intensities has been given by Weedman, [33] who argues that this supports both the contention that QSOs are (simply) distant Seyferts, at $z>0.1$, and a cosmological interpretation of the red shifts. Optically the first distinguishing property of a Seyfert galaxy is its extremely bright nucleus, followed by the breadth of its emission lines. It now seems clear that a third feature of Seyfert galaxy is their powerful x-ray emission and it would be surprising if it were found that this was not associated with the Seyfert nucleus. The observed variability for several Seyferts supports this view.

Elvis et al [34] have made a detailed comparison of 15 x-ray Seyferts with their optical, infra-red and radio emission. They find the x-ray luminosity to be correlated with the infra-red power (at 3.5μ and 10μ), the optical continuum power and with the luminosity in the $H\alpha$ emission line, but not significantly with the radio power or the luminosity in the forbidden optical emission lines. Elvis et al suggest that these correlations strongly support the view that the x-ray emission originates in the dense inner nuclear region, of extent <0.1 pc, which is peculiar to Type I Seyferts. In passing, it seems clear that the high success rate of identifying x-ray Seyferts in the 2A catalogue is due, in part, to the x-ray to optical continuum correlation, in that the detected x-ray Seyferts are also among the brightest known.

Of the 25 x-ray Seyferts listed in Table 3, it is interesting that 4 were found to be Type I Seyfert galaxies as a result of an optical search of 2A source error boxes. The first new Seyfert to be discovered in this way, MCG 8-11-11, previously listed as a barred spiral galaxy, was found by Ward et al [35] to be by far the brightest extragalactic object in the 0.093 sq. deg. error box of 2A 0551+466, and a Type I Seyfert. A continuing search of the error boxes of unidentified 2A sources by the same group, at the Anglo-Australian Observatory, the Royal Greenwich Observatory and Leicester, has now led to the discovery [36] of 3 further Type I Seyferts, viz ESO 113-IG45, MCG-2-58-22 and ESO 141-G55. Notably, ESO 113-IG45 is optically the most luminous (and the brightest) in the Southern Hemisphere, comparable with a low luminosity QSO, while MCG-2-58-22 has Balmer lines broader than any other known Seyfert, except 3C 382. The identification of ESO 113-IG45 with 2A 0120-591 was also proposed, independently, by Ricker [37] on the basis of positional coincidence and the Seyfert-like properties of this galaxy reported by Fairall [38].

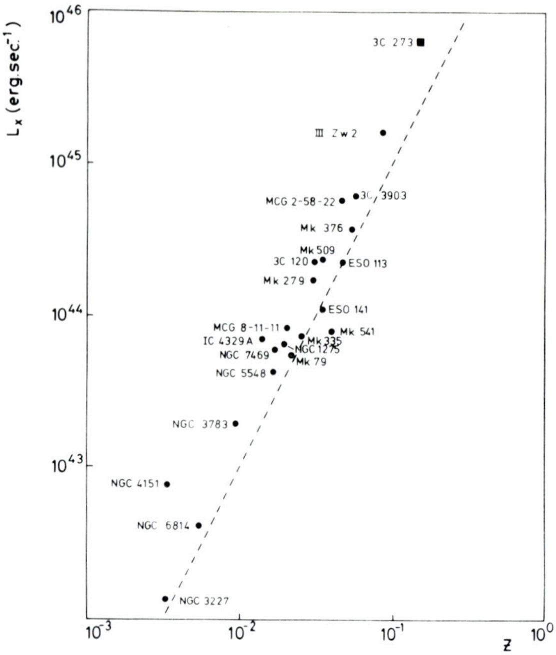

Fig. 8

Recently, several additional x-ray Seyferts have been identified, 2 with refined source positions in the 4U catalogue [4], 2 from SAS-3 observations, 1 from a PST search of Uhuru data [32], and 1 further 2A source identification (2A 1326-310 ≡ MCG-6-30-15) on the basis of a substantially improved error box obtained by SAS-3. An improved SAS-3 x-ray position has also confirmed the 2A identification of IC 4329 A, distinguishing this Seyfert from the nearby cluster of galaxies, bringing the current total of identified x-ray Seyferts (Table 3) to 25.

It is interesting to note that if the correlation between optical continuum and x-ray power extends to much fainter objects, future x-ray surveys may provide the most certain method of detecting new active galaxies, particularly distant QSOs, since optical searches are notably difficult at $M > 16$ or 17 mag, while only a fraction of all QSOs are also strong radio sources.

Variability in Seyfert X-radiation

Establishing variability in the intensity of faint x-ray sources is difficult, particularly when such evidence is based on a comparison of data from different instruments. Nevertheless, there is now no doubt that some x-ray Seyferts do vary on quite short time-scales and it can be anticipated that over the next few years observations of greater precision will show this to be a feature common to the class. Charles et al [39] first showed an apparent factor of four variation for 3C 390.3 between Uhuru (1971) and Copernicus (1973) measurements. In the same way Ives et al [30] found evidence for a factor 2.5 increase in the 2-6 keV emission of NGC 4151 from 1971 to 1974. More recently, Elvis [31] reported in Ariel V SSI observations of NGC 4151 a flare with a factor 1.7 ± 0.2 increase in less than 3 days during August 1975. Also from the SSI data, Ward et al [34] have found MCG 8-11-11 to vary by at least a factor 3 on a time scale of 30 days or less. A comparison of Uhuru and Ariel V SSI data on MK 279 [32] further indicates variability on a time scale $\lesssim 3$ years.

Most dramatic, is the recent report from Tananbaum et al [32] of flares in the x-radiation from NGC 4151 at an implied rate of 1 or 2 per day, with one event showing an order-of-magnitude change on a time possibly as short as 730 seconds. If confirmed, such violent events would have substantial implications for models of the emitting region while, in any case, the now established variability on time scales of days clearly implies that the x-rays arise within the innermost region of the nuclei of Seyfert galaxies. Related evidence for this view is included in the summary of x-ray Seyfert properties (Table 4).

TABLE 3 Seyfert Galaxies Identified with X-ray Sources

3U catalogue identifications	(3)	NGC 4151, 1275$^{(a)}$, 3C 390.3$^{(b)}$
2A catalogue identifications	(10)	NGC 3783, 5548, 6814, 7469 IC 4329A MK 79, 279, 376, 509 MCG 8-11-11$^{(c)}$
2A source identifications - post catalogue	(4)	MCG 2-58-22$^{(c)}$ ESO 113-IG45$^{(c)}$ ESO 141-G55$^{(c)}$ MCG -6-30-15$^{(d)}$
SAS-3 sources	(2)	III Zw 2, 3C 120
4U sources	(2)	MK 335, 541
Ariel V SSI PST search	(1)	NGC 3227$^{(e)}$
Uhuru PST search	(3)	AKn 120, VV 144, MK 304$^{(f)}$
Total	= 25	x-ray Seyferts

Notes

(a) Resolved from Perseus cluster in Fabian et al [19]
(b) Confirmed by Copernicus position [39]
(c) Discovered after x-ray detection [35] [36] [37]
(d) Identification from improved SAS-3 position
(e) Elvis et al [34]
(f) Tananbaum et al [32]

TABLE 4 Summary of Properties of X-ray Seyferts

Lx (2-10 keV) covers range $10^{42.5}$ - 10^{45} erg s^{-1}

Variability on time-scale ~1 month common, NGC 4151 varies over days and possibly ~12 min.

Correlation of Lx with
- optical continuum power
- 2.2μ, 3.5μ and 10μ power
- luminosity of Hα emission line
- FWZI of Hβ

and not with
- radio power
- forbidden opt. em. lines

support the view that the x-radiation arises in the dense core of a Type 1 Seyfert galaxy of extent <0.1 pc.

Luminosity function yields (non-evolving) contribution to background of ~6% at 2-10 keV.

OTHER ACTIVE GALAXIES

The 3U catalogue [2] lists a handful of other active galaxies identified as powerful x-ray emitters. These included the nearby radio galaxy Cen A, the bright quasar 3C 273, and (although with a very large error box) the irregular galaxy M 82. Again, subsequent observations have considerably extended this list. Table 5 summarises the present non-Seyfert identifications in this active galaxy category.

TABLE 5 Active Galaxies Identified with X-ray Sources

Radio galaxies (?)	(2)	Cen A (NGC 5128) M 87$^{(a)}$
QSOs	(3)	3C 273 MR 2251-178 ≡ 2A 2251-179$^{(b)}$ QSO 0241+622$^{(c)}$
BL Lac-type	(3)	MK 421 MK 501 I Zw 186$^{(d)}$
High excitation emission line galaxies	(7)	NGC 5506$^{(e)}$ NGC 7582$^{(f)}$ NGC 1365$^{(g)}$ M 82 MCG -5-33-16$^{(h)}$ NGC 2110$^{(i)}$
Total	= 15	

Notes

(a) Separation of M87 from Virgo cluster (Lawrence [23]).
(b) Identification based on improved SAS-3 position [42]
(c) Apparao et al [43].
(d) Shulman [44].
(e) Ricker et al [45].
(f) Ward et al [36].
(g) Elvis M. Private communication.
(h) Schnopper et al [46].
(i) Bradt et al [47].

Radio Galaxies

Although several active galaxies combine powerful x-ray and radio emission (eg. NGC 1275, 3C 390.3, 3C 273) the majority of extragalactic x-ray sources are, at most, weak radio emitters. In fact, only for one object, Cen A, is a physical relation between x-radiation and radio emission established to date. Although Cen A is tentatively listed in Table 5 under the sub-head 'radio galaxies', the major x-ray emission arises from the nucleus of the central, giant elliptical galaxy NGC 5128 [50]. X-ray observations spanning almost a decade - a more complete 'light curve' than at any other wavelength - have shown the ~2-10 keV emission to vary by an order-of-magnitude, with factor-of-two changes within a few days. A compilation of data up to 1977 is given by Lawrence et al [51], and additional OSO-8 observations are reported by Mushotsky et al [52]. The latter authors also report radio flare changes in phase with the x-ray variability and argue for a dominant x-ray production by Compton scattering of the radio photons by a single relativistic electron population. X-radiation has recently been detected also from the large radio lobes of Cen A [53]. Again, the emission is likely to be by the so-called inverse Compton process, though in these extended regions the scattered photons are more likely to be of the microwave background. The derived magnetic field, of 0.7μG, is close to the equipartition value and indicates Cen A is merely the first crude example of an x-ray map of extended radio sources that will be obtained over the next few years.

X-radiation from QSOs

Until recently 3C 273 was distinguished as the nearest and brightest QSO and also the only certain QSO identified as an x-ray source. A PST search of 65 radio-bright QSOs using the Ariel V sky survey data produced only one marginal additional detection [54]. Upper limits to the x-ray luminosity ranged from ~0.15 to ~50 times that of 3C 273.

Now, a refined SAS-3 position of the source 2A 2251-179 has led to its unambiguous identification with a bright (B~16) quasi-stellar object [42]. Given the preliminary designation MR 2251-178, this object - at a red shift of 0.068 - has an x-ray luminosity varying (at 2-10 keV) from 5×10^{44} to 1.6×10^{45} ergs s^{-1} and which exceeds the optical luminosity by a factor of ~10. Very recently, a third x-ray QSO has been reported, again from the optical search of a small SAS-3 RMC error box. As yet un-named, the quasi-stellar object identified with the x-ray source 2S 0241+622 (=4U 0241+61) is, at a red shift of 0.0438, the nearest known QSO and represents a further example of the increasing number of active galaxies whose outstanding properties have been recognised only after the discovery of an adjacent x-ray source.

Although the number of identified x-ray QSOs is still insufficient to allow a useful study of the relation of the x-ray emission to other properties - or to provide a useful limit to the likely evolutionary effects - the increased sensitivity of the HEAO-1 sky survey may do so, and certainly HEAO-B will.

BL Lacertids

The first positive detection of x-radiation from an object of this class came with the identification of the bright BL Lac-type object, MK 421, with the x-ray source 2A 1101+38. The identification was first proposed by Ricketts et al [40], on the basis of positional coincidence, albeit with an x-ray error box 0.15 deg^2, and highly variable x-ray emission. Subsequent Ariel V observations have shown the x-ray source to persist at a generally faint level (~0.8 Ariel count s^{-1}) with a further flaring in late December 1977. No simultaneous optical observations have yet been obtained, but the rapidly improving location (Fig. 9) has substantially strengthened the identification with MK 421. Very recently the revised Uhuru (4U) catalogue has been issued [4] and contains a source, 4U 1651+39, identified with a second BL Lac-type object, MK 501. It is interesting to note that MK 421 and 501 are in the brightest (in V) 3 or 4 in the 32 BL Lacertae objects listed by Stein et al [49]. Early analysis of HEAO-A1 data has produced a third candidate, I Zw 186 [44]. The ill-understood nature of BL Lac-type objects, with their compact appearance, featureless blue continuum and high optical variability, possibly lying somewhere between Type I Seyfert galaxies and QSOs, makes the establishment of powerful x-ray emission especially interesting.

High Excitation, Narrow Emission Line Galaxies

Also of interest is the rapidly increasing number of x-ray identifications with active galaxies having high excitation forbidden line spectra, but without the broad permitted lines characteristic of Type I Seyferts. This number, now 7 if M 82 is included, suggests strongly that powerful x-ray emission is also a feature of these objects. Perhaps the prototype of this kind of galaxy is NGC 5506 [55]. This galaxy has long been a candidate identification for the x-ray source 3U 1410-03 (=2A 1410-029) [56], and this has now been confirmed by SAS-3 RMC observations [45] which produced a 1.1 arcmin radius error circle (90% confidence) centred only ~12 arcsec north of the galaxy nucleus (Fig. 10).

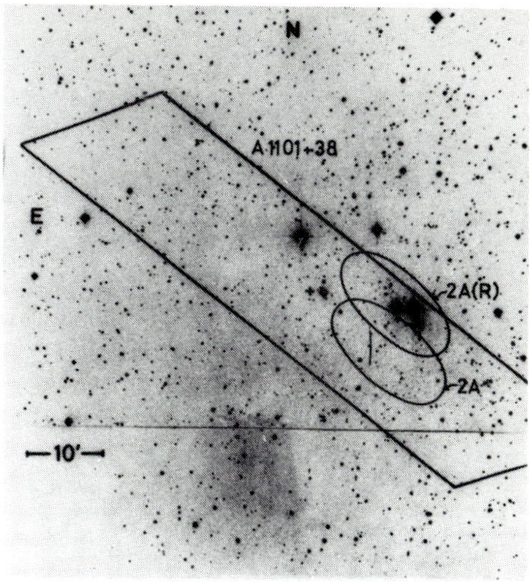

Fig. 9

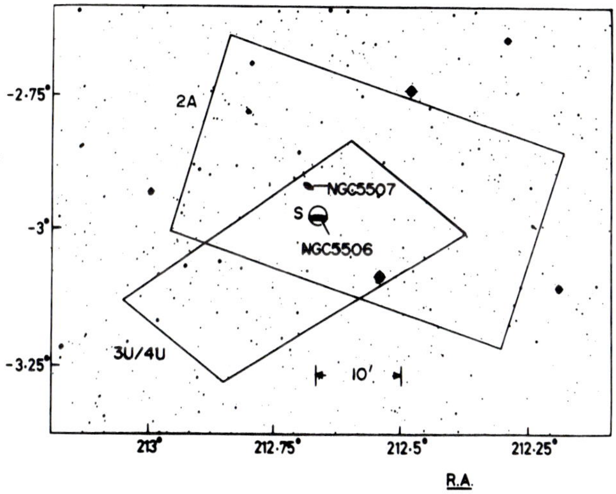

Fig. 10

The optical search of 2A error boxes by Ward et al [36] produced two other emission line galaxies, NGC 7582 and NGC 2992 and gave the first indication of a new class of x-ray emitter being established. A resulting PST search of the Ariel V sky survey data in the vicinity of 25 other bright emission line galaxies revealed a further positive detection, NGC 1365. The nearby galaxy M 82 is morphologically very similar to NGC 5506 [55] and shows emission lines in places, although its nucleus is obscured by dust. The SSI error box of 2A 0904+700 includes M82 and makes it a reasonable additional candidate in this class.

Finally, two recent additions have been reported from SAS-3 observations. These are MCG-5-23-16 [46] and NGC 2110 [47], both discussed elsewhere in this volume. Together they bring the number of emission line galaxies identified as x-ray emitters to 7, if M 82 is included. Several also appear to be variable and the major importance of the x-ray observations may well be in providing the first evidence that these objects, too, possess a compact 'quasar-like' region deep in their nucleus.

The x-ray luminosity of the emission line galaxies discovered to date lies in the range 10^{42} - 10^{43} erg s^{-1} except for that of M 82 which is only 5 x 10^{40} ergs s^{-1}. They thus occur at the lower end of the Seyfert galaxy range. Warwick [41] has shown how the inclusion of the emission line galaxies and QSOs, respectively, at the low and high luminosity ends, extends the x-ray luminosity function of Type I Seyferts to give a total, 'active galaxy' luminosity function, covering the range $L_X \sim 10^{42} - 10^{46}$ erg s^{-1}. Nevertheless, he finds the contribution from all known active galaxies to the (2-10 keV) diffuse x-ray background to still not exceed 10-15%. Thus, the sum of both major groups of extragalactic source, clusters and active galaxies, account for $<30\%$ of the background. Of course, as several authors have pointed out, a modest cosmological evolution of x-ray Seyferts, an up-turn in the low luminosity end of the active galaxy luminosity function (eg. many more, weaker emission line galaxies)or the discovery of a new class of luminous extra-galactic source, could easily account for the remainder. Thus, there is still no strong case for a substantial diffuse component of the x-ray background. The forthcoming HEAO-B mission could well decide.

REFERENCES

1. R. Giacconi et al. (1972). Astrophys. J. 178, 281.

2. R. Giacconi et al. (1974). Astrophys. J. (Supp) 27, 37.

3. M. J. Rees. This volume.

4. W. Forman et al. (1978). Astrophys. J. (Supp)(in press).

5. B.A. Cooke et al. (1978). M.N.R.A.S. 182, 455.

6. G. Villa et al. (1976). M.N.R.A.S. 176, 609.

7. C. Jones and W. Forman (1978). Astrophys. J. 224, 1.

8. M.J. Ricketts, (1978). M.N.R.A.S. 183, 51P.

9. N. A. Bahcall, (1977). Astrophys. J. 217, L77.

10. N. A. Bahcall, (1977). Astrophys. J. 218, L93.

11. I. McHardy, (1978). M.N.R.A.S. 184, 783.

12. D. Tytler and N.V. Vidal, (1978). M.N.R.A.S. 182, 33P.

13. M.G. Watson et al. (1978). M.N.R.A.S. 184, 79P.

14. R.J. Mitchell et al. (1976). M.N.R.A.S. 176, 29P.

15. R.F. Mushotzky et al. (1978). Astrophys. J. 225, 21.

16. R.J. Mitchell et al. (1977). M.N.R.A.S. 181, 25P.

17. A.C. Fabian and J.E. Pringle, (1977). M.N.R.A.S. 181, 5P.

18. E. Kellogg and S. Murray, (1974). Astrophys. J. 193, 2.

19. A.C. Fabian et al. (1974). Astrophys. J. 189, L59.

20. W. Cash et al. (1976). Astrophys. J. 209, L111.

21. P. Gorenstein et al. (1978). Astrophys. J. 224, 718.

22. M.P. Ulmer and J.G. Jernigan, (1978). Astrophys. J. 222, L85.

23. A.X. Lawrence, (1978). M.N.R.A.S. (in press).

24. G. de Vaucouleurs, (1961). Astrophys. J. (Supp). 6, 213.

25. P. Davison, (1978). M.N.R.A.S. 183, 39P.

26. P. Gorenstein et al. (1977). Astrophys. J. 216, L95.

27. R. E. Griffith et al. (1974). Nature, 250, 471.

28. P. Gorenstein. This volume.

29. V. Schönfelder, (1978). Nature 274, 344.

30. J.C. Ives et al. (1976). Astrophys. J. 207, L159.

31. M.S. Elvis (1976). M.N.R.A.S. 177, 7P.

32. H. Tanabaum et al. (1978). Astrophys. J. 223, 74.

33. D. W. Weedman, (1976). Astrophys. J. 208, 30.

34. M.S. Elvis et al. (1978). M.N.R.A.S. 183, 129.

35. M.J. Ward et al. (1977). Astron. and Astrophys. 56, 219.

36. M.J. Ward et al. (1978). Astrophys. J. 223, 788.

37. G.R. Ricker (1978). Nature 271, 334.

38. A.P. Fairall (1977). M.N.R.A.S. 180, 391.

39. P.A. Charles et al. (1975). M.N.R.A.S. 170, 17P.

40. M.J. Ricketts et al. (1976). Nature 259, 546.

41. R.S. Warwick (1979). Proc. R. Soc. (to be published).

42. G. R. Ricker et al. (1978). Nature 271, 35.

43. K.M.V. Apparao et al. (1978). Nature 273, 450.

44. S. Shulman. This volume.

45. G. R. Ricker et al. (1978). Astrophys. J. (in press).

46. H.W. Schnopper et al. (1978). CFA preprint.

47. H.V. Bradt et al. This volume.

48. J.C. Ives and P.W. Sanford (1976). M.N.R.A.S. 176, 13P.

49. W.A. Stein et al. (1976). Ann. Rev. Astron. Astrophys. 14, 173.

50. J.P Delvaille et al. (1978). Astrophys. J. 219, L81.

51. A. Lawrence et al. (1977). M.N.R.A.S. 181, 43P.

52. R.F. Mushotzky *et al*. (1978). *Astrophys. J*. *220*, 790.

53. B.A. Cooke *et al*. (1978). *M.N.R.A.S*. *182*, 661.

54. G.J. White and M.J. Ricketts (1977). *M.N.R.A.S*. *181*, 435.

55. A.S. Wilson *et al*. (1976). *M.N.R.A.S*. *177*, 673.

56. J.N. Bahcall *et al*. (1975). *Astrophys. J*. *199*, L9.

THE X-RAY SPECTRA AND VARIABILITY OF ACTIVE GALAXIES

J. P. Stark

Mullard Space Science Laboratory, University College London, Holmbury St. Mary, Dorking, Surrey, U.K.

INTRODUCTION

The results of the Ariel V Sky Survey (SSI) [1] have indicated that extragalactic X-ray astronomy may, broadly speaking be divided into two categories: X-ray emission from clusters of galaxies and X-ray emission from the nuclei of active galaxies and in particular from Seyfert galaxies.

The spectral data obtained by Ariel V Experiment C, on six Seyfert galaxies are summarised. These data are compared and contrasted with observations of NGC 5128 (a radio galaxy), obtained during three periods, spanning 4 years.

SEYFERT GALAXIES

A total of approximately 20 Seyfert galaxies have now been identified as X-ray sources. Experiment C on Ariel V [2] has observed a total of six of these, namely NGC 4151, 3C 390.3, NGC 5548, Mkn 279, Mkn 506 and Mkn 509. The identification of most of these with Seyfert galaxies appears fairly certain [3] and may therefore be taken, with some caution, as a typical group within the Seyfert class of X-ray emitters. The spectra for all these are shown in Fig. 1; the spectral parameters are summarised in Table 1.

Table 1: Spectral Parameters of Seyferts

Source	α	N_H	L_X(2-10 keV)
NGC 4151	1.53	$3\text{-}18.10^{22}$	$2\text{-}12.10^{42}$
NGC 5548	1.9	10^{22}	4.10^{43}
Mkn 509	1.7	10^{22}	2.10^{44}
Mkn 506	1.3	10^{23}	2.10^{44}
Mkn 279	1.2	2.10^{22}	1.10^{44}
3C390.3	1.6	10^{22}	$2\text{-}9.10^{44}$

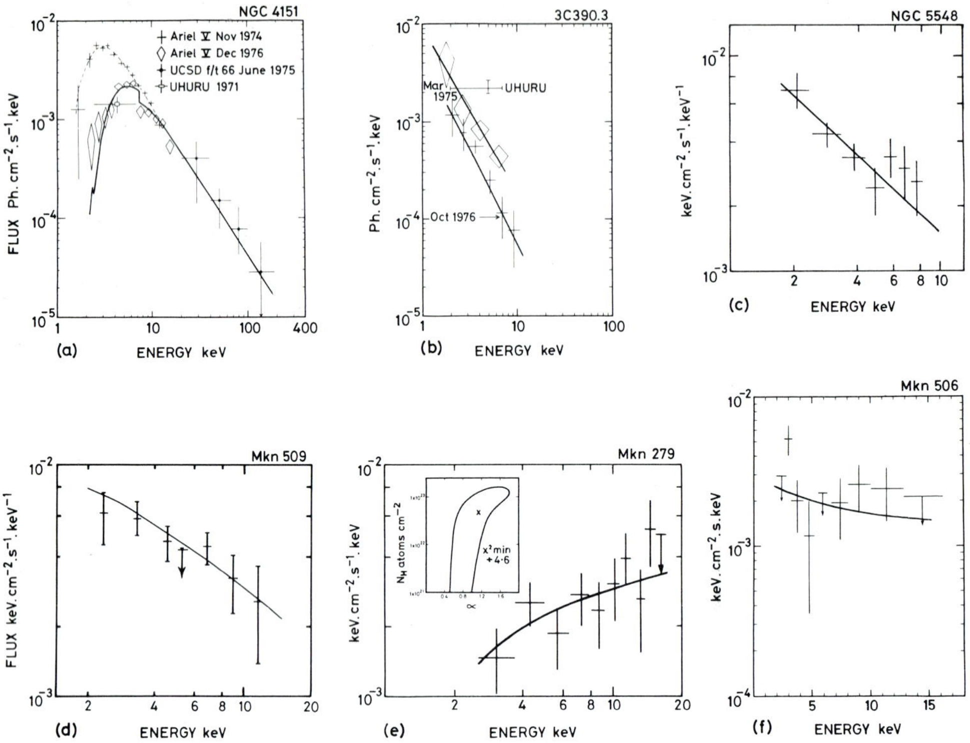

Fig 1: The X-ray spectra of six seyfert galaxies

For a detailed discussion of the spectra the reader may refer to [3, 4, 5]; however, it is apparent from a cursory glance that there is a wide range of spectral parameters noted within this class. These data provide the first catalogue of X-ray spectra for Seyfert galaxies. Clearly for the majority of these sources the flux at the Earth is very small, and thus the range of acceptable spectral parameters is considerable. However, it is of interest to note if there are any spectral trends within this group of sources. This might indicate possible associations of emission from these galaxies in different wavebands, and thus provide further information concerning the emission region. The work of Elvis et al. [6] has already shown that there is a strong correlation between X-ray emission and Seyfert type, namely that X-ray emission occurs in the optically brightest Type 1 Seyferts. The only exception to this is NGC 1275, whose classification is in some doubt.

In Fig. 2 the best fitting spectral index observed in the 2 - 10 keV band is plotted as a function of the U-B colours for the six sources. The error bars are 1σ errors noted from the fitting procedure. From this there does indeed appear to be a trend in the sense that the lower the ultra-violet excess, the harder the X-ray spectrum becomes. The linear correlation coefficient for this data is 0.9. Considerable caution should be exercised however as the optical data were taken at different epochs from the X-ray data and variation in the value of U-B has been observed in some Seyferts.

NGC 5128

The first detailed examination of the spectrum of this source [7] showed that when the source was at its peak intensity in January 1975 the best fitting power law was 1.79 ± 0.02; the column density at this time was $1.35 \pm 0.02 \times 10^{23}$ H atoms cm^{-2}.

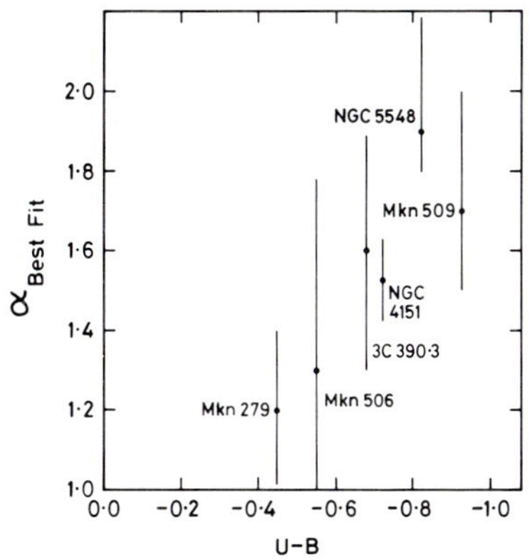

Fig 2: Best fitting spectral index: U - B.

The spectral data obtained for this period, together with those obtained in December 1975 and January 1976 is plotted in Fig. 3. Although the intensity for the source had decreased by 40% of its original value for these two later observations the spectral parameters for the source had not varied significantly.

A re-analysis of all these data, together with new data obtained in March 1978 has recently been performed, adopting the heavy element absorption cross section of Fireman [8]. The spectral results for the three sightings of this source are displayed in Fig. 4. In this the χ^2_{min} + 4.6 contours are plotted showing the values for combinations of power law index and hydrogen column density which are acceptable for these data. This indicates that although it is not necessary to postulate variation in spectral index, it is required that between January 1975 and March 1978 a change in the amount of material in the line of sight occurred. The 2 - 10 keV flux for these periods are: 1.14×10^{-9}, 6.8×10^{-10} and 4.9×10^{-10} erg cm^{-2}/s for the January 1975, December 1975/ January 1976 and March 1978 respectively, thus the latest observations indicate the source to have decreased by 60% of its January 1975 level. One of the simplest interpretations for these observations is that gross changes in the intensity of an underlying source occur; these changes are not correlated with spectral variations but are indicative of fluctuations in the activity of the source. However, superimposed on this variability is an additional change in the extinction of the low energy X-ray flux, possibly caused by fast filaments of gas passing through the line of sight to the source.

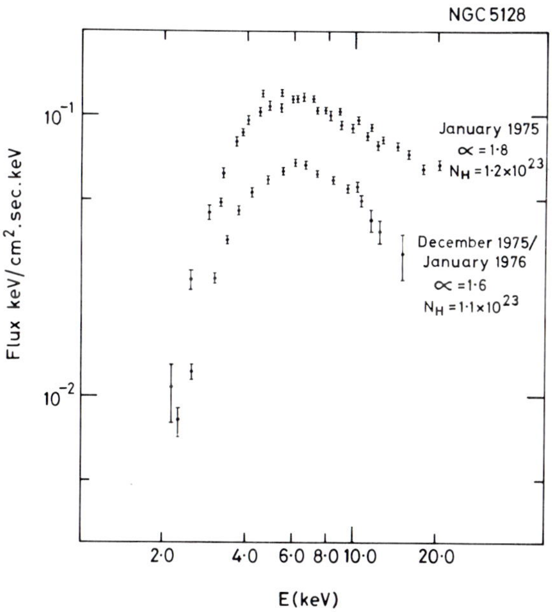

Fig 3: Spectra of NGC 5128 in January 1975 and December 1975/January 1976.

DISCUSSION

The scenario depicted above for NGC5128 is broadly similar to the X-ray emitting region of NGC 4151 [5], the most extensively studied Seyfert galaxy. For both these sources a decrease in X-ray luminosity has been correlated with an increase in hydrogen column density, implying the existence of fast filamentary motions. The recent observations of NGC 4151 by OSO-8 [9] have also shown that variability in the X-ray flux may occur which is not correlated with any change in spectral

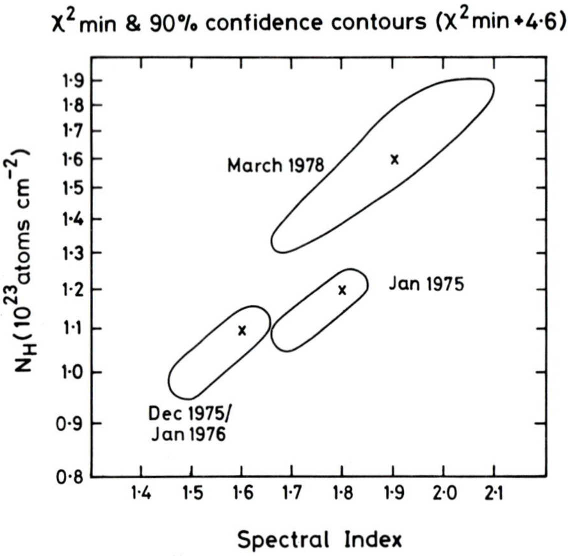

Fig 4: χ^2 grid of spectral index: Hydrogen-column density.

parameters, thus similar to the situation which had already been noted in NGC 5128. It is apparent that single component model satisfies these observations in both sources; indeed, the failure to observe a change in spectral index correlated with intensity variability strongly suggest a single emission region.

REFERENCES

1. Cooke, B.A., Ricketts, M.J., Maccacaro, T., Pye, J.P., Elvis, M., Watson, M.G., Griffiths, R.E., Pounds, K.A., McHardy, I., Maccagni, D., Seward, F.D., Page, C.G., Turner, M.J.L., 1978. M.R.A.S., 182, 489.

2. Ives, J.C. and Sanford, P.W., 1976. M.N.R.A.S., 176, 13.

3. Stark, J.P., 1978. Proc. Roy. Soc., in press.

4. Barr, P., White, N.E., Sanford, P.W. and Ives, J.C., 1977. M.N.R.A.S., 181, 43.

5. Stark, J.P., Bell-Burnell, J., Culhane, J.L., 1978. M.N.R.A.S., 182, 23.

6. Elvis, M., Maccacaro, T., Wilson, A.S., Ward, M.J., Penston, M.V., Fosbury, R.A.E. and Perola, G.C., 1978. M.N.R.A.S., 183, 129.

7. Stark, J.P., Bell-Burnell, J., Culhane, J.L., 1978. M.N.R.A.S., 174, 35.

8. Fireman, E.L., Astrophys. J., 187, 7.

9. Mushotzky, Holt and Serlemitsos, 1978, Preprint.

OBSERVATION OF 2A1415+255 (NGC 5548) BY OSO-7

R. E. Rothschild*, W. A. Baity*, R. F. Mushotzky**, W. A. Wheaton*** and L. E. Peterson*

*University of California, San Diego, La Jolla, CA 92093
**NASA Goddard Space Flight Center, Greenbelt, MD 20771
**Massachusetts Institute of Technology, Cambridge, MA 02139

INTRODUCTION

The active galaxy NGC 5548 is one of the bright, blue, nearby Seyferts that also exhibit strong optical emission line spectra. Mushotzky [1] predicted that such objects and this one in particular, would produce radio, optical, and X-radiation by means of a synchrotron self-Compton process. The existence of a flat X-ray spectrum could indicate that the underlying emission mechanism is of such a non-thermal origin. Stark, Burnell, and Culhane [2] reported observations of NGC 5548 in the 2-8 keV range that indicated a flat X-ray spectrum. In addition they noted a possible factor of two variability with respect to UHURU observations. Elvis et al. [3] suggest, too, that X-radiation from Seyfert galaxies is strongly variable.

We report here the results of the UCSD OSO-7 observations of NGC 5548 over the range 5-350 keV. The UCSD X-ray telescope aboard OSO-7 scanned the sky for 600 days from its launch in September 1971 until the tape recorder failed in May of 1973. The actively shielded NaI (Tl) detector was sensitive to X-rays from 4->600 keV, had 64 cm^2 effective area, and a 6.5° (FWHM) field of view [4]. The detector scanned a great circle on the sky every two seconds, and pulse height plus time of arrival data formed the telemetered information.

In 1975 the UCSD OSO-7 Skymap program was begun to combine the 20 months of scanning data into a map of the sky with one degree solid angle bins. A background correction was performed on an orbit-by-orbit basis with only those bins free from known source contributions forming the background region. At present the UCSD/Skymap data base contains 7×10^5 seconds of live time and covers over 90% of the sky. This presents 40% of the usable scanning data.

Hale Bradt [5] has reported that the RMC on SAS-3 confirms the indentification of 2A1415+255 with NGC 5548. The HEAO A2 scanning data also confirm the identification.

OBSERVATION

The position of NGC 5548 was scanned intermittently for 47 days between January 3, 1972 and November 17, 1972. This resulted in a total of 860 seconds of live time on this source. The resulting X-ray counts spectrum was fit by folding analytic models through the instrument response function and performing a chi-squared comparison. The best fit power law had a number index of $1.78 \pm .25$ for the data from 5.4 - 100 keV. Above 100 keV only upper limits are seen. The inferred incident spectrum is shown in Fig. 1 along with the data points from the Ariel 5 observation (Stark, Burnell and Culhane [2]). The error bars shown were used in the spectral fitting process and represent the formal uncertainties due to statistical fluctuations in the number of source and background counts and uncertainties due to the live time correction. The present measurement of the spectral index is in agreement with the MSSL measurement of 1.9 ± 0.3, although there is a factor 10 difference in intensities. Our best fit model was used to calculate the 2-10 keV flux assuming no low energy absorption. The result is $4.2 \pm 2.1 \times 10^{-10}$ ergs/cm^2 sec and can be compared to other observations in Table 1.

An attempt to fit the data to a thermal bremsstrahlung model yielded a worse fit than for a power law and was highly insensitive to temperature as evidenced by the fact that the one sigma variation yielded values of kT from 29 to 115 keV.

DISCUSSION

Prior to this observation only two extragalactic X-ray sources, NGC 4151 and Cen A, had been seen above 30 keV. With the addition of NGC 5548 the list of three is composed entirely of compact active galaxies. This would tend to suggest that most compact extragalactic X-ray sources have flat X-ray spectra extending to 100 keV and perhaps beyond.

The observed variability implies an active region whose extent is of the order of a light year. This is small, on a galactic scale, and implies that the bulk of the emission originates at the core of the galaxy. This would support a synchrotron self-Compton model which requires relatively high electron and photon fluxes to produce observable X-ray flux. The only place such densities exist in this galaxy is in the core.

Sandage [9] has pointed out a relationship between the position of a galactic nucleus on a B-V vs U-B plot and the strength of the non-thermal component. Mushotzky [1] has used a synchrotron self-Compton model to explain the relationship between radio-optical and X-ray components of extragalactic X-ray sources. Thus, he pointed out that position on the (B-V) vs (U-B) plot might also be suggestive of the probability of X-ray emission. Of the ~ 120 Seyferts listed by Weedman [10], B-V and U-B values were given for about sixty; these are shown in Fig. 2. We feel that this demonstrates that the bluer type 1 Seyferts are brighter X-ray sources than the redder type 1 Seyferts, or the type 2 ones for that matter. Using this plot as a guide for searching for as yet undetected X-ray emitting Seyferts, we would suggest NGC 985, II Zw 136, Mkn 478, I ZW 1, and Mkn 64 as prime candidates. Number 19 in this figure is MR2251-178, the new X-ray QSO mentioned by George Ricker [11] earlier in this conference.

It's position on this plot is consistent with X-ray emitting Seyferts, and thus suggests that either it, too, is a type 1 Seyfert or that the same processes are at work in this object as are in type 1 blue Seyferts.

TABLE 1. Fluxes from NGC 5548

Date	2-10 keV flux (ergs/cm^{2}sec)	Experiment	Fraction of UCSD flux	Reference
Jan'72-Nov'72	$4.2 \pm 2.1 \times 10^{-10}$	UCSD OSO-7	1.00	Present Paper
Jan'77	$3.7 \pm 0.3 \times 10^{-11}$	MSSL Ariel-5	$0.088 \pm .044$	Stark, Burnell & Culhane [2]
Oct'74-Oct'76	$4.1 \pm 0.3 \times 10^{-11}$	SSI Ariel-5	$0.098 \pm .049$	Cooke et al. [6]
Dec'70-Mar'73	$8.5 \pm 1.0 \times 10^{-11}$	UHURU (4u)	$0.202 \pm .104$	Forman et al. [7]
	$2.2 \pm 0.8 \times 10^{-11}$	UHURU (PST)	$0.052 \pm .032$	Forman [8]

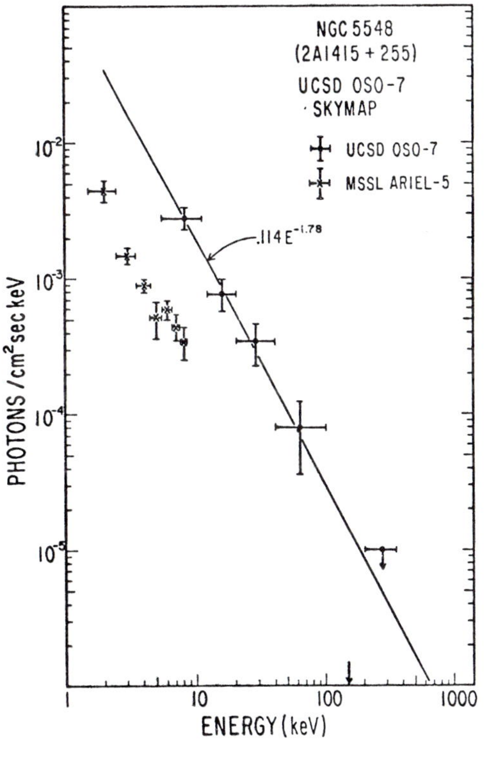

Fig. 1. Spectrum of NGC 5548 (2A1415 + 255)

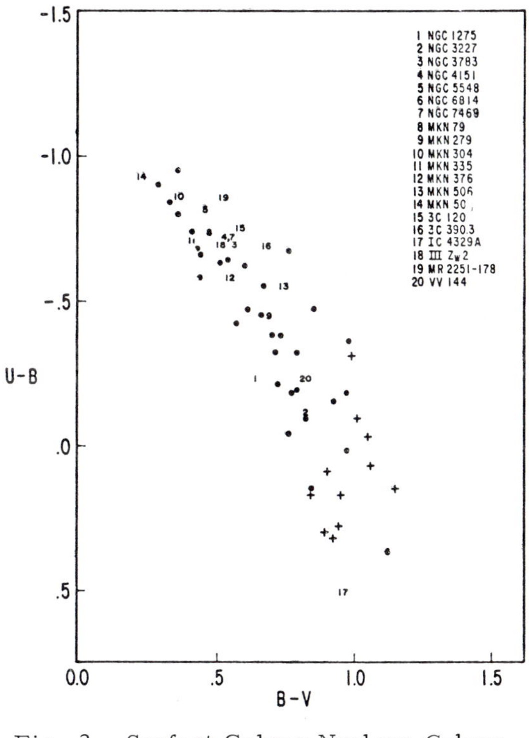

Fig. 2. Seyfert Galaxy Nuclear Colors. Sixty Seyferts from Weedman [10] plotted as ". " (type 2 without detected X-ray emission), or as a number corresponding to the X-ray source list on the right of the figure.

REFERENCES

1. R. F. Mushotzky, Nature 265, 225 (1977).

2. J. P. Stark, J. B. Burnell, and J. L. Culhane, M.N.R.A.S., 182 23 (1978)

3. M. Elvis, T. Maccacaro, A. S. Wilson, M. J. Ward, M. V. Penston, R. A. E. Fosbury and G. C. Perola, M.N.R.A.S., 183, 129 (1978)

4. L. E. Peterson, Ann. Rev. Astron. Astrophys., 13, 423 (1975).

5. H. Bradt, 1978 private communication.

6. B.A. Cooke, M.J. Ricketts, T. Maccacaro, J.P. Rye, M. Elvis, M.G. Watson, R.E. Griffiths, K.A. Pounds, I. McHardy, D. Maccagni, F.D. Seward, C.G. Page, and M.J.L. Turner. M.N.R.A.S. (1978)

7. W. Forman, C. Jones, L. Caminsky, P. Julien, S. Murray, G. Peters, H. Tananbaum, ed. R. Giacconi. Preprint, 1977.

8. W. Forman, 1977 private communication.

9. A. Sandage. Astrophys. J., 180, 687 (1973)

10. D.W. Weedman. Ann. Rev. Astron. Astrophys. 15, 69 (1977)

11. G. Ricker, 1978 private communication.

HIGH ENERGY X-RAY OBSERVATIONS OF EXTRAGALACTIC OBJECTS

W. A. Baity*, D. E. Gruber*, J. L. Matteson*, F. K. Knight*, P. L. Nolan*, A. Scheepmaker**, ***, W. A. Wheaton**, J. A. Hofman**, F. A. Primini**, W. H. G. Lewin** and L. E. Peterson*

**University of California at San Diego, La Jolla, CA, USA*
***Massachusetts Institute of Technology, Cambridge, MA, USA*
****Cosmic Ray Working Group, Huygens Lab, Leiden, The Netherlands*

INTRODUCTION

Extragalactic X-ray emitters excite the imagination because of their prodigious luminosities. Certain of them are inferred to be very compact, due to intensity variations on time scales of days, hours or even minutes. Two of these, Cen A and NGC 4151, have been observed with the hard X-ray instrument (A-4) on HEAO-1. This instrument comprises three sets of detectors: the Low Energy Detectors (LED) cover the range ~15-170 keV with two crossed fields of view of 1.6° x 20° FWHM and 206 cm^2 effective total area; the Medium (MED) and High (HED) Energy Detectors extend the overlapping energy coverage to 1 and 10 MeV respectively, with larger, circular fields of view [1]. These preliminary results pertain only to scanning data from the LEDs. Similar studies, using only 30% to 50% of the available scans from the first six months, have so far yielded only upper-limits consistent with previous results or extrapolations for M87, 3C390.3, 3C273, BL Lac, MK421, MK501 and other objects. The use of all scans, as well as pointings, will be required to overcome the low fluxes at hard X-ray energies and the small detector areas available.

CEN A, NGC 5128

The active galaxy NGC 5128 has been studied from radio to gamma-ray wavelengths, although the nucleus is dust-obscured in the visual. Figure 1 shows our data from five days scanning in January 1978, together with fits to previous work above 10 keV from OSO-7 [2] and OSO-8 [3].

Clearly the source rose in intensity from 1972 to 1975 (c.f. [4]), with spectral softening after early 1973, and has since decreased in intensity while keeping the $E^{-1.7}$ photon power law. This corresponds well with the 2-6 keV behavior as summarized by Lawrence, Pye and Elvis in 1977 (c.f. [5] and references therein).

As has been pointed out by Mushotzky et al [3], the total Cen A emission can be accounted for by a model in which the X-rays are due to a synchroton self-Compton source embedded in a cold H_2 cloud. This model indicates that a spectral break should be seen just beyond our last LED point, above 150 keV. Data from the MEDs is currently being analysed and will be reported elsewhere.

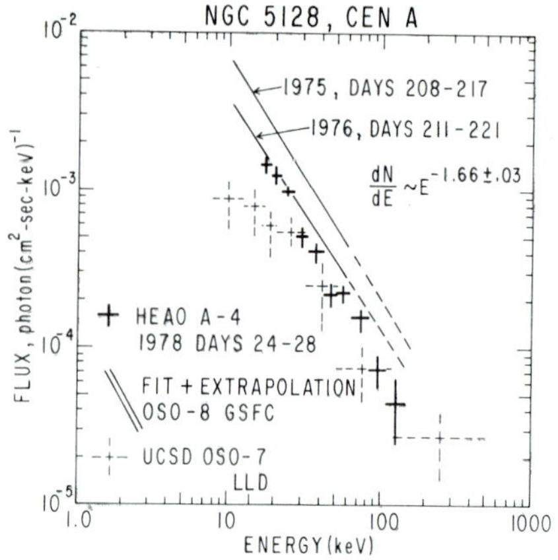

Fig. 1.

The 10-150 keV X-ray spectrum of NGC 5128. The UCSD OSO-7 results are from July, 1972 [2] and are best fit by an $E^{-1.16 \pm 0.23}$ power law.

NGC 4151

This object is frequently taken to be the archtypical Seyfert Type 1 X-ray emitter, but it is actually one of the nearest and intrinsically weakest of that class, which presently numbers some 25 objects [6]. Figure 2 is a color-color plot of the nuclei of various Seyferts indicating that NGC 4151 might not be entirely typical, a caveat to be recalled when calculating the contribution of Seyfert galaxies to the diffuse (cosmic) background.

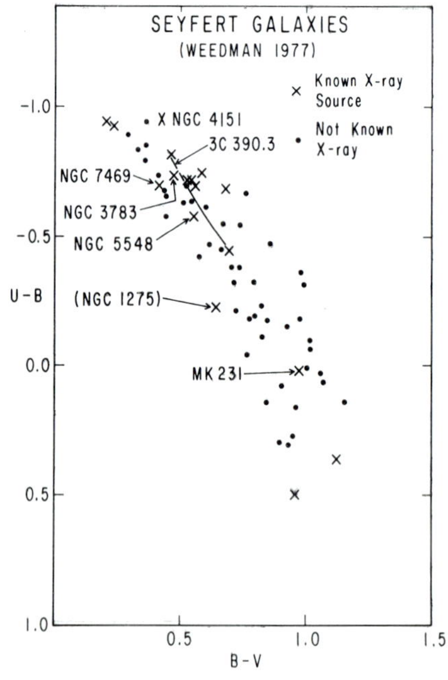

Fig. 2.

Color-color plot for the nuclei of Seyferts listed by Weedman [7], from [8] and private communication from R. Rothschild. The Seyfert character of NGC 1275 is questionable. 3C390.3 is also classified as an N galaxy.

As NGC 4151 is known to vary over time scales down to 3 days [9,10] or even 12 minutes [11], we summed data into 24-hour accumulations in the 17-39 keV range, where the signal to noise ratio is most favorable (Fig. 3). No significant daily variations were found between 1977 days 337 through 349.

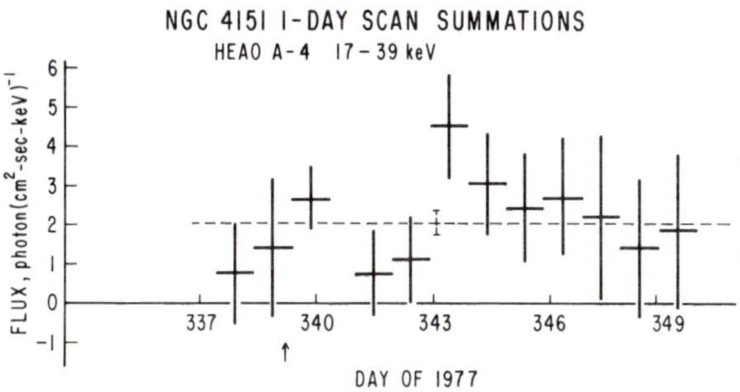

Fig. 3.

Daily variations of NGC 4151 in our most favorable energy range. $\chi^2 = 0.72$ for the fit to the average value (dashed line).

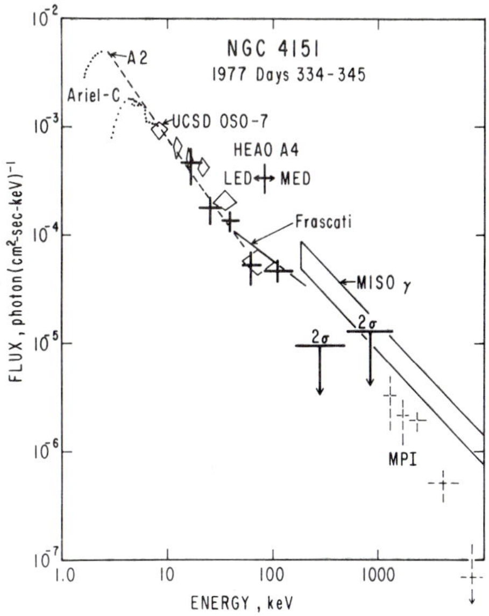

Fig. 4.

Spectrum of NGC 4151, 10 keV-10 MeV, with indications of variable absorption below 10 keV.

The hard X-ray spectrum for this series of scans is shown in Fig. 4, together with previous work in this energy range and somewhat higher from UCSD OSO-7 [12], the Frascati group [13] and the Milano-Southampton (MISO) group [14] balloon results, and the double-Compton telescope results of the Munich (MPI) group [15]. The latter consider NGC 4151 to be the only source in the field of view, and thus these preliminary points might also be taken as upper limits. Auriemma et al [12] have given a number of points between 40 and 200 keV roughly consistent with our results although the best fit line would have a flatter slope ($\sim E^{-0.9}$) than ours ($\sim E^{-1.4}$). It is possible to draw an $E^{-1.0}$ line through the A2, A4 and MPI data. This falls a factor of three below the MISO results, which would indicate either variability or some difficulty in their conversion of excess counting rate >200 keV into γ-ray flux. There was reportedly an X-ray flare shortly before the MISO observations (G. Bignami, private communication).

A connection between the NGC 4151 spectrum and the diffuse cosmic γ-ray spectrum has been frequently made [10,15,16,17]. It appears that Seyferts may supply 6-25% of the 2-10 keV diffuse flux while their summed contribution at energies above ~100 keV is estimated at 50-100%. Adding cosmological evolution effects may increase the contribution from the nuclei of active galaxies to 100% at 2-10 keV [18].

SUMMARY

Preliminary HEAO A4 studies at 15-100 keV have shown spectra consistant with previous observations for Cen A and NGC 4151. The addition of all the scanning data will decrease the error bars on the average measured fluxes by a factor of three.

In the difficult 100-1000 keV (MED) range, we have just begun the analysis. The exact spectral shape here will be a test of the self-Compton source model. In addition, the spectral shapes and their relation to the cosmic background may help set an upper limit to the amount of cosmological evolution of the active nucleus galaxies. In particular, for NGC 4151, if the A-4 final results between 200 keV and 1 MeV as high as our preliminary MED results shown in Fig. 4 and if the MPI results are not upper limits, then the (required) spectral break may have been found at 3 MeV, and it must break to $\sim E^{-3}$ to conform to the observed isotropic diffuse background [15]. We must reserve judgment on this point pending further analysis of A-4 data.

ACKNOWLEDGMENTS

We would like to acknowledge the support work by P.Campbell, R.Pegg and G. Huszar which made these studies possible and discussions with R. Rothschild, B. Cooke, and J. Doty. This work was supported by NASA contract NAS 8-27974.

REFERENCES

1. J.L. Matteson, Proceedings of the AIAA, Paper 78-35 (1978).
2. R.F. Mushotzky et al., Ap. J. (Letters) 206, L45 (1976).
3. R.F. Mushotzky et al., Ap. J. 220, 790 (1978).
4. J.H. Beall et al., Ap. J. 219, 836 (1978).
5. A. Lawrence et al., MNRAS 181, 93 (1977).
6. K.A. Pounds this conference, Paper A.6.1, Pergamon, Oxford (1979).
7. D.W. Weedman, Ann. Rev. Astron. Astrophys. 15, 69 (1977).
8. R.F. Mushotzky, Nature 265, 225 (1977).
9. J.C. Ives et al., Ap. J. (Letters) 207, L159 (1976).
10. M.S. Elvis, MNRAS 177, 7 (1976).
11. H. Tananbaum, et al., Ap. J. 223, 74 (1978).
12. W.A. Baity, et al., Ap. J. (Letters) 199, L5 (1975).
13. G. Auriemma, et al., Ap. J. (Letters) 221, L7 (1978).
14. G. DiCocco, et al., Nature 270, 319 (1977).
15. V. Schoenfelter, Nature 274, 344 (1978).
16. D. Schwartz et al., X-Ray Astronomy 359, (1973).
17. J.E. Grindlay, Nature 273, 211 (1978).
18. Y. Avni, Preprint 77/52-Ph, Weizmann Institute (1977).

FAR-ULTRAVIOLET SPECTRUM OF THE SEYFERT GALAXY NGC 4151

Arthur F. Davidsen* and George F. Hartig

Department of Physics, The Johns Hopkins University, Baltimore, Maryland 21218, U.S.A.

INTRODUCTION

NGC 4151, one of the most extensively studied Seyfert galaxies, has been observed from radio to γ-ray wavelengths. It was the first Seyfert detected as an x-ray source [1] and is now merely the brightest of a rapidly growing list of Seyfert x-ray emitters [2] . We have obtained the first far ultraviolet spectrum of NGC 4151 and present here a brief description of the emission and absorption lines detected.

OBSERVATIONS

The Faint Object Telescope sounding rocket payload [3] , previously used to observe the far ultraviolet spectrum of 3C 273 [4] , was launched aboard a Black Brant VB vehicle from White Sands Missile Range on 1978 February 10 at 0530 UT in order to observe the spectrum of NGC 4151 in the range $\lambda\lambda$ 1160-1760 A. Both the sensitivity and the resolution (~8A) were improved over the version flown previously. Circular apertures of 36 arc second diameter were employed for both the target and background channels, and the pointing jitter of the telescope was limited to a small fraction of this. Nearly 330 seconds of data were accumulated, of which 235 were spent on target. The remaining data was used to cross-calibrate the target and background channels. By far the major portion of the background is due to the geocoronal Lyman α emission line. The raw target spectrum contains some 36,000 counts, of which ~17,000 are due to the Ly α background, confined to a 60A band centered at λ1216 A.

The ultraviolet spectrum of NGC 4151, corrected for background and reduced to absolute flux units, is shown in Figure 1. The instrument was calibrated both before and after the observation by comparison with a reference photomultiplier tube which is itself periodically compared with an NBS-calibrated photodiode. The absolute fluxes are believed to be accurate to about $\pm$14 per cent. The wavelength scale is accurate to $\pm$3A due to uncertainty in the position of the target within the spectrograph entrance aperture. The horizontal bar near Ly α in Figure 1 indicates the region over which the background correction was substantial. The

*Alfred P. Sloan Foundation Research Fellow

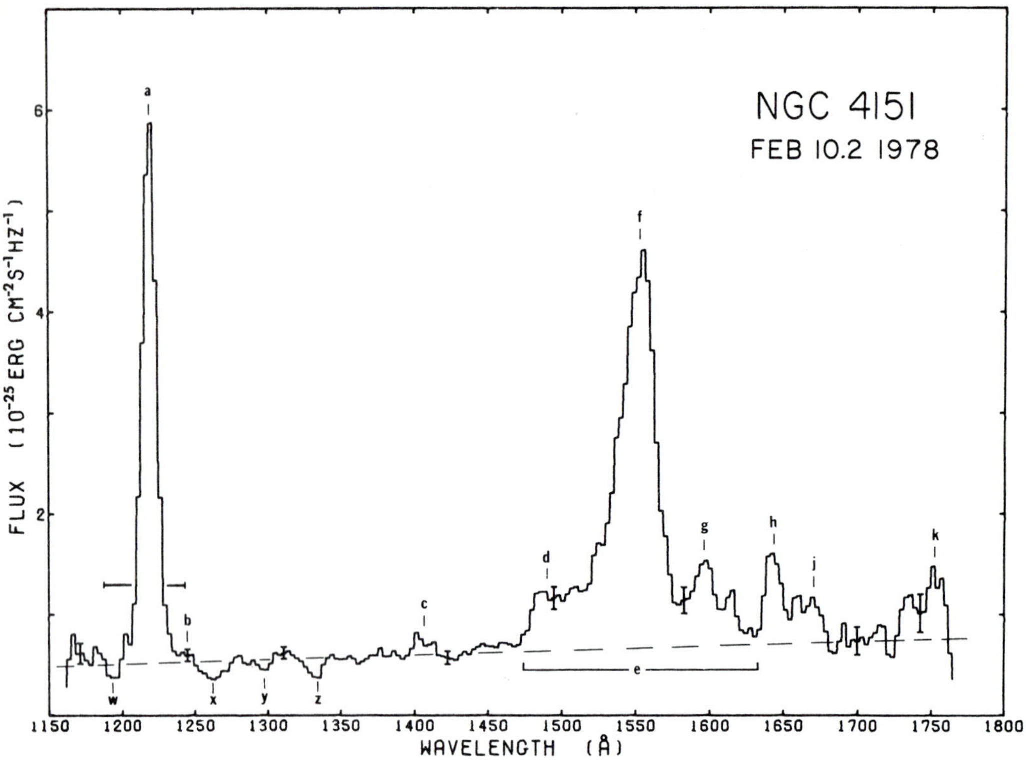

Fig. 1. Far-ultraviolet Spectrum of NGC 4151

dashed line is the assumed continuum $f_\nu \propto \nu^{-1.0}$. Statistical error bars ($\pm 1\sigma$) are indicated in several places. Emission and absorption features are designated by small letters, and possible contributors to the emission features, equivalent widths, and intensities relative to Ly α are given in Table 1. Data for several absorption features are presented in Table 2.

RESULTS AND DISCUSSION

As expected both theoretically and from previous observations of quasars, the strongest uv emission line is Ly α, with a flux of 1.38 x 10^{-11} erg $cm^{-2}s^{-1}$. However, relative to the optical emission line strengths, Ly α is much weaker than predicted by theoretical calculations of radiative recombination in the photoionized gas in a Seyfert nucleus [5] . The Hβ flux measured by Osterbrock and Koski [6,7] is 6.53 x 10^{-12} erg $cm^{-2}s^{-1}$, with an uncertainty of ±20 per cent. Thus I(Ly α)/I(Hβ) = 2, even smaller than the ratio observed in 3C 273 [4] and similar to that deduced for quasars generally [8] . Resonance scattering and absorption by dust within the Milky Way cannot affect the observed Ly α flux very significantly. Thus the reduced Ly α/Hβ ratio must be an intrinsic feature of the Seyfert galaxy.

CIV λ1550 is the next strongest line, also similar to the case of high redshift quasars. However, the CIV feature is extremely broad and asymmetrical and is surrounded by emission extending from λ1470 to λ1630. Much of this broad feature may be due to permitted FeIII emission, perhaps resulting from continuum

TABLE 1. Emission Features

Label	Possible Contributors λ(Å)	ion	W_λ [1,2] (Å)	Intensity (Lα=100) [2,3]	Comments
a	1215.7	HI	133(+17,-24)	100.	
b	1238.8,1242.8	NV	<3.2	<2.4	uncertain due to large geo-coronal Ly α correction
c	1402-1413	OIV]	3.9(+1.8,-0.7)	2.6	
	1393.8,1402.8	SiIV			
	1394-1396	FeIII			
d	1486	NIV]	<12.7	<7.3	contaminated by broad (FeIII?) feature
e	1465-1625	FeIII	103:	60:	very uncertain, dependent on relative strengths of >20 multiplets
f	1548.2,1550.8	CIV	119.(+69):	71:	contaminated by broad (FeIII?) feature
g	1584-1591	?	9.1(+14):	5.3:	contaminated by broad (FeIII?) feature
h	1640.5	HeII	18.2(+4.4,-4.5)	10.2	
j	1661.2,1666.5	OIII]	9.2(+3.9,-2.8)	5.1	
k	1747-1754	NIII]	?	?	at edge of spectral range, part of feature may be cut off
	1735-1753	MgII			
	1722-1749	MgIII			
	1722-1748	FeIII			

[1] equivalent width error estimates in ( ) reflect limits to which the continuum may be reasonably drawn.

[2] particularly uncertain estimates indicated by a colon.

[3] I(Ly α) = 1.38×10^{-11} erg cm^{-2} s^{-1}.

excitation of the more than 20 multiplets concentrated in this wavelength region. In obtaining the CIV line strength given in Table 1 we have assumed the FeIII multiplets are present with the relative strengths given by Kelly and Palumbo [9] and have synthesized the spectrum in this wavelength region. In this way we find I(CIV)/I(Ly α) = 0.7, somewhat higher than the ratio typically found in high redshift quasars [8] . However, the relative contributions to the feature due to CIV and FeIII are not well-determined. If the broad feature arises from CIV alone, velocities up to ±15,000 km s^{-1} are implied.

Another significant line which is very well-detected is HeII λ1640. We find I(HeII λ1640)/I(Ly α) = 0.10, similar to the ratio found in several high redshift quasars by Bahcall and Oke [10] . Here again the uv line is much weaker than expected relative to the optical lines, however. Comparison with the HeII λ4686 flux [6] yields I(λ1640/I(λ4686) = 1.3, whereas ~7 is expected from radiative recombination theory [11] . It is quite clear that some mechanism, perhaps extinction by dust, is very effective at reducing the uv line strengths relative to the optical in quasars and in Seyfert nuclei. Detailed models incorporating this effect are needed in order to deduce information on abundances and physical conditions in the line emitting gas.

The ultraviolet spectrum of NGC 4151 also contains at least 2 very strong absorption features, labelled x and z in Figure 1. These are ascribed to SiII (and possibly SII) at λ1260 and to CII at λ1335. The two other absorption

TABLE 2. Absorption Features

Label	Possible Contributors λ(Å)	Ion	W_λ (Å)	Comments
w	1190-1197	SiII	2.2:	shape and strength highly uncertain due to geocoronal Ly α correction
x	1260-1265	SiII	5.4(+2.1,-1.5)	
	1250-1260	SII		
y	1304.4	SiII	1.5(+0.8,-0.7)	detected at about 2σ level
	1302	OI		
z	1334.5,1335.7	CII	3.8(+0.5,-0.9)	

features listed in Table 2 are of marginal significance. The strength of the CII and SiII absorption features suggests a very large column density along the line of sight to the continuum source, consistent with the observation of strong x-ray absorption corresponding to $N_H \approx 10^{23}$ cm^{-2} in NGC 4151 [12, 13] .

REFERENCES

1. H. Gursky *et al.*, *Astrophys. J. (Letters)*, 165, L43 (1971).

2. M. Elvis *et al.*, *M.N.R.A.S.*, 183, 129 (1978).

3. A. F. Davidsen, W. G. Fastie and G. F. Hartig, in: *New Instrumentation for for Space Astronomy*, Pergamon Press, New York, 1978, p. 57.

4. A. F. Davidsen *et al*, *Nature*, 269, 203 (1977).

5. G. A. Shields, *Astrophys. J.*, 191, 309 (1974).

6. D. E. Osterbrock and A. T. Koski, *M.N.R.A.S.*, 176, 61P (1976).

7. D. E. Osterbrock, private communication.

8. J. A. Baldwin, *M.N.R.A.S.*, 178, 67P (1977).

9. R. L. Kelly and L. J. Palumbo, Naval Research Lab. Rep. No. 7599, 1973.

10. J. N. Bahcall and J. B. Oke, *Astrophys. J.*, 163, 235 (1971).

11. M. Brockelhurst, *M.N.R.A.S.*, 153, 471 (1971).

12. J. C. Ives, *et al.*, *Astrophys. J. (Letters)*, 207, L159 (1976).

13. P. Barr *et al.*, *M.N.R.A.S.*, 181, 43P (1977).

X-RAY EMISSION FROM CLUSTERS OF GALAXIES AND GALACTIC NUCLEI

A. C. Fabian and M. J. Rees

Institute of Astronomy, Madingley Road, Cambridge, England

ABSTRACT

Three main topics are discussed and reviewed: (i) the implications of the X-ray data on clusters of galaxies, and models for cluster evolution; (ii) the nature of the compact X-ray sources in quasars and active galactic nuclei; and (iii) the likely impact of HEAO-B observations on extragalactic X-ray astronomy and cosmology.

INTRODUCTION

In this review, we attempt to discuss the types of extragalactic X-ray sources so far discovered, to summarise what can already be inferred, and to indicate how our understanding may soon be enlarged by future observations. The first section deals with compact X-ray sources in galactic nuclei and quasars, followed by a section dealing with clusters of galaxies. In the last section we mention other possible classes of source which may be discovered by HEAO-B and the next generation of X-ray telescopes, and emphasise the importance of optimum observing strategies and complementary programmes in other wavebands.

COMPACT X-RAY SOURCES IN GALACTIC NUCLEI AND QUASARS

General Remarks

The 2-10 keV X-ray luminosities of galactic nuclei range from $L_x \lesssim 10^{36}$ erg s^{-1} for our own Galactic nucleus to $L_x \sim 10^{46}$ erg s^{-1} for quasars. The X-ray spectra often appear hard and may be considered as quasi-power law continua, similar to the continuum spectra observed in the radio and optical wavebands. The hard X-ray and γ-ray luminosities may exceed the total emission from the nucleus at other wavebands, as is the case with the nearby galaxy Centaurus-A (NGC 5128) in which the spectrum has been confirmed above 1 MeV [1] (see also NGC 4151, [2,3]). X-ray variability has been observed on timescales of days or less (e.g. Lawrence et al. [4] and references therein) and possibly down to $\sim$ 12 min in the Seyfert 1 galaxy NGC 4151 [5]. Such high luminosities, coupled with rapid variations, strongly suggests that the X-rays originate close to the power source in these galactic nuclei.

The broad optical emission lines may be interpreted as indicating mass motions at speeds up to $\sim$ 0.03c in 3C 390.3 and 3C 382 (both are luminous X-ray sources [6-8]. Shocks created by cloud collisions heat the gas to temperatures $T_s \gtrsim 10^9$ K; it then predominantly radiates X-rays, either directly by thermal bremsstrahlung, or indirectly by comptonising soft photons. Shocks and magnetic fields may produce relativistic electrons which Compton scatter soft photons to X-ray energies. It might then be expected that the amount of gas present (which, when cold, is in some way proportional to the width of the broad emission lines) is related to the X-ray luminosity. This is borne out by the correlation observed between FWZI (Hβ) and L_x [9]. Osterbrock, Koski and Phillips [10] report variations in the wings of the broad optical emission lines in 3C 390.3. It would be interesting to discover whether such variations correlate with changes in the X-ray luminosity.

Model-Independent Limits

Luminous outbursts of radiation L erg s^{-1}, cannot occur on timescales, Δt, much shorter than the light crossing time of the emitting region. Moreover, if the electron scattering optical depth, τ_{es}, through this region exceeds unity, the photons diffuse out on a timescale τ_{es} times larger than the light crossing time. Considering a spherical region of radius R, containing particle density n, we see that

$$\tau_{es} \simeq n\sigma_T R \tag{1}$$

and

$$\Delta t \gtrsim (\tau_{es} + 1)R/c \tag{2}$$

Cavallo and Rees [11] have discussed a general constraint depending on efficiency, η, with which matter is converted to energy ($\eta < 1$):

i.e. $$L\Delta t \simeq \eta Mc^2, \tag{3}$$

where M is the mass directly involved in the energy production. The density n must then satisfy

$$n \gtrsim \frac{3L\Delta t}{4\pi R^3 \eta m_p c^2} \tag{4}$$

(the inequality allows for the possibility that more matter resides in the source than is actually involved in the outburst). Δt is a minimum when the source has a radius such that $\tau_{es} = 1$.

Thus

$$L \lesssim 2 \times 10^{41} (\eta/0.1)\Delta t \text{ erg s}^{-1}. \tag{5}$$

Storing the energy in, for example, magnetic fields and then releasing it does not overcome this limit unless the processes take place in a vacuum. Matter is needed to generate and confine the magnetic fields and this itself will present some opacity. Limit (5) is such that direct variability on timescales much less than a day ought not to be observed in the most luminous quasars, unless relativistic outflow occurs. (Note that the $L^{-2.5}$ shape of the X-ray Seyfert luminosity function implies that this limit ought to be observed most readily in the most luminous sources.)

A second limit applies to the hard X-ray and γ-ray radiation due to photon-photon collisions [12, 13]. Photons above a threshold energy E_s may collide and produce electron-positron pairs with harder photons of energy E_γ provided that

$$\frac{E_s}{m_e c^2} \frac{E_\gamma}{m_e c^2} \gtrsim 2 \qquad (6)$$

The optical depth to photon-photon collisions,

$$\tau_{\gamma\gamma} \simeq n_x \sigma_T R \qquad (7)$$

where n_x is the number density of photons of within a bandwidth $\sim 2\ E_s$ centred around $2\ E_s$ [13]. γ-rays of energy E_γ are not therefore expected if $\tau_{\gamma\gamma}(E_\gamma) \gtrsim 1$. This limit (7) and the efficiency limit (5) are both shown in Fig. 1, where a power law hard X-ray spectrum of photon index α has been assumed. The hard X-rays and γ-rays are both taken to originate from the same region, of size $c\Delta t$. A distinct emission line at 510 keV may appear if the pairs are brought to rest before annihilating.

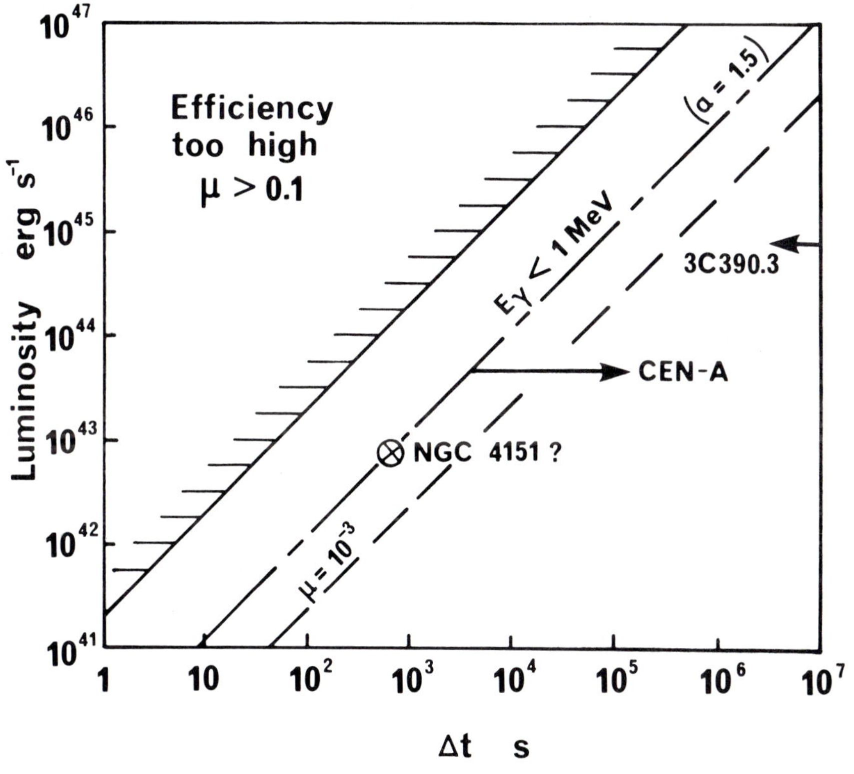

Fig. 1. Constraints on luminosity variations on a timescale Δt seconds.

If a compact source has an X-ray spectrum $P(\nu)d\nu \propto \nu^{-\alpha}$ (i.e. differential photon number spectrum $\propto \nu^{-(\alpha+1)}$), then the optical depth for γ-rays of energy $E_\gamma (> m_e c^2)$ is proportional to E_γ^{α}(for $\alpha > 0$): $\tau_{\gamma\gamma}$ increases with E_γ because the more energetic γ-rays can interact with less energetic (and more numerous) X-ray photons.

Interesting constraints can be placed on any powerful source which is known to emit γ-rays and X-rays. For instance, Cos B observations [14] have recently shown that 3C 273 emits $\sim 2 \times 10^{46}$ erg s^{-1} in the form of $\gtrsim 100$ MeV γ-rays. This source also emits $\sim 5 \times 10^{45}$ erg s^{-1} in the form of X-ray photons with energies

$\gtrsim$ 5 keV which can interact with these γ-rays. If the γ-ray source size is R_γ and the X-rays come from within this region then τ_γ would exceed 1 unless $R_\gamma \gtrsim 10^{18}$ cm. This means that the γ-rays cannot come from a small central source (or display large-amplitude variability on timescales $\lesssim$ 1 yr) unless relativistic beaming effects occur.

For a given r_γ, we expect that the spectrum of γ-rays from a compact X-ray source should cut off above a γ-ray energy at which $\tau_\gamma \simeq 1$.

X-ray Emission Mechanisms

A schematic diagram of emission mechanisms is shown in Fig. 2. The two major reservoirs of energy that can be directly radiated are hot gas and relativistic electrons, both of which may be supplied by shocks etc. These can radiate by various pathways, the dominate X-ray mechanism being decided by the relative emission timescales. Processes such as shocks, supernovae etc. produce hot gas with a higher energy efficiency than they create relativistic electrons. This tends to imply that the dominant radiation observed is due to thermal processes. It is difficult to attribute all the X-rays and γ-rays in Cen-A to non-thermal processes (involving relativistic electrons, for example) without questioning the efficiency with which those particles have been produced. The thermal particles may, of course, have put most of their energy into adiabatic expansion, or have been accreted into a massive black hole, but as we shall discuss, the cooling time-scale for thermal electrons is expected to be very short. Both of these alternatives imply a low value of η in equation (3).

Thermal bremsstrahlung. This occurs on a timescale

$$t_{br} \simeq \frac{2 \times 10^{11} T^{\frac{1}{2}}}{n} \text{ s} \tag{8}$$

where T is the gas temperature ($\sim 10^9$K). Particle densities $\gtrsim 6.10^{10}$ cm^{-3} are clearly needed if decay times of less than an hour are observed. Under these conditions τ_{es} through the emission region (equation (1)) is then high:

$$\tau_{es} \simeq 30\, L_{42}\, t_3^{-1} \tag{9}$$

where $L = 10^{42} L_{42}$ erg s^{-1} and $\Delta t = 10^3 t_3$s. Lightman, Giacconi and Tananbaum [15] have pointed out that Comptonisation of the bremmstrahlung photons will then occur, resulting in an emergent Wien spectrum if $\tau_{es} \gtrsim 10 (T \sim 10^9$K).

Comptonisation. This involves Compton scattering of radiation by non-relativistic electrons. The electrons cool on a timescale

$$t_c \simeq \frac{2 \times 10^7}{\mathcal{E}_{rad}} \text{ s} \tag{10}$$

where $\mathcal{E}_{rad}$ is the energy density in photons to be Comptonised. $\mathcal{E}_{rad} \simeq 3\mathcal{L}_{42} R_{15}^{-2}$ erg cm^{-3} where $L_p = 10^{42}\mathcal{L}_{42}$ erg s^{-1} is the photon luminosity before Comptonisation (we assume $L >> L_p$) from a region containing the electron source but of size $10^{15} R_{15}$ cm. (the photons contributing to L_p need not be due to this source). It is clear that short timescales of the order of hours can result if $\mathcal{E}_{rad} >> 10^4$ erg cm^{-3} which may be reasonable if $R_{15} << 1$ and the photon source and electron source are intimately connected. Due to the competition between escape and scattering probabilities, Comptonisation leads to a power law spectrum extending between the energy of the input photons and that associated with the energy of the electrons kT_e, where $1 < \tau_{es} (kT_e/m_e c^2)^{\frac{1}{2}} < 4$ [16-20]. In view of the chaotic and inhomogeneous situation expected in galactic nuclei, the prevalence of power law spectra and

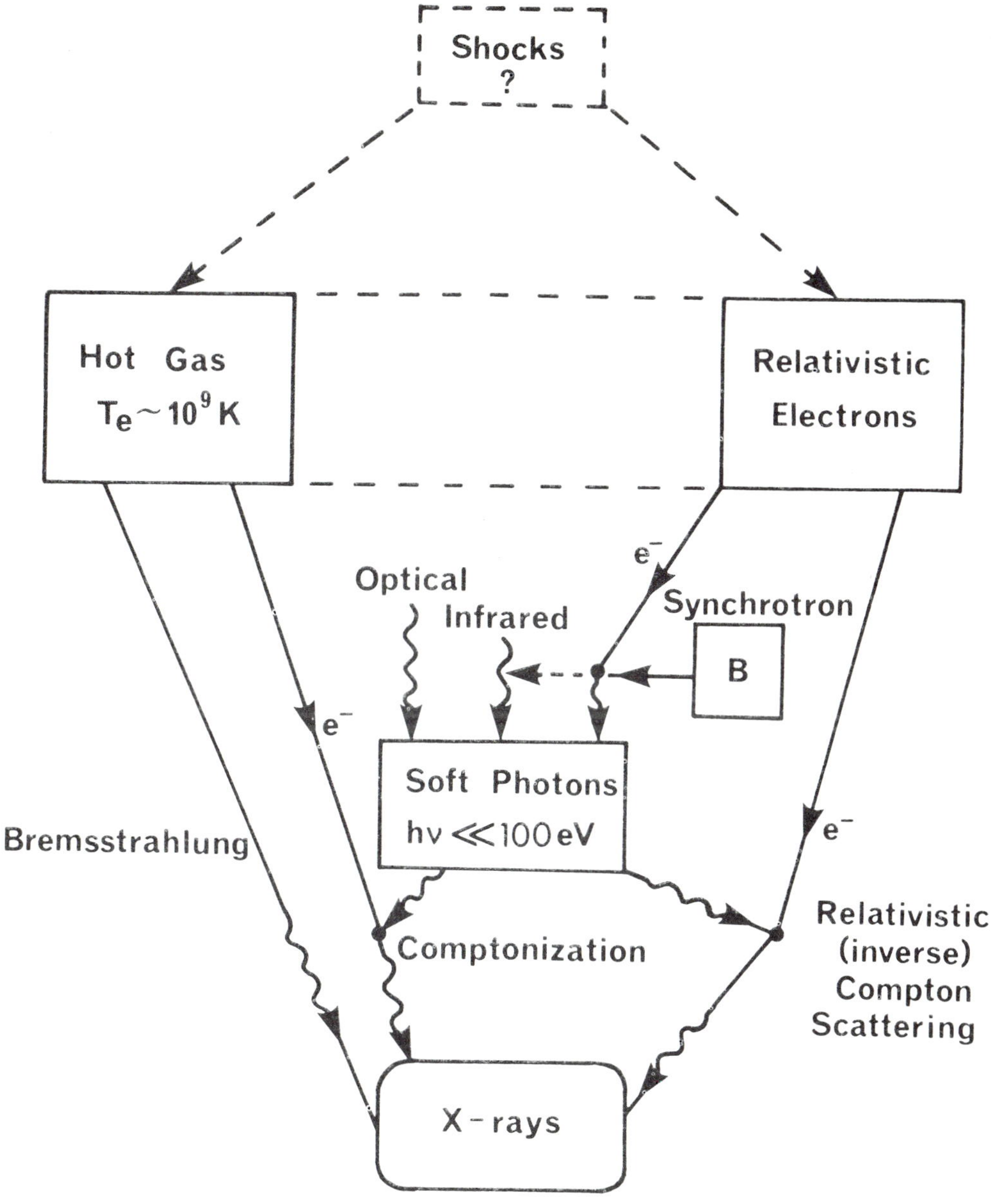

Fig. 2. Possible pathways for X-ray emission

$(kT_e/m_ec^2) \simeq 1$ seem to require τ_{es} to lie in a somewhat unlikely narrow band. This problem may be resolved if the photon source lies outside the hot cloud and what we observe are the reflected photons. Note that photons emerging with low energies are scattered fewer times than those of higher energy and are thus expected to diffuse out of the source faster [15].

Gas clouds shocked at velocities $\gtrsim 0.03c$ reach proton temperatures $T_p \gtrsim 10^{10}$K. The electron temperature may not much exceed $T_e \simeq 5.10^9$K however. This is due to the high cooling rate of mildly relativistic electrons when electron-electron bremsstrahlung becomes important [21] and pairs are produced, thereby enhancing the cooling rate [22, 23]. Conditions in a galactic nucleus are likely to be very complicated but radiation by Comptonisation and bremsstrahlung from shocked gas clouds may give the required spectrum and variability. The soft photons scattered up to form the power law may originate in the outer cooler regions of shocked clouds. Those reflected back from the hot inner core of such clouds are rapidly upscattered, but scatterings on the passage through the cooler outer regions do not result in large energy shifts since

$$\frac{\Delta E}{E} = (kT_e/m_ec^2) \qquad (11)$$

We pursue this problem in greater detail elsewhere [24].

Relativistic Compton scattering. (Inverse) Compton scattering of radiation by relativistic electrons gives the same relative energy shift as (11), except that (kT_e/m_ec^2) is replaced by $\gamma^2 = (1 - v_e^2/c^2)^{-1}$. Many observations of variable radio or infrared emission (e.g. Cen-A; [25]) indicate that relativistic electrons occur in galactic nuclei, probably emitting at those low frequencies by the synchrotron process. The lifetime of relativistic electrons to inverse Compton scattering is a factor γ^{-1} shorter than (10):

$$t_{ic} \simeq \frac{2 \times 10^7}{\gamma \mathcal{E}_{rad}} \text{ s} \qquad (12)$$

Lorentz factors $\sim 10^3$-10^5 are needed to produce X-rays from infrared or radio photons, respectively, giving rise to the possibility of decay timescales $\sim$ 1 hr if $\mathcal{E}_{rad} \sim 1$ erg s^{-1}. Since the X-ray energy density generated by scattering, $\mathcal{E}_x$, probably exceeds the initial $\mathcal{E}_{rad}$ due to radio (or infrared) photons, second-order Compton scattering may be important, extending the spectrum up to $\sim \gamma m_ec^2$ (above which the Klein-Nishina cross section effectively truncates this process). Multiple scattering can present a serious problem (the so-called 'Compton Catastrophe' [26]) in sources of the kind under discussion in which $\mathcal{E}_x >> \mathcal{E}_{rad}$ [27]. This may be avoided if the electron spectrum exists only over a short range, but note that a synchrotron or other cut-off observed at low frequencies need not imply a similar cut-off in the electron spectrum. Most of the energy is then radiated as photons of energy $>> mc^2$ (mean photon energy $\sim$ mean electron energy). Photon-photon collisions reduce small emission regions to a fireball [11] of interacting pairs, (see section on model-independent limits) which ultimately radiate by bremsstrahlung etc.

Many authors [27-30] have considered (inverse) Compton scattering of relativistic electrons on synchrotron photons produced by the same electrons in an ambient magnetic field, B. The energy loss timescale for the electrons is similar to (12) with $\mathcal{E}_{rad}$ replaced by the magnetic energy density, $B^2/8\pi$. The usual approach is to assume that a low frequency cutoff at ν_m (where the observed flux is F_m) is due to synchrotron self-absorption; then

$$B = 7 \times 10^{-60} \nu_m^5 \theta^4 F_m^{-2} \text{ G} \qquad (13)$$

where θ is the angular size of the source. Relating this to the observed X-ray flux, F_x, we obtain

$$F_x \propto \theta^{-8} \nu_m^{-6.5} F_m^{5} \tag{14}$$

where a typical photon spectral index (radio, X-ray) of 1.5 has been assumed. It is clearly not difficult to obtain a fit to F_x! Many estimates of θ are, however, inconsistent with variability on timescales $\Delta t <$ weeks. Appealing to source expansion [31] severely reduces the efficiency η as defined in (10). The observations usually restrict θ fairly well, and although small sizes related to an unobserved cut-off may be involved, these probably involve small values of B. Once again, second (or higher) order Compton scattering may prove more important ($\mathcal{E}_x >> B^2/8\pi$) and few relativistically Compton-scattered X-rays may actually be observed.

Accretion onto Massive Black Holes

Our discussion has, so far, involved no specific assumption about the actual power source, generally considered to be gravitational in origin. The collapse of stars with a dense star cluster (giving rise to supernovae and pulsars), the collapse and spin-up of supermassive spinars, or accretion into a massive black hole, are all possibilities that have been considered (see Rees [32] for references). Both the core of a dense star cluster and a spinar are likely to evolve into massive black holes, however. The vast reservoir of energy in, for example, the Cen-A radio lobes indicates $\sim 10^7$ $M_\odot$ of 'spent' matter if the energy originated in the nucleus and was produced from matter with an efficiency $\lesssim 0.1$. Furthermore the large increases in flux observed from many of these nuclei argues for a coherent potential well deep enough to confine large quantities of matter at X-ray temperatures. If confinement does not occur, then most of the energy release goes into source expansion. All these arguments strongly suggest, on the basis of current theory, that accretion onto a central black hole is responsible for the high luminosities of active nuclei.

Many authors have followed the suggestion that massive black holes accreting gas [33] or stars [34] power galactic nuclei. A major problem with 'standard' accretion disc models [35,36] is that they are predominantly opaque to radiation and have large surface areas which increase with the black hole mass, M_{bh}. Consequently the temperature of the dominant radiation emitted is low:

$$T_{disc} \lesssim 10^8 (M_{bh}/M_\odot)^{-\frac{1}{4}} \text{ K} \tag{15}$$

This problem presents itself even in the case of Cyg X-1 where a quasi-power law hard spectrum is observed up to $\sim$ 1 MeV (see Mandrou et al. [37]) and in which, for example, thermal instabilities [38] and/or Comptonisation has been invoked [19, 38, 39]). The production of thermal X-rays from massive black holes requires optically thin conditions and, of course, high temperatures.

'Standard' two-dimensional accretion discs are less likely to occur in galactic nuclei than in stellar-mass binary systems. Directed transfer of matter via a stream may take place in the latter, but the conditions in a galactic nucleus are likely to be far more chaotic. Even when a disc forms, matter is likely to join it at all radii from the z-direction, just because no one directed stream is responsible for setting up the disc. A flattened mass oriented perpendicular to the direction of the mean incoming angular momenta seems the best that can be attained. The ratio of X-ray emission to lower energy emission depends on the fraction of matter striking the disc at velocities $> 10^8$ cm s^{-1} (i.e. $r < 10^4$ r_s, where r_s is the Schwarzschild radius of the black hole). Dominant X-ray emission suggests that this fraction is large, and we are in the regime of quasi-spherical accretion [40, 41]. It is important in flows close to spherical symmetry that the gas is heated to high temperatures, thereby enhancing the pdV work available, and also that the gas can cool before entering the black hole.

McCray and others [42, 43] have emphasised the importance of heating instabilities on accretion flows. Gas initially falling in can be heated by Compton recoil or photoionization to temperatures such that it is forced to expand away from the central mass. This may in part be responsible for some of the variability in galactic nuclei. Radiation pressure may be important (especially when $L \simeq L_{Edd}$). Some of the differences between ordinary Seyferts/quasars and BL Lac objects may be due to the observer's orientation with respect to an outflow of material. Indeed we may be viewing BL Lac objects along a beam of relativistic particles [44]. The beam itself may disperse any gas that would potentially absorb the X-rays, and scattering within it might subtly change the X-ray spectrum (note Mkn 501).

The presence of gas with a range of temperatures, 10^4K - 10^9K, in galactic nuclei suggests the possibility of observing Fe emission lines. Payne and Eardley [45] have hown that a characteristic edge due to the maximum blueshift, may be observed if the gas is accreting onto a black hole. Changes in X-ray absorption, such as reported by Barr et al. [46] for NGC 4151, may provide us with a handle on the cooler gas in the nuclei. Rapid changes in absorption, on timescales less than the observation integration time may be interpreted as spectra with a positive energy power law slope over the 2 - 10 keV range.

It seems likely that X-ray observations of galactic nuclei probe closest to the power house involved. Further extended spectral observations covering the hard X- and γ-ray bands are needed. These, combined with variability timescales, may allow many possible emission mechanisms to be eliminated. Highly variable sources seem most likely to emit via non-relativistic Compton scattering, the timescales for this process being shorter than for bremsstrahlung from the same electrons. Relativistic Compton scattering seems unable to produce short timescales without multiple scattering being important, thereby creating an energy problem and predicting dominant γ-rays. The bulk of the material in the emitting region may be mildly relativistic ($T_e = 10^9 - 10^{10}$ °K). Observations of X-rays from galactic nuclei should stimulate much needed study of processes such as:

Radiation processes in 'transrelativistic' plasmas where electron-electron bremsstrahlung and relativistic corrections are important.

Production of $e^+ - e^-$ pairs, and their effects on cooling and opacity.

Effects of major differences between electron and ion temperatures.

Comptonisation of spectra in inhomogeneous gas clouds.

Acceleration of ultrarelativistic particles by mildly relativistic shocks.

CLUSTERS OF GALAXIES

The Present State of Cluster Sources

In clusters such as Coma, there now seems little doubt that most of the X-rays below $\sim$ 10 keV result from thermal bremsstrahlung. The X-ray emission comes predominantly from the core of the cluster, since the emissivity per unit volume is proportional to n_e^2. The following three inferences about conditions in the core seem fairly uncontroversial:

(i) The electron temperature T_e is close to the virial temperature T_V, so that (sound speed) ≈ (virial velocity).

(ii) The gas mass in the core amounts to only ∿ 2 per cent of the total virial mass (which is partly contributed by the galaxies, but is largely in the form of non-gaseous 'dark matter').

(iii) The data are consistent with an iron abundance in the hot gas which is ∿ ½ the solar value.

Statements (i) and (ii) imply a cooling time such that $t_{cool} \simeq t_{Hubble} >> t_{dynamical}$. As a first approximation, the present state of the intracluster gas can thus be fitted by a quasi-static model, evolving slowly due to infall, cooling, conduction, mass loss (and energy input) from galaxies, etc.

Note that - in order that they should indeed be 'uncontroversial' - the above statements (i) - (iii) are worded in a cautious and limited way: they refer only to the cores of rich clusters at the present epoch. We know much less about the much larger amount of gas residing in the outer parts of clusters, because its emissivity is low. This gas may not be in equilibrium (it could be flowing either in or out), and its temperature and composition are very uncertain. It is unclear to what extent the same statements apply to small groups (where T_V is lower), or to rich clusters at earlier times. However, the uncertainties are not quite as basic as those that faced us in the context of galactic nuclei (see section on compact X-ray sources in galactic nuclei and quasars): in the case of clusters, we do know what radiation mechanism predominates (at least at photon energies ≲ 10 keV), and can therefore proceed to worry about the quantitative details.

Some relevant questions can be summarised under the following headings:

A. What is the present role of heating and cooling?

a) Is there evidence for inflow into central massive galaxies? Silk [47] suggested that giant CD galaxies might have gained mass by infalling of cooling material. Cowie and Binney [80] and Fabian and Nulsen [48] have discussed the gas dynamics of this inflow, and considered its possible relevance to (e.g.) NGC 1275.

b) Is T_e (or the core radius for the gas distribution) significantly different from that implied by the distribution of galaxies? This is not of course a straightforward question if the cluster is non-spherical [49]; nor is it obvious that the galaxies should be accurate tracers for the distribution of the missing mass [50, 51]. This question is however relevant to the balance between cooling and heating for the core gas as discussed by Lea and Hohlman [52].

c) Are the X-ray properties of clusters correlated with their radio properties? If so, this might indicate a heat input or dynamical effect from relativistic particles [52]; or, conversely, the lifetime or luminosity of diffuse non-thermal radio sources might depend on the external gas pressure, which is in turn correlated with the bremsstrahlung X-ray emissivity. Note that inverse Compton scattering of the microwave background (by relativistic electrons associated with current or past radio outbursts) may be a significant contribution to the X-ray flux from clusters above 10 keV even if it is swamped by bremsstrahlung at lower photon energies.

d) What are the temperatures, densities and gas flows in the outer parts of clusters? This question could in principle be answered if X-ray maps with adequate angular resolution were available in several energy bands (but note Forman et al. [53]). Relevant but indirect information may come from searches for the Sunyaev-

Zel'dovich [54] 'microwave dip', or from the orientation of radio trail sources on the edges of clusters.

e) What are the effects of non-sphericity, bulk motions in gas, etc? High resolution X-ray maps are obviously crucial here. Tentative evidence for inhomogeneities has been reported by Gorenstein elsewhere in these proceedings.

B. How much Fe is there, and where did it come from?

a) Is Fe restricted to the cores of clusters? The required amount could then have been plausibly supplied by 'sweeping' of processes material lost from stars in the individual galaxies [55, 56].

b) Can 'sedimentation' concentrate Fe in cores? Fabian and Pringle [57] point out that it is marginally possible (cf. Rephaeli [58]) for heavy ions to sink preferentially into the central gravitational well. This could happen only if the magnetic field configuration was such that thermal conductivity was also high, and in the absence of large-scale mixing motions.

C. What correlations exist between the X-ray and optical properties of clusters?

a) Little can at present be deduced from comparisons of X-ray luminosity with velocity dispersion, owing to the large measurement errors (see Culhane, these proceedings). Similarly, it is not yet clear just how the correlations of luminosity with cluster richness and central galaxy density [59-63] are to be interpreted (see Smith et al. [64]). What controls the balance of mass and energy input? The authors cited above have also shown that X-ray luminosity correlates with a deficiency in core spiral galaxies, as is to be expected by ram stripping arguments, but note that cluster richness may inhibit the formation of spirals [65-67].

b) Does the radiative cooling of cluster gas produce or enhance the appearance of central CD galaxies? Are there other galaxies like NGC 1275 due to accretion of such gas?

D. Is there a high energy tail to the X-ray spectrum from clusters?

Apart from the possibility of inverse Compton X-rays, some gas may be heated to temperatures > T_V by explosive outbursts from galaxies. Clusters may conceivably contain a population of unresolved point sources [68, 69].

In view of all these uncertainties, particularly those listed under A and B, it is unlikely that either a simple 'isothermal' or an 'adiabatic' model will adequately reproduce the spatial or spectral properties of X-ray clusters as observations improve.

Evolution of Gas Content of Clusters

Some consideration of how clusters evolve is a prerequisite not only for understanding their present properties, but also for interpreting the data on large-redshift clusters that HEAO-B should soon provide.

Two contrasting hypotheses have been explored:

A. Galaxies (and dark matter) may have condensed from primordial gas before clusters assemble. Galaxies then gradually aggregate into clusters in a manner that can be simulated by gravitational N-body computations. The present gas in clusters then results from either: a) Infall of the small fraction of material ($\sim$ 10%) that

escaped incorporation into the first generation of gravitationally-bound systems.
b) Ejection from galaxies, via stellar winds and planetaries, supernovae, 'sweeping' of interstellar matter by ram pressure of pre-existing cluster gas [55].

(Note that the high inferred Fe abundance, though suggestive of alternative (b), might be compatible with (a) if a hypothetical 'Population III' of early stars (or massive objects), now perhaps constituting the 'dark mass', indulged in pre-galactic nucleosynthesis).

B. Protoclusters (i.e. gas clouds of $\sim 10^{14}$ $M_{\odot}$) may have been the first objects to have condensed out of the expanding universe; they then cool and fragment into galaxies. In this scheme, advocated particularly by the Moscow group [70], the present intracluster gas represents material that has not (yet) cooled and condensed.

Obviously a type-A hypothesis yields clusters whose X-ray luminosities <u>increase</u> with time; on the other hand, hypothesis B would predict that clusters were brightest where they had just virialised but were still predominantly gaseous, but that the X-ray luminosity would thereafter decline as the initial gas gradually cooled and condensed into galaxies.

The volume of work based on B [71-75] is outweighed by that predicated on A [56, 76-81]. One problem with either hypothesis involving strong cooling is that substantial atmospheres at X-ray temperatures ($\gtrsim 3.10^7$K) cannot have cooled in clusters without contradicting observations of the present day clusters, or of the X-ray background (Fabian and Nulsen, these proceedings). This can be overcome if the gas is dispersed or if it never attained X-ray temperatures.

HEAO-B observations of distant clusters offer our best hope of discriminating between these two basic schemes. In practice, a complete range of schemes between the extremes of A and B seem possible within the limits of current observation. Most of this has been ill-explored theoretically, but the next generation of X-ray data ought to provide the necessary impetus and narrow down the range of possibilities.

In a detailed study of hypothesis A, Perrenod [56] has considered the X-ray emission from the gas that accumulates in a realistically-evolving cluster potential. He considers two alternative origins for the gas (t_{cool} is in general $\gtrsim H_0^{-1}$): either

(i) The gas was distributed like the galaxies at the epoch of cluster 'turnaround', and falls in (being shock-heated) as the cluster virialises, <u>or</u>

(ii) The gas results from mass-loss from galaxies. In this case, stellar-evolutionary considerations suggest that the loss rate would have been higher in the past, going roughly as t^{-1} [66].

Perrenod presents numerous alternative models in these two categories. The general findings are that L_X may increase by a factor up to ~ 10 between the epoch corresponding to $z \simeq 1$ and the present. This is a combined consequence of the cumulative build-up of gas, and the deepening of the cluster potential well as virialisation proceeds.

Even though the evolutionary effects are large, Perrenod points out that there may be grounds for optimism about the prospects of using X-ray clusters for 'geometrical' cosmology. This is because the X-ray evolution involves physical processes that seem more straightforward and tractable than those bedevilling optical and radio tests. (Perrenod [56] claims that once a cluster has turned around, its evolution depends solely on its dynamical age, which is related to z in a manner that depends on q_o. Uncertainties on the heat and mass input from galaxies, and their dependence on time, would complicate the picture, so that prospects of clear cosmological tests seem remote).

The models that Perrenod and others have investigated do, however, cover only a small fraction of the tenable options. For instance, any of the following three schemes would lead to higher values of L_x in the past.

(i) According to hypothesis B [70], a newly-virialised gaseous protocluster could have $t_{cool} < t_{Hubble}$, and could then radiate much of its binding energy on a timescale $\sim t_{cool}$. (The actual amount of energy radiated would depend on the fraction of gas which was shock-heated to X-ray temperatures rather than being squeezed into dense cool 'pancakes').

(ii) Even in hypothesis A, infalling gas might overshoot and reach $T_e \simeq T_{escape} \simeq 10\, T_{virial}$ before bouncing and establishing a quasi-static distribution.

(iii) If cluster gas is ejected from galaxies via supernova-driven winds, the velocities and temperatures may be so high that only a small fraction of the gas remains gravitationally confined even with a cluster (the bulk may escape and constitute an ultra-hot intercluster medium).

DEEP X-RAY SURVEYS

In separate work ([82] and these proceedings) we have estimated how many distant clusters, active galaxies and quasars may be detected in a typical HEAO-B field. The form of the expected X-ray source counts will depend on at least three factors: (i) the X-ray luminosity function for various classes of source at the current epoch; (ii) the evolutionary effects for clusters (as discussed above), and (iii) the evolution of L_x/L_{opt} for quasars. As a corollary, data on the content of typical HEAO-B deep fields will help to pin down these three important uncertainties. HEAO-B may, however, reveal new categories of cosmologically-interesting X-ray sources, and we mention two such possibilities here.

Compton Scattering in Remote Radio Sources

Inverse Compton scattering of microwave background photons by relativistic electrons was long ago suggested as a contributor to the X-ray background [83]. This process would tend to be more efficient at large redshifts, for two reasons [84, 85]: the energy density of the microwave background varies as $(1 + z)^4$; and the many suppliers of relativistic electrons - radio source, quasars, etc. - were more prolific at early epochs. As a corollary of this effect, Compton scattering may 'snuff out' extended sources at large redshifts [86-88].

X-rays observed in the range 1 - 10 keV are generated predominantly by electrons with Lorentz factors in the range $(1 - 3) \times 10^3$. The radio sources with the highest inverse Compton X-ray luminosities are therefore those with the largest energy content in the form of such electrons. These sources will not necessarily be those with the highest radio luminosities: the maximum stored energy is inferred to exist in very extended 'giant' sources and in the low surface-brightness 'bridges' joining the components of some strong double sources. On the assumption of equipartition, the energy stored in relativistic electrons in NGC 6251 and 3C 236 is $\gtrsim 10^{60}$ erg [89, 90]: and the energy is indeed mainly in electrons with γ of $\sim 10^3$, these being the values corresponding to the low-frequency end of the radio (synchrotron) spectrum. If the magnetic field is weaker than its equipartition strength, the inferred energy stored in relativistic electrons is even larger.

A radio source with a typical spectra index ~ 0.7 containing $10^{60}\mathcal{E}_{60}$ erg of relativistic electrons with $\gamma \gtrsim 10^3$, and at redshift z, will emit an inverse Compton X-ray power of $\sim 3 \times 10^{43} (1 + z)^4 \mathcal{E}_{60}$ erg/s. We do not know the appropriate values of $\mathcal{E}_{60}$, but it would seem quite probably that there may be diffuse objects at $z \simeq 2$, a few hundred kpc in extent, emitting $\sim 10^{46}$ erg of 1 - 10 keV X-rays. These sources would generally not be particularly powerful radio sources (objects such as Cygnus A or 3C 9 are powerful sources because they have a strong magnetic field, rather than because they have an exceptional energy content), but they may feature in deep radio surveys.

Young galaxies. According to some theories, galaxies pass through a bright early phase when the rate of star formation, supernova outbursts, etc. is ~ 100 times higher than in a present-day galaxy [91, 92]. The properties of such systems, and their potential detectability, depend on many uncertainties; but we can readily see that a young galaxy where supernova outbursts were frequent could give rise to thermal X-rays with a hard spectrum.

A supernova ejects debris at speeds $\sim 10^4$ km/s. When the ejecta from separate supernovae collide, shocks raise the kinetic temperature to $\sim 10^9$ K (~ 100 keV). This hot material would be expelled from the galaxy as a wind, provided that radiative cooling is unimportant on the outflow time-scale. Further shock-heating would occur when the winds from individual 'young galaxies' in a cluster interacted. This process would generate a patchy distribution of thermal X-rays with a spectrum characterised by temperatures exceeding T_V (see discussion in section on model-independent limits).

The X-ray background. Rich clusters of galaxies cannot alone produce the 2-6 keV X-ray background, even with evolution ([82] and Schwartz, these proceedings). Any candidate source for the background must be more numerous than 10^6 sr^{-1} and weaker than 2×10^{-3} Uhuru count/s. The HEAO-B detectors are just capable of detecting any sources at this limit in 10^5 s, although there would then be ~ 500 per field if they constituted the X-ray background. Extrapolations of existing source counts with a -1.5 slope down to $\sim 6 \times 10^{-5}$ Uhuru count/s, might give the entire background, but this requires $\sim 10^4$ quasars per square degree. HEAO-B may clarify the various contributions of quasars, clusters and otner types of sources, but is unlikely to image the X-ray background completely.

The 2-6 keV X-ray background may therefore not be due primarily to point-like sources. A total number of sources $\gtrsim 1$ per cent of that of all galaxies is required, unless they are all at $z >> 1$. At energies > 10 keV inaccessible to HEAO-B the main contributors may be active nuclei and quasars (which appear to have hard spectra), extended radio galaxies at large redshifts and possibly very hot gas in clusters. These considerations suggest that it is important to improve limits on, and estimates of, fluctuations in both 2-6 keV and > 10 keV X-ray backgrounds. If the relative contributions of different source populations depend on energy, it is unlikely that the X-ray background spectrum is smooth and featureless.

An Observing Strategy, and Complementary Programmes

Ten (or more) separate samples with the IPC of 10^5s observing time each, such as suggested in the HEAO-B consortium observing programme [93], with HRI images of the same fields, should reveal a significant number of clusters of galaxies (10-20) and active galaxies or quasars ($\lesssim 100$). The majority of these objects will be at small redshifts ($z \lesssim 0.5$): although a few will appear at $z \simeq 1$. Spending more time on each field does not increase the number of distant clusters in proportion to the investment in observing time. Clusters may be studied in some detail, but it would seem more advantageous to spend that time studying the much brighter clusters found by previous scanning instruments.

Objective prism surveys, followed by detailed optical study, are important in obtaining cluster and active galaxy redshifts. Clusters of galaxies evolving through a high-luminosity phase may be detected in some of the HEAO-B fields. Optical recognition of these events may prove fruitless, but a search for microwave dips [54] may be worthwhile. On the Rayleigh-Jeans part of the background spectrum, the fractional dip amounts to

$$\frac{\Delta T}{T} = -2\,\tau_{es}\,(kT_e/m_e c^2), \tag{16}$$

so a particle density of $\sim 0.1\ cm^{-3}$ in the core of a cluster implies an electron scattering optical depth $\tau_{es} \simeq 0.05$. $\Delta T/T$ is thus ~ 0.02 if $T = T_{escape}$, or 0.002 if $T = T_V$. Such a core subtends 1 arcmin at $z \simeq 2$-3 and thus fills about one-quarter of the beam of a 60-m dish operated at 3 cm. Dips of 5×10^{-2} to 5×10^{-3} are thus expected [74, 82], and may be as readily detected as those from nearby rich clusters which have smaller τ_{es}, and where the dip may already have been discovered [94, 95]. Failure to observe the predicted dips might indicate that the emitting gas was patchy. Alternatively, extended X-ray emission detected in the IPC without variation, and yet not in the HRI, may be due to inverse Compton radiation in radio lobes. A sensitive radio map of the region can test this. (Note that the study of microwave dips in nearby rich clusters provides information on the gas content and distribution complementary to that given by the X-ray observations themselves.)

Unless internal photoelectric absorption causes attenuation at photon energies as high as ~ 4 keV, most of the active galaxies detected by HEAO-B should be at $z \lesssim 0.4$ and discernible on optical Sky Survey plates. Only if there are active galaxies in which L_x/L_{opt} greatly exceeds that in known sources can no immediate optical candidate be expected.

On the basis of present information we expect the following possibilities:

1) Sources extended over several Imaging Proportional Counter (IPC) pixels are probably clusters of galaxies with $z \lesssim 0.2$. This could be checked on optical Sky Survey plates.

2) A source detected in one IPC pixel but not in High Resolution Imager (HRI) 6-arcsec pixel, with no evidence for variability, could be one of the following:

a) If there were no associated microwave dip, nor a single radio source at the same position, it would probably be a more distant cluster with $z \lesssim 1$. Optical techniques [96] may confirm this.

b) If a microwave dip were present, it would be most likely a protocluster (see section on model-independent limits). A patchy distribution may be expected in the HRI, especially since any protoclusters detected may be well above the sensitivity limit.

c) If a single radio source (perhaps itself extended) were observed, the X-rays could then result from the inverse Compton mechanism (see section on the present state of cluster sources). An estimate of magnetic field is then possible.

3) Any sources observed in the HRI 6-arcsec pixels are expected to be observed in the IPC unless they are variable:

a) If there were an associated galaxy on optical Sky Survey plates, it would probably be an active galaxy with $z \lesssim 0.4$.

b) An optical counterpart that appeared stellar with emission lines would be a quasar, radio emission then being possible.

c) Absence of any optical counterpart would indicate a new class of source with an abnormally high ratio L_x/L_{opt}.

ACKNOWLEDGEMENTS

We are grateful for helpful discussions with Bill Forman, Christine Jones and Paul Nulsen. ACF acknowledges the support of a Radcliffe Fellowship.

REFERENCES

1. R. D. Hall et al., Astrophys. J., 210, 631 (1976).

2. G. Di Cocco et al., Nature, 270, 319 (1978).

3. G. Auriemma et al., Astrophys. J. Lett., 221, L7 (1978).

4. A. Lawrence, J. P. Pye and M. Elvis, Mon. Not. R. astr. Soc., 181, 93P (1977).

5. H. Tananbaum et al., preprint (1978).

6. P. A. Charles, M. S. Longair and P. W. Sanford, Mon. Not. R. astr. Soc., 170, 17P (1975).

7. M. Ricketts, private communication (1978).

8. R. E. Marshall et al., preprint (1978).

9. M. Elvis et al., Mon. Not. R. astr. Soc., 183, 129 (1978).

10. D. E. Osterbrock, A. T. Koski and M. M. Phillips, Astrophys. J., 197, L41 (1975).

11. G. Cavallo and M. J. Rees, Mon. Not. R. astr. Soc., 183, 359 (1978).

12. J. Jelley, Nature, 211, 472 (1966).

13. K. Herterich, Nature, 250, 311 (1974).

14. B. N. Swanenburg et al., preprint (1978).

15. A. P. Lightman, R. Giacconi and H. Tananbaum, Astrophys. J., 224, 373 (1978).

16. A. F. Illarionov and R. A. Sunyaev, Sov. Astron., 16, 45 (1972).

17. J. E. Felten and M. J. Rees, Astr. Astrophys., 17, 226 (1972).

18. S. A. Colgate, J. D. Colvin and A. G. Petschek, Astrophys. J. Lett., 197, L105, (1975).

19. D. M. Eardley, A. P. Lightman and S. L. Shapiro, Astrophys. J. Lett., 199, L15, (1975).

20. J. I. Katz, Astrophys. J., 206, 910 (1976).

21. S. Maxon, Phys. Rev., A5, 1630 (1972).

22. G. A. Bisnovatyi-Kogan, Y. B. Zel'dovich and R. A. Sunyaev, Sov. Astron., 15,17 (1971).

23. W. R. Stoeger, Astr. Astrophys., 61, 659 (1977).

24. A. C. Fabian and M. J. Rees, in preparation.

25. J. H. Beall et al., Astrophys. J., 219, 836 (1978).

26. F. Hoyle, G. E. Burbidge and W. L. W. Sargent, Nature, 209, 751 (1966).

27. M. J. Rees, Mon. Not. R. astr. Soc., 137, 444 (1967).

28. T. W. Jones, S. L. O'Dell and W. A. Stein, Astrophys. J., 188, 353 (1974).

29. J. E. Grindlay, Astrophys. J., 199, 49 (1975).

30. R. F. Mushotzky, Nature, 265, 225 (1977).

31. R. F. Mushotzky et al., Astrophys. J., 220, 790 (1978).

32. M. J. Rees, Ann. N. Y. Acad. Sci., 302, 613 (1977).

33. D. Lynden-Bell, Nature, 233, 690 (1969).

34. J. Hills, Nature, 254, 295 (1975).

35. J. E. Pringle and M. J. Rees, Astr. Astrophys., 21, 1 (1972).

36. S. S. Shakura and R. A. Sunyaev, Astr. Astrophys., 24, 337 (1973).

37. P. Mandrou, M. Neil, G. Vedrenne and A. Dupont, Astrophys. J., 219, 288 (1978).

38. J. E. Pringle, M. J. Rees and A. G. Pacholczk, Astr. Astrophys, 129, 179 (1973).

39. D. M. Eardley, A. P. Lightman, D. G. Payne and S. L. Shapiro, Astrophys. J., 224, 53 (1978).

40. A. C. Fabian, D. Maccagni, M. J. Rees and W. J. Stoeger, Nature, 260, 683 (1976).

41. P. Meszaros and J. Silk, Astr. Astrophys., 55, 289 (1977).

42. J. P. Ostriker, R. McCray, R. Weaver and A. Yahil, Astrophys. J., 208, L61 (1977).

43. L. L. Cowie, J. P. Ostriker and A. A. Stark, Astrophys. J. in press (1978).

44. R. Blandford and M. J. Rees, in: Proceedings of Pittsburgh BL Lac Conference, (1978).

45. D. G. Payne and D. M. Eardley, Astrophys. Letters, 19, 39 (1978).

46. P. Barr, N. E. White, P. W. Sanford and J. C. Ives, Mon. Not. R. astr. Soc., 131, 43P (1977).

47. J. Silk, Astrophys. J., 208, 646 (1976).

48. A. C. Fabian and P. E. J. Nulsen, Mon. Not. R. astr. Soc., 180, 479 (1977).

49. J. Binney and O. Strimpel, preprint (1978).

50. S. D. M. White, Mon. Not. R. astr. Soc., 177, 717 (1976).

51. S. D. M. White and M. J. Rees, Mon. Not. R. astr. Soc., 183, 341 (1978).

52. S. M. Lea and G. D. Hohlman, Astrophys. J., 222, 29 (1978).

53. W. Forman, C. Jones, S. S. Murray and R. Giacconi, preprint (1978).

54. R. A. Sunyaev and Y. B. Zel'dovich, Comm. Astrophys. Sp. Phys., 4, 173 (1972).

55. F. Takahara and S. Ikeuchi, Prog. Theor. Phys., 58, 1728 (1977).

56. S. C. Perrenod, Astrophys. J. in press (1978).

57. A. C. Fabian and J. E. Pringle, Mon. Not. R. astr. Soc., 181, 5P (1977).

58. Y. Rephaeli, preprint (1978).

59. N. A. Bahcall, Astrophys. J. Lett., 217, L77 (1977).

60. N. A. Bahcall, Astrophys. J. Lett., 218, L93 (1977).

61. C. Jones and W. Forman, Astrophys. J., 224, 1 (1978).

62. I. McHardy, Mon. Not. R. astr. Soc. in press (1978).

63. R. F. Mushotzky et al., preprint (1978).

64. B. W. Smith, R. F. Mushotzky and P. J. Serlemitsos, preprint (1978).

65. J. R. Gott and T. X. Thuan, Astrophys. J., 204, 649 (1976).

66. G. R. Gisler, Astr. Astrophys., 51, 137 (1976).

67. G. R. Gisler, preprint (1978).

68. J. I. Katz, Astrophys. J., 207, 25 (1976).

69. A. C. Fabian, J. E. Pringle and M. J. Rees, Nature, 263, 301 (1976).

70. A. C. Doroshkevitch, R. A. Sunyaev and Y. B. Zel'dovich, in: Confrontation of Cosmological Theories and Observation, Reidel, Holland, 1974, p. 213, and references cited therein.

71. L. Maraschi and G. C. Perola, Astr. Astrophys., 40, 387 (1975).

72. E. Kellogg, Astrophys. J., 218, 582 (1977).

73. T. W. Hartquist, Astrophys. J., 217, 3 (1977).

74. A. Cavaliere, L. Danese and G. De Zotti, Astrophys. J., 217, 6 (1977).

75. A. C. Fabian and P. E. J. Nulsen, preprint (1978).

76. J. E. Gunn and J. R. Gott, Astrophys. J., 176, 1 (1972).

77. S. F. Gull and K. J. E. Northover, Mon. Not. R. astr. Soc., 173, 585 (1975).

78. S. M. Lea, Astrophys. J., 203, 569 (1976).

79. F. Takahara, S. Ikeuchi, N. Shibazaki and R. Hoshi, Prog. Theo. Phys., 56, 1093, (1976).

80. L. L. Cowie and J. Binney, Astrophys. J., 215, 723 (1977).

81. L. L. Cowie and S. C. Perrenod, Astrophys. J., 219, 354 (1978).

82. A. C. Fabian and M. J. Rees, Mon. Not. R. astr. Soc., to be published (1978).

83. J. E. Felten and P. Morrison, Astrophys. J., 146, 686 (1966).

84. R. Bergamini, P. Londrillo and G. Setti, Nuovo Cimento, 52B, 495 (1969).

85. J. E. Felten and M. J. Rees, Nature, 221, 924 (1969).

86. M. J. Rees and G. Setti, Nature, 217, 326 (1968).

87. M. Rowan-Robinson, Mon. Not. R. astr. Soc., 150, 389 (1970).

88. P. A. G. Scheuer, in: Radio astronomy and cosmology, Reidel, Holland, 1977, p. 343.

89. P. C. Waggett, P. J. Warner and J. E. Baldwin, Mon. Not. R. astr. Soc., 81, 465, (1977).

90. A. G. Willis, R. G. Strom and A. S. Wilson, Nature, 250, 625 (1974).

91. J. P. Ostriker and T. X. Thuan, Astrophys. J., 202, 353 (1975).

92. D. Meier, Astrophys. J., 207, 343 (1976).

93. R. Giacconi and E. J. Schreier, Center for Astrophysics document. HEAO-B Consortium Observing Programme.

94. G. Lake and R. B. Partridge, Nature, 270, 502 (1977).

95. S. F. Gull and K. J. E. Northover, Nature, 263, 572 (1976).

96. J. E. Gunn and J. B. Oke, Astrophys. J., 195, L55 (1975).

HIGH RESOLUTION X-RAY OBSERVATIONS OF THE COMA CLUSTER

M. Johnson*, R. G. Cruddace**, G. Fritz**, S. Shulman** and H. Friedman**

*Department of Astronomy, University of Pennsylvania, Philadelphia, PA, U.S.A.
**E. O. Hulburt Center for Space Research, Naval Research Laboratory, Washington, D.C, U.S.A.

INTRODUCTION

During the last few years X-ray instruments of steadily improving angular resolution have made observations, which suggest that isothermal gas spheres provide at best an oversimplified and at worst an inaccurate description of X-ray emission from clusters of galaxies. The Perseus cluster is a prominent example. Although most of the emission is concentrated around the cluster center, about 35% extends along a protrusion to the west [1], following a line of galaxies, and the central emission region appears to be elongated [2] along an axis roughly aligned with the northerly extension of the radio source 3C84B. Finally, there is a bright region of emission about 3' in radius and centered on NGC 1275 [3]. Measurements of the extent of the X-ray emission in different energy bands [4] reveal that the size becomes greater as the photon energy increases, a result quite unexpected as a result of the emphasis given to adiabatic and isothermal models. This result could be due to cooling near the cluster center, a process which may have been precipitated by accretion of intracluster gas onto NGC 1275.

Another example is the Virgo cluster. Observations by Ariel 5 have turned up evidence that the X-ray emission is not centered on M87, but has a more complex structure which may be separated into two emission regions, a cool ($\sim 10^7$ °K) one around M87 and a hotter one ($\sim 10^8$ °K) concentrated around the cluster center near M84 [5]. There is also evidence for cooling in the M87 region, because the isothermal core radius in the soft X-ray band (15.3'$\pm$1.7', 0.5-1.5 keV [6]) is significantly less than the UHURU result in the 2-6 keV band (25' $\pm$ 4' [7]).

The Coma cluster would seem to have a better chance of fulfilling the predictions of the early theoretical models. The galaxy distribution approaches spherical symmetry and there is no sign of the violent activity exhibited by M87 and NGC 1275. However there is an elongation of the galaxy distribution [7], which at radii less than 30' has an E-W orientation. This and the E-W orientation of the line joining the giant elliptical galaxies NGC 4889 and 4874, should produce gravitational equipotential surfaces near the cluster center which are roughly ellipsoidal. We might expect to see this reflected in the distribution of intracluster gas and hence of the X-ray surface brightness. Further, it may not be true that the cluster contains no active galaxy. NGC 4874 is a radio source, which at 5 GHz has been resolved into two components separated by about 20" [8]. With these considerations in mind, in this paper we describe the results of an observation of the Coma cluster, using an X-ray instrument which has both high angular resolution and a response over a wide photon energy band (0.1 - 15 keV).

RESULTS

The observations were made by a rocket payload, which was launched at 1710 UT on 1977 February 15. The payload contained two large proportional counters equipped with 'Kimfol' windows 2 microns thick. Each viewed the sky through a collimator having a field of view of 10.3' x 4° (FWHM). The total effective aperture of the instrument was 754 cm^2. Details of the collimator assembly and alignment have been described in an earlier paper [9]. The estimated error in defining the pointing direction of the collimator, taking all uncertainties into account, is 1.2' along the narrow dimension of the collimator field.

The Coma cluster was scanned in three directions, having position angles of 105°, 150° and 20° (Fig. 1), at a controlled rate of 0.02 deg s^{-1}. The X-ray count-rate in the 0.5-5 keV energy band during these scans is shown in Fig. 2. These data have been fitted by an isothermal gas sphere model containing four parameters, the core radius, the time at which the collimator axis passed over the emission center, the central gas density and the background count-rate. A stretch of data between 90 and 120s (Fig. 2) was chosen as a background sample, which the model was forced to fit at large values of the cluster radius. Although isothermal conditions probably are not sustained in the cluster gas, the model fits the data sufficiently well to allow comparisons with the results of other investigations to be made readily.

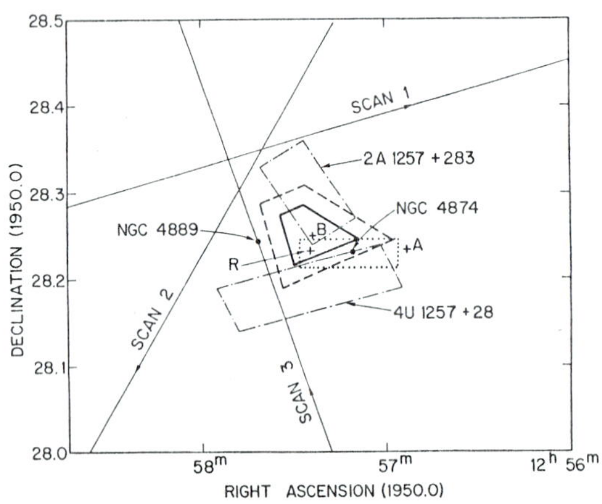

Fig. 1. A map of the central region of the Coma cluster, showing the tracks of the center of the collimator field during each scan and our 90% confidence error box for the center of the diffuse X-ray emission. The heavily lined box was obtained from fits to the data, and the dashed lines show the enlargement necessary as a result of a pointing uncertainty of 1.2'. The crosses show the centers of the clusters determined from galaxy counts, as measured by Bahcall (B; [12]) Rood et al (R; [13]), and Abell (A; [20]). The dotted rectangle is an error box containing the center of mass of the galaxies, as determined by Thompson and Gregory [14].

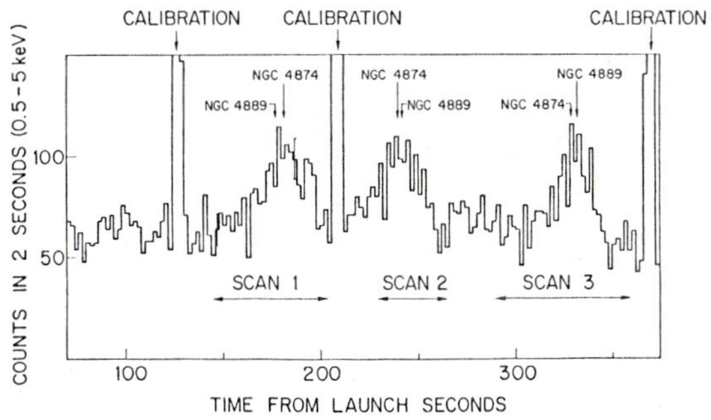

Fig. 2 - The X-ray count-rate during a rocket flight devoted to three scans of the Coma cluster,using detectors equipped with 10' x 4° (FWHM) collimators.

We have obtained 90% confidence limits for the parameters of interest, the position of the center and the core radius, using an increment $\Delta\chi^2$ to the minimum value of χ^2 which is appropriate to a single parameter fit. Following the prescription described by Avni [10] for projecting a χ^2-surface onto a subspace of the parameters, we have performed simulations to verify the correctness of the procedure for this particular non-linear model. Calculations were made of the count-rate expected from the scan of the collimator across an isothermal sphere. The core radius, sphere position, central gas density and background rate were defined. Then the predicted count-rate was perturbed using Poisson statistics and fitted to the model in two ways. The first allowed all parameters to vary and the second fixed a subset q of the parameters at the defined values. We calculated the χ^2-values of the best fits, χ^2_{min} and $\chi^2(q)$ respectively, and then tested whether $\Delta\chi^2 = \chi^2(q) - \chi^2_{min}$ was less than the χ^2 appropriate to q degrees of freedom. One hundred tests were made and the number of successes recorded. This simulation was performed for q = 1, 2, 3 and 4, and the results are summarized in Table I. The numbers of successes are close to 90%, and the average values of χ^2 are close to 1, 2, 3 and 4 for the one-, two-, three- and four-parameter fits respectively. We conclude that in estimating 90% confidence limits of one parameter, a χ^2 limit of χ^2_{min} + 2.71 may be employed safely.

TABLE 1. Result of Simulations of Fitting Isothermal Models to X-ray Data Obtained Using a One-Dimensional Scanning Collimator

Number of parameters, q, in the subset of interest	4	3	2	1
Number of successes* in 100 tries	94	90	89	91
Average value of $\chi^2(q) - \chi^2_{min}$	3.86	3.24	2.14	1.10

From the measurements of the emission center on each scan, we have constructed the 90% error box shown in Fig. 1. The enlargement shown by the dashed line allows for the 1.2' pointing uncertainty. This error box intersects both the Uhuru and Ariel 5 error boxes. Although it does just include NGC 4874 (and just excludes NGC 4889), it seems more likely that the center of the emission lies between the two giant ellipticals. This portion of the error box includes the centers of the galaxy distribution determined by Bahcall [11] and by Rood et al [12], and the eastern end of the error box drawn by Thompson and Gregory [13], in which they place the center of mass of the cluster. They state that the eastern end contains the mass center of the inner regions (radius $< 1^{\circ}$), and that the center moves west toward NGC 4874 as galaxies ever more distant from the center are considered.

In order to search for signs of asymmetry in the X-ray surface brightness, the 0.5-5 keV data for each scan have been fitted separately to an isothermal model. The results are shown in Fig. 3a, in the form of 90% confidence contours in a plot of core radius against background count-rate. The χ^2 for the best fits are 0.54, 1.07, and 1.02 per degree of freedom, and the core radii are $16.3^{+6.7}_{-6.4}$, $18.7^{+5.9}_{-4.3}$, and $9.7^{+4.3}_{-3.4}$ arc min for scans 1, 2 and 3 respectively. The results for scans 2 and 3 do not overlap, and a comparison of scans 1 and 2 with scan 3 provides us with evidence for some elongation of the surface brightness contours, along an axis which is approximately E-W. This feature is evident also in maps, not yet ready for publication, which have been constructed from the scan data.

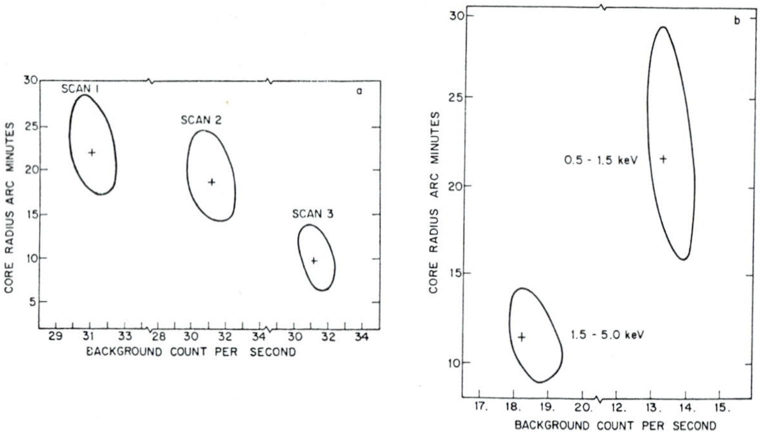

Fig. 3 - a) The results of fitting an isothermal model to the 0.5-5 keV data for each scan separately. The crosses represent the best fits, and the lines are 90% confidence contours for a 1-parameter fit ($\chi^2 + 2.71$). b) The results of fitting an isothermal model to all three scans simultaneously, using data in two photon energy bands. The crosses and contours have the same meaning as described in a) above.

The search for temperature gradients has been pursued using data divided into two energy bands, 0.5-1.5 and 1.5-5 keV. In each band separately, all three scans were fitted simultaneously to an isothermal model, giving the results shown in Fig. 3b. The 90% confidence limits for the "isothermal" core radii are $21.6^{+7.4}_{-5.6}$ arc min and $11.5^{+2.8}_{-2.5}$ arc min for the 0.5-1.5 and 0.5-5 keV bands respectively. The corresponding χ^2 of the two fits are .79 and .96 per degree of freedom for the 0.5-1.5 and 1.5-5.0 keV bands respectively. We conclude that the "isothermal" core radius increases as the photon energy band is lowered, and therefore that there is a marked departure from isothermal conditions.

DISCUSSION

The results of a number of measurements of the isothermal X-ray core radius of the Coma cluster are summarized in Table 2. In a band of moderately high photon energy, approximately 1.5-5 keV, our results agree reasonably well with those obtained by Lea et al. (Uhuru data [14]), Griffiths et al (Copernicus data [15]) and Malina et al (rocket data [16]). These measurements overlap in a band between 13' and 14'. In the softer energy band (0.5 - 1.5 keV), we find ourselves in better agreement with the Copernicus result [15] than with the result reported by Gorenstein et al [6], and suggest that the "isothermal" core radius is greater than 19' in this band.

TABLE 2. Summary of Measurements of the X-ray Isothermal Core Radius of the Coma Cluster

Source of Measurement	Energy Band keV	Core Radius arc minutes
Uhuru [14]	2 - 6	16 ± 3
Copernicus [15]	1.5 - 4.6	15 ± 3
UCB rocket [16]	.5 - 3.2	10^{+8}_{-5}
Present work	1.5 - 5	$11.5^{+2.8}_{-2.5}$
Copernicus [15]	.5 - 1.5	>19.5
SAO rocket [6]	.5 - 1.5	<15
Present work	.5 - 1.5	$21.6^{+7.4}_{-5.6}$

We conclude that isothermal conditions do not obtain in the Coma intracluster gas, and further that there is no sign of substantial cooling. The latter may be demonstrated, for example, by calculating the surface brightness distribution in the cooling model proposed by Cowie and Binney [17]. The effect of cooling in a cluster at a temperature of 10^8 ^oK is to make the brightness distribution significantly broader in the 1.5-5 keV band than the 0.5-1.5 keV band. Conditions in the intracluster gas appear to be between isothermal and adiabatic, which agrees with earlier results. Lea [18] obtained a best fit to the Uhuru data using a polytropic exponent of 1.35, and Malina et al. [19] obtained 90% confidence limits of 1.15 and 1.48 after fitting rocket data.

This conclusion is supported by other evidence revealed by our observations, which places the emission center between the two giant ellipticals and suggests an elongation of the emission region along a roughly E-W axis. The distribution of X-ray emission reflects, within the uncertainties of our results, the probably shape of the gravitational equipotential surfaces in the central regions (radii$<1^o$). We have found no evidence* for concentrated emission around NGC 4889 or 4874, neither in reconstructed X-ray images nor in the temperature distribution in the gas. The implication is that accretion and cooling ground giant ellipticals have not reached significant levels in the Coma cluster, whereas in the Perseus and Virgo clusters they have, and may have precipitated the violent activity in NGC 1275 and M87.

REFERENCES

1. W. Cash, et al., Ap. J., 209, L111, 1976.
2. R.S. Wolff, et al., Ap. J., 208, 1, 1976.
3. H. Helmken, et al., Ap. J., 221, L43, 1978.
4. M.P. Ulmer and J.G. Jernigan, Ap. J., 222, L85, 1978.
5. P.J.N. Davison, Mon. Not. R.A.S. 183, 39P, 1978.
6. P. Gorenstein, et al., Ap. J., 183, L57, 1973.
7. S.M. Lea, et al., Ap.J., 184, L105, 1973.
8. L. Schipper and I.R. King, Ap. J. 220, 798, 1978.
9. W. Jaffe and G.C. Perola, Astron. and Astrophys., 31, 223, 1974.
10. R.G. Cruddace, et al., Ap. J., 222, L95, 1978.
11. Y. Avni, Ap. J., 210, 642, 1976.
12. N. Bahcall, Ap. J., 183, 783, 1973.
13. H.J. Rood, et al., Ap. J., 175, 627, 1972.
14. L.A. Thompson, and S.A. Gregory, Ap. J., 220, 809, 1978.
15. R.E. Griffiths, et al., Nature, 250,471, 1974.
16. R. Malina, et al. Ap. J., 209, 678, 1976.
17. L.L. Cowie and J. Binney, Ap. J., 215, 723, 1977.
18. S.M. Lea, Astrophys. Letters, 16, 141, 1975.
19. R. Malina, et al., Ap. J., 219, 795, 1978.
20. G.O. Abell, Ap. J. Suppl., No. 31, Vol III, 211, 1958.

*Note added in proof: Results of subsequent analyses are leading us to reconsider this conclusion. However it is certainly true that any Coma galaxy X-ray source is less conspicuous than M87 and NGC 1275.

X-RAY OBSERVATIONS OF THE LUNAR OCCULTATION OF ABELL 401

M. P. Ulmer*, R. Kinzer**, R. G. Cruddace**, K. Wood**, E. T. Byram**, T. A. Chubb** and H. Friedman**

**Department of Physics and Astronomy, Northwestern University, Evanston, IL, U.S.A.*
***Naval Research Laboratory, Washington, D.C. 20375, U.S.A.*

INTRODUCTION

The A401 cluster of galaxies, classified as distance class 3 and richness class 2 by Abell [1], is one of the more distant clusters (z = .0746; Hintzen et al. [2]) identified as a source of X-ray emission. However, neither of the established error boxes shown in Fig. 1, 4U0254+13 (Forman et al., [3]) or 2A0255+132 (Cooke et al., [4]), encloses the center of the cluster, irrespective of whether one selects the Abell center, which coincides with a prominent cD galaxy in the cluster, or the center obtained by Bahcall [5]. Therefore there remains some uncertainty about its identification as a simple source centered on the cluster. Some support for the identification is given by the implied intracluster gas temperature, $7.8^{+2.4}_{-1.5} \times 10^8$ ^oK (Mitchell et al. [6], Mushotzky et al. [7]) and the three-dimensional cluster velocity dispersion (2410 km s^{-1} [2]), which fit well the correlation established using measurements of many X-ray clusters [6,7]). Radio observations of the cluster (Slingo, [8]) indicate that the extent of the intracluster gas is large. 4C13.17A is a source with an unusually long tail (7' or 1.2 Mpc, H_o = 50 km/sec/Mpc), and 4C13.17C is a double source in which the components are separated by only 50" (140 kpc). Therefore at radii as large as 20' (3.3 Mpc), the distance of 4C13.17C from the center, the intracluster gas appears to be dense enough to constrain the expansion of plasma from radio galaxies. The purpose of this paper is to present new information on the position and extent of the X-ray source, using the results of an observation made by the HEAO A-1 instrument during a lunar occultation of the cluster.

RESULTS

A401 was occulted by the Moon during two satellite orbits on February 14, 1978. As a result of the orbital motion of the spacecraft during this period two occultations were observed. The position of the Moon observed from the spacecraft is indicated by the two tracks shown in Fig. 1. The detector was a large proportional counter equipped with a 2.5μ Mylar window and filled with a mixture of 77.5% xenon and 22.5% methane at a total pressure of 1.4×10^4 n/m^2. It was equipped with a $1^o \times 4^o$ (FWHM) collimator having an aperture of 2350 cm^2. For a bremsstrahlung spectrum with a temperature of 8×10^8 ^oK, the mean efficiency of each detector is .33 in the 1-10 keV band. The limit-cycle of the spacecraft pointing system allowed a dither of a few tenths of a degree, which in this observation permitted the center of the field of view to wander over a region about 0.75^o in extent and containing both the cluster center and the Ariel error box.

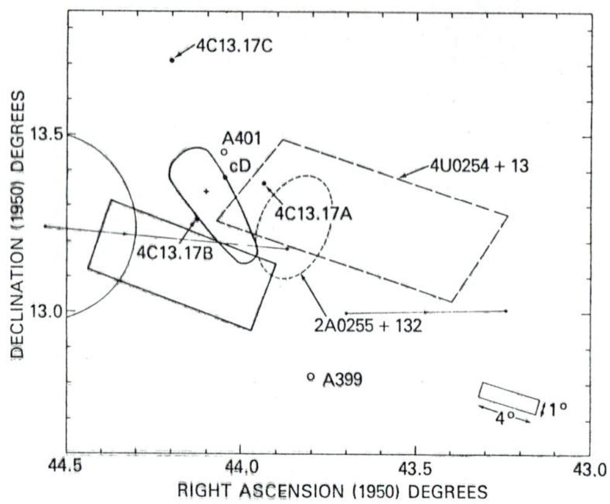

Fig. 1 A map of the region of sky containing the clusters of galaxies Abell 401 and 399. The open circles are the cluster centers, as determined by Abell [1] for A399 and by Bahcall [5] for A401. Abell places the center of A401 very close to the cD galaxy. The dashed lines show the Uhuru and Ariel error boxes for the center of the X-ray source, and the heavily-lined contour defines our error region. The light lines and semi-circle describe the tracks of the center of the moon during two occultations of A401, and the size of the moon. The small sketch in the lower right corner, which is not drawn to scale, shows the orientation of the detector collimator.

In Figure 2 we show the detector count-rate in the 1-10 keV energy band during selected periods of the two occultations. The detector background count-rates, measured before source acquisition, have been subtracted. These stretches of data were chosen for analysis, because the particle background, as measured by the detector coincidence count-rate, was known to be steady. The data were tested further for variable background effects by changing the length of the data sample by $\pm 30\%$ and forcing the fitting program to include a variable background term, taken to be a function of the coincidence count rate. In no case was there a significant change from the results obtained with a constant background. The background count-rate was 125 cts/5.12 sec, which is the cause of the large fluctuations evident in Fig. 2. Other data obtained during this observation but not used in the final analysis, showed significant changes in background. The presence of an X-ray source is evident both during the initial acquisition of the target at the beginning of the observation, and during the second occultation (Figure 2b), when there is a decrease in count-rate as the Moon covers the source. No motion of the collimator can explain this modulation. We observed no sudden change in count-rate, and therefore can conclude that most of the X-rays come from an extended source.

The data have been fitted to models in which the X-ray source is a point, a uniform disk, a uniform disk with a point at the center, a uniform disk and a point, and a uniform disk with a point located at the center of A399. The purpose of the latter two models is to investigate the implications of assuming some other emission region besides that centered on A401. Such a possibility is evident in the location of the published error boxes (Figure 1) and was mentioned by Cooke et al. (1978). A point source is chosen for convenience in estimating limits to the intensity of such emission and the corresponding size of the region around

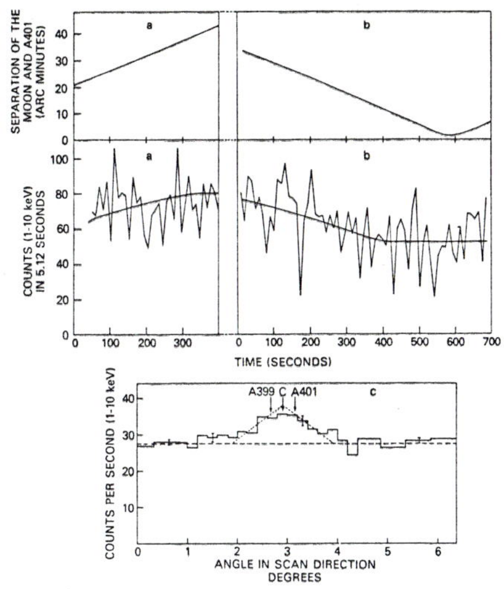

Fig. 2 The X-ray count-rate during each of the two lunar occultations of A401. The statistical fluctuations are large because a relatively large background level has been subtracted from the signal. The smooth curves in the lower graphs are the best fits of a uniform disk model (Table 1) to the data.

A401. With the data at hand, a convergent solution was not obtained when the model contained two extended sources. The parameters in these fits were the position and intensity of the point and disc sources, and the radius of the disc. The results are summarized in Table I, in which uncertainties are 90% confidence levels for 3-5 parameter fits, the number depending on the model.

Table I

Summary of Results of Fitting Models to Data Obtained during the Lunar Occultation of A401

Model	Best fit positions, source center R.A.	Dec	No. of degrees of freedom	χ^2 per degree of freedom	Confidence in the fit +	Source* intensity counts s^{-1}	Disc radius arc minutes	Point source intensity / Disc source intensity
Point Source	44.0	13.4	208	1.62	10^{-11}	5.9	-	-
Uniform Disc Source	44.10	13.33	207	1.20	.03	13.4±4.5	25.5±4.4	-
Uniform disc source + central point source	44.10	13.35	206	1.20	.035	11.9±5.9	$30.2^{+22.0}_{-10.2}$	$.13^{+.20}_{-.13}$
Uniform disc source + point source	44.10 42.38	13.43 11.98	204	1.18	.04	$11.9^{+17.3}_{-8.6}$	$25.9^{+9.3}_{-5.9}$	$.9^{+.3}_{-.9}$
Uniform disc source + point source at A399	44.10 43.80	13.40 12.82	206	1.19	.04	$9.3^{+12.2}_{-6.1}$	$13.4^{+9.1}_{-3.4}$	$.43^{+.05}_{-.21}$

+Probability of finding a larger value of χ^2 if the model is correct.
*To convert counts s^{-1} to energy flux in ergs cm^{-2} s (1-10 keV), multiply by 6.4×10^{-12}.

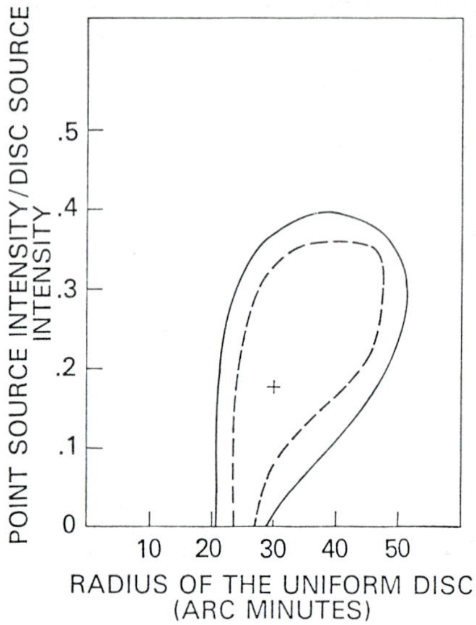

Fig. 3 Contours of 90% (solid line) and 67% (dashed line) confidence for the fit of a uniform disc and point source model (model 3 in Table I) to the data. The cross marks the best fit. The χ^2-values for these contours are those appropriate to a 4-parameter fit.

Using the results for the uniform disc and central point source model, in Fig. 3 we have projected the χ^2-surfaces for 90% and 67% confidence levels onto a plane in which the co-ordinates are disc radius and the ratio of the point and disc source intensities. Finally we have used the results to calculate a 90% confidence error region (Fig. 1) for the position of the center of a uniform disk representing the X-ray source.

DISCUSSION

The results we have obtained are more consistent than earlier results (Fig. 1) with the hypothesis that the X-ray emission is concentric with the distribution of galaxies, especially if we adopt Abell's result placing the center at the cD galaxy. The measured luminosity of the source, 1.9×10^{45} ergs s^{-1} (1-10 keV) agrees well with earlier values of 2.2×10^{45} ergs s^{-1} (2-10 keV) [6,7]. Although Figure 3 places no stringent limits on the strength of any point X-ray source, it does set a lower limit to the radius of the extended source, which is surprisingly large. The radius of a uniform disc is only 10% larger than the isothermal core radius, so that the X-ray core radius is greater than 18' (Fig. 3) or 2.4 Mpc. This is at least four times greater than the core radii of the Perseus and Coma clusters (400-600 kpc, or 10'-15' c.f. Malina, Lampton and Bowyer, [9]), which implies that the central gas density is substantially lower, by perhaps a factor of 3 or more. The quality of the best fit to the data was not changed when a second source was introduced (models 4 and 5 in Table I). The best fits for these cases suggest that this source, if it exists, is to the south of A401, and has an intensity no more than one half of the total.

The extent of the X-ray emission is similar to the extent of the intracluster gas deduced from radio observations, which place a head-tail galaxy 3.3 Mpc from the cluster center[8]. This is not the first evidence for existence of intracluster gas over a megaparsec from the cluster center. For the Perseus cluster, using data obtained in a higher X-ray energy band, 5.5 - 15.5 keV, Ulmer and Jernigan [10] derived a core radius of 40^{+6}_{-16} arc min, or $1.2^{+0.2}_{-0.5}$ Mpc. Further, the radio-tail galaxies IC310 and NGC1265 are located 0.85 and 1.2 Mpc, respectively, from the Perseus cluster center.

In summary, we have confirmed the Uhuru and Ariel 5 identification of an X-ray source with A401, and we have shown that the extent of the emission is at least 18' or 2.4 Mpc. We suggest that these extensive haloes of intracluster gas, having diameters of several magaparsecs, may be common, and should be emphasized in future studies and observations of clusters.

REFERENCES

1. G.O. Abell, Ap. J. Suppl.,No. 31, Vol. III, 211,(1958).
2. P. Hintzen et al., Ap. J., 212, 8,(1977).
3. W. Forman et al., to be published in Ap. J. Suppl.
4. B.A. Cooke et al., M.N.R.A.S., 182, 489,(1978).
5. N.A. Bahcall, Ap. J., 198, 249,(1975).
6. R.J. Mitchell et al., M.N.R.A.S., 181, 25P,(1977).
7. R.F. Mushotzky et al., NASA Technical Memorandum 78097, (1978).
8. A. Slingo, M.N.R.A.S., 168, 307, (1974).
9. R. Malina, et al., Ap. J., 209, 678,(1976).
10. M.P. Ulmer et al., Ap. J. Letters, 222, L85, (1978).

X-RAY STRUCTURE OF CLUSTERS OF GALAXIES

Paul Gorenstein and Daniel Fabricant

Harvard-Smithsonian Center for Astrophysics, 60 Garden Street, Cambridge, Massachusetts 02138, U.S.A.

INTRODUCTION

The X-ray structure of three clusters of galaxies have been observed with a two-dimensional imaging X-ray telescope in a series of sounding rocket flights. Considerable structure has been detected in the X-ray surface brightness distribution with particular effects concentrated in the vicinity of massive members of the cluster. Some results pertaining to M87 in Virgo and the Perseus cluster of galaxies have already been reported [1,2]. A few additional results on these objects are reported here, plus some works in progress on the Coma cluster of galaxies.

M87

Spectral observations on the Virgo cluster from OSO-8 have revealed a thermal component with a temperature of about 25 x 10^6 K and a relative abundance of iron that is close to the cosmic value [3,4]. For these conditions one would expect that the X-ray spectrum near 1 keV is dominated by a complex of emission lines due principally to Fe XXVII-XXIV. The intensity of the complex should be considerably greater than the Fe XXV-XXVI lines found in the 6-7 keV range. The focal plane detector was a position sensitive proportional counter that had an energy resolution of about 50% FWHM at 1 keV. With this in mind, we analyzed the spectral data from the rocket flight on M87 from the standpoint of detecting the line complex. Furthermore, the spatial resolution of the imaging system could be used to search for gradients in the temperature and/or iron abundance of the hot gas in the halo of M87.

Ariel 5 [5,6] reports of a new component in Virgo some 2° in extent should not have a significant effect upon the interpretation of our results. Within a $12\overset{'}{.}5$ radius of M87 this new component is expected to make no more than a ten percent contribution to the total flux. In any case, our results refer to the total spectrum of Virgo in three bins at different radial distances from M87.

Plasma emission spectra were calculated from the Raymond and Smith (RS) [7] code for a range of temperatures and abundances of heavier elements. Three sets of emission spectra files were generated for use in this analysis: one in which the Fe/H ratio was varied between 10^{-7} and 10^{-4} while holding the temperature fixed at 2.55 x 10^7 K, and two in which the temperature was varied between 1.6 and 4 x 10^7 K while keeping the Fe/H ratio constant at 2.5 x 10^{-5}. The energy range and statistical precision of our data were insufficient to allow T and Fe/H to be determined simultaneously. After including N_H as a free parameter to account for interstellar absorption, these models were convolved with the instrumental response and compared to the observed pulse height spectra for the radial bins, 0' to 5', 5' to 10', 10' to 25', as well as for the entire data set 0' to 25'.

A good fit to all of the data within 0' to 25' region around M87 is shown in Fig. 1. Fe/H was fixed at 4.0 x 10^{-5} and the following values were determined for T and N_H in a RS model spectrum (90% confidence): T = (2.6 $\pm$ 0.5) x 10^6K, and N_H = (2.5 $\pm$ 1.4) x 10^{20}/cm^2. The values of T and N_H are consistent with the soft component for Virgo found by Mushotsky [4] from data at higher energy. The value of N_H is in agreement with the column density of hydrogen as derived from tables of 21 cm radiation of Tolbert [8] and Heiles [9]. The best fit of a bremsstrahlung plus gaunt factor model, without lines (dashed curve) is not acceptable for two reasons: the minimum chi-square is too high and the column density of hydrogen, at 6.5 x 10^{20} cm^2, is unreasonably large.

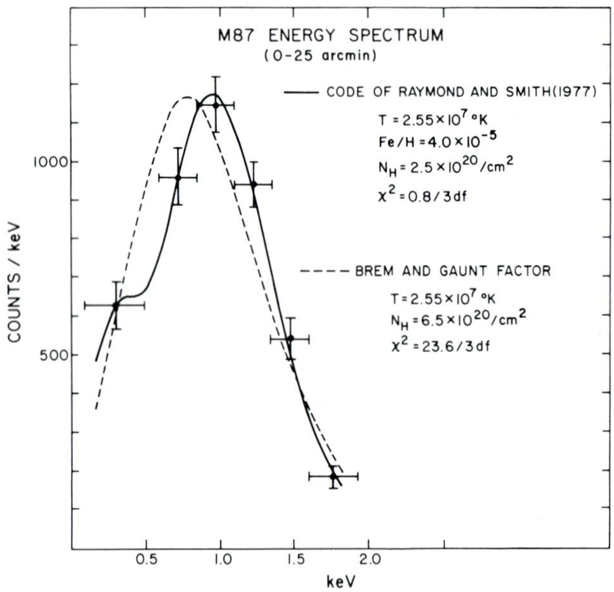

Fig. 1. Points are observed pulse height spectrum of M87 (r < 25'). The solid curve is the result of convolving the spectrum of Raymond and Smith [7] for a 2.55 x 10^7K plasma containing a cosmic elemental abundance with the instrument response and allowing the column density of hydrogen, N_H, to be a free parameter. The dashed curve is the best fit of a 2.55 x 10^7K bremsstrahlung plus gaunt factor spectrum. On the basis of a chi-square test the fit of the solid curve to the data is acceptable while that of the dashed curve is not.

Tables 1 and 2 show the results of the spectral analyses carried out on the radial bins 0' to 5', 5' to 10', and 10' to 25' (as well as 0' to 25'). In obtaining Table 1, T is fixed at 2.55 x 10^7 K, while Fe/H and N_H are allowed to vary freely. For Table 2, Fe/H is fixed at 4 x 10^{-5}, while T and N_H are varied. Results are also given in both tables for the case where N_H is fixed at 2.5 x 10^{20} /cm^2 over the M87 region. This is probably a more realistic situation than allowing N_H to vary freely between 1.0 and 4.0 x 10^{20}/cm^2. The Fe/H ratio is comparable to the cosmic value of 4.0 x 10^{-5} [10] in all of the bins. The temperature ratios of the three bins to the entire data are given in Table 3. All three are consistent with a temperature ratio of one, although the statistical error is large in the 10' to 25' bin.

THE COMA CLUSTER

A picture of the 0.15 - 2.0 keV X-ray emission from the Coma cluster of galaxies binned in 2' x 2' cells over a region of $\sim 1^{o}$ diameter is shown in Fig. 2. As the resolution of the instrument is no better than 4', the picture has been smoothed by taking the final number of counts in each cell to be one-fourth of the sum of the original number of counts in that cell and three adjacent cells. Several features are notable. In general there are significant variations in the surface brightness. While the intensity is largest in the vicinity of the center of the picture, the actual centroid of the flux is at a local minimum. Within an aspect error of no more than 2', the centroid of the X-ray emission is coincident with the "supergiant" elliptical galaxy NGC 4874. Fig. 3 shows the distribution of counts in a strip that is 6' wide in declination and 2' wide in right ascension as a function of right ascension. The dashed line represents the best fit to all of the data assuming azimuthal symmetry by integrating over ϕ and fitting the radial distribution of surface brightness. The chi-square between the best fit and the 9 data points in the center is 23 which is unacceptably high, whereas the fit to all of the data (20 degrees of freedom) is acceptable. We interpret this as indicative of significant variations in surface brightness in the central region of the Coma cluster. Clumping of the gas is probably caused in part by strong local variations of the gravitational potential. The region north of the center (Fig. 2) is characterized by a larger flux than the region south of the center. However, the north-south excess occurs entirely in the 0.15 - 0.28 keV band, where the contribution from the galaxy is most pronounced. The excess is equal to the galactic flux we have detected in other pointing positions. The enhanced number of counts along the border is an artifact originating in saturation effects of the position sensing circuits.

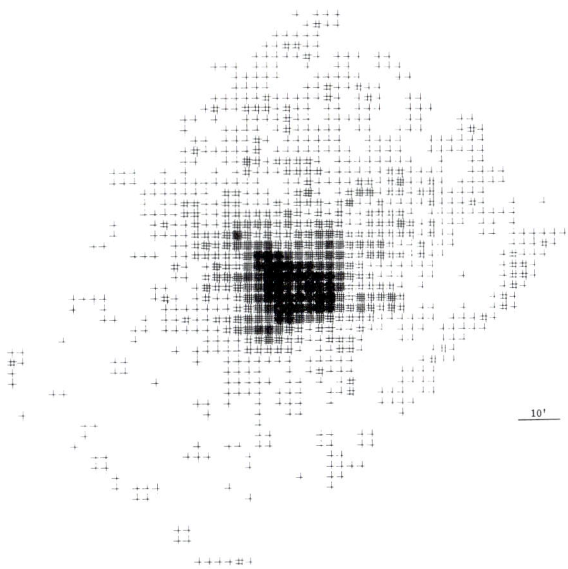

Fig. 2. X-ray map of the Coma cluster of galaxies for the 0.15 - 2.0 keV band. Cells are 2' x 2'. North is at the top and east at the left of the picture. Data have been smoothed by averaging four adjacent cells before plotting. The excess in the northern half of the picture occurs primarily in the 0.15 - 0.28 keV range and may have a galactic origin.

Ignoring the local variations in surface brightness we assume azimuth symmetry and plot the radiation distribution of surface brightness in Fig. 4. The solid curve is a thermal hydrostatic model and the dash curve is an adiabatic model. Both are consistent with the data.

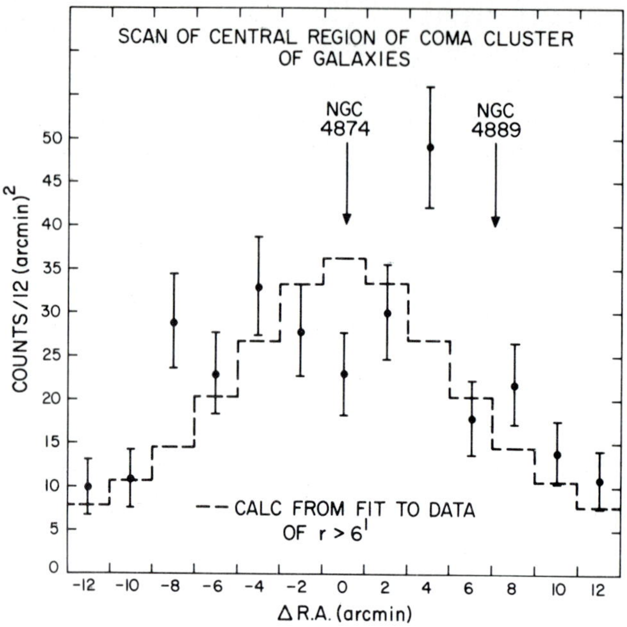

Fig. 3. X-ray intensity in central region of the Coma cluster in strips of 2' (R.A.) x 6' (DEC) centered on DEC = 28° 14'. The dashed histogram is calculated from the best fit to all of the data. The positions of NGC 4874 and NGC 4889 are indicated.

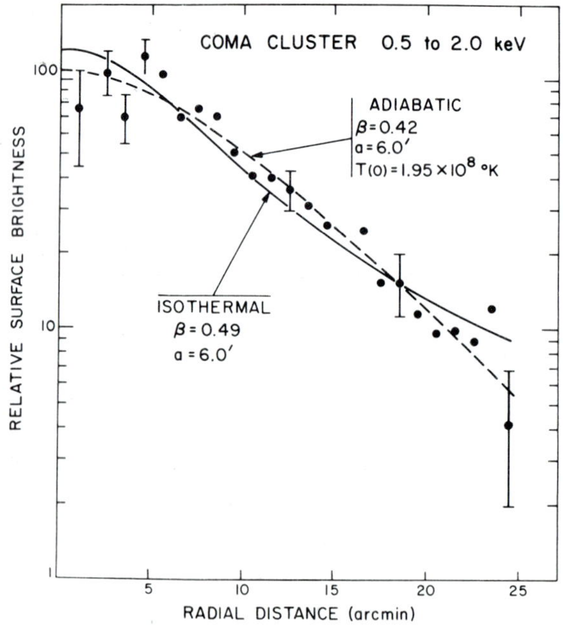

Fig. 4. Radial distribution of X-ray surface brightness of the Coma cluster of galaxies. Best fits of an isothermal hydrostatic and adiabatic hydrostatic model are shown.

CONCLUSIONS

The X-ray structure of clusters of galaxies exhibits considerable variations in surface brightness. Significant enhancements in regions of ~ 100 kpc above that of the cluster sources as a whole have been detected at M87 in the Virgo cluster and NGC 1275 in the Perseus cluster. The central region of the Coma cluster exhibits local variations in surface brightness or clumpiness. The centroid of the X-ray emission is near the supergiant elliptical galaxy NGC 4874 and is a local minimum in intensity.

The ellipticity of the extended X-ray sources in clusters can be investigated and on advanced missions should reveal the true distribution of mass in the cluster. From rocket data we can detect a (30 ± 10)% East-West enhancement in the Perseus cluster. This is much less than the amount derived from the distribution of bright galaxies. We detect a marginal E-W enhancement within the central region of the Coma cluster that is consistent with the galaxy distribution.

We see positive evidence for the Fe XXVII-XXIV complex of lines in the spectrum of M87. The most straightforward interpretation of our spatially resolved spectral data is that the gas in the halo of M87 is isothermal and contains an iron abundance that is uniformly close to the cosmic value.

Table 1. Fe/H Relative Atomic Abundance as a Function of Radius for M87

$T = 2.55 \times 10^7$ K

Radial Range (arcmin)	Fe/H* $(1.0 < N_H < 4.0 \times 10^{20})$	Fe/H* $(N_H = 2.5 \times 10^{20})$
0 - 5	$> 0.9 \times 10^{-5}$	$> 2.8 \times 10^{-5}$
5 - 10	$> 0.2 \times 10^{-5}$	$(3.9 \pm 2.8) \times 10^{-5}$
10 - 25	$> 10^{-7}$	$> 1.2 \times 10^{-5}$
0 - 25	$> 1.0 \times 10^{-5}$	$(4.3 \pm 2.6) \times 10^{-5}$

90% confidence

Table 2. Temperature as a Function of Radius for M87

$Fe/H = 4.0 \times 10^{-5}$

Radial Range (arcmin)	Temperature* $(1.0 < N_H < 4.0 \times 10^{20}$	Temperature* $(N_H = 2.5 \times 10^{20})$
0 - 5	$(2.75 \pm 0.75) \times 10^7$ K	$(2.45 \pm 0.35) \times 10^7$ K
5 - 10	$(2.55 \pm 0.5) \times 10^7$ K	$(2.65 \pm 0.45) \times 10^7$ K
10 - 25	$> 2.1 \times 10^7$ K	$(2.9 \pm 0.7) \times 10^7$ K
0 - 25	$(2.7 \pm 0.5) \times 10^7$ K	$(2.55 \pm 0.30) \times 10^7$ K

90% confidence

Table 3. Temperature Ratios for Various Radial Bins Around M87

$Fe/H = 4.0 \times 10^{-5}$
$N_H = 2.5 \times 10^{20}$

Radial Range (arcmin)	T in Radial Range/T (0' - 25')*
0 - 5	0.96 ± 0.16
5 - 10	1.04 ± 0.21
10 - 25	1.14 ± 0.30

90% confidence

ACKNOWLEDGEMENTS

We would like to thank F.R. Harnden Jr. and K. Topka of the Harvard-Smithsonian Center for Astrophysics for their contribution to the analysis of sounding rocket data that is the basis for this paper. We would also like to thank J. Raymond for the use of his computer code in the analysis of spectral data. This work was supported by the National Aeronautics and Space Administration under grant NSG-5138.

REFERENCES

1. P. Gorenstein et al., The Astrophysical Journal (Letters), 216, L95 (1977).
2. P. Gorenstein et al., The Astrophysical Journal, in press (1978).
3. P.J. Serlemitsos et al., The Astrophysical Journal (Letters), 211, L63 (1977).
4. R.F. Mushotzky et al., The Astrophysical Journal, in press (1978).
5. P.J.N. Davison, Monthly Notices of the Royal Astronomical Society, 183, 39p (1978).
6. A. Lawrence, Monthly Notices of the Royal Astronomical Society, submitted (1978).
7. J.C. Raymond and B.W. Smith, Astrophysical Journal Supplement, 35, 419 (1977).
8. C.R. Tolbert, Astronomy and Astrophysics (Supplement), 3, 349 (1971).
9. C. Heiles, Astronomy and Astrophysics (Supplement), 20, 37 (1975).
10. C.W. Allen, Astrophysical Quantities, 3rd Edition, University of London (1973).

HEAO-1: LOW ENERGY X-RAY OBSERVATIONS OF CLUSTERS OF GALAXIES

S. M. Lea*, K. O. Mason** †, G. Reichert**
and G. Riegler***

*Astronomy Department and Space Sciences Laboratory, University of California, Berkeley, California 94720, USA
**Space Sciences Laboratory, University of California, Berkeley, California 94720, USA
***Jet Propulsion Laboratory, 4800 Oak Grove Drive, Pasadena, California 91103, USA

INTRODUCTION

The discovery of iron line emission at 6.7 keV (e.g. [1], [2]) in the X-ray spectra of several clusters has provided convincing evidence for a thermal origin of the bulk of the emission, and for an abundance of iron within a factor of 2 of the solar value. This abundance determination provides strong constraints on the origin and evolution of the gas (e.g. [3], [4]). We report here preliminary results for clusters of galaxies from the low energy detectors (LED's) of the HEAO-1 A-2 experiment[1], including the detection of iron line emission at 1.1 keV.

OBSERVATIONS

Of the three strongest cluster X-ray sources, Coma, Perseus, and Virgo, both Coma and Perseus are within 3° of low energy stellar X-ray sources, HZ 43 and Algol, respectively. HZ 43 is stronger than the Coma cluster below 1 keV, and precludes our making any statement about the Coma spectrum without an extensive analysis. Algol is much weaker and is not within the field of view during the last 3 days of the Perseus observation. This discussion will concentrate on our results for Perseus and Virgo.

The Perseus Cluster

The Perseus cluster was observed in October of 1977. Our best fit spectrum indicates thermal bremsstrahlung emission from a plasma at temperature $2(\pm1) \times 10^7$ K, with an absorbing hydrogen column of $2(\pm1) \times 10^{21}$ cm^{-2}. This column is in agreement with previous determinations (3,1) and with the 21 cm hydrogen column of 1.6 10^{21} cm^{-2} (6,7). The temperature is lower than the previously measured value of $\sim$7.8 10^7 K (1,3). Our data are inconsistent with this latter temperature at more than 99% confidence. No line emission is evident in our data. The spectrum is marginally consistent, at the 3 sigma level, with emission from a plasma having

[1]The A-2 experiment of HEAO-1 is a collaborative effort led by E. Boldt of GSFC and G. Garmire of CIT with collaborators at CIT, GSFC, UCB and JPL.
†Miller Research Fellow

the iron abundance determined by OSO-8 (1).

The Virgo Cluster

The Virgo Cluster was scanned in December, 1977. We observed a source centered on the position of M87. The raw count rates in three broad band energy channels are shown in Fig. 1. Also shown is the response of the detector to a point source (the actual response to the star HZ 43 has been plotted). The source is extended in all three channels with a FWHM (for an assumed Gaussian source) of 1.1^{o}. We see no evidence for a change in size with energy in the range .2 - 2 keV. Our measured size is consistent with previous measurements, but probably not with recent low energy measurements (9). The size is nevertheless still much smaller than the cluster as a whole. The discrepancy between the two measurements in approximately the same energy range may be due to a region of somewhat lower surface brightness to which the HEAO detectors are more sensitive because of their larger field of view. It suggests that caution be used when comparing sizes at different energies when the observations are made with dissimilar detectors. We have also detected a broad, low surface brightness extended emission region around M87. We believe that this region originates in our galaxy. Morphologically, it seems to be associated with the north polar spur, while its spectrum indicates an absorbing column $< 10^{19}$ cm^{-2}, which is inconsistent with the observed galactic 21 cm column of (1.8 - 2.8) x 10^{20} cm^{-2} observed in the direction of the Virgo cluster [7,10]. However, this emission would prohibit us from observing a low surface brightness extension similar to the one discussed by Lawrence [11], if such existed in our energy range.

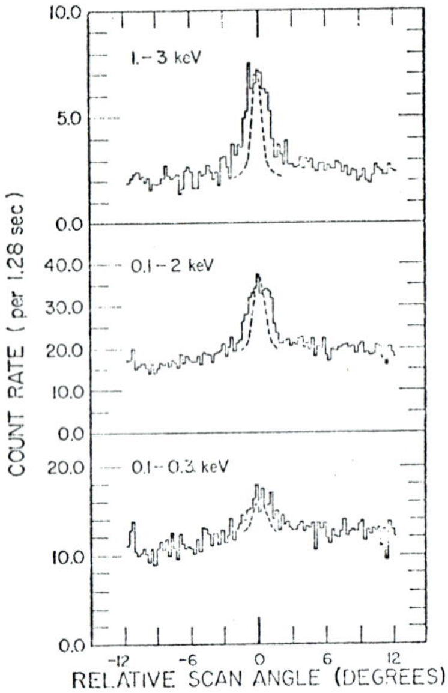

Fig. 1 Scan across the Virgo cluster. The detector response to a point source is shown as a dashed line.

TABLE 1 Spectral Fits to Virgo Data

MODEL	Degrees of freedom	χ^2	Log T	Log N_H	Prob.	Other parameters
H, He Plasma	24	35.8	7.6	20.5	.04	---
Thermal plus	21	24.7	6.1 3.2*	21.0 21.7	.25	---
Thermal bremsstrahlung	22	21.8	7.6	20.4	.5	E_{line} = 1.15 keV EW = .176 keV
Plasma, cosmic abundance	24	22.9	7.64	20.3	.5	
Plasma, variable	23	22.8	7.60	20.3	.5	F_e = .78 solar

*Power law index

Our spectral results for the source at M87 are shown in Table 1. We could not obtain an acceptable fit for a single temperature thermal bremsstrahlung model or for a power law. The best fit bremsstrahlung spectrum is shown in Fig. 2a. A two component model could be fit. However, neither component is consistent with the spectrum seen at higher energies (e.g. [1]), while both components require absorbing columns in excess of 10^{21} cm^{-2}, implying absorption intrinsic to the source. This requirement is inconsistent with our observation of finite source extent. A more realistic fit was obtained for a thermal bremsstrahlung model with an emission line. The best fit was for a line centered at 1.15 keV, the expected position of the iron line complex at a temperature of 4×10^{7} K. An equally good fit was obtained for emission from a plasma of cosmic composition at 4×10^{7} K [12]. This spectrum is shown in Fig. 2b. As suggested by the single line fit, the only feature which we see in our data with any significance is the 1.15 keV iron line. This line is a blend of transitions in Fe XXIII and Fe XXIV. The abundance which we determine for iron is $3.1\ (\pm^{7}_{2.3}) \times 10^{-5}$ relative to hydrogen, e.g., the abundance is within a factor of 3 or so of the "cosmic" value [13].

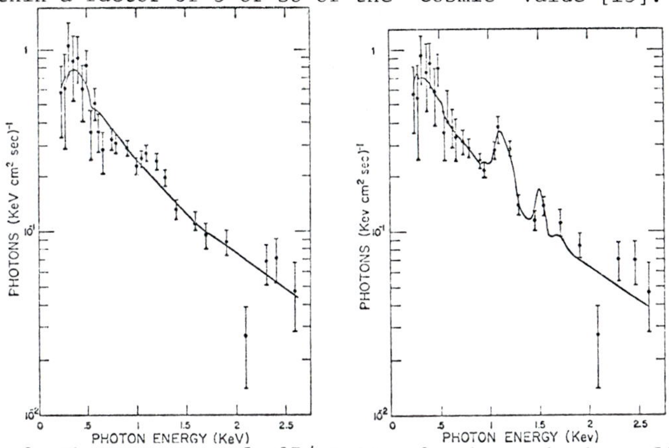

Fig. 2 The spectrum of M87/VIRGO. 2a shows the best fit bremsstrahlung spectrum; 2b shows the best fit model for emission from a cosmic plasma.

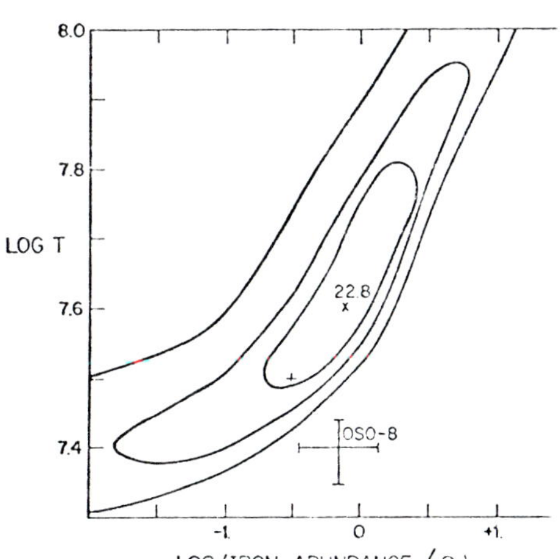

In Fig. 3 we show the χ^2 contours in (temperature, abundance) parameter space. The temperature determination is based both on the continuum shape and on the position of the line which moves to lower energy as the temperature drops.

Fig. 3 Contours of constant χ^2. The best fit point is shown by an X. Contours are drawn at 68%, 90% and 99% confidence. Also shown is a point corresponding to the low temperature component found by OSO-8 [1].

Using our model and the theoretical calculations of Raymond and Smith [12], we predict the intensity of the 6.7 keV iron line to be $0.0003^{+.006}_{-.0026}$ photons $cm^{-2} sec$. This result should be compared with the observed value of 0.0021 ± .0006 photons $cm^{-2}sec^{-1}$[14].

DISCUSSION

The main discrepancy between our results and those of previous observers is the measurement of temperature. For Perseus, our low temperature may indicate that we are observing a region of more extended, cooler gas not seen by the higher energy detectors. The source is not obviously extended in our scans, but with a collimator of 1.5° FWHM sources up to ∿ 500 kpc in radius would appear as point sources at the distance of the Perseus cluster. Such an extended region may also have been missed by X-ray telescopes [15] if the surface brightness were low. The lack of line emission indicates that this region consists of lower abundance (primordial?) material than is found in the cluster center.

For Virgo, we observe a higher temperature than either OSO-8 [1] or Ariel V [16]. The Virgo spectrum is complex above 2 keV requiring at least 2 components to obtain a good fit [1,16]. In view of the uncertainties in such multi-component fits, and especially since no line emission (other than the 7 keV iron line) was included by these authors, we feel that our data are not necessarily incompatible with the higher energy results. Our data points fit well with the Ariel V points [16] at 2 keV.

It is obvious from these data that the simple models which have been used to describe clusters of galaxies are inadequate. For Virgo at least two components are indicated by both the spectral and spatial data. Perhaps one source is associated with M87, and one with the cluster as a whole. A similar situation may hold for the NGC 1275/Perseus cluster source. We hope that further insight may be obtained when all the data from the A-2 experiment (LED, MED and HED) are combined into a single spectrum for each source, covering the range .2 - 50 keV.

Observation of the 1.1 keV iron line in more distant sources by HEAO-B may give us some important clues as to the evolution of clusters and galaxies within clusters, by allowing us to measure the abundance of iron as a function of time. It is unfortunate that this line is strongest in the cooler (and hence less luminous [1]) cluster sources.

REFERENCES

1. R.F. Mushotzky et al., Astrophys. J. (in press, 1978).
2. R.J. Mitchell et al., Mon. Not. R. Ast. Soc 176, 29P (1978).
3. R.F. Malina et al., Astrophys. J. 219, 795 (1978).
4. D.S. De Young, Astrophys. J. 223, 47 (1978).
5. S. Perrenod, Astrophys. J. (in press, 1978).
6. M.P. Hughes et al., Astrophys. J. (Suppl.) 23, 223 (1971).
7. C. Heiles, Astr. and Astrophys. Suppl. 20 37, (1976).
8. E. Kellogg et al., Astrophys. J. (Letters) 174, L65 (1972).
9. P. Gorenstein et al., Astrophys. J. (Letters) 216, L95 (1977).
10. C. Heiles et al.,Aust. J. Phys. Astrophys. Suppl. (submitted 1978).
11. A. Lawrence, preprint (1978).
12. J.C. Raymond et al., Astrophys. J. (Suppl.) 35, 419 (1977).
13. C.W. Allen, Astrophysical Quantities, The Athlone Press (3rd Edition), 1973.
14. P.J. Serlemitsos et al., Astrophys. J. (Letters) 211, L63 (1977).
15. P. Gorenstein et al., Astrophys. J., (in press, 1978).
16. P.J.N. Davison, Mon. Not. R. Astr. Soc., (in press, 1978).

Unidentified Extragalactic Sources and Diffuse Component

FAINT EXTRAGALACTIC OBJECTS AND THE UNIDENTIFIED HIGH LATITUDE X-RAY SOURCES

S. Shulman

E. O. Hulburt Center for Space Research, Naval Research Laboratory, Washington, D.C. 20375, U.S.A.

INTRODUCTION

Since high sensitivity satellite X-ray observations were begun with the launch of Uhuru in 1970, a large number of relatively faint, high galactic latitude sources have been detected. A substantial fraction of these now have been identified with extragalactic optical objects, but a certain number of X-ray sources do not appear to coincide with cataloged objects. In this paper, I examine the classes of extragalactic objects which have been identified with X-ray sources in the Fourth Uhuru (4U) Catalog [1] and the Ariel V (2A) Catalog [2] and discuss preliminary results from the HEAO A-1 instrument. In making this survey, I have used only the data presented in the available versions of the catalogs and have not made any attempt to update this information for individual sources which have since been discussed in the literature. However, as the results are statistical in nature, these few sources should not seriously affect the conclusions.

In comparing the 4U and 2A catalogs, one must exercise caution because of the different source location capabilities of the Uhuru and Ariel V survey instruments. These differences are illustrated in Fig. 1, which shows the number of sources located by each satellite at galactic latitudes $|b| \geq 20^{\circ}$ as a function of the localization or error box size.

The error box distributions are distinctly different for the two catalogs, with half the 2A sources located to better than 0.25 sq. deg. while almost half the 4U sources have error boxes exceeding 1.0 sq. deg. in area. Also shown in Fig. 1 are the number of accidental associations with X-ray sources expected in searching a catalog with an average source density of 1 source per 10 sq. deg. (approximately the density of the Abell cluster catalog [3]). The number of chance identifications as a function of error box size indicates that identifications of X-ray sources with error boxes less than 1.0 sq. deg. are reasonably secure while larger error boxes cannot provide reliable identifications.

Based on the above conclustions, it is necessary to restrict the samples from the two catalogs further in order to insure similar location precision and to minimize the probability of including chance identifications. In the following discussions then, I will restrict the samples to $|b| \gtrsim 20^{\circ}$ and error boxes less than 1.0 sq. deg. In addition, only declinations above -30° will be considered as it is only in the northern sky that the optical catalogs are reasonably complete.

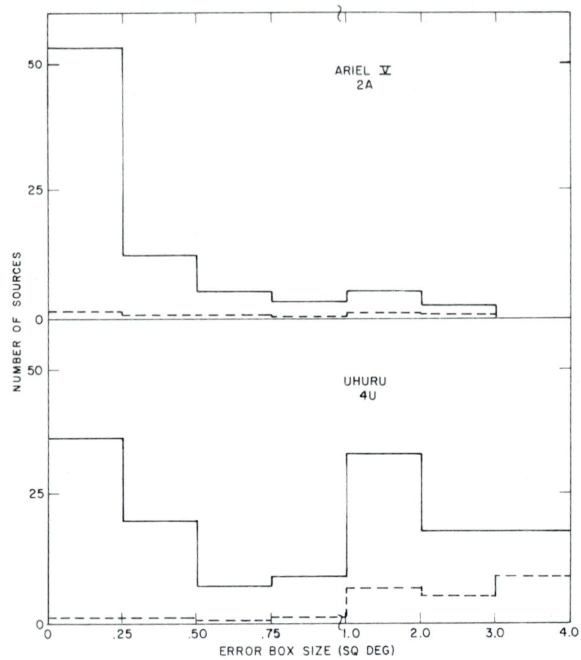

Fig. 1 Distribution of error box sizes for the 2A and 4U catalogs. The dashed lines show the accidental identifications expected from a catalog containing randomly distributed sources with an average density of 1 per 10 sq. deg.

IDENTIFIED SOURCES

The 2A and 4U catalogs have been searched for all sources satisfying the following criteria: (a) $\delta \geq -30^{\circ}$, (b) $|b| \geq 20^{\circ}$, and (c) error box 1.0 sq. deg. All sources which are definitely known to be galactic such as Her X-1, AM Her, and Sco X-1, have been excluded. The 2A catalog contains 51 sources and the 4U contains 49 sources which meet all of the above criteria. These are listed in Table 1 along with the suggested identifications ("X" indicates no suggested identification). Also listed in parentheses are any catalog entries which have error boxes larger than 1.0 sq. deg. but which correspond to sources located in the other catalog to better than 1.0 sq. deg.

From the entries in Table 1, the following statements can be made:

1) A total of 81 sources, presumably mostly extragalactic have been located.
2) 40 sources or 78 percent of the 2A sources are identified.
3) 32 sources of 65 percent of the 4U sources are identified.
4) From a total of 52 sources identified in both catalogs, the major classes of X-ray sources are clusters of galaxies (30, 58%), and Seyferts (10,19%). Other non-Seyfert active galaxies comprise another large group (6, 12%).
5) 29 sources (36 percent of the total of 81) remain unidentified.

Clusters of galaxies constitute the most numerous class of extragalactic sources. These have been discussed in detail by various authors [4], [5], [6]. The next

Table 1

Ariel V (2A) and Uhuru (4U) Source Lists*

2A	4U	Identification
	0005+20	Seyfert: Mkn 335
	0028+22	x
0039-096	0037-10	Cluster: Abell 85
	0041+36	x
0054-015	(0050-01)	Cluster: Abell 119
0102-222	0103-21	Cluster: Abell 133
0102-242		Cluster: Abell 140
	0148+36	Cluster: Abell 262
	0223+31	x
	0228-13	Cluster: Abell 358
0252+060		x
0255+132	0254+13	Cluster: Abell 401
0311-227	(0302-22)	x
	0334-30	x
0335+096	(0344+11)	Group
0349-139		x
0411+103	0410+10	Cluster: Abell 478
0431-136	(0431-12)	Cluster: Abell 496
	0432+05	Seyfert: 3C120
	0506-03	x
	0518-26	x
0710+456		Seyfert: Mkn 376
0738+498		Seyfert: Mkn 79
0906-095	0900-09	Cluster: Abell 754
0943-140	(0937-12)	Galaxy: NGC 2992/2993
0954+700	(1004+70)	Galaxy: M82
1033-270	1033-26	Cluster: Abell 1060
1058-226	1057-21	Cluster: Abell 1146
1102+384		BL Lac: Mkn 421
(1141+199)	1143+19	Cluster: Abell 1367
1150+720		x
	1153-11	x
	1203-06	Supercluster (?)
1207+397	1206+39	Seyfert: NGC 4151
1219+305		x
	1221-08	x
1225+022	1226+02	QSO: 3C273
1228+125	1228+12	Cluster: Virgo, M87
	1232+07	Galaxy: IC 3576
1251-290	1249-28	Cluster: SC 1251-288
	1253-00	x
1257+283	1257+28	Cluster: Abell 1656, Coma
1306-012		x
	1308+86	x
1346+266	(1348+25)	Cluster: Abell 1795
1347-300		Cluster, Seyfert: IC4329A
1348+700		Seyfert:Mkn 279
1410-029	1410-03	Galaxy: NGC 5506/5507
1415+255	(1414+25)	Seyfert: NGC 5548
1418+485		Cluster: Abell 1904
	1438-18	x
	1455+19	Cluster: Abell 1991
	1505+57	Supernova: 1954C (?)
1508+062		Cluster: Abell 2029
1519+082		Galaxy: NGC 5920
1556+274	1556+27	Cluster: Abell 2142
1600+164	1601+15	Cluster: Abell 2147
	1627-09	x
1626+396	1627+39	Cluster: Abell 2199
	1644+69	x
	1651+39	BL Lac: Mkn 501
	1651-06	x
	1652+63	x
1659+337		x
1704+241	1700+24	x
1705+786	1707+78	Cluster: Abell 2256
	1715+02	x
	1722+11	x
	1847+78	Seyfert: 3C390.3
1854+683	(1859+69)	Cluster: Abell 2312
2040-115		Seyfert: Mkn 509
2220-022		Cluster: Abell 2440
2237-256		x
2251-179		Group
2259+085	(2300+08)	Seyfert: NGC 7469
2302-088	(2305-07)	x
2318-272		x
2322+166	(2315+15)	Cluster: Abell 2589
2344-285	2344-27	Cluster: Klemola 44
	2345+27	Cluster: Abell 2666
	2358+21	x

*Sources listed have $\delta \gtrsim -30^{\circ}$, $|b| \gtrsim 20^{\circ}$, and are not definitely identified with galactic objects, and have error boxes $\lesssim$ 1.0 sq. deg in at least one catalog. Sources identified in a second catalog with an error box >1.0 sq deg. are listed in parentheses (e.g. 4U0050-01).

largest class found to date is the Seyfert galaxies. A great deal has been learned about these sources also [7], [8], including observations of variability on time scales of hours [7], days [8], and months [9]. These time scales demonstrate that the X-ray emission is produced in the bright energetic nuclei of the Seyferts. By analogy, such variability is also anticipated from X-ray sources associated with other types of active galaxies such as QSO's and BL Lacertae type objects.

In identifying and classifying extragalactic X-ray sources there are some pitfalls due to selection effects. First, the X-ray observer is limited to searching catalogs which are readily available in an appropriate form (usually computer readable). Second, catalogs must be large enough to yield a reasonable chance of success, and yet small enough that the relatively large errors in X-ray position do not produce a high probability of spurious identifications. Third, the objects in the catalog must have properties which make them likely X-ray emitters. This last selection effect leads to some obvious biases. For instance, once clusters and Seyferts were identified as a large fraction of extragalactic X-ray sources, a great deal of effort went into refining the positions of known X-ray sources near such objects and in attempting to find significant X-ray emission from such objects where none had been found before. In this way, the percentage of X-ray sources identified with clusters and Seyferts tends to be biased somewhat higher than its true value.

Another bias of X-ray survey catalogs is in favor of steady sources over variable ones. As most of the extragalactic sources are found near the limiting sensitivity of the survey instrument, even modest variations of a factor of two may change the source intensity from above to below the detection threshold of the instrument. This feature can be demonstrated from the sources identified in the list in Table 1. Of the 32 sources identified in both catalogs, 21, or 66 percent, are clusters of galaxies while only 3, or 9 percent, are Seyferts. Stated another way, 21 out of 30 clusters, or 70 percent, are identified in both catalogs while only 3 out of 10 Seyferts, or 30 percent, are identified in both. Although it is very difficult to compare the absolute limiting sensitivities of two surveys for a particular source, this statistical comparison indicates a fundamental difference between these two classes of extragalactic X-ray source which is most simply explained by variability of the Seyferts.

UNIDENTIFIED SOURCES

In addition to the identified sources, there is still a substantial fraction of high latitude X-ray sources which have not been identified with objects cataloged in other parts of the spectrum. The history of such unidentified X-ray sources is worth sketching as it may provide some clues to their nature. In analyzing the data contained in the 3U catalog, Maltisky et al. [10] pointed to the existence of a substantial number of unidentified high galactic latitude sources as evidence for a class of X-ray galaxies with large values of L_x/L_{opt}. The analysis was made, however, before a systematic search of all catalogs had been made and before any systematic surveys of the southern sky had even been made for clusters of galaxies. Since they did not distinguish between regions of the sky visible or not visible with ground based instruments in the northern hemisphere, the number of unidentified objects was overestimated.

The number of unidentified 3U sources had declined from 45 to 18 by the time Culhane [4] reviewed their status. Since then, the 4U catalog has appeared which withdraws 7 of the remaining source detections and presents possible identifications for 7 more. This process of attrition from 45 to 4 suggests that no new class of extragalactic is needed to explain X-ray sources identified at the limiting sensitivity of current instruments. Although the unidentified high latitude sources from the 3U catalog have largely disappeared, the higher sensitivity of the 2A and 4U catalogs has brought forth an entirely new list of such objects. Those fitting the declination and location criteria are listed in Table 1 with an "x" in the identification column. They still represent a significant fraction of all high latitude sources, and the lack of an identification carries increased weight due to the additional time and effort that has gone into attempts to identify them. Ultimately, the nature of these sources can best be determined by locating them within error boxes of 1 sq. arc minute so that unequivocal optical identifications can be made. However, until the appropriate instruments are available to make such precise measurements, it is important to draw all possible inferences from data obtained with existing instruments.

One simple observation can be drawn from Table 1. Of the 28 unidentified sources, only 3 are listed in both catalogs. As noted in the previous section, this characteristic also distinguishes the Seyferts from the cluster sources, and the difference is attributed to the known X-ray variability of the Seyferts. If one makes the conservative assumption that the unidentified sources fall into the same classes as identified sources, then their variability suggests that they are largely Seyferts or other active galaxies. Other classes of objects are also consistent with the variability of the unidentified sources. In particular, nearby low luminosity galactic sources as suggested by Holt et al. [11] might also exhibit variability. However, the preponderance of extragalactic objects and the lack of instances in which galactic stars are known to have been misidentified as extragalactic objects among the identified high latitude sources make this suggestion seem artificial.

Another possibility is that the unidentified sources are simply spurious artifacts produced by detector, electronic, telemetry, or orbital background problems. This would account for the fact that unidentified sources are usually not seen in both surveys. However, the history of the 3U unidentified sources suggests that at most 15-20 percent of the objects can be accounted for in this way. Furthermore, the lack of overlap in the Seyferts listed in the two catalogs presents the possibility of a simple alternative explanation.

If the unidentified objects are extragalactic, then the lack of an optical counterpart seems strange since almost all of the identified extragalactic X-ray sources were well known optical and radio objects before they were detected in X-rays. However, the sensitivity of the X-ray surveys is now sufficiently high that this sequence of "optical-then-X-ray-identification" may be reversed. Several examples now exist in the literature where the error box of a faint extragalactic X-ray source contained no previously cataloged optical object, but was later found to contain an active galaxy. Two new Seyfert galaxies, MGC8-11-11 and ESO 141-G55, have been found by searching X-ray error boxes [8], and two new QSO's, 4U0241+61 [12] and 2A2251-179 [13] have also been found in this way.

The primary reason that these optical objects were not previously cataloged is that their apparent magnitudes are below the completeness limits of optical surveys. For Seyfert galaxies, Huchra and Sargent [14] have estimated that the Markarian lists are essentially complete for magnitudes brighter than $m_p \sim 14.0$ but are less than 30 percent complete for $m_p \gtrsim 15.5$. Some deeper surveys over small areas of the sky have been made in search of strong emission line QSO's [15],[16]. These surveys may be complete to a limiting magnitude of 17-18 but in most of the sky the cataloging may be incomplete at much brighter levels.

Given the uncertainties in the completeness of optical surveys, it is worth considering the hypothesis that a substantial fraction of the unidentified sources are active galaxies belonging to well known classes such as Seyferts, BL Lacertae type objects, and QSO's, and that the optical counterparts have not yet been cataloged. For the known Seyferts, the ratio of L_x/L_{opt} is estimated to be in the range 1-5 [7] although for one object, 3C390.3, $L_x/L_{opt} \simeq 10$. At the upper end of this range, X-ray sources with intensities of 1 Uhuru ct/sec might be expected to have optical counterparts with visual magnitudes below 15. At this level, the incompleteness of the Seyfert and BL Lacertae surveys, and the poor area coverage of the QSO searches leave us with the possibility that these types of active galaxy are the unidentified X-ray sources. For instance, if all the 2A/4U unidentified sources were Seyferts, then the current catalogs must be lacking at least 29 such objects as opposed to the 10 X-ray objects which are cataloged. This would suggest a completeness factor of about 25 percent. As some of these would presumably be in parts of the sky not yet surveyed, this percentage is not incompatible with the 30 percent completeness estimates of Huchra and Sargent [14] assuming that most of the optical counterparts are fainter than $m_p \sim 15.5$.

The arguments given above and the successful searches for Seyferts and QSO's suggest that active galaxies are the most likely candidates for the unidentified high latitude X-ray sources. The uncertainty will not be resolved until the error boxes of a substantial number have been searched down to a limiting magnitude of $m_p \sim 16$-17.

HEAO A-1 PRELIMINARY RESULTS

The first 21 days of production data from the Naval Research Laboratory experiment (A1) on board the HEAO 1 spacecraft have now been analyzed, and preliminary results on faint extragalactic objects from the 10 percent of the sky survey can be compared with the above analysis of the Ariel V and Uhuru catalogs. This comparison is particularly useful because of the increased sensitivity of the HEAO A-1 instrument over previous surveys. The relevant parameters of the A1 detectors are given in Table 2.

Table 2

HEAO A1 Instrument Parameters

DETECTORS	FIELD OF VIEW (FWHM)	AREA	LIMITING-SENSITIVITY (0.5-10 keV)
1,2,3,4,	$1^\circ \times 4^\circ$	6400 cm^2	5×10^{-4} cts/cm^2 sec
5	$1^\circ \times 0.5^\circ$	1400	5×10^{-3}
6	$1^\circ \times 0.5^\circ$	1400	5×10^{-3}
7	$8^\circ \times 2^\circ$	2000	2×10^{-2}

The principal detectors used in the search for faint extragalactic objects are the four $1^\circ \times 4^\circ$ detectors with a combined area of 6400 cm^2. The detectors scan a great circle perpendicular to the ecliptic plane which advances 1° per day as the spacecraft spin axis follows the sun. Sky coverage for the first 3 weeks is shown in Fig. 2.

The production data for the first 21 days of the HEAO 1 mission have been searched for faint, previously undetected high latitude sources in the intensity range 2-7.5×10^{-3} cts/cm^2 sec (approximately 0.4 - 1.5 Uhuru cts/sec). A total of 27 such sources have been found. An estimate of the expected number of sources in this intensity range can be obtained from the Uhuru number intensity distri-

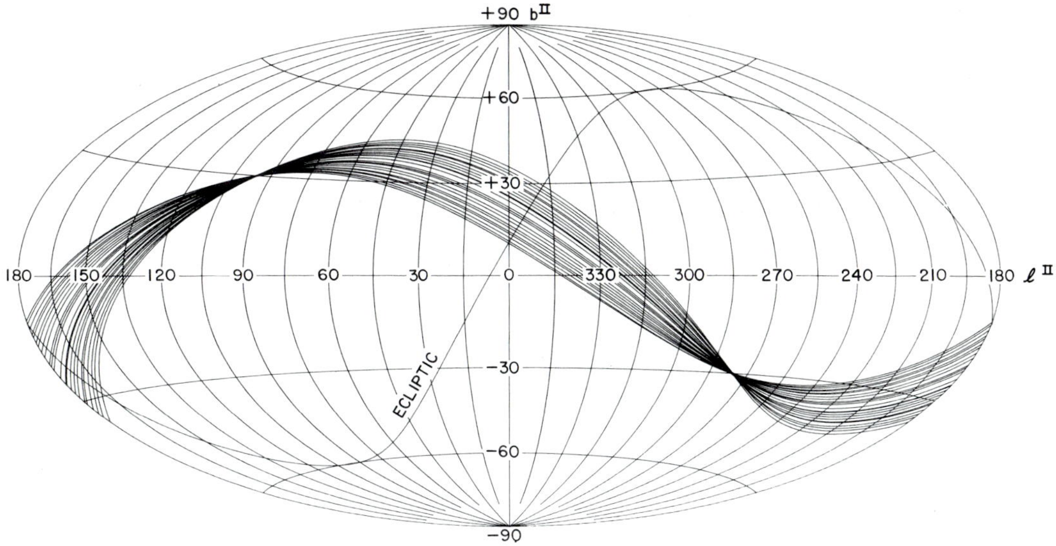

Fig. 2 Sky coverage, in galactic coordinates, of the HEAO A-1 experiment from 19 August to 10 September 1977.

bution [17], $n(s) = (21 \pm 7)\ s^{-5/2}$ sources/steradian where "s" is the intensity in Uhuru cts/sec. This distribution predicts that the A1 experiment should see approximately 60±14 sources in the 10 percent of the sky covered. Thus, we have probably found only about half the sources that are present in the data. This is due to source confusion problems which are not readily unraveled by the data analysis programs. Figure 3 shows the superposition of 4 days of scanning data for successive groups of days. Only half of the scan is shown over the northern ecliptic hemisphere. Previously known sources are labeled, and many new sources are also marked. Source confusion problems can be seen in several places such as scan azimuths 16°, 37°, 57° (Her X-1), and 61° (A2199, Mkn 501).

To compare the preliminary HEAO A-1 results with those from Ariel V and Uhuru, the same criteria must be placed on the source locations and error box size. Of the 27 new sources detected, 18 have $|b| \gtrsim 20^{o}$, $\delta \gtrsim -30^{o}$ and error boxes smaller than 1.0 sq. deg. They are listed in Table 3 along with intensities, error box sizes, and tentative identifications. Of the 18 new sources listed, 7 have been tentatively identified with extragalactic objects. As expected, the clusters of galaxies dominate the list of identified objects, and at this stage the fraction of unidentified sources is quite large. Six of the unidentified sources have error boxes smaller than 0.32 sq. deg., and one has an error box less than 0.1 sq. deg. so that optical identifications are likely to be made.

The new QSO and BL Lacertae objects are of interest because these classes of X-ray source as yet contain only a few members each. The QSO is relatively bright, $m_v = 16$, with a redshift of Z = 0.13. Based on just these two identifications, it seems clear that the A1 survey will add substantial numbers of such objects to the X-ray catalogs.

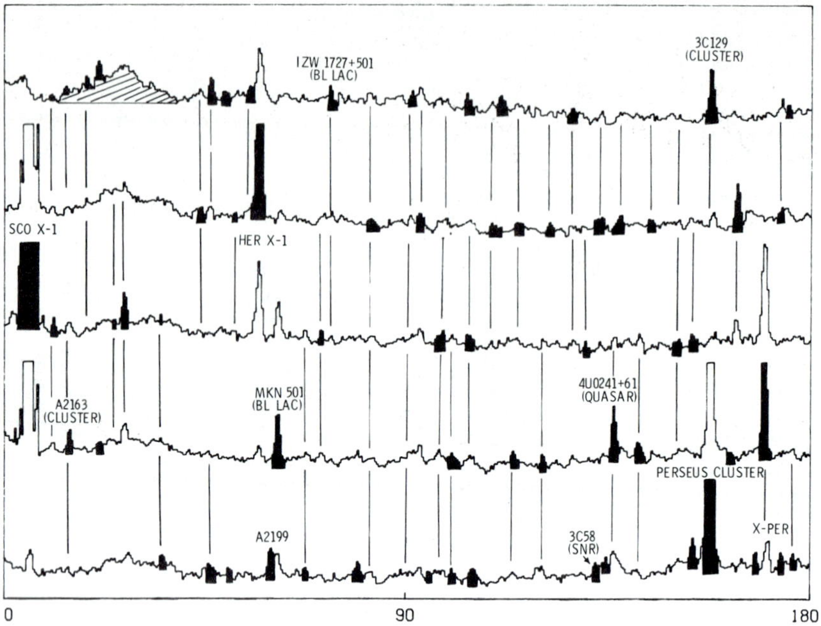

Fig. 3 Superposition of successive 4 day intervals of HEAO A-1 scan data. Only half of the full scan is shown in the northern ecliptic hemisphere. A scan azimuth of 90° corresponds to the north ecliptic pole.

Table 3

HEAO A-1, New High Latitude Sources

Source (R.A., Dec)	Intensity ($x10^{3}$ cts/cm^{2} sec)	Error Box Size (sq. deg)	Identification
0405+149	2.6	.99	x
0450-040	7.2	.27	x
0406-081	7.2	.30	x
0436-165	2.1	.76	x
1615-097	2.7	.25	x
1642-111	2.5	.068	x
1632-069	4.2	.43	x
1611-069	5.0	.066	?Cluster:A2163 (.2° outside)
1637+137	4.0	.71	Cluster: A2224
1658+458	2.5	.58	x
1714+488	2.6	.32	x
1728+502	2.3	.28	BL Lac: IZW186
1805+677	2.3	.46	QSO:1803+676
1815+693	2.9	.34	Cluster: A2304
1810-696	3.0	.15	?Cluster:A2301 (.1° outside)
1848+725	1.8	.54	x
1934+732	4.5	.32	Cluster: A2321
2130+817	1.9	.24	x

HEAO A-1 RESULTS ON UNIDENTIFIED SOURCES

Because only a small sample of the A-1 survey has been analyzed, no substantial results can be drawn from a statistical analysis of the objects found. However, the increased sensitivity makes it possible to obtain information on the location and variability of unidentified sources in the 2A and 4U catalogs. A list of the unidentified sources from Table 1 which were within the band of sky surveyed during the first 21 days of the HEAO-1 mission is given in Table 4.

Table 4

HEAO A1 Observations of 2A/4U Unidentified Sources

Source	2A/4U Intensity** (2A or 4U cts/sec)	A1 Intensity* ($X10^3$ cts/cm^2 sec)
4U0506-03	1.4 ± 0.3	4.6 ± 0.5
4U0518-26	1.55 ± 0.4	$\lesssim 0.5$
4U1627-09	1.89 ± 0.4	$\lesssim 0.5$
4U1651-06	1.32 ± 0.6	0.9 ± 0.3
2A1659+337	0.6 ± 0.1	4.2 ± 0.3
{2A1704+241} {4U1700+24}	{0.6 ± 0.1} {4.2 (variable)}	4.9 ± 0.2

*Approximate conversion 1.0 Uhuru ct/sec = 0.4 Ariel ct/sec = 5.0×10^{-3} cts/cm^2 sec (A1)

** Uhuru intensities are taken from the updated 4U list [1]

There are two contributions from the increased sensitivity of HEAO A-1. The first is to make definitive statements about the variability of some of the unidentified sources. Clear variability of factors of 5-10 are seen for 4U0518-26, 4U1627-09, and 4U1651-06. Combined with the Uhuru report of at least a factor of 2 variability in 4U1700+24, at least 4 out of the 6 entries in Table 4 show significant variability.

The second A-1 contribution is in refining the positions for those unidentified sources which were sufficiently intense to be located precisely. Error boxes, (90 percent confidence contours) obtained for the 4 sources detected by the A-1 instruments (Table 4) are shown in Fig. 4. In two cases (4U0506-03, 4U1700+24), the A1 error box does not overlap those of the other surveys. No likely candidates have been found in the HEAO error boxes, although a variable star UX Ori is within the box of the source nearest 4U0506-03. Because the A1 intensity for 4U1651-06 is only 0.9×10^{-3} cts/cm^2 sec (0.2 Uhuru cts/sec), the error box is quite long in a direction perpendicular to the scan direction. However, the position is consistent with the Uhuru error box. The A1 intensity for 2A1659+337 shows no significant variations from the Ariev V value. The improved HEAO error box indicates that Abell 2244 is the most likely candidate for the X-ray source, and this identification is consistent with the lack of variability. In the 2A catalog, Abell 2244 was noted as lying just outside the edge of the Ariel error box.

The preliminary A1 results regarding the 2A/4U unidentified sources confirm two of the conclusions reached on the basis of the previous catalogs alone. First, they indicate that a certain number of identifications will be forthcoming simply

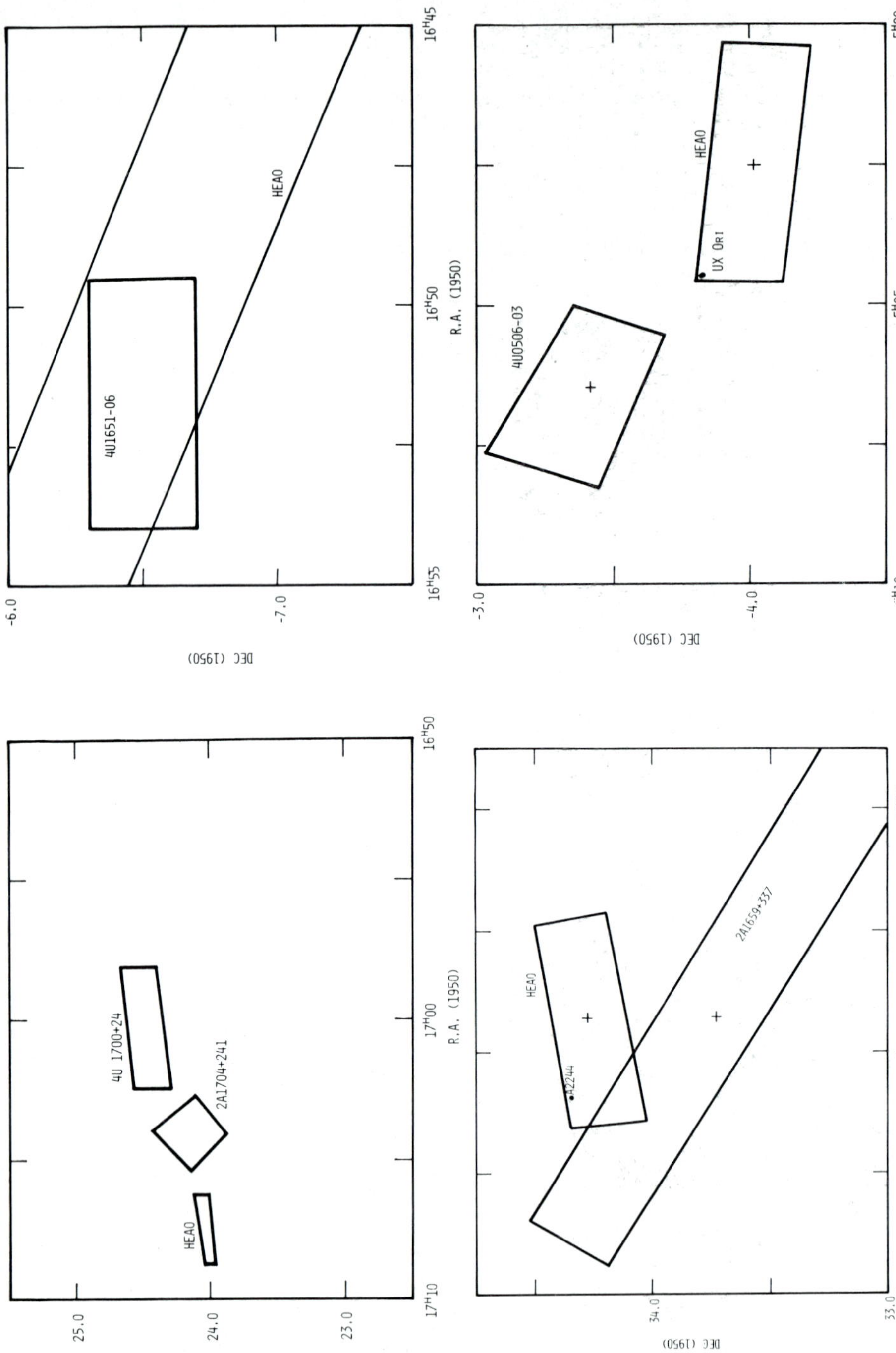

Fig. 4 HEAO A-1 error boxes for 4 Arid/Uhuru unidentified high latitude sources.

as a result of improved positions. This process has been responsible in large measure for the attrition of the original 3U unidentified sources. Second, the A-1 results also show that a substantial fraction of unidentified sources are highly variable and thus support the suggestion that they may be identified with active galaxies.

The HEAO A-1 experiment will contribute to the study of extragalactic X-ray sources in several important ways. It will provide a substantial number of new detections which will greatly increase our knowledge of such relatively rare X-ray sources as QSO's and BL Lacertae type objects and will continue to expand the already familiar classes such as clusters and Seyferts. It will provide information on source variability over a range not previously accessible. Finally, it will provide improved positions for previously unidentified sources.

ACKNOWLEDGEMENTS

I would like to thank Kent Wood, Arthur Davidsen and Ray Cruddace for many helpful discussions of various aspects of extragalactic astronomy. The HEAO A-1 experiment is the product of many people's efforts at the Naval Research Laboratory: the principal investigator is Dr. H. Friedman, the hardware development was directed by Dr. T.A. Chubb and Mr. E.T. Byram, the data reduction software development is directed by Dr. K. Wood, and much of the data analysis is due to the combined efforts of W. Evans, R. Kinzer, J. Johnson, W.N. Johnson, D.P. McNutt, J.F. Meekins R. Morin, L. Scott, G. Share, H. Smathers, W. Snyder and D. Yentis. The HEAO program at the Naval Research Laboratory is supported by NASA under contract H87023A.

REFERENCES

1. Forman, W., et al., Ap. J. Supplement, to be published 1978.
2. Cooke, B.A., et al., M.N.R.A.S., 182, 489, 1978.
3. Abell, G.O. Ap. J. Supplement, 3, 211, 1958.
4. Culhane, J.L. Quart. J.R.A.S., 18, 1, 1977.
5. Jones, C. and Forman W., Ap. J., 224, to be published 1978.
6. McHardy, I. submitted to M.N.R.A.S., 1978.
7. Tananbaum, H., et al., Ap. J., 223, 74, 1978.
8. Elvis, M., et al., M.N.R.A.S., 183, 129, 1978.
9. Ward, M.J. et al., Astr. and Ap., 56, 219, 1977.
10. Matilsky, T., et al., Ap. J., 181, 753, 1973.
11. Holt, S.S. et al., Ap. J. (Letters), 188, L97 1974.
12. Apparao, K.M.V., et al., Nature, 273, 450, 1978.
13. Ricker, G.R. et al., Nature, 271, 35 1978.
14. Huchra, J. and Sargent, W.L.W., Ap. J., 186, 433, 1973.
15. Smith, M.G. Ap. J., 202, 591, 1975.
16. Osmer, P.S. Ap. J. (Letters), 218, L89, 1977.
17. Murray, S.S. preprint, 1977.

LOCATION OF UNIDENTIFIED WEAK X-RAY SOURCES BY THE HEAO-1 SCANNING MODULATION COLLIMATOR

D. A. Schwartz*, H. V. Bradt**, U. Briel*, R. Dower**, R. Doxsey**, G. Fabbiano*, R. Griffiths*, H. Gursky*, M. Johnston**, R. Leach*, W. Liller*, A. Ramsey** and J. Schwarz*

**Harvard-Smithsonian Center for Astrophysics, Cambridge, Massachusetts, U.S.A.*
***Center for Space Research, Massachusetts Institute of Technology*

THE SCANNING MODULATION COLLIMATOR EXPERIMENT ON HEAO-1

This paper discusses the results on four of the first weak sources for which we obtained precise scanning modulation collimator (MC) positions. Our results are summarized in the table below. The experiment has been described previously [1,2,3].

Source	b		Size (Gaussian FWHM)	Flux (UFU)*
4U 0446+44	0°.2	Cluster of galaxies	6'	5.4 (HEAO/A-2)
2A 1052+606	51°	K-type sub-giant	< 1.7'	2.7 (variable x4)
2A 0526-328	-31°	Unidentified	< 1.7'	1.8
2A 0922-317	13°	Unidentified	< 40"	4.9 (variable x2)

* 1 Uhuru Flux Unit = 1.7 x 10^{-11} ergs/cm^{2}s, 2-6 keV; $\simeq 1.6\,\mu$Jy at 3.6 keV.

AN EXTRAGALACTIC SOURCE IN THE GALACTIC PLANE

The source 4U 0446+44 (= 2U 0447+44) was originally suggested as arising in a cluster of galaxies [4]. Since the source is in the galactic plane in a region of absorption $A_V \sim 3$, evidence for the existence of the cluster is based on the appearance of only two galaxies, the optical counterparts of the radio sources 3C 129 and 3C 129.1. These galaxies have a measured redshift z = 0.0213 [5] (D = 128 Mpc, for H_0 = 50). Most convincing is the existence of a radio tail on 3C 129 [6,7]. Tails have been associated with radio sources in clusters [8], and can be explained in terms of ram pressure confinement by gas of density $\sim 10^{-3}$ to 10^{-4} particles/cm^3 at temperatures of the order of 10^8 °K. Such temperatures and densities are just those appropriate to expl: a the X-ray emission in terms of thermal bremsstrahlung from intracluster gas.

Figure 1 shows the radio contours [6] of 3C 129 and 3C 129.1. The tail of 3C129 can be seen extending about 20 arcmin (740 kpc) to the west. The modulation collimator position is given in Fig. 1, and clearly shows, for the first time, that the emission is not from either of the radio galaxies. The location shown is the only possible one within the 0.0074 deg^2 uncertainty region given by the 4U catalog [9], and within the HEAO/A-1 LASS position.

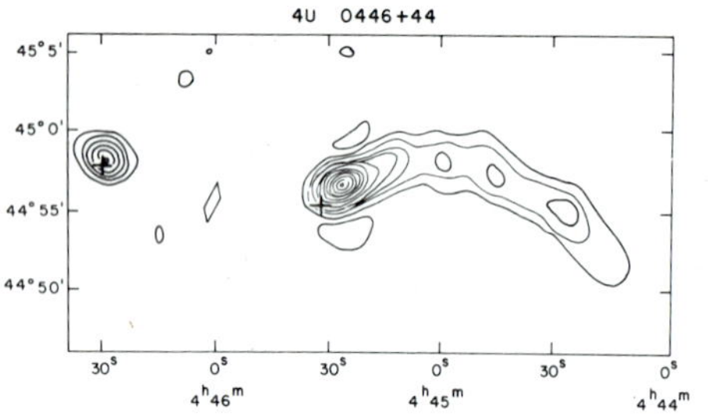

Fig. 1 The 408 MHz contours [6] for the radio sources 3C129 and 3C129.1 are shown, bracketing the 90% uncertainty error box (diamond) obtained for the X-ray source 4U0446+44 by the HEAO-1 modulation collimator (MC) experiment. The MC error box has an area 0.0006 deg^2. Positions of the optical galaxies are indicated by the crosses.

Our result of 2.0 ± 0.8 UFU (3.6σ) in the 30" FWHM collimator (MC1) and 2.4 ± 1.2 UFU (2.6σ) in the 120" FWHM collimator (MC2) seemed surprisingly low at first, compared with the Uhuru [9] and OSO-8 [10] fluxes of 5.0 and 5.7 respectively. If the source were variable it could not be a cluster. The MC experiment would detect less than 0.5 UFU if its angular extent were similar to the 15 arcmin measured for the Coma and Perseus clusters [11], which are at similar distances. The definitive data were provided by the GSFC HEAO/A-2 experiment [12]. They measured an equivalent flux of 5.4 ± 0.3 UFU, 2–6 keV, during the same time as our modulation collimator observations.

Simulations indicate that a 6 arcmin FWHM Gaussian source would explain our measured fluxes. An approximate 90% confidence range is 3 to 12 arcmin. The X-ray core radius is ~3 times smaller, and the surface brightness ~twice that of the Coma cluster. Our result is only marginally consistent with the relatively high ratio of MC1 to MC2 fluxes, which may indicate the existence of a more compact core to the surface brightness distribution. Huchra [13] has examined 2 of the 3 brightest stars in the error box, and found them to be normal F and A stars. They are therefore unlikely candidates. The midpoint of 3C129 and 3C129.1 falls just outside the location uncertainty.

A GALACTIC SOURCE AT HIGH GALACTIC LATITUDE

The source 2A 1052+606 was listed in the preprint version of the Ariel V Sky Survey Instrument (SSI) catalog [14] with a flux of 0.7 UFU, and it is in fact confirmed by analysis of the first 20,000 Ariel V orbits [15]. The error box size was 1.3 deg^2, and contained several NGC galaxies although none was suggested as a likely candidate. The modulation collimator data from 1977 November 5 to 9 gives equivalent fluxes of 2.4 ± 0.8 and 3.1 ± 0.9 UFU, in MC1 and MC2 respectively. Our flux measurement clearly proves variability on a time scale of a few years, due to the absence of the source from the 4U [9] and 2A [14] catalogs.

We immediately noticed a very curious result (Fig. 2): two SAO stars were inside or very near to one of our error boxes. The star SAO 015355 was observed by McCrosky and Liller to be a G-type dwarf, and therefore not a likely candidate.

An echelle spectrograph of SAO 015338 showed it to have an early K-type spectrum, but with H_α emission [16]. This is quite an unusual feature for a K star at such a high galactic latitude, and indicates a significant gaseous component external to the photosphere of the star. This suggests that the star is a very strong candidate for the X-ray emitting system. Further observation showed the star to be a sub-giant [17], and to have a complex spectrum

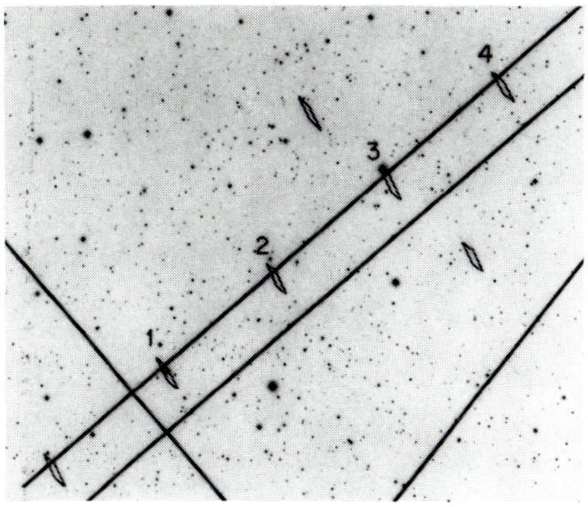

Fig. 2 The Ariel V location uncertainty [14, preprint] for the source originally designated 2A 1052+606 is shown (large box), along with the band of position given by the HEAO/A-1 LASS experiment. Four HEAO/A-3 error boxes (total area 5.4 arcmin^2) lie within the intersection of the Ariel V SSI and HEAO/A-1 locations. Bright SAO stars fall within HEAO/A-3 boxes #1, SAO 015355, and #3, SAO 015338. The latter is the candidate identification.

with characteristics of an RS CVn system [18]. The cataloged visual magnitude is $m_V = 8.8$. Since $m_V \sim 3.2$ for a KO IV star [19], we estimate a distance of 130 pc and an X-ray luminosity 5×10^{31} ergs s (2-6 keV). If the identification is correct, it is a previously unknown type of low luminosity hard X-ray emitting system. An optical search for periodicities is needed to show if SAO 015338 is a binary system such as produces hard X-rays in AM Her [20] or SS Cyg [21]. This X-ray source may be related to the RS CVn systems which have recently been shown to be a prominent class of soft X-ray emitters [22].

TWO UNIDENTIFIED HIGH GALACTIC LATITUDE SOURCES

Analysis of about 1/2 of the sky to a level of about 2 UFU, results in three sources at high galactic latitude, ($|b| > 10^\circ$) for which we have precise positions but no candidate identification. These are 2A0526-326, 2A0922-317 = 4U0923-31, and 2A2151-316 = H2157-303. We will discuss the first two, since some attempts have been made at the optical identifications. For 2A0922-317, the 2A and 4U error boxes are disjoint. The NRL LASS line lies between these two, and we originally took the two MC positions nearest the 2A and 4U boxes. Subsequently, SAS-3 has indicated that the position near the 2A box is the correct one [23]. Figure 3 shows an enlargement of this region. Liller [24] studied the four brightest stars just outside our location and found nothing unusual. Our fluxes are 4.8 ± 0.5 UFU for MC1 and 5.0 ± 0.8 UFU for MC2. The agreement of these two fluxes indicates an angular size of less than 40 arcsec. The Uhuru point summation technique gives a 2.2 UFU flux, and the 2A catalog gives 2.8 ± 0.5 UFU. We therefore have clear evidence for variability.

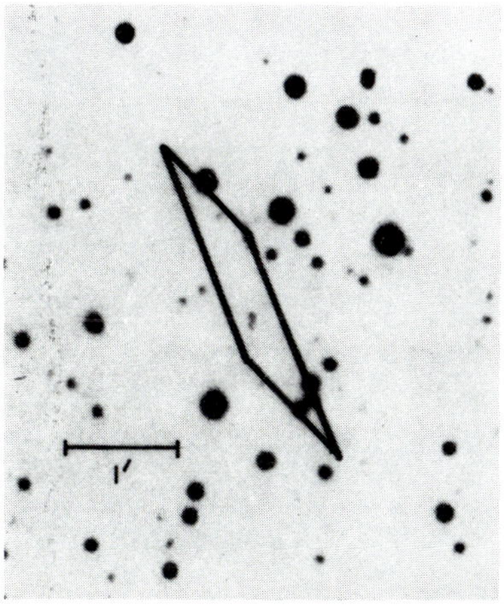

Fig. 3 The HEAO/A-3 MC error diamond at the N end of the 2A0922-317 location. The four brightest stars just outside this region have been studied by Liller [24], but give no indication of being likely candidates for the optical counterpart. The objects in the box are fainter than about $m_V = 18$.

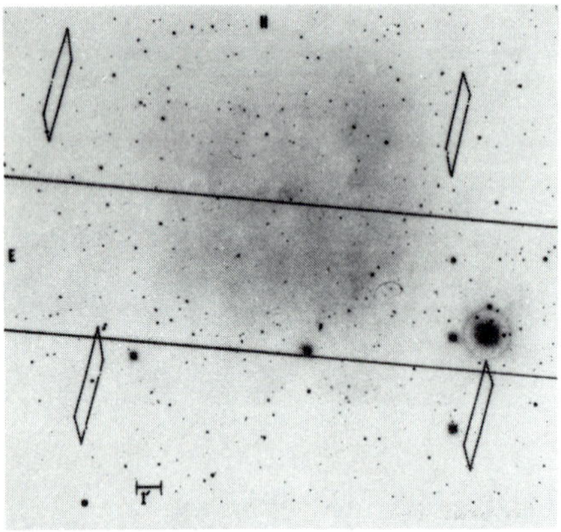

Fig. 4 The only four HEAO/A-3 MC error diamonds for 2A0526-328 which are inside the 2A location are shown superimposed on a CTIO 4m red plate taken by W. Liller. The HEAO LASS line is also shown.

Figure 4 is a reproduction of a CTIO 4m red plate taken by W. Liller. Liller [24] has observed the three brightest stars outside the SW box, and the brightest star in the SE box, to have ordinary spectra. Our HEAO/A-3 fluxes are 1.4 ± 0.4 UFU and 2.3 ± 0.5 UFU in MC1 and MC2 respectively. Their average agrees with the equivalent 2A flux of 1.7 UFU. The ratio of counting rates in the two modulation collimators indicates an angular size less than 1.7 arcmin FWHM. Two distant clusters appear in Fig. 4, but each is several arcmin from a possible location. One 19th magnitude galaxy is near the center of the SW error box.

What are these unidentified objects? They are not clusters of galaxies, and not active galaxies with $m_V \lesssim 19$. Most likely they are either galactic stars, or quasars. In any case, they have $m_V \gtrsim 15$ and $L_x/L_{op} \gtrsim 1$. Therefore, if the objects are extragalactic, they may be called X-ray galaxies as first suggested by Giacconi et al. [25].

REFERENCES

1. H. Gursky et al., Astrophys. J., 223, 973 (1978).
2. A. Roy et al., IEEE Trans. Nuc. Sci., vol. NS-24, no. 1, 804 (1977).
3. D. Schwartz et al., Proceedings AIAA 16th Aerospace Sciences, 78-34 (1978).
4. E. Kellogg et al., Astrophys. J. (Letters), 185, L13 (1973).
5. H. Spinrad, Astrophys. J. (Letters) 199, L 1 (1975).
6. R.H. MacDonald et al., MNRAS, 138, 259 (1968).
7. J.M. Hill and M.S. Longair, MNRAS, 154, 125 (1971).
8. J.P. Vallee and A.S. Wilson, Nature, 259, 451 (1976).
9. W. Forman et al., Astrophys. J. (Suppl.), in press (1978).
10. R.F. Mushotzky et al., Astrophys. J., (in press) (1978).
11. S.M. Lea et al., Astrophys. J. (Letters), 184, L105 (1973).
12. R. Mushotzky, private communication from GSFC HEAO/A-2 group (1978).
13. J. Huchra, private communication (1978).
14. B.A. Cooke et al., MNRAS, 182, 489 (1978).
15. K. Pounds, private communication (1978).
16. W. Liller, IAU Circ. No. 3176 (1978).
17. G. Wallerstein, private communication to W. Liller (1978).
18. D. Crampton, private communication (1978).
19. C.W. Allen, Astrophysical Quantities, The Athlone Press, London (1963).
20. J.H. Swank et al., Bull. AAS, 9, 591 (1977).
21. G. Fabbiano et al., Nature, in press (1978).
22. P. Charles et al., this volume.
23. H. Schnopper, colloquium at Harvard College Observatory, 4 May (1978).
24. W. Liller, private communication (1978).
25. R. Giacconi et al., Bull. AAS, 3, 477 (1971).

ARIEL V MEASUREMENTS OF THE X-RAY BACKGROUND IN THE 2-18 keV ENERGY RANGE

R. S. Warwick and J. P. Pye

X-ray Astronomy Group, Physics Department, University of Leicester, Leicester LE1 7RH, England

INTRODUCTION

We use measurements of small scale fluctuations in the intensity of the x-ray background to determine the number-flux relation for x-ray sources at a flux level well below present survey limits. Using the 2A catalogue [1], Warwick and Pye [2] have previously shown that the source counts of x-ray sources at high galactic latitude ($|b|>10^{\circ}$) are consistent with the relation

$$N(S)dS = 5 \pm 1\ S^{-2.5}\ dS$$

where N(S)dS is the number of sources per steradian with fluxes in the range S to S + dS (with S in Sky Survey Instrument (SSI) count s^{-1} where 1 SSI count $s^{-1} \approx 2.7$ Uhuru count s^{-1}).

DATA ANALYSIS

Figure 1 shows part of a typical observation by the SSI on Ariel V [1, 3] in which data recorded over many satellite orbits have been summed to improve sensitivity. A number of discrete sources including the Coma cluster, are detected in this observation. The source responses are superimposed upon a large background count, which results from both the diffuse x-ray background and from cosmic ray interactions with the detector.

The counts recorded in a particular sector due to cosmic rays depend upon the experiment on-time, which is constant for all sectors. By comparison, the counts due to the diffuse x-ray background depend upon the exposure time (ie the time the detector is actually viewing the sky), which varies with sector number due to Earth blocking of the detector field of view [3]. It is therefore possible to determine the cosmic ray count per sector by fitting the total background count per sector against the computed exposure time, and then linearly extrapolating to zero exposure time.

The analysis of SSI background measurements is based on a total of 17 separate summed data sets. The cosmic ray count was determined for each data set and then subtracted from the total background count. Background count rates were then calculated with the data summed into 12 sector bins, allowing 4 sector gaps between adjacent bins to eliminate overlap. Bins centred within 20° of the galactic plane were rejected (to avoid contamination by galactic sources), as were those within 20° of Sco-X1. In addition to allow for the fact that many regions of sky were scanned more than once, the data from a particular area of sky were weighted inversely according to the number of times that particular region of sky was scanned.

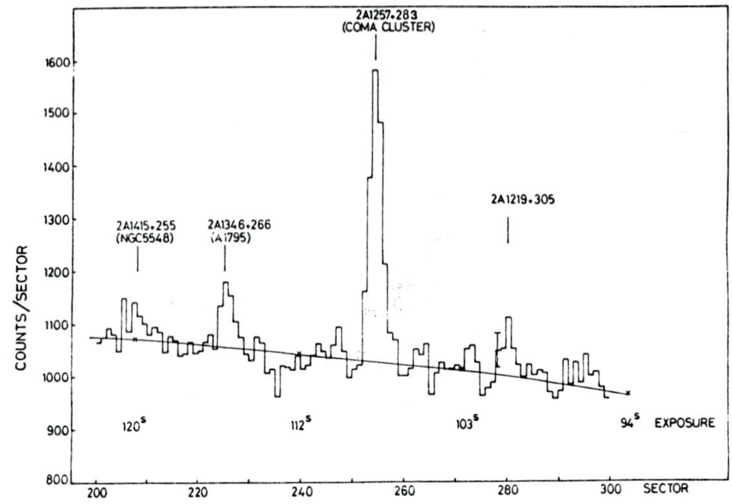

Fig. 1 Part of a summed-orbit data set from the SSI in which the Coma cluster and a number of other discrete sources are detected. The source responses are superimposed on a large background count. The slope of the background is due to a changing exposure time across the scan.

Fig. 2 The observed distribution of background count rates. The theoretical fit to the data predicted from the differential source count relation, $N(S)dS = 5\ S^{-2.5}dS$, is also shown. The inset shows the χ^2 contour map (68% and 90% confidence intervals).

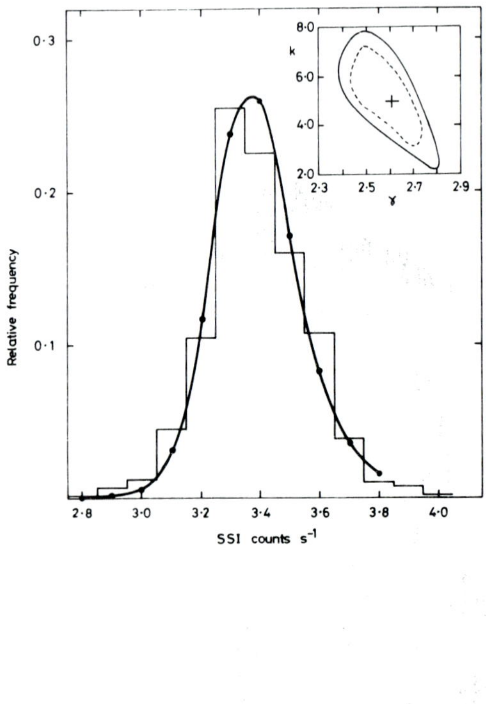

BACKGROUND FLUCTUATIONS AND THE X-RAY SOURCE COUNTS

The observations have been analysed by fitting a theoretical fluctuations distribution (often termed the P(D) relation) to the observed distribution of background count rates (Fig. 2). This approach has previously been used by Fabian [4] in a study of fluctuations in the x-ray background as measured by the Uhuru satellite, and by Wall and Cooke [5] in a study of the radio source counts at 2.7 and 5.0 GHz.

The form of the P(D) relation depends upon the effective beamwidth of the instrument used in the observations ($0.75^{\circ} \times 10.6^{\circ}$ FWHM for the SSI) and upon the normalisation and slope of the differential source counts [6, 7]. The observed distribution is wider than can be explained by counting statistics, and Fig. 2 shows the fit obtained using the source count relation derived by Warwick and Pye [2] for sources in the 2A catalogue. The theoretical curve is clearly an excellent fit to the observational data. The range of acceptable fits is shown in the inset where contours of constant χ^2 are plotted against the normalisation, K, and slope, γ, of the differential source counts. A minimum value of χ^2 is obtained at $K = 5$, $\gamma = 2.6$ which is consistent with the results of Fabian [4].

The 68% confidence contour is translated to a plot of the normalised differential source counts in Fig. 3, which also shows the results derived from the 2A catalogue [2]. Scheuer [7] has shown that the background fluctuations are due primarily to sources at flux levels above that corresponding to one source per effective beamwidth. This criterion defines the lower flux limit of the shaded region in Fig. 3. The upper flux limit is given by the count rate (above the mean value) at which the χ^2 fitting of the theoretical and observed distributions was truncated. This analysis of the SSI background data indicates that the x-ray source count relation derived for 2A sources can be extrapolated to fluxes at least an order of magnitude fainter than the 2A catalogue limit.

DISCUSSION

The observed source count relation may now be compared with forms of the source counts calculated using the luminosity functions of the two main classes of identified extragalactic x-ray source, namely clusters of galaxies and active galaxies. The luminosity functions which were used in this calculation are derived from Ariel V observations of Abell clusters [8] and active galaxies (ie emission line galaxies, Seyfert galaxies and QSOs) [9]. The luminosity range for the clusters of galaxies was taken to be 10^{36} to 3×10^{38} watts, and that for the active galaxies to be 10^{34} to 10^{39} watts (using a linear extrapolation of the luminosity function for active galaxies at the low luminosity end). An Einstein-de Sitter world model with no evolution was assumed. In addition the clusters of galaxies were assumed to have thermal bremsstralung spectra with a temperature of 10 keV and the active galaxies to have power law spectra with an energy index of -0.5. The predicted forms of the differential source counts for these two classes of source are shown in Fig. 3.

At high fluxes, clusters of galaxies and active galaxies comprise approximately 70% of the total counts, which is in agreement with the identification statistics for 2A sources. However, at a flux level of 0.1 counts s^{-1} the contribution of these two classes of source (with no evolution) has fallen to 40% of the total.

A number of interpretations are possible at this stage. If, for example, the unidentified sources have a steep source count, then this would compensate for the decreasing contribution of clusters of galaxies and active galaxies. However, if this is not the case, then other possibilities are:

i) The counts of clusters of galaxies or active galaxies (or both) may be higher than the predicted values as a result of cosmological evolution. For example the forms of density evolution which have been invoked to account for the diffuse x-ray background in terms of the integrated x-ray emission from Seyfert galaxies [10, 11], if applicable to active galaxies as a whole, would give a flat source count relation for these sources over the range of present measurements.

ii) The observed fluctuations in the background count rate from the SSI may be due in part to an excess population of very faint sources (ie the source counts may actually follow the predicted form over the present range but then turn up sharply at fainter fluxes).

iii) The luminosity functions used in the calculations may not accurately describe the source population at a flux level of ~0.1 SSI counts s^{-1}.

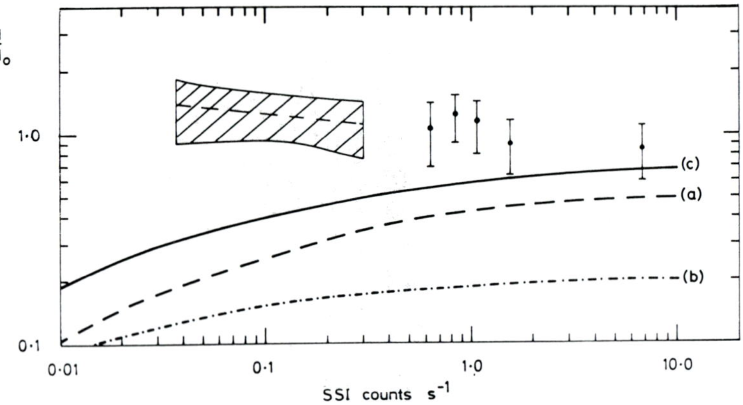

Fig. 3 The differential source counts normalised to the relation $N_o(S)dS=5S^{-2.5}dS$. The individual points refer to 2A sources. The curves show the predicted forms of the source counts for a) clusters of galaxies, b) active galaxies, c) clusters of galaxies and active galaxies combined.

Clearly further progress will be possible when 2A source identifications are more complete and when data are available on discrete x-ray sources at fluxes well below the level of the 2A catalogue.

REFERENCES

1. B.A. Cooke et al., Mon. Not. R. astr. Soc. 182, 489 (1978).
2. R.S. Warwick and J.P. Pye, Mon. Not. R. astr. Soc. 183, 169 (1978).
3. G. Villa et al., Mon. Not. R. astr. Soc. 176, 609 (1976).
4. A.C. Fabian, Mon. Not. R. astr. Soc. 172, 149 (1975).
5. J.V. Wall and D.J. Cooke, Mon. Not R. astr. Soc. 171, 9 (1975).
6. J.J. Condon, Astrophys. J. 188, 279 (1974).
7. P.A.G. Scheuer, Mon. Not R. astr. Soc. 166, 329 (1974).
8. I. McHardy, Mon. Not. R. astr. Soc. in press.
9. R.S. Warwick, Proc. Roy. Soc. To be published.
10. M.S. Elvis et al., Mon. Not. R. astr. Soc. 183, 129 (1978).
11. Y. Avni, Astr. Astrophys. 63, L13 (1978).

THE COSMIC X-RAY BACKGROUND: I. THE DISCRETE SOURCE COMPONENT

E. Boldt, F. Marshall*, R. Mushotzky*, S. Holt, R. Rothschild** and P. Serlemitsos

Laboratory for High Energy Astrophysics, NASA/Goddard Space Flight Center, Greenbelt, Maryland 20771, USA
**NAS/NRC Research Associate*
***Now at University of California (San Diego)*

THE X-RAY BACKGROUND EXPERIMENT (A2)[†] ON HEAO-1

Although X-ray astronomy has certainly flourished during the last decade, research on the diffuse sky, particularly the hard component, has been relatively dormant. Satellite experiments since 1970 have focused our attention on individually discernible sources, particularly compact objects of high astrophysical interest, such as neutron stars. However, there was also a practical reason for this restriction during the initial decade or so of X-ray astronomy with orbiting observatories. The reason is that all X-ray detectors in orbit have a significant background arising from other than the X-ray sky. Some of the extraneous causes are cosmic rays, ambient electrons, locally generated gamma rays and induced radioactivity. And, for discrete source measurements, even the "true" diffuse X-ray sky is an undesirable contamination. Hence, the standard approach has been to make the detector's field of view as small as practical and to measure the increase in signal as the detector scans across the discrete source of interest.

Our strategy for extracting the diffuse flux of celestial X-rays from the overall signal with the A-2 experiment on HEAO-1 is summarized as follows:

1) The most fundamental aspect of a diffuse flux is that it increases linearly with solid angle. Hence, we employ two different fields of view per detector; one of them is always 3° x 3° (FWHM), the second is 3° x 1½° for some detectors and 3° x 6° for others.
2) Ambient electrons can be a problem. For our high-energy gas proportional counters, the main detector is separated from the entrance window by a charged particle veto layer (propane filled) relatively transparent to hard X-rays. For our low energy detectors this is impossible and deflection magnets are used.
3) We detect X-rays via the photoelectric effect, which is a strong function of energy. To exploit this characteristic we have three types of multi-layer gas proportional counters, optimized for low, medium and high energy X-rays, designated LED, MED and HED, respectively. For soft X-rays (0.15 - 3 keV) we use propane, for medium energy X-rays (1.5 - 15 keV) we use argon and for hard X-rays (3 - 60 keV) we use xenon.

[†]The A2 experiment on HEAO-1 is a collaborative effort led by E. Boldt of GSFC and G. Garmire of CIT, with collaborators at GSFC, CIT, JPL and UCB.

4) The electrons emerging from a photo-electrically excited atom are confined to a small region near the atom; hence we employ multi-anode veto to remove more penetrating charged particle events.
5) Finally, since the scan period of HEAO involves much of the orbit around the Earth we separate spatial and temporal effects with some counters offset by only 6° along the scan path.

Figure 1 is a photograph of the medium energy detector (MED) in the lab. The other detectors look about the same. X-rays enter a mechanical collimator on top, traverse a thin window and are detected in a multi-anode multi-layer argon gas proportional chamber. The dual collimator is matched to the internal multi-celled structure of the proportional chamber; odd numbered cells are aligned with the 3° x 3° collimator, while even numbered cells are aligned with the 3° x 1½° collimator. Figure 2 shows the dual collimator itself in greater detail in order to convey how this scheme has been implemented.

Fig. 1. Laboratory photograph of the medium energy detector (MED); overall length is about one meter.

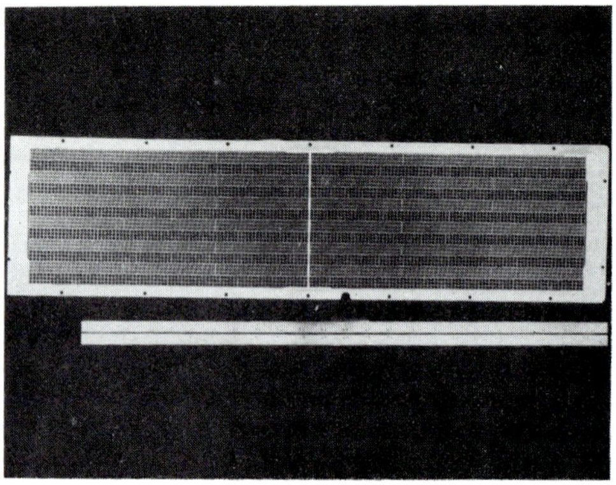

Fig. 2. Dual collimator for MED showing 3° x 3° and 3° x 1½° cell arrangement, as viewed from below, on axis.

The A-2 experiment consists of six proportional chambers, two LED's for low energies, one MED for midband coverage and three HED's for hard X-rays. To get some idea of how this experiment scans the sky let us concentrate on any one of the six detectors and examine its scan path on the celestial sphere. This is shown in Figure 3. Every half hour this dual collimator detector scans a complete great circle on the celestial sphere in an angular band 3° wide. At any instant of time it views an angular region $\pm\ \theta$ along the scan path with the detector half corresponding to the large field of view collimation as well as an included angular region half this size with the compliment-ary portion of the detector corresponding to the small field of view. Each day the spin axis moves 1° along the ecliptic equator so that the entire sky is scanned in 6 months.

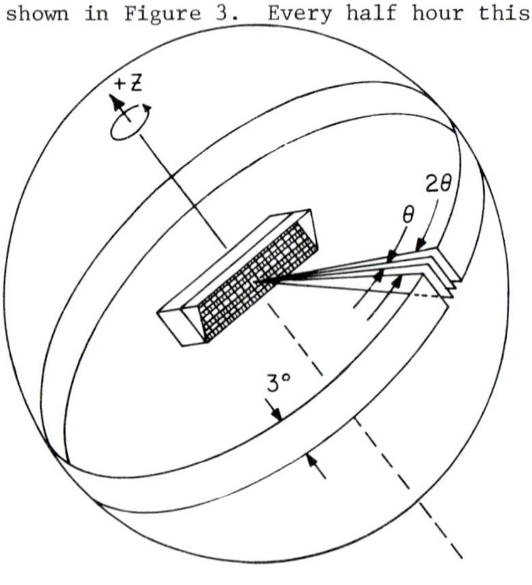

Fig. 3. A schematic representation of the celestial scan. The z direction points from HEAO to the sun and provides the axis of scanning. $\theta = 1\frac{1}{2}°$ for LED #1, MED and HED #3. $\theta = 3°$ for LED #2, HED #1 and HED #2.

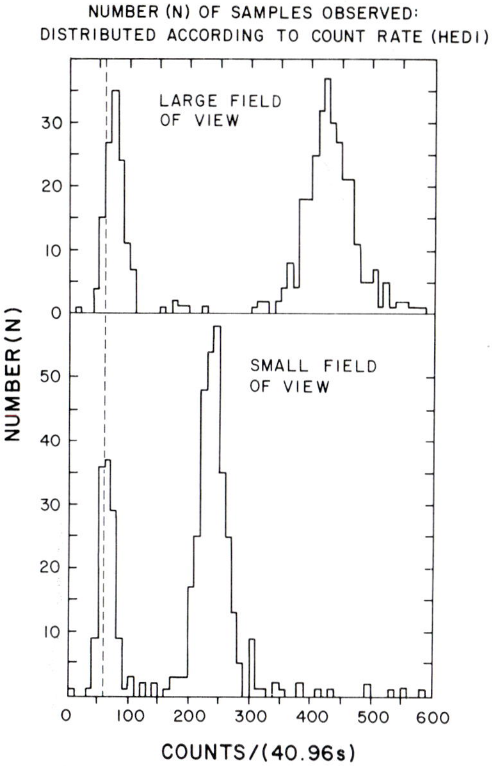

Fig. 4. Histograms of observed samples sorted according to count per telemetry major frame (40.96s) for HED #1, classified according to field of view, 3° x 6° or 3° x 3°.

How does the dual collimation scheme work out in practice? The in-orbit performance of HED #1 relative to this question is exhibited in Figure 4, where the effectiveness of the scheme is quite evident. The two histograms displayed give the observed population of samples sorted according to the total accepted X-ray count per telemetry major frame (i.e. 40.96s). They are classified only as regards field of view, the top histogram for counts associated with the 3° x 6° collimation and the bottom for counts associated with the 3° x 3° collimation. These histograms are based on data accumulated over many scan cycles regardless of what was in the field of view, be it the Earth or celestial sources. The histogram for each of the fields of view exhibits two clearly separated peaks, the high one attributed to exposures dominated by the sky. If there were no extraneous sources of background the two histograms would scale as the ratio of solid angles (i.e. both the Earth and the sky represent essentially isotropic sources). In fact, the Earth is a relatively weak X-ray source, even in the hard X-ray band (~ 3-60 keV) considered here, and most of the signal when the earth fills the field of view arises from background internal to the detector (e.g. Compton collisions of gamma rays). More extensive data bear out the qualitative indication in Fig. 4 that the internal background to be associated with the two fields of view are equal. Furthermore, a comparison of the internal background derived from the two peaks associated with the diffuse sky has been shown to be the same as that derived from the two peaks associated with the diffuse earth, both in magnitude and spectral shape. As shown in Fig. 4 by a dashed line, the internal background for HED #1 represents an average contamination of ~ 14% for the large field of view and ~ 25% for the small field of view, when the full energy bandwidth (~ 60 keV) of this detector is included.

SPECTRA OF EXTRAGALACTIC SOURCES AND BACKGROUND

After confirming the proper performance of our experiment in-orbit, we felt ready to start our analysis of the spectrum of the diffuse X-ray background, and a pilot study has been carried out by Marshall et al.(1). To exclude effects of internal detector background, the pulse height spectrum to be associated with the sky background was obtained in each detector by simply subtracting the raw spectrum observed with the small field of view from the raw spectrum observed with the large field of view. Then we tested some models for incident spectra of interest by folding them through the calibrated response function of the detector, normalizing to the total number of counts and comparing observed counts per channel with that predicted. The response function for this procedure was verified, in orbit, on the Crab Nebula. At this stage of data analysis we find a "best-fit" thermal spectrum characterized by a value for kT that is within about 5 keV of 45 keV (for more details and models see Marshall et al.(1)). The spectrum is remarkably structureless, at least below about 20 keV. For the absolute magnitude of the flux, we agree quite well with estimates by Schwartz(2) based on a compilation of pre-HEAO measurements by various experiments over different portions of the spectrum.

Now we can compare this background spectrum with the spectra of specific extragalactic objects. Most of those identified so far are clusters of galaxies(3,4); the brightest extragalactic X-ray object in the sky is the Perseus Cluster, which has a thermal spectrum as shown in Fig. 5. The list of 20 such spectra now known from our OSO-8 experiment(5) should increase to about 60 when we complete the HEAO analysis of clusters from an all-sky survey. In this case, the thermal emission is characterized by kT = 6.8 keV for the continuum and an equivalent width of 400 eV for a line at 6.7 keV, indicative of a near-cosmic abundance of iron in collisional equilibrium at this temperature. These are reasonably typical parameters for X-ray clusters. In other words, they are much cooler than the overall background, and they do have spectral structure.

The next most common extragalactic X-ray source category is that of the Seyfert galaxies. Although there were about 14(6) identified prior to HEAO-1, we knew the spectrum of only one, namely NGC4151. We expect to obtain spectra for all those so far identified and already have several besides NGC4151. Fig. 6 shows the spectrum of an object at about twenty times the distance of NGC4151. This extreme Seyfert N-galaxy 3C390.3 was in a low state during the accumulation of these data,

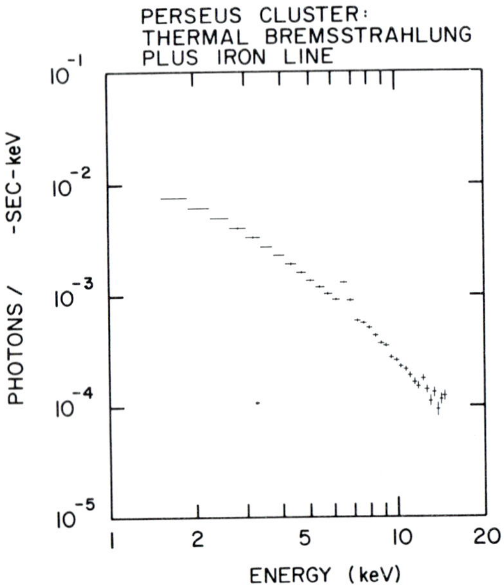

Fig. 5. Incident photon number spectrum for the Perseus Cluster inferred from data obtained with GSFC Cosmic X-Ray Spectroscopy experiment aboard OSO-8(5). The spectrum is unfolded from the data on the basis of a model of thermal bremsstrahlung (with iron line emission.

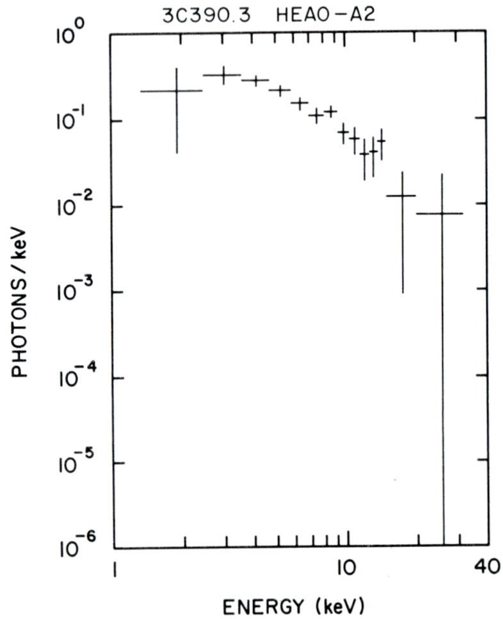

Fig. 6. Incident photon number spectrum for 3C390.3 inferred from data obtained with HED #3 from the A2 experiment aboard HEAO-1. The spectrum is unfolded from the data on the basis of a power-law spectrum absorbed by cool matter of cosmic abundance.

with an intensity of about 0.8 UFU (Uhuru flux units). The incident spectrum inferred is based on a best-fit power law of number index 1.7 and absorption by 5×10^{22} atoms cm^{-2}. In summary, the spectrum is not thermal and it is highly absorbed, similar to that of NGC4151.

Now that the A1 and A3 experiments on HEAO-1 have isolated the X-ray source associated with the BL Lac object Markarian 501(7), we have obtained its spectrum(8), shown in Fig. 7. It is a remarkably flat spectrum, much like that of Markarian 421(8), with a photon number spectral index of 1.2 and no apparent absorption.

We are now in a position to compare the various major components of the extragalactic hard X-ray sky. First, we have the spectrum of the diffuse background (i.e., the spectrum for the composite emission of all unresolved sources). For resolved sources we already have a good idea of the typical spectrum to be associated with clusters of galaxies. As we have just seen, we also have some idea of the highly absorbed spectra to be associated with relatively compact objects such as Seyferts and N-galaxies, and of the very flat unabsorbed spectra to be associated with BL Lac objects such as Markarian 501. From previous surveys, such as those conducted with Uhuru and Ariel 5, we know how the number of such objects decreases with source intensity over about a decade; that is, we know something about the Log N-Log S curve for observations in the 2-10 keV band(9,10). Hence, we have what we need to estimate the spectrum for the composite flux from almost all extra-galactic sources that could have been resolved prior to HEAO. This information is summarized in Figure 8.

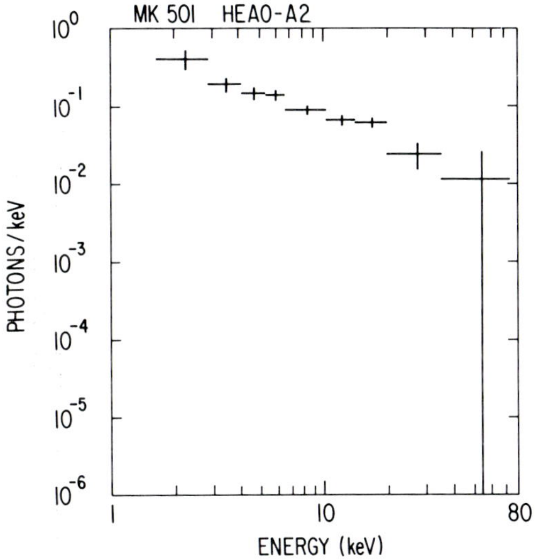

Fig. 7. Incident photon number spectrum for Markarian 501 inferred from data obtained with HED #1 from the A2 experiment aboard HEAO-1. The spectrum is unfolded from the data on the basis of a power-law spectrum.

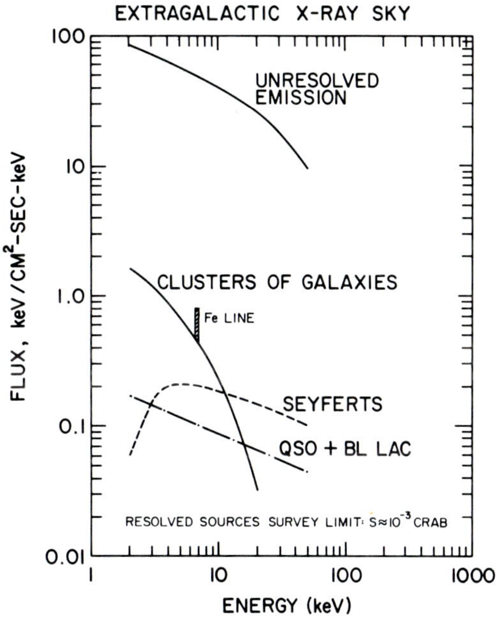

Fig. 8. The X-ray flux from the extragalactic sky as it would be viewed by an omnidirectional (4π) spectrometer (see text). Resolved source limit is 1 UFU.

Figure 8 shows the spectral characteristics of the total flux from the extragalactic X-ray sky as would be viewed by an omnidirectional (4π) spectrometer. The curve labeled "unresolved emission" is the best fit thermal continuum for the diffuse background; as discussed before, this corresponds to kT ≈ 45 keV. The curve labeled "clusters of galaxies" represents a thermal continuum with kT ≈ 6 keV and iron line emission at ~ 6.7 keV spread out in energy over a bandwidth corresponding to a spread in redshifts of 0.09 since the most distant cluster with detected line emission (Abell 478) has that redshift. The curves labeled "Seyferts" and "QSO + BL Lac" are very approximate in shape since spectra have been measured for only a few sources in each instance, although the absolute 2-10 keV level is based on larger samples, especially for Seyferts. From the most distant source of each class positively identified we can crudely estimate what the multiplication factor is for getting from these flux values shown here to ones corresponding to the integral of all such sources in the universe. For clusters of galaxies that factor is less than about 5. For Seyferts the factor is approximately 10. For "QSO + BL Lac" the factor would be less than 10, if we really believed we could ignore evolution for such sources. On this basis, we conclude that at no energy do discrete sources of the type indicated contribute more than about a tenth or so of the unresolved emission spectrum shown here, with clusters making their major contribution at the low energy end and Seyferts at the high end. In particular, the iron line emission from clusters is so pronounced that the resulting discontinuity in the unresolved emission as observed in this experiment would be close to 1%; anything much more than this would have been detected in the limited data sample already analyzed(1).

REFERENCES

1. F. Marshall, E. Boldt, S. Holt, R. Miller, R. Rothschild, and P. Serlemitsos, Ap. J. (to be submitted), 1978.
2. D. Schwartz, In X-rays in Space, ed. E. Venkatesan (Calgary: University of Calgary), p. 1096, 1975.
3. I. McHardy, Mon. Not. R. Astr. Soc., in press, 1978.
4. C. Jones and W. Forman, Center for Astrophysics, Ap. J. 224, 1, 1978.
5. R. Mushotzky, P. Serlemitsos, B. Smith, E. Boldt, and S. Holt, Ap. J. 225, 21, 1978
6. M. Elvis, T. Maccaro, A. Wilson, M. Ward, M. Penston, R. Fosbury, and G. Perola, Mon. Not. R. Astr. Soc., 183, 129, 1978.
7. D. Schwartz, H. Bradt, R. Doxsey, R. Griffith, H. Gursky, M. Johnston, and J. Schwartz, Ap. J. 224, L103, 1978.
8. R. Mushotzky, E. Boldt, S. Holt, S. Pravdo, P. Serlemitsos, J. Swank, and R. Rothschild, Ap. J., in press, 1978.
9. S. S. Murray, Center for Astrophysics, preprint No. 680, 1977.
10. R. Warwick and J. Pye, Mon. Not. R. Astr. Soc., 183, 169, 1978.

THE COSMIC X-RAY BACKGROUND: II. FLUCTUATIONS

E. Boldt, A. Stottlemyer*, R. Schafer**, S. Holt, R. Rothschild*** and P. Serlemitsos

Laboratory for High Energy Astrophysics, NASA/Goddard Space Flight Center, Greenbelt, Maryland 20771, USA

**Computer Sciences Corp.*
***Also Dept. of Physics and Astronomy, Univ. of Maryland*
****Now at University of California (San Diego)*

INTRODUCTION

Most of the X-ray sky involves presently unresolved radiation, and there is mounting evidence that this is dominated by genuinely diffuse emission. For soft X-rays (i.e. $\lesssim$ 0.3 keV) this diffuse flux is highly anisotropic, associated mainly with nearby regions of our galaxy. For hard X-rays (i.e. > 3 keV) the fluctuations in surface brightness are relatively small and the global isotropy is sufficiently good to indicate extragalactic origins that are extremely remote. The A-2 experiment on HEAO-1[†] is the first specifically developed to investigate the diffuse radiation of the entire X-ray sky over a wide band of photon energies (i.e. 0.15 - 60 keV), covering both of these regimes. In this paper we concentrate on the spectrum of the extragalactic component of the diffuse background as compared to that for the fluctuations in surface brightness.

DEVIATIONS FROM ISOTROPY

We are confronted with the situation that it is extremely difficult, if not impossible, to account for most of the unresolved extragalactic X-ray sky in terms of extragalactic X-ray objects like those already observed(1). In fact, the spectral shape suggests thermal emission from a hot plasma at an effective temperature of a half-billion degrees(2). If this plasma is spatially uniform, we can apply the analysis of Field and Perrenod(4) to conclude that the plasma mass is about half that required for closure of the universe. However, there are serious problems with this picture, as pointed out by these authors and others(5). For example, the existence of diffuse neutral hydrogen clouds in intergalactic space indicates that the hot plasma under consideration can not pervade all intergalactic space for if it did these clouds could not persist. One way to get around this is to postulate that the intergalactic plasma is clumped. A model discussed by Field and Perrenod(4) has this matter in isothermal spheres of a type described by Chandresekhar(6) many years ago. If one assumes that the total mass in such hot clouds is less than the closure mass, then the number of such objects must be less than 2×10^5. However, upper limits on background fluctuations, such as obtained

[†] The A2 experiment on HEAO-1 is a collaborative effort led by E. Boldt of GSFC and G. Garmire of CIT, with collaborators at GSFC, CIT, JPL and UCB.

by Schwartz et al.(7), Fabian and Sanford(8) and Shafer et al.(10), require that the number of discrete sources for the background exceed 2×10^5 objects by at least a factor of 2. However, these analyses were basically for point sources, while the hot spheres discussed by Field and Perrenod(4) could have radii larger than 100 Mpc and present an angular size greater than a degree. For radii > 20 Mpc the mass is more than that usually associated with galaxies. That's the situation as it exists now. Defining some key characteristics of this background necessary for solving this puzzle is a major goal of our elaborate program of data analysis. The results discussed in the preceeding paper(1) on the spectrum of the background represent less than 5% of the data we hope to have from this mission. Research on extragalactic sources with HEAO-1 is just beginning, and we expect to define the X-ray properties of Seyferts, quasars, BL Lac objects and other active galactic nuclei much as our OSO-8 experiment has already done for clusters of galaxies(3). What are we leaving out? The answer resides in weaker sources and fluctuations.

For the fields of view employed by this experiment, source confusion is a problem at a level $\lesssim$ 1 UFU, even for extragalactic sources. Yet we would like to extend our spectroscopy for extragalactic sources down below that level for a complete, unbiased sample, independent of source identification. To do this we ignore source confusion and measure the energy spectrum of the brightness enhancements characteristic of the extragalactic sky and compare this directly with the spectrum of the overall diffuse background. We have made some progress on this, and some preliminary results are shown in Fig. 1. The experiment is described in the preceeding paper(1).

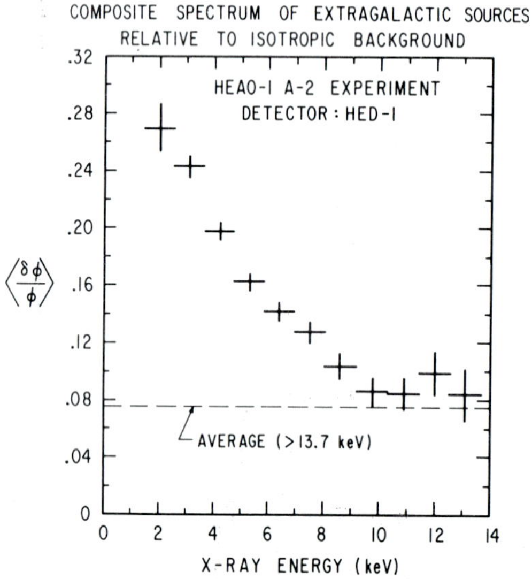

Fig. 1. The average enhancement ($\delta\phi$) measured with the 3° x 3° portion of HED-1 relative to the diffuse background flux (ϕ) is plotted as a function of photon energy for data accumulated into those cells (3° in ecliptic latitude by 1° in longitude) defined as "enhanced" by data from the 3° x 3° portion of HED-3.

Figure 1 shows the average enhancement in surface brightness relative to the diffuse background as a function of energy for HED #1. The data correspond to 47 days of accumulation and were obtained from regions more than 20° away from the galactic plane that are devoid of strong sources. Enhanced cells (up to a fluctuation of 100%) were defined on the sky by HED #3, which is offset from HED #1 by 6° along the scan. The complimentary experiment was also performed in which HED #1 was used to define enhanced cells from which HED #3 spectral data were selected; the results were similar to those shown in Fig. 1. The enhancement plotted here is the average for those cells selected; in each of the two experiments the cells selected amounted to ~ 16% of the total sample examined. The smallest detectable fluctuation in surface brightness was about 5% for this analysis. The weakest detectable source corresponding to this fluctuation is equivalent to ~ 0.5 UFU. The positive fluctuations at the level considered here are dominated by discrete extragalactic sources. If the composite spectrum of these sources were of the same spectral shape as the diffuse background, the fractional enhancements plotted in Fig. 1 would be consistent with a constant value (i.e. independent of energy).

Although the fractional enhancements at energies above ~ 14 keV are consistent with the average value shown by the horizontal dashed line in Fig. 1, a gradual increase with energy can not be ruled out statistically at this time; an order of magnitude more data will be used for further analysis. Discrete sources such as BL Lac objects have harder spectra than the diffuse background at these higher energies and would contribute to such an increase. Below 10 keV the observed spectrum of the enhancements (i.e. composite spectrum for discrete extragalactic sources) as shown in Fig. 1 is clearly softer than the diffuse background, as expected from the theoretical composite spectrum for clusters of galaxies(1). For this portion of the plot dominated by clusters of galacies, the characteristic temperature for a thermal enhancement corresponds to $kT < 5$ keV, assuming that other components of the enhancement have spectra no softer than the diffuse background. This upper limit to kT is less than the average $kT \gtrsim 6$ keV obtained for individually studied clusters(3), but might be indicative of the larger average redshift in this sample and/or a small detection bias in the sample of identified clusters. The contribution of clusters to the enhanced cells at $\gtrsim 2$ keV, as exhibited in Fig. 1, is consistent with the theoretical estimate inferred from Fig. 8 of the preceeding paper(1) when the Log N - Log S relation is extended down to 0.5 UFU (i.e. $S \approx 5 \times 10^{-4}$ Crab).

We have considered as a subsample the smallest fluctuations among the enhancements previously described and also separately considered negative fluctuations. While the relatively poor statistics for these samples do not permit any significant spectral analysis of the type exhibited in Fig. 1, we have attempted a "two color" classification for obtaining some preliminary indication of a trend. The two colors considered for this spectral analysis are 2-10 keV and 10-16 keV. As before, the threshold for detectable fluctuations is about 5%. The subsample of small enhancements was obtained by removing cells with cataloged sources or fluctuations $\gtrsim$ 10%. We define a spectral hardness ratio R as follows:

$$R \equiv \frac{\left\langle\frac{\delta\phi}{\phi}\right\rangle_{(10\,-\,16\ \mathrm{keV})}}{\left\langle\frac{\delta\phi}{\phi}\right\rangle_{(2\,-\,10\ \mathrm{keV})}} .$$

For fluctuations having the same spectrum as the diffuse background, $R = 1$. For the full sample of enhancements previously considered $R = 0.55 \pm 0.03$, using the results of both HED 1 and HED 3. For the subsample of small enhancements, $R = 1.18 \pm 0.23$. For negative fluctuations, $R = 0.89 \pm 0.39$. The full sample of enhancements is clearly softer than the diffuse background in this representation as well as that previously discussed. However, as we restrict these fluctuations to values that are closer to our threshold, the spectrum is probably no longer as soft and, within statistics, it is consistent with that of the diffuse background. This indication that the contribution of clusters such as we know diminishes for the weakest fluctuations is reinforced by preliminary results from a more complete statistical treatment of fluctuations (down to zero) now underway by Shafer et al.(10). As demonstrated in the preceeding paper(1), clusters of galaxies dominate the presently resolved discrete source component of the extragalactic sky below ~ 10 keV. By examining the fluctuations in sky surface brightness separately for these two regimes we will, in effect, be setting constraints on the Log N - Log S relation for these two classes of objects down to a level that is an order of magnitude lower than previously considered.

To obtain statistically independent samples of the sky relative to point objects it is sufficient to require that there be no overlap in the fields of view. For extended objects it's more complicated. In this case we are using an autocorrelation analysis. This is particularly important for establishing some characteristic parameters of a possibly clumpy intergalactic hot plasma.

We already know from the work of de Vaucouleurs(11) and others that the local region of the universe within ~ 20 Mpc is not isotropic in optically observable galaxies and that this "local supergalaxy" is likely to be anisotropic in X-rays as well. However, the rather pronounced fluctuations intrinsic to the X-ray sky impose a noise on such measurements of surface brightness variations that amounts to several per cent over resolution elements of a few square degrees. To get around this problem S. Pravdo (private communication) is first trying a global strategy whereby he will integrate diffuse X-ray data over supergalactic longitude as a function of supergalactic latitude. Later on he will try correlations with galaxy counts.

Since the proper frame of the extragalactic background is not likely to be anchored to the solar system we should be able to detect our velocity relative to this preferred frame. In particular, observers of the microwave background(12) have already detected a net velocity of ~ 300 km/sec. The limitation to measuring a velocity of such magnitude in X-rays is that the corresponding anisotropy amplitude is only about 0.4% and we have relatively large fluctuations intrinsic to the sky to deal with. However, these fluctuations can be minimized by rejecting data contaminated with resolved sources. By rejecting sources at the level of 0.5 UFU the ultimate sensitivity corresponds to the detection of an amplitude of 0.4% at a statistical level somewhat better than 4σ. For the full HED bandwidth (3-60 keV) of this experiment, we have already established an upper limit of about 1% based on a small fraction of the available data.

Finally, there should be some truly diffuse galactic X-ray background since we know that cosmic-ray electrons colliding with ambient microwave photons and starlight do produce hard X-rays. Here again, as in the possible smooth anisotropy of the extragalactic background, we must get around the problem of brightness fluctuations intrinsic to the extragalactic sky. In other words, model testing and correlations with other data are required. D. Worrall (private communication) is working on this by considering models for the galactic confinement of relativistic electrons that might be tested, and we are investigating correlations with the galactic radio background.

ACKNOWLEDGEMENTS

We thank our collaborators G. Garmire, F. Marshall, R. Mushotzky, S. Pravdo, L.A. Rose, B. Smith, A. Szymkowiak and D. Worrall for valuable discussions.

REFERENCES

1. E. Boldt, F. Marshall, R. Mushotzky, S. Holt, R. Rothschild and P. Serlemitsos (this symposium), 1978.
2. F. Marshall, E. Boldt, S. Holt, R. Miller, R. Rothschild and P. Serlemitsos, Ap. J. (to be submitted), 1978.
3. R. Mushotzky, P. Serlemitsos, B. Smith, E. Boldt, and S. Holt, Ap. J. 225, 21, 1978
4. G. Field and S. Perrenod, Ap. J. 215, 717, 1977.
5. J. Bergeron and J. Gunn, Ap. J. 217, 892, 1977.
6. S. Chandresekhar, Principles of Stellar Dynamics (Chicago: University of Chicago Press), p. 231, 1942.
7. D. Schwartz, E. Boldt, S. Holt, P. Serlemitsos, and R. Bleach, Nature Phys. Sci. 233, 110, 1971.
8. A. Fabian and P. Sanford, Nature Phys. Sci. 234, 21, 1971.
9. R. Shafer, E. Boldt, S. Holt, and P. Serlemitsos, Bull. AAS 9, 616, 1977.
10. R. Shafer, E. Boldt, L.A. Rose, S. Holt, R. Rothschild, and P. Serlemitsos, Bull. Am. Phys. Soc. 23, 527, 1978.
11. G. de. Vaucouleurs, Pub. A.S.P. 83, 113, 1971
12. G. Smoot, M. Gorenstein, and R. Muller, Phys. Rev. Lett. 39, 898, 1977.

DISCRETE SOURCE CONTRIBUTIONS TO THE DIFFUSE COSMIC X-RAYS

Daniel A. Schwartz

Harvard - Smithsonian Center for Astrophysics, Cambridge, Massachusetts 02138, U.S.A.

INTRODUCTION

In this review I will concentrate on the contribution of discrete sources to the diffuse background. The observational features of the isotropy [1,2,3], the spectral shape [4,5], and the intensity [6,7,8] of the diffuse X-ray background in the 2-200 keV region have been well established for many years. A surfeit of theories regarding the mechanism of the origin exists [8,9,10,11], yet none gives a definitive explanation of the background. The difficulty is to relate the energy source to other observed quantities. The procedure has been to postulate such effects as cosmic evolution [12,13], production of metagalactic cosmic ray electrons [14,15], cosmological heating of an intergalactic plasma containing an amount of mass of order required for $q_o = 1/2$ [16], or a cosmological epoch of enhanced supernova activity [17]. The observed intensity is calculated from the relation [5]:

$$I(E) = \frac{R_H}{4\pi} \int_0^{z_{max}} \frac{B\,[E(1+z);\,z]\,dz}{(1+z)^5\,\sqrt{1+2\,q_o z}} \qquad (1)$$

where I(E) is the measured spectral intensity (keV/cm^{2}s keV ster), R_H is the Hubble radius (H_o = 50 km/s Mpc, R_H = 600 Mpc are used here), and B(E) is the spectral emissivity (keV/cm^3 keV) per co-moving coordinate volume. We see in Fig. 1 that even in a uniform emissivity model with no evolutionary effects (i.e., $B = B_o\,(1+z)^3$) about 20% of the emission arises at a

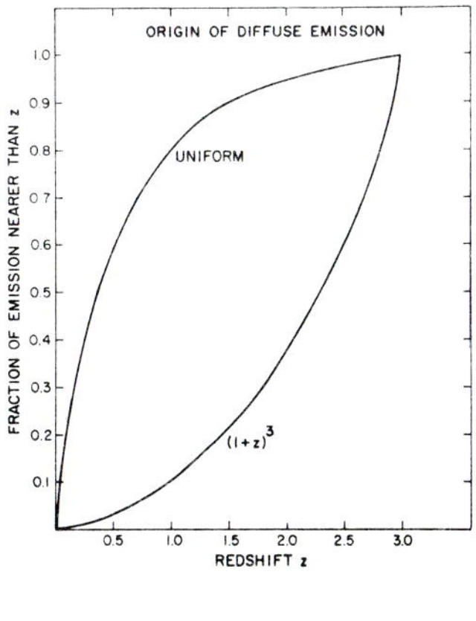

Fig. 1 Fraction of the X-ray background which is produced at redshifts nearer than the abscissa. Two models are shown: emissivity which is uniform (in co-moving coordinates) throughout the universe (upper, left curve) and emissivity which evolves by an additional factor $(1+z)^3$ (lower, right curve). In both cases we assume q_o = 0.5, the X-ray energy spectral index is -0.5, and the emission is cut off at z = 3.

redshift greater than 1. In this case we may set $z_{max} = \infty$, since the integral in equation (1) converges. In general we might anticipate a cosmological evolution of the form $B = [B_o (1+z)^3] (1+z)^n$. Thermal bremsstrahlung from a metagalactic plasma could give $n \approx 3$. For $n \geq 1.5$ we must truncate the emission to avoid a divergence. A value $z_{max} = 3$ is assumed in Fig. 1. Figure 1 also shows that at redshifts of order 0.1 the volume emissivity of a population of such evolving sources need only be a few percent of that of a non-evolutionary population in order to give the same integrated diffuse intensity. This fact has been presented by Rowan-Robinson and Fabian [18] to stress that evolution of known extragalactic X-ray sources is restricted to be less than that of quasars (which have $n \approx 6$ [19]) to avoid exceeding the measured background intensity.

MEASURED SPECTRAL INTENSITY

Figure 2 shows selected measurements of the spectral energy intensity [20-27]. Extensive references to other measurements appear in previous reviews [5,6,28,29]. The points (shown as open triangles) from the GSFC portion of the HEAO/A-2 diffuse background

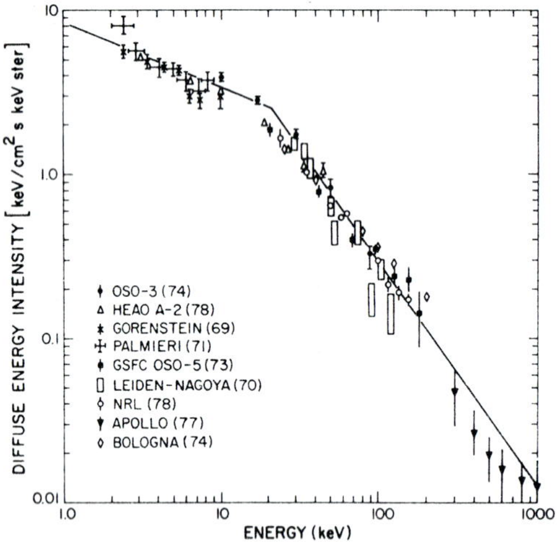

Fig. 2 Selected measurements of the spectral energy intensity. Recent measurements shown as HEAO/A-2 [31], NRL [26], Apollo [27], and Bologna [25], confirm the shape (solid lines) chosen as a numerical representation of selected older data shown as OSO-3 [20], Gorenstein [21], Palmieri [22], GSFC OSO-5 [23], and Leiden-Nagoya [24].

experiment [30,75] are of special interest. Their agreement with the average of previous measurements (represented as the solid lines) is surprising if one takes the viewpoint of the previous IAU X-ray symposium [28] which expressed a healthy skepticism of the observational techniques. Such a skeptical view is no longer tenable: the measurements now clearly show that the spectrum steepens between 2 and 200 keV.

The HEAO/A-2 results [75] in the 3-50 keV range accurately fit a thermal bremsstrahlung spectrum, with gaunt factor, of effective temperature $kT = 45 \pm 5$ keV [30,31]. At energies less than 1 keV it is now well known that galactic emission and absorption dominate the background, so that we do not expect the isotropic diffuse spectrum to extrapolate down to this region. In the region above 100 keV, a temperature of $kT = 95$ keV must be used to fit the background [26], and above about 1/2 MeV another emission component must be evoked. As we will discuss later, correction of the spectrum below 10 keV flattens it significantly.

As a numerical representation of the background a power law is extremely convenient because it can be easily integrated over redshift. A good representation of the background can now be obtained by lowering the solid curve of Fig. 2 by about 10%:

$$I(E)\ [\text{keV/keV cm}^2\text{s ster}] = \begin{cases} 7.7\ E^{-0.4} & 1 < E < 21\ \text{keV} \\ 152\ E^{-1.38} & 21 < E \end{cases} \quad (2)$$

Using equation (1) and assuming a uniform (no evolution) source distribution leads to an integrated 2-6 keV volume emissivity of

$$\overline{B}\ [\text{ergs/Mpc}^3\text{s}] = \begin{cases} 1.4 \times 10^{39} & q_o = 1/2 \\ 0.85 \times 10^{39} & q_o = 0 \end{cases} \quad (3)$$

This value of $\overline{B}$, calculated for $H_o = 50$, compares with the value 2.2×10^{39} quoted by Schwartz and Gursky [5] using $H_o = 75$, and $q_o = 0.5$. Schwartz and Gursky incorrectly state that their $\overline{B}$ is integrated from 2 to 10 keV: the numerical calculation was actually from 2 to 6 keV.[1] Recent evidence favors a value of the decelleration parameter, q_o, near zero [32], so we will adopt $\overline{B} = 9 \times 10^{38}$ ergs/Mpc^{3}s in this review.

CONTRIBUTIONS FROM KNOWN CLASSES OF DISCRETE SOURCE

There are two relatively well-known classes of extragalactic source: clusters of galaxies and Seyfert Type I galaxies. Figure 3 shows estimates of their volume luminosity functions dN/dL = (number of objects) Mpc^{-3} (10^{44} erg/s)$^{-1}$. The solid dots are the Uhuru estimates

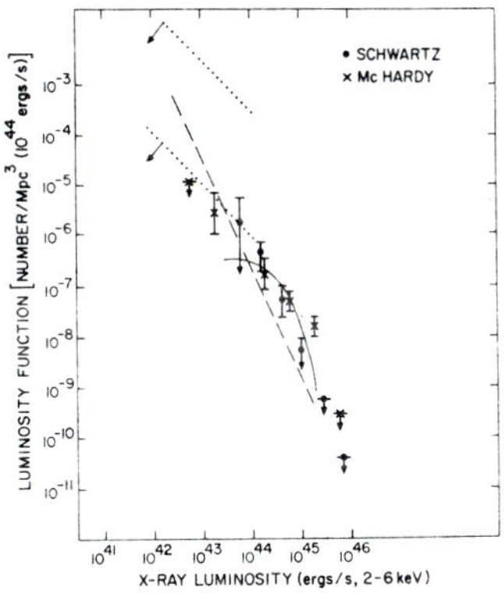

Fig. 3 Estimates of the differential luminosity function, number/Mpc^{-3} (10^{44} ergs/s)$^{-1}$, of 2-6 keV X-ray sources. Dots and crosses: clusters of galaxies. Dashed line: Seyfert galaxies. The dotted lines give the limits imposed by the total densities of all Seyferts (upper) or Abell clusters (lower).

[33] fcr clusters of galaxies, corrected to use the point summation technique fluxes. These agree well with the Ariel V estimates due to McHardy [34], shown as crosses. The solid curve represents an empirical fit

$$dN/dL = (3.5 \times 10^{-7})\ e^{-L44/3} \quad (4)$$

(where L_{44} is in units of 10^{44} erg/s, 2-6 keV), which seems to agree with the Uhuru data for $0.3 < L_{44} < 30$. This leads to an estimated contribution of 35% of the background, for $q_o = 0$, or 22% for $q_o = 1/2$. The dashed line represents the fit to the Seyfert luminosity function [35,36], corrected from the reported bandwidth of 2-10 keV to the bandwidth of 2-6 keV. The data fit a curve [35]

$$dN/dL = (2.5 \times 10^{-7})\ L^{-2.3} \quad (5)$$

in the range $0.03 < L_{44} < 10$. The integrated emissivity over this luminosity range is

[1] We thank A. Wilson for first pointing out this error.

2×10^{38} ergs/s Mpc^3s, or 22% of the background for $q_o = 0$ (15% for $q_o = 1/2$). To significantly increase the volume emissivity

$$B = \int_{L_o}^{L_1} L\left(\frac{dN}{dL}\right) dL \tag{6}$$

by integrating to much lower luminosities L_o, we obviously must have dN/dL increase more rapidly than L^{-2} as L decreases. For Abell clusters this clearly is not the case based on the estimated data. Also, the total volume density of X-ray clusters would exceed that of all Abell clusters at $L_o = 0.1 \times 10^{44}$ ergs/s.

In the case of Seyferts, the data in the range $0.1 < L_{44} < 10$ could be fit as well to $dN/dL = 4 \times 10^{-7} L^{-3}$ within the large uncertainties noted [35,36]. Extrapolation of this function down to 3×10^{42} ergs/s formally provides the diffuse emissivity. There are other constraints which do not allow this extrapolation. First, the normalization of the log N vs log S curve would be greatly exceeded, as discussed later in this paper. Second, the total density of X-ray Seyferts of luminosity above 3×10^{42} ergs/s would be 3×10^{-4} Mpc^{-3}, which is at least as great as generous estimates of the volume density of all Seyferts [37]. Third, there would need to be a very sharp cutoff below $L_{44} = 0.03$ to avoid exceeding the volume emissivity of the background radiation.

Physically, the X-ray source in Seyfert galaxies may be the same as in the newly recognized class of emission line galaxies [38,39,40] and or as in the giant radio source Cen A. It must be stressed that the observed emission line galaxies are at distances 30-50 Mpc and luminosities 1 to 30×10^{42} ergs/s, and therefore are not themselves the dominant class of source which contributes to the background. It must be assumed that they are the "tip of the iceberg" of an appropriate luminosity function, with sources $\lesssim 10^{42}$ actually making up the background. Cen A is the only identified extragalactic source in the proper luminosity range. When it is associated with the class of weak radio galaxies [41], the expected volume density is 3×10^{-5} Mpc, and only a few percent of the background is produced. However, if we consider that its distance [42] is 5 Mpc, we estimate a volume density 10^{-3} Mpc^{-3}, although the uncertainty with only this one object is large enough that 3×10^{-5} Mpc^{-3} is still allowed. M82 is also a possible prototype object.

Normal galaxies have been estimated [5] to provide 6×10^{37} ergs/s Mpc^3 emissivity due to stellar X-ray sources. They would provide 6% of the background with the revised emissivity estimate for $q_o = 0$ (or 4% with $q_o = 1/2$).

There are now five known X-ray emitting quasi-stellar objects [43-48]. Three "classical" quasars are within a redshift of 0.16 and emit 0.3 to 5×10^{45} ergs/s; two BL Lac objects at a distance of about 200 Mpc emit about 3×10^{44} ergs/s. Considering only these objects at their estimated distances gives rough estimates of the equivalent local volume density of 10^{-7} for the BL Lac objects and 3×10^{-9} for the quasars. These give estimates of 3 and 1 percent contributions, respectively. The numbers of unidentified X-ray sources might allow the above density estimates to be increased only by a factor of 3 to 10.

In summary, no single class of X-ray source provides the total luminosity in the 2-6 keV region. A new viewpoint arising from our increased knowledge of extragalactic discrete sources is as follows: The sum of all classes, with the most generous estimates for each and with q_o near zero, could formally produce the measured intensity. Summing the contributions from these classes does not solve the problem of producing the measured spectral shape, however. What is significant is that each type of source produces at least a few percent of the background, and therefore, according to Fig. 1, could easily produce the entire background emissivity with either density or luminosity evolution less than the amount actually observed for quasars. Our present understanding of the X-ray production processes predicts that some evolution must take place [49,50]. The key feature of any theory of the X-ray background must be to predict an appropriate evolution of the X-ray spectral shape.

THE NET BACKGROUND SPECTRUM

Figure 4 shows the spectral volume emissivity plotted as a function of energy. The contribution marked clusters normalizes their 2-6 keV emissivity to be that estimated from their luminosity function of Fig. 3. The midpoints denote the "typical" cluster spectrum of 7 keV

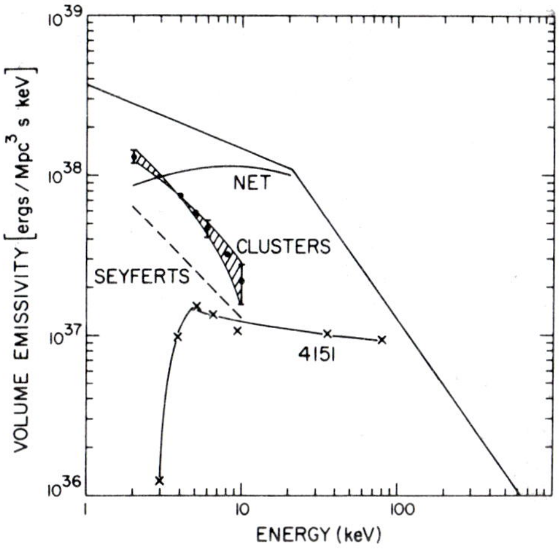

Fig. 4 Differential volume emissivity function. The solid curve is a representation of the data in Fig. 2, as derived from equation 1. We have assumed $q_0 = 0$. The shaded region is an estimate of the contribution due to clusters, and the dashed line of that due to Seyferts. The average spectrum of all Seyferts is poorly known, and might be nearer to that measured for the object NGC 4151. The line marked "net" is the residual background emissivity after subtracting the predicted contributions of clusters and Seyferts, and is the curve which must be explained by a successful theory of the origin of the background.

and the shaded region denotes the range of temperatures 5-10 keV which fit most of the measured X-ray spectra of clusters [51]. The dashed line represents the Seyfert volume emissivity obtained by integrating the luminosity function in Fig. 3 from 7×10^{43} to 10^{45} ergs/s, and by assuming an energy spectrum proportional to E^{-1}.

This particular shape is obtained from taking the power law number index of $2.8 \pm .8$ quoted by Tananbaum et al. [35], and subtracting the quoted error to try to mimic the flatness of the measured spectrum. The original result was obtained by lumping together the spectra of the 14 Seyferts observed by Uhuru, and is not claimed to be a fair average spectrum weighted by the luminosity function. Figure 4 includes actual spectral data [52,53] on NGC 4151, arbitrarily transformed to a volume emissivity by multiplying by a nominal density of 3×10^{-4} Mpc^{-3} for all Seyferts. Clearly this spectrum is everywhere flatter than the background. The Seyfert 3C390.3 is reported to have an energy index 0.7 [31], approximately equal to the background below 20 keV, and much flatter above. The difficulty with these two objects is their relatively high cutoffs at 3-4 keV. The heavy curve labelled "net" from 2 to 20 keV is obtained by subtracting the cluster and Seyfert emissivity curves from that due to the original measured intensity. I conclude that Seyfert galaxies can be hypothesized to have a distribution of spectral indices ranging at least from that of NGC 4151 to that of the lumped spectra of Tananbaum et al. [35] and further to have a range of low energy cutoffs sufficient to mimic the "net" spectrum from 2 to 20 keV.

Even if we subtract a contribution from clusters alone, the net curve would be flatter than the gaunt factor $E^{-.3}$ dependence of a thermal bremsstrahlung curve of any temperature. If we use this factor to constrain the amount of hot intergalactic plasma we obtain $(C^{1/2}\,\Omega) < .3$, where Ω is the ratio of the plasma density to closure density, and C is the "clumpiness" factor [54]. (To obtain this limit we, of course, started from the value 4×10^{38} ergs/Mpc^{3}s keV at 2 keV which would be allowed if $q_o = 1/2$.)

Above 20 keV very few extragalactic sources have been studied, and we cannot say with any accuracy what fraction of the background remains unexplained, nor a fortiori, what is its spectral shape. Brecher and Morrison [55,56,57] first stressed that a range of power law indices and break frequencies similar to those seen in radio galaxies could simulate the apparent diffuse X-ray spectrum from 1 keV to 1 MeV. Their argument was in the very different context of explaining the diffuse background by inverse Compton scattering on the microwave background of the same electrons responsible for the radio emission.

We may speculate that a similar argument is applicable in a new context. The flatness of the net spectrum in Fig. 4 is similar to that of the class of compact radio sources. Many of these sources have been observed to have milliarcsec sizes in the gHz region and to show cutoffs which can be interpreted as due to synchrotron self-absorption. They would then be expected to emit X-rays by the inverse Compton scattering on the radio photons of the same high energy electrons which produce the radio emission [58,59]. Conversely, synchrotron self-Compton models have been evoked to explain the observed X-ray emission from specific objects [59,60,61,45]. The synchrotron and inverse Compton processes produce identical spectra from a given population of electrons. A measurement of the BL Lac type object, Markarian 501, indeed shows an X-ray energy index of 0.2, consistent with the radio index of 0.1 [31]. The measurement also shows no low energy cutoff, which is especially encouraging for producing the 2-5 keV X-ray background. An appropriate distribution and evolution of radio source spectra must be postulated.

FLUCTUATIONS AND THE LOG N VS. LOG S CURVE

If the luminosity function dN/dL is independent of distance, then those sources much nearer than a redshift of z = 0.1 will follow the classical log N vs log S relation

$$n(>S) = K\,S^{-3/2} \quad \text{sources/ster} \tag{7}$$

where

$$K = \frac{1}{16\pi^{3/2}} \int_0^\infty L^{3/2} \left(\frac{dN}{dL}\right) dL \tag{8}$$

or

$$K = (3.7 \times 10^6) \int_0^\infty L_{44}^{3/2} \left(\frac{dN}{dL}\right) dL_{44} \tag{9}$$

and where S is in UFU, (dN/dL) = number Mpc^{-3} $(10^{44}$ ergs/s$)^{-1}$, K is the number of sources stronger than 1 UFU per steradian, and L_{44} is in units of 10^{44} ergs/s, (2-6 keV). The number K can be estimated in two ways, through direct counts of X-ray sources [62, 63] after applying appropriate corrections, and independently through measurement of the small scale fluctuations [64,3] via the relation

$$p^2 = (6.3 \times 10^{-5})\,K \tag{10}$$

where $p^2 = \text{var } I / I^2$ is the percentage fluctuation squared. The numerical constants in equation (10) depend on Uhuru parameters, in particular on the effective solid angle $\Omega_e = 0.004$ ster. Strickly speaking, equation (10) sets only an upper limit to K since there may be other sources of fluctuation which contribute to p^2 as the sum of squared quantities.

In Fig. 5 we show independent measurements of K plotted as horizontal error bars, and of p plotted as vertical error bars. These are centered on a straight line of unit slope and the scales are adjusted so that this line represents equation (10). I have adjusted the value of

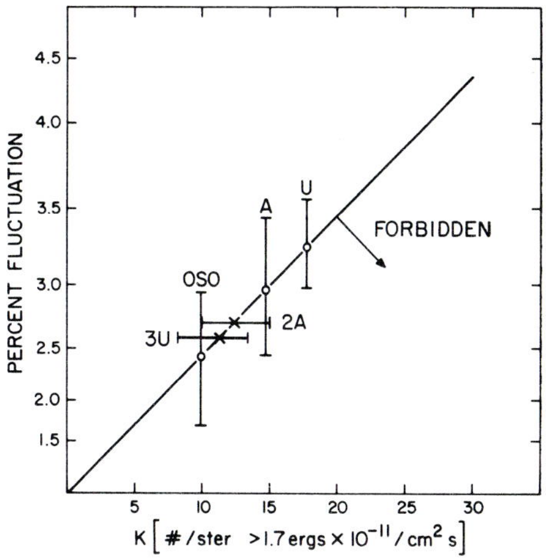

Fig. 5 Measurement of the number of sources per steradian stronger than 1 UFU (horizontal bars and scale) and of the percentage fluctuation in the background in an equivalent 0.004 steradian beam (vertical bars and scale). The solid line is given by equation (10), assuming that all fluctuations arise from the unresolved sources of population $KS^{-3/2}$. The area to the right of this line is forbidden, but to the left is allowed if there are additional contributions to the fluctuations.

K labelled "2A" as quoted by Warwick and Pye [63] by a factor of 0.84, which represents a conversion of 2.4 UFU per Ariel V ct s^{-1}. The curve labelled "3U" is based on the value $K = 21 \pm 5$ obtained by Murray [62], after his corrections for source confusion. It is now known that the fluxes in the 3U catalog and preprint version of the 4U catalog are based on averaging only positive fluctuations, and overestimate the average intensity of a given source. For the 4U catalog [44], I have computed the ratio of PST intensities to the original 4U catalog entry, and find a ratio of 0.65. This ratio applies to 117 sources at galactic latitudes $|b| > 20^o$, and excludes sources identified with galactic objects and sources with greater than 2% chance of being spurious. This leads to a correction faction $(0.65)^{1.5} = 0.52$, or $K = 11 \pm 2.6$, bringing it into very good agreement with the Ariel V result. Therefore, with the constant K in the range 7-15 and with intrinsic sky fluctuations [3,65, 66] of 1.5 to 3.2 percent on an equivalent angular scale of 0.004 degrees, we have a self-consistent picture.

The luminosity functions of equations (4) and (5) may be used in equation (9) to predict contributions of $K_c = 15$ due to clusters and $K_s = 5$ due to Seyferts. These numbers show that clusters and Seyferts can account statistically for all the high latitude sources. Of course, the absolute normalizations of the luminosity functions are subject to an uncertainty which could be at least a factor of three. If anything, they must have been overestimated, and therefore their contribution to the X-ray background is overestimated. I will adopt an upper limit of $K_x \leq 10$ as the normalization constant in equation (7) for any additional class of extragalactic X-ray emitter.

Note that if the sum $K = K_c + K_s + K_x$ takes on its maximum value, then additional fluctuations intrinsic to the universe, and not due merely to additional unresolved sources [64], cannot be more than 1%. However, if K is as low as 7, then there may be a residual 2.4%

fluctuation. These numbers are very relevant to the models in which evolution of source density and/or luminosity produces the background. From Fig. 1 we see that in such a case we are roughly observing a shell of sources near the cutoff distance. In this model we must have a minimum of 1.5×10^5 (8.6×10^5) sources/ster if the additional fluctuations are indeed 2.4% (1%). This in turn implies equivalent local volume densities of a minimum of 7.8×10^{-6} (1% fluctuations) and 1.4×10^{-6} Mpc^{-3} (for 2.4% fluctuations). The density calculation is based on $q_o = 0$, and would be a factor of a few larger for $q_o = 1/2$.

As stressed recently by Fabian and Rees for the 1% fluctuation case [67], the local volume density of Abell clusters is $\sim 10^{-6}$ Mpc^{-3}, and therefore luminosity evolution to produce the background would violate the isotropy. On the other hand, we do not expect an Abell cluster to disappear within several times 10^{10} years of its formation, therefore density evolution also can be ruled out.

Let us return to the question of whether a new population of discrete sources comprises the background. We know that locally such sources obey $n(>S) = K_x S^{-3/2}$ with $K_x \leq 10$. To obtain the diffuse X-ray surface brightness we must have

$$3.4 \times 10^{-8}\ (\text{ergs/cm}^2\ \text{ster}) = \int_{S_{min}}^{\infty} S \left(\frac{dN}{dS}\right) dS \qquad (11)$$

or

$$2000\ (\text{UFU per ster}) = \int_{S_{min}}^{\infty} \frac{K_x}{1.5}\ f(z)\ S^{-3/2}\ dS \qquad (12)$$

where $f(z) \equiv (dN/dS)/(dN/dS)_E$ is defined as the cosmological correction factor which multiplies the Euclidean relation. The factor f depends on S implicitly through the red-shift z. Figure 6 shows a plot of f for the two different cases $q_o = 0$, and $q_o = 1$. An X-ray energy spectral index of -0.5 is assumed. The curves are very similar down to a flux

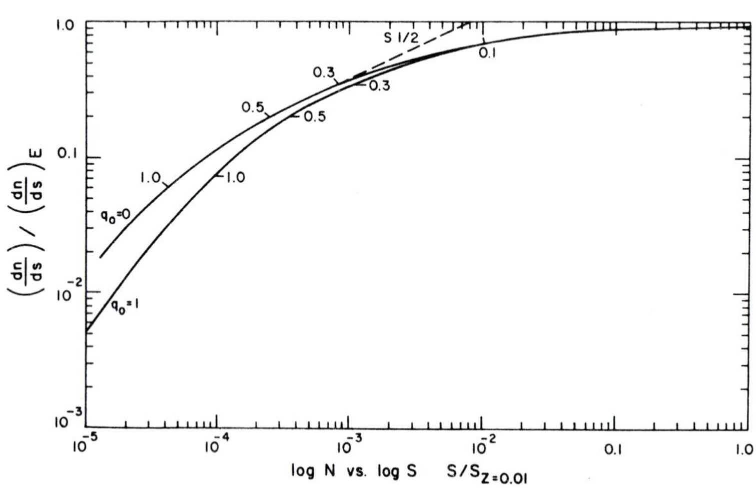

Fig. 6 Correction factor to the Euclidean relation due to relativistic geometry. Decrease of this correction more steeply than $S^{1/2}$ for sources more distant than about z = 0.3 resolves Olbers paradox.

level 10^{-4} times the strength which the same source would appear to have at a distance of 60 Mpc (z = 0.01), although the redshift labels of particular points may vary. To explain the integrated background we see from Fig. 6 that S_{min} in equation (12) should be considered to be at a redshift of z = 0.3, where $S_{min} = 10^{-3}$ times $S_{z=0.01}$. Since f(z) is proportional to $S^{1/2}$ at this redshift, the integrated contribution in equation (12) from all sources weaker than this S_{min} will converge. This can also be seen in Fig. 1, where about 40% of the background arises from z less than 0.3.

To explain 80% of the background it is necessary to assume that the sources which are at 60 Mpc give a flux of about 0.1 UFU, which implies an average luminosity L (2-6 keV) = 7 x 10^{41} ergs/s, and an average volume density of 10^{-3} Mpc^{-3}. The luminosity is an upper limit, and the volume density a lower limit, since we have taken the upper limit K_x = 10/ster for this hypothetical class of source.

The emission line galaxies recently discovered to be X-ray sources cannot be this class of object. The table shows that their luminosities, except for M82, significantly exceed the upper limit of the allowed average. This is a strong inconsistency since we have only 8

Galaxy	Distance (Mpc)	L_X (10^{42} ergs/s; 2-6 keV)	Reference
NGC 1365	34	1.3	[39]
NGC 2110	45	13	[68]
NGC 2992	44	13	[39]
MGC -5-23-16	49	30	[40,69]
M 82	3.1	0.05	[39]
NGC 5506	50	7	[70]
NGC 7582	30	2.7	[39]
Cen A	5	0.4 to 2.5	

objects, of strength above 1 UFU, in a sample of roughly 1/4 of the sky. Thus, K is actually less than 2 for these objects, S at 60 Mpc would be 0.004 UFU, and the emission line galaxies with flux 1 UFU should be at 4 Mpc and have luminosities 3 x 10^{40} ergs/s, if they constituted the background. M 82 is the single object consistent with these properties.

ISOTROPY OF THE BACKGROUND

Figure 7 shows a representation of the X-ray isotropy as measured by Uhuru. The data are sorted into 10 degree bins along the scan circle of any one day, and the accumulated counts and exposure times are dumped into bins of 10 x 10 degrees in galactic coordinates. Data are plotted only if the net exposure time was greater that 100 seconds, so that photon statistics contribute only about 25% as much to the scatter as do the intrinsic sky fluctuations. Bins are excluded if they contain a deflection more than 3 counts/s above the all sky mean [3]. Figure 7 shows some large scale effects which are significant at about 5 times the intrinsic sky fluctuation noise. I interpret these as a galactic population, and also an excess of sources along the supergalactic equator.

We analyze the data for a supergalactic component, taking a model of an oblate spheroid of major axis 30 Mpc and minor axis 6 Mpc [71]. Figure 8 shows the data plotted as a function of path length. The point at 5 Mpc may contain some contamination from the Large Magel-

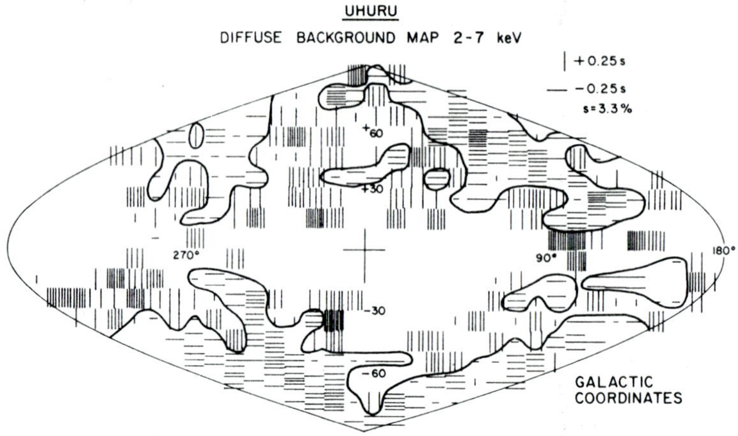

Fig. 7 Spatial distribution of the 2-7 keV background as measured by Uhuru. The shading of each box is proportional to the number of rms intrinsic fluctuations from the mean over the whole sky. Blank areas indicate no coverage, due to the proximity of discrete sources. Residual contribution from these sources leads to positive fluctuations near the galactic and local supercluster planes.

lanic Cloud sources. The other data is consistent with no supercluster emission, but allows a contribution of up to 1.5 UFU/50 Mpc of path length. This would correspond to a volume emissivity 10 times that of the X-ray background, and gives an upper limit of 2×10^{44} ergs/s for the luminosity of the local supercluster, compared to 3×10^{45} ergs/s for the X-ray superclusters discovered by Murray et al. [72].

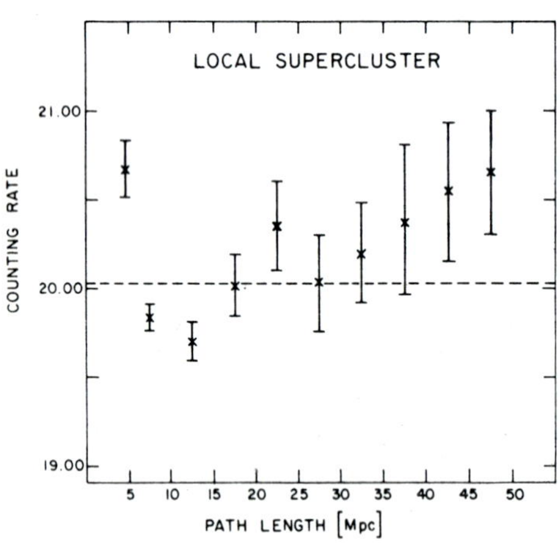

Fig. 8 Total detector counting rate (3.50 counts/s is non-X-ray background), as a function of path length viewed through the local supercluster [71]. The dashed line is the mean over the whole sky, and is consistent with all the data except the 5 Mpc point. An upper limit to the local supercluster contribution is 0.03 counts/(s Mpc).

CONCLUSIONS

We conclude that no class of discrete X-ray source, whether or not it is presently recognized, can constitute the X-ray background without evolutionary effects. Within this restriction, many theories are possible, and none are particularly favored on purely observational grounds.

I list the following hypotheses in the order in which they may be amenable to verification observationally:

1. Evolution of the luminosity and/or space density of Seyferts and other active galaxies. The observational tests are for HEAO/A-2 to show that their combined spectra can mimic the net background spectrum of Fig. 4, and for HEAO-B to show via the $< V/V_m >$ test [73] that their volume emissivity at a redshift $z = 0.5$ (10^{-3} UFU for $L = 10^{43}$ ergs/s) is enhanced by a factor of 3 relative to their local contribution.

2. Quasars (including BL Lac type objects). The observational tests are to observe enough objects to conclude that their evolution is typical of quasars as a class, and to show that their combined spectrum can simulate that of the net background. Note that the quasars which comprise the bulk of the background must have an average luminosity less than 6×10^{44} ergs/s to be consistent with the fluctuation constraint.

3. Evolution of luminosity or space density of clusters of galaxies, along with evolution of their effective temperatures. Observationally, pure luminosity evolution implies greatly increased angular sizes to avoid the fluctuation constraint discussed above, and density evolution implies the existence of X-ray clusters in which the galaxies have not formed to avoid exceeding the density of all Abell clusters. The observational test is again to determine, via $< V/V_m >$, that the density of objects above a given luminosity limit increases with redshift. HEAO-B can detect the clusters of 2×10^{44} ergs/s luminosity to a redshift $z = 1$.

4. A hot intergalactic plasma. The observational tests are that the above three tests must fail, and the discrete source contributions in Fig. 5 must be overestimated so that the net spectrum is very close to the measured spectrum. A combination of hypotheses 3 and 4 has recently been suggested by Kellogg [74].

5. Inverse Compton scattering. If the direct X-ray astronomy measurements do not reveal the source of the X-ray background, we must appeal to radio astronomy observations for the sources of metagalactic cosmic ray electrons with an appropriate spectrum and intensity to match the net diffuse background.

ACKNOWLEDGEMENTS

I am grateful to E. Boldt, F. Marshall, and R. Schafer for communication of results from OSO-8 and HEAO/A-2 on the spectrum and small scale fluctuations of the background. I thank A. Fabian and R. Mushotzky for discussion of discrete source contributions. I thank several members of the High Energy Astrophysics Division, Harvard-Smithsonian Center for Astrophysics, for comments on an early version of this talk which have influenced the viewpoint presented here. H. Gursky has collaborated on studied of the Uhuru X-ray background results. I especially thank J. Peritz for careful preparation of this manuscript in camera-ready form.

REFERENCES

1. D.A. Schwartz, Astrophy. J., 162, 439 (1970).
2. A.C. Fabian, Nature (Phys. Sci.), 327, 19 (1972).

3. D.A. Schwartz et al., Astrophys. J., 204, 315 (1976).
4. D.A. Schwartz et al., Astrophys. J., 162, 431 (1970).
5. D.A. Schwartz and H. Gursky, in: X-ray Astronomy, Reidel, Dordrecht, 359 (1974).
6. D.A. Schwartz, in: X-rays in Space, U. of Calgary Press, Calgary, 1096 (1975).
7. R. Giacconi et al., Phys. Rev. Lett., 9, 439 (1962).
8. J. Silk, Space Sci. Rev., 11, 671 (1970).
9. J. Silk, A. Rev. Astron. Astrophys., 11, 269 (1973).
10. J.E. Felten, in: IAU Symposium No. 55 on Non-Solar X-ray and Gamma-Ray Astronomy, Reidel, Dordrecht, 258 (1973).
11. M.J. Rees, in: IAU Symposium No. 55 on Non-Solar X-ray and Gamma-Ray Astronomy, Reidel, Dordrecht, 250 (1973).
12. J. Silk, Astrophys. J. (Letters), 151, L19 (1968).
13. Y. Avni, preprint WIS-77/52-Ph. (1978).
14. J.E. Felten and P. Morrison, Astrophys. J., 146, 686 (1966).
15. K. Brecher and P. Morrison, Astrophys. J. (Letters), 150, L61 (1967).
16. G.B. Field and S.C. Perrenod, Astrophys. J., 215, 717 (1977).
17. C. Hogan and D. Layzer, Astrophys. J., 212, 360 (1977).
18. M. Rowan-Robinson and A.C. Fabian, MNRAS, 170, 199 (1975).
19. M. Schmidt, Astrophys. J., 176, 273 (1972).
20. D.A. Schwartz and L.E. Peterson, Astrophys. J., 190, 297 (1974).
21. P. Gorenstein et al., Astrophys. J., 156, 315 (1969).
22. T.M. Palmieri et al., Astrophys. J., 169, 33 (1971).
23. B.R. Dennis et al., Astrophys. J., 186, 97 (1973).
24. J.A.M. Bleeker and A.J.M. Deerenberg, Astrophys. J., 159, 215 (1970).
25. E. Horstman-Moretti et al., Il Nuovo Cimento, 6B, 68 (1971).
26. R.L. Kinzer et al., Astrophys. J., 222, 370 (1978).
27. J.I. Trombka et al., Astrophys. J., 212, 925 (1977).
28. T. Pal, in: IAU Symposium No. 55 on Non-Solar X-ray and Gamma-Ray Astronomy, Reidel, Dordrecht, 279 (1973).
29. H.M. Horstman et al., Riv. Nuovo Cimento, 5, 255 (1975).
30. F. Marshall, private communication (1978).
31. E.A. Boldt, preprint NASA TM-78106 (1978).
32. J.R. Gott et al., Astrophys. J., 194, 543 (1974).
33. D.A. Schwartz, Astrophys. J., 220, 8 (1978).
34. I. McHardy, preprint (1978).
35. H. Tananbaum et al., Astrophys. J., 223, 74 (1978).
36. M. Elvis et al., MNRAS, 183, 129 (1978).
37. J. Huchra and W.L.W. Sargent, Astrophys. J., 186, 433 (1973).
38. J.N. Bahcall et al., Astrophys. J. (Letters), 199, L9 (1975).
39. M.J. Ward et al., Astrophys. J., 223, 778 (1978).
40. F. Pineda and H.W. Schnopper, I.A.U. Circ. No. 3190 (1978).
41. M. Ryle, Q. Jl. Astron. Soc., 11, 429 (1970).
42. G.R. Burbidge and E.M. Burbidge, Astrophys. J., 129, 271 (1959).
43. M.J. Ricketts et al., Nature, 259, 546 (1976).
44. W. Forman et al., Astrophys. J. (Suppl), in press (1978).
45. D.A. Schwartz et al., Astrophys. J. (Letters), 224, in press (1978).
46. E. Kellogg et al., Astrophys. J. (Letters), 165, L49 (1971).
47. K.M.V. Apparao et al., Nature, 273, 453 (1978).
48. G. Ricker, this volume (1979).
49. L. Cowie and S. Perrenod, Astrophys. J., 219, 354 (1978).
50. G. Garmire, private communication (1978).

51. R.F. Mushotzky et al., Astrophys. J., in press (1978).
52. D. Koch, private communication (1973).
53. W.A. Baity et al., Astrophys. J. (Letters), 199, L5 (1975).
54. G.B. Field, in: IAU Symposium No. 63 The Confrontation of Cosmological Theories with Observational Data, Reidel, Dordrecht, 13 (1974).
55. K. Brecher and P. Morrison, Phys. Rev. Letters, 23, 802 (1969).
56. K. Brecher and G.R. Burbidge, Comments on Astrophys. and Space Sci., 2, 75 (1970).
57. K. Brecher, Astron. Astrophys., 23, 105 (1973).
58. G.R. Burbidge et al., Astrophys. J., 193, 43 (1974).
59. R.F. Mushotzky et al., Astrophys. J. (Letters), 206, L45 (1976).
60. J.E. Grindlay, Astrophys. J., 199, 49 (1975).
61. H.W. Schnopper et al., Astrophys. J. (Letters), 215, L7 (1977).
62. S.S. Murray, Astrophys. J., in press (1978).
63. R.S. Warwick and J. Pye, MNRAS, 183, 1169 (1978).
64. P.A.G. Scheuer, Proc. Camb. Phil. Soc., 53, 764 (1957).
65. R.A. Schafer et al., Bull. A.A.S., 9, 616 (1977).
66. J.P. Pye et al., Bull. A.A.S., 9, 558 (1977).
67. A.C. Fabian and M.J. Rees, preprint (1978).
68. H. Bradt et al., preprint (1978).
69. M. Davis et al., I.A.U. Circ. No. 3203 (1978).
70. G.R. Ricker et al., preprint CSR-P-77-33 (1977).
71. G. de Vaucouleurs, Astron. J., 63, 253 (1958).
72. S.S. Murray et al., Astrophys. J. (Letters), 219, L89 (1978).
73. M. Schmidt, Astrophys. J., 151, 393 (1968).
74. E.M. Kellogg, Astrophys. J., 218, 582 (1977).
75. E.A. Boldt et al., this volume (1979).

X-RAY EMISSION FROM SUPERCLUSTERS

A. Cavaliere*, L. Danese*, G. De Zotti*,
R. Fusco Femiano** and M. Salvati**

**Istituto di Astronomia, Università di Padova*
***Laboratorio Astrofisica Spaziale CNR di Frascati*

OBSERVATIONS

Evidence of large X-ray emission from regions of high density of galaxy clusters has been reported by Murray et al. [1] : luminosities $L=6$ to 10×10^{45} erg s^{-1} (2-10 keV, $H_o=50$ Km s^{-1} Mpc^{-1}) are apparently associated in the 4U Catalog with 3 out of 12 such regions falling in the sky area covered to 2 ct s^{-1}; these comprise respectively 6 to 7 rich clusters with $D>4$ within a radius $\simeq30$ Mpc ("superclusters", henceforth SC). However, the more conservative PST fluxes from these sources correspond to $L\simeq2$ to 4×10^{45} erg s^{-1}, consistent with the upper limits by Ricketts [2] ; Pravdo et al. [3] set a still lower limit for one of these sources. This contrasting evidence poses anyway the interesting problem of the effects of superclustering on the cluster X-ray emission.

MODES OF EMISSION

The 4U observations have been discussed in terms of thin Bremsstrahlung from hot gas pervading the SCs with $T\simeq2\times10^{8}$°K and $n\simeq10^{-5}$ cm^{-3} (implying an overdensity $\simeq3/\Omega$, Ω being the density parameter); the mass of the emitting gas is then $M\simeq2\times10^{16}$ $M_\odot$. If that were the equilibrium temperature of a virialized system, a total mass $\gtrsim 5\times10^{16}$ $M_\odot$ would be required [1] . However, a nearly uniform condition is unlikely in a system whose sound crossing time $t_s=3\times10^{10}$ $r_{50Mpc}/T_8^{\frac{1}{2}}$ yr matches or exceeds the Hubble time. A gravitational system virializing now must show a much larger density contrast $\rho_{vo}/\bar{\rho}_o\simeq177/\Omega$ $(H_o t_o)^2$ [4] . If the SCs are neither virialized nor appreciably recollapsed it is not easy to account for the huge thermal energy postulated, 10^{66} $(M_{gas}/10^{16}$ $M_\odot)$ erg. Finally, Kellogg [5] finds no evidence of intercluster emission in a number of small nearby SCs.
We argue that any SC emission must be composed of the combined emissions associated with the clusters included. The luminosity function of Abell clusters has been derived by Schwartz [6] : $P_1(>L)=\exp(-L/\bar{L})$ with $\bar{L}=2\times10^{44}$ erg s^{-1}. We find that the probability of a SC luminosity synthesized with summed emissions associated with n component clusters is

$$P_n(>L) = \exp(-L/\bar{L}_{SC}) \sum_{k=0}^{n-1} (L/\bar{L}_{SC})^k/k! \quad . \qquad (1)$$

Eq. (1) is tabulated, for values of interest here, in Table 1.

TABLE 1 A Tabulation of the Integral Luminosity Function (Eq. (1); $x=L/\bar{L}_{SC}$) for 6, 7, and 8-Component Clusters

x	1	2	3	4	5	6	7	8	9	10	14
P_6	.9994	.983	.916	.785	.616	.446	.301	.191	.116	.067	.0076
P_7	.99992	.996	.967	.889	.762	.606	.450	.313	.207	.130	.0163
P_8	.99999	.999	.988	.949	.867	.744	.599	.452	.324	.220	.0340

This shows that $P_7(>L)\simeq 3/12$ requires an average luminosity of the superclustered clusters $\bar{L}_{SC}>\bar{L}$ if $L>1.7\times10^{45}$ erg s^{-1}.

ENHANCED EMISSION

Enhanced emission may result in two ways. If the mass fraction surviving the process of galaxy formation in diffuse form is $\varepsilon\gtrsim10\%$, one expects in the forming SCs a large density and pressure of the intercluster gas; then a considerable luminosity enhancement obtains. The external pressure acts as a boundary condition on a static distribution of hot gas in the cluster potential wells [7] . To compare an isolated object (labelled 1) structured like the Coma cluster with an analogous object (labelled 2), we have computed the ratio L_2/L_1 from detailed polytropic models. In the absence of inward heat flow ($T_{ext}\lesssim10^7$°K or long diffusion time of the emitting volume) we find the results given in Table 2.

Consider then the dynamical stage of infall of external gas. Using the gravitational instability picture as a guide for magnitudes, we find an overdensity $\rho/\bar{\rho}\simeq2\div3$ at $z\simeq4$ ($\Omega=0.1$); this parameter governs the local deceleration parameter q_{eff}, while the fraction $\varepsilon\rho$ left over in diffuse matter constitutes the infalling gas mass, whose dynamical stress also adds to the containment. The infall process gives, according to the numerical computations of Cowie and Perrenod [8] and Perrenod [9] , $L_2/L_1\gtrsim5$ at $z\simeq0$, in conditions where $\varepsilon>10\%$.

If, instead, the process of galaxy formation were so efficient as to incorporate $\gtrsim$ 90% of the mass, then most of the intracluster gas must originate from mass shed by the internal galaxies; but then, even in a well advanced SC the effects of the residual gas would be little. The X-ray luminosity of a superclustered cluster should not exceed that of an isolated object with the same richness R. (Note however that, in the presence of external gas of low pressure but high temperature, superclustered clusters could also be underluminous on account of heat inflow which would lower their central density while increasing their temperature.)

In fact, as the cluster emission is apparently correlated with R [10] , the expected SC outputs depend now on the statistics of richness distribution in SCs. We may evaluate an upper limit to $\bar{L}_{SC}$

TABLE 2 Values of L_2/L_1 as a Function of $y=p_2/p_1$, Computed from Detailed Hydrostatic Models of Gas Distribution in Clusters[(a)]

γ	T_o ($\times 10^8$ °K)	n_o ($\times 10^{-3}$ cm^{-3})	n T cm^{-3} °K	y	L_2/L_1
1.33	2.07	1.23	3	1	1.0
1.33	2.10	1.79	9	3	2.1
1.33	2.15	1.82	30	10	2.6
1.33	2.19	1.90	60	20	3.2
1.27	1.88	1.48	10	1	1.0
1.27	1.93	2.00	30	3	2.0
1.27	2.00	2.20	100	10	3.1
1.27	2.06	2.35	200	20	4.2
1.19	1.66	1.88	50	1	1.0
1.19	1.73	2.65	150	3	2.5
1.19	1.83	3.45	500	10	5.6
1.19	1.90	4.22	1000	20	9.4

(a) The basic models (y=1, n T is the external density x temperature at the boundary) fit the X-ray brightness distribution in Coma Cluster with $\chi^2=17.7$ to 19.5 vs. 23 degrees of freedom [7].

$$\bar{L}_{SC} = \bar{L}(F_1+rF_2)/(f_1+rf_2) \lesssim 2.7\times10^{44} \text{ ergs s}^{-1} , \qquad (2)$$

using the percentages $f_1=73\%$ and $f_2=27\%$ of clusters of R=1 and R$\gtrsim$2 in the Abell sample [11], and the corresponding percentages in SCs: $F_1=60\%$ and $F_2=40\%$ [12]; the upper bound r<9 for the ratio of the average luminosities of the R=1 and R$\gtrsim$2 clusters reported in Jones and Forman [10] has also been inserted. After Table 1 this value of $\bar{L}_{SC}$ implies an average SC luminosity $L\lesssim2.5\times10^{45}$ erg s^{-1}. Even if $F_1=F_2\cong50\%$, $L<3\times10^{45}$ erg s^{-1} obtains.

CONCLUSIONS

The SC emission may constitute an intercluster density probe. A SC emission $L<3\times10^{45}$ erg s^{-1} can be explained in terms of the richness effect only. On the other hand, should values $L>3\times10^{45}$ erg s^{-1} be confirmed, they would imply the presence of large amounts of gas in SCs acting on the member clusters; the direct emission from such intercluster medium, however, is likely to be swamped by that induced as cluster-enhanced emission, provided that the thermal state of the intracluster gas is not too close to an isothermal (polytropic exponent $\gamma>1.2$, c.f. Table 2).
On the observational side, the total SC emission can be fixed by HEAO-1 and the luminosity of individual clusters in SCs can be gauged by HEAO-B. The physical conditions in SCs can also be checked selectively with measurements of $-\Delta T/T$ in the microwave background [13]. Any diffuse medium would produce $-\Delta T/T<10^{-4}$ over a large area, still redu-

ced by effect of beam subtraction with the throws limited to 15' by atmospheric disturbances. Overluminous clusters would also produce $-\Delta T/T \simeq 0.7\times10^{-4}\ L_{44}^{2} > 2\times10^{-4}$. Clusters of R=2 and 3 in SCs should be the prime targets for such measurements.

Work supported by Consiglio Nazionale delle Ricerche.

REFERENCES

1. S.S. Murray, W. Forman, C. Jones, R. Giacconi, *Astrophys. J. (Letters)* 219, L89 (1978).
2. M.J. Ricketts, *Monthly Notices Roy. Astron. Soc.* 183, 51P (1978).
3. S.H. Pravdo, R.F. Mushotzky, R.H. Becker, E.A. Boldt, S.S. Holt, P.J. Serlemitsos, J.K. Swank, *Nature* 270, 158 (1977).
4. J.R. Gott, M.J. Rees, *Astron. and Astrophys.* 45, 365 (1975).
5. E.M. Kellogg, *Astrophys. J. (Letters)* 220, L63 (1978).
6. D.A. Schwartz, *Astrophys. J.* 220, 8 (1978).
7. A. Cavaliere, R. Fusco-Femiano, *Astron. and Astrophys.* (1978), to be published.
8. L.L. Cowie, S.C. Perrenod, *Astrophys. J.* 219, 354 (1978).
9. S.C. Perrenod, *Ph.D. Thesis*, Harvard University (1977).
10. C. Jones, W. Forman, *Astrophys. J.* 224, 1 (1978).
11. G.O. Abell, *Astrophys. J. Suppl.* 3, 211 (1958).
12. S.C. Perrenod, *Nature* 274, 39 (1978).
13. R.A. Sunyaev, Ya.B. Zeldovich, *Comments Astrophys. Space Phys.* 4, 173 (1972).

COOLING GAS IN CLUSTERS OF GALAXIES AND THE X-RAY BACKGROUND

A. C. Fabian and P. E. J. Nulsen

Institute of Astronomy, Madingley Road, Cambridge CB3 0HA, England

Most previous work on the evolution of hot gas in clusters of galaxies has been based upon numerical hydrodynamic calculations [1 - 5] and has investigated only a small region of parameter space. This may seem reasonable when it is considered that the cooling time of the gas in observed clusters, $t_c \gtrsim H_o^{-1}$. Such a situation is, however, likely to occur in any gas remaining after a cooling phase. We have therefore attempted to estimate the cooling evolution of a cluster by a simple method, which can then indicate the range of possible histories more readily than any detailed numerical study.

We assume that the gas is initially in the form of a hydrostatic atmosphere, characterized by a central density, n_o, central temperature, T_o, and polytropic index γ (see e.g. [1]) in a potential well provided by

$$\rho(r) = 2.4 \times 10^{-25}[1 + (r/240 \text{ kpc})^2]^{-3/2} \text{ g cm}^{-3}.$$

Cooling occurs first in the core, then proceeds outward with little change in the outer atmosphere owing to the r^3 increase in volume. The luminosity evolution is then, at least, that due to the gas radiating all its thermal energy when its cooling time is reached:

$$L_x(t) = \alpha \frac{dM}{dt}\Big|_{r=r_c} \frac{kT_c}{\mu M_H}$$

where the subscript c refers to cooling and

$$\frac{dM}{dt}\Big|_{r=r_c} = \frac{4\pi r_c^2 M_H n(r_c)}{(dt_c/dr)\big|_{r=r_c}}.$$

α takes account of the flow details and is roughly $3 < \alpha < 10$. The results for $\gamma = 5/3$ and a total elapsed time of H_o^{-1} are plotted in Fig. 1. Further details are found in Fabian and Nulsen 6 .

Most clusters of galaxies have present X-ray luminosities $< 10^{44}$ erg s^{-1}, showing from Fig. 1 that n_o can never have been $\gtrsim 10^{-3}$ at the base of any atmospheres unless T_o was small (and thus so was the atmosphere). Hardening of the cluster potential well as the cluster itself evolved, energy release due to gravitational infall and the factor α all contribute to making the curves in Fig. 1 lower limits. The best

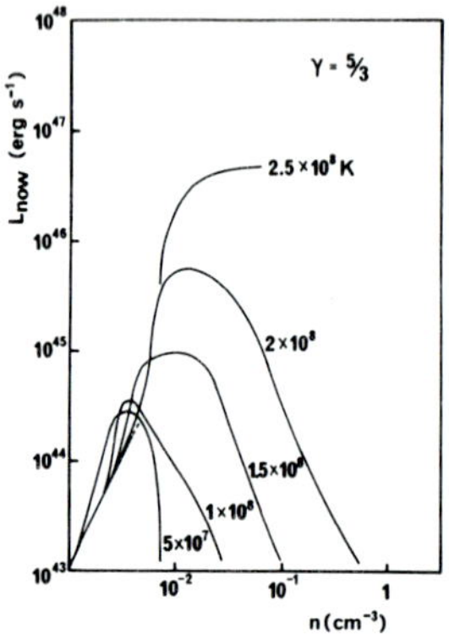

Fig. 1. X-ray luminosity now (divided by α) as a function of initial central density for a set of adiabatic atmospheres in the gravitational potential of the mass distribution $2.4 \times 10^{-25} [1 + (r/240 \text{ kpc})^2]^{-1.5}$ gm cm $^{-3}$. Curves are labelled with their initial central temperature. The quantity α allows for the details of the cooling process, but for most cases the flow proceeds approximately hydrostatically and then $\alpha \gtrsim 5$, while at the peaks $\alpha \simeq 10$. The simple model described in the text would give $L_X = 0$ before the core starts to cool. The figure is partially corrected (see [6]) for this by including a core luminosity until $t = t_c$.

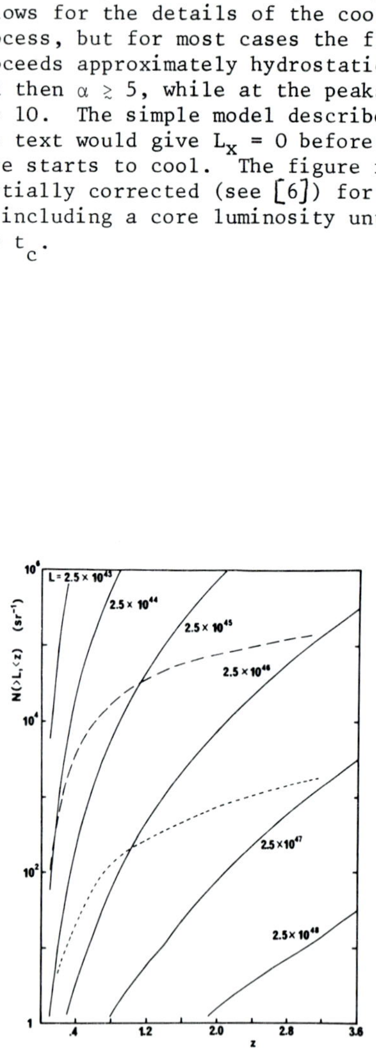

Fig. 2. Limits on the integral X-ray luminosity function versus redshift. The solid curves give the maximum number of sources sr^{-1} (with kT = 10 keV) with redshift less than z and luminosity greater than L (in erg s^{-1}) compatible with observations of the X-ray background. The upper dashed curves are the total number of Abell clusters sr^{-1} normalized to 4×10^5 $(c/H_0)^{-3}$. The lower dashed curves give the maximum number of clusters which can have passed through a cooling phase since z (see text).

defined point on the curves is the peak which is due to the cooling of the core.

We can also place constraints on X-ray cooling in clusters by comparing with observations of the X-ray background and source counts. These lead to (Fabian and Rees [7]):

$$N(>S) \lesssim \begin{cases} 5\ S^{-3/2} & 0.6 < S \\ 3.8\ S^{-2} & 2.10^{-3} \lesssim S \lesssim 0.6 \\ 2.2 \times 10^{3}\ S^{-1} & S \lesssim 2.10^{-3} \end{cases}$$

where S is in Uhuru ct s^{-1}, and N is sr^{-1}. These convert to Fig. 2 in which limits on the integrated luminosity function N(> L, < Z) are shown for q = 0.5 (for details see [6]).

The above observational constraints and theoretical arguments show that few rich clusters can have undergone strong cooling since Z = 3. This indicates that the inner regions of rich clusters of galaxies contained less than a few percent of their binding mass which as since cooled at X-ray temperatures. This conclusion can only be avoided if the gas cooled at much lower temperatures (such as when the cluster core was in the process of formation), or if the gas was ejected by some explosive events before it cooled.

We have studied the effects of widespread cooling of gas in clusters of galaxies. The range of central particle densities and temperatures that can have occurred in the history of such clusters is severely limited by theoretical arguments and by current observations. We find that there are three possible solutions involving: (i) little or no widespread cooling; (ii) catastrophic cooling throughout the entire cluster and (iii) cooling during a transient hydrodynamic phase.

ACF thanks the Radcliffe Trust for support and PEJN acknowledges receipt of a Hackett Studentship from the University of Western Australia.

REFERENCES

1. S. M. Lea, Astrophys. J., 203, 569 (1976).

2. S. F. Gull and K. J. E. Northover, Mon. Not. R. astr. Soc., 173, 585 (1975).

3. F. Takahara et al., Prog. Theo. Phys., 56, 1093 (1976).

4. L. L. Cowie and S. C. Perrenod, Astrophys. J., 219, 354 (1978).

5. S. C. Perrenod, preprint (1978).

6. A. C. Fabian and P. E. J. Nulsen, Mon. Not. R. astr. Soc., in press. (1978).

7. A. C. Fabian and M. J. Rees, Mon. Not. R. astr. Soc., in press. (1978).

HUBBLE CONSTANT FROM X-RAY AND MICROWAVE OBSERVATIONS OF CLUSTERS OF GALAXIES

A. Cavaliere, L. Danese and G. De Zotti

Istituto di Astronomia, Università di Padova

OUTLINE OF THE METHOD

Sunyaev and Zeldovich [1] pointed out that if hot intergalactic gas is the dominant contributor to the X-ray emission from cluster of galaxies, it would also produce measurable local distortions of the cosmic background radiation spectrum (Sunyaev-Zeldovich effect). In the Rayleigh-Jeans region, such distortions amount to a relative diminution $\Delta T/T$ of the radiation temperature:

$$\frac{\Delta T}{T} = -2.1\times10^{-4}\left(\frac{T_o}{10^8\,{}^oK}\right)^{3/4}\left(\frac{F_x}{Uct/s}\right)^{\frac{1}{2}}\left(\frac{3'}{\theta_c}\right)^{\frac{1}{2}}\left(\frac{D_L}{10^3 Mpc}\right)^{\frac{1}{2}}(1+z)\eta^{-\frac{1}{2}}\zeta \qquad (1)$$

where T_o is the central gas temperature, F_X (in Uhuru counts/s; 1 Uct/s = 2.4×10^{-11} erg cm^{-2} s^{-1}) is the 2-10 keV X-ray flux, θ_c is the angular core radius of the galaxy distribution, η and ζ are shape factors (see [2]) and D_L is the luminosity distance.

Now $\Delta T/T$ can already be measured ([3], [4], [5], [6]); accurate evaluations of T_o and some good determinations of the gas density and temperature distributions which may permit in principle an *observational* definition of the shape factors will soon become available with the sensitivity of the experiments on board of HEAO-1, and the spectral and spatial resolutions of those of HEAO-B. Then, in principle, from eq. (1) we can obtain D_L in terms of observable quantities; if $z \ll 1$ this amounts to an *absolute* determination of the Hubble constant (since $D_L \simeq cz/H_o$) obtained through combined X-ray, microwave and optical observations on a *single* cluster (observations on a second cluster at $z \simeq 1$ would also allow an evaluation of the deceleration parameter q_o).

In practice, however, one must beware of various entangling effects. On one side Compton scattering of relativistic electrons on the microwave photons and discrete sources can contribute significantly to the cluster X-ray luminosity; of course good spectral and structural data could isolate the bremsstrahlung share. On the other side cluster radiosources, background confusion and emission from "tepid" (10^4-10^7 °K) gas [7] could blur somewhat the microwave "dip". However measurements of ΔT at different frequencies (e.g. at 10.6 GHz and at 33 GHz, as already used by Gull and Northover [3] and by Lake and Partridge [4]) can single out the Sunyaev-Zeldovich effect since, in the Rayleigh-

Jeans region, its amplitude scales as ν^2 or as $\sim\nu^3$ as compared with that of the contribution of tepid gas and of radiosources, respectively. In particular measurements at the wavelength $\lambda=3.8/T$ mm (T=thermodynamic temperature of the cosmic background radiation), where the Sunyaev-Zeldovich effect vanishes, could constitute useful checks of the amplitude of "spurious" contributions.
A comparison of the measurements by Gull and Northover [3] and by Lake and Partridge [4] shows no evidence of $\Delta T(10.6\text{ GHz})>\Delta T(33\text{ GHz})$ as it would be expected if there is any relevant emission. Also Owen [8] lists 299 Abell clusters (out of a total of 503 observed) not detected at 1400 MHz above 0.1 Jansky; under conservative assumptions this corresponds, at 33 GHz, to a radiosource signal <0.16 mK, to be contrasted with measured $\Delta T\simeq -1$ mK ([5], [6]).

REDUCING THE INFORMATION NEEDED

The amount of structural information needed can be greatly reduced resorting to reasonable theoretical models. The intergalactic gas distribution is likely to be static at least for $z\lesssim 0.4$ ([9], [10]). The cores of static models are characterized by the ratio $\tau\equiv\mu\, m_H\langle v_{\shortparallel}^2\rangle/kT_o$ of the galaxy to the gas scale-heights (μ= mean molecular weight, $\langle v_{\shortparallel}^2\rangle^{\frac{1}{2}}$= line-of-sight velocity dispersion of the galaxies at the cluster center). The detailed distribution of a polytropic gas ($T\propto n^{\gamma-1}$) is efficiently approximated by ([9], [11], [12], [13]):

$$n/n_o = \begin{cases} (1+x^2)^{-3\tau/2} & \gamma=1 \\ \left[1-\frac{3}{2}\frac{\gamma-1}{\gamma}\tau\ \ln(1+x^2)\right]^{1/(\gamma-1)} & 1<\gamma\leq 5/3 \end{cases} \qquad (2)$$

in the central region $x\lesssim 20$. Fits to the Coma cluster data have been produced by Cavaliere and Fusco-Femiano [12] who found $1.3\times10^8\leq T_o\leq 2.1\times10^8$ °K $(0.39\leq\tau\leq 0.63)$, $1.1\leq\gamma\leq 1.4$ and by Malina et al. [14] who found $0.8\times10^8\leq T_o\leq 2\times10^8$ °K $(0.40\leq\tau\leq 1.0)$, $1.15\leq\gamma\leq 1.48$ (90% confidence levels). Of course, the parameters cannot vary independently within those ranges: in polytropic models higher T_o are associated with higher γ, the correspondence being biunivocal in the particular model adopted by Malina et al.. To be conservative we use here the union of the allowed regions in the parameter space. Also, we consider the Coma cluster structure (with the additional constraint cooling time > Hubble time) to be typical in the present absence of detailed structural information on more distant objects.
We have computed the shape factor $\phi=\tau^{3/2}\eta\zeta^{-2}$ for a number of cluster at different redshifts. If $z<0.05$ the variation of ϕ within the allowed region is mainly due to the beam throw effect: since atmospheric irregularities limit the beam separation to 10'÷15' for ground based observations, the reference beam seems a region still inside the gas distribution. On the other hand, at $z\gtrsim 0.2$ the dilution effect in the main beam becomes important as τ approaches unity (gas scale-height close to the galaxy scale-height). In the intermediate z range $(0.05\leq z<0.2)$ the variation of ϕ within the allowed region is generally $\lesssim$ 20% provided that the effect of the X-ray instrumental window are not significant. These conditions amount to select objects in the Abell distance classes D=3 to 6; richness classes R=2,3 are obvious prime candidates.

INFORMATION FROM EXISTING OBSERVATIONS

At present, F_X and ΔT are not both accurately known for any object; with this data situation, still simpler arguments can be used to derive bounds on H_0.
For Coma cluster, $F_X=(3.6\pm0.07)\times10^{-10}$ erg cm^{-2} s^{-1}(2-10 keV, [15]). The measurement of ΔT turns out to be difficult because of the presence of extended and relatively intense radiosources near the centre. The most recent result reads $\Delta T=+0.25\pm0.2$ mK for a beam separation of 9' [16], this corresponds to a 2σ limit $\Delta T \gtrsim -0.2$ mK. This value, since the maximum allowed values of ϕ is, in this case, $\phi_{max}\simeq6.6$, would imply the formal bound $H_0\gtrsim45$ km s^{-1} Mpc^{-1}. Another object that gives significant limits to H_0 is A2319 ($\Delta T=-0.13\pm0.41$ mK, [3]) associated with the X-ray source 4U 1919+44 ($F_X=(9.4\pm0.7)\times10^{-11}$ erg cm^{-2} s^{-1}, [15]). If A2319 is one cluster, these values imply $H_0\gtrsim60$ km s^{-1} Mpc^{-1}; if, instead, it is two clusters superposed along the line of sight (as discussed in [17]) and the same X-ray source is associated with A2319A, then the bound is $H_0\gtrsim15$ km s^{-1} Mpc^{-1}.
Preliminary values of ΔT have been measured [5] in the direction of three very rich and distant clusters of galaxies (A1689, A2125 and A2218) which however have not been observed to date in X-rays down to $\simeq$0.5 Uct/s. Very accurate measurements by Birkinshaw et al. [6] have confirmed the presence of the "dip" in the direction of A2218 with $\Delta T=-1.08\pm0.18$ mK. An inspection of the values of ϕ appropriate for these objects shows that they have an absolute minimum ($\phi_{min}=0.93$). Hence an upper limit to H_0 independent of structural information, can be derived. We find for A2218:

$$H_0 \lesssim 80(T_0/3\times10^8\ °K)^{3/2}(\bar{r}_c/r_c)(Uct\ s^{-1}/0.5) \qquad (3)$$

where $\bar{r}_c=0.25$ $(H_0/50)$ Mpc is the average core radius of the galaxy distribution in a sample of 15 rich and regular clusters [18].

CONCLUSIONS

The method here outlined has merits and drawbacks associated with an absolute measurement: it bypasses the conventional chain of distance indicators, but it requires precise measurements of the relevant observables. The sources of uncertainties, however, do not appear to be forbidding provided that a set of "reliable" object is selected: even within the present experimental situation, we have argued that the set is populated. The rapidly evolving data accumulation conceivably can lead in the near future to a determination of H_0 with an uncertainty comparable to, or less than, the overall indetermination associated with the classical methods.

We are indebted with R.B. Partridge for communicating to us his recent results, and to G. Setti for critical reading. We wish also to thank Mrs. M. Tomas Zanon for carefully typing the manuscript. Work supported by C.N.R..

Note

During this Symposium we have been told that a similar idea for determining H_0 was mentioned by J.E. Gunn at the Saas-Fee Spring Course "Observational Cosmology" (April 1978), proceedings to be published by

Geneva Observatory.

REFERENCES

1. R.A. Sunyaev and Ya.B. Zeldovich, *Comments Astrophys. Space Phys.* 4, 173 (1972).
2. A. Cavaliere, L. Danese and G. De Zotti, submitted to *Astron. Astrophys.* (1978).
3. S.F. Gull and K.J.E. Northover, *Nature* 263, 572 (1976).
4. G. Lake and R.B. Partridge, *Bull. A.A.S.* 8, 354 (1976).
5. G. Lake and R.B. Partridge, *Nature* 270, 502 (1977).
6. M. Birkinshaw, S.F. Gull and K.J.E. Northover, submitted to *Monthly Notices Roy. Astron. Soc.* (1978).
7. J.C. Tarter, *Astrophys. J.* 220, 749 (1978).
8. F.N. Owen, *Astron. J.* 79, 427 (1974).
9. S.M. Lea, in: *Highlights of Astronomy* 4, 329 (1977).
10. S.C. Perrenod, submitted to *Astrophys. J.* (1978).
11. A. Cavaliere and R. Fusco-Femiano, *Astron. Astrophys.* 49, 137 (1976).
12. A. Cavaliere and R. Fusco-Femiano, *Astron. Astrophys.*, to be published (1978).
13. J.N. Bahcall and C.L. Sarazin, *Astrophys. J. Suppl.* 34, 451 (1977).
14. R.F. Malina, S.M. Lea, M. Lampton and C.S. Bowyer, *Astrophys. J.* 219, 795 (1978).
15. C. Jones and W. Forman, *CFA* preprint n. 857 (1977).
16. G. Lake and R.B. Partridge, private communication (1978).
17. S.M. Faber and A. Dressler, *Astron. J.* 82, 187 (1977).
18. N.A. Bahcall, *Astrophys. J.* 198, 249 (1975).

THE CONTENT OF A TYPICAL HEAO-B FIELD ($|b| > 10°$)

A. C. Fabian and M. J. Rees

Institute of Astronomy, Madingley Road, Cambridge, UK

INTRODUCTION

The advent of orbiting imaging X-ray telescopes such as HEAO-B will allow deep X-ray surveys to be made for clusters of galaxies, active galaxies and quasars. We have combined information from the observed X-ray source counts and identifications to predict the numbers of sources of each class that might be expected in a typical HEAO-B one degree field at high galactic latitude.

PREDICTED SOURCE COUNTS

Observed X-ray source counts and identifications allow the luminosity functions of various classes of objects to be calculated. In this paper we concentrate upon the extragalactic sources in clusters of galaxies and galactic nuclei. The luminosity functions of these two classes of objects (see e.g. Warwick & Pye, these proceedings) are remarkably similar, and may be combined with a cosmological model and telescope response to predict source counts for any X-ray instrument. We have done this for the HEAO-B telescope using the updated sensitivities [1] and method explained in Fabian and Rees [2]. Figures 1 and 2 show the results for an observing time of 10^5s in the Imaging Proportional Counter (IPC) and for 4×10^4s in the High Resolution Imager (HRI).

The effective area of the HEAO-B detectors were approximated by the following:

$$A_{IPC} = \begin{cases} 100 & \\ 50 \ \text{cm}^2 & \\ 130 & \end{cases} \qquad \begin{matrix} 0.2 < E < 0.3 \\ 0.6 < E < 1.0 \ \text{keV} \\ 1.0 < E < 3.0 \end{matrix}$$

and

$$A_{HRI} = \begin{cases} 15 & \\ 15 \ \text{cm}^2 & \\ 7.5 & \end{cases} \qquad \begin{matrix} 0.2 < E < 0.3 \\ 0.5 < E < 0.8 \ \text{keV} \\ 0.8 < E < 2.5 \end{matrix}$$

We required sources fluxes to exceed 5σ above detector background (IPC: 3×10^{-3} ct arcmin^{-2} s^{-1}; HRI: 5×10^{-3} ct arcmin^{-2} s^{-1}) and used a pixel size of 1 arcmin for the IPC and 12" for the HRI. Clusters were assumed to be extended to a radius of 240 kpc. The same luminosity function was used for both clusters of galaxies and active galaxies [2]:

$$n(L)dL_{44} = 1.7 \times 10^5 \, L_{44}^{-2.5} \, (c/H_o)^{-3},$$

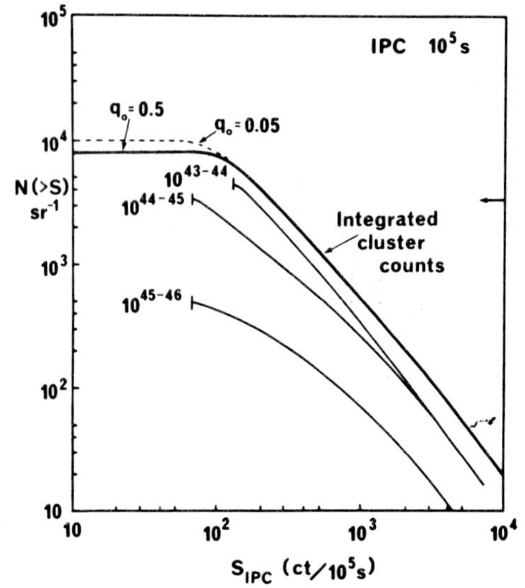

Fig. 1. Integrated IPC counts for clusters of galaxies in 10^5 seconds observing time. Active galaxies give a similar result unless their spectra are steep or absorbed.

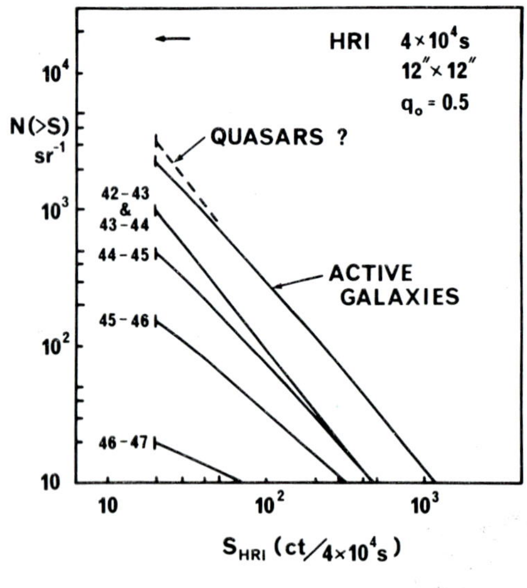

Fig. 2. Integrated HRI counts in 4×10^4 s and a pixel size of 12 x 12 arcsec. Arrow denotes value of N corresponding to 1 source/HRI field. See text for further explanation.

truncated at 10^{43} erg s^{-1} for clusters and 10^{42} erg s^{-1} for active galaxies.

A thermal bremsstrahlung spectrum, of kT = 10 keV and no low energy cutoff, has been assumed. Any soft X-ray excess increases the predicted source counts; a low energy cutoff (such as is evident in several galactic nuclei) reducing the counts.

We assume a Friedmann cosmological model and find that the effect of varying the deceleration parameter, q_o, from 0.05 to 0.5 is slight. Source evolution is a major uncertainty. Clusters of galaxies may have been less luminous in the past if they follow the evolution discussed by Perrenod [3]. Alternatively, they may have been very luminous at redshifts of 2 and 3 (see Fabian and Rees, these proceedings). We are similarly ignorant of the evolution of galactic nuclei, although quasar observations may indicate that such objects were more luminous in the past. We indicate in Fig. 2 the possible contribution of quasars assuming that L_x/L_{opt} is fixed at the value appropriate to 3C 273 and evolves similarly with redshift for all quasars.

Galactic stellar objects at $|b| > 20^o$ are expected to contribute much smaller numbers than clusters or active galaxies above 2 keV. This is partly because they are only minor contributors to the Uhuru and Ariel V source counts but is also due to the flat slope of star counts. Below 2 keV, soft galactic sources may be more numerous. Typical RS CVn systems such as UX Ari (see Charles et al. these proceedings) may be detectable in the IPC down to $\sim 14^m$. The counts for these sources would, however, have flattened off because at this magnitude the corresponding distance exceeds the disc scale height. At 14^m there are [4] about 100 stars per square degree ($|b| > 20^o$). These systems are unlikely to constitute more than $\sim 10^{-4}$ of all stars however, suggesting that fewer than $\sim 10^{-2}$ per square degree will be detected.

It is hoped that these predictions may enable more exotic objects, such as discussed in Fabian and Rees (these proceedings), to be identified more easily.

REFERENCES

1. R. Giacconi, E. Schreier, W. Forman and C. Jones, Center for Astrophysics document CFA/HEA 78-214. 1978.

2. A. C. Fabian and M. J. Rees, Mon. Not. R. astr. Soc., in press, (1978).

3. S. Perrenod, preprint (1978).

4. C. W. Allen, Astrophysical Quantities, 3rd edition, 1972.

Observations of Transient X- and γ-Ray Cosmic Sources

NEW RESULTS ON HERCULES X-1 CYCLOTRON LINES

W. Voges*, W. Pietsch*, C. Reppin*, J. Trümper*, E. Kendziorra** and R. Staubert**

**Max-Planck-Institut für Physik und Astrophysik, Institut für extraterrestrische Physik, 8046 Garching, Germany*
***Astronomisches Institut der Universität Tübingen, Germany*

After its discovery by Uhuru [1] the eclipsing binary X-ray source Her X-1 has been widely studied in the whole X-ray band [2-12]. During a balloon observation in May 1976 we detected the 1.24 sec pulsations in the energy range 30 - 70 keV for the first time, and discovered a strong line feature in its spectrum at $\sim$ 58 keV. Since an atomic or nuclear origin of this feature appeared impossible for intensity reasons, we interpreted it in terms of quantized electron cyclotron line emission originating in the hot and highly magnetized plasma created at the magnetic poles of a rotating neutron star [13, 14]. The corresponding magnetic field strength is 5.3×10^{12} G. Our data were equally well fitted by an absorption feature at $\sim$ 40 keV leading to a magnetic field strength of $\sim 4 \times 10^{12}$ G [15].

To study the spectrum of Her X-1 in more detail and with better statistical significance, a larger (by a factor of 6.5) detector was built and flown successfully three times during the fall of 1977 from Palestine, Texas [16]. In this paper we shall discuss the energy dependent pulse profiles and the spectrum of Her X-1 obtained during the flight on September 3, 1977. The new detector system is described by C. Reppin et al. [17].

Figure 1 shows the 1.24 sec light curves of Her X-1 for different energy intervals in comparison with the results of May 3, 1976 obtained with the smaller prototype system. The improvement in statistical significance is obvious. For deriving the energy spectrum of Her X-1 we have developed the following method. Using our pulse profiles one can define a pulse and an off-pulse region as indicated in Fig. 2a. Treating the off-pulse region as a "background" region, a P-OP spectrum is defined as the count rate in the pulse region minus the count rate in the off-pulse region. In the case of the P-OS spectrum we have subtracted the background count rate as obtained from the off-source measurements from the count rate in the pulse region (Fig. 2b).

Figure 3 shows the so derived raw data count rate spectra. The error bars indicate $\pm$ 1 σ statistical deviations, the upper limits are at 2 σ. The dashed lines represent the best-fitting exponential spectrum with a Gaussian line, taking into account the atmosphere absorption, detection efficiency, energy resolution and K-iodine escape.

$$I = \frac{I_B}{E} \exp(-E/kT) + \frac{I_L}{\sigma\sqrt{2\pi}} \exp\left(-\frac{(E-E_L)^2}{2\sigma^2}\right)$$

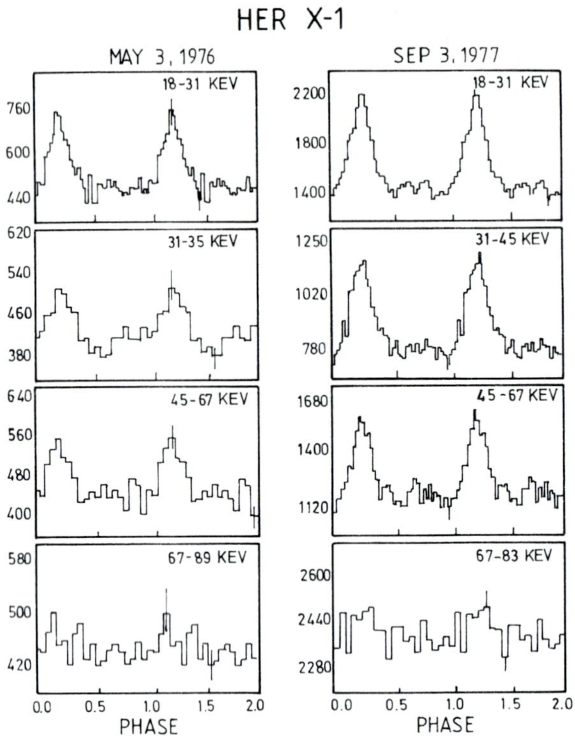

Fig. 1 Her X-1 pulse profiles on May 3, 1976 and Sept 3, 1977 for four different energy intervals. Error bars are 1 σ statistical deviations.

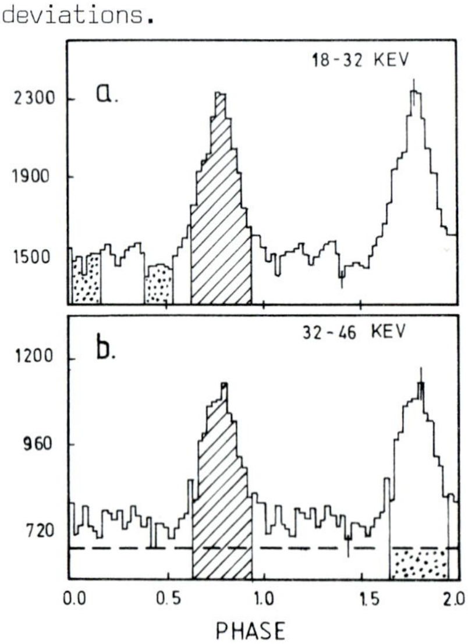

Fig. 2a P-OP (pulse minus off-pulse) regions.
Fig. 2b P-OS (pulse minus off-source) regions.

Both spectra show similar features.

a) In the energy range from 45 to 70 keV a clear excess is present.
b) At about 40 keV there appears to be a sharp dip.
c) Above 60 keV the count rate decreases quite rapidly.

These features are evident during the whole observation time and in all spectra, obtained separately by each of the four detector subsystems.

In a preliminary spectral analysis we have applied different fits to the data. Single power laws, thermal bremsstrahlung or black-body spectra yield large χ^2, χ is typically ~ 9 per degree of freedom, resulting mainly from bad fitting in the region of the line feature. It also appears difficult to fit the spectral excess by a second continuous spectral component. A much better fit (χ ~ 2 per degree of freedom) is provided by a thermal bremsstrahlung spectrum with a Gaussian emission line.

In Fig. 4 both deconvoluted P-OP and P-OS spectra are shown. The corresponding spectral parameters are listed in Table 1.

We also used a thermal bremsstrahlung law plus absorption line and obtained an equally good fit to the data (E_L = 38.5 keV). Therefore, on the basis of our preliminary spectral analysis, we cannot distinguish between either alternative.

The comparison of our results obtained on September 3, 1977 with the ones of May 3, 1976 can be summarized as follows:

a) The existence of the line feature is confirmed with large statistical signifi-

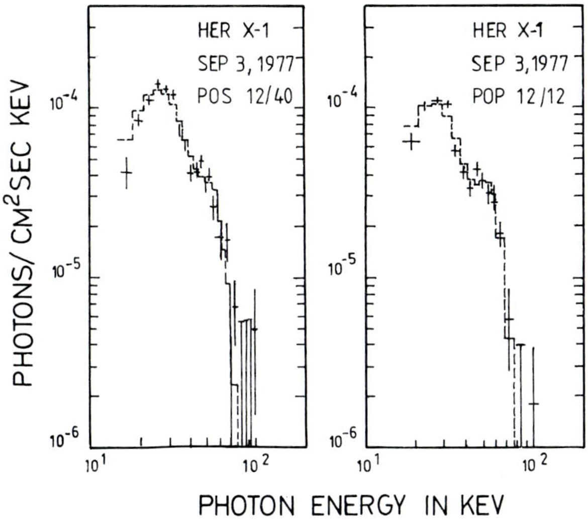

Fig. 3 Pulsed count rate spectra. The dashed line represents a thermal bremsstrahlung + Gaussian line fit through the data.

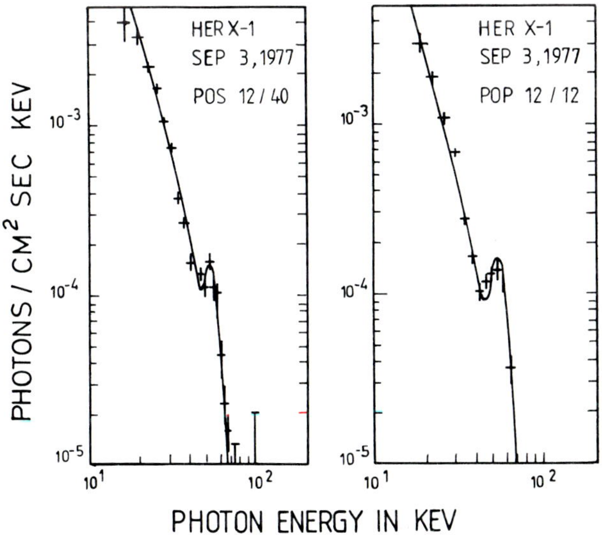

Fig. 4 Deconvoluted X-ray spectra of the Her X-1 pulses. Solid line, best fitting bremsstrahlung spectrum with a Gaussian line to the data points.

cance ($\geq 10\ \sigma$). It may be fitted by an emission ($\sim$ 55 keV) or absorption line ($\sim$ 40 keV).

b) The continuum flux and the line flux are a factor $\sim$ 2.5 lower during the September 1977 observation.

c) We find no second harmonic line feature in the September 1977 data.

The corresponding conservative upper limit ($3\ \sigma$) for the line flux at 110 keV is 1.5×10^{-4} /cm² sec for P-OP and P-OS spectra. The line intensity ratio is

$$\frac{I_2\ (110\ \text{keV})}{I_1\ (55\ \text{keV})} \leqslant 0.1$$

as compared with

$$\frac{I_2}{I_1} \approx 1 \text{ in May 1976.}$$

The spectral variability indicated by this observation can be explained in terms of a positive correlation between the continuum luminosity and source temperature [18].

TABLE 1 Hercules X-1 Spectral Parameters

Count Rate Spectrum	Fitted Spectrum	Continuum I_B	Parameters kT (keV)	Line Parameters Energy (keV)	Width (keV) FWHM	Intensity (10^{-3} cm^{-2} s^{-1})
P-OP	Brems-strahlung+ Gaussian line pro-file	0.60	8.7	55.4	11.5	1.7
P-OS	same	0.61	9.2	54.8	10.6	1.4

REFERENCES

1. R. Giacconi et al., Ap. J. 184, 227 (1973)
2. M. P. Ulmer et al., Ap. J. (Letters) 181, L33 (1973)
3. R. Doxsey et al., Ap. J. (Letters) 182, L25 (1973)
4. S. S. Holt et al., Ap. J. (Letters) 190, L109 (1974)
5. S. H. Pravdo, GSFC preprint X-661-76-280 (1976)
6. S. Shulman et al., Ap. J. (Letters) 199, L101 (1975)
7. R. L. Catura et al., Ap. J. (Letters) 202, L5 (1975)
8. J. Middleditch et al., Ap. J. 208, 567 (1976)
9. D. Gerend et al., Ap. J. 209, 562 (1976)
10. C. Jones et al., Ap. J. (Letters) 208, L131 (1976)
11. V. S. Iyengar et al., Nature 251, 292 (1974)
12. J. D. Kurfess et al., Nature Phys. Sci., 245, 116 (1973)
13. E. Kendziorra et al., Ap. J. (Letters) 217, L93 (1976)
14. J. Trümper et al., Ann. N.Y. Acad. Sci. 302, 538 (1976)
15. J. Trümper et al., Ap. J. (Letters) 219, L105 (1978)
16. W. Voges et al., IAU-Telegram 3184 (1978)
17. C. Reppin et al., ESA SP-135, 293 (1978)
18. C. Reppin et al., to be published

THE 35 DAY VARIATION OF HERCULES X-1 IN HARD X-RAYS

R. Staubert*, E. Kendziorra*, W. Pietsch**, C. Reppin**, J. Trümper** and W. Voges**

**Astronomisches Institut der Universität Tübingen, Germany*
***Max-Planck-Institut für Physik und Astrophysik, Institut für Extraterrestrische Physik, Garching, Germany*

The general picture of the 35 day modulation of Hercules X-1 at low energies is mainly known from extensive studies by Uhuru [1] . The light curve is characterized by a 12 day 'on' state, followed by what was believed at first to be a 23 day 'off' state. It was later discovered that this 'off' is interrupted by another 'on' with less intensity and a duration of about five days [2] . During this so-called 'short on' state, all the normal features such as the 1.24 sec pulsation, the eclipses with binary period, as well as the intensity dips are present.

With respect to this time scale we have performed four balloon observations of Hercules X-1 in the hard X-ray range of 15-150 keV: the first during the 'on' in May of 1976 [3] which led to the discovery of the cyclotron line [4] , and a series of three flights in the Fall of 1977 which covered the 'on', 'short on' and 'off'. A summary of observations of Hercules X-1 for phases outside the normal 'on' state is given in Fig. 1. Only three of these observations detected the 1.24 sec pulsation: Uhuru and our two balloon observations. In the 'short on' it is the first detection above 15 keV, and in the 'off' it is the first detection of pulsation at all.

The light curves are shown in Fig. 2 and a summary of observational data are given in Table 1. The main results are:

- the 1.24 sec pulsations are present at all three phases of the 35 day cycle (the statistical significance of the main pulse is 30, 7 and 4 standard deviations respectively),
- the relative total intensity in the 20-50 keV range scales like 1 : 0.25 : 0.1

for 'on' : 'short on' : 'off',

- a pulse/interpulse structure is present in the 'on' and 'short on' (Fig. 2),
- the intensity ratio of pulse to interpulse changes: it is 12 : 1 for 'on', 2.5 : 1 for 'short on', or in other words: the main pulse is reduced by a factor of five whereas the interpulse stays essentially the same.

This is consistent with lately reduced Uhuru data in which a continuing decrease of the main pulse is observed towards the end of the 'on' state [5].

TABLE 1 Balloon Observations of Hercules X-1 in hard X-rays

	'on'	'short on'	'off'
date of observation	3 Sep 77	20 Sep 77	18 Oct 77
phase of 35 day cycle (with respect to 'turn on')	0.12	0.61	0.41
time on source (hours)	2.3	2.1	1.0
phase of binary cycle (with respect to center of eclipse)	0.11-0.23	0.12-0.22	0.54-0.61
phase of 'dip'-cycle (with respect to leading edge)	0.34-0.45	0.60-0.70	0.91-0.98
total relative intensity (20-50 keV)	1.0	0.25	0.10
statistical significances of pulses: main pulse / interpulse (st. dev.)	30/6	7/4.5	4/-
percent flux in pulses: main pulse / interpulse	46/4	31/13	-
absolute flux in pulses: main pulse / interpulse $(cm^2 \text{ sec keV})^{-1}$	$(46/4)\ 10^{-5}$	$(8.9/3.6)\ 10^{-5}$	-

This behaviour may have interesting implications for models of the 35 day modulation mechanism. It is clear that the hard X-rays must come from regions very close to the stellar surface. There are two possible beam geometries - the pulse and interpulse are either generated on the two magnetic poles or at one pole in different beams. The 35 day variation is widely believed to be produced by variable obscuration of the radiation by a wobbling accretion disc. Such a disc has a radius of about 10^{11} cm, it's inner radius is given by the Alfven surface which in the case of Hercules X-1 is about 10^8 cm. Interpreting our data with this model, there must be a different attenuation for pulse and interpulse or for the different beams, which means that the thickness of the occulting accretion disc must not be larger than a few neutron star diameters. An alternative picture, consistent with our data, is that the modulation is being produced by the precession of the neutron star itself.

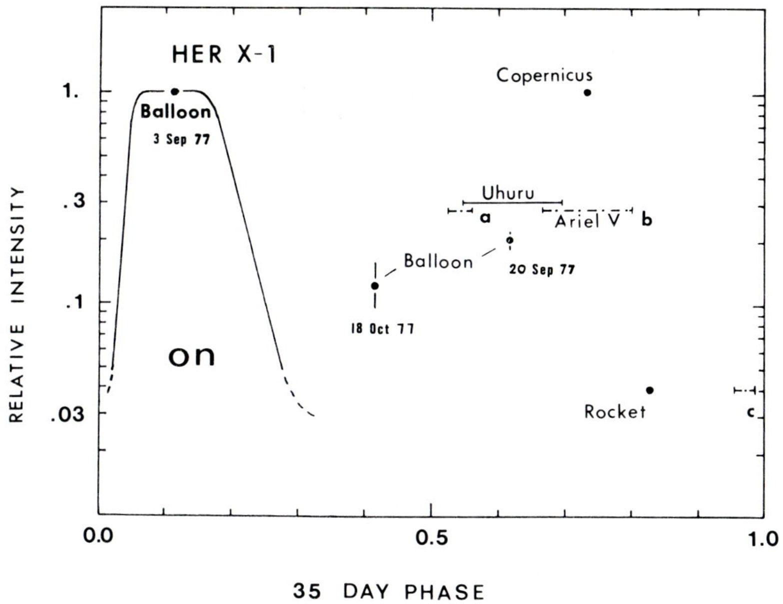

Fig. 1 Summary of X-ray observations of Hercules X-1 outside the 'on' state
Ariel 5: a) 2.4-19.8 keV [7], b) 2-12 keV [8], c) 1.3-13 keV [9]
Uhuru : 2-6 keV [2] Copernicus: 4-12 keV [6] Rocket: .18-.28 keV [10]
Balloon: 20-50 keV, AIT/MPE [this work]

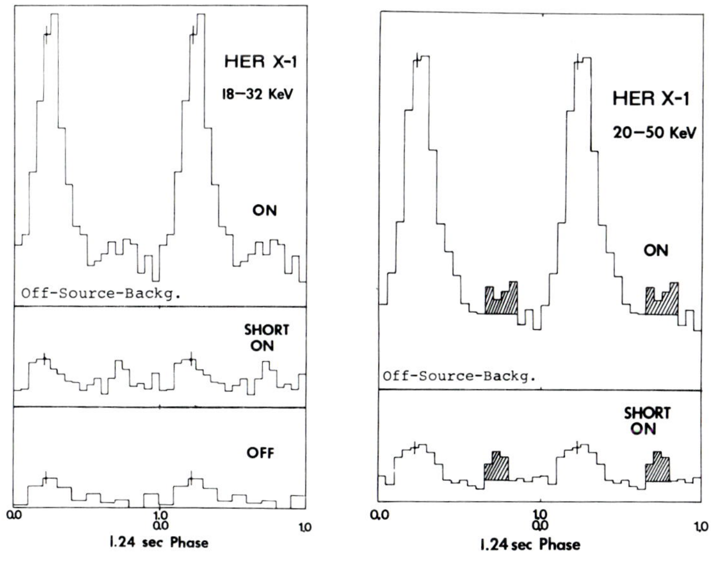

Fig. 2 The 1.24 sec light curve of Hercules X-1 at
18-32 keV for three different phases in the 35 day cycle (left)
and 20-50 keV for the 'on' and 'short on' state - the shaded
area is taken as the 'interpulse' - (right)

REFERENCES

1. R. Giacconi et al., Astrophys. J. 184, 227 (1973)

2. C. Jones et al., Astrophys. J. (Letters) 209, L131 (1976)

3. E. Kendziorra et al., Astrophys. J. (Letters) 217, L97 (1978)

4. J. Trümper et al., Astrophys. J. (Letters) 219, L105 (1978)

5. P.C. Joss et al., preprint 1977

6. A.C. Fabian et al., Nature 244, 212 (1973)

7. B.A. Cooke et al., Nature 256, 712 (1975)

8. S.S. Holt et al., Nature 263, 484 (1976)

9. P.J.N. Davison et al., M.N.R.A.S. 178, 1P (1977)

10. G. Fritz et al., Astrophys. J. (Letters) 207, L29 (1976)

HEAO-1 OBSERVATIONS OF HER X-1 AT ENERGIES ABOVE 13 keV IN FEBRUARY 1978

J. L. Matteson*, D. E. Gruber*, R. E. Rothschild* and J. A. Hoffman**

*Department of Physics, University of California, San Diego, U.S.A.
**Department of Physics and Center for Space Research, Massachusetts Institute of Technology. Present address: Astronaut Office, Johnson Space Center

INTRODUCTION

The cornucopia of periodicities in the Her X-1 X-ray spectrum includes a 1.24 sec pulsation, 1.7 day binary eclipse, an unexplained 35 day modulation, and intensity dips of a few hours duration [1]. These periodicities are in general accompanied by spectral variability [2,3]. Although rather thoroughly observed from 1-10 keV, Her X-1 has only recently been observed above 30 keV because of a very steeply declining energy spectrum [4]. In May 1976 a Max Planck/Tübingen balloon observation of Her X-1 sensitive above ~20 keV (hereafter MPT-76) confirmed the spectral steepening and revealed a pulsed spectral excess at ~58 keV which was interpreted as cyclotron line emission in the ~10^{12} gauss field of the neutron star in the Her X-1 system [5]. A line flux of (2.9 + 3.6, - 1.2) x 10^{-3} ph/cm^2-sec was required to fit the observation. Emission interpreted as the second harmonic of the cyclotron line was detected at 3.3σ significance at ~110 keV with a flux of ~2.6 x 10^{-3}ph/cm^2 sec. A UCSD balloon observation in June 1975 placed an upper limit on line emission at 57 keV of 1.1 x 10^{-3}ph/cm^2-sec (95% confidence) [6]. In order to confirm the reported line emission, clarify its variability and extend the observations of Her X-1 at high energies to additional intensity phases, a series of HEAO-1 pointing observations of Her X-1 were conducted in February 1978 with the UCSD/MIT Hard X-ray and Low Energy γ-ray Experiment, A4. In this paper we present preliminary results based on quick-look data from these observations.

RESULTS

Table I presents details of the observations. The day 59 observation was chosen to be as close as possible to the phases of the MPT-76 observation. All observations were straight pointings at Her X-1 with no off-source data taken during the observation. Observed light curves at three energies of interest are shown in Fig. 1. Pulsed emission is clearly indicated up to pulse heights equivalent to ~64 keV and includes the energy range of the reported cyclotron line. However, at ~110 keV, the energy expected for the cyclotron harmonic, no pulsation is apparent. That the day 59 pulsed emission is weaker than that of the other days is likely due to an intensity dip in Her X-1 since scanning data (30-min period) indicates the average intensity around this time was ~0.75 of that on days 55 and 58. Additional analysis of scanning data will be required to verify the presence of an intensity dip.

Pulse height spectra of the pulsed emission were obtained as the difference between the "pulsed" and "non-pulsed" parts of the light curve (see Fig. 1). The resulting spectra are shown in Fig. 2. Multiparameter fits to each day's spectrum were performed with a model consisting of a power law below a break energy, an exponential above the break and a spectral line with intrinsic width. Corrections for detector efficiency, energy resolution, K X-ray escape and absorption in the overlying material were included in the fitting. The same model was used in the MPT-76 analysis. Our analysis corresponds to the Pulse minus Off-Pulse MPT-76 analysis. Best fit values were estimated from the method of Bevington [7].

The model and the results of the fitting are given in Table II. Chi-square, computed for 20 bins of ~1/2 FWHM each, was entirely unacceptable for fits without the line term, but acceptable with it. The best-fit line intensity on days 55 and 58 is in agreement with the average for these days, $1.37 \pm 0.11 \times 10^{-3}$ ph/cm^2-sec. The day 59 spectrum has a fitted line flux of $(0.75 \pm 0.19) \times 10^{-3}$ ph/cm^2-sec, a factor of two below that measured one day earlier. On all days the fitted line energy is consistent with the average value of 55.1 ± 0.5 keV. The intrinsic width of the line is poorly determined because of the relative insensitivity of χ^2 to it, which is partly due to the low flux in the line. The fitted value of kT on all days is consistent with the average value of 6.8 ± 0.3 keV. The exponential spectral component at 25 keV is constant on days 55 and 58, but is a factor 2.0 lower on day 59. No overtone line at 110 keV was observed. Upper limits of 0.2×10^{-3} ph (cm^2-sec)$^{-1}$ for an unresolved line were obtained for the three observations.

If this same model is fit, but with the line term in absorption at a lower energy ~42 keV, the best-fit chi-squares are even somewhat lower than for the emission line model. Best fit values for kT are here about 12 keV. The intrinsic width of the line in absorption is somewhat better defined by the data, with upper limits $\Delta E/E_0$~30% (95% confidence).

DISCUSSION

Significant high-energy variability of Her X-1 is indicated by these observations. The fitted continuum and line decreased a factor ~2 in one day. This is not surprising, in view of the short time scales (hours) of the intensity dips of Her X-1. The fitted kT remained constant through the variation, suggesting that the mechanism responsible for the apparent temperature is uncoupled from that producing the luminosity changes. Our observation of the constancy of the 20 to 40 keV spectrum shape is in agreement with previous observations [3]. The MPT-76 fitted line strength is $(2.9 + 3.6, - 1.2) \times 10^{-3}$ ph/cm^2-sec, a factor 2.6 above our values for day 55 and 58, the fitted kT is 7.3 + 2.4, - 3.8 keV in agreement with our observation. When combined with our results, these indicate a change in fitted line strength from 2.9×10^{-3} (MPT-76) to $<1.1 \times 10^{-3}$ ph/cm^2-sec (day 59, 95% confidence) while the exponential component's intensity varied in approximate proportion to the 55 keV flux and the kT value remained essentially constant.

CONCLUSION

The HEAO-1 observations reported here clearly indicate pulsed emission from Her X-1 at energies above 50 keV which is in excess of the pulsed spectrum extrapolated from lower energies. This excess is fitted by spectral line emission at 55.1 ± 0.5 keV with a flux of $(1.37 \pm 0.11) \times 10^{-3}$ ph/cm^2-sec on 1978 day 55 and 58 and $(0.75 \pm 0.19) \times 10^{-3}$ on day 59. The data do not require a spectral line to explain the excess; other spectral shapes are possible, provided they have an upward concavity somewhere between 35 and 50 keV, and a 12 keV continuum with a narrow line in absorption at 42 keV is particularly suitable. The fitted continuum below 50 keV was observed to decrease a factor 2.0 in one day and the fitted line decreased in approximate proportion and the fitted kT remained constant at 6.8 ± 0.3 keV. Analysis of observations planned for August and September 1978 should clarify the spectral variability of Her X-1.

ACKNOWLEDGEMENTS

We gratefully acknowledge the assistance of many people in the work reported here. B. Baity, F. Knight, and P. Nolan contributed at UCSD under the direction of L. Peterson, experiment Principal Investigator. B. Cooke, J. Doty, F. Primini, G. Tsiang and W. Wheaton contributed at MIT under the direction of W. Lewin, experiment Co-Principal Investigator.

This work was supported by Contract NAS 8-27974.

REFERENCES

1. Giacconi, R., et al. Ap. J., 184, 227 (1973).
2. Pravdo, S.H., et al. Ap. J., (Letters), 216, L23 (1977).
3. Becker, R.H., et al. Ap. J., 214, 879 (1977).
4. Holt, S.S., et al. Ap. J., (Letters), 190, L109 (1974).
5. Trumper, J., et al. Ap. J., (Letters), 219, L105 (1978).
6. Paciesas, W.S., Ph.D. Thesis, UCSD SP 78-02 (1978).
7. Bevington, P.R."Data Reduction and Error Analysis for the Physical Sciences," McGraw-Hill pp 153-154 (1969).

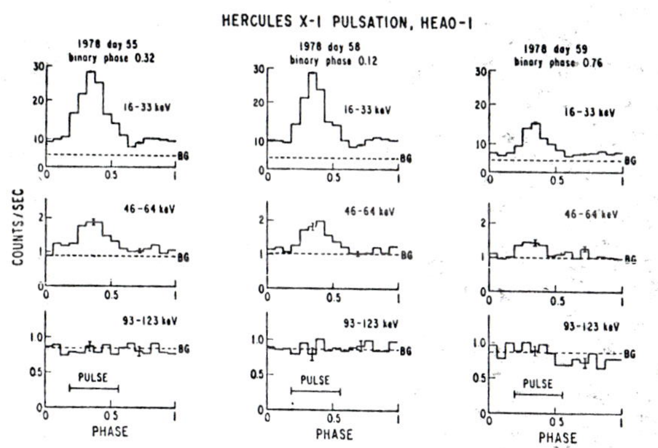

Figure 1

Observed light curves of Her X-1. The pulsed spectra, Fig. 2, are formed from the difference of the average spectra of the "Pulse" and "non- Pulse", i.e. off-Pulse, phases. 1σ error bars are indicated. The dashed lines are background rates from 1 day earlier or later when HEAO-1 was in the same geomagnetic environment and scanning source-free regions. These rates were not used in the analysis reported here.

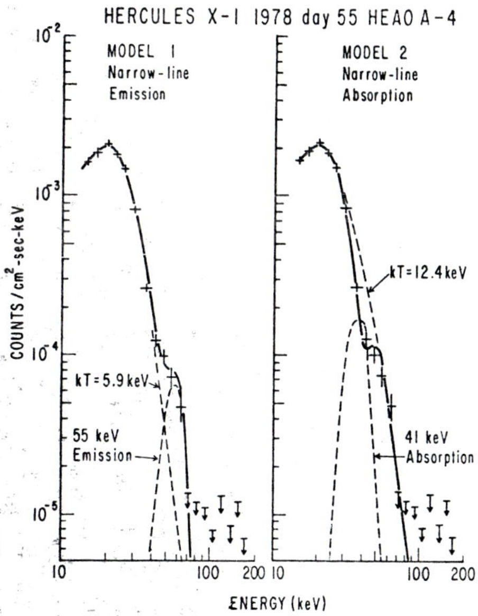

Figure 2

Representative pulsed spectrum in one detector of Her X-1 & two model fits. 1σ error bars and 2σ upper limits are indicated. The heavy dots represent the best fit model after folding through the instrument's response. They are connected for convenience. Either fit is satisfactory, but not without the line term.

TABLE I - HEAO-1 Pointing Observations of Her X-1

Day of 1978	Pointing Interval (UT)	1.7 Day Phase	Days After 34.9 Day Turn-on*
55	1452-1809	0.30-0.34	3.9
58	1547-1845	0.09-0.14	6.9
59	1620-1937	0.74-0.77	8.1

TABLE II - Best Fit Parameters from Model* Fitting

	Day 55	Day 58	Day 59
Line Parameters:			
I_o (10^{-3} photons/cm^2-sec)	1.44 ± 0.15	1.28 ± 0.18	0.75 ± 0.19
E_o (keV)	58.0 ± 1.1	54.1 ± 0.6	57.6 ± 2.2
FWHM† (keV)	$15 \pm^{30}_{15}$	$10 \pm^{30}_{10}$	$3 \pm^{50}_{3}$
Continuum Parameters:			
$N^{‡}$ (10^{-3} photons/cm^2-sec-keV)	2.68 ± 0.08	2.60 ± 0.07	1.50 ± 0.09
$kT^{‡}$ (keV)	6.9 ± 0.4	6.6 ± 0.5	6.7 ± 0.8
E_b (keV)	27.1 ± 0.5	26.7 ± 0.6	26.9 ± 1.0
$S^{‡}$	0.6 ± 0.1	0.5 ± 0.1	0.4 ± 0.3

$^*\Phi(E) = N_o \cdot (E/25 \text{ keV})^{-S}$, $E \leq E_b$

$N_1 \cdot \exp(-E/kT)/E + I_o \cdot G(E, E_o, \text{FWHM})$, $E \geq E_b$,

Where $N_1 = N_o \cdot E_b \cdot \exp(E_b/kT) \cdot (E_b/25 \text{ keV})^{-S}$

and G is the gaussian line profile

† χ^2 weakly dependent on this parameter.

‡ these parameters correlated.

CYCLOTRON LINE EMISSION FROM BINARY X-RAY SOURCES

V. S. Iyengar, T. M. K. Marar, K. Kasturirangan
and U. R. Rao

ISRO Satellite Centre, Bangalore 560 058, India

Introduction

With the discovery by Trümper et al [1] of a hard x-ray electron cyclotron emission line in the spectrum of Her X-1, we can now glean a wealth of new information on the morphology of the x-ray production zone and the strength and temporal variations of the magnetic field associated with it. In this paper we show that the observed spectral shape and intensity constrain thermal x-ray production to annular hot spots of $\sim$0.15 km^2 area at the neutron star surface around the magnetic poles and that extended radiative shocks of the type described by Inoue [2] and Basko and Sunyaev [3] may be ruled out for Her X-1. Further we show evidence for magnetic field changes $\sim 2\text{-}3 \times 10^{12}$ gauss in the hot spots over short time periods. Lastly, we enquire into some promising candidates, among x-ray binaries, for the observation of x-ray cyclotron emission.

In the 1 to $\sim$20 keV energy band the photon spectrum of Her X-1 is well represented by $N(E)=CE^{-0.9}$, for the period after turn on and outside the dips. For energies $>$ 20 keV, a continuum spectrum best fitted by $N(E)=(C_1/E^{1.4})\exp(-E/kT)$ with $kT \sim 8$ keV, on which is superposed a narrow (FWHM$\sim$12 keV) line of intensity 3.4 (+2.0,-1.7)$\times 10^{-3}$ photons $cm^{-2}s^{-1}$ centred at 58 keV, has been observed by Trümper et al. This gives a total source luminosity $L_X \sim 10^{37}$ erg/s. Table 1 summarises presently available observational data which suggest the presence of hard x-ray emission lines in Her X-1 spectra obtained by various groups. The data have been normalised so as to yield the same total continuum intensity in the 20-40 keV band as Trümper et al under the assumption that variations in intensity are only due to electron scattering of the intrinsic source spectrum.

A SELF-CONSISTENT MODEL FOR X-RAY PRODUCTION IN HER X-1

The cyclotron emission line and the x-ray thermal bremsstrahlung continuum must arise from the same hot plasma. The temperature of this plasma is 93×10^6 °K (kT$\sim$8 keV), as deduced from the 20-46 keV continuum. The observed cyclotron line intensity of 3.4×10^{-3}photons

X.R.A.—R

cm^{-2} s^{-1} at E_c=58 keV corresponds to a luminosity of $5.7x10^{35}$erg/s for isotropic emission, for a source distance of 4 Kpc. Since $E_c \gg kT$, the cyclotron absorption co-efficient for kT=8keV, given by $K_c(\theta) = 6.0x10^{21}N_e(1+Cos^2\theta)$, in the usual notation, is large and the cyclotron emission line intensity reaches the blackbody limit. Using the blackbody emission rate of $(4\pi h \nu_c^3/c^2)(exp(h\nu_c/kT)-1)^{-1}$ergs $cm^{-2}s^{-1}$ over 2π steradians, the observed cyclotron luminosity must arise from the source subtending an area of 0.15 km^2. Following Basko and Sunyaev [4] we assume that the accreted matter flows to the magnetic pole of the neutron star in a narrow hollow cone of radius, $a \sim 1$ km and thickness $d_o \sim 50$ metres.

Two possible scenerios for the deceleration of the infalling matter are (i) Stoppage at the surface of the neutron star primarily by nucleon-nucleon collisions with a mean free path $\sim 50g\ cm^{-2}$ [4] perhaps aided in part by radiation pressure; and (ii) Shock transition within the accretion funnel, above the neutron star surface, due to radiation pressure [2,3] . In the former case a hot spot of area $2\pi a d_o \sim 0.3\ km^2$ develops on the surface of the neutron star which is comparable to the emission area for cyclotron radiation deduced earlier. The thermal bremsstrahlung rate for a plasma with cosmic abundances of elements is $P_B=2.4x10^{-27}T^{\frac{1}{2}}N_e^2$ ergs $cm^{-3}s^{-1}$. Assuming a hot spot thickness $t \sim 1$ cm, the emission volume is $\sim 3x10^9 cm^3$ and $L_x \sim 10^{37}$erg s^{-1} requires, for $kT \sim 8$ keV, a hot spot electron density $Ne \simeq 1.2x10^{25} cm^{-3}$ which is easy to accommodate. The free-free absorption co-efficient for this plasma is $K_{ff}(\nu)=5.2x10^8 N_e^2(1-exp(-h\nu/kT))g/T^{\frac{1}{2}}\nu^3$ cm^{-1} and the Thomson scattering co-efficient $K_{sc}=6.65x10^{-25}N_e$ cm^{-1}. The plasma is optically thick when $\tau_{ff}(\nu)\tau_{sc} \equiv K_{ff}K_{sc}t^2 \geq 1$. The transition from optically thick to optically thin conditions occur at $h\nu_{t1} \sim 16$ keV for the parameters deduced earlier. Below 16 keV the source spectrum should therefore flatten and approach that of a blackbody. This is consistent with the observations of the thermal bramsstrahlung above ~ 20 keV.

If the hot plasma resides in the accretion column, as suggested in the latter model above, the observational constraint on the emission area for cyclotron line emission ($A \sim 0.15\ km^2$) restricts the emission zone to a slice of the column of height $h_1 \sim 75$ m. The hot plasma then has a volume of $2\pi a d_o h_1 \simeq 2.3x10^{13}$ cm^3 and N_e required for $L_x \sim 10^{37}$ erg s^{-1} is then $\sim 1.4x10^{23}$ cm^{-3} at kT=8 keV. The plasma would then be optically thick below $h\nu_{t2}$ 80 keV. Since the emission spectrum below 80 keV is that of a blackbody no cyclotron line can stand out, which contradicts the observations.

Further, as shown by Inoue [2] and Basko and Sunyaev [3], the hot zone electron density for $\dot{M} \sim 10^{17}g\ s^{-1}$ is only 10^{22} cm^{-3} and the hot plasma of the nearly stopped infalling matter must extend to a height much greater than 10^5 cm above the neutron star surface. As shown by Trümper et al [1] the cyclotron line width is compatible with thermal doppler broadening and even if the entire line width is ascribed to the radial magnetic field inhomogenity in a dipole field, the radial extent of the line emission region cannot exceed 700 metres.

We thus conclude that all observational features support x-ray emission from the localised hot spots at the neutron star surface around the magnetic poles, but not the formation of an extended shock heated zone in Her X-1.

TABLE-1

HARD X-RAY SPECTRA OF HER X-1
WITH EVIDENCE FOR CYCLOTRON LINE EMISSION

OBSERVER AND DATE OF EXPT.	35-DAY PHASE	LINE ENERGY (keV)	NORMALISED LINE INTENSITY (10^{-3} photons $cm^{-2}s^{-1}$)	SIGNIFICANCE	DERIVED MAG.FIELD (10^{12}Gauss)	RELATIVE DISTANCE FOR DIPOLAR FIELD (R/R_T)
TRÜMPER et al [1] (MAY 3, 1976)	$T_{ON}+5$ d	58	$3.4^{+2.0}_{-1.7}$	5 σ	5.8	1.00
COE et al [7] (FEB.4-6 & 9-13,1977	$T_{ON}+1$ to 3d & $T_{ON}+6$ to 10d	64	22±9.3	2.4 σ	6.4	0.96
BECKER et al [5] (AUG.28,1975)	MID TURN ON $\sim T_{ON}+1$ h	36	21.5±9.0	2.4 σ	3.6	1.17
	POST TURN ON $\sim T_{ON}+2$ h to 14 h	53	10.6±7.1	1.5 σ	5.3	1.03
	DIP $\sim T_{ON}+14.5$ to 16.5 h	44	12.3±12	1 σ	4.4	1.10
ULMER et al [6] (DEC.8-10,1971)	$T_{ON}+2.7$ d to 4.7 d	25	15.4±30	-	2.5	1.32
MAY 29-JUNE 1 1972)	$T_{ON}-1$d to $T_{ON}+3$d	23.5	59.6±33.6	1.8 σ	2.3	1.35

POSSIBLE TEMPORAL VARIATIONS IN THE POLAR MAGNETIC FIELD

Excess photon counting rates over the best fit continuum indicative of line emission have been observed in Her X-1 spectrum above 20 keV by several groups as shown in Table 1. The line energy varies considerably from one measurement to the other, the most notable being the variations in the 17-hour observation of Becker et al [5]. The deduced magnetic field ($H=(2\pi mc\ E_c/eh)$ corrected for a 10% gravitational redshift, and the radial distance of the line emitting region computed on the assumption of a dipolar magnetic field normalised to the radial distance R_T of the observation of [1] are given in the table. If $R_T \sim 10$ km, the emitting region must move radially as much as 3.5 km. under this hypothesis and must therefore lie in the accretion funnel, a hypothesis that contradicts the results of the previous section. The line intensities shown in the table are best fit by a blackbody emission profile of kT=8±0.4keV, with an emission area of 0.15 km^2, suggesting that the x-ray lines observed by the different groups are produced in the same hot spot on the surface of the neutron star, and that temporal changes in the magnetic field at the location of the hot spot vary the cyclotron emission energy, moving the line intensity along the Planck curve. If the spectral features at 36, 53, 44 keV in the mid-turn-on, post-turn-on, and dip spectra of [5] are indeed genuine cyclotron lines, local magnetic fields at the hot spot must vary by factor ~ 2 in few hours, making cyclotron line spectroscopy in Her X-1 a powerful means of studying temporal variations in the magnetic field associated with the line emission region. The available data (Table 1) are too scanty to draw any conclusions regarding the correlation of the hot spot magnetic field intensity with the orbital or 35-day periods. It is presumable that localised ring currents within the magnetosphere near the hot spot give rise to the magnetic field fluctuations.

CONCLUDING REMARKS

The published hard x-ray spectra of other binary x-ray sources do not contain any firm evidence for a cyclotron emission line. SMC X-1 is believed to emit $\sim 3\times10^{38}$ erg/s from an optically thick extended radiative shock zone [3]. Positive detection of a cyclotron line in the intermediate x-ray band from SMC X-1 would make this model untenable. Cen X-3 and Vela XR-1 are also likely candidates which may show evidence for cyclotron line emission between 20 and 30 keV.

ACKNOWLEDGEMENTS

This work was supported by the Department of Space, Govt. of India.

REFERENCES

1. Trumper, J., Pietsch, W., Reppin, C., Voges, W., Staubert, R., and Kendziorra, E., 1978, Ap.J (letters) 219, L105.

2. Inoue, H., 1975, Publ. Astron. Soc. Japan 27, 311.

3. Basko, M.M., and Sunyaev, R.A., 1976, M.N.R.A.S.175, 395.

4. Basko, M.M., and Sunyaev, R.A., 1975, Astron & Astrophys.42, 311.

5. Becker, R.H., Boldt, E.A., Holt, S.S., Pravdo, S.H., Rothschild, R.E., Serlemitsos, P.J., Smith, B.W., & Swank, J.H., 1977 Ap.J. 214, 879.

6. Ulmer, M.P., Baity, W.A., Wheaton, W.A., and Peterson, L.E., 1973 Ap.J (letters) 181, L33.

7. Coe, M.J., Engel, A.R., Quenby, J.J., and Dyer, C.S., 1977 Nature, 268, 508.

19-SECOND VARIATION OF CYGNUS X-1 HARD X-RAYS

M. Nakagawa, H. Sakurai and M. Uchida

Department of Physics, Osaka City University, Osaka, Japan

INTRODUCTION.

In the soft x-ray region, Cyg X-1 is known to exhibit complex time variation [1,2]. On the other hand, in the hard(20keV) x-ray range, data on the short-term variation of Cyg X-1 are still meagre.

We have been studing Cyg X-1 in the hard x-ray region(30keV - 70 keV) with our balloon-borne detectors for several years. We present here the results of Fourier-analysing Cyg X-1 data obtained in 1972.

EXPERIMENT.

A balloon payload launched on 7 October 1972 from Sanriku Balloon Center reached its ceilling altitude at 0730 U.T. and floated between 35.4 km and 36.7km.

The x-ray detector comprises two scintillation counter units each consisting of a NaI(Tl) crystal and a plastic scintillator mounted between the crystal and a photomultiplier tube. Both scintillators are 76 mm in diameter; one is 3 mm thick, the others 30 mm. A cellular collimator with 4°(north- south) x 20°(east-west) FWHM field of view is placed on the top of each counter. It's axis is inclined by 4° with respect to the zenith. A graded shield of lead and tin surrounds the whole counter assembly. The charged particle background is removed by means of the difference in the rise time of the pulse coming from the NaI(Tl) crystal and the plastic scintillator. The aspect of the detector system is controlled to point alternately to the north (OFF-mode) and to the south (ON-mode) by ground command. During the ON-mode, Cyg X-1 lies in the field of view of the detector, and no x-ray sources are in the field of view during the OFF-mode.

RESULT.

300 second of data around the meridian transit of Cyg X-1 for the ON-mode and another 300 seconds of data for the OFF-mode are analysed with a FFT algorithm and a folding method. We assume that the background is stationary and isotropic for the examination of the time variation of Cyg X-1 comparing the ON-mode data with the OFF-mode. For the ON- and OFF-modes, Table 1 and Fig. 1 show the counting rates for 4.5 sec intervals and the statistical fluctuations, respectively.

Table 1 TIME VARIATION OF CYG X-1

COUNTING RATE / 4.5 sec

	ON - B	OFF - B	CYG X-1
$\bar{N}$ = $\overline{\text{COUNTS}}$	29.1	22.8	⟨6.3⟩
$\sqrt{N}$	5.4	4.8	
S = $\sqrt{\text{VARIANCE}}$	6.8	4.4	
$S/\sqrt{\bar{N}}$	1.3	0.9	

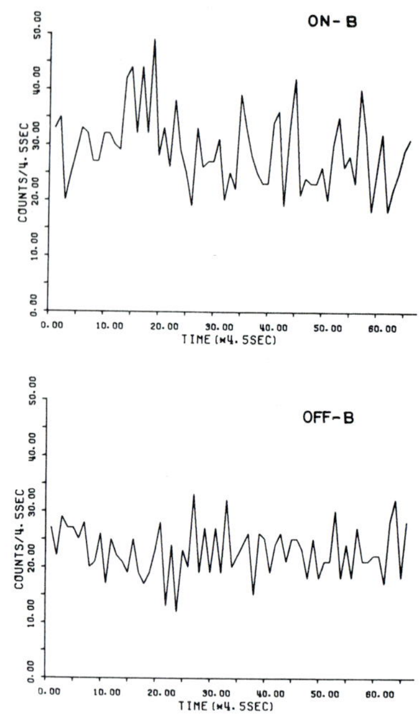

Fig. 1 The counting rates for 4.5 sec intervals.

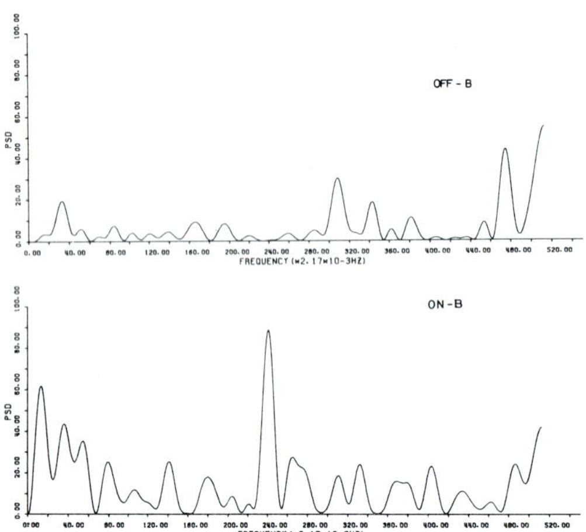

Fig. 2 PSD for the ON- and OFF-modes.

From these time series of the counting rates, the power spectrum density (PSD) is obtained using an FFT algorithm with application of the Hanning window. Fig. 2 shows the PSD for the OFF- and ON-modes. For the ON-mode, the PSD exhibits a significant peak at $(5.2 \pm 0.3) \times 10^{-2}$ Hz (period= 19.2 sec), but not for the OFF-mode. The PSDs of other frequencies are similar for the ON- and OFF-modes. Statistical significance of this peak is estimated to be 99.7 %.

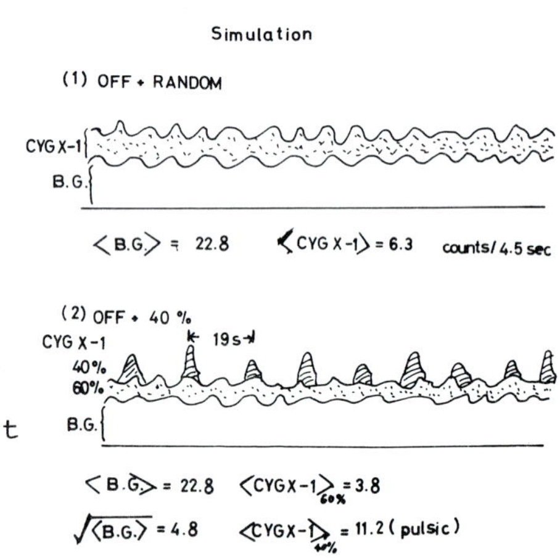

Fig. 3 Schematic diagram of the simulation.

In order to test the significance of this peak we conducted two simple simulations (1,2) according to the method as shown in Fig. 3. The first simulation is as follows; a time series was generates by adding the counting rates obtained by randum number sampling with a Poisson distribution to the OFF-mode data at each interval. The average value, λ, is assumed to be the mean counting rate of Cyg X-1 obtained by the differnce between the

ON-mode and the OFF-mode counts for 300 seconds. Fig. 4(a) shows the PSD of this simulation. The next simulation is performed by adding two components to the OFF-mode data. One is computed in the same manner to the above, except the average value is 60 % of λ and the other uses an average value of 40 % of λx(19.2/4.5) and having 19.2 sec periods intermittently as shown in Fig. 3(b). Fig. 4(b) shows the PSD of this simulation. In this Fig., the magnitude of the PSD around the frequency of 5.2 x 10^{-2} Hz is in agreement with the result of the ON-mode. Whereas in Fig. 4(a) the PSD exhibits no peak around this frequency and is similar to the result of the OFF-mode.

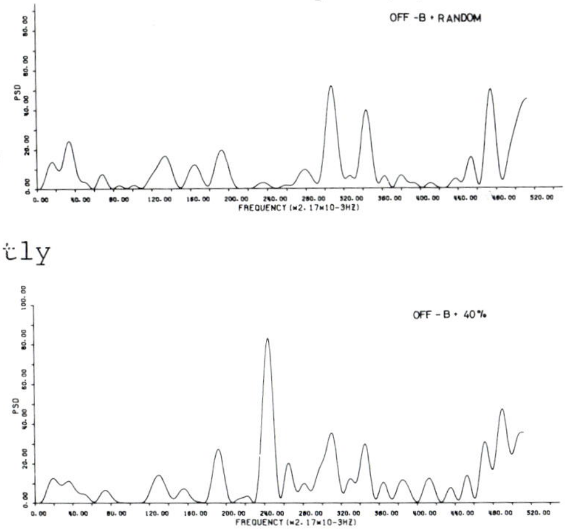

Fig. 4 Result of the PSD for the simulation.

To check the PSD analysis, we applied a folding method to these data. The results of this method as shown in Fig. 5. Fig. 5(a) shows the result with a 19.2 sec folding period for the ON-mode, and Fig. 5(b) is the difference between the ON-mode folding and the OFF-mode folding, namely Cyg X-l's folding. The latter is like a sine curve.

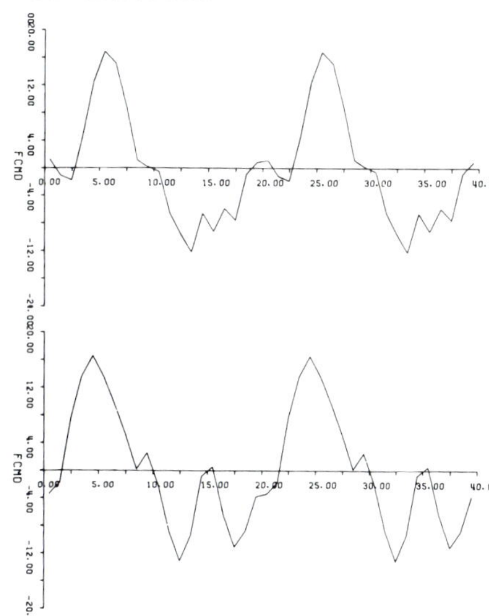

Fig. 5 Result of the folding method with a 19.2 sec folding period.

DISCUSSION.

If the gondola were to swing like a pendulam during the ON-mode but not in the OFF-mode, the period of the pendulum motion for our gondola whould be 7.3 sec. But during the ON- and OFF-modes the data obtained by the flight level and the geomagnetic aspect sensor were constant within the accuracy of the system, namely ±0.8 % of the Cyg X-l counts.

In order to examine this peak at 5.2 x10^{-2} Hz, the folding method is applied to the data of the ON- and OFF-modes varing the periods from 17.8 sec to 20.1 sec with 0.1 sec interval. Lack of similar modulation at these periods proves the existence of the 19.2 sec period obtained from the power spectrum analysis, and suggests that 24 % of the Cyg X-l intensity is variable with the 19.2 sec period. Due to the short series of dataapplied to the PSD analysis, 5 minutes, where the best S/N value of the ON-mode is obtained, it is not clear that this periodicity is continuous.

Frontera et al reported that the PSDs show a peak at a frequency of 5.75 x10^{-2} Hz in the hard x-ray region in 1972. Our total counts and signal to noise ratio are comparable to theirs. So, it is attractive that the similar period has been analysed by other groups in the same year.

The authors are very grateful to professor J.Nishimura and members of the balloon flight group of Institute of Space and aeronautical Science, University of Tokyo for the balloon flight. We wish also to thank prof. Y.Tanaka, University of Tokyo and prof. S.Osaki, Osaka City University for their useful advices.

REFERENCES.

1) R.E.Rothschild, et al 1974, Ap.J.(Letters). 189, L13.
2) Y.Ogawara et al 1977, ISAS RN 26.
3) J.B.Bendat and A.G.Piersol 1971, "Randum Data: Analysis and Measurement Procedures" in John Wiley & Sons, Inc.
4) F.Frontera and F.Fuligni 1975, Ap.J.(Letters), 198, L105.

HARD X-RAY TRANSIENT SOURCE OBSERVED DURING THE 1976 TRANSATLANTIC FLIGHT

F. Fuligni, W. Dusi, F. Frontera, E. Morelli and G. Ventura

Laboratorio TE.S.R.E./C.N.R., Via de'Castagnoli 1, 40126 Bologna, Italy

INTRODUCTION

Transatlantic balloon flights offer a unique low cost opportunity to observe a number of times, at periods of about one day, the same sky zones, thus offering a very suitable tool for the study of long term variations of cosmic X-ray sources. We report here preliminary results of the analysis of data, collected during the 1976 mission, specifically intended to search for low intensity hard X-ray sources not immediately evident above background counts. The balloon was launched from Trapani (Sicily, Lat. 38°), on July 29, with on board, among other experiments, two identical NaI(Tl) detectors having a total effective area of 525 cm^2 and an aperture of 14° FWHM, obtained by semiactive collimation. The nominal working range was between 20 and 300 keV, with relatively poor energy resolution. The gondola was unstabilized and both detectors were mounted with axes of view looking along the vertical . In this way a sky belt centered approximately around 38° declination was scanned three times during the 75 hour flight. The available bit rate (96 bps) allowed us to obtain continuously data for both detectors in two energy channels with a time resolution of 0.83 sec. In addition, 60- channel spectra for each detector were transmitted alternately with a time resolution of 106.6 sec.

DATA ANALYSIS AND RESULTS

There was a continuous increase of the background counts due to latitude effect. To this there were superimposed slow oscillating variations, related to changes in the ceiling altitude. These last were accounted for by knowledge of counting rate versus pressure, obtained by best fitting an exponential curve to the data collected during the ascent and then checking the parameters at different geomagnetic latitudes, made possible by the sudden large increases in altitude at the balloon dawns.
A reliable complete history of the motion of the balloon was obtained, despite omega system malfunctions, by interpolations, checked with the help of the data concerning Cyg X-1 and dawns. In this way an overall behaviour of the latitude effect was obtained for all energy channels. Corrections were then performed by using the overflow (> 300 keV) counting rate as an indicator of the geomagnetic cut-off.
Results of this analysis, referred to a sky belt from 03^h10^m to 09^h00 R.A., are summarized in Fig. 1, where excess counts (20-150 keV) above background versus U.T.

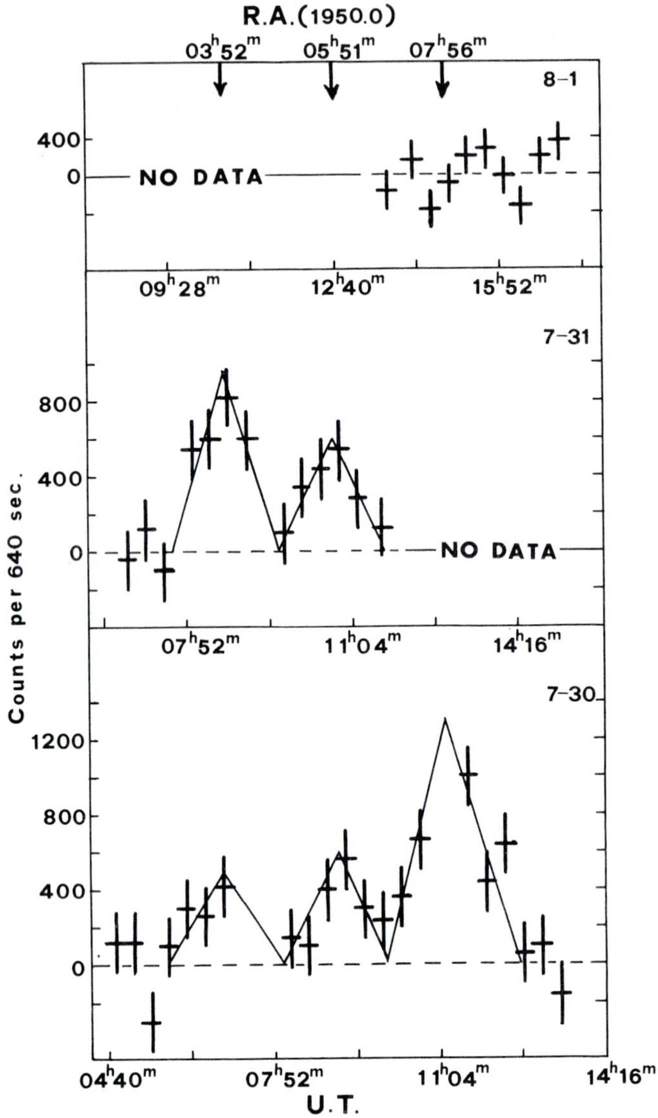

Fig. 1. Excess counts above background for three successive scans of the region between 03^h10^m and 09^h00 R.A. centered around 39° declination. The energy range is 20-150 keV. Lacking data are due to telemetry noise. Fits to the angular response refer to X-Per, MCG 8-11-11 and a source with hypotetical position 07^h56^m R.A. and 39° D., respectively. In all three scans the average pressure was about 3.8 mb.

are reported for the three available successive scans. Unfortunately telemetry noise has not allowed, as it can be seen from lacking data in the figure, a complete observation of the entire region during its subsequent passages. In the scan of July 30, three sources are visible. Only very preliminary results are available, at this stage of analysis, for the two less intense ones.

The first object, at 03^h50^m R.A., can be identified with the interesting system X-Per/3U0352 + 30. The observation of July 31 shows that its intensity has varied by a factor about 2 in approximately one day. This variation is immediately evident in the figure by comparison with the second source which appears stable. A rough evaluation of the energy spectrum suggests a power law with spectral index $\alpha \simeq 2$, in agreement with the results of Mushotzky *et al.* [1].

The second source is compatible in position with the Seyfert galaxy MCG 8-11-11 already detected by Ariel at lower energies. Further study is in progress to check the plausibility of this attractive association. No data are available for these sources during the passage of Augst 1, because of telemetry problems.

We discuss now the third object, for which the analysis has been almost completed. It was not observed on July 31 due to lacking of data, and its flux resulted below the sensitivity of the instrument during the third passage on August 1. This corresponds to a variation in intensity by a factor of at least 6, which implies, in the assumption of an exponential decay, an e-folding time $\lesssim$ 1 day. No previous sources have been reported, even at lower energy, in the region where it can be located, i.e. $07^h56^m \pm 15^m$ R.A. and $39° \pm 6°$ D. ($l^{II} = 182° \pm 6°$, $b^{II} = 30° \pm 5°$). (The limits in declination were set by studying the width of the angular response, which shows that the object was very near the axis of view). These considerations definitely demonstrate the transient or highly variable nature of the source we have detected.

Rough spectral information has been obtained. Both power law with $\alpha \simeq 1.7$ and thermal Bremsstrahlung with KT $\simeq$ 50 keV are acceptable.

A preliminary search for the optical counterpart does not show the presence of any peculiar object inside the error box, except for the type 1 Seyfert galaxy MKN 382, which is located very close to the centroid position. This coincidence is very suggestive and worthy of some speculation. As it is well known, a class of transient sources is characterized by having decay times in the range 1-8 wks and a spatial distribution very close to the galactic plane $|b| \lesssim 5°$ (see e.g. [2]). Both the rapid decay of our source and its relatively high galactic latitude ($b \simeq 30°$) in the general direction of the anticenter seem to exclude it as a member of the mentioned class and suggest rather an association with the "weak, high-latitude transients", reported by Ricketts *et al.* [3]. These are characterized, in addition to their location, by relatively low flux levels and rapid decay times ($\lesssim$ 1 wk). 15 Seyfert galaxies are presently known as X-ray emitters in the low energy band with typical powers of $10^{42.5}$ - $10^{44.5}$ erg/sec, and a few of them exhibit flux variations [4].

Association of our object with MKN 382 gives a flux density of the order of that of X-Per (reference to Fig. 1 must take into account the difference in exposed area). Assuming a Bremsstrahlung spectrum with KT = 50 keV we obtain a flux, in 20-150 keV, of 8.9×10^{-10} erg/cm^2 sec corresponding to a radiated power of 4.4×10^{45} erg/sec. Extrapolation of this spectrum to lower energies gives a value of 10^{45} erg/sec for the radiated power in 2-10 keV. An upper limit to the X-ray emission from MKN 382 was set by Ariel in the latter energy range, at 8.7×10^{43} erg/sec [4]. This would imply a variation in X-ray luminosity by more than a factor of ten, which is not completely unusual, as it is shown by the behaviour of A 1103+38 [3], having a suggested association with the faint elliptical galaxy MKN 421, probably a BL Lac type object.

As a conclusion, we can say that detection of a transient or highly variable object in the general direction of the galactic anticenter around b=30° has been firmly established. The suggested association with the type 1 Seyfert galaxy MKN 382 appears, in some respects plausible and is certainly attractive, especially for the many implications concerning the energy requirements for production of such tremendous powers in very restricted regions (in our case $\lesssim 10^{-3}$ pc). More extended observations of the region are necessary and simultaneous ground based measurements desirable.

We thank Mssr G. Baravelli, A. Brighenti, E. Rossi, A. Rubini, S. Silvestri, S. Tarlazzi, M. Trifoglio and O. Zerlotin for their help in solving the many technical problems. Particular thanks to Prof. D. Brini for his continuous interest and encouragement.

REFERENCES

1. R.F. Mushotsky, D.H. Roberts, W.A. Baity and L.E. Peterson, *Astrophys. J.* 211, L129 (1977).

2. L.J. Kaluziensky Ph. D., Thesis, University of Maryland (1977).

3. M.J. Ricketts, B.A. Cooke and K.A. Pounds, *Nature,* London 259, 546 (1976).

4. M. Elvis, T. Maccacaro, A.S. Wilson, M.T. Ward, M.V. Penston, R.A.E. Fosbury and G.C. Perola, *Mon. Not. R. Astr. Soc.* 183, 129 (1978).

RECENT OBSERVATIONS OF GAMMA RAY BURSTS FROM THE SIGNE 2MP AND SIGNE 3 SATELLITE EXPERIMENTS

G. Chambon*, K. Hurley*, M. Niel*, G. Vedrenne*,
I. V. Estuline** and A. S. Melioransky**

**Centre d'Etude Spatiale des Rayonnements,
31400 Toulouse, France
**Institute for Cosmic Research, Moscow, U.S.S.R.*

In 1977, as part of a long term program of scientific cooperation between France and the Soviet Union, two X and gamma ray experiments were put into orbit. The first, called Signe 3, was in a French satellite launched by a Soviet rocket into a low altitude earth orbit, in June. The second, called Signe 2 MP, was aboard the Soviet Prognoz 6 satellite, launched into a highly eccentric earth orbit in September. Both satellites are spin stabilized. The scientific objectives of the two experiments include the study of solar and cosmic X and gamma ray bursts, and of cosmic X ray sources. A complete description of these two instruments will appear in the near future; here we describe only those features which pertain to the gamma burst detection capabilities.

The Signe 3 experiment consists of a 12 cm^2, 3 cm thick NaI(Tl) crystal, in a CsI(Na) anticoincidence well 2.5 cm thick with a total surface area of about 950 cm^2; both have burst triggering systems and rapid memories associated with them. When triggered, the central detector measures the integral count rates above 20 keV, with 32 ms resolution, for 8 seconds; the anticoincidence well measures the integral count rates above 60 keV with 8 ms resolution for 4 seconds.

The Signe 2MP experiment consists of a 3.7 cm thick NaI(Tl) detector with about 63 cm^2 surface area, and two 1.4 cm thick lateral NaI(Tl) detectors with about 12 cm^2 surface area, in a 1.25 cm thick CsI(Na) anticoincidence well with about 750 cm^2 surface area (200 cm^2 form factor). A total of seven rapid memories are associated with the above 4 detectors; when a burst is detected, part or all of the following data may be transmitted, depending upon which detectors triggered :

- integral count rates above 80 keV, with a time resolution of 2 ms or 16 ms, for 2 or 16 sec, respectively
- 6 channel energy spectra, every 250 ms or every 4 sec, over the energy range 20-280 keV or 80-1000 keV, respectively.

In 1977, Signe 2 MP and Signe 3 detected two confirmed gamma ray bursts up to energies of about 1 MeV; these were the events of 20 October and 10 November, for which confirmation was provided by the Helios 2 gamma burst experiment {1} and the HEAO A-4 experiment {2} . Correlation studies are now underway which will eventually lead to precise arrival direction information; we have used the approximate, preliminary arrival directions to estimate the effective detector surface areas and thus the energy reaching the detectors. These are 2×10^{-4} $ergs/cm^2$ and 6.5×10^{-5} $ergs/cm^2$ for the 20 October and 10 November events, respectively, above an energy of about 160 keV. These must clearly be regarded as lower limits to the total energy, since it is known that a substantial fraction of the total energy may be present in lower energy photons {3} and also because the bursts may last longer than our observations indicate.

Fig. 1 shows the 20 October burst as seen in the 63 cm^2 detector of Signe 2 MP at 250 ms time resolution. A multiple-peaked structure is evident (cf. {3}) with characteristic (e-folding) rise and fall times of about 700 ms for the large peaks; significant structure may be present on smaller time scales, however.

Fig. 2 shows the time structure of the 10 November burst, which was seen by Signe 2 MP and Signe 3. To date, not all of the Signe 2 MP data has been recovered; the figure shows a 1 sec long "prememory" for the Signe 2 MP upper detector, followed by a 250 ms time resolution histogram from one of the slower memories. Superimposed upon the latter curve is the 8 ms time history from the anticoincidence of Signe 3. As observed by either Signe 2 MP or Signe 3, the total duration of the event is about 0.7 sec. In an inset, two uncorrected spectra are shown. "U" refers to the upper (63 cm^2) detector of Signe 2 MP, while "L" refers to a lateral detector; the latter spectrum turns over at low energies due to absorption by the anticoincidence crystal, and perhaps also by surrounding material in the spacecraft.

Finally, Fig. 3 shows an unconfirmed event observed only by Signe 3, on 12 December, in both detectors. In time structure and total energy, it resembles the burst of 10 November.

In October and November of 1978, three detectors similar to the upper detector of Signe 2 MP will be launched. Two of these will be aboard two Soviet Venus probes, and one will be aboard a Prognoz satellite. Data will be transmitted from Moscow to Toulouse by means of a telephone line ; data reduction will allow arrival directions to be determined rapidly with precisions of the order of 10's of arc minutes.

REFERENCES

1. T. Cline, private communication, 1977

2. L.E. Peterson, private communication , 1977

3. S.R. Kane and G.H. Share, Astrophys. J., 217, 549, 1977

4. S.R. Kane and K.A. Anderson, Astrophys. J., 210, 875, 1976

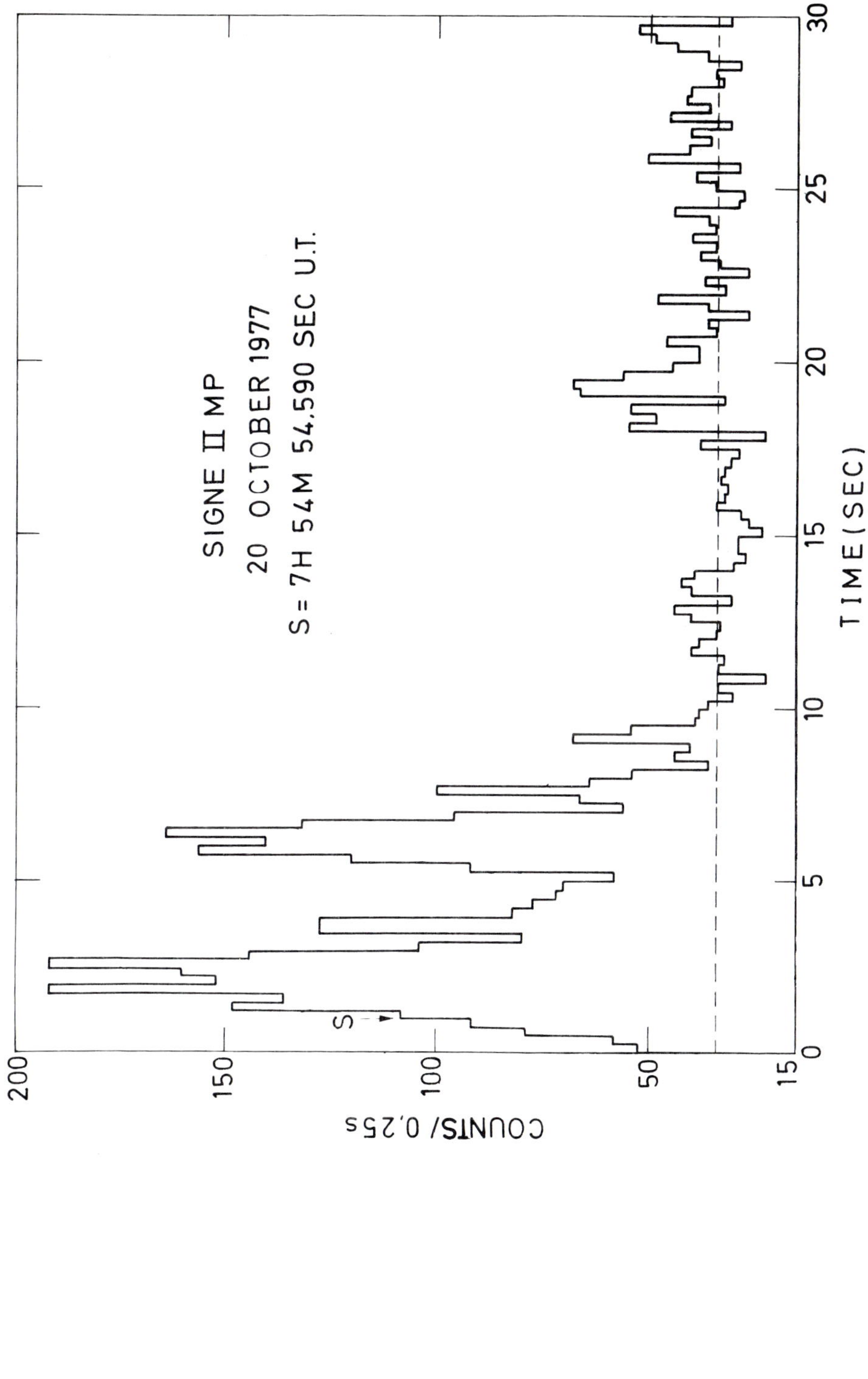

Fig. 1 The 20 October 1977 event as seen by the 63 cm^2 detector of Signe 2 MP, at energies above 115 keV. The time resolution is 250 ms.

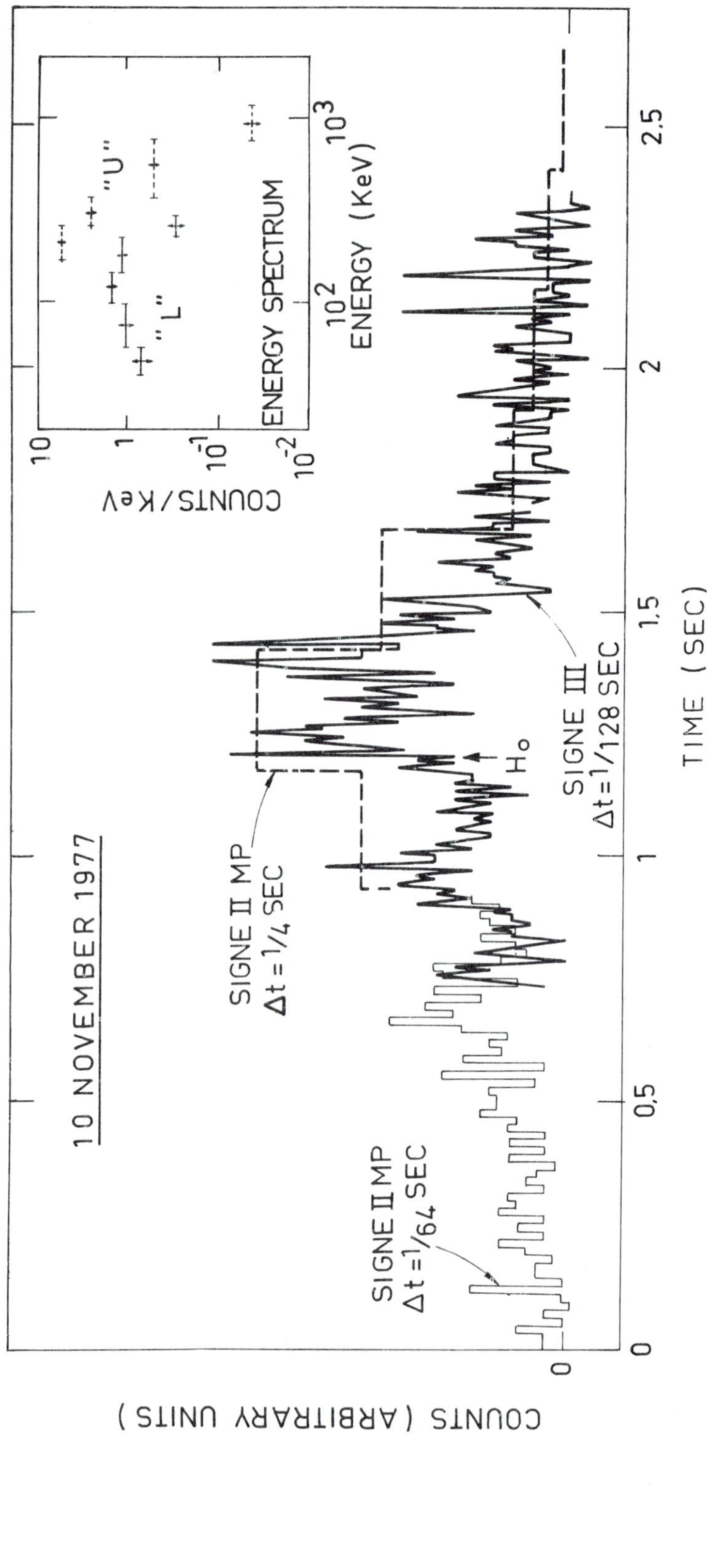

Fig. 2 The 10 November event as seen by Signe 2 MP and Signe 3. Inset :raw spectra measured by Signe 2 MP in the 63 cm^2 ("U") and 12 cm^2 ("L").

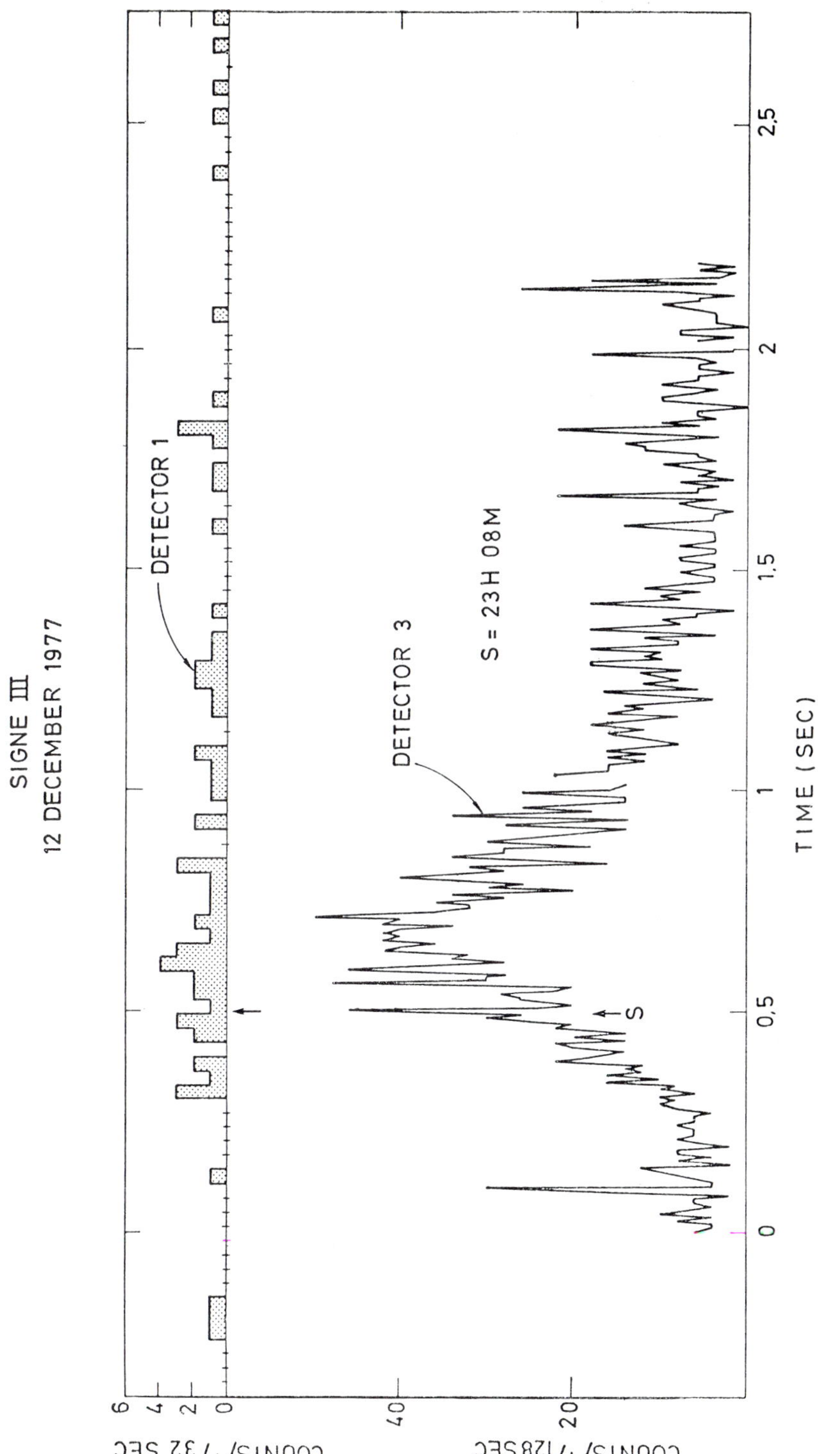

Fig. 3 An unconfirmed burst observed by Signe 3 on 12 December 1977.

SEARCH FOR GAMMA RAY BURSTS — RESULTS FROM A BALLOON EXPERIMENT

P. C. Agrawal, S. V. Damle, G. S. Gokhale,
S. Naranan and B. V. Sreekantan

Tata Institute of Fundamental Research, Bombay 400005, India

ABSTRACT

A 1 m square, 5 cm thick plastic scintillation detector, having a sensitivity for Gamma Ray Bursts of size $S > 8 \times 10^{-8}$ ergs/cm^2, was successfully flown on balloon flights from Hyderabad, India, on March 31, 1977 and on April 13, 1978. Twelve hours of observations do not show any Gamma Ray Burst event. This gives an upper limit of 1.5×10^3 events per year for a burst of $> 8 \times 10^{-8}$ ergs/cm^2. This is consistent with an S^{-1} curve for the size distribution, indicating a galactic origin for Gamma Ray Bursts.

INTRODUCTION

We report here investigations carried out in a search for Cosmic Gamma-Ray Bursts (of the type discovered by Klebesadel et al[1] in a series of two balloon flights from Hyderabad (India). The present investigation was aimed mainly at the detection of weaker gamma-ray bursts of total energy $S > 10^{-7}$ erg/cm^2 or less. Measurements of the frequency distribution of weaker gamma-ray bursts is of great significance in order to decide whether such bursts are of galactic or extragalactic origin. Almost all the 50 or so Gamma Ray Bursts detected so far from satellite borne observations are relatively intense bursts, of energy content (or size) $S \gtrsim 10^{-5}$ ergs/cm^2. A few bursts of less than 10^{-5} ergs/cm^2 have also been observed in balloon borne experiments [2-12]. From the number distribution of these bursts over a limited size range, it is, at present difficult to distinguish between $S^{-3/2}$ and S^{-1} distributions. An experimental determination of the frequency distribution of the weaker bursts, $< 10^{-5}$ erg/cm^2, is therefore crucial from the viewpoint of deciding between the $S^{-3/2}$ and S^{-1} distributions.

THE EXPERIMENT

The gamma-ray burst detector used consisted of four plastic scintillation counters with a total area of 10,000 cm^2. Each scintillation counter was 50 cm x 50 cm, and 5 cm thick, and was viewed by diffuse reflection from a millipore-paper coated cone by two 12.5 cm diameter photomultipliers. The two phototubes were operated in mutual coincidence to minimize contributions of the noise to the counting rates. The detectors were calibrated prior to flight using the Compton edges of 0.67 MeV and 1.17 and 1.3 MeV gamma-rays from a Cs^{137} and Co^{60} radioactive source respectively. The four detectors were operated in pairs of two for the purpose of pulse height analysis, each pair having an independent high voltage supply and electronics. The two detectors in a pair were operated in mutual anticoincidence with a time resolution of 4 microseconds for rejecting air shower events. The counting rates of each pair of detectors were sorted in three energy intervals of 100-400 keV, 400-1000 keV and 1000-4000 keV, and all events which resulted in energy deposition of more than 4 MeV were counted in the integral channel. The counting rates in each energy window were read out every 12.8 milliseconds and then telemetered to the ground using a PCM/FM telemetry link. The sensitivity of the detector was such that a gamma-ray burst of intensity 10^{-7} erg/cm^2 would be detectable at a significance level of 6σ.

Two balloon flights have been made with this detector, one on 31 March 1977 and the other on 13 April 1978. The payload floated at a level of 5.7 g/cm^2 of residual atmosphere in the first flight and 6.1 g/cm^2 level in the second flight. The total observation time in both flights at the float altitude was 12 hours. The instrument performed well in both flights.

RESULTS AND DISCUSSION

A very preliminary analysis of the data from both flights has been carried out using counting rates integrated over 1.28 sec intervals. Detailed analysis of the data with a finer time resolution of 12.8 msec is in progress. The criterion that we adopted for calling any counting rate increase a burst was that there should be at least 3σ increases in the total counting rates of the three energy channels in each of the two pairs of detectors. The probability of seeing a 3σ increase, due to statistical deviations, simultaneously in two independent detector pairs is only about 9×10^{-6} per 1.28 sec frame of observation time. Not a single event which satisfied the above criterion was detected in the preliminary analysis.

Based on our failure to observe a single gamma-ray burst in 12 hours of observing time, we can place an upper limit on the frequency of these gamma-ray bursts. In calculating this upper limit we have assumed that the energy spectrum observed by Metzger et al [13] for the Apollo event is a typical spectrum for gamma-ray bursts. Our upper limit is shown in Fig. 1, along with the data points of other balloon and satellite experiments. The burst size used for our points in Fig. 1 is based on the integration of 1.28 sec. used for the burst search. Our upper limit is closer to the S^{-1} curve. If we take this result at its face value, the galactic origin of Gamma Ray Burst events is indicated. Our result seems to rule out the $S^{-3/2}$ law for the size distribution of Gamma Ray Bursts.

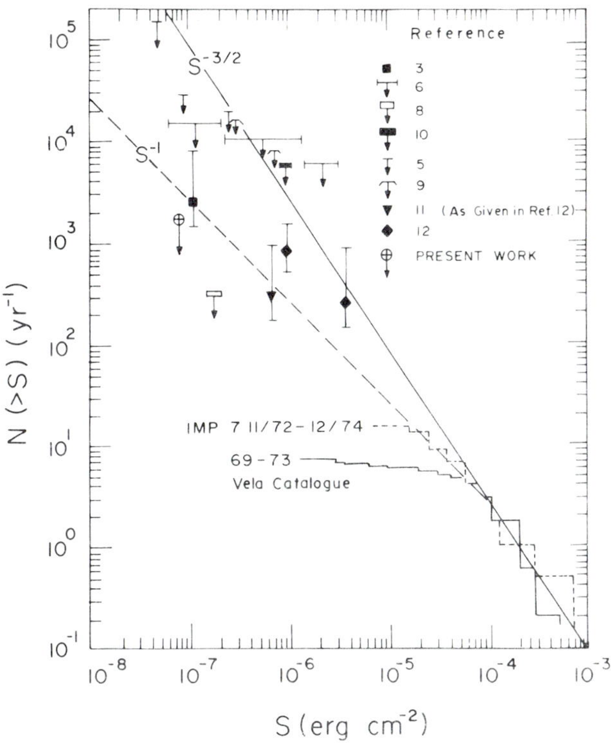

Fig. 1. Gamma Ray Burst Size spectrum showing our three standard deviation upper limit and data points from other balloon and satellite experiments.

Because of the high geomagnetic cut-off of 16.8 GV over Hyderabad, the problem of possible precipitation of electrons (which might be important for high latitude flights) is not very serious, and therefore the experiment does not have the disadvantage of non-simultaneous balloon flights. However, a longer exposure time, with possibly more sensitive instruments, may be required to arrive at a definitive conclusion. (This aspect is clearly demonstrated in the work of White et al [12], where, in their first 24 hour exposure, they recorded 3 bursts while a further 41 hour exposure recorded no events.)

ACKNOWLEDGEMENTS

We gratefully acknowledge the excellent support rendered by Messers R. V. Kundapurkar, A. T. Kothare, Neville Cor-rio and J. G. Deodhekar in the development and construction of the experiment. We thank Balloon Flight Group of TIFR under Shri R. T. Redkar for the balloon launches and acknowledge interesting discussions with Dr. R. K. Manchanda.

REFERENCES

1. Klebesadel R. W., Strong I. B., and Olson R. A., Ap. J. (Letters), 182, L85 (1973).

2. Ogawara et al., Astrophysics and Sp. Sci., 42, 271 (1976).

3. Bewick A., Coe M. J., Mills J. S. and Quenby J. J., Nature, 258 686 (1975).

4. Koga R., Simnett G. and White R. S., Ap. J. (Letters), 203, L115 (1976).

5. Cline T. L., Desai U. D., Astrophysics & Sp. Sci., 42, 17 (1976).

6. Johnson W. N., Kurfess J. D. and Bleach R. D., Astrophysics and Sp. Sci., 42, 35 (1976).

7. Herzo D., Dayton B., Zych A. D. and White R. S., Ap. J. (Letters), 203, L115 (1976).

8. Carter J. C., Dean, A. J., Manchanda R. K. and Ramsden D., Nature, 266, 750 (1977).

9. Cline T. L., Desai U. D., Schimdt W. K. H. and Teegarden B. J., Nature, 266, 694 (1977).

10. Cline T. L. and Schmidt W. K. H., Nature, 266, 749 (1977).

11. Nishimura J. et al., Nature, 272, 337 (1978).

12. White R. S., Ryan J. M., Wilson R. B., Zych A. D. and Dayton W. B., Nature, 271, 635 (1978).

13. Metzger A. E., Parker R. K., Gilman D., Peterson L. E. and Trombka J. I., Ap. J. (Letters), 194, L19 (1974).

HARD X-RAY SHOULDERS IN THE SPECTRA OF THREE SCOX-1 LIKE OBJECTS AS OBSERVED BY ARIEL V

M. C. Coe*, J. G. Greenhill*, S. J. Bell-Burnell**
and K. T. Strong**

**The Blackett Laboratory, Imperial College, London, U.K.*
***Mullard Space Science Laboratory, University College London, Holmbury St. Mary, Dorking, U.K.*

ABSTRACT

The energy spectra of three ScoX-1 like sources (ScoX-1, GX349+2, and GX17+2) have been measured in the energy range 2 to ~100 keV by detectors on Ariel V. The data in the 2-15 range are fitted by thermal bremsstrahlung spectra with kT $\sim$ 5 keV. At energies above about 30 keV, ScoX-1 and GX349+2 show a marked excess flux above the thermal bremsstrahlung continuum. There are similar indications for GX17+2 although the data are of limited statistical significance. The high energy data can in all three cases be fitted by thermal spectra with kT $\gtrsim$ 30 keV.

Large-Scale Distribution of Cosmic γ-Ray Emission

THE SKY IN THE LIGHT OF GAMMA RAYS

K. Pinkau

Max-Planck-Institut für Physik und Astrophysik,
Institut für extraterrestrische Physik, 8046 Garching, West Germany

INTRODUCTION

Gamma ray observations have now provided us with results at energies above about 1 MeV that allow us to paint a picture of the sky with a rather broad brush (angular resolution $\geqslant 1^{\circ}$ above several 100 MeV, about 10° between 1 and 10 MeV) and few colours (spectral resolution typically $\leqslant$ 8% FWHM between 1 - 10 MeV, 50% FWHM at 100 MeV).

In this picture, the milky way stands out at high energies as a narrow band shining brilliantly in the direction towards the center and more dimly in the anticenter direction, where individual gamma ray sources dominate the appearance of the galaxy.

Gamma radiation from extragalactic space appeared to be of a diffuse nature - the extension of the diffuse extragalactic X-radiation into the gamma domain. The spectral shape of that diffuse radiation now appears to be well known. However, the first extragalactic objects have been observed in the light of gamma rays with the likely identifications as NGC 4151 (a Seyfert galaxy) and 3C273 (a Quasar), and these discoveries may be important also for the interpretation of the diffuse radiation.

At low energies, the Compton scattering technique has clearly made its impact on the state of the art. At energies above the SAS-II range, it is the ESA gamma ray observatory Cos-B which has greatly advanced our knowledge. This satellite records celestial gamma rays using a wire spark chamber technique in connection with a CsI spectrometer of 4.7 radiation lengths thickness since August 1975. It is still operating and could continue to operate until the end of 1980 if ESA finds the money to record the telemetry signals.

GALACTIC GAMMA RAYS

Large Scale Distribution

The milky way shines brightly as a narrow band. Figure 1 shows the line intensity of gamma rays within a latitude interval $|b^{II}| < 6^{\circ}$. That this line source is very narrow can be seen from the latitude profiles shown in Fig. 2.

Let us perform a scan through the galactic plane. First, we note that the region of high emission towards the inner part of the galaxy is not symmetric; with respect to the center it extends from just beyond 280° in longitude to about 55°; in the spiral model of our galaxy by Georgelin and Georgelin [1] from high excitation

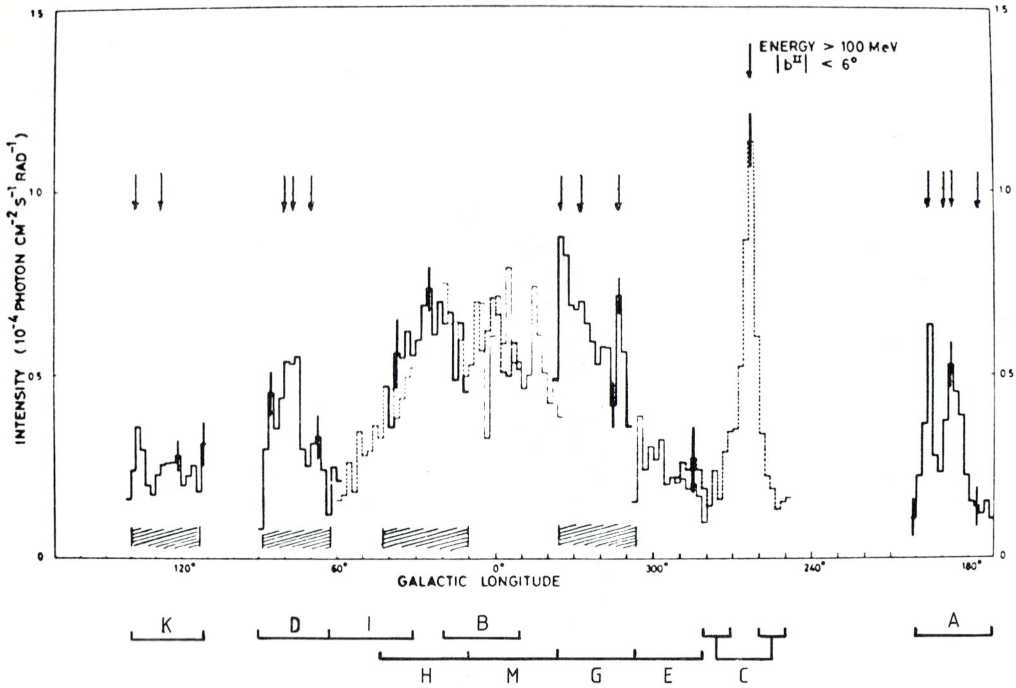

Fig. 1 Longitude profile of the intensity of gamma rays of energy > 100 MeV. The arrows indicate the positions of point sources. Capital letters denote the longitude intervals for which latitude profiles are given in Fig. 2 and the shaded areas those intervals for which the average energy spectra are shown in Fig. 8. (After Bennett et al. [2])

parameter HII regions, this is just the inner region bounded by the Sagittarius-Carina arm.

Then, as we closely investigate the region between $l^{II} = 280^0$ and 60^0 we see maxima of local intensity. These maxima appear to be correlated to spiral features of the galaxy, as has been pointed out before (Fichtel et al. [3], and further references therein, Paul et al. [4]).

The agreement in detail is striking; it clearly brings out the asymmetric nature of the milky way with a skewness of about 25^0 towards the side of large galactic longitudes. The maxima identified above in the Cos-B results do agree with the intensity maxima in neutral hydrogen and the radio continuum.

On the other hand, Fig. 1 shows clearly that a number of point sources has already been discovered, and their contribution must be taken into account.

Turning our attention now towards the outer galaxy, the Cygnus region is highly structured just as in the radio domain, for example. Its peak emission is close to the star γ-Cygni and not associated with the X-ray source Cyg X-3.

Beyond Cygnus, the general galactic emission is very low, apart from some point sources, of which the Crab complex with two new sources at $l^{II} = 189^0$ and 195^0, and the Vela point source at $l^{II} = 263^0$ are the most prominent.

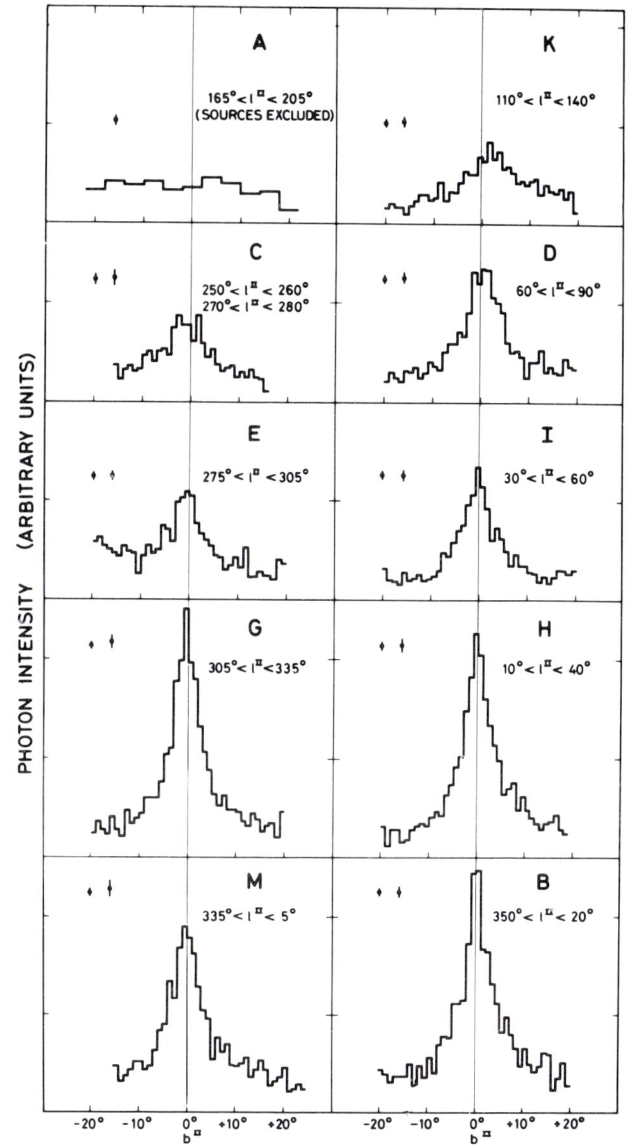

Fig. 2 Latitude profiles of the intensity of gamma rays of energy > 100 MeV, averaged over the longitude intervals indicated. Typical statistical errors are for the values in the wings (left) and the peak (right). (After Bennett et al. [2])

Point Sources

In our scan through the galaxy, the importance of localized sources has become apparent. This can most easily be recognized in the anticenter direction, because the limited power of angular resolution suffices to isolate and investigate them in the quiet environment.

Cos-B has brought the list of localized sources to a total of 13 (Fig. 3). In addition, it is very likely that PSR 1822 has been identified as a 14th source due to its time variation (and not, as in the other cases, as a local maximum of intensity).

Source	$l^{II}(^\circ)$	$b^{II}(^\circ)$	Relative intensity	Remarks
CG64+0	64.5±0.5	0.0±1.0	$0.30^{+0.10}_{-0.05}$	
CG75+0	75.0±1.0	0.0±0.5	$0.45^{+0.15}_{-0.10}$	
CG78+1	78.5±0.5	+1.5±0.5	$1.00^{+0.30}_{-0.20}$	
CG121+3	121.0±1.0	+3.5±1.0	$0.20^{+0.10}_{-0.05}$	
CG135+1	135.5±1.0	+1.5±1.0	$0.30^{+0.15}_{-0.10}$	
CG176−7	176.0±1.0	−7.0±1.0	0.50±0.10	
CG185−5	185.3±0.5	−5.6±0.5	1.00	PSR0531+21 at l^{II} = 184.6, b^{II} = −5.8
CG189+1	189.0±1.5	+1.0±1.5	0.40±0.10	
CG195+4	195.9±1.0	+4.5±0.5	0.90±0.20	195+5 (SAS−2)
CG263−2	263.7±0.3	−2.6±0.3	3.30±0.50	PSR0833−45 at l^{II} = 263.6, b^{II} = −2.8
CG312−1	312.0±1.0	−1.5±1.0	0.45±0.15	
CG327−0	327.5±1.0	−0.5±1.0	0.35±0.15	
CG333+0	333.5±1.0	0.0±1.0	0.70±0.20	

Fig. 3 Gamma ray sources detected by Cos-B. (After Hermsen et al.[5])

The intensities of these localized sources cover a range from 0.2 (CG 121+3) to 3.3 times the Crab (Vela, CG 263-2). Among many interesting objects are CG 78+1 (in Cygnus) and CG 195+5, which both are about as intense as the Crab. CG 195+5 has not been seen as an interesting object in other wavelength regions.

It is most interesting to study the Crab and Vela, and to compare their properties in the gamma ray range with those at other wavelengths (Figs. 4,5).

The Crab has been known to pulse in phase at all energies. The gamma ray domain makes no exception, and the pulse shape and height does not change with energy. In contrast to this, the Vela pulsar is very similar to the Crab in showing a main pulse and an interpulse in gamma rays, while the pulse shape and phase is different in the optical, and again different in the radio region.

The similar behaviour in gamma rays, yet different appearance in the X-ray, optical and radio range between the two pulsars is also apparent from their flux data over this very large energy range. Both pulsars have similar values in the unpulsed flux over the entire range of observations. Their pulsed fluxes are also similar in the radio and gamma ray domain, but show very large differences in the optical and X-ray region.

Within the gamma ray region, the Crab pulsed flux shows a power law energy spectrum that connects well to measurements at lower energies. Vela has no power law energy spectrum but becomes flatter at lower energies. Only upper limits exist in the X-ray and no measurements at all in the low energy gamma ray region for the Vela pulse.

Models of the production of pulsed radiation will not be discussed here. The large degree of similarity of the Crab and Vela in the gamma ray region may take us very close here to the actual pulsar production mechanism.

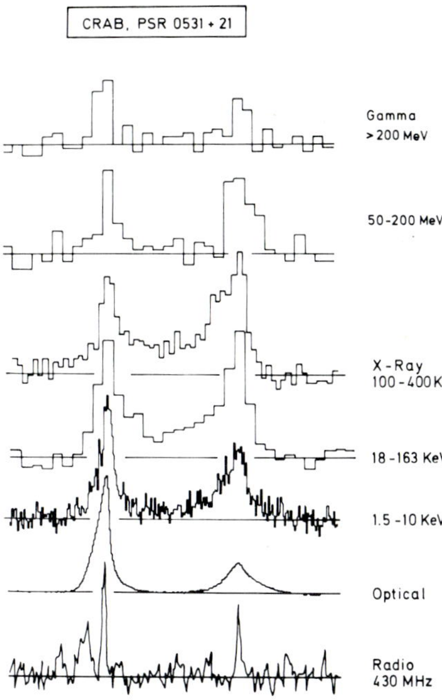

Fig. 4 Light curves of the Crab pulsar in various parts of the electromagnetic spectrum. (Taken from Pinkau [6])

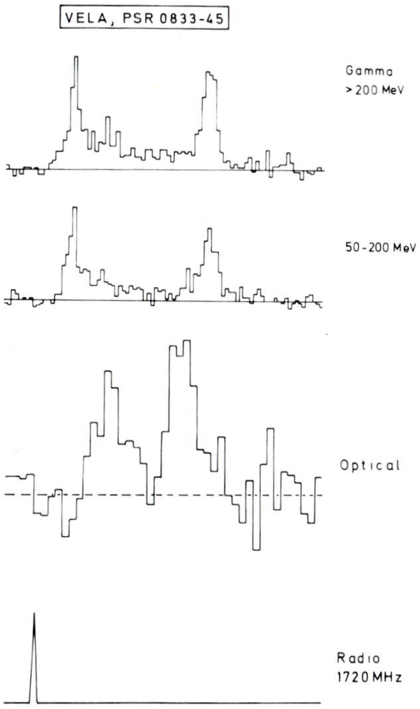

Fig. 5 Light curves of the Vela pulsar in four parts of the electromagnetic spectrum where it has been seen. (Taken from Pinkau [6])

At high energies, both sources are practically 100% pulsed (Fig. 6).

Energy range	PSR 0531 + 21	PSR 0833 − 45
Pulsed emission*	photon $cm^{-2}s^{-1}$	photon $cm^{-2}s^{-1}$
>50 MeV	(5.9 ± 0.8) 10^{-6}	(15.6 ± 0.8) 10^{-6}
>200 MeV	(1.3 ± 0.3) 10^{-6}	(5.3 ± 0.4) 10^{-6}
Continuous emission, 2−σ upper limit	photon $cm^{-2}s^{-1}$	photon $cm^{-2}s^{-1}$
>50 MeV	1.7 × 10^{-6}	1.6 × 10^{-6}
>200 MeV	0.5 × 10^{-6}	0.7 × 10^{-6}
Pulsed fraction, 2−σ lower limit		
>50 MeV	78%	91%
>200 MeV	72%	89%
Contribution to total pulsed flux >50 MeV		
first pulse	(50 ± 9) %	(32 ± 3) %
interregion	<15 %	(30 ± 4) %
second pulse	(50 ± 9) %	(38 ± 4) %

*Only statistical errors included

Fig. 6 Results on the Crab and Vela gamma ray point sources. (After Kanbach et al. [7])

The upper limits for the unpulsed flux derived from Cos-B can be improved in the future, but they already now put constraints on models for the production of the nucleonic component of cosmic rays in supernova remnants. If these cosmic rays were produced in the explosive event itself or shortly afterwards when the pulsar slowed down fast, they also must have diffused away quickly from the Vela source so that the present gamma ray emission has no continuous component.

Nature of the Galactic Emission

From the longitude distribution of gamma rays one has tried to derive information on their mode of production. The flat-topped extended ridge has been interpreted to mean that the gamma ray emissivity is large in a ring-like area extending from about 4 to 7 kpc around the galactic center. In different schools of thought (Fichtel et al. [8]; Stecker [9], [10]) this in turn has been attributed either to the fact that cosmic ray and matter density are very high in the spiral arms thus making them very bright in gamma rays or, alternatively, that the radio data predict a ring of clouds of molecular hydrogen to exist at about 6 kpc, and that the assumed high flux of cosmic rays there would produce gamma rays by collisions with this high density gas. In both models it had been argued that gamma ray point sources are of relatively little importance, and that the gamma rays would mainly be produced by the π^0-production process.

The room for these additional modes of production was generated by the early observation that the gamma radiation from the center region was much more (about 5 times) intense than could be understood by the interaction of the nominal cosmic ray intensity (solar vicinity) and neutral hydrogen.

While the observations of Cos-B have shown surprising coincidence with galactic spiral arm structures, it has also become obvious that one must be careful with assumptions on cylindrical symmetry in the galaxy, and that point sources must be taken into account (cf. Fig. 1, Fig. 3).

That point sources must carefully be considered is most clearly seen in the anticenter direction. If point sources are excluded, the gamma ray emission profile from the galactic disc is much lower than would be expected if cosmic rays extended all the way out to where neutral hydrogen is found. It has thus been concluded that a cosmic ray gradient must exist in the galaxy with the cosmic ray intensity decreasing beyond our distance from the galactic center (Fig. 7).

Estimates have been made on the contribution of pulsars to the gamma ray flux from the galactic plane, and some authors arrived at a possible contribution between 5 and 20%, while others think it could well be as high as 50% (see, for example, Kniffen et al. [11]). The problem is to establish a correlation between gamma ray luminosity on the one hand, and pulsar period and its first derivative on the other. Furthermore, it is not clear if all pulsars have been found in the radio surveys, especially in the regions of large continuum radiation, and, finally, if radio pulsars are the only class of gamma ray objects possible.

Indeed, the correlation between the Cos-B localized sources and radio pulsars is very weak; only two of the 13 were previously known to be pulsars (the Crab and Vela).

Therefore, the question whether the visibility of the galactic structure is due to a diffuse process, like the coincident concentration of matter and cosmic rays, or due to the fact that the gamma ray sources are located in the spiral arms, must remain an unsolved problem as yet. Unfortunately, the angular dimensions of galactic features are of the same order of magnitude as the angular resolution of the gamma ray instrument.

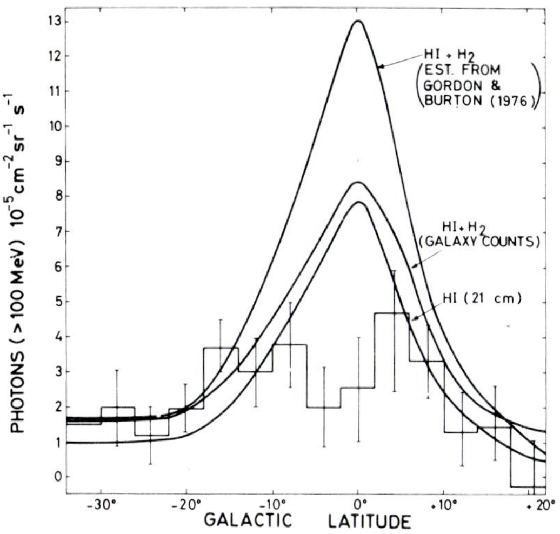

Fig. 7 Latitude distribution of gamma rays of measured energy > 100 MeV observed by Cos-B and expected distribution for cosmic ray/gas interactions for a spatially uniform cosmic ray density. The estimates are based on (a) total gas from galaxy counts (b) HI gas from 21 cm data (c) total gas from 21 cm data and CO data. (After Strong et al. [12], Gordon & Burton [13])

The average energy spectra (Fig. 8) of the galactic emission in various regions are not due to a pure π^0-generation process but are compatible with a large contribution of bremsstrahlung from electrons, as has been computed by Kniffen et al. [14]. However, it turns out that the Vela pulsed energy spectrum of gamma rays is of exactly the same shape as the average galactic emission. Thus, it is not clear at present whether these average spectra are due to a large number of point sources or to a diffuse emission process by cosmic ray nucleons and electrons.

EXTRAGALACTIC GAMMA RAYS

In the spectrum of the "diffuse" extragalactic gamma radiation, the SAS-II group could successfully eliminate a galactic contribution to the extragalactic flux at high latitudes. The spectrum (Fig. 9) on which most experimental groups agree, extends from the X-ray domain with a 2.3 differential power law, shows a shoulder between 1 and 10 MeV and continues beyond 10 MeV with a 2.85 differential power law index. The nature of the shoulder has been the subject of many investigations, and it has been noted that cosmological considerations could play a role because space is very transparent to gamma rays of this energy. On the other hand, the possible contribution of other galaxies may be important, and this field is now developing rapidly. Graml et al. [15] have presented the Compton telescope measurement of the energy spectrum of NGC 4151 which shows a sharp break at 3 MeV (Fig. 10). Schönfelder [16] has pointed out that X-ray Seyfert galaxies could explain the observed diffuse gamma ray flux if nearly all of them had an energy spectrum and a gamma ray luminosity like that of NGC 4151. This luminosity for NGC 4151 is 10^{45} ergs/sec below 3 MeV and above 1 MeV.

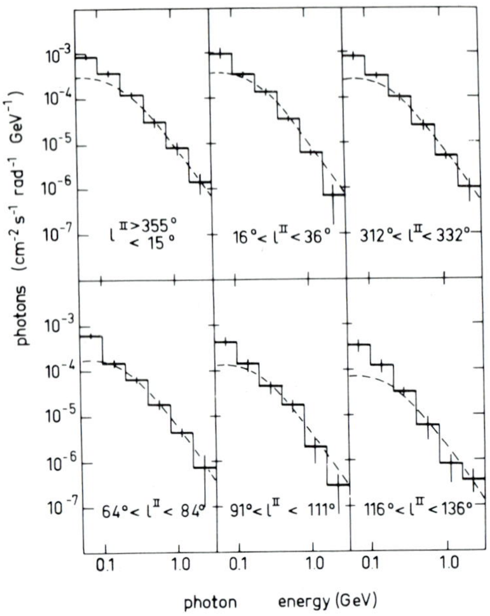

Fig. 8 Energy spectra of the gamma radiation from all the regions of the galactic disc listed in Fig. 1. The error bars represent both statistical uncertainties and fluctuation of the background. The dashed lines are the π^0-spectrum of Badhwar and Stephens [17] normalized at 400 MeV. (After Paul et al. [18]).

At the same time, Swanenburg et al. [19] have reported observing an extragalactic gamma ray source with Cos-B. The Quasar 3C273 is within the error box and the flux spectrum multiplied by E^2 from the radio to the gamma ray range for this object is shown in Fig. 11. Its luminosity in the gamma ray range from 50 to 500 MeV is $\sim 2 \times 10^{46}$ ergs/sec for H_0 = 60 km s^{-1} Mpc^{-1}.

With the discovery of extragalactic gamma ray sources a new chapter has begun. Once again the cosmos has taught us that there are more surprises for us in store than we had anticipated.

SUMMARY

On the galactic emission, structures of the galaxy appear in high contrast, showing a deviation from symmetry about the galactic center. The contribution of localized sources to the high value of gamma ray emission is yet open. Many gamma ray point sources have been discovered including pulsars. The light curves of Vela and Crab show double pulses. Cosmic rays probably do not fill the region beyond our distance from the galactic center because the emission from the plane is very low.

The extragalactic diffuse gamma ray spectrum is very steep with a shoulder between

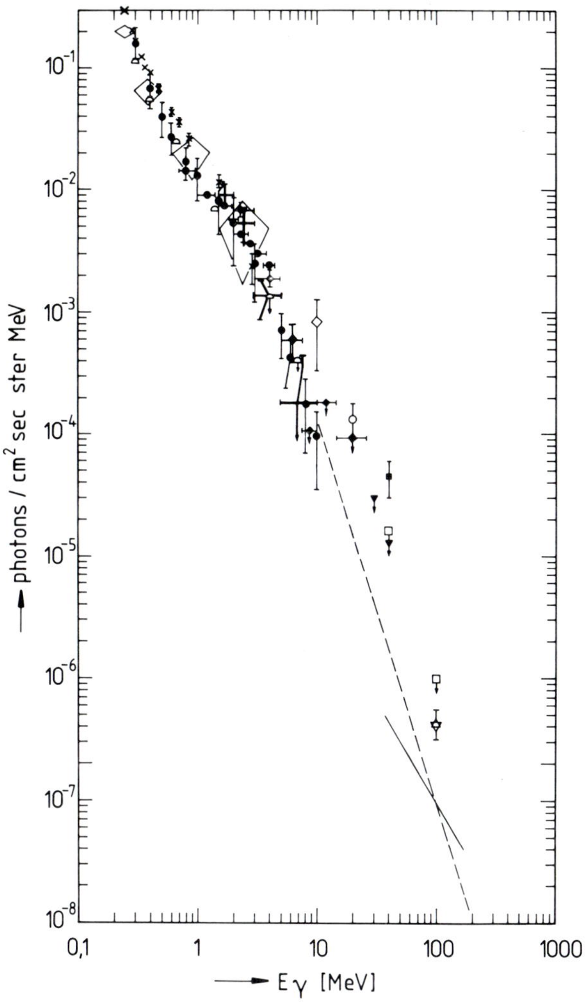

Fig. 9 Spectra of the diffuse gamma radiation (adapted from Schönfelder et al. [20] where the authors of the various measurements are referenced. At high energies, the total diffuse flux has been subdivided into a galactic contribution (full line) and an extragalactic part (dashed line). (After Fichtel et al. [21])

1 and 10 MeV. Extragalactic point sources have now been discovered, NGC 4151 being a rather safe, 3C273 a probable identification.

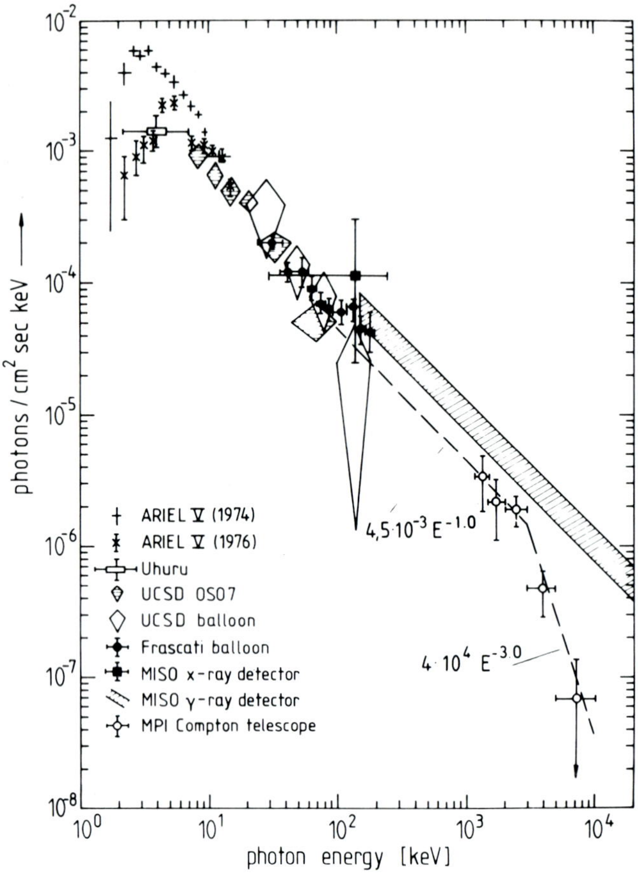

Fig. 10 Energy spectrum of the Seyfert galaxy NGC 4151 from 2 keV to 10 MeV. (After Graml et al. [15] and Schönfelder [16])

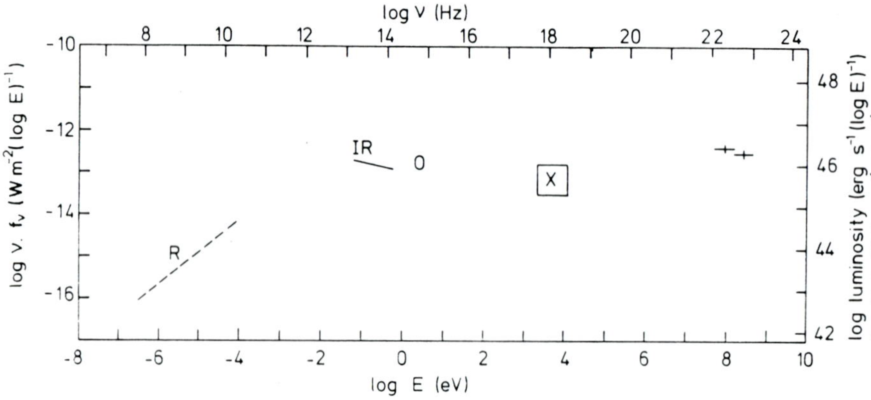

Fig. 11 Flux spectrum of 3C273 from the radio to the gamma ray range. (After Swanenburg et al. [19])

REFERENCES

Frascati 1977: Proceedings of 12th ESLAB Symposium "Recent Advances in Gamma Ray Astronomy", ESA SP-124.

1. Y.M. Georgelin and Y.P. Georgelin, Astron. Astrophys. 49, 74 (1976).
2. K. Bennett et al., Frascati, A 12 (1977).
3. C.E. Fichtel et al., Astrophys. J. 198, 163 (1975).
4. J. Paul et al., Astrophys. J. 207, 62 (1976).
5. W. Hermsen et al., Frascati, A 2 (1977).
6. K. Pinkau, Proc. 15th Intern. Cosmic Ray Conf., Plovdiv, Bulgaria, Rapporteur Paper (1977).
7. G. Kanbach et al., Frascati, A 3 (1977).
8. C.E. Fichtel et al., Frascati, A 13 (1977).
9. F.W. Stecker, in: Origin of Cosmic Rays (eds. J.L. Osborne and A.W. Wolfendale), Reidel, Dordrecht, 1975, p. 267.
10. F.W. Stecker, in: The Structure and Content of the Galaxy and Galactic Gamma Rays, Goddard Space Flight Center, X-662-76-154, 1976, p. 378.
11. D.A. Kniffen et al., Frascati, A 6 (1977).
12. A.W. Strong et al., Frascati, A 26 (1977).
13. M.A. Gordon and W.B. Burton, Astrophys. J. 208, 346 (1976).
14. D.A. Kniffen et al., Frascati, A 16 (1977).
15. F. Graml et al., Proc. Esrange Symposium, Ajaccio, 24-29 April 1978, ESA SP-135, 1978, p. 287.
16. V. Schönfelder, Nature 274, 344 (1978).
17. G.D. Badhwar and S.A. Stephens, Proc. 15th Intern. Cosmic Ray Conf., Plovdiv, 1, 198 (1977).
18. J.A. Paul et al., Astron. Astrophys. 63, L31 (1978).
19. B.N. Swanenburg et al., Nature (1978) in press.
20. V. Schönfelder et al., Astrophys. J. 217, 306 (1977).
21. C.E. Fichtel et al., Frascati, A 29 (1977).

X-RAY EMISSION IN THE VICINITY OF HIGH-ENERGY γ-RAY SOURCES

G. H. Share, K. Wood, R. L. Kinzer, S. Shulman, J. Samimi*, W. N. Johnson, D. Yentis, J. Meekins, W. Evans, D. McNutt, E. Byram, T. Chubb and H. Friedman

Naval Research Laboratory, Washington, D.C., U.S.A.
**Department of Physics, Ferdowsi University, Iran*

Thirteen localized sources of high-energy γ-rays have been identified by the COS B and SAS 2 satellites and various balloon-borne experiments [1]. Only two of these sources, the Crab nebula and Vela pulsar, have been positively identified with counterparts observed at other wavelengths. These γ-ray sources lie close to the galactic plane ($|b^{II}|<7^{o}$). The angular resolution of these observations is no better than about $\pm 1^{o}$; therefore some of the sources may in fact be due to extended galactic features. Two recent publications have described searches for X-ray counterparts of the unidentified sources using data from Uhuru, OSO-7 and SAS-3 [2,3]. X-ray sources have been found within the error circles of CG 78+1, CG135+1 and CG312-1.

During the first three weeks of its operation, from 18 Aug. to 10 Sept. 1977, the NRL large area survey instrument on HEAO-1 (the A1 experiment) scanned through the positions of 4 of the 11 unidentified sources: CG135+1, CG312-1, CG327-0 and CG333+0. These sources were observable for ~10 sec at intervals of about 30 minutes, with the exception of times when they were Earth occulted or when the satellite passed through high radiation zones. Four of the seven A1 modules were used. Each one has an area of 1600 cm^2 and a field of view of 1° x 4° FWHM (the 1° aperture is in the scan direction). As the spin axis of the satellite follows the sun, the sources could be observed for ~8 days. From these observations we have identified five new X-ray sources in the vicinity of the four COS B positions. These sources are relatively weak; their intensities are $\lesssim 5 \times 10^{-3}$ cts/cm^2-sec (~1 Uhuru count) in the 0.5 to 20 keV range.

<u>CG135+01</u>

The weak Uhuru source 4U0241+61 was recently identified by SAS-3 and optical and radio co-workers [4] with an extragalactic object, despite its low galactic latitude. The object is a highly luminous quasar, 0241+622, with a redshift of 0.04. It was observable by HEAO on single scans. Upon summing these individual scans over one-day intervals at least one additional source, H0235+60, became evident (a second weak source was present but its position could not be adequately determined). Its intensity is ~3.5×10^{-3} cts/cm^2 sec. A preliminary position for this new source is shown in Fig. 1 along with the positions of the quasar and γ-ray source. Also shown is the location of a highly variable galactic radio source, GT0236+610 which flared on 26 Aug. 1977 from 50 to 250 mJy [5], just after it left the field of view of the A1 experiment. This radio source lies near the edge of the error box of the new X-ray source. If the new X-ray source is

GT0236+610, then it is definitely galactic, unlike 4U0241+61.

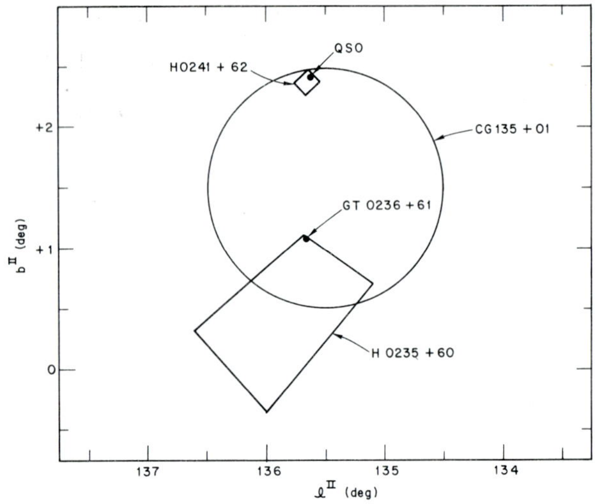

Fig. 1. X-ray sources and other interesting objects in the vicinity of CG135+1.

The X-ray intensity of the quasar was monitored for about nine days. Its daily average intensities are shown in Fig. 2. During the six days (233-238) in which it

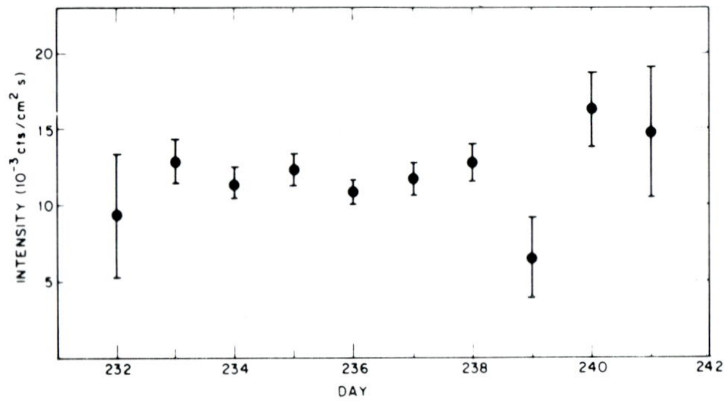

Fig. 2. Averaged daily intensities of X-ray emission from 0241+62.

was centered in the field of view, its intensity varied by $\lesssim 10\%$. Systematic effects due to reduced observations are likely to have caused the fluctuation seen near day 239.

CG312-1

As described in Ref. 2 this is a confused region containing several sources. During our observation the X-ray source 4U1416-62, which lies within the γ-ray error box, was not detected; however A1354-64, which lies ~1° from the periphery of this box, was seen. This is the only one of the four γ-ray sources discussed here for which no new X-ray sources have been isolated. It is possible that flux from fainter sources has been detected but that they have not been resolved; there is a diffuse enhancement throughout the region which is consistent with unresolved sources. Similar enhancement is seen at other crossings of the galactic plane, including regions containing the three other γ-ray sources discussed here.

CG327-0 and CG333+0

This is a vicinity of the galactic plane which is rich in X-ray sources and other potential counterparts of γ-ray sources, such as pulsars and supernova remnants. During our period of observation two strong sources, 4U1608-52 and 4U1624-49, dominated this region of the plane. 4U1608-52 was in the decay phase of a flare which reached its maximum in July. These strong sources made it difficult to detect and position weak sources using the $1^{\circ} \times 4^{\circ}$ collimator. Despite this difficulty, four new sources have been detected. Their preliminary positions are shown in Fig. 3 in relationship to the γ-ray sources and other objects in the

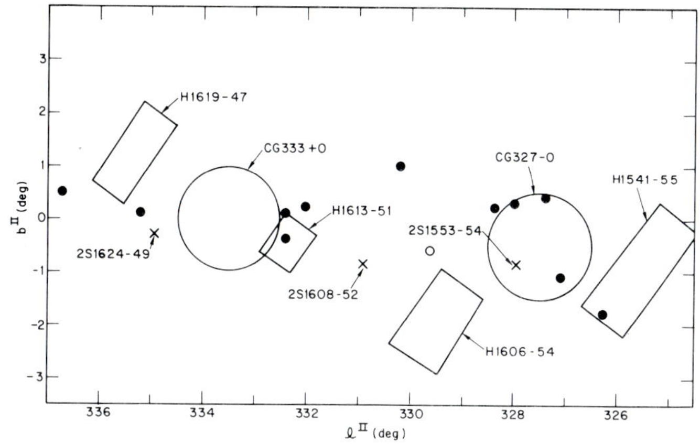

Fig. 3. X-ray sources and other interesting objects in the vicinity of CG327-0 and CG333+0. • supernova remnant; o pulsar.

region. The variable source 2S1553-54, a possible counterpart of CG327-0 was not detected during this observation period. The positions of several supernova remnants [6] are shown.

Discussion

Five new X-ray objects have been found in a high sensitivity survey of four of the eleven high-energy γ-ray sources using the HEAO scan modules. In future work we hope to improve on the positions given for these sources in this report.

The growth in the number of detectable objects as the detectable flux limit decreases means that more than one candidate X-ray counterpart will typically be found in the vicinity of the relatively large COS B error boxes. We cannot exclude the possibility that the fainter X-ray objects are the correct identifications, anymore than we can exclude the possibility that the γ-ray sources are X-ray quiet. In lieu of improved error boxes for the γ-ray sources, it appears that distinctive temporal or spectral characteristics will be needed to establish their true counterparts.

References

1. W. Hermsen et al., Nature 269, 494 (1977).

2. L. Maraschi et al., Nature 272, 679 (1978).

3. P.F. Julien and H.F. Helmken, Nature 272, 699 (1978).

4. K.M.V. Apparao et al., Nature 273, 450 (1978).

5. P.C. Gregory and A.R. Taylor, Nature 272, 704 (1978).

6. D.H. Clark and J.K. Caswell, Mon. Not. R. Astr. Soc. 174, 267 (1976).

PERFORMANCE OF A GAMMA RAY CAMERA FOR ASTROPHYSICS

H. Horstman, E. Horstman Moretti, F. Fuligni, G. Di Cocco, W. Dusi, F. Frontera, E. Morelli, and A. Spizzichino

Laboratorio TE.S.R.E./C.N.R., via de'Castagnoli 1, 40126 Bologna, Italy

INTRODUCTION

Accurate positions for X and γ sources, important for identification purposes, are particularly difficult to obtain in the energy range of the hard X rays and soft γ rays. On the other hand in the soft and medium X-ray range imaging devices (pinhole cameras, grazing incidence telescopes, rotating modulation collimators, etc.) have been in use for some time. These instruments can be divided roughly into two types, large angular aperture - survey detectors, and small field of view-dedicated detectors. The former are used to search and identify unknown sources while the latter are used to study in detail already well known objects.
Recently we have been studying possible applications of the Anger [1] or gamma-camera to X-ray astronomy. This instrument, well known in medical applications, with a suitable coded mask could apparently be used either to localize γ-ray burst sources in a wide field use or to measure positions and possibly angular extents of already known sources. The energy range given by the limits of the detecting system covers rougly the interval between 50 and 300 keV, thus extending the range of other "imaging" devices. We briefly list the advantages of the gamma camera:

- No modulation of the incoming flux is necessary for localization purposes, hence position and timing information are independent as is highly desirable for rapidly varying sources.
- Half of the active area is always available for source observation.
- The spurious background is measured simultaneously with the source.
- Energy resolution should be almost as good as in a normal NaI detector.

THE GAMMA CAMERA

The original gamma camera by Anger [1] had a single circular NaI crystal with seven phototubes mounted at a certain distance behind the crystal in a hexagonal array and was used as a pinhole camera. The number of phototubes for the modern medical gamma camera is usually 19 or 37 (or more). The outputs (V_i) of the phototubes for a pulse are weighted by the position (x_i, y_i) of the phototube to find the position centroids of each X-ray absorbed in the crystal:

$$\langle x\rangle = \Sigma\ V_i x_i\ /\ \Sigma\ V_i$$
$$\langle y\rangle = \Sigma\ V_i y_i\ /\ \Sigma\ V_i \qquad (1)$$

The length of the light pipe affects both the linearity and regularity of the position determination and the resolution. If the light pipe is too short, the response of the detector will be extremely non uniform, with the positional resolution being as large as the diameter of the phototube. On the other hand with a long light guide all the phototubes would give identical responses and the apparent average position of any interaction would be the center of the crystal. A measure of this effect can be had from the derivative dx'/ds between the calculated position (x') and the actual point of interaction. There will be an optimum intermediate length of the light pipe which will give dx'/dx roughly constant over the active crystal surface, and as large as possible. We sought to optimize this parameter, among others, with Monte Carlo simulations whose results are given in Fig. 1 for two cases, 7 and 19 phototubes (mounted on a surface larger than the crystal to avoid possible double solutions). The root mean square deviation of the detected position at about 50 keV is also shown for point scintillation sources. Both parameters indicate that no satisfactory solution can be found with 7 phototubes while the best light guide length is 0.3 crystal radii for 19 phototubes. Both uniformity and resolution would be improved going to 37 phototubes or more, as the above parameters are roughly proportional to the size of the phototube up to the limits given by light dispersion and the existence of multiple interactions in the crystal from single X-ray photons. In a complete model of the position determination the multiple events and also the optics (changing index of refraction) of the system should be taken into account. In the model used in Fig. 1 the walls and bottom of the light guide were chosen to be absorbing, while a diffuse reflector, which improved the resolution, was allowed for above the crystal. Typical resolutions for a 20 cm diameter crystal are 7 to 8 mm at 50 keV, and 4 mm at 200 keV, with the very conservative assumption of 10 photons/keV and 20% conversion efficiency. The actual performance is in fact expected to be better by a factor of 2.

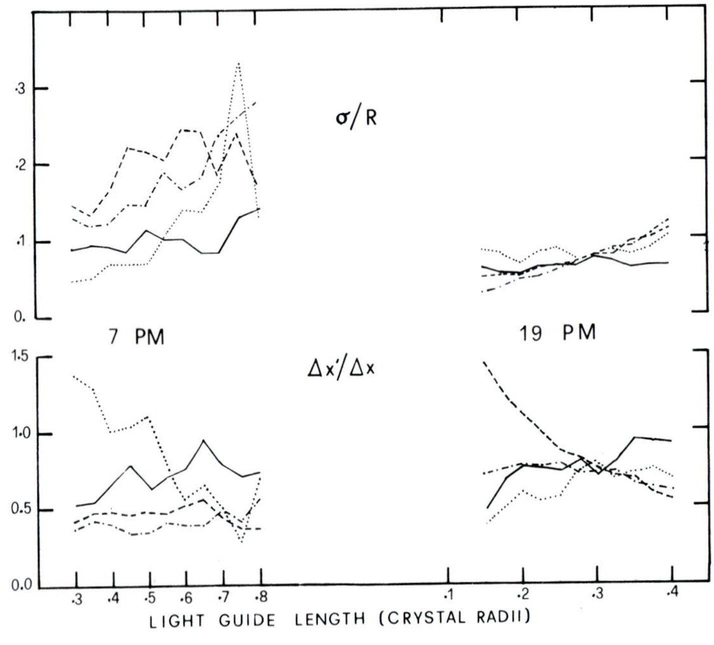

Fig. 1. Top: the linear resolution (1σ at 50 keV) in crystal radii is given as a function of light guide length for the cases of 7 and 19 phototubes, for several positions in the crystal. Bottom: the conversion factor Δx'/Δx between calculated (x') and original (x) scintillation coordinates is shown for the same points. (—— center; .... half the radius of the crystal; ---- and —.— on the perimeter of the crystal, separated by 30° in azimuth).

γ-RAY BURST DETECTION

The problem of localizing γ-ray burst sources is well known. The distribution in space is not known, but it seems as if the present evidence tends toward a local--galactic distribution. Source locations are usually defined using two or more distinct detectors well separated in space. Single detector localizations are few and among the recent observations we may mention Nishimura *et al.* [2] and Sommer and Müller [3]. Both were balloon flights and while the former used a modulation collimator, the latter used single crystals inclined in different directions, so as to have different projected areas for a burst. This idea is recurrent in a number of different proposals [4] or in attemps at γ-ray burst observations [5]. Multiple scattering in the atmosphere will have modified the response of the modulation collimator, giving a new higher background proportional to the time variation of the burst, while the wide open response of the inclined crystals leads to counting rates that cannot be interpreted simply by the application of projected areas. In one case [3], in fact, some detectors were completely "in the shadow" but finite counting rates resulted all the same. This problem for inclined detectors does not disappear even when the detectors are flown above the atmosphere in satellites. For near-earth orbits the earth is almost always in the field of view, and there is usually the structure of the spacecraft nearby that can contribute in practically unpredictable ways. A gamma camera open about 2π steradians and with a set of 4σ rings in front of it to define the direction from their shadow (such as roughly sketched in Fig. 2) will see the scattered flux as an increased background level which should not vary appreciably with position over the crystal surface. It will still see the direct photons in contrast against this enhanced background and be able to separate out the spectrum. An instrument of this kind has been proposed for a possible Italian satellite and for GRO (while we plan to fly a bigger version on a transatlantic balloon flight in collaboration with K. Hurley of the CESR of Toulouse). The sensitivity of the satellite instrument has been evaluated for different spectra, since the spectral hardness determines the number of photons in our range (50 to 300 keV) for a given total intensity (the harder the spectrum the poorer the sensitivity). A number spectrum varying as E^{-1}, breaking to E^{-2} at 200 keV, which can be considered a worst case,

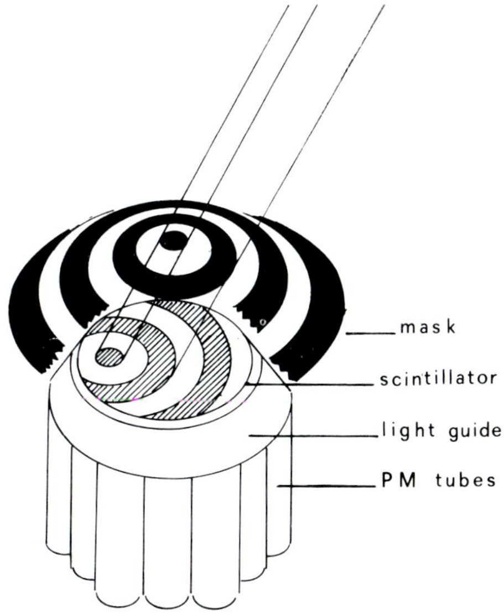

Fig. 2. Schematic view of the instrument.

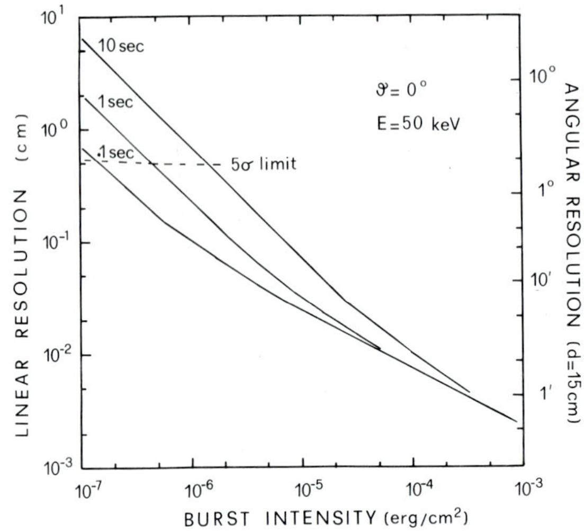

Fig. 3. Linear and angular resolution for impulsive event localization.

would give a 5σ sensitivity of about $4\times10^{-7}\sqrt{T}/\cos\Theta$, where Θ is the incident angle and T the duration of the burst in seconds. The angular sensitivity will go roughly as $\Delta x \cos\Theta/d$, where Δx is the positional error in the shadow determination and d the distance to the mask. Δx will in general decrease with improved statistics, down to about 1/10 or 1/15 of the crystal positional resolution, after which systematic errors will predominate. If the mask were at a distance of only 15 cm, the angular sensitivity to γ-ray bursts (in orbit) would be as sketched in Fig. 3.

SOURCE DETECTION

For source observation it is only necessary to put the ring mask (possibly a Fresnel zone plate in this case) further away, and to collimate to reduce the number of sources in the field of view. Supposing that one can distinguish a point source from an extended source to at least the level of 0.25σ of the positional accuracy and putting the mask at 2 meters gives an angular resolving power for extended sources of about 3 arcminutes. Positional accuracy for a single point source will instead be only limited by the integration time and the signal to noise ratio, down to the limit of systematic errors.

As mentioned in the introduction, it should be noted that half the area of the detector is sensitive to the source, which is equivalent to a full area on-off observation. More important, the spurious background, instrumental, diffuse, or even source-generated, is measured simultaneously with the source. Hence, while the time and energy structure information is preserved, the derivation of intensity and spectrum may be "cleaned" for spurious effects.

REFERENCES

1. H.O. Anger, *Rev. Sci. Instrum.* 29, 27 (1958).
2. J. Nishimura *et al.*, *Nature*, London 272, 337 (1978).
3. M. Sommer and D. Müller, *Astrophys. J. Lett.* 222, L17 (1978).
4. T.L. Cline *et al.*, *Bull. Am. Phys. Soc.* 22, 556 (1977).
5. G.J. Fishman *et al.*, *Astrophys. J. Lett.* 223, L13 (1978).

A SEARCH FOR GAMMA-RAY SPECTRAL LINES FROM THE GALACTIC CENTER BY HEAO-1

J. L. Matteson, P. L. Nolan and L. E. Peterson

University of California at San Diego, La Jolla, CA, USA

INTRODUCTION

Gamma-ray lines are the signature of nuclear deexcitation, radiative capture and positron annihilation, processes which result from extreme temperatures and non-thermal particle populations in astrophysical sites. Such line emission has already been detected from solar flares due to accelerated nuclei interacting in the ambient flare plasma [1]. The much lower line fluxes expected from cosmic sources has made their detection difficult. Balloon observations with NaI(Tl) detectors [2,3] have indicated lines from the galactic center (0.511, 1.6 and 4.6 MeV) and Centaurus A (1.6 and 4.5 MeV) at fluxes of 1 to 3 x 10^{-3} ph/cm^2-sec and a statistical significance of $\sim 3.5\sigma$. A recent balloon observation with a cooled Ge detector [4] confirms the 0.511 MeV line from the galactic center at a flux of $(1.21 \pm 0.22) \times 10^{-3}$ ph/cm^2-sec. The UCSD/MIT Hard X-Ray and Low Energy Gamma-Ray Experiment of HEAO-1 provides the first opportunity for extended observations of the sky with a collimated instrument capable of detecting line fluxes with a sensitivity typically better than that achieved in balloon observations to date. Here we report preliminary HEAO-1 results from a search for gamma-ray lines from the galactic center in the 1.5 to 8 MeV range.

INSTRUMENTATION

The instrument, shown in Fig. 1, has been described earlier [5] and will be briefly described here. It consists of seven NaI(Tl)/CsI(Na) phoswich scintillators surrounded by a massive CsI(Na) shield. Three NaI(Tl) detector geometries provide a 1.7° x 20° FWHM field of view from 10-200 keV (0.3cm thick NaI(Tl), 206cm^2 total effective area), 16° circular FWHM from 80-2000 keV (2.5 cm thick NaI(Tl), 168 cm^2) and 43° circular FWHM from 0.3-10 MeV (7.5 cm thick NaI(Tl), 120 cm^2). Data from the latter detector is presented here. It is shielded by an average of $\sim$12 cm of CsI(Na) and has an energy resolution of $\sim$6% FWHM at 2.6 MeV. NaI events which satisfy the anticoincidence condition are pulse-height-analyzed into 512 channels. Gain variations of $\sim$5% occur on time scales of $\sim$15 min due to the changing orientation of the detector photomultiplier in the geomagnetic field as HEAO-1 executes its great circle scan each 30 min. These variations are monitored by continuous internal gain calibration and corrected in data analysis. The background spectrum is shown in Fig. 2. From 0.2 to 2 MeV it is dominated by spallation and neutron capture products of I^{127}. The greatest effect is the β-decay continuum of I^{128}. Lines from electron capture decays of Te^{121}, I^{124}, I^{125} and I^{126}

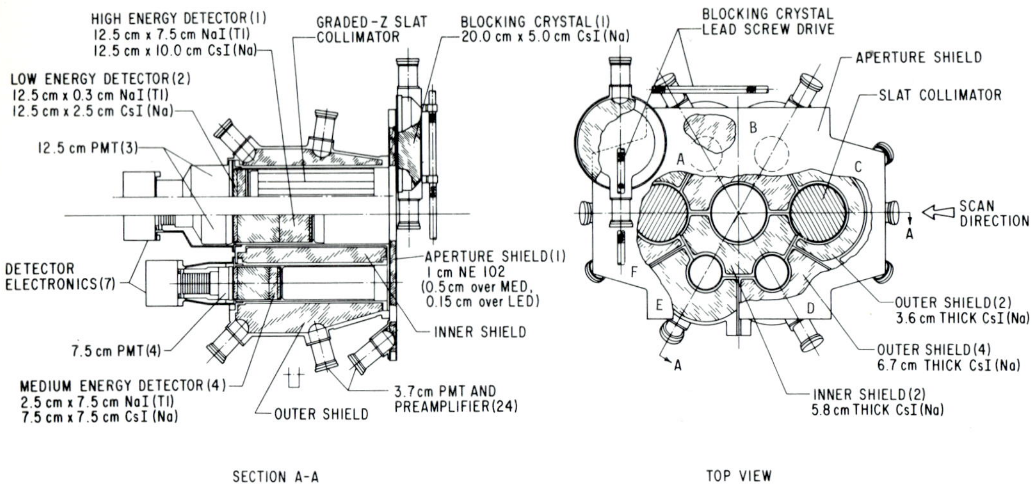

Fig. 1. UCSD/MIT Hard X-Ray and Low Energy Gamma-Ray Instrument of HEAO-1

are also present. Above 2 MeV the background is due to relatively short lived spallation effects with $T_{1/2} \leq 10$ min, and gamma-ray leakage through the shield. A steeply falling continuum results which is strongly modulated by variations in L during the orbit as is indicated in Fig. 2.

DATA AND ANALYSIS

The results reported here are based on data taken during 16 days in September 1977 when the HEAO-1 scan plane was passing over the galactic center at a rate of $\sim 1^{\circ}$/day. For each day only the 8 hours of data most remote from previous encounters with the South Atlantic Anomaly (SAA) were utilized. This allowed the radioactivity produced in the detectors by the SAA protons to have decayed to a level where further decay was acceptably small in the 30 min scan period. Data when the earth was outside the field-of-view were gain corrected and sorted into 18 scan azimuth bins for 4 L-ranges. A clear L-dependence of the background was indicated by this analysis, but within each L-range the data were smoothly distributed about an average value, indicating that uncorrected systematic background effects were not a significant problem. The observed pulse height spectrum of the galactic center was calculated using the two lowest L-ranges, $L < 1.3$, and taking all the non-galactic center data as background, excluding the Crab Nebula which was on the scan plane. The resulting pulse height spectrum is presented in Fig. 3. Visual inspection shows that no obvious statistically significant spectral lines are present. The total flux in the 1.5-7.3 MeV range is $(3 \pm 7) \times 10^{-4}$ ph/cm^2-sec. Correction from measured counts to cosmic gamma-rays was done using the instrumental response as given by the UCSD Monte Carlo computer program. This uses a 9-element geometry to model the instrument in sufficient detail to give good agreement with pre-launch calibrations. The predicted response at 4.43 MeV indicated in Fig. 3 was calculated in this manner. Upper limits to spectral lines at various energies are given in Table 1, column 2. At 4.43 MeV our 3σ limit is well below the reported 3σ detection from balloons [2]. However, the balloon observation [2] indicated a broadened feature and in this case our upper limit is also below the flux reported from the balloon observation.

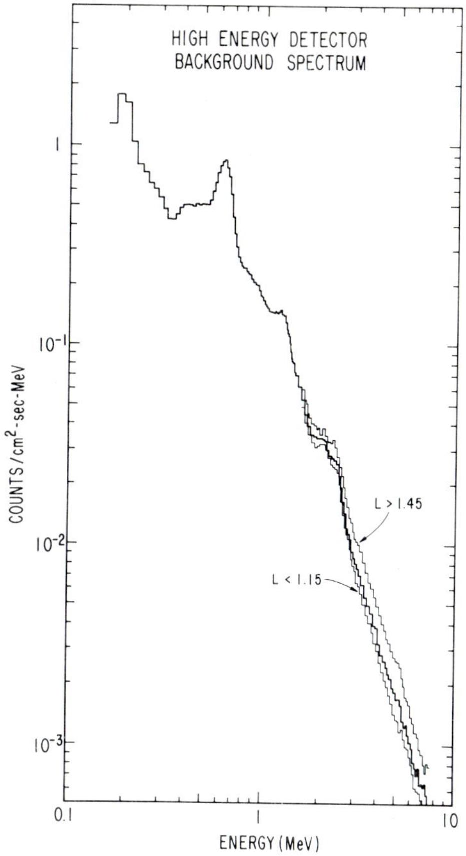

Fig. 2.

High energy detector background spectra at times well away from South Atlantic Anomaly encounters. The L-modulation is a factor $\sim$Z above 3 MeV but <10% below 1 MeV. The heavy line is the average over L.

CONCLUSION

The HEAO-1 observations reported here have failed to confirm the reported [2] 4.4 MeV flux of $(9.5 \pm 2.7) \times 10^{-4}$ ph/cm^2-sec from the galactic center. Our 3σ upper limits are 0.35 of the reported flux if it is due to a narrow line and 0.60 of it for a line at the reported [2] width of 700 keV. Therefore, the existence of the reported feature at 4.4 MeV cannot be strongly rejected by the present results. However, its flux must be somewhat less than the reported value. When the total data base of three sets of galactic center scans is analyzed the sensitivity will improve by a factor 2.5 to that indicated in Table 2, column 3.

At the oral presentation of this paper a spectrum of the galactic center presented by us was based on an analysis of five days of data which included all L-values and did not correct for L-variation of the background. The near synchronization of the L-induced background variations and the galactic center sightings, nearly all at high L-values, produced an apparently strong source at the galactic center. This "source" had a continuum spectrum as well as an indication of a spectral line at 4.4 MeV with a flux of $\sim 10^{-3}$ ph/cm^2-sec. These are entirely due to the uncorrected background effects and should be disregarded. In the analysis presented here the synchronization effect described above was searched for by inspection of the scan azimuth dependence of the rates in several energy bands for all four L-ranges. No indication of the effect was apparent.

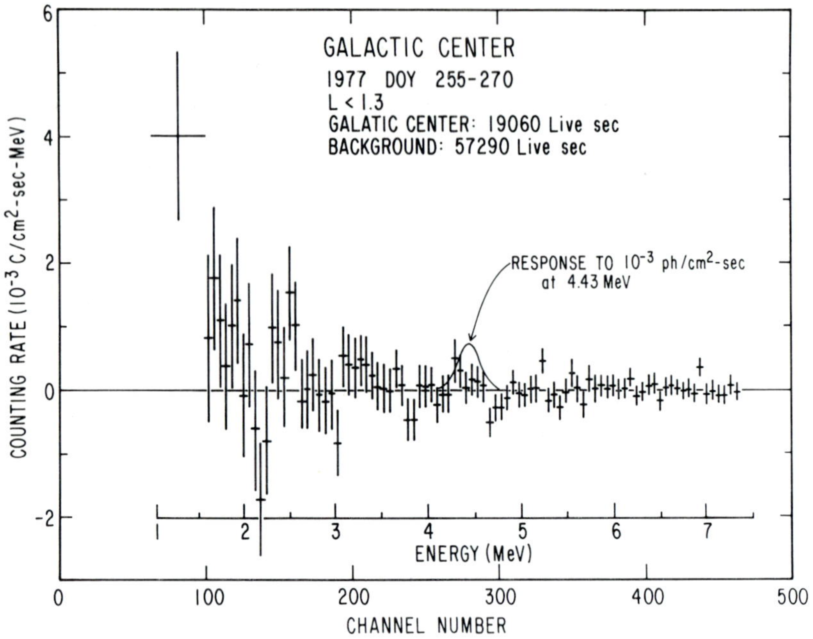

Fig. 3. The measured photon spectrum from the galactic center. The lowest energy point has significantly less livetime due to the instruments' telemetry priorities. The other points are averages over 4 PHA channels.

ACKNOWLEDGMENTS

It is a pleasure to acknowledge the contributions of G.L. Huszar to the extensive computer coding required for this work,which was supported by NASA NAS 8-27974.

Table 1. Sensitivity to Spectral Lines from the Galactic Center*

Energy (MeV)	3σ Upper Limits (10^{-4}ph/cm^2-sec)	
	16 days scanning	3 complete all sky scans
1	-	2.9
2.2	6.6	2.6
4.43	3.3	1.3
8	2.2	0.9
4.43	5.7	2.2

*For unresolved narrow lines except for last row which is for a 700 keV line width.

REFERENCES

1. E.L. Chupp et al., Nature, 241, 333, (1973).
2. R.C. Haymes et al., Ap. J., 201, 593, (1975).
3. R.D. Hall et al., Ap. J., 210, 631, (1976).
4. M. Leventhal et al., Ap. J. (Letters), 225, L11, (1978).
5. J.L. Matteson, 1978, Proceedings of AIAA 16th Aerospace Sciences Meeting, Huntsville, Alabama, January 16-18, 1978, paper 78-35.

SEARCH FOR GAMMA RAY LINES IN THE DIRECTION OF THE GALACTIC CENTER

F. Albernhe*, D. Boclet**, J.M. de Costa***,
Ph. Durouchoux**, J. F. Le Borgne*, R. Rocchia**
and G. Vedrenne*

**Centre d'Etude Spatiale des Rayonnements - 31400 Toulouse, France*
***DPh/EP/ES, Centre d'Etudes Nucléaires de Saclay - 91190 Gif-sur-Yvette, France*
****Instituto de Pesquisas Espaciais - Sao Jose Dos Campos, SP Brésil*

INTRODUCTION

A high resolution gamma-ray spectrometer aimed at stellar measurements has been jointly developed at Saclay and Toulouse. The instrument {1} consists of a large Ge(Li) crystal (140 cm^3) cooled at liquid nitrogen temperature. The Ge(Li) detector is surrounded by an anticoincidence NaI scintillator system, which is made up of a 55 mm wide tube of NaI(Tl) crystal, and divided into two optically separated halves along its axis and of two half circular disks, 50 mm in diameter, installed below the Ge(Li) detector. The whole assembly is shown in Fig. 1

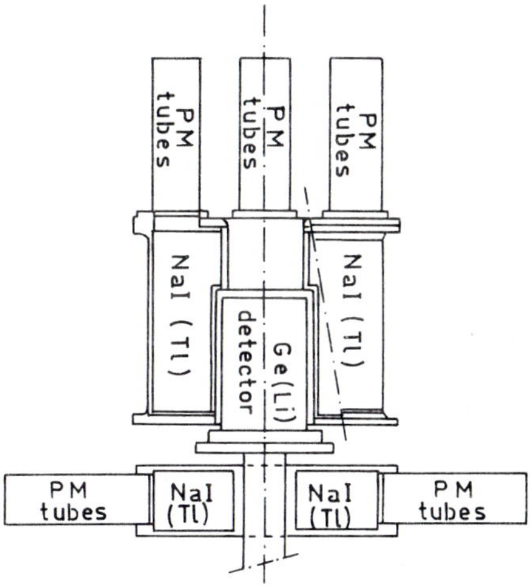

Fig. 1 View of the spectrometer

BALLOON FLIGHTS

Two flights (with 350 000 m^3 balloons) of the spectrometer have been successfully conducted in Brasil in Feb. 1977 (Guaratingueta, Sao Paulo, L - 45° 12 W, 1 = 22° 50 S, 12 GV rigidity cut-off). We accumulated data for 8.30 hours on the galactic center and 4.40 hours of background observation at a mean pressure of 3 gm/cm^2. The detector was oriented vertically and the observation made by transit of the source through the collimator.

Unfortunately the 3.4 keV spectrometer resolution rapidly broadened up to 8 keV during the ascent of both balloon flights and remained at this value during all the observations. In spite of this failure, we resolve several discrete peaks, as will be shown.

During each flight, a radioactive calibration source of ^{88}Yttrium (0.898 MeV - 1.836 MeV) allowed us to continuously check the gain stability of the electronics. The total count rate in the central detector was about 12 counts s^{-1} at float altitude in the 0.1 - 2.3 MeV energy range.

ATMOSPHERIC AND INDUCED GAMMA-RAY LINES

On the overall time observation (48805 s) the energy loss spectrum (Fig. 2) exhibits a total of 14 lines which contain greater than 3 σ excess.

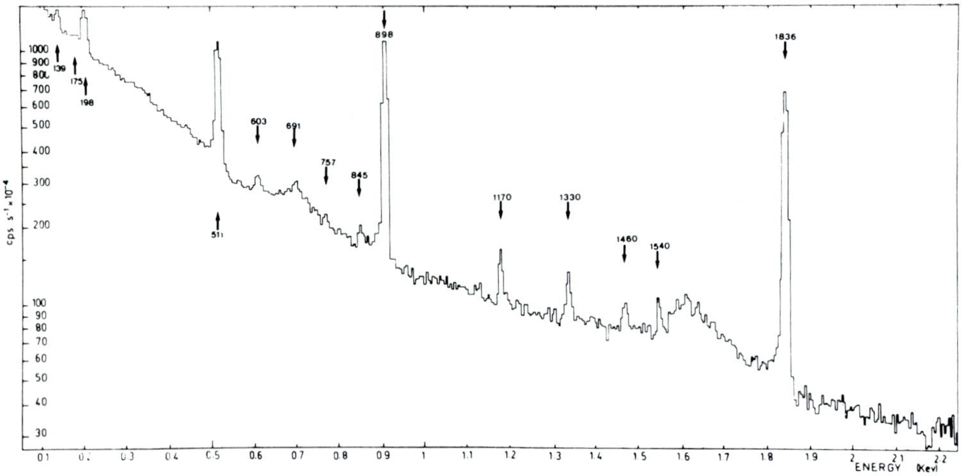

Fig. 2 Energy loss spectrum accumulated during 48 805 s.

The strongest peaks (898 keV and 1836 keV) correspond to the onboard inflight Y^{88} calibration sources, with slight contribution of Co^{60} impurities.
In addition to the 511 keV positron-electron annihilation line, the other lines listed on table 1 in majority can be explained in terms of cosmic ray and neutron induced nuclear reactions in the detector, and gondola.

TABLE 1

Photon energy (MeV)	σ	Flux (counts $cm^{-2}s^{-1}$)	
0.139	14.7	$6.83\ 10^{-3}$	Ge^{74} (n,γ) Ge^{75}
0.175	7.1	$1.7\ 10^{-3}$	Ge^{70} (n,γ) Ge^{71}
0.198	26.5	$1.15\ 10^{-3}$	Ge^{74} (n,γ) Ge^{75}
0.511	46.5	$3.47\ 10^{-2}$	$e^+ + e^-$
0.603	11		Ge^{74} (n,n') Ge^{74}
0.691	7.1		Ge^{72} (n,n') Ge^{72}
0.757	3		
0.845	6.7		Al^{27}(n,n')Al^{27}+Fe^{56}(p,p')Fe^{56}
1.460	4.8		K^{40}
1.540	6.3		

We dit not detect significant lines in the direction of the galactic center region and give in Table 2 some 3 σ upper limits at the energies lines predicted.

With the method of observation used (transit of the galactic center through the field of view, the altitude variation between 2 and 4 g/cm^2 dot not permit a direct evaluation of the contribution from the galactic center at 511 keV. However, after correction for the variation in pressure altitude and with the two flights results we can give a 3 σ upper limit for the 511 keV line from the direction of galactic center of 2.3 10^{-3} photons/cm^2.sec. This value is in agreement with the positive measurement of 1.21 ± 0.22 10^{-3} photons/cm^2.s. given by Leventhal {2}.

Even with the nominal resolution (3 keV) the sensitivity is much too low and an order of magnitude better sensitivity seems necessary to detect gamma lines from astrophysical sites with good confidence.

TABLE 2

Energy (MeV)	Origine	3σ upper limit ΔE = 12 keV counts $cm^{-2}s^{-1}$	Reference
431	$^4He(\alpha,n)^7Be^*$	2.3 10^{-3}	{5}
478		2.4 10^{-3}	{3}
847	$^{56}Fe(p,p)^{56}Fe^*$	1.7 10^{-3}	{4}
1.156	^{44}Ti	8.3 10^{-4}	{4}
1.173	^{60}Co	1.2 10^{-3}	{4}
1.238	$^{56}Fe(p,p')^{56}Fe^*$	1.1 10^{-3}	{5}
1.275	^{22}Ne	8.5 10^{-4}	{5}
1.369	$^{24}Mg(p,p')^{24}Mg^*$	9.3 10^{-4}	{5}
1.611		1.3 10^{-3}	
1.634	$^{20}Ne(p,p')^{20}Ne$	1.5 10^{-3}	{5}
1.779	$^{28}Si(p,p')^{28}Si^*$	1.5 10^{-3}	{5}
1.809	^{26}Al	1.2 10^{-3}	{4}
2.223	$^1H(n,\gamma)^2H^*$	7.0 10^{-4}	
2.313	$^{14}N(,p,p')$	7.0 10^{-4}	{5}

REFERENCES

1. F. Albernhe et al., Nucl. Instrum. Meth. in press

2. M. Leventhal, C.J. Mc Callum, P.D. Stang, G.S.F.C. Gamma Ray Spectroscopy in Astrophysics, 1978

3. R.C. Haymes, G.D. Walraven, C.A. Meegan, R.D. Hall, F.T. Djuth and D.H.Shelton, Astrophysics. J. 201, 593, 1975

4. R. Ramaty, R.E. Lingenfelter, G.S.F.C. Report No X-660-284

5. R.E. Lingenfelter, R. Ramaty, Diffuse Galactic gamma ray lines, report.

FAR UV SPECTRUM OF THE SKY BACKGROUND AND GEOCORONA OBSERVED BY GALACTIKA ABOARD THE PROGNOZ-6 SATELLITE

C. T. Hua*, P. Cruvellier*, G. Courtes*, A.M. Zvereva**, L. V. Granitsky**, A. B. Severny**

**Laboratoire d'Astronomie Spatiale, CNRS, 13012 Marseille, France*
***Crimean Observatory, 334413 Crimee, U.S.S.R.*

INTRODUCTION

A preliminary report is given about recent data obtained by means of a scanning spectrometer GALACTIKA λλ 1100-1900 a French-Soviet cooperative experiment, aboard the Prognoz-6 satellite, launched on September 22, 1977. The progressive decay of the Lyman α intensity with increasing distances from the Earth leads to the conclusion that previous observations of the sky background continuum near λ 1500 Å could be largely contaminated by scattered light from the geocoronal emission, when measured at low altitudes. The very high apogee of Prognoz-6 permits a much better evaluation of the galactic and extragalactic sky background.

Because of absorption by the interstellar medium, observations of the sky background can only begin at $\gtrsim$ 1000 Å. The ultraviolet spectrum is the result of the superimposition of the general galactic and extragalactic radiation, combined with scattering phenomena by the interplanetary and interstellar media.
The difficulty in measuring the background continuum, when observed for instance with a wide field scanner, is to properly discriminate its very low intensity from that of bright sources in the field and in the foreground. This is particularly the case for low altitude satellites and rocket.

Furthermore, there remain great difficulties in the precise evaluation linked to absolute calibration in the ultraviolet range; there is always uncertainty about the standard source or detector set-up. In this respect, observationnal data to date could be reasonably given confidence within a factor ∿ 2.
The highly elongated orbit (apogee 200 000 km) of Prognoz-6 offers the possibility of measuring the spectral distribution from parts of the sky (6°x6°) both inside and outside the Geocorona. The change of the Lyman α intensity with distance from the Earth also permits us to eliminate the influence of this radiation on the background intensity, i.e. the radiation from stars and galaxies.

OBSERVATIONS

The "GALACTIKA" photoelectric scanner was used to record the spectrum from 1100 Å to 1900 Å in 53 Å steps with a band width of 200 Å, in the antisolar direction (Fig. 1). More than 4000 scans were obtained, including the dark sky region in Pisces, bright regions in the Milky Way, the Pleiades and a bright B star ζ Tau. Examples of data at different distances from the Earth are given in Fig. 2, where the upper part represents the background outside Lyman α, at apogee, with a larger scale of intensities. Figure 3 shows the typical change of the Lyman α intensity with the distance from the Earth for ascending and descending parts of the orbit. Near the apogee, the emission becomes constant at about 170 Rayleigh (in our calibration scale 1 count.s^{-1}= 1.7 Rayleigh (R)). We remark that the Geocorona extends up to 90-150 x 10^3 km from the Earth. Near the perigee, the emission reaches ∿ 400 R, but for distances less than 40 x 10^3 km, the emission can reach 1200 R. Furthermore, the measurements at apogee show a decrease of ∿ 2 when approaching the Milky Way, which corresponds to earlier observations by many authors.

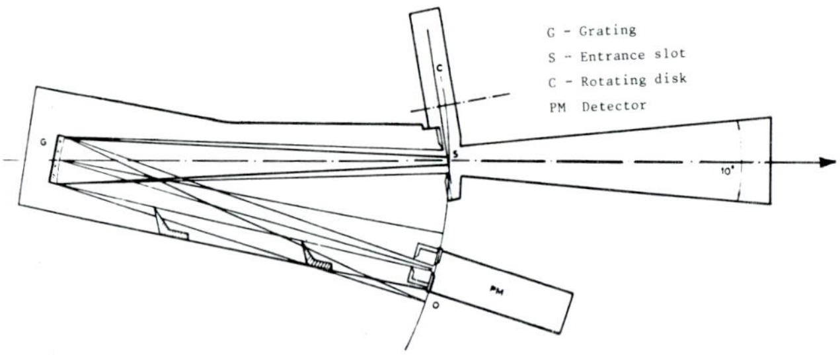

Fig. 1 Schematic view of the spectrometer "Galactika"

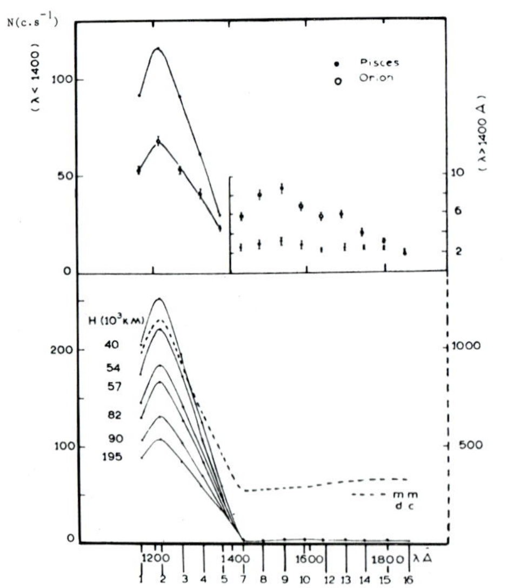

Fig. 2 Top. Intensity distribution of the sky background for the dark region (filled dots) and the Orion region (open circles)

Bottom. Changes in Lyman α intensity versus wavelength and distance to Earth. The dashed line corresponds to the dark current when approaching the radiation belt (right side scale)

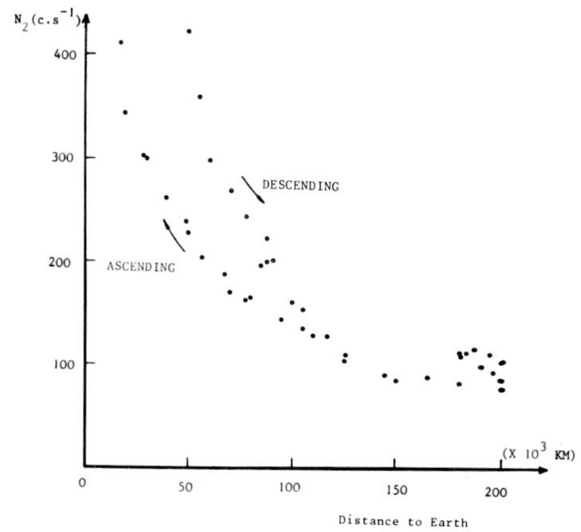

Fig. 3 Intensity of Lyman α during ascending and descending branches in orbit

To obtain the background emission, it is necessary to eliminate light scattered inside the spectrometer from the bright Lyman α emission. We found that the most efficiency way to do this is to plot the values of intensities for a given wavelength against the Lyman α intensity (I_2) near apogee. This dependence is well fitted by a straight line $I\lambda = K\lambda\ I_2 + b\lambda$, where $b\lambda$ obviously, would represent the true background intensity if Lyman α were absent $(I_2 = 0)$. (See Fig. 4).

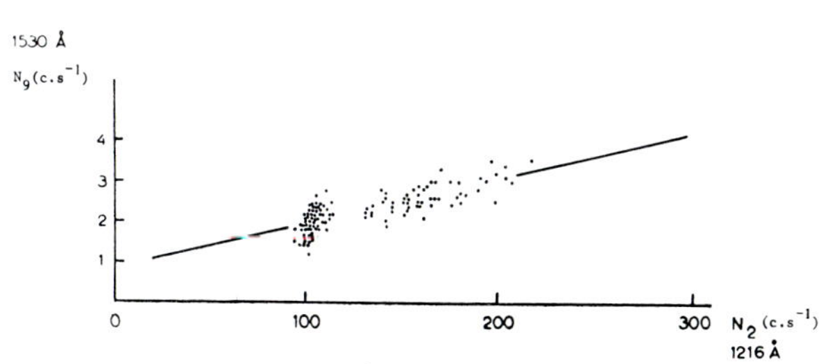

Fig. 4 Influence of scattered light

The distribution of b_λ for different wavelengths is presented in Fig. 5. These values are corrected for the photometric error of the field and reduced to the absolute intensity scale of observations by Henry et al. and TD1 [1, 2]. The dashed curve is the contribution of B and A stars in this dark region (coinciding with the DS region of Henry et al. 3). If we subtract the stellar component (as computed by Henry et al. loc.cit.) from our distribution of b_λ, we get the distribution clearly showing the hump near λ 1500 Å. Result which is in very good agreement with that obtained by Henry et al. [3]. This hump can be attributed to the redshifted emission of the HeII λ 304 from hot, dense intergalactic plasma (loc.cit. Fig. 11).

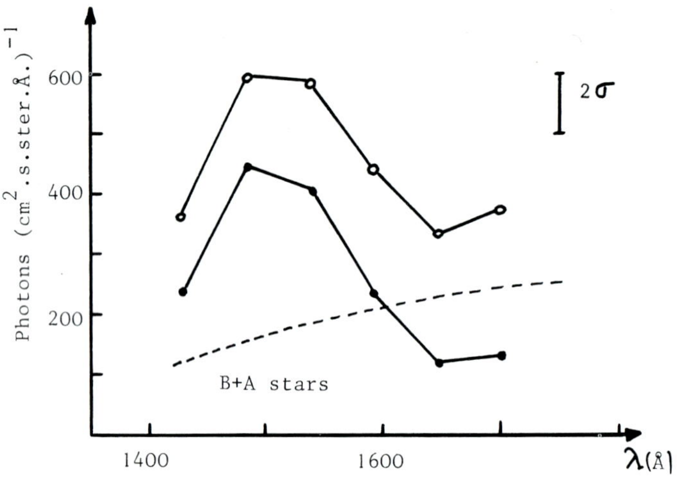

Fig. 5 1. The raw intensity of the sky background for the Pisces region as observed by Galactika I spectrometer (upper curve).
2. The contribution of stars as calculated by Henry et al. [3] (dashed line).
3. Lower curve: intensity of the sky background obtained after subtraction of the stellar contribution.

SUMMARY

Recent data of the sky background in the spectral range λλ 1000-1900 are presented, and comparison made with previous observations.

REFERENCES

1. R.C. Henry, A. Weinstein, P.D. Feldman, W.G. Fastie and H.W. Moos, Astrophys. J. 201, 613 (1975).

2. F. Beeckmans, Astron. Astrophys. 49, 263 (1976).

3. R.C. Henry, A. Weinstein, P.D. Feldman and W.G. Fastie, Astrophys. J. preprint (July 15, 1978).

INDEX